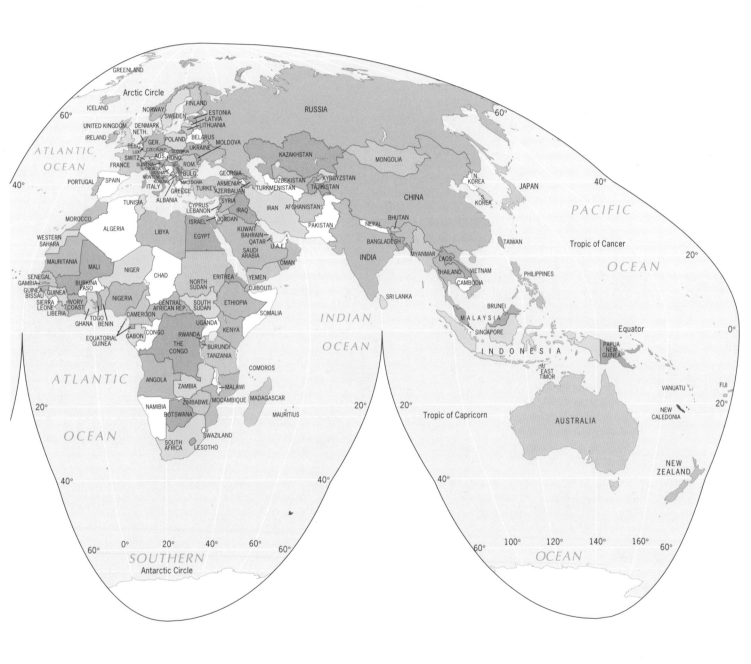

GREENLAND

Arctic Circle

RUSSIA

ICELAND
NORWAY
FINLAND
SWEDEN
ESTONIA
LATVIA
LITHUANIA

UNITED KINGDOM
DENMARK
NETH.
GER.
POLAND
BELARUS
UKRAINE
MOLDOVA
KAZAKHSTAN
MONGOLIA

IRELAND
BELG.
LUX.
CZECH REP.
SLOVAKIA

ATLANTIC
OCEAN

SWITZ.
FRANCE
AUS.
HUNG.
SLOVENIA
CROATIA
BOSNIA
MONTENEGRO
KOSOVO
ROM.
BULG.
GEORGIA

N.
KOREA
JAPAN

PORTUGAL
SPAIN
ITALY
MACEDONIA
GREECE
ALBANIA
TURKEY
ARMENIA
AZERBAIJAN
UZBEKISTAN
TURKMENISTAN
TAJIKISTAN
KYRGYZSTAN

S.
KOREA

CHINA

PACIFIC

TUNISIA
CYPRUS
LEBANON
SYRIA
IRAQ
IRAN
AFGHANISTAN

MOROCCO
ISRAEL
JORDAN
PAKISTAN
NEPAL
BHUTAN

TAIWAN

Tropic of Cancer

WESTERN
SAHARA
ALGERIA
LIBYA
EGYPT
KUWAIT
BAHRAIN
QATAR
U.A.E.
SAUDI
ARABIA
OMAN
BANGLADESH
INDIA
MYANMAR

LAOS

OCEAN

MAURITANIA
MALI
NIGER
CHAD
ERITREA
YEMEN
THAILAND
VIETNAM
PHILIPPINES

SENEGAL
GAMBIA
GUINEA
BISSAU
GUINEA
BURKINA
FASO
NIGERIA
NORTH
SUDAN
DJIBOUTI
SRI LANKA
CAMBODIA

SIERRA
LEONE
LIBERIA
IVORY
COAST
GHANA
TOGO
BENIN
CENTRAL
AFRICAN REP.
SOUTH
SUDAN
ETHIOPIA
SOMALIA
BRUNEI

INDIAN

EQUATORIAL
GUINEA
CAMEROON
GABON
CONGO
THE
CONGO
UGANDA
RWANDA
BURUNDI
KENYA
MALAYSIA

SINGAPORE

Equator

PAPUA
NEW
GUINEA

OCEAN

TANZANIA
COMOROS

INDONESIA

EAST
TIMOR

ATLANTIC
ANGOLA
ZAMBIA
MALAWI
MADAGASCAR
MAURITIUS

VANUATU
FIJI

NAMIBIA
ZIMBABWE
MOÇAMBIQUE
BOTSWANA

Tropic of Capricorn

NEW
CALEDONIA

OCEAN
SOUTH
AFRICA
SWAZILAND
LESOTHO

AUSTRALIA

NEW
ZEALAND

SOUTHERN
Antarctic Circle

OCEAN

60°
40°
20°
0°
20°
40°
60°

To Will Pesce

With our deep appreciation for your many years
of sustained confidence and support

WILEY PLUS +

www.wileyplus.com

WileyPLUS is a research-based online environment for effective teaching and learning.

WileyPLUS builds students' confidence because it takes the guesswork out of studying by providing students with a clear roadmap:

- what to do
- how to do it
- if they did it right

It offers interactive resources along with a complete digital textbook that help students learn more. With *WileyPLUS*, students take more initiative so you'll have greater impact on their achievement in the classroom and beyond.

GEOGRAPHY
REALMS, REGIONS, AND CONCEPTS

H. J. de Blij
John A. Hannah Professor of Geography
Michigan State University

Peter O. Muller
Professor, Department of Geography and Regional Studies
University of Miami

Jan Nijman
Professor, Department of Geography and Regional Studies
University of Miami

WILEY
John Wiley & Sons, Inc.

VICE PRESIDENT AND EXECUTIVE PUBLISHER	Jay O'Callaghan
EXECUTIVE EDITOR	Ryan Flahive
SENIOR DEVELOPMENTAL EDITOR	Ellen Ford
EDITORIAL PROGRAM ASSISTANT	Marcus Van Harpen
EDITORIAL ASSISTANT	Darnell Sessoms
SENIOR CONTENT MANAGER	Micheline Frederick
SENIOR PRODUCTION EDITOR	Janet Foxman
MARKETING MANAGER	Margaret Barrett
CREATIVE DIRECTOR	Harry Nolan
SENIOR DESIGNER	Wendy Lai
SENIOR PHOTO EDITOR	Jennifer MacMillan
SENIOR ILLUSTRATION EDITOR	Anna Melhorn
SENIOR MEDIA EDITOR	Lynn Pearlman
MEDIA SPECIALIST	Aaron Gass
PRODUCTION SERVICES	Jeanine Furino/ Furino Production

Cover photographs by Zachary Woodward; © Jan Nijman.

This book was set in 10.5/12 Adobe Garamond by Aptara®, Inc., and printed and bound by Courier/Kendallville. The cover was printed by Courier/Kendallville.

This book is printed on acid-free paper. ∞

Founded in 1807, John Wiley & Sons, Inc. has been a valued source of knowledge and understanding for more than 200 years, helping people around the world meet their needs and fulfill their aspirations. Our company is built on a foundation of principles that include responsibility to the communities we serve and where we live and work. In 2008, we launched a Corporate Citizenship Initiative, a global effort to address the environmental, social, economic, and ethical challenges we face in our business. Among the issues we are addressing are carbon impact, paper specifications and procurement, ethical conduct within our business and among our vendors, and community and charitable support. For more information, please visit our website: *www.wiley.com/go/citizenship*.

Library of Congress Cataloging-in-Publication Data

De Blij, Harm J.
    Geography: realms, regions and concepts/H. J. de Blij, Peter O. Muller & Jan Nijman.—15th ed.
        p. cm.
    Includes index.
    ISBN 978-1-118-09360-3 (acid-free paper)
 1. Geography—Textbooks.   I. Muller, Peter O.   II. Nijman, Jan.   III. Title.
    G128.D42 2012
    910—dc23
                         2011024076

    Main Book ISBN 978-1-118-09360-3
    Binder-Ready Version ISBN 978-1-118-12320-1

Printed in the United States of America

10 9 8 7 6 5 4 3 2 1

For the past four decades, *Geography: Realms, Regions, and Concepts* has reported (and sometimes anticipated) trends in the discipline of Geography and developments in the world at large. In fourteen preceding editions, *Regions,* as the book has come to be called, has explained the contemporary world's **geographic realms** (the largest regional entities on the face of the Earth) and their natural environments and human dimensions. In the process, this book has become an introduction to Geography itself, the discipline that links the study of human societies and natural environments. We look at the ways people have organized their living space, have adapted to changing social as well as environmental circumstances, and continue to confront forces beyond their control ranging from climate change to globalization. From old and still relevant concepts to new and untested ideas, *Regions* provides geographic perspective on our transforming world.

The book before you, therefore, is an information highway to geographic literacy. The first edition appeared in 1971, at a time when school geography in the United States (though not in Canada) was a subject in decline. It was a precursor of a dangerous isolationism in America, and geographers foresaw the looming cost of geographic illiteracy. Sure enough, the media during the 1980s began to report that polls, public surveys, tests, and other instruments were recording a lack of geographic knowledge at a time when our world was changing ever faster and becoming more competitive by the day. Various institutions, including the National Geographic Society, banks, airline companies, and a consortium of scholarly organizations, mobilized to confront an educational dilemma that had resulted substantially from a neglect of the very topics this book is about.

Before we can usefully discuss such commonplace topics as our "shrinking world" and our "global village," or deal with popular misconceptions such as the "flat world" and the "death of distance," we should know the different parts of the world, the interrelated components that together make up the whole. This is not just an academic exercise. You will find that much of what you encounter in this book is of immediate, practical value to you—as a citizen, a consumer, a traveler, a voter, a job-seeker. North America is a geographic realm with ever-intensifying global interests and involvements. Those interests and involvements require countless, often instantaneous decisions, whether they are in international business, media, affairs of state, disaster relief, and so forth. Such decisions must be based on the best possible knowledge of the world beyond our continent. That knowledge can be gained by studying the layout of our world, its environments, societies, resources, policies, traditions, and other properties—in short, its regional geography.

REALMS, REGIONS . . . AND CONCEPTS

This book begins with an introductory chapter followed by 12 sets of chapters that each cover one of the world's major geographic realms and their constituent regions. The Introduction discusses the world as a whole, outlining the physical stage on which the human drama is being played out, providing environmental frameworks, demographic data, political background, and economic-geographical context. In the remaining chapters, we have divided the world into 12 major realms and these are in turn subdivided into smaller regions. We explain why the realms and regions are delineated the way they are, and we describe their prevailing geographic features; we also place recent events around the world in their proper geographic context.

Along the way, we introduce more than 150 geographic concepts placed in their regional settings. Most of these concepts are indeed primarily geographical, but others are ideas about which, we believe, students of geography should have some knowledge. Although such concepts are listed on the opening page of every chapter, we have not, of course, enumerated every geographic notion used in that chapter. Many colleagues, we suspect, will want to make their own realm-concept associations, and as readers will readily perceive, the book's organization is quite flexible. It is possible, for example, to focus almost exclusively on substantive regional material or, alternatively, to concentrate mainly on conceptual issues.

NEWS ABOUT THE FIFTEENTH EDITION

New editions of *Regions* have a reputation for comprehensive and detailed revision, and this is one of the most substantial revisions yet. To be sure, the world has undergone some momentous changes over the past two years. Some of the most striking developments include the ever-expanding influence of China in many of the world's geographic realms, from Southeast Asia to South America; India's continuing economic rise; the impact of the global commodity boom on economies in the less developed world; Europe's financial troubles and their consequences; and the remarkable, ongoing political upheavals throughout the Arab World, initially dubbed the 'Arab Spring' by many observers. These events, and others, have reverberated throughout the world and are transforming their respective realms today. We will detail many of these changes in their geographic context as we proceed through our global regional survey.

This fifteenth edition has also benefited from the perspectives and insights of new co-author Jan Nijman, Professor Emeritus of Geography at the University of Miami and Professor of Urban Studies at the University of Amsterdam. His international scholarly experience and his long-time affiliation with the National Geographic Society, including worldwide travels on behalf of the Society, make him an ideal partner in the writing of *Regions*. He contributed to revisions throughout the book.

New Structure

In previous editions of *Regions*, we devoted one full chapter to each of the 12 realms. The first part of each chapter treated the realm at large and the second part focused on the individual regions within the realm—but this separation was often informal. In this edition, we decided to divide the material more formally into separate **A** and **B** chapters. As the table of contents shows, each of the world's geographic realms (with the exception of the Austral and Pacific Realms) is now treated in a *pair* of chapters: **A** chapters discuss broad geographic features at the scale of the entire realm whereas **B** chapters provide a more detailed analysis at the scale of the realm's constituent regions and countries.

This more formalized structure has two advantages. First, it increases flexibility for the user because it offers the option of 'customizing' reading assignments. As anyone who has taught or taken a course in World Regional Geography knows, the subject is complex, the world is still a very large and highly diverse place, and semesters are short. In principle, each of the **A** or **B** chapters can be read in isolation. If users prefer to reduce reading assignments on one or more realms, they might concentrate on the **A** chapters; the opposite approach focused on the **B** chapters is possible as well. Neither scale (realm or region) can be said to be more important than the other. They are most revealing, of course, in terms of their complementary nature: one informs the other—from the general to the specific, from the whole to its constituent parts, from the global to the local.

The second advantage is more substantive and is anchored in current disciplinary trends. *Scale,* along with *region,* is one of the most crucial general concepts in Geography. Understanding geographic events almost always demands a multi-scalar perspective. Using the **A-B** structure, we consistently discuss processes at two different scales. First we employ the broader scale of the realm to appreciate, for example, matters of geopolitics or economic integration or core-periphery relationships—developments that unfold and project across an entire realm or major parts of it. Subsequently, we "zoom in" on the regions nested within the realm to consider developments at a finer resolution. Typically, **B** chapters facilitate a closer look at the role of individual countries and national cultures (and, where necessary, their internal subdivisions).

The **A-B** structure should not be regarded as a rigid mechanism that separates geographic reality into two 'fixed' scales. Geography is a dynamic field; there is continuity in some ways, but change abounds. Geographic space is continuously reorganized and reshaped. The study of realms and regions is more based on "sliding" scales, from the level of the realm downward. **A** chapters typically deal with larger-scale issues; **B** chapters provide a more detailed analysis. Thus international issues that do not always involve the entire realm may be discussed in either **A** or **B** chapters, or both (e.g., India-Pakistan relations in the South Asian realm).

Voices From the Region

The fifteenth edition also introduces a set of boxed vignettes entitled *Voice From the Region*. This feature provides a platform to let people from each geographic realm speak directly to the reader—a local voice, unfiltered, that presents a reflection on their corner of the world as seen through their own eyes. Located in each **B** chapter (plus Chapters 11 and 12), these vignettes, accompanied by the name and a photo of the individual providing the local perspective, focus on important current events ranging from undocumented immigration in Arizona to the still-unfolding political upheaval in Egypt. Each *Voice* was solicited by the authors specifically for this purpose. We believe this new feature will help students identify more closely with developments and people across the spectrum of world regions.

A Changing World

The past two years have witnessed exceptional change in several parts of the world. One of the most fascinating things about Geography is that it exudes a sense of permanence in certain respects (e.g., the natural environment), but in reality everything changes over time—sometimes quite slowly and sometimes abruptly and dramatically. To cite just a few examples: (1) today's periphery of the European Union is a very different one from just a few years ago; (2) the geopolitics of the Indian Ocean, with China and India vying for dominance, has swiftly assumed global importance; (3) world market demand for raw materials continues to alter landscapes in a host of countries from Mongolia to Chile; and (4) what began as the 'Arab Spring' at the end of 2010 has sparked profound changes in many countries across North Africa and Southwest Asia, reconfigured relationships between them, and forced external powers like the United States to reexamine their strategic involvement in that realm.

Occasionally the world changes so rapidly that interpretations based on earlier understandings are overtaken by new developments—forcing us to rethink particular realms, regions, and countries, or even the global system that binds them together. In the early 2010s, the penetrating influence of China around the world and the enormous ramifications of the Arab revolutions are cogent examples. Thus new editions of *Regions* not only contain myriad valuable updates but also reappraisals of the fundamental nature of the world's geography.

What to Look For

With a revision manuscript totaling tens of thousands of words, countless map modifications, and a large number of new photos and Field Note boxes (many from new co-author Jan Nijman), no summary can adequately encapsulate all that this new edition contains. Nevertheless, here are some high points.

In the **Introduction**, your attention will be drawn to a pair of new maps and accompanying text that illustrate the process of globalization while underscoring the growing inequalities in our world, a phenomenon visible at every scale from the global to the local. Ever wider core-periphery disparities mark individual realms, regions, and countries on a small planet with a steadily growing population, intensifying risks to the very order and stability the world will need to make progress.

In **Chapters 1A and 1B**, contemporary Europe is discussed within a core-periphery framework, an approach that could not be more relevant given today's economic circumstances. Europe's grand experiment in unification and coordination now faces its mightiest challenge as core-periphery contrasts and disagreements threaten six decades of progress toward integration. The financial crises in countries such as Greece, Ireland, Portugal, and Italy have put enormous pressure on the Union while political imperatives for expansion to the east stand in sharp contrast to their staggering cost and the possible weakening of internal cohesion. And look for the new NATO map that highlights the different political and economic logics that operate inside the European realm.

Chapters 2A and 2B, which cover the Russian realm, continue to emphasize Russia's growing difficulties in dealing with Muslim populations along its southern periphery as well as the specter of terrorism within its borders. Climate change is likely to have a major impact on this gigantic country's northern regions, yet Siberia and the Far East are rapidly losing people now that their stay is no longer enforced by the government. Russian leaders proclaim a desire to revive their country's stature as a global power, but representative government, human rights, and legal institutions are among the casualties of this process.

Chapters 3A and 3B on North America, present new material on the economic recession that continued into 2011 with little sign of a recovery, including a discussion of the U.S. housing crisis and a new map of foreclosures. These chapters also discuss such recent events as the impact of the BP oil spill disaster on the Gulf Coast, Canadian efforts to protect its northern boreal forests, and how Arizona's new laws affect immigration via Mexico.

In **Chapters 4A and 4B** on Middle America, the emphasis continues on the realm's giant, Mexico, where the drug wars and accompanying violence pose a long-term challenge to national cohesion. New detailed maps of the drug conflict show international trafficking, the carving up of Mexican turf by rival cartels, and the violent killings that have occurred. Elsewhere, new material is introduced on the predicament of small-island developing economies that prevail in the Caribbean Basin. Tiny in size and often insular, most Middle American countries, including those on the Central American land bridge, face imposing challenges arising from their geography.

The big new story in South America, discussed in **Chapters 5A and 5B**, is the impact of the global commodities boom. China's apparently insatiable demand for raw materials (and to a lesser extent, India's), has led to the rapid expansion of primary economic sectors (mining, agriculture) in countries like Brazil, Chile, and Peru. This has changed the landscape, sometimes with serious environmental consequences, but also boosted government revenues while stimulating specific industries. Brazil, considered one of the world's four big "emerging" economies (BRICs) alongside Russia, India, and China, has grown spectacularly. By the end of 2012, an enormous new port will open on the Brazilian coast north of Rio de Janeiro, providing "a highway to China" in the bulk exporting of raw materials.

Subsaharan Africa, too, is under the influence of China and the rising global demand for minerals and other commodities. As **Chapters 6A and 6B** show, of the 10 fastest-growing economies on Earth, no less than six are located in this realm, including Angola and Nigeria. However, these countries are starting their growth from deep levels of poverty, and far greater development will be needed to lead them to prosperity. Moreover, as the text indicates, economic growth that is one-sidedly based on natural resources tends to be concentrated in particular locales and in the hands of a few. In these chapters, also look for the background story on the world's newest country: Equatorial Africa's South Sudan.

Chapters 7A and 7B, on North Africa/Southwest Asia, are rewritten to include the revolutionary events that began unfolding in late 2010. Tunisia's "Jasmine Revolution" triggered an explosion of simmering discontent across an Arab World in which young people, especially, took to the streets to confront autocratic governments and demand democratic reform. In Tunisia and Egypt, longtime despots resigned; a full-scale civil war erupted in Libya; ongoing upheaval continues to plague Syria and Yemen, and the rest of the realm (and the world) watched in anxiety. No matter how things eventually turn out in the different countries of the realm, they will never be the same. Chapter **7A** provides an extensive description and explanation of what began as the "Arab Spring," whereas Chapter **7B** offers a more detailed coverage of events in individual countries through mid-2011. In addition, look for a new map on Afghanistan's ethnic mosaic and a new section on the rise and fall of Usama bin Laden.

Chapters 8A and 8B on South Asia bear the strongest imprint of co-author Jan Nijman, whose field research has repeatedly taken him to that realm over the past fifteen years. Look for many new photos and Field Notes, new additions to our boxed series of "Great Cities of the Realm," and new maps such as the one on Naxalite terrorist activity in India. New text was added on Indian Ocean geopolitics, gender issues in several countries, demographic concerns, economic developments (especially India's IT sector), and the mounting challenges that make Pakistan such a difficult place to govern.

Chapters 9A and 9B cover another incredibly dynamic realm, supported by a number of new maps. If China's influence can now be felt all over the world, its dominance within East Asia is overwhelming. China's own economic geography is quickly changing as well, with the rapid westward expansion of the country's Coastal Core region into the Interior. In the meantime, the gap between rich and poor continues to widen and the Chinese government faces increasing demands from its workforce for higher wages. Whereas China's rise is indisputable, Japan's retreat stands in sharp contrast. Look for new text on Japan's "lost two decades" and the disastrous earthquake-tsunami-nuclear catastrophe that rocked the country in March 2011. This pair of chapters also contains updated material on the strained relations between the Koreas and between the People's Republic of China and Taiwan.

Chapters 10A and 10B on Southeast Asia provide new and updated coverage on Singapore's world-city status, on Vietnam's stop-and-go liberalization, and on Indonesia's emergence as the realm's dominant power. Also look for a new section on the rapid expansion of oil palm plantations on the Malay Peninsula and Borneo.

Chapters 11 and 12, because of their limited size, were retained as single chapters without the **A/B** split. In **Chapter 11**, the Austral Realm, we continue our focus on three ongoing concerns: the environment (Australia has suffered both dreadful droughts and death-dealing fires); aboriginal peoples and their circumstances in the modern, globalizing societies of Australia and New Zealand; and Australia's cultural diversification as immigration continues to alter its social geography. Australia has enormous resources and its future could be very bright indeed. In some ways, Australia is also becoming an Asian country, but the process is not a smooth one.

Chapter 12, Pacific Realm and Polar Futures, concentrates on the effects of a growing human presence in (and designs on) both polar zones, where environmental change is facilitating entry for tourists, tens of thousands of whom are cruising Antarctic waters and landing on the Earth's southernmost shores. Climate change may also be opening long-icebound waterways in the Arctic Basin, where Russians, Canadians, Americans, and others dream of northern navigation routes and extracting seafloor resources such as oil and natural gas. Our final map in this chapter reveals these opportunities in the Arctic Basin, but what it cannot show is the threat this invasion of the planet's last frontier poses to the fragile ecosystems prevailing there. And in the enormous water hemisphere between the poles, residents of the Pacific Realm face other consequences of climate change as rising sea level, inextricably bound to events in the Arctic and Antarctic, becomes an existential threat to everyone living on its low-lying shores.

This latest edition of *Regions* reflects our continuing commitment to bring to our readers our geographic

perspectives on a fast-changing world and its dynamic regional components. Since the previous edition, the author team has spent much time traveling, researching, and lecturing in all 12 realms as well as in dozens of countries around the globe. The authors' international involvements are invaluable in the continuation, innovation, and strengthening of *Regions*, which after more than 40 years is still one of the most exciting projects in the domain of college geography textbooks. We trust that you will find this Fifteenth Edition as informative and challenging as you have our earlier ones.

Data Sources

Numerous print and Internet sources were consulted during the updating of the book. For all matters geographical, of course, we consult *The Annals of the Association of American Geographers, The Professional Geographer, The Geographical Review, The Journal of Geography*, and many other academic journals published regularly in North America—plus an array of similar periodicals published in English-speaking countries from Scotland to New Zealand.

All quantitative information was updated to the year of publication and checked rigorously. Hundreds of other modifications were made, many in response to readers' and reviewers' comments. New spellings of geographic names continue, and we pride ourselves in being a reliable source for current and correct usage.

The statistical data that constitute Appendix B are derived from numerous sources. As users of such data are aware, considerable inconsistency marks the reportage by various agencies, and it is often necessary to make informed judgments about contradictory information. For example, some sources still do not reflect the rapidly declining rates of population increase or life expectancies in AIDS-stricken African countries. Others list demographic averages without accounting for differences between males and females in this regard.

In formulating Appendix B we have used among our sources the United Nations, the Population Reference Bureau, the World Bank, the Encyclopaedia Britannica *Books of the Year*, *The Economist* Intelligence Unit, the *Statesman's Year-Book*, and *The New York Times Almanac*. The urban population figures—which also entail major problems of reliability and comparability—are mainly drawn from the most recent database published by the United Nations' Population Division. For cities of less than 750,000, we developed our own estimates from a variety of sources. At any rate, the urban population figures used here are estimates for 2012 and they represent *metropolitan-area totals* unless otherwise specified.

Pedagogy

We continue to devise ways to help students learn important geographic concepts and ideas, and to make sense of our complex and rapidly changing world. Special features include the following:

Two-Part (A/B) Regional Chapter Organization. As explained above in the *New Structure* section of this preface, **A** chapters focus on the featured geographic realm as a whole whereas **B** chapters concentrate on that realm's major regional components. This subdivision scheme is followed for Chapters 1–10; Chapters 11 and 12, because they are much shorter, preserve the new structure within a single chapter. The opening Introduction chapter, which treats global perspectives, does not adhere to the scheme.

Opener Maps. A comprehensive map of the realm opens each **A** chapter. These maps were specially created according to our specifications, and they have been updated for this Fifteenth Edition; as in the previous edition, these maps are assigned the first figure number in each **A** chapter. A map of the realm's constituent regions opens each **B** chapter, and it too is assigned the first figure number in that chapter.

Concepts, Ideas, and Terms. Each **A** and **B** chapter begins with a numbered listing of the key geographic concepts, ideas, and terms that appear in that chapter. When these items are introduced in the text, they appear in orange-colored boldface type accompanied by numbers that correspond to those in the chapter-opening list.

List of Regions. On the opening page of each **B** chapter, a list of the regions within that particular realm provides a preview and helps to organize the chapter.

Major Geographic Qualities (Boxes). Near the beginning of each **A** chapter (except in the Introduction), we list, in boxed format, the major geographic qualities that best summarize that portion of the Earth's surface.

Sidebar Boxes. Certain topics and issues of interest are highlighted in boxes offset from the main text. Occasionally, these boxes will include a place that merits the reader's closer attention.

Voices From the Region (Boxes). As elaborated on page viii above, these boxed vignettes appear in each **B** chapter (plus Chapters 11 and 12) and provide a platform for a native of each realm to describe an aspect of local geography in her or his own words.

Points to Ponder (Boxes). This boxed feature at the end of each chapter offers a few brief statements about things of geographic import happening (or likely to happen) in a realm or region that may give one pause.

Regional Issues (Boxes). In the coverage of each realm (in either the **A** or **B** chapter), a controversial issue has been singled out for debate by two opposing voices. These commentaries are drawn from the senior author's field notes and correspondence, and are intended to stimulate classroom discussion. Individual readers may "vote" their opinions by e-mail, as indicated at the bottom of each box.

Among the Realm's Great Cities (Boxes). This feature reflects the ever-expanding process and influence of urbanization worldwide. More than 30 profiles of the world's leading cities are presented, each accompanied by a specially drawn map.

Major Cities of the Realm Population Tables. Near the beginning of every **B** chapter, we have included a table reporting the most up-to-date urban population data (based on 2012 estimates drawn from the sources listed earlier).

Using Geography (Boxes). At the end of each **B** chapter, two or three thought questions are posed that ask readers to call upon their accumulating knowledge of geographic concepts and apply them to important regional issues affecting the realm just covered.

From the Field Notes (Boxes). Many of the photos in this book were taken by the authors while doing fieldwork. By linking them to extended captions entitled *From the Field Notes*, valuable insights are provided into how a geographer observes and interprets information in the field. Each Field Note box is also linked to the ConceptCaching website (www.conceptcaching.com), a pedagogical feature described in the Ancillaries section below.

Table of Area and Demographic Data for the World's States. This seven-page table in Appendix B (inside the back cover) displays the most important geographic and population data for each of the world's countries, updated to 2012. Definitions and brief discussions of each data category in the table are found at the end of Appendix B.

Appendices, Glossary, and Index. In addition to the data table (Appendix B), we feature Appendix A that lists metric and customary units as well as their conversions; in the text, we consistently use metric measurements with British units provided in parentheses. The extensive Glossary and then the Index follow the Appendices. Four additional sections that appeared in the back matter of earlier editions, which should not be overlooked, now appear exclusively on the book's website (www.wiley.com/college/deblij): (1) *Using the Maps*, a guide to map interpretation; (2) *Opportunities in Geography*, an essay describing advanced study in the discipline of Geography to provide access to a wide range of geography-related professional careers; (3) a *Pronunciation Guide*; and (4) a detailed bibliography—under the title *References and Further Readings*, and organized by chapter—that introduces the wide-ranging literature of the discipline and world regional geography.

Ancillaries

A broad spectrum of print and electronic ancillaries are available to accompany the Fifteenth Edition of *Regions*. (Additional information, including prices and ISBNs for ordering, can be obtained by contacting John Wiley & Sons.) These ancillaries are described next.

THE TEACHING AND LEARNING PACKAGE

This Fifteenth Edition of *Geography: Realms, Regions, and Concepts* is supported by a comprehensive supplements package that includes an extensive selection of print, visual, and electronic materials.

Wiley/National Geographic
College Atlas of the World 2e

EDITED BY H. J. DE BLIJ AND ROGER DOWNS

Wiley is proud to offer the *College Atlas of the World 2e* to your students. State-of-the-art cartographic technology plus award-winning design and content make this affordable, compact, yet comprehensive atlas the ultimate resource for every geography student. The past decade has seen an explosion of innovative and sophisticated digital mapping technologies—and the Society has been at the forefront of developing these powerful new tools to create the finest, most functional, and most informative atlases available anywhere. This powerful resource can be purchased individually or at a significant discount when packaged with any Wiley textbook.

Resources That Help Teachers Teach

MEDIA-ENHANCED PowerPoint Presentations. This fresh collection of PowerPoint presentations will help you bring the striking photographs, precise maps, and vivid illustrations from *Regions* into your classroom. We understand that each professor has unique needs for their classroom presentations, so we offer three different types of presentation for each chapter:

• PowerPoints with just the text art.
• PowerPoints with text art and presentation notes.
• Media-Integrated PowerPoints with links to videos and animations.

Videos and Podcasts. We have created a series of streaming video resources and podcasts to support *Geography: Realms, Regions, and Concepts 15e.* Our streaming videos provide short lecture-launching video clips that can be used to introduce new topics, enhance your presentations, and stimulate classroom discussion. To help you integrate these resources into your syllabus, we have developed a comprehensive library of teaching resources, study questions, and assignments available in WileyPLUS.

Image Gallery. We provide online electronic files for the line illustrations and maps in the text, which the instructor can customize for presenting in class (for example, in handouts, overhead transparencies, or PowerPoints).

In addition to the text images, you also have access to ***ConceptCaching***, an online database of photographs that explores what a region looks like and what it feels like to live in that region. Photographs and GPS coordinates are "cached" for viewing by core geographical concept and by region. Professors can access the images or submit their own by visiting www.conceptcaching.com.

Instructor's Manual and Test Bank. This manual includes a test bank, chapter overviews and outlines, and lecture suggestions. The *Test Bank* for *Regions* contains over 3000 test items, including multiple-choice, fill-in, matching, and essay questions. It is distributed via the secure Instructor's website as electronic files, which can be saved into all major word processing programs. The *Instructor's Manual* is also available on the password-protected Instructor's website.

Overhead Transparency Set. All of the book's maps and diagrams are available as print-on-demand transparencies in beautifully rendered, four-color format, and have been resized and edited for maximum effectiveness in large lecture halls.

WileyPLUS. Online course management of the educational learning package is also available to accompany *Regions*. WileyPLUS combines videos, interactivities, automated grading of many questions, and grades that flow automatically into a grade book—including an electronic version of the textbook—in an all-in-one easy-to-use system. It has been used successfully in both small and large classes, and is particularly well suited as a stand-alone platform for online classes. WileyPLUS is flexible. Its features include:

1. Map testing: develop students' spatial knowledge with interactive map tests of countries, cities, and physical features.
2. Ask the questions specific to your course by creating questions to add to any assignment. For instance, you could create your own series of questions about migration to your State or employment shares in your Province.
3. Set your own policies: give students from one to five (or even unlimited) attempts at each question, give hints after a missed attempt, set due dates, assign point values, and give individual students extensions. You can set up the assignments so that students lose some percentage of the possible points if they don't give the correct answer by the *n*th attempt.
4. Make announcements or post files for the class.
5. Draw from the instructor resources for discussion questions or PowerPoint files of graphics.
6. Organize and manage class rosters and grades. You can create rosters or allow students to pay and self-enroll online.
7. The test banks are included among the instructor resources for each chapter, and you can use these to create quizzes and tests for your students to take online. These can also be automatically graded and will flow into the grade book.

Your investment of time to develop a good course using WileyPLUS can be carried over to subsequent semesters and shared with other instructors in your university.

Wiley Faculty Network. This peer-to-peer network of faculty is ready to support your use of online course management tools and discipline-specific software/learning systems in the classroom. The WFN will help you apply innovative classroom techniques, implement software packages, tailor the technology experience to the needs of each individual class, and provide you with virtual training sessions led by faculty for faculty.

The Regions 15e Instructors' Site. This comprehensive website includes numerous resources to help you enhance your current presentations, create new presentations, and employ our pre-made PowerPoint presentations. These resources include:

- A complete collection of PowerPoint presentations. Three for each chapter.
- All of the line art from the text.
- Access to photographs.
- A comprehensive test bank.

Resources That Help Students Learn

Student Website www.wiley.com/college/deblij. This easy-to-use and student-focused website provides additional resources that complement the textbook and enhance your students' understanding of geography and improve their grade.

Student Study Guide. This popular Study Guide supports the book content and is packed with useful study and review tools. For each chapter in the textbook, the Study Guide gives students and faculty access to chapter objectives, content questions-and-answers, outline maps of each realm, sample tests, and more. For further information, please contact your local Wiley sales representative.

WileyPLUS. For the student, WileyPLUS offers a number of enhanced features—including an affordable price. There are a number of bonus features like map testing, flash cards, and area and demographic data for countries around the world. The links to related web pages are live and instantaneous. Students can answer questions online and submit electronically to receive immediate feedback. When answering a question, students can click on links to refer to the text or media resources. These resources include:

- **Videos** provide a first-hand look at life in other parts of the world.
- **Interactive Animations and Exercises** allow you to explore key concepts from the text.
- **Virtual Field Trips** invite you to discover different locations through photos and help you better understand how geographers view the world.

- **Podcasts, Web Cams, and Live Radio** let you see and hear what is happening all over the world in real time.
- **Flashcards** offer an excellent way to drill and practice key concepts, ideas, and terms from the text.
- **Map Quizzes** help students master the place names that are crucial to their success in this course. Three game-formatted place-name activities are provided for each chapter.
- **GeoDiscoveries Modules** allow students to explore key concepts in greater depth using videos, animations, and interactive exercises.
- **Chapter Review Quizzes** provide immediate feedback to true/false, multiple-choice, and additional short-answer questions.
- **Concepts, Ideas, and Terms Interactive** Flash-cards help students review and quiz themselves on the concepts, ideas, and terms discussed in each chapter.
- **Interactive Drag-and-Drop Exercises** challenge students to correctly label important illustrations from the textbook.
- **Audio Pronunciation** is provided for over 2000 key words and place names from the text.
- **Annotated Web Links** put useful electronic resources into context.
- **Area and Demographic Data** are provided for every country and world realm (see Appendix B).

ACKNOWLEDGMENTS

Over the 41 years since the publication of the First Edition of *Geography: Realms, Regions, and Concepts*, we have been fortunate to receive advice and assistance from literally hundreds of people. One of the rewards associated with the publication of a book of this kind is the steady stream of correspondence and other feedback it generates. Geographers, economists, political scientists, education specialists, and others have written us, often with fascinating enclosures. We make it a point to respond personally to every such letter, and our editors have communicated with many of our correspondents as well. Moreover, we have considered every suggestion made and many who wrote or transmitted their reactions through other channels will see their recommendations in print in this edition.

Student Response

A major part of the correspondence we receive comes from student readers. We would like to take this opportunity to extend our deep appreciation to the several million students around the world who have studied from our books. In particular, we thank the students from more than 140 different colleges across the United States who took the time to send us their opinions. Students told us they found the maps and graphics attractive and functional. We have not only enhanced the map program with exhaustive updating but have added a number of new and updated maps to this

Fifteenth Edition as well as making significant changes in many others. Generally, students have told us that they found the pedagogical devices quite useful. We have kept the study aids the students cited as effective: a boxed list of each chapter's key concepts, ideas, and terms (numbered for quick reference in the text itself); a box summarizing each realm's major geographic qualities; and an extensive and still-expanding Glossary.

Faculty Feedback

In assembling the Fifteenth Edition, we are indebted to the following people for advising us on a number of matters:

MICHELE BARNABY, *Pittsburg State University (Kansas)*
PATRICIA BOUDINOT, *George Mason University (Virginia)*
SEAN CHENOWITH, *University of Louisiana-Monroe*
ELLEN HANSEN, *Emporia State University (Kansas)*
EDWARD JACKIEWICZ, *California State University, Northridge*
JAMES LEONARD, *Marshall University (West Virginia)*
ALAN MARCUS, *Towson University (Maryland)*
CRISTIAN MELO, *Florida International University*
DALTON W. MILLER, JR., *Mississippi State University*
BIMAL PAUL, *Kansas State University*
CRAIG REVELS, *Central Washington University*
JOELIA ROBINSON, *Houston Community College*
BETTY SHIMSHAK, *Towson University (Maryland)*
LEON YACHER, *Southern Connecticut State University*

In addition, several faculty colleagues from around the world assisted us with earlier editions, and their contributions continue to grace the pages of this book. Among them are:

JAMES P. ALLEN, *California State University, Northridge*
STEPHEN S. BIRDSALL, *University of North Carolina*
J. DOUGLAS EYRE, *University of North Carolina*
FANG YONG-MING, *Shanghai, China*
EDWARD J. FERNALD, *Florida State University*
RAY HENKEL, *Arizona State University*
RICHARD C. JONES, *University of Texas at San Antonio*
GIL LATZ, *Portland State University (Oregon)*
IAN MACLACHLAN, *University of Lethbridge (Alberta)*
MELINDA S. MEADE, *University of North Carolina*
HENRY N. MICHAEL, *late of Temple University (Pennsylvania)*
CLIFTON W. PANNELL, *University of Georgia*
J. R. VICTOR PRESCOTT, *University of Melbourne (Australia)*
JOHN D. STEPHENS, *University of Washington*
CANUTE VANDER MEER, *University of Vermont*

Faculty members from a large number of North American colleges and universities continue to supply us with vital feedback and much-appreciated advice. Our publishers arranged several feedback sessions, and we are most grateful to the following professors for showing us where the text could be strengthened and made more precise:

MARTIN ARFORD, *Saginaw Valley State University (Michigan)*
DONNA ARKOWSKI, *Pikes Peak Community College (Colorado)*
GREG ATKINSON, *Tarleton State University (Texas)*
CHRISTOPHER BADUREK, *Appalachian State University*
DENIS BEKAERT, *Middle Tennessee State University*
THOMAS L. BELL, *University of Tennessee*
DONALD J. BERG, *South Dakota State University*
JILL (ALICE) BLACK, *Missouri State University*

KATHLEEN BRADEN, *Seattle Pacific University*
DAVID COCHRAN, *University of Southern Mississippi*
JOSEPH COOK, *Wayne County Community College (Michigan)*
DEBORAH CORCORAN, *Southwest Missouri State University*
MARCELO CRUZ, *University of Wisconsin at Green Bay*
WILLIAM V. DAVIDSON, *Louisiana State University*
LARRY SCOTT DEANER, *Kansas State University*
JEFF DE GRAVE, *University of Wisconsin, Eau Claire*
JASON DITTMER, *Georgia Southern University*
JAMES DOERNER, *University of Northern Colorado*
STEVEN DRIEVER, *University of Missouri-Kansas City*
ELIZABETH DUDLEY-MURPHY, *University of Utah*
DENNIS EHRHARDT, *University of Louisiana-Lafayette*
BRYANT EVANS, *Houston Community College*
WILLIAM FLYNN, *Oklahoma State University*
WILLIAM FORBES, *Stephen F. Austin State University (Texas)*
BILL FOREMAN, *Oklahoma City Community College*
ERIC FOURNIER, *Samford University (Alabama)*
GARY A. FULLER, *University of Hawai'i*
RANDY GABRYS ALEXSON, *University of Wisconsin-Superior*
WILLIAM GARBARINO, *Community College of Allegheny County (Pennsylvania)*
HARI GARBHARRAN, *Middle Tennessee State University*
CHAD GARICK, *Jones County Junior College*
JON GOSS, *University of Hawai'i*
DEBRA GRAHAM, *Messiah College*
RICHARD J. GRANT, *University of Miami*
JASON B. GREENBERG, *Sullivan University*
MARGARET M. GRIPSHOVER, *University of Tennessee*
SARA HARRIS, *Neosho County Community College (Kansas)*
JOHN HAVIR, *Ashland Community & Technical College*
JOHN HICKEY, *Inver Hills Community College (Minnesota)*
SHIRLENA HUANG, *National University of Singapore*
INGRID JOHNSON, *Towson State University (Maryland)*
KRIS JONES, *Saddleback College (California)*
UWE KACKSTAETTER, *Front Range Community College, Westminster (Colorado)*
CUB KAHN, *Oregon State University*
J. MIGUEL KANAI, *University of Miami*
THOMAS KARWOSKI, *Anne Arundel Community College (Maryland)*
ROBERT KERR, *University of Central Oklahoma*
ERIC KEYS, *University of Florida*
JACK KINWORTHY, *Concordia University-Nebraska*
MARTI KLEIN, *Saddleback College (California)*
CHRISTOPHER LAINGEN, *Kansas State University*
HEIDI LANNON, *Santa Fe College (Florida)*
UNNA LASSITER, *Stephen F. Austin State University (Texas)*
RICHARD LISICHENKO, *Fort Hays State University (Kansas)*
CATHERINE LOCKWOOD, *Chadron State College (Nebraska)*
GEORGE LONBERGER, *Georgia Perimeter College*
CLAUDE MAJOR, *Stratford Career*
STEVE MATCHAK, *Salem State College (Massachusetts)*
CHRIS MAYDA, *Eastern Michigan University*
TRINA MEDLEY, *Oklahoma City Community College*
DALTON W. MILLER, JR., *Mississippi State University*
ERNANDO F. MINGHINE, *Wayne State University (Michigan)*
VERONICA MORMINO, *Harper College (Illinois)*
TOM MUELLER, *California University (Pennsylvania)*
IRENE NAESSE, *Orange Coast College (California)*
VALIANT C. NORMAN, *Lexington Community College (Kentucky)*
RICHARD OLMO, *University of Guam*
PAI YUNG-FENG, *New York City*
J. L. PASZTOR, *Delta College (Michigan)*
IWONA PETRUCZYNIK, *Mercyhurst College (Pennsylvania)*
PAUL E. PHILLIPS, *Fort Hays State University (Kansas)*
ROSANN POLTRONE, *Arapahoe Community College (Colorado)*
JEFF POPKE, *East Carolina University*

DAVID PRIVETTE, *Central Piedmont Community College (North Carolina)*
JOEL QUAM, *College of DuPage (Illinois)*
RHONDA REAGAN, *Blinn College (Texas)*
PAUL ROLLINSON, *Missouri State University*
RINKU ROY CHOWDHURY, *Indiana University*
A. L. RYDANT, *Keene State College (New Hampshire)*
JUSTIN SCHEIDT, *Ferris State University (Michigan)*
KATHLEEN SCHROEDER, *Appalachian State University*
NANCY SHIRLEY, *Southern Connecticut State University*
DMITRII SIDOROV, *California State University, Long Beach*
DEAN SINCLAIR, *Northwestern State University (Louisiana)*
RICHARD SLEASE, *Oakland, North Carolina*
SUSAN SLOWEY, *Blinn College (Texas)*
DEAN B. STONE, *Scott Community College (Iowa)*
JAMIE STRICKLAND, *University of North Carolina at Charlotte*
RUTHINE TIDWELL, *Florida Community College at Jacksonville*
IRINA VAKULENKO, *Collin County Community College (Texas)*
CATHY WEIDMAN, *Austin, Texas*
KIRK WHITE, *York College of Pennsylvania*
THOMAS WHITMORE, *University of North Carolina*
KEITH YEARMAN, *College of DuPage (Illinois)*
LAURA ZEEMAN, *Red Rocks Community College (Colorado)*
YU ZHO, *Vassar College (New York)*

We also received input from a much wider circle of academic geographers. The list that follows is merely representative of a group of colleagues across North America to whom we are grateful for taking the time to share their thoughts and opinions with us:

MEL AAMODT, *California State University, Stanislaus*
GILLIAN ACHESON, *Southern Illinois University, Edwardsville*
WILLIAM V. ACKERMAN, *Ohio State University*
W. FRANK AINSLEY, *University of North Carolina, Wilmington*
SIAW AKWAWUA, *University of Northern Colorado*
VICTORIA ALAPO, *Metropolitan Community College*
DONALD P. ALBERT, *Sam Houston State University*
TONI ALEXANDER, *Louisiana State University*
R. GABRYS ALEXSON, *University of Wisconsin-Superior*
KHALED MD. ALI, *Geologist, Dhaka, Bangladesh*
NIGEL ALLAN, *University of California, Davis*
JOHN L. ALLEN, *University of Wyoming*
KRISTIN J. ALVAREZ, *Keene State College (New Hampshire)*
DAVID L. ANDERSON, *Louisiana State University, Shreveport*
KEAN ANDERSON, *Freed-Hardeman University*
JEFF ARNOLD, *Southwestern Illinois College*
JERRY R. ASCHERMANN, *Missouri Western State College*
JOSEPH M. ASHLEY, *Montana State University*
PATRICK ASHWOOD, *Hawkeye Community College*
THEODORE P. AUFDEMBERGE, *Concordia College (Michigan)*
JAIME M. AVILA, *Sacramento City College*
EDWARD BABIN, *University of South Carolina-Spartanburg*
ROBERT BAERENT, *Randolph-Macon College*
MARVIN W. BAKER, *University of Oklahoma*
GOURI BANERJEE, *Boston University*
MICHELE BARNABY, *Pittsburg State University (Kansas)*
J. HENRY BARTON, *Thiel College (Pennsylvania)*
STEVEN BASS, *Paradise Valley Community College (Arizona)*
RENE BATAILLE, *Nashville State Technical Community College*
THOMAS F. BAUCOM, *Jacksonville State University (Alabama)*
KLAUS J. BAYR, *Keene State College (New Hampshire)*
DENIS A. BEKAERT, *Middle Tennessee State University*
JAMES BELL, *Linn Benton Community College (Oregon)*
KEITH M. BELL, *Volunteer State Community College-(Tennessee)*
WILLIAM H. BERENTSEN, *University of Connecticut*
DONALD J. BERG, *South Dakota State University*
ROYAL BERGLEE, *Morehead State University (Kentucky)*

RIVA BERLEANT-SCHILLER, *University of Connecticut*
LEE LUCAS BERMAN, *Southern Connecticut State University*
RANDY BERTOLAS, *Wayne State College*
THOMAS BITNER, *University of Wisconsin-Marshfield/Wood County*
WARREN BLAND, *California State University, Northridge*
DAVIS BLEVINS, *Huntington College (Alabama)*
HUBERTUS BLOEMER, *Ohio University*
S. BO JUNG, *Bellevue College (Nebraska)*
R. DENISE BLANCHARD-BOEHM, *Texas State University*
MARTHA BONTE, *Clinton Community College (Idaho)*
GEORGE R. BOTJER, *University of Tampa (Florida)*
R. LYNN BRADLEY, *Belleville Area College (Illinois)*
KEN BREHOB, *Elmhurst College (Illinois)*
JAMES A. BREY, *University of Wisconsin-Fox Valley*
ROBERT BRINSON, *Santa Fe Community College (Florida)*
REUBEN H. BROOKS, *Tennessee State University*
PHILLIP BROUGHTON, *Arizona Western College*
LARRY BROWN, *Ohio State University*
LAWRENCE A. BROWN, *Troy State-Dothan (Alabama)*
ROBERT N. BROWN, *Delta State University (Mississippi)*
STANLEY D. BRUNN, *University of Kentucky*
RANDALL L. BUCHMAN, *Defiance College (Ohio)*
MICHAELE ANN BUELL, *Northwest Arkansas Community College*
DANIEL BUYNE, *South Plains College*
MICHAEL BUSBY, *Murray State University (Kentucky)*
DEAN BUTZOW, *Lincoln Land Community College (Illinois)*
MARY CAMERON, *Slippery Rock University (Pennsylvania)*
MICHAEL CAMILLE, *University of Louisiana at Monroe*
MARY CARAVELIS, *Barry University (Florida)*
DIANA CASEY, *Muskegon Community College (Michigan)*
DiANN CASTEEL, *Tusculum College (Tennessee)*
BILL CHAPPELL, *Keystone College*
SEAN CHENOWETH, *University of Louisiana-Monroe*
STANLEY CLARK, *California State University, Bakersfield*
JOHN E. COFFMAN, *University of Houston (Texas)*
DWAYNE COLE, *Cornerstone University (Michigan)*
JERRY COLEMAN, *Mississippi Gulf Coast Community College*
JONATHAN C. COMER, *Oklahoma State University*
BARBARA CONNELLY, *Westchester Community College (New York)*
FRED CONNINGTON, *Southern Wesleyan University*
WILLIS M. CONOVER, *University of Scranton (Pennsylvania)*
OMAR CONRAD, *Maple Woods Community College (Missouri)*
ALLAN D. COOPER, *Otterbein College*
WILLIAM COUCH, *University of Alabama, Huntsville*
BARBARA CRAGG, *Aquinas College (Michigan)*
GEORGES G. CRAVINS, *University of Wisconsin-La Crosse*
ELLEN K. CROMLEY, *University of Connecticut*
JOHN A. CROSS, *University of Wisconsin-Oshkosh*
SHANNON CRUM, *University of Texas at San Antonio*
WILLIAM CURRAN, *South Suburban College (Illinois)*
KEVIN M. CURTIN, *University of Texas at San Antonio*
JOSE A. DA CRUZ, *Ozarks Technical Community College*
ARMANDO DA SILVA, *Towson State University (Maryland)*
MARK DAMICO, *Green Mountain College (Vermont)*
DAVID D. DANIELS, *Central Missouri State University*
RUDOLPH L. DANIELS, *Morningside College (Iowa)*
SATISH K. DAVGUN, *Bemidji State University (Minnesota)*
WILLIAM V. DAVIDSON, *Louisiana State University*
CHARLES DAVIS, *Mississippi Gulf Coast Community College*
JAMES DAVIS, *Illinois College*
JAMES L. DAVIS, *Western Kentucky University*
PEGGY E. DAVIS, *Pikeville College*
ANN DEAKIN, *State University of New York, College at Fredonia*
KEITH DEBBAGE, *University of North Carolina at Greensboro*
MOLLY DEBYSINGH, *California State University, Long Beach*
DENNIS K. DEDRICK, *Georgetown College (Kentucky)*
JOEL DEICHMANN, *Bentley College*

STANFORD DeMARS, *Rhode Island College*
TOM DESULIS, *Spoon River College*
THOMAS DIMICELLI, *William Paterson College (New Jersey)*
MARY DOBBS, *Highland Community College, Wamego*
SCOTT DOBLER, *Western Kentucky University*
D. F. DOEPPERS, *University of Wisconsin-Madison*
JAMES DOERNER, *University of Northern Colorado*
ANN DOOLEN, *Lincoln College (Illinois)*
STEVEN DRIEVER, *University of Missouri-Kansas City*
WILLIAM DRUEN, *Western Kentucky University*
ALASDAIR DRYSDALE, *University of New Hampshire*
KEITH A. DUCOTE, *Cabrillo Community College (California)*
ELIZABETH DUDLEY-MURPHY, *University of Utah*
WALTER N. DUFFET, *University of Arizona*
MIKE DUNNING, *University of Alaska Southeast*
CHRISTINA DUNPHY, *Champlain College (Vermont)*
ANTHONY DZIK, *Shawnee State University (Kansas)*
DENNIS EDGELL, *Firelands BGSU (Ohio)*
RUTH M. EDIGER, *Seattle Pacific University*
JAMES H. EDMONSON, *Union University (Tennessee)*
M. H. EDNEY, *State University of New York-Binghamton*
HAROLD M. ELLIOTT, *Weber State University (Utah)*
JAMES ELSNES, *Western State College*
ROBERT W. EVANS, *Fresno City College (California)*
EVERSON DICKEN, *California State University, San Bernardino*
MICHAEL FERBER, *Regent University*
JEFF FESPERMAN, *Illinois Valley Community College*
DINO FIABANE, *Community College of Philadelphia*
G. A. FINCHUM, *Milligan College (Tennessee)*
IRA FOGEL, *Foothill College (California)*
RICHARD FOLEY, *Cumberland College*
ROBERT G. FOOTE, *Wayne State College (Nebraska)*
RONALD FORESTA, *University of Tennessee*
ELLEN J. FOSTER, *Texas State University*
G. S. FREEDOM, *McNeese State University (Louisiana)*
EDWARD T. FREELS, *Carson-Newman College*
PHILIP FRIEND, *Inver Hills Community College*
JAMES FRYMAN, *University of Northern Iowa*
OWEN FURUSETH, *University of North Carolina at Charlotte*
RICHARD FUSCH, *Ohio Wesleyan University*
GARY GAILE, *late of University of Colorado-Boulder*
EVELYN GALLEGOS, *Eastern Michigan University & Schoolcraft College*
GAIL GARBRANDT, *Mount Union College, University of Akron*
DALE GARRET, *Evangel University*
RICHARD GARRETT, *Marymount Manhattan College*
JERRY GERLACH, *Winona State University (Minnesota)*
MARK GISMONDI, *Northwest Nazarene University*
LORNE E. GLAIM, *Pacific Union College (California)*
SHARLEEN GONZALEZ, *Baker College (Michigan)*
DANIEL B. GOOD, *Georgia Southern University*
GARY C. GOODWIN, *Suffolk Community College (New York)*
S. GOPAL, *Boston University*
MARVIN GORDON, *Lake Forest Graduate School of Business*
ROBERT GOULD, *Morehead State University (Kentucky)*
MARY GRAHAM, *York College of Pennsylvania*
GORDON GRANT, *Texas A&M University*
PAUL GRAY, *Arkansas Tech University*
DONALD GREEN, *Baylor University (Texas)*
GARY M. GREEN, *University of North Alabama*
STANLEY C. GREEN, *Laredo State University (Texas)*
RAYMOND GREENE, *Western Illinois University*
MARK GREER, *Laramie County Community College (Wyoming)*
WALT GUTTINGER, *Flagler College (Florida)*
JOHN E. GYGAX, *Wilkes Community College*
LEE ANN HAGAN, *College of Southern Idaho*
RON HAGELMAN, *University of New Orleans*
W. GREGORY HAGER, *Northwestern Connecticut Community College*

RUTH F. HALE, *University of Wisconsin-River Falls*
JOHN W. HALL, *Louisiana State University-Shreveport*
PETER L. HALVORSON, *University of Connecticut*
DAVID HANSEN, *Pennsylvania State University, Harrisburg & Schuylkill*
MERVIN HANSON, *Willmar Community College (Minnesota)*
ROBERT C. HARDING, *Lynchburg College (Virginia)*
MICHELLE D. HARRIS, *Bob Jones University (South Carolina)*
SCOTT HARRIS, *CGCS Drury University*
ROBERT J. HARTIG, *Fort Valley State College (Georgia)*
SUZANNA HARTLEY, *Shelton State Community College*
TRUMAN A. HARTSHORN, *Georgia State University*
CAROL HAZARD, *Meredith College*
HARLOW Z. HEAD, *Barton College*
MARK HEALY, *William Rainey Harper College*
DOUG HEFFINGTON, *Middle Tennessee State University*
JAMES G. HEIDT, *University of Wisconsin Center-Sheboygan*
CATHERINE HELGELAND, *University of Wisconsin-Manitowoc*
NORMA HENDRIX, *East Arkansas Community College*
JAMES E. HERRELL, *Otero Junior College (Colorado)*
JAMES HERTZLER, *Goshen College (Indiana)*
JOHN HICKEY, *Inver Hills Community College (Minnesota)*
THOMAS HIGGINS, *San Jacinto College (Texas)*
EUGENE HILL, *Westminster College (Missouri)*
LARISSA HINZ, *Eastern Illinois University*
MIRIAM HELEN HILL, *Indiana University Southeast*
SUZY HILL, *University of South Carolina-Spartanburg*
ROBERT HILT, *Pittsburg State University (Kansas)*
SOPHIA HINSHALWOOD, *Montclair State University (New Jersey)*
PRISCILLA HOLLAND, *University of North Alabama*
MARK R. HOOPER, *Freed-Hardeman University*
R. HOSTETLER, *Fresno City College (California)*
ERICK HOWENSTINE, *Northeastern Illinois University*
LLOYD E. HUDMAN, *Brigham Young University (Utah)*
JANIS W. HUMBLE, *University of Kentucky*
TERESA HUTCHINSON, *Columbus State Community College*
JUANA IBANEZ, *University of New Orleans*
TONY IJOMAH, *Harrisburg Area Community College (Pennsylvania)*
WILLIAM IMPERATORE, *Appalachian State University (North Carolina)*
EDWARD JACKIEWICZ, *California State University, Northridge*
RICHARD JACKSON, *Brigham Young University (Utah)*
MARY JACOB, *Mount Holyoke College (Massachusetts)*
GREGORY JEANE, *Samford University (Alabama)*
SCOTT JEFFREY, *Catonsville Community College (Maryland)*
JERZY JEMIOLO, *Ball State University (Indiana)*
NILS I. JOHANSEN, *University of Southern Indiana*
DAVID JOHNSON, *University of Southwestern Louisiana*
INGRID JOHNSON, *Towson State University (Maryland)*
KAREN JOHNSON, *North Hennepin Community College (Minnesota)*
RICHARD JOHNSON, *Oklahoma City University*
SHARON JOHNSON, *Marymount College (New York)*
JEFFREY JONES, *University of Kentucky*
KRIS JONES, *Saddleback College (California)*
MARCUS E. JONES, *Claflin College (South Carolina)*
TAMOURA JONES, *Atlanta Technical College (Georgia)*
MOHAMMAD S. KAMIAR, *Florida Community College, Jacksonville*
MELINDA KASHUBA, *Shasta College (California)*
MATTI E. KAUPS, *University of Minnesota-Duluth*
JO ANNE W. KAY, *Brigham Young University, Idaho*
PHILIP L. KEATING, *Indiana University*
DAVID KEELING, *Western Kentucky University*
COLLEEN KEEN, *Gustavus Adolphus College (Minnesota)*
ARTIMUS KEIFFER, *Wittenberg University (Ohio)*
GORDON F. KELLS, *Mott Community College (Michigan)*
KAREN J. KELLY, *Palm Beach Community College, Boca Raton (Florida)*
RYAN KELLY, *Lexington Community College (Kentucky)*
VIRGINIA KERKHEIDE, *Cuyahoga Community College and Cleveland State University (Ohio)*

Tom Kessenger, *Xavier University (Ohio)*
Masoud Kheirabadi, *Maryhurst University*
Susanne Kibler-Hacker, *Unity College (Maine)*
Changjoo Kim, *Minnesota State University*
James W. King, *University of Utah*
John C. Kinworthy, *Concordia College (Nebraska)*
Albert Kitchen, *Paine College (Georgia)*
Ted Klimasewski, *Jacksonville State University (Alabama)*
Robert D. Klingensmith, *Ohio State University-Newark*
Lawrence M. Knopp, Jr., *University of Minnesota-Duluth*
Lynn Koehnemann, *Gulf Coast Community College*
Terrill J. Kramer, *University of Nevada*
Joe Krause, *Community College of Indiana at Lafayette*
Debra Kreitzer, *Western Kentucky University*
Brender Krekeler, *Northern Kentucky University; Miami University (Ohio)*
Arthur J. Krim, *Cambridge, Massachusetts*
Michael A. Kukral, *Rose-Hulman Institute of Technology*
Chris Laney, *Berkshire Community College (Massachusetts)*
Elroy Lang, *El Camino Community College (California)*
Richard L. Langill, *Saint Martin's University (Washington)*
Christopher Lant, *Southern Illinois University*
A. J. Larson, *University of Illinois-Chicago*
Paul R. Larson, *Southern Utah University*
Larry League, *Dickinson State University (North Dakota)*
David R. Lee, *Florida Atlantic University*
Wook Lee, *Texas State University*
Joe Leeper, *Humboldt State University (California)*
Yechiel M. Lehavy, *Atlantic Community College (New Jersey)*
Scott Leith, *Central Connecticut State University*
James Leonard, *Marshall University (West Virginia)*
Elizabeth J. Leppman, *St. Cloud State University (Minnesota)*
John C. Lewis, *Northeast Louisiana University*
Dan Lewman Jr., *Southwest Mississippi Community College*
Caedmon S. Liburd, *University of Alaska-Anchorage*
T. Ligibel, *Eastern Michigan University*
Z. L. Lipchinsky, *Berea College (Kentucky)*
Allan L. Lippert, *Manatee Community College (Florida)*
Richard Lisichenko, *Fort Hays State University (Kansas)*
John L. Litcher, *Wake Forest University (North Carolina)*
Lee Liu, *Central Montana State University*
Li Liu, *Stephen F. Austin State University (Texas)*
William R. Livingston, *Baker College (Michigan)*
Catherine M. Lockwood, *Chadron State College (Nebraska)*
George E. Longenecker, *Vermont Technical College*
Cynthia Longstreet, *Ohio State University*
Jack Looney, *University of Massachusetts, Boston*
Tom Love, *Linfield College (Oregon)*
K. J. Lowrey, *Miami University (Ohio)*
James Lowry, *Stephen F. Austin State University (Texas)*
Max Lu, *Kansas State University*
Robin R. Lyons, *University of Hawai'i-Leeward Community College*
Susan M. Macey, *Texas State University*
Michael Madsen, *Brigham Young University, Idaho*
Ronald Magden, *Tacoma Community College (Washington)*
Christiane Mainzer, *Oxnard College (California)*
Laura Makey, *California State University, San Bernardino*
Mike Makowsky, *Midland College*
Harley I. Manner, *University of Guam*
Anthony Paul Mannion, *Kansas State University*
Gary Manson, *Michigan State University*
Charles Manyara, *Radford University (Virginia)*
James T. Markley, *Lord Fairfax Community College (Virginia)*
Sister May Lenore Martin, *Saint Mary College (Kansas)*
Kent Mathewson, *Louisiana State University*
Patrick May, *Plymouth State College (New Hampshire)*
Dick Mayer, *Maui Community College (Hawai'i)*

Sara Mayfield, *San Jacinto College, Central (California)*
Dean R. Mayhew, *Marine Maritime Academy*
J. P. McFadden, *Orange Coast College (California)*
Bernard McGonigle, *Community College of Philadelphia*
Molly McGraw, *Southeastern Louisiana University*
Paul D. Meartz, *Mayville State University (North Dakota)*
Dianne Meredith, *California State University-Sacramento*
David Merwin, *Framingham State College (Massachusetts)*
Gary C. Meyer, *University of Wisconsin-Stevens Point*
Judith L. Meyer, *Southwest Missouri State University*
Mark Micozzi, *East Central University (Oklahoma)*
John Milbauer, *Northeastern State University*
Dalton W. Miller, Jr., *Mississippi State University*
Raoul Miller, *University of Minnesota-Duluth*
Roger Miller, *Black Hills State University (South Dakota)*
James Mills, *State University of New York, College at Oneonta*
Ines Miyares, *Hunter College, CUNY (New York)*
Bob Monahan, *Western Carolina University*
Keith Montgomery, *University of Wisconsin-Madison*
Dan Morgan, *University of South Carolina, Beaufort*
Debbie Morimoto, *Merced College (California)*
John Mortron, *Benedict College (South Carolina)*
Anne Mosher, *Syracuse University*
Barry Mowell, *Broward Community College (Florida)*
Tom Mueller, *California University (Pennsylvania)*
Donald Myers, *Central Connecticut State University*
Robert R. Myers, *West Georgia College*
Gary Nachtigall, *Fresno Pacific University (California)*
Yaser M. Najjar, *Framingham State College (Massachusetts)*
Katherine Nashleanas, *University of Nebraska-Lincoln*
Jeffrey W. Neff, *Western Carolina University*
David Nemeth, *University of Toledo (Ohio)*
Robert Newcomer, *East Central University (Oklahoma)*
William Nieter, *St. Johns University (New York)*
William N. Noll, *Highland Community College*
Valiant C. Norman, *Lexington Community College (Kentucky)*
Joseph A. Naumann, *University of Missouri, St. Louis*
Raymond O'Brien, *Bucks County Community College (Pennsylvania)*
Patrick O'Sullivan, *Florida State University*
Diane O'Connell, *Schoolcraft College*
Nancy Obermeyer, *Indiana State University*
John Odland, *late of Indiana University*
Doug Oetter, *Georgia College and State University*
Anne O'Hara, *Marygrove College (Michigan)*
Patrick Olsen, *University of Idaho*
Joseph R. Oppong, *University of North Texas*
Lynn Orlando, *Holy Family University (Pennsylvania)*
Mark A. Oumette, *Hardin-Simmons University (Texas)*
Richard Outwater, *California State University, Long Beach*
Cissie Owen, *Lamar University (Texas)*
Mary Ann Owoc, *Mercyhurst College (Pennsylvania)*
Eugene J. Palka, *U.S. Military Academy (New York)*
Steve Palladino, *Ventura College (California)*
Bimal K. Paul, *Kansas State University*
Selina Pearson, *Northwest-Shoals Community College (Alabama)*
Mauri Pelto, *Nichols College (Massachusetts)*
James Penn, *Southeastern Louisiana University*
Nick Petropouleas, *Daytona State College (Florida)*
Linda Pett-Conklin, *University of St. Thomas (Minnesota)*
Diane Philen, *Lower Brule Community College (South Dakota)*
Paul Phillips, *Fort Hays State University (Kansas)*
Michael Phoenix, *ESRI, Redlands, California*
Jerry Pitzl, *Macalester College (Minnesota)*
Brian Plaster, *Texas State University*
Armand Policicchio, *Slippery Rock University-(Pennsylvania)*
Rosann Poltrone, *Arapahoe Community College (Colorado)*
Billie E. Pool, *Holmes Community College (Mississippi)*

GREGORY POPE, *Montclair State University (New Jersey)*
JEFF POPKE, *East Carolina University*
WILLIAM PRICE, *North Country Community College*
VINTON M. PRINCE, *Wilmington College (North Carolina)*
GEORGE PUHRMANN, *Drury University (Missouri)*
DONALD N. RALLIS, *Mary Washington College (Virginia)*
RHONDA REAGAN, *Blinn College (Texas)*
DANNY I. REAMS, *Southeast Community College (Nebraska)*
JIM RECK, *Golden West College (California)*
ROGER REEDE, *Southwest State University (Minnesota)*
JOHN RESSLER, *Central Washington University*
JOHN B. RICHARDS, *Southern Oregon State College*
DAVID C. RICHARDSON, *Evangel University (Missouri)*
GRAY RINGLEY, *Virginia Highlands Community College*
SUSAN ROBERTS, *University of Kentucky*
CURT ROBINSON, *California State University, Sacramento*
AMY ROCK, *John Carroll University*
WOLF RODER, *University of Cincinnati*
JAMES ROGERS, *University of Central Oklahoma*
PAUL A. ROLLINSON, *Southwest Missouri State-University*
JAMES C. ROSE, *Tompkins/Cortland Community College (New York)*
THOMAS E. ROSS, *Pembroke State University (North Carolina)*
THOMAS A. RUMNEY, *State University of New York, College at Plattsburgh*
GEORGE H. RUSSELL, *University of Connecticut*
BILL RUTHERFORD, *Martin Methodist College*
RAJAGOPAL RYALI, *Auburn University at Montgomery (Alabama)*
PERRY RYAN, *Mott Community College*
JAMES SAKU, *Frostburg State University (Maryland)*
DAVID R. SALLEE, *University of North Texas*
RICHARD A. SAMBROOK, *Eastern Kentucky University*
EDUARDO SANCHEZ, *Grand Valley State University (Michigan)*
JOHN SANTOSUOSSO, *Florida Southern College*
GINGER SCHMID, *Texas State University*
BRENDA THOMPSON SCHOOLFIELD, *Bob Jones University (South Carolina)*
ADENA SCHUTZBERG, *Middlesex Community College (Massachusetts)*
ROGER M. SELYA, *University of Cincinnati*
RENEE SHAFFER, *University of South Carolina*
WENDY SHAW, *Southern Illinois University, Edwardsville*
SIDNEY R. SHERTER, *Long Island University (New York)*
HAROLD SHILK, *Wharton County Jr. College (Texas)*
NANDA SHRESTHA, *Florida A&M University*
WILLIAM R. SIDDALL, *Kansas State University*
DAVID SILVA, *Bee County College (Texas)*
STEVEN SILVERN, *Salem State College (Massachusetts)*
JOSE ANTONIO SIMENTAL, *Marshall University (West Virginia)*
MORRIS SIMON, *Stillman College (Alabama)*
ROBERT MARK SIMPSON, *University of Tennessee at Martin*
KENN E. SINCLAIR, *Holyoke Community College (Massachusetts)*
ROBERT SINCLAIR, *Wayne State University (Michigan)*
JIM SKINNER, *Southwest Missouri State University*
BRUCE SMITH, *Bowling Green State University (Ohio)*
EVERETT G. SMITH, JR., *University of Oregon*
PEGGY SMITH, *California State University, Fullerton*
RICHARD V. SMITH, *Miami University (Ohio)*
JAMES SNADEN, *Charter Oak State College (Connecticut)*
DAVID SORENSON, *Augustana College (Illinois)*
SISTER CONSUELO SPARKS, *Immaculata University (Pennsylvania)*
CAROLYN D. SPATTA, *California State University, Hayward*
M. R. SPONBERG, *Laredo Junior College (Texas)*
DONALD L. STAHL, *Towson State University (Maryland)*
DAVID STEA, *Texas State University*
ELAINE STEINBERG, *Central Florida Community College*
D. J. STEPHENSON, *Ohio University Eastern*
HERSCHEL STERN, *Mira Costa College (California)*
REED F. STEWART, *Bridgewater State College (Massachusetts)*
NOEL L. STIRRAT, *College of Lake County (Illinois)*
JOSEPH P. STOLTMAN, *Western Michigan University*

DEAN B. STONE, *Scott Community College*
WILLIAM M. STONE, *Saint Xavier University, Chicago*
GEORGE STOOPS, *Minnesota State University*
DEBRA STRAUSSFOGEL, *University of New Hampshire*
JAMIE STRICKLAND, *University of North Carolina at Charlotte*
WAYNE STRICKLAND, *Roanoke College (Virginia)*
PHILIP STURM, *Ohio Valley College, Vienna*
PHILIP SUCKLING, *Texas State University*
ALLEN SULLIVAN, *Central Washington University*
SELIMA SULTANA, *University of North Carolina at Greensboro*
RAY SUMNER, *Long Beach City College (California)*
CHRISTOPHER SUTTON, *Western Illinois University*
T. L. TARLOS, *Orange Coast College (California)*
WESLEY TERAOKA, *Leeward Community College*
MICHAEL THEDE, *North Iowa Area Community College*
DERRICK J. THOM, *Utah State University*
CURTIS THOMSON, *University of Idaho*
BEN TILLMAN, *Texas Christian University*
CLIFF TODD, *University of Nebraska, Omaha*
STANLEY TOOPS, *Miami University (Ohio)*
RICHARD J. TORZ, *St. Joseph's College (New York)*
HARRY TRENDELL, *Kennesaw State University (Georgia)*
ROGER T. TRINDELL, *Mansfield University (Pennsylvania)*
DAN TURBEVILLE, *East Oregon State College*
NORMAN TYLER, *Eastern Michigan University*
GEORGE VAN OTTEN, *Northern Arizona University*
GREGORY VEECK, *Western Michigan University*
C. S. VERMA, *Weber State College (Utah)*
KELLY ANN VICTOR, *Eastern Michigan University*
SEAN WAGNER, *Tri-State University (Indiana)*
GRAHAM T. WALKER, *Metropolitan State College of Denver*
MONTGOMERY WALKER, *Yakima Valley Community College (Washington)*
DEBORAH WALLIN, *Skagit Valley College (Washington)*
MIKE WALTERS, *Henderson Community College (Kentucky)*
LINDA WANG, *University of South Carolina, Aiken*
J. L. WATKINS, *Midwestern State University (Texas)*
DAVID WELK, *Reedley College (California)*
KIT W. WESLER, *Murray State University (Kentucky)*
PETER W. WHALEY, *Murray State University (Kentucky)*
MACEL WHEELER, *Northern Kentucky University*
P. GARY WHITE, *Western Carolina University (North Carolina)*
W. R. WHITE, *Western Oregon University*
GARY WHITTON, *Fairbanks, Alaska*
REBECCA WIECHEL, *Wilmington College (North Carolina)*
MARK WILJANEN, *Eastern Kentucky University*
GENE C. WILKEN, *Colorado State University*
FORREST WILKERSON, *Texas State University*
P. WILLIAMS, *Baldwin-Wallace College (Ohio)*
STEPHEN A. WILLIAMS, *Methodist College (North Carolina)*
DEBORAH WILSON, *Sandhills Community College (Nebraska)*
MORTON D. WINSBERG, *Florida State University*
ROGER WINSOR, *Appalachian State University (North Carolina)*
ELIZABETH WINTERNITZ-RUSSELL, *Maui Community College (Hawai'i)*
WILLIAM A. WITHINGTON, *University of Kentucky*
A. WOLF, *Appalachian State University (North Carolina)*
JOSEPH WOOD, *University of Southern Maine*
RICHARD WOOD, *Seminole Junior College (Florida)*
GEORGE I. WOODALL, *Winthrop College (North Carolina)*
STEPHEN E. WRIGHT, *James Madison University (Virginia)*
DAWN WROBEL, *Moraine Valley Community College*
LEON YACHER, *Southern Connecticut State University*
KYONG YUP CHU, *Bergen Community College (New Jersey)*
FIROOZ E. ZADEH, *Colorado Mountain College*
DONALD J. ZEIGLER, *Old Dominion University (Virginia)*
YU ZHOU, *Bowling Green State University (Ohio)*
ROBERT C. ZIEGENFUS, *Kutztown University (Pennsylvania)*

PERSONAL APPRECIATION

For assistance with the map of North American indigenous people, we are greatly indebted to Jack Weatherford, Professor of Anthropology at Macalester College (Minnesota); Henry T. Wright, Professor and Curator of Anthropology at the University of Michigan; and George E. Stuart, President of the Center for Maya Research (North Carolina). The map of Russia's federal regions could not have been compiled without the invaluable help of David B. Miller, then Senior Edit Cartographer at the National Geographic Society, and Leo Dillon of the Russia Desk of the U.S. Department of State. The map of Russian physiography was updated thanks to the suggestions of Mika Roinila of the State University of New York, College at New Paltz. Special thanks also go to Charles Pirtle, Professor of Geography at Georgetown University's School of Foreign Service for his advice on Chapter 4A; to Bilal Butt (post-doctoral fellow, University of Wisconsin-Madison) for his significant help with Chapter 6A; to Charles Fahrer of Georgia College and State University for his suggestions on Chapter 6B; and to C. Frederick Lohrengel II of Southern Utah University for his advice on Chapter 10A. Enormous gratitude goes to Beth Weisenborn, Virtual Course Coordinator of the Department of Geography at Michigan State University, and all the instructors of Geo 204 (World Regional Geography) for their invaluable suggestions to improve the book. And we are most grateful to Tanya de Blij, who holds a graduate degree in geography from Florida State University and whose computer and editing skills have helped keep our projects on track.

We also record our appreciation to those geographers who ensured the quality of this book's ancillary products. Elizabeth Muller Hames, D.O. (as well as M.A. in Geography, University of Miami) co-authored the original *Study Guide*. At the University of Miami's Department of Geography and Regional Studies, we are most grateful for the advice and support we received from faculty colleagues Richard Grant, Miguel Kanai, Mazen Labban, Shouraseni Sen Roy, and Ira Sheskin as well as GIS Lab Manager Chris Hanson. We also take this opportunity to reconfirm our appreciation of the work done on the ancillaries for earlier editions by Eugene J. Palka, Professor of Geography at the United States Military Academy, West Point, NY, whose vision and enthusiasm contributed importantly to the achievement of our objectives.

We are privileged to work with a team of professionals at John Wiley & Sons that is unsurpassed in the college textbook publishing industry. As authors we are acutely aware of these talents on a daily basis during the crucial production stage, especially the outstanding coordination and leadership skills of Senior Production Editor Janet Foxman, Senior Developmental Editor Ellen Ford, Senior Photo Editor Jennifer MacMillan, and Illustration Editor Anna Melhorn. Others who played a leading role in this process were Senior Designer Wendy Lai, copyeditor Betty Pessagno, and Don Larson (assisted by Ann Kennedy) of Mapping Specialists, Ltd. in Fitchburg, Wisconsin. For more than a decade, we have much appreciated the generosity and leadership of Executive Geosciences Editor Ryan Flahive, always the prime mover in launching and guiding this book, and who was so diligently assisted throughout the preparations for this latest edition by Darnell Sessoms. For this latest revision, we have especially benefitted from the managerial skills and savvy of our College Marketing Manager, Margaret Barrett, whose boundless enthusiasm and energy perfectly complement the talents of the *Regions* team. We also thank Veronica Armour, Lynn Pearlman, Bridget O'Lavin, and Harry Nolan for their help and support. Beyond this immediate circle, we acknowledge the long-time support and encouragement we continue to receive from many others at Wiley including Vice-President for Production Ann Berlin and Publisher Jay O'Callaghan.

If the authors were to give out an MVP award, it would unanimously go to Jeanine Furino of Furino Production, one of the most gifted textbook-production professionals we have ever encountered. In less than six months (!), not only did she oversee and coordinate every aspect of the hugely complicated process of creating this attractive volume out of a mountain of word and graphic files, email attachments, sketches, design layouts, and myriad rounds of preliminary pages—she made us feel that we were the only authors she was working with while handling a sizeable workload of other books.

Finally, and most of all, we express our gratitude to our spouses—Bonnie, Nancy, and Jannie—for seeing us through the constantly challenging schedule of this latest edition of *Regions*.

H. J. DE BLIJ
Boca Grande, Florida

PETER O. MULLER
Coral Gables, Florida

JAN NIJMAN
Coral Gables, Florida

May 9, 2011

GEOGRAPHY
REALMS, REGIONS, AND CONCEPTS

WORLD REGIONAL
GEOGRAPHY: Global Perspectives

FIGURE G-1

© H. J. de Blij, P. O. Muller, and John Wiley & Sons, Inc.

© H. J. de Blij

Searing social contrasts mark the cityscape
of Mumbai, India.

What are your expectations as you open this book? You have signed up for a course that will take you around the world to try to understand how it functions today. Hopefully, you will also discover how interesting and unexpectedly challenging the discipline of geography is. We hope that this course, and this book, will open new vistas and

The limestone-tower landscape of South China's Li River Valley.
© Jan Nijman

bring new perspectives, and help you navigate our increasingly complex and often daunting world.

You could not have chosen a better time to be studying geography. The world is changing on many fronts, and so is the United States. Still the most formidable of all countries, the United States remains a great power capable of influencing nations and peoples, lives and livelihoods from pole to pole. That power confers on Americans the responsibility to learn as much as they can about those nations and livelihoods, so that the decisions of their government representatives are well-informed. But in this respect, the United States is no superpower. Geographic literacy is a measure of international comprehension and awareness, and Americans' geographic literacy ranks low among countries of consequence. For the world, that is not a good thing, because such geographic fogginess tends to afflict not only voters but also the representatives they elect, from the school board to the White House.

So you are about to strike a blow against geographic illiteracy, and at the end of this term you will find yourself among the still-small minority of Americans who have spent even just one semester studying the world at large. When you get into discussions about global affairs or faraway places, you will be able to bring to the table a knowledge and awareness not often heard. Enjoy!

A WORLD ON MAPS

Just a casual glance at the pages that follow reveals a difference between this and other textbooks: there are almost as many maps as there are pages. Geography is more closely identified with maps than any other discipline, and we urge you to give as much (or more!) attention to the maps in this book as you do to the text. It is often said that a picture is worth a thousand words, and the same or more applies to maps. When we write "see Figure XX," we really mean it . . . and we hope that you will get into the habit. We humans are territorial creatures, and the boundaries that fence off our 200 or so countries reflect our divisive ways. But other, less visible borders—between religions, languages, rich, and poor—partition our planet as well. When political and cultural boundaries are at odds, there is nothing like a map to summarize the circumstances. Just look at the maps of Afghanistan and Iraq in Chapters 7A and 7B of this book. How clear were the implications of these maps to those who made the decisions to send American troops into war back in 2001 and 2003?

Maps in Our Minds

All of us carry in our minds maps of what psychologists call our *activity space*: the apartment building or house we live in, the streets nearby, the way to school or workplace, the general layout of our hometown or city. You will know what lane to use when you turn into a shopping mall, or where to park at the movie theater. You can probably draw from memory a pretty good map of your hometown. These mental maps [1] allow you to navigate your activity space with efficiency, predictability, and safety. When you arrived as a first-year student on a college or university campus, a new mental map will have started forming. At first you needed an online or hard-copy map to find your way around, but soon you dispensed with that because your mental map was sufficient. And it will continue to improve as your activity space expands.

If a well-formed mental map is useful for decisions in daily life, then an adequate mental map is surely indispensable when it comes to decision-making in the wider world. You can give yourself an interesting test. Choose some part of the world, beyond North America, in which you have an interest or about which you have a strong opinion—for example, Israel, Taiwan, Afghanistan, North Korea, or Venezuela. On a blank piece of paper, draw a map that reflects your impression of the regional layout there: the country, its neighbors, its internal divisions, major cities, seas (if any), and so forth. That is your mental map of the place. Put it away for future reference, and try it again at the end of this course. You will have proof of your improved mental-map inventory.

The Map Revolution

The maps in this book show larger and smaller parts of the world in various contexts. Some depict political configurations; others display ethnic, cultural, economic, or environmental features. *Cartography* (the making of maps) has undergone a dramatic technological revolution—a revolution that continues. Earth-orbiting satellites with *remote sensing* ability (special on-board scanners and television cameras) transmit remotely sensed information to computers on the surface, recording the expansion of deserts, the shrinking of glaciers, the depletion of forests, the growth of cities, and myriad other geographic phenomena. Earthbound computers possess ever-expanding capabilities not only to organize this information but also to display it graphically. This allows geographers to develop *geographic information systems (GISs)*, presenting on-screen information within seconds that would have taken months to assemble just a few decades ago.

Nonetheless, satellites—even spy satellites—cannot record everything that occurs on the Earth's surface. Sometimes the borders between ethnic groups or cultural sectors can be discerned by satellites, for example, in changing types of houses or religious shrines, but this kind of information tends to require on-the-ground verification through field research and reporting. No satellite view of Iraq could show you the distribution of Sunni and Shia Muslim adherents. Many of the boundaries you see on the maps in this book cannot be seen from space because long stretches are not even marked on the ground. So the maps you are about to "read" have their continued uses. They summarize complex situations and allow us to begin forming durable mental maps of the areas they represent.

There is one other point we should make that is especially important when it comes to *world* maps: never forget that the world is a sphere, and to project it onto a two-dimensional flat surface must necessarily entail some very significant distortions. Try peeling an orange and flattening the entire peel on a surface—you will have to tear it up and try to stretch it in places to get the job done. Have a look at Figure G-1 and note how the Atlantic Ocean and other parts are interrupted. You can produce a map like this in many different ways, but you will always end up distorting things. When studying world maps, there is nothing like having a *globe* at hand to remind you of our three-dimensional reality!

GEOGRAPHY'S PERSPECTIVE

Geography is sometimes described as the most interdisciplinary of disciplines. That is a testimonial to geography's historic linkages to many other fields, ranging from geology to economics and from sociology to political science. And, as has been the case so often in the past, geography is in the lead on this point. Today, *interdisciplinary* studies and research are more prevalent than ever. The old barriers between disciplines are breaking down.

This is not to suggest that college and university departments are no longer relevant; they are just not as exclusive as they used to be. These days, you can learn some useful geography in economics departments and some good economics in geography departments. But each discipline still has its own particular way of looking at the world.

A Spatial Perspective

Most disciplines focus on one key theme: economics is about money; political science is about power; psychology is about the mind; biology is about life. Geography, then, is about *space* on the Earth's surface. More specifically, geographers are interested in the *organization* of that space. Social space (cities, buildings, political boundaries, etc.) as well as natural space (climates, terrain, water bodies, etc.) are not randomly configured. Instead, there generally

From the Field Notes . . .

"On the descent into Tibet's Lhasa Gongga Airport, I had a great view of the Yarlung Zangbo Valley, its braided stream channels gently flowing toward the distant east. The Yarlung Zangbo is the highest major river on Earth, running from the Tibetan Plateau into northeastern India where it joins the mighty Brahmaputra River that continues on to Bangladesh where it empties into the Indian Ocean. It was mid-October and the water levels were low. The landing strip of the airport can be seen in the center-right of the photo, on the south bank. The airport is quite far from Lhasa, the Tibetan capital, about 62 kilometers (40 mi) to its southwest. Despite major road and tunnel construction, it is still more than an hour's drive. The airport had to be built away from the city and in this widest part of the valley because it allows relatively easy landings and take-offs. It lies at 3700 meters above sea level (12,100 ft), one of the highest airports in the world."

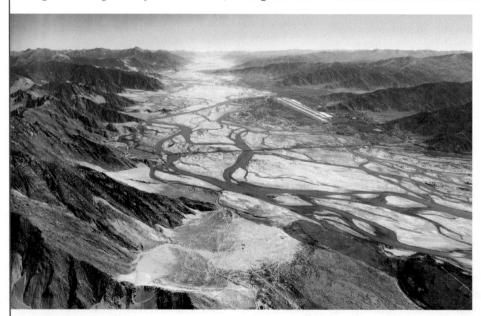

© Jan Nijman

www.conceptcaching.com

Concept Caching

prevails a particular order, regularity, even predictability about the ways in which space is organized. Sometimes it is the deliberate work of human beings and sometimes it is the work of nature, but very often there are particular patterns. Geographers consider these spatial patterns and processes as not only interesting but also crucial to how we live and how we organize our societies. The **spatial perspective [2]** has defined geography from its beginnings.

Environment and Society

There is another glue that binds geography and has done so for a very long time: an interest in the relationships between human societies and the natural (physical) environment. Geography lies at the intersection of the social and natural sciences and integrates perspectives from both, being the only discipline to do so explicitly. This perspective comes into play frequently: environmental change is in the news on a daily basis in the form of worldwide climate change, but this current surge of global warming is only the latest phase of endless atmospheric and ecological fluctuation. Geographers are involved in understanding current environmental issues not only by considering climate change in the context of the past, but also by looking carefully at the implications of global climate change for human societies.

More generally, think of this relationship between humans and their environment as a two-way street. On one hand, human beings have always had a transformative effect on their natural surroundings, from the burning of forests to the creation of villages. On the other hand, humans have always been heavily dependent on the natural environment, their individual and collective behaviors very much a product of it. There are so many examples that it is hard to know where to begin or when to end: we eat what nature provides and traditional diets vary regionally; rivers allow us to navigate and connect with other peoples—or they serve as natural boundaries like the Rio Grande; wars are fought over access to water or seaports; landlocked countries seem to have different cultures from those of islands; and so on.

At times we are faced with the interrelationship between humans and their environment. For example, humans modify the environment through escalating carbon dioxide emissions (the so-called greenhouse effect) and are subsequently confronted with the need to adjust to rising sea levels. We will always be part of nature, no matter how far technology advances.

Spatial Patterns

Geographers, therefore, need to be conversant with the location and distribution of salient features on the Earth's surface. This includes the natural (physical) world, simplified in Figure G-1 as well as the human world, and our inquiry will view these in temporal (historical) as well as spatial perspective. We take a penetrating look at the

overall geographic framework of the contemporary world, the still-changing outcome of thousands of years of human achievement and failure, movement and stagnation, stability and revolution, interaction and isolation. The spatial structure of cities, the layout of farms and fields, the networks of transportation, the configurations of rivers, the patterns of climate—all these form part of our investigation. As you will find, geography employs a comprehensive spatial vocabulary with meaningful terms such as area, distance, direction, clustering, proximity, accessibility, and many others we will encounter in the pages ahead. For geographers, some of these terms have more specific definitions than is generally assumed. There is a difference, for example, between *area* (surface) and *region*, between *boundary* and *frontier*, and between *place* and *location*. The vocabulary of geography holds some surprises, and what at first may seem to be simple ideas turn out to be complex concepts.

Scale and Scope

One prominent item in this vocabulary is the term **scale [3]**. Whenever a map is created, it represents all or part of the Earth's surface at a certain level of detail. Obviously, Figure G-1 displays a very low level of detail; it is little more than a general impression of the distribution of land and water as well as lower and higher elevations on our planet's surface. A few prominent features such as the Himalayas and the Sahara are named, but not the Appalachians or the Gobi Desert. At the bottom of the map you can see that one inch at this scale must represent about 1650 miles of the real world, leaving the cartographer little scope to insert information.

A map such as Figure G-1 is called a *small-scale* map because the ratio between map distance and real-world distance, expressed as a fraction, is very small at 1:103,750,000. Increase that fraction, and you can represent less territory—but also enhance the amount of detail the map can represent. In Figure G-2, note how the fraction increases from the smallest (1:103,000,000) to the largest (1:1,000,000). Montreal, Canada is just a dot on Map A but an urban area on Map D. Does this mean that world maps like Figure G-1 are less useful than larger-scale maps? It all depends on the purpose of the map. In this chapter, we often use world maps to show global distributions as we set the stage for the more detailed discussions to follow. In later chapters, the scale tends to become larger as we focus on smaller areas, even on individual countries and cities. But whenever you read a map, be aware of the scale because it is a guide to its utility.

The importance of the scale concept is not confined to maps. Scale plays a fundamental role in geographic research and in the ways we think about geographic problems: scale in terms of *level of analysis*. This is sometimes referred to as *operational scale*, the scale at which social or natural processes operate or play out. For instance, if you want to investigate the geographic concentration of

EFFECT OF SCALE

FIGURE G-2

© H. J. de Blij, P. O. Muller, and John Wiley & Sons, Inc.

wealth in the United States, you can do so at a range of scales: within the neighborhood, the city, the county, the State,* or at the national level. You choose the scale that is the most appropriate for your purpose, but it is not always that straightforward. Suppose you had to study patterns of ethnic segregation: what do you think would be the most relevant scale(s)?

In this book, our main purpose is to understand the geography of the world at large and how it works, and so, inevitably, we must deal with large spatial entities. Our focus is on the world's realms and on the main regions within those realms, and in most cases we will have to forego analyses at a finer scale. For our purposes, it is the big picture that matters most.

---

*Throughout this book we will capitalize *State* when this term refers to an administrative subdivision of a country: for example, the U.S. *State* of Ohio or the Australian *State* of New South Wales. Since this term is also synonymous with country (e.g., the state of Brazil), we use the lower case when referring to such a national *state.*

WORLD GEOGRAPHIC REALMS

Ours is a globalized, interconnected world, a world of international trade and travel, migration and movement, tourism and television, financial flows and Internet traffic. It is a world that, in some contexts, has taken on the properties of a "global village"—but that village still has neighborhoods. Their names are Europe, South America, Southeast Asia, and others familiar to us all. Like the neighborhoods of a city or town, these global neighborhoods may not have sharply defined borders, but their persistence, after tens of thousands of years of human dispersal, is beyond doubt. Geographers call such global neighborhoods **geographic realms** [4]. Each of these realms possesses a particular combination of environmental, cultural, and organizational properties.

These characteristic qualities are imprinted on the landscape, giving each realm its own traditional attributes and social settings. As we come to understand the human and environmental makeup of these geographic realms, we learn not only *where* they are located but also *why they are located where they are* (a key question in geography), how they are constituted, and what their future is likely to be in our fast-changing world. Figure G-3, therefore, forms the overall framework for our investigation in this book.

Criteria for Geographic Realms

The existence and identification of world geographic realms depends on a combination of factors. Our world offers a highly complex and variable environment of large and small continents, enormous oceans and countless waterways, innumerable islands, diverse habitats and cultures, and intricate political geographies. What makes a realm depends on the circumstances, but we can still identify three main sets of criteria:

- *Physical and Human* Geographic realms are based on sets of spatial criteria. They are the largest units into which the inhabited world can be divided. The criteria on which such a broad regionalization is based include both physical (that is, natural) and human (or social) yardsticks. On the one hand, South America is a geographic realm because physically it is a continent and culturally it is comprised of comparable societies. The realm called South Asia, on the other hand, lies on a Eurasian landmass shared by several other geographic realms; high mountains, wide deserts, and dense forests combine with a distinctive social fabric to create this well-defined realm centered on India.

- *Functional* Geographic realms are the result of the interaction of human societies and natural environments, a *functional* interaction revealed by farms, mines, fishing ports, transport routes, dams, bridges, villages, and countless other features that mark the landscape. According to this criterion, Antarctica is a continent but not a geographic realm.

- *Historical* Geographic realms must represent the most comprehensive and encompassing definition of the great clusters of humankind in the world today. China lies at the heart of such a cluster, as does India. Africa constitutes a geographic realm from the southern margin of the Sahara (an Arabic word for desert) to the Cape of Good Hope and from its Atlantic to its Indian Ocean shores.

Figure G-3 displays the 12 world geographic realms based on these criteria. As we will show in more detail later, waters, deserts, and mountains as well as cultural and political shifts mark the borders of these realms. We will discuss the position of these boundaries as we examine each realm.

Delineating Realms: Boundaries and Transition Zones

Oceans and seas are the most common natural boundaries of the world's realms, such as the South Atlantic to Subsaharan Africa's west, or the North Atlantic to North America's

east. But where two geographic realms meet, **transition zones [5]**, not sharp boundaries, often mark their contacts.

We need only remind ourselves of the border zone between the geographic realm in which most of us live, North America, and the adjacent realm of Middle America. The line in Figure G-3 coincides with the boundary between Mexico and the United States, crosses the Gulf of Mexico, and then separates Florida from Cuba and the Bahamas. But Hispanic influences are strong in North America north of this boundary, and the U.S. economic influence is strong south of it. The line, therefore, represents an ever-changing zone of regional interaction. Again, there are many ties between South Florida and the Bahamas, but the Bahamas resemble a Caribbean more than a North American society. Miami has so many Cuban or Cuban-American inhabitants that it is sometimes considered the second-largest Cuban city after Havana.

In Africa, the transition zone from Subsaharan to North Africa is so wide and well defined that we have put it on the world map; elsewhere, transition zones tend to be narrower and less easily represented. Early in this second decade of the twenty-first century, such countries as Belarus (between Europe and Russia) and Kazakhstan (between Russia and Muslim Southwest Asia) lie in inter-realm transition zones. Remember, over much (though not all) of their length, borders between realms are zones of regional change.

Transition zones are fascinating spaces: it is almost as if they rebel against a clear ordering of the world's geography. They remind us that the world is a restless and contested place with shifting boundaries and changing geographic fortunes. They challenge the geographer's mapping skills and, above all, they underscore just how complex the study of geography is. As you will see, transition zones are often places of tension and/or conflict.

Geographic Realms: Dynamic Entities

Had we drawn Figure G-3 before Columbus made his voyages (1492–1504), the map would have looked different: Amerindian states and peoples would have determined the boundaries in the Americas; Australia and New Guinea would have constituted one realm, and New Zealand would have been part of the Pacific Realm. The colonization, Europeanization, and Westernization of the world changed that map dramatically. During the four decades after World War II relatively little change took place, but since 1985 far-reaching realignments have again been occurring.

Two Varieties of Realms

The world's geographic realms can be divided into two categories. The first are *monocentric* realms that are dominated by a single major political entity, in terms of territory and/or population. North America (United States), Middle America (Mexico), East Asia (China), South Asia (India), Russia, and the Austral Realm (Australia) are all

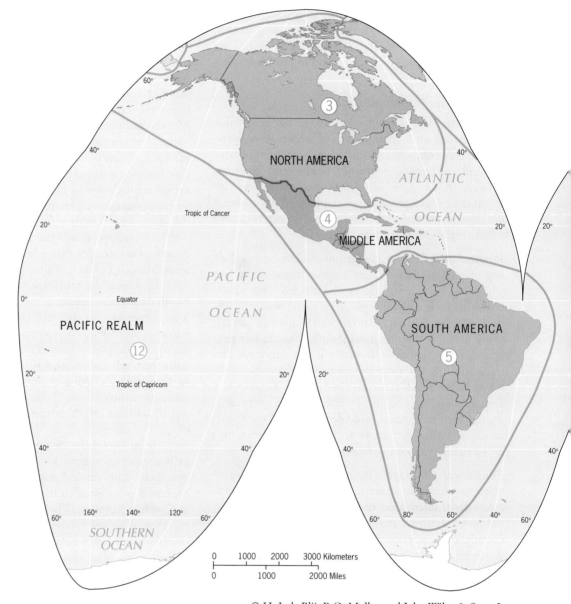

WORLD GEOGRAPHIC REALMS

1. EUROPE
2. RUSSIA
3. NORTH AMERICA
4. MIDDLE AMERICA
5. SOUTH AMERICA
6. SUBSAHARAN AFRICA
7. NORTH AFRICA/ SOUTHWEST ASIA
8. SOUTH ASIA
9. EAST ASIA
10. SOUTHEAST ASIA
11. AUSTRAL REALM
12. PACIFIC REALM

FIGURE G-3

© H. J. de Blij, P. O. Muller, and John Wiley & Sons, Inc.

monocentric realms. They are, in their entirety, heavily influenced by the presence of that one country. It is as if the realm is organized around them.

The second type of realm is *polycentric* in nature. There, the appearance, functioning, and organization of the realm are dispersed among a number of more or less equally influential regions or countries. Europe, North Africa/Southwest Asia, Subsaharan Africa, and the Pacific Realm are in this category. Polycentric realms can be more volatile in some ways, their development determined by the sum of many different parts.

Two of the world's realms are a bit more difficult to categorize. Southeast Asia is a dynamic realm that contains nearly a dozen countries, some of them regarded as emerging economies. Arguably, Indonesia is becoming the most influential power, but it may be premature to label this a monocentric realm. The other realm that seems to be in-

between is South America. Here it is Brazil that has the biggest population and increasingly the largest and most influential economy. South America, more emphatically than Southeast Asia, seems to be moving toward a monocentric organization of its realm.

Of course, some of the dominant powers in the monocentric realms influence events beyond their realm and truly demonstrate a global reach. The United States has dominated world events in an unprecedented manner since the Second World War, but more recently it has had to make way for newly emergent powers. Nowadays, the rise of China is a hot topic of debate. We will see, for example, that China's role in South America has grown rapidly while that of the United States has waned. Our discussion of the various realms will give due consideration to the influence of global trends and outside powers.

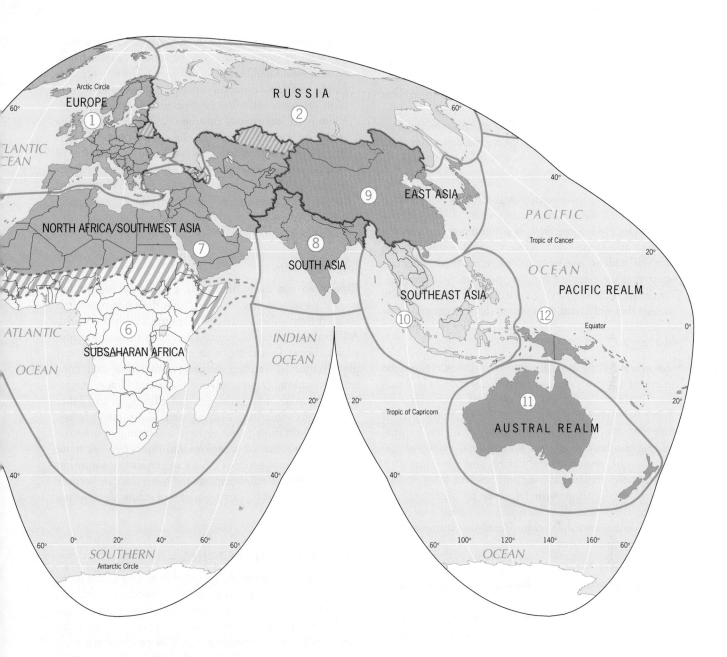

REGIONS WITHIN REALMS

The spatial division of the world into geographic realms establishes a broad global framework, but for our purposes a more refined level of spatial classification is needed. This brings us to an important organizing concept in geography: the **regional concept** [6]. To establish regions within geographic realms, we need more specific criteria.

Let us use the North American realm to demonstrate the regional idea. When we refer to a part of the United States or Canada (e.g., the South, the Midwest, or the Prairie Provinces), we employ a regional concept—not scientifically but as part of everyday communication. We reveal our *perception* of local or distant space as well as our mental image of the region we are describing.

But what exactly is the Midwest? How would you draw this region on the North American map? Regions are easy to imagine and describe, but they can be difficult to

outline on a map. One way to define the Midwest is to use the borders of States: certain States are part of this region, others are not. You could use agriculture as the chief criterion: the Midwest is where corn and/or soybeans occupy a certain percentage of the farmland. Look ahead to Figure 3B-6, where you will notice that a different name for this region is used—the Heartland—because of the differing (agricultural) criteria that define it. Each method results in a different delimitation; a Midwest based on States is different from a Midwest based on farm production or on industrial activity. Therein lies an important principle: regions are devices that allow us to make spatial generalizations, and they are based on selected criteria to help us construct them. If you were studying the geography behind politics, then a Midwest region defined by State boundaries would make sense. If you were studying agricultural distributions, you would need a different definition.

Criteria for Regions

- *Area* Given these different dimensions of the same region, we can identify properties that all regions have in common. To begin with, all regions have *area*. This observation would seem obvious, but there is more to this idea than meets the eye. Regions may be intellectual constructs, but they are not abstractions: they exist in the real world, and they occupy space on the Earth's surface.

- *Boundaries* It follows that regions have *boundaries*. Occasionally, nature itself draws sharp dividing lines, for instance along the crest of a mountain range or the margin of a forest. More often, regional boundaries are not self-evident, and we must determine them using criteria that we establish for that purpose. For example, to define a citrus-growing agricultural region, we may decide that only areas where more than 50 percent of all farmland stands under citrus trees qualify to be part of that region.

- *Location* All regions also possess *location*. Often the name of a region contains a locational clue, as in Amazon Basin or Indochina (a region of Southeast Asia lying between India and China). Geographers refer to the absolute location [7] of a place or region by providing the latitudinal and longitudinal extent of the region with respect to the Earth's grid coordinates. A more useful measure is a region's relative location [8], that is, its location with reference to other regions. Again, the names of certain regions reveal aspects of their relative locations, as in *Mainland* Southeast Asia and *Equatorial* Africa.

- *Homogeneity* Many regions are marked by a certain *homogeneity* or sameness. Homogeneity may lie in a region's human (cultural) properties, its physical (natural) characteristics, or both. Siberia, a vast region of northeastern Russia, is marked by a sparse human population that resides in widely scattered, small settlements of similar form, frigid climates, extensive areas of permafrost (permanently frozen subsoil), and cold-adapted vegetation. This dominant uniformity makes it one of Russia's natural and cultural regions, extending from the Ural Mountains in the west to the Pacific Ocean in the east. When regions display a measurable and often visible internal homogeneity, they are called formal regions [9]. But not all formal regions are visibly uniform. For instance, a region may be delimited by the area in which, say, 90 percent or more of the people speak a particular language. This cannot be seen in the landscape, but the region is a reality, and we can use that criterion to draw its boundaries accurately. It, too, is a formal region.

- *Regions as Systems* Other regions are marked not by their internal sameness but by their functional integration—that is, the way they work. These regions are defined as spatial systems [10] and are formed by the areal extent of the activities that define them. Take the case of a large city with its surrounding zone of suburbs, urban-fringe countryside, satellite towns, and farms. The city supplies goods and services to this encircling zone, and it buys farm products and other commodities from it. The city is the heart, the *core* of this region, and we call the surrounding zone of interaction the city's hinterland [11]. But the city's influence wanes on the outer periphery of that hinterland, and there lies the boundary of the functional region of which the city is the focus. A functional region [12], therefore, is usually forged by a structured, urban-centered system of interaction. It has a core and a periphery.

Interconnections

Even if we can easily demarcate a particular region and even if its boundaries are sharp, that does not mean it is isolated from other parts of the realm or even the world. All human-geographic regions are more or less *interconnected*, being linked to other regions. As we shall see, globalization is causing ongoing integration and connections among regions around the world. Trade, migration, education, television, computer linkages, and other interactions sometimes blur regional identities. Interestingly, globalization tends to have a seemingly paradoxical effect: in some ways, regions and places become more alike, more homogeneous (think of certain consumption patterns), but in other respects the contrasts can become stronger (for example, a reassertion of ethnic or religious identities).

THE PHYSICAL SETTING

This book focuses on the geographic realms and regions produced by human activity over thousands of years. But we should not forget the natural environments in which all this activity took place because we can still recognize the role of those environments in how people make their living. Certain areas of the world, for example, presented opportunities for plant and animal domestication that other areas did not. The people who happened to live in those favored areas learned to grow wheat, rice, or root crops and to domesticate oxen, goats, or llamas. We can still discern those early *patterns of opportunity* on the map in the twenty-first century. From such opportunities came adaptation and invention, and thereby arose villages, towns, cities, and states. But people living in other kinds of environments found it much harder to achieve this organization. The Americas, for instance, had no large animals that could be domesticated except llamas. This meant that societies created agricultural systems that did not involve ploughing as there were no draught animals. When Europeans introduced cows, horses, and other livestock, this change completely revolutionized the environments and cultural systems of the Western Hemisphere. The modern map carries many such imprints of the past.

From the Field Notes . . .

© AP/Wide World Photos

"Flying over Iceland's volcanic topography is to see the world in the making: this is some of the youngest rock on the planet, and even at rest you can sense its impermanence. Here nature shows us what mostly goes on deep below the surface along the mid-oceanic ridges, where tectonic plates pull apart and lava pours out of fissures and vents. When that happens on dry land, the results can be catastrophic. In the 1780s, an Icelandic volcano named Laki, in a series of eruptions, killed tens of thousands and caused a global ecological crisis. In 2010, the eruption of this far smaller volcano, Eyjafjallajökull, disrupted air travel across much of the Northern Hemisphere."

www.conceptcaching.com

Natural (Physical) Landscapes

The landmasses of Planet Earth present a jumble of **natural landscapes [13]** ranging from rugged mountain chains to smooth coastal plains (Fig. G-1). Certain continents are readily linked with a dominant physical feature—for example, North America and its Rocky Mountains, South America with its Andes and Amazon River Basin, Europe with its Alps and Rhine and Danube river basins, Asia with its Himalaya Mountains and numerous river basins, and Africa with its Sahara and Congo River Basin. Physical features have long influenced human activity and movement. Mountain ranges form barriers to movement but also channeled the spread of agricultural and technical innovations. Large deserts similarly form barriers, as do rivers, although rivers also permit accessibility and connectivity between people. River basins in Asia still contain several of the planet's largest population concentrations: the advantages of fertile soils and ample water supplies that first enabled clustered human settlement now sustain hundreds of millions in crowded South and East Asia.

As we study each of the world's geographic realms, we will find that physical landscapes continue to play significant roles in this modern world. That is one reason why the study of world regional geography is so important: it puts the human map in environmental as well as regional perspective.

Geology and Natural Hazards

Our planet may be 4.6 billion years old, but it is far from placid. As you read this chapter, Earth tremors are shaking the still-thin crust on which we live, volcanoes are erupting, storms are raging. Even the continents themselves are moving measurably, pulling apart in some areas, colliding in others. Hundreds of thousands of human lives are lost to natural calamities of this sort in almost every decade, and such calamities have at times altered the course of history.

About a century ago a geographer named Alfred Wegener, a German scientist, used spatial analysis to explain something that is obvious even from a small-scale map like Figure G-1: the apparent, jigsaw-like fit of the landmasses, especially across the South Atlantic Ocean. He concluded that the landmasses on the map are actually pieces of a supercontinent that existed hundreds of millions of years ago (he called it *Pangaea*) that drifted away when, for some reason, that supercontinent broke up. His hypothesis of **continental drift [14]** set the stage for scientists in other disciplines to search for a mechanism that might make this possible, and much of the answer to that search proved to lie in the crust beneath the ocean surface. Today we know that the continents are "rafts" of relatively light rock that rest on slabs of heavier rock called **tectonic plates [15]** (Fig. G-4) whose movement is propelled by giant circulation cells in the red-hot magma below (when this molten magma reaches the surface through volcanic vents, it is called lava).

Inevitably, moving tectonic plates collide. When they do, earthquakes and volcanic eruptions result, and the physical landscape is thrown into spectacular relief. Compare Figures G-4 and G-5, and you can see the outlines of the tectonic plates in the distribution of these hazards to human life. The January 12, 2010 earthquake adjacent to Port-au-Prince, Haiti measured 7.0 on the Richter scale. Although a shallow quake, its epicenter was located in a very densely populated area. More than 300,000 people died, a similar number were injured, and 1.3 million were

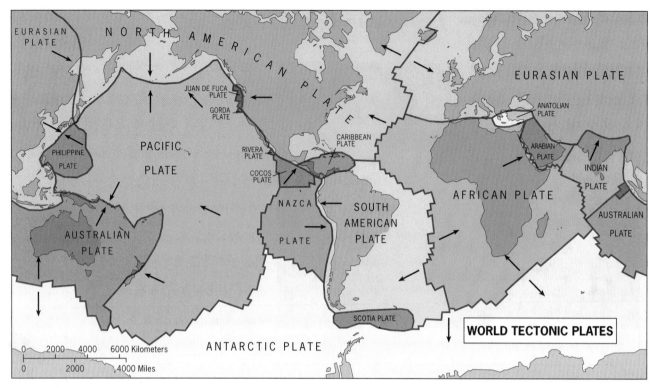

FIGURE G-4

© H. J. de Blij, P. O. Muller, and John Wiley & Sons, Inc.

made homeless and destitute. One of the Earth's oceans is almost completely encircled by active volcanoes and earthquake epicenters. Appropriately, this is called the **Pacific Ring of Fire [16]**.

It is useful to compare Figure G-5 to Figure G-3 to see which of the world's geographic realms are most susceptible to the hazards inherent in crustal instability. Russia, Europe, Africa, and Australia are relatively safe; in other realms the risks are far greater in one sector than in others (western as opposed to eastern North and South America, for instance). As we will find, for certain parts of the world the details in Figure G-5 present a clear and present danger. Some of the world's largest cities (e.g., Tokyo, Mexico City) lie in zones that are highly vulnerable to sudden disaster, as indeed occurred with the March 2011 earthquakes and tsunamis a bit north of Tokyo.

FIGURE G-5

© H. J. de Blij, P. O. Muller, and John Wiley & Sons, Inc.

Climate

The prevailing climate constitutes a key factor in the geography of realms and regions (in fact, some regions are essentially *defined* by climate). But climates change: those dominating in certain regions today may not have prevailed there several thousand years ago. Thus any map of climate, including the maps in this chapter, is but a still-picture of our always-changing world.

Climatic conditions have swung back and forth for as long as the Earth has had an atmosphere. Periodically, an ice age [17] lasting tens of millions of years chills the planet and causes massive ecological change. One such ice age occurred while Pangaea was still in one piece, between 250 and 300 million years ago. Another started about 35 million years ago, and we are still experiencing it. The current epoch of this ice age, on average the coldest yet, is called the *Pleistocene* and has been going on for nearly 2 million years.

In our time of global warming this may come as a surprise, but we should remember that an ice age is not a period of unbroken, bitter cold. Rather, an ice age consists of surges of cold, during which glaciers expand and living space shrinks, separated by warmer phases when the ice recedes and life spreads poleward again. The cold phases are called glaciations [18], and they tend to last longest, although milder spells create some temporary relief. The

truly warm phases, when the ice recedes poleward and mountain glaciers melt away, are known as interglacials [19]. We are living in one of these interglacials today. It even has a geologic name: the *Holocene*.

Imagine this: just 18,000 years ago, great icesheets had spread all the way to the Ohio River Valley, covering most of the Midwest; this was the zenith of a glaciation that had lasted about 100,000 years, the *Wisconsinan Glaciation* (Fig. G-6). The Antarctic Icesheet was bigger than ever, and even in the tropics, great mountain glaciers pushed down valleys and onto plateaus. But then Holocene warming began, the continental and mountain glaciers receded, and ecological zones that had been squeezed between the advancing icesheets now spread north and south. In Europe particularly, where humans had arrived from Africa via Southwest Asia during one of the milder spells of the Wisconsinan Glaciation, living space expanded and human numbers grew.

Global Climate Change

Today we are living in an era of global climate change [20], particularly natural global warming that has been accelerated by anthropogenic (human-source) causes. Since the Industrial Revolution, we have been emitting gases that have enhanced nature's *greenhouse effect* wherein the sun's radiation becomes trapped in the Earth's atmosphere.

FIGURE G-6

© H. J. de Blij, P. O. Muller, and John Wiley & Sons, Inc.

This is leading to a series of climate changes, especially the overall warming of the globe. In 2007, the Intergovernmental Panel on Climate Change (IPCC) released an important series of documents that present indisputable scientific evidence on global climate change and the possible implications of these changes. The experts predict an increase of 2–3°C (3.6–5.4°F) overall for the globe, but with significant regional variability (e.g., more at higher latitudes, less at lower latitudes). Precipitation patterns are predicted to become more variable, particularly in regions where they are already seasonal.

This change in temperature may seem small but is expected to have significant impacts on global climate patterns, agricultural zones, and the quality of human lives. The full ramifications are not known, but scenarios are being modeled so that societies can confront the changes that are coming. Leaders of some countries are more skep-

tical than others, and some have already made greater adjustments than others. One of the most significant consequences of global climate change is that the Arctic icecap is melting faster than even recent models predicted, with environmental and geopolitical implications. We pick up this issue in Chapters 2A and 12.

Climate Regions

We have just learned how variable climate can be, but in a human lifetime we see little evidence of this variability. We talk about the ***weather*** (the immediate state of the atmosphere) in a certain place at a given time, but as a technical term climate [21] defines the aggregate, total record of weather conditions at a place, or in a region, over the entire period during which records have been kept.

Figure G-7 may appear very complicated, but this map is useful even at a glance. Devised long ago by Wladimir

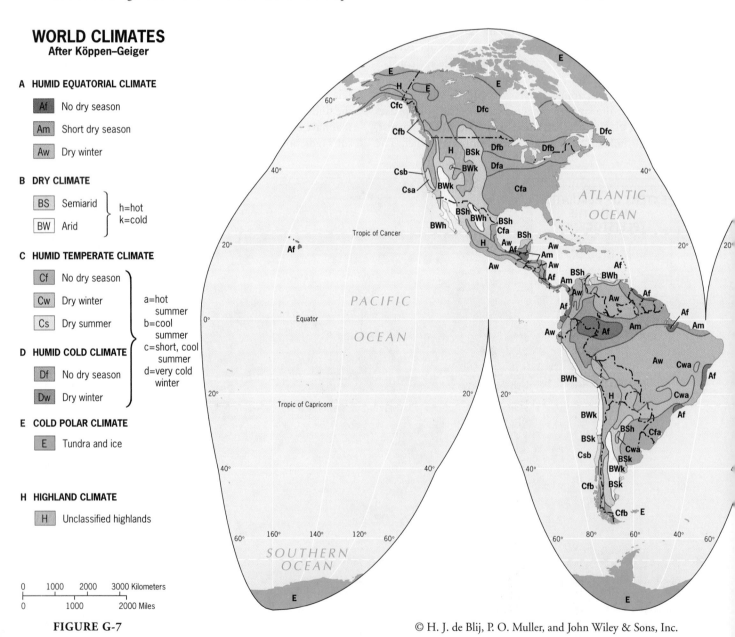

WORLD CLIMATES
After Köppen–Geiger

A HUMID EQUATORIAL CLIMATE

Af	No dry season
Am	Short dry season
Aw	Dry winter

B DRY CLIMATE

| BS | Semiarid |
| BW | Arid |

h=hot
k=cold

C HUMID TEMPERATE CLIMATE

Cf	No dry season
Cw	Dry winter
Cs	Dry summer

D HUMID COLD CLIMATE

| Df | No dry season |
| Dw | Dry winter |

a=hot summer
b=cool summer
c=short, cool summer
d=very cold winter

E COLD POLAR CLIMATE

| E | Tundra and ice |

H HIGHLAND CLIMATE

| H | Unclassified highlands |

0   1000   2000   3000 Kilometers
0   1000   2000 Miles

FIGURE G-7

© H. J. de Blij, P. O. Muller, and John Wiley & Sons, Inc.

Köppen and later modified by Rudolf Geiger, it represents climatic regions through a combination of colors and letter symbols. In the legend, note that the *A* climates (rose, orange, and peach) are equatorial and tropical; the *B* climates (tan, yellow) are dry; the *C* climates (shades of green) are temperate, that is, moderate and neither hot nor cold; the *D* climates (purple) are cold; the *E* climates (blue) are frigid; and the *H* climates (gray) prevail in highlands like the Himalayas and the Andes.

Figure G-7 merits your attention because familiarity with it will help you understand much of what follows in this book. The map has practical utility as well. Although it depicts climatic regions, daily weather in each color-coded region is relatively standard. If, for example, you are familiar with the weather in the large area mapped as *Cfa* in the southeastern United States, you will feel at home in Uruguay (South America), Kwazulu-Natal (South Africa), New

South Wales (Australia), and Fujian Province (China). Let us look at the world's climatic regions in some detail.

Humid Equatorial (A) Climates The humid equatorial, or tropical, climates are characterized by high temperatures all year and by heavy precipitation. In the *Af* subtype, the rainfall arrives in substantial amounts every month; but in the *Am* areas, the arrival of the annual wet **monsoon** (the Arabic word for "season" [see Chapter 8A]) marks a sudden enormous increase in precipitation. The *Af* subtype is named after the vegetation that develops there—the tropical rainforest. The *Am* subtype, prevailing in part of peninsular India, in a coastal area of West Africa, and in sections of Southeast Asia, is appropriately referred to as the monsoon climate. A third tropical climate, the savanna (*Aw*), has a wider daily and annual temperature range and a more strongly seasonal distribution of rainfall.

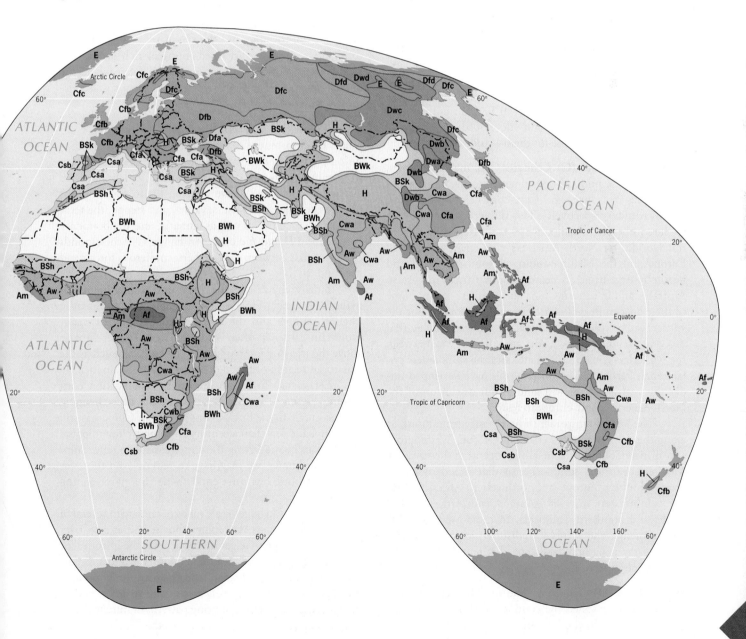

Savanna rainfall totals tend to be lower than those in the rainforest zone, and savanna seasonality is often expressed in a "double maximum." Each year produces two periods of increased rainfall separated by pronounced dry spells. In many savanna zones, inhabitants refer to the "long rains" and the "short rains" to identify those seasons; a persistent problem is the unpredictability of the rain's arrival. Savanna soils are not among the most fertile, and when the rains fail hunger looms. Savanna regions are far more densely peopled than rainforest areas, and millions of residents of the savanna subsist on what they cultivate. Rainfall variability is their principal environmental problem.

Dry (B) Climates Dry climates occur in both lower and higher latitudes. The difference between the *BW* (true *desert*) and the moister *BS* (semiarid *steppe*) varies but may be taken to lie at about 25 centimeters (10 in) of annual precipitation. Parts of the central Sahara in North Africa receive less than 10 centimeters (4 in) of rainfall. Most of the world's arid areas have an enormous daily temperature range, especially in subtropical deserts. In the Sahara, there are recorded instances of a maximum daytime shade temperature of more than 50°C (122°F) followed by a nighttime low of less than 10°C (50°F). Soils in these arid areas tend to be thin and poorly developed; soil scientists have an appropriate name for them—aridisols.

Humid Temperate (C) Climates As the map shows, almost all these mid-latitude climate areas lie just beyond the Tropics of Cancer and Capricorn (23.5° North and South latitude, respectively). This is the prevailing climate in the southeastern United States from Kentucky to central Florida, on North America's west coast, in most of Europe and the Mediterranean, in southern Brazil and northern Argentina, in coastal South Africa, in eastern Australia, and in eastern China and southern Japan. None of these areas suffers climatic extremes or severity, but the winters can be cold, especially away from water bodies that moderate temperatures. These areas lie midway between the winterless equatorial climates and the summerless polar zones. Fertile and productive soils have developed under this regime, as we will note in our discussion of the North American and European realms.

The humid temperate climates range from moist, as along the densely forested coasts of Oregon, Washington, and British Columbia, to relatively dry, as in the so-called Mediterranean (dry-summer) areas that include not only coastal southern Europe and northwestern Africa but also the southwestern tips of Australia and Africa, central Chile, and Southern California. In these Mediterranean environments, the scrubby, moisture-preserving vegetation creates a natural landscape different from that of richly green western Europe.

Humid Cold (D) Climates The humid cold (or "snow") climates may be called the continental climates, for they seem to develop in the interior of large landmasses, as in the heart of Eurasia or North America. No equivalent land areas at similar latitudes exist in the Southern Hemisphere; consequently, no *D* climates occur there.

Great annual temperature ranges mark these humid continental climates, and cold winters and relatively cool summers are the rule. In a *Dfa* climate, for instance, the warmest summer month (July) may average as high as 21°C (70°F), but the coldest month (January) might average only 11°C (12°F). Total precipitation, much of it snow, is not high, ranging from about 75 centimeters (30 in) to a steppe-like 25 centimeters (10 in). Compensating for this paucity of precipitation are cool temperatures that inhibit the loss of moisture from evaporation and evapotranspiration (moisture loss to the atmosphere from soils and plants).

Some of the world's most productive soils lie in areas under humid cold climates, including the U.S. Midwest, parts of southern Russia and Ukraine, and northeastern China. The winter dormancy (when all water is frozen) and the accumulation of plant debris during the fall balance the soil-forming and enriching processes. The soil differentiates into well-defined, nutrient-rich layers, and substantial organic humus accumulates. Even where the annual precipitation is light, this environment sustains extensive coniferous forests.

Cold Polar (E) and Highland (H) Climates Cold polar (*E*) climates are differentiated into true icecap conditions, where permanent ice and snow keep vegetation from gaining a foothold, and the tundra, which may have average temperatures above freezing up to four months of the year. Like rainforest, savanna, and steppe, the term *tundra* is vegetative as well as climatic, and the boundary between the *D* and *E* climates in Figure G-7 corresponds closely to that between the northern coniferous forests and the tundra.

Finally, the *H* climates—the unclassified highlands mapped in gray (Fig. G-7)—resemble the *E* climates. High elevations and the complex topography of major mountain systems often produce near-Arctic climates above the tree line, even in the lowest latitudes such as the equatorial section of the high Andes of South America.

Let us not forget an important qualification concerning Figure G-7: this is a still-picture of a changing scene, a single frame from an ongoing film. Climate is still changing, and less than a century from now climatologists are likely to be modifying the climate maps to reflect new data. Who knows: we may have to redraw even those familiar coastlines. Environmental change is a never-ending challenge.

You will find larger-scale maps of climate in several of the regional chapters that follow, but it is useful to refer back to this Köppen-Geiger map whenever the historical or economic geography of a region or country is under discussion. The world climatic map reflects agricultural opportunities and limitations as well as climatic regimes, and as such helps explain some enduring patterns of human distribution on our planet. We turn next to this crucial topic.

REALMS OF POPULATION

Earlier we noted that population numbers by themselves do not define geographic realms or regions. Population distributions, and the functioning society that gives them common ground, are more significant criteria. That is why we can identify one geographic realm (the Austral) with less than 30 million people and another (South Asia) with more than 1.6 billion inhabitants. Neither population numbers nor territorial size alone can delimit a geographic realm. Nevertheless, the map of world population distribution shows some major clusters that are part of some specific realms (Fig. G-8).

Before we examine these clusters in some detail, remember that the world's human population now rounds off at 7.1 billion—confined to the landmasses that constitute less than 30 percent of our planet's surface, much of which is arid desert, rugged mountain terrain, or frigid tundra. (Remember too that Fig. G-8 is another still-picture of an ever-changing scene: the rapid growth of humankind continues.) After thousands of years of slow growth, world population during the nineteenth and twentieth centuries grew at an increasing rate. That rate has recently been slowing down, even imploding in some parts of the world. But consider this: it took about 17 centuries after the birth of Christ for the world to add 250 million people to its numbers; now we are adding 250 million about every three years.

Major Population Clusters

One way to present an overview of the location of people on the planet is to create a map of **population distribution** [22] (Fig. G-8). As you see in the map's legend, every dot represents 100,000 people, and the clustering of large numbers of people in certain areas as well as the near-emptiness of others is readily evident. There is a technical difference between population distribution and **population density**, which is another way of showing where people live. Density maps reveal the number of persons per unit area, requiring a different cartographic technique.

- **South Asia** The *South Asia* population cluster lies centered on India and includes its populous neighbors, Pakistan and Bangladesh. This huge agglomeration of humanity focuses on the wide plain of the Ganges River (**A** in Fig. G-8). South Asia has just become the world's largest population cluster, overtaking East Asia in 2010. A larger percentage of the people remain farmers here, although pressure on the land is greater, while agriculture is less efficient than in East Asia.

- **East Asia** Now surpassed by South Asia, the second-ranked *East Asia* population cluster lies centered on China and includes the Pacific-facing Asian coastal zone from the Korean Peninsula to Vietnam. Not long ago, we would have reported this as a dominantly rural, farming population, but rapid economic growth and associated urbanization have changed the picture. In the interior river basins of the Huang (Yellow) and Chang/Yangzi (**B** and **C** on the map), and in the Sichuan Basin between these two letters, most of the people remain farmers—and as Appendix B shows, farmers still outnumber city-dwellers in China as a whole. But the great cities of coastal and near-coastal China are attracting millions of new inhabitants, and interior cities are growing rapidly as well. Before 2020, the East Asia cluster will become more urban than rural.

- **Europe** The third-ranking population cluster, *Europe*, also lies on the Eurasian landmass but at the opposite end from China. The European cluster, including western Russia,

From the Field Notes . . .

"One early January morning in northern Vietnam, just outside the city of Ninh Binh, a girl came by on her bicycle, pulling an ox on a long rope. When I asked, through an interpreter, where she was taking the ox she replied that she had to take it to her uncle who still worked his land. The girl's own family was moving on to more urban lifestyles, with jobs in manufacturing or services. Vietnam has been changing and modernizing in recent years. Some sectors of the economy are growing vigorously and urbanization has proceeded apace. Sometimes people move to the cities, and sometimes the city comes to them (see the high-rise of the encroaching city in the left background). To the girl, it all seemed for the better. 'I like the city,' she said. She did not have much eye for the ox, but she proudly pointed to her new bicycle. To her, the future looked good."

© Jan Nijman

www.conceptcaching.com

FIGURE G-8

© H. J. de Blij, P. O. Muller, and John Wiley & Sons, Inc.

counts over 700 million inhabitants, which puts it in a class with the two larger Eurasian concentrations—but there the similarity ends. In Europe, the key to the linear, east-west orientation of the axis of population (**D** in Fig. G-8) is not a fertile river basin but a zone of raw materials for industry. Europe is among the world's most highly urbanized and industrialized realms, its human agglomeration sustained by factories and offices rather than paddies and pastures.

The three world population concentrations just discussed (South Asia, East Asia, and Europe) account for almost 4 billion of the world's 7-plus billion people. No other cluster comes close to these numbers. The next-ranking cluster, *Eastern North America*, is only about one-quarter the size of the smallest Eurasian concentrations. As in Europe, the population in this area is concentrated in major metropolitan complexes; the rural areas are now relatively sparsely settled. Geographic realms and regions, therefore, display

varying levels of **urbanization [23]**, the percentage of the total population living in cities and towns. Some regions are urbanizing much more rapidly than others, a phenomenon we will explain as we examine the world's realms.

REALMS OF CULTURE

Imagine yourself in a boat on the Nile River, headed upstream (south) from Khartoum, North Sudan. The desert sky is blue, the heat is searing. You pass villages on the shore that look much the same: low, square, or rectangular dwellings, some recently whitewashed, others gray, with flat roofs, wooden doors, and small windows. The minaret of a modest mosque may rise above the houses, and you get a glimpse of a small central square. There is very little vegetation; here and there a hardy palm tree stands in a courtyard. People on the paths wear long white robes and headgear, also white, that looks like a baseball cap without

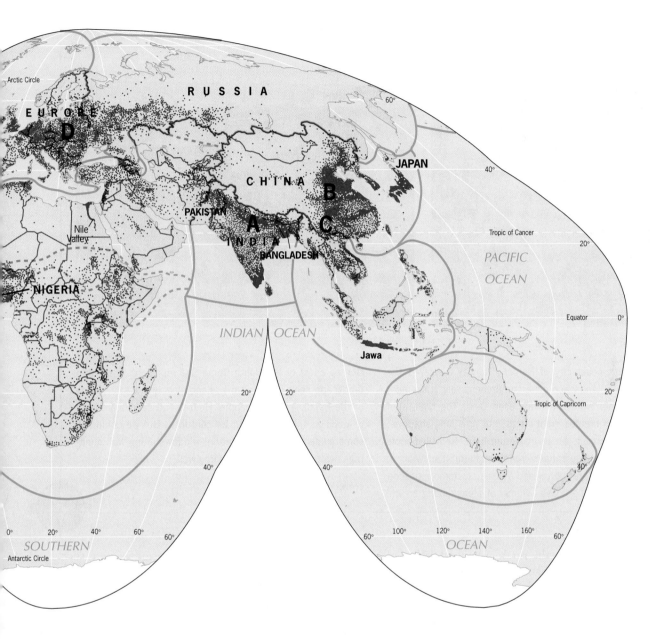

the visor. A few goats lie in the shade. Along the river's edge lie dusty farm fields that yield to the desert in the distance. At the foot of the river's bluff lie some canoes.

All of this is part of central North Sudan's rural **cultural landscape [24]**, the distinctive attributes of a society imprinted on its portion of the world's physical stage. The cultural landscape concept was initially articulated in the 1920s by a University of California geographer named Carl Sauer, who stated that "a cultural landscape is fashioned from a natural landscape by a culture group" and that "culture is the agent; the natural environment the medium." What this means is that people, starting with their physical environment and using their culture as their agency, fashion a landscape that is layered with forms such as buildings, gardens, and roads, and also modes of dress, aromas of food, and sounds of music.

Continue your journey southward on the Nile, and you will soon witness a remarkable transition. Quite suddenly, the square, solid-walled, flat-roofed houses of North

Sudan give way to round, wattle-and-thatch, conical-roofed dwellings of South Sudan. You may note that clouds have appeared in the sky: it rains more here, and flat roofs will not do. The desert has given way to green. Vegetation, natural as well as planted, grows between houses, flanking even the narrow paths. The villages seem less orderly, more varied. People ashore wear a variety of clothes, the women often in colorful dresses, the adult men in shirts and slacks, but shorts when they work the fields, although you see more women wielding hoes than men. You have traveled from one cultural landscape into another, from Arabized, Islamic Africa to animist or Christian Africa. You have crossed the boundary between two geographic realms.

No geographic realm, not even the Austral Realm, has just one single cultural landscape, but cultural landscapes help define realms as well as regions. The cultural landscape of the high-rise North American city with its sprawling suburbs differs from that of Brazil and South America; the organized terraced paddies of Southeast Asia are unlike

From the Field Notes . . .

© H. J. de Blij

"The Atlantic-coast city of Bergen, Norway displayed the Norse cultural landscape more comprehensively, it seemed, than any other Norwegian city, even Oslo. The high-relief site of Bergen creates great vistas, but also long shadows; windows are large to let in maximum light. Red-tiled roofs are pitched steeply to enhance runoff and inhibit snow accumulation; streets are narrow and houses clustered, conserving warmth . . . The coastal village of Mengkabong on the Borneo coast of the South China Sea represents a cultural landscape seen all along the island's shores, a stilt village of the Bajau, a fishing people. Houses and canoes are built of wood as they have been for centuries. But we could see some evidence of modernization: windows filling wall openings, water piped in from a nearby well."

www.conceptcaching.com

anything to be found in the rural cultural landscape of neighboring Australia. Variations of cultural landscapes *within* geographic realms, such as between highly urbanized and dominantly rural (and more traditional) areas, help us define the world's regions.

The Geography of Language

Language is the essence of culture. People tend to feel passionately about their mother tongue, especially when they believe it is threatened in some way. In the United States today, movements such as "English Only" and "English First" reflect many people's fears that the primacy of English as the national language is under threat as a result of immigration. As we will see in later chapters, some governments try to suppress the languages (and thus the cultures) of minorities in mistaken attempts to enforce national unity, provoking violent reactions.

In fact, languages emerge, thrive, and die out over time, and linguists estimate that the number of lost languages is in the tens of thousands—a process that continues. One year from the day you read this, about 25 more languages will have become extinct, leaving no trace. Just in North America, 115 native languages were lost during the past half-century. Some major ones of the past, such as Sumerian and Etruscan, have left fragments in later languages. Others, like Sanskrit and Latin, live on in their modern successors. At present, about 7000 languages remain, half of them classified by linguists as endangered. Some of the "hot spots" are the Amazon, Siberia, northern Australia, and the Andes. By the end of this century, the bulk of the world's population will be speaking just a few hundred languages, which means that many millions will no longer be speaking their ancestral mother tongues.

Scholars have tried for many years to unravel the historic roots and branches of the "language tree," and their debates continue. Geographers trying to map the outcome of this research keep having to modify the pattern, so you should take Figure G-9 as a work in progress, not the final product. At minimum, there are some 15 so-called *language families*, groups of languages with a shared but usually distant origin. The most widely distributed language family, the Indo-European (shown in yellow on the map), includes English, French, Spanish, Russian, Persian, and Hindi. This encompasses the languages of European colonizers that were carried and implanted worldwide, English most of all. Today, English serves as the national or official language of many countries and outposts, and remains the

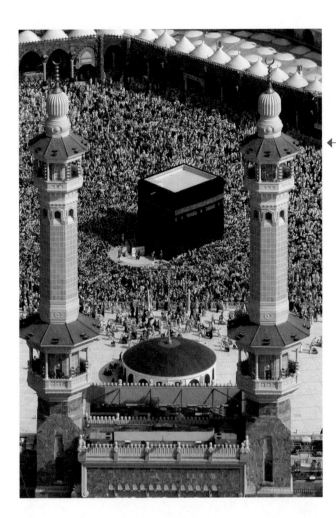

The *hajj* is the yearly pilgrimage of Muslims to the holy city of Mecca in Saudi Arabia. The pilgrimage is referred to as the fifth "pillar" of Islam, the obligation of every able-bodied Muslim to worship Allah in this holiest of sites at least once in their lifetime. This is the Grand Mosque of Mecca on November 17, 2010, as more than two million Muslim pilgrims launched into the final rituals of this largest religious pilgrimage in the world.

Mustafa Ozer/AFP/GettyImages, Inc.

gion are increasingly diffuse and dynamic. But today, still, we find that geographic realms are often dominated by a single religion or family of religions: Christianity in Europe and the Americas, Islam in North Africa/Southwest Asia, Hinduism in South Asia, and Buddhism in Southeast Asia. But the boundaries tend not to be very sharp and usually take the form of transition zones (e.g., between North and Middle America or between North and Subsaharan Africa).

A WORLD OF STATES

Take a look at a map of the countries of the world today, such as the one just inside this book's front cover, and it looks like a giant jigsaw puzzle of about 200 pieces. As noted earlier in this chapter, although we often refer to these political entities on the world map as *countries*, the appropriate geographic term for them is **states [25]**.

Getting familiar with the states of the world is not as daunting as it may seem at first because you are likely to have a head start. Territorially, Russia is the largest of all. Australia occupies a whole continent. Brazil covers nearly half of South America. It's at the other end of the scale, where small countries seem to vie for space, that recognition gets tougher. But size is no dependable criterion when it comes to importance. Think Switzerland and its banks; think Israel and its security; think Kuwait and its oil.

The Modern State

Ours is a world of states. The political territorial organization of the world in a system of states hinges on the notion of **sovereignty [26]**. It is a concept from international law which means that the government of a state rules supreme within its borders. Normally, states recognize each other's sovereignty, but this becomes a matter of contention at times of conflict and war.

In the tens of thousands of years of human history, the modern state is a relatively recent invention, and so is the international system of which it forms the cornerstone. The modern state emerged from other kinds of politico-territorial organization that have existed since the beginnings of complex civilizations. In the study of ancient history, scholars sometimes use the term "polity" or "proto-state" to indicate the difference. Ever since farm surpluses enabled the growth of large and prosperous towns, this was accompanied by the more sophisticated and centralized exercise of power and political organization. From these origins, the earliest states took shape.

lingua franca (common second language) of government, commerce, and higher education in many multicultural societies (see Fig. G-9 inset map). In the postcolonial era, English became the chief medium of still another wave of ascendancy now in progress: globalization.

But even English may eventually go the way of Latin, morphing into versions you will hear (and learn to use) as you travel, forms of English that may, generations from now, be the successors that Italian and Spanish are to Latin. In Hong Kong, Chinese and English are producing a local "Chinglish" you may hear in the first taxi you enter. In Lagos, Nigeria, where most of the people are culturally and ethnically Yoruba, a language called "Yorlish" is emerging. No map can keep up with the continuing evolution of language.

Landscapes of Religion

Religion played a crucial part in the emergence of ancient civilizations and has shaped the course of world history. Hinduism, for example, was one of the earliest religions that helped shape an entire realm (South Asia). Later, Buddhism, Christianity, and Islam emerged as major belief systems, often splitting up into various branches stretching across realms and regions. Figure G-10 shows the current distribution of world religions. Our world has become a more complicated place in recent times, and its patterns of reli-

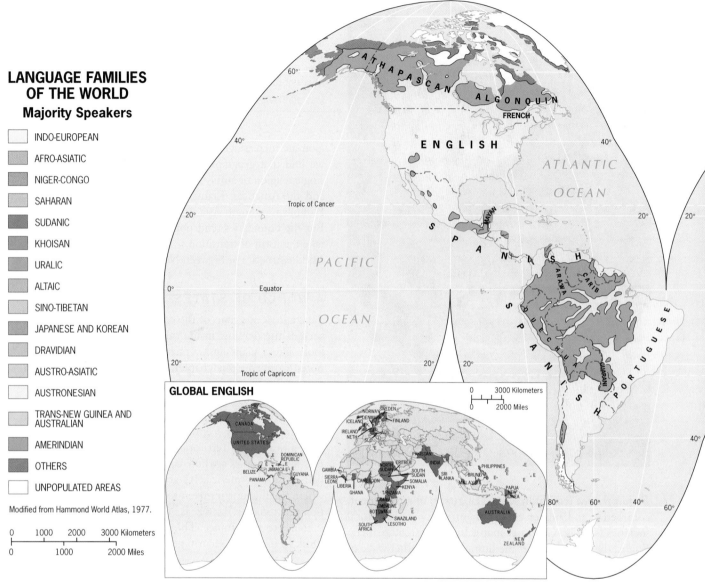

FIGURE G-9

© H. J. de Blij, P. O. Muller, and John Wiley & Sons, Inc.

Although ancient states such as the Greek city-states and the Roman Empire exhibited several qualities of modern states, it was not until the seventeenth century that European rulers and governments began to negotiate treaties that defined the state in international law. That is why the modern state is often described as based on the **European state model [27]**, with definitions of nationality and sovereignty. Often, the model assumed that state and nation were ideally conterminous, so that a **nation-state** would enclose an ethnically and culturally homogeneous people within a national boundary. That was never truly the case (even France, the "model of models," had its minorities), and today the ideal state is defined as a clearly and legally defined territory inhabited by a citizenry governed from a capital city by a representative government. As we shall discover in Chapters 1A and 1B, not even in Europe itself are all governments truly representative, but

the European state model has, for better or worse, been adopted throughout the world.

So the modern state is a historical phenomenon, and there are also signs that it may not last forever. The state system is challenged "from below" by ethnic minorities and regional secessionist movements (e.g., Kurds in Turkey, Quebec in Canada). But it is also challenged "from above" through increasingly powerful international organizations such as the European Union. There, states voluntarily transfer some of their power to "Brussels," mainly because they think it will be to their economic advantage.

Even though many states find themselves negotiating these challenges through the decentralization of authority to regional governments or through the transfer of part of their authority to international bodies, it is important to keep in mind that they do so, almost always, with the capacity to retain their powers and to seize control at their discretion.

Subdivisions of the State

For all our efforts to cooperate diplomatically (the United Nations), economically (the European Union), strategically (the North Atlantic Treaty Organization), and in other ways, it is the state and its government—not regions or realms—that holds the ***power*** and the authority to make decisions in the global arena. This is still a world of states.

Subdivisions of the State

Meanwhile, we are all too well aware that states contain subdivisions. Even the smallest states are partitioned in this manner. As all Americans—as well as Mexicans, Brazilians, and Australians—know, some larger states call their subdivisions *States*: the State of Virginia, the State of Chihuahua, the State of Minas Gerais, the State of Victoria. (As pointed out earlier, a state denotes a sovereign country whereas a [capitalized] State signifies a subdivision.) The subdivisions of other states have alternate names: provinces (Canada), regions (Italy), Autonomous Communities (Spain), Federal Districts (Russia), Divisions (Myanmar). And some of these subnational political units are becoming increasingly assertive, occasionally making their own decisions about their economic or social policies whether the central (state) government likes it or not. When that happens—in Scotland, in Catalonia, in California—we should pay even closer attention to the map.

Our analysis of the world's regional geography requires data, and it is crucial to know the origin of these data. Unfortunately, we do not have a uniformly sized grid that we can superimpose over the globe: we must depend on the world's 190-plus countries to report vital information (think of this information as "state-istics"). Fortunately, all large and populous countries tend to also provide information on each of these subdivisions when they conduct their census. So at least some data are available at a finer scale.

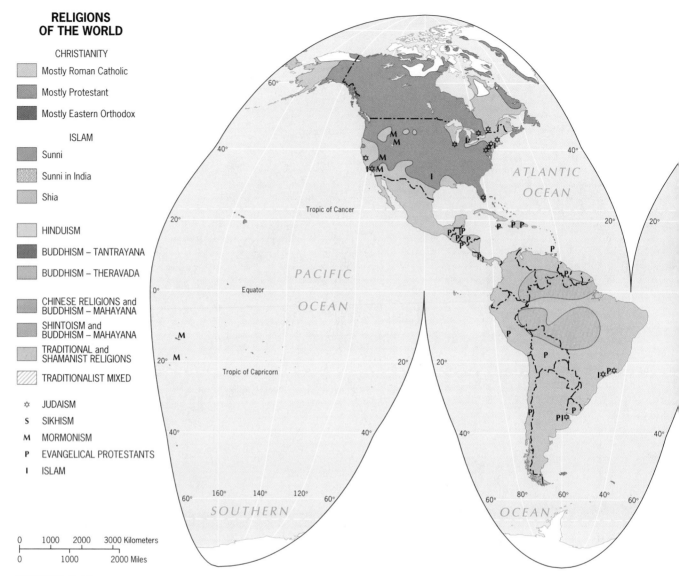

RELIGIONS OF THE WORLD

CHRISTIANITY
- Mostly Roman Catholic
- Mostly Protestant
- Mostly Eastern Orthodox

ISLAM
- Sunni
- Sunni in India
- Shia

- HINDUISM
- BUDDHISM – TANTRAYANA
- BUDDHISM – THERAVADA
- CHINESE RELIGIONS and BUDDHISM – MAHAYANA
- SHINTOISM and BUDDHISM – MAHAYANA
- TRADITIONAL and SHAMANIST RELIGIONS
- TRADITIONALIST MIXED

- ☆ JUDAISM
- S SIKHISM
- M MORMONISM
- P EVANGELICAL PROTESTANTS
- I ISLAM

| 0 | 1000 | 2000 | 3000 Kilometers |
| 0 | | 1000 | 2000 Miles |

FIGURE G-10

© H. J. de Blij, P. O. Muller, and John Wiley & Sons, Inc.

Geopolitics and the State

As we shall observe in the pages that follow, states vary not only in terms of their dimensions, relative location, domestic resource base, productive capacity, and other physical and cultural properties, but also in terms of their influence in world affairs. Napoleon once remarked that "the politics of the state lies in its geography." There is no doubt that geography is vital to state affairs and to the relations among states within realms and regions. But the actual influence of geography is far from simple and difficult to measure. There can be many different aspects to a country's geography, and sometimes they are hard to separate from cultural or economic factors.

Take size, for example. Big countries tend to be more powerful than small countries, but this is hardly a perfect relationship, and economic prowess also counts for a lot. Some relatively small countries carry a lot of weight in international affairs, while other larger and more populous ones are much less influential. The United Kingdom, for instance, is about the size of the U.S. State of Oregon and has

a population smaller than a half-dozen of the States of India, but on the global stage it is a major player and its capital, London, is one of the world's leading cities. Russia, on the other hand, has more than twice the population of the UK and is nearly double the territorial size of the United States—but its role in the world has far less impact than Britain's.

When gauging the role of geography in international affairs we must also remember that while physical geography may be "permanent" in certain ways, the *meaning* ascribed to those geographic features can change over time. Think of the impact of transport and communication technology on relative distance, for example. And it is important that we consider the role of geography in historical context as well. For instance, the origins of the European Union must be understood in the context of World War II, particularly the role of Germany during and after the war.

In the upcoming chapters we will pay due attention to the role of the United States in those realms. The United States is still the most influential state in the world, reaching far beyond its Atlantic and Pacific shores. In terms of population and territorial size, the United States may not

Adapted from E.H. Fouberg et al., *Human Geography*, 9e, based on several data sources.

be the biggest, but its economy is and its military might is unrivalled. The question, in today's highly interdependent and increasingly multipolar world, is how useful that military power really is. In the end, economic prowess and diplomatic maneuvering may matter the most. Thus America's involvement around the world will continue to be based on geopolitical considerations, especially in countries and regions deemed critical to U.S. interests.

States, Realms, and Regions

As Figures G-3 and G-11 suggest, geographic realms are mostly assemblages of states, and the borders between realms frequently coincide with the boundaries between countries—for example, between North America and Middle America along the U.S.-Mexico border. But a realm boundary can also cut *across* a state, as does the one between Subsaharan Africa and the Muslim-dominated realm of North Africa/Southwest Asia. Here the boundary takes on the properties of a wide transition zone, but it still divides states such as Nigeria, Chad, and Ethiopia. The transformation of the margins of the former Soviet

Union is creating similar cross-country transitions. Recently independent states such as Belarus (between Europe and Russia) and Kazakhstan (between Russia and Muslim Southwest Asia) lie in transition zones of regional change.

Most often, however, geographic realms consist of groups of states whose boundaries also mark the limits of the realms. Look at Southeast Asia, for instance. Its northern border coincides with the political boundary that separates China (a realm practically unto itself) from Vietnam, Laos, and Myanmar (Burma). The boundary between Myanmar and Bangladesh (which is part of the South Asian realm) defines its western border. Here, the state boundary framework helps delimit geographic realms.

The global boundary framework is even more useful in delimiting regions *within* geographic realms. We shall discuss such divisions every time we introduce a Regions of the Realm (B) chapter, but an example is appropriate here. In the Middle American realm, we recognize four regions. Two of these lie on the mainland: Mexico, the giant of the realm, and Central America, which consists of the seven comparatively small states located between Mexico

STATES AND ECONOMIES OF THE WORLD, 2009

- High income economies
- Upper-middle income economies
- Lower-middle income economies
- Low income economies
- Data unavailable

Data source: World Bank.

FIGURE G-11

© H. J. de Blij, P. O. Muller, and John Wiley & Sons, Inc.

and the Panama-Colombia border (which also marks the boundary with the South American realm).

Political Geography

To our earlier criteria of physical geography, population distribution, and cultural geography, we now add **political geography** as a shaper of world-scale geographic regions. In doing so, we should be aware that the global boundary framework continues to change. Sometimes, new boundaries are created, as between Serbia and Kosovo when Kosovo declared its independence in 2008, or between North Sudan and South Sudan in 2011. Occasionally, boundaries are eliminated as was the case between former West and East Germany in 1990. And then, of course, the meaning of boundaries can change because the countries involved agree on new political relationships; this is especially clear in the case of the European Union. But the overall system has endured.

In Chapter 12, we will discuss a recent development in boundary-making: the extension of boundaries onto and into the oceans and seas. This process has been dividing up the last of the Earth's open frontiers, with uncertain consequences.

GEOGRAPHIES OF DEVELOPMENT

Finally, as we prepare for our study of world regional geography, it is all too clear that realms, regions, and states do not enjoy the same level of prosperity. The field of **economic geography** focuses on spatial aspects of the ways people make their living and deals with patterns of production, distribution, and consumption of goods and services. As with all else in this world, these patterns reveal much variation. Individual states report the nature and value of their imports and exports, farm and factory output, and many other economic data to the United Nations

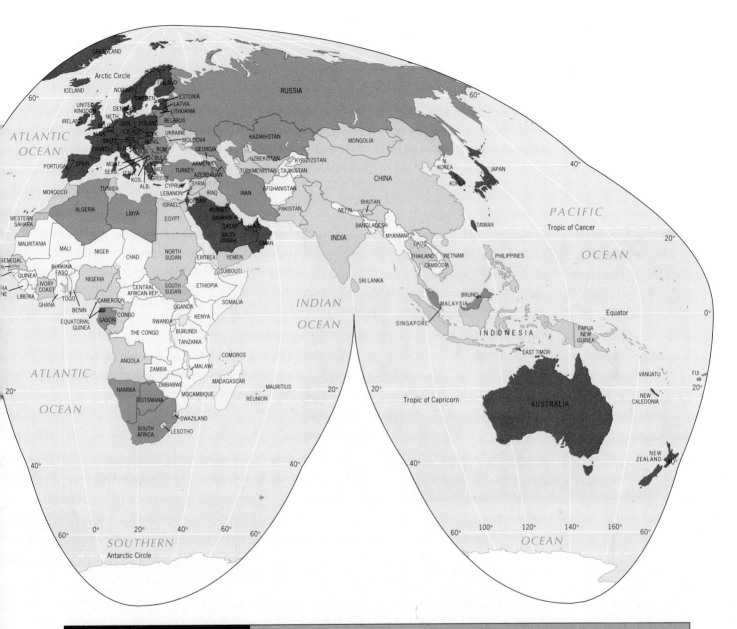

The Gini Coefficient

ECONOMIC GEOGRAPHERS STUDY regional disparities and their causes, including variations of income. But it can be difficult to determine who is earning how much across sizeable populations. Enter Corrado Gini, an Italian statistician, who made pioneering contributions—including a mathematical formula to measure the degree of dispersion of a phenomenon through a population, including economic gains. His name is forever linked to an index that reveals what proportion of a population is sharing in the wealth, and who is not.

This index ranges from 0.0 (no differences at all; everyone earns the same amount) to 1.0 (one earner takes all). A country in which a few tycoons control all the wealth and everyone else labors for a pittance will have a "GC" of close to 1; but a country with a more equitable spread of income will be much closer to 0.

As important as the actual number is the way the GC is changing. When China was under strict communist rule and before its modern economic boom began, its GC was low (nobody, of course, actually measured it). By 1993, however, it was reported to be 0.41, and today it is approaching 0.5. China's incomes are increasingly concentrated in the country's wealthier Pacific Rim. India's GC, probably already underestimated at 0.38, may be rising even faster than China's. But Brazil, long exhibiting one of the world's highest (if not the highest) GC approaching 0.6, is showing signs of a decrease partly as a result of social programs we discuss in Chapter 5B. The GC for some states is unavailable, even as an estimate. Certain governments prefer not to let the Gini out of the bottle.

From the Field Notes . . .

"February 1, 2003. Thanks to a Brazilian intermediary I was allowed to enter and spend a day in two of Rio de Janeiro's hillslope *favelas*, an eight-hour walk through one into the other. Here live millions of the city's poor, in areas often ruled by drug lords and their gangs, with minimal or no public services, amid squalor and stench, in discomfort and danger. And yet life in the older *favelas* has become more comfortable as shacks are replaced by more permanent structures, electricity is sometimes available, water supply, however haphazard, is improved, and an informal economy brings goods and services to the residents. I stood in the doorway of a resident's single-room dwelling for this over-view of an urban landscape in transition: satellite-television disks symbolize the change going on here. The often blue cisterns catch rainwater; walls are made of rough brick and roofs of corrugated iron or asbestos sheeting. No roads or automobile access, so people walk to the nearest road at the bottom of the hill. Locals told me of their hope that they will some day have legal rights to the space they occupy. Brazilian government at times expresses support for these claims, but it is complicated. As the photo shows, people live quite literally on top of one another, and mapping the chaos will not be simple (but will be made possible with geographic infor-mation systems). This would allow the gov-ernment to tax residents, but it would also allow residents to obtain loans based on the value of their *favela* properties, and bring millions of Brazilians into the formal econ-omy. The hardships I saw on this excursion were often dreadful, but you could sense the hope for and anticipation of a better future."

© H. J. de Blij

www.conceptcaching.com

and other international agencies. From such information, economic geographers can measure the comparative well-being of the world's countries (see box titled "The Gini Coefficient"). The concept of **development [28]** is used to gauge a state's economic, social, and institutional growth.

Statistics: A Caution

The concept of development as measured by data that reflect totals and averages for entire national populations entails some risks of which we should be aware from the start. When a state's economy is growing as a whole, and even when it is "booming" by comparison to other states, this does not automatically mean that every citizen is better off and the income of every worker is rising. Averages have a way of con-cealing regional variability and local stagnation. In very large states such as India and China, it is useful to assess regional, provincial, and even local economic data to discover to what extent the whole country is sharing in "development." In the case of India, we should know that the State of Maharashtra (containing the second-largest city, Mumbai) is far ahead of most others when it comes to its share of the national econ-omy. In China, the coastal provinces of the Pacific Rim far outstrip those of the interior. In Spain, the people of the Autonomous Community of Catalonia (focused on Barce-lona) delight in telling you that theirs is the most productive entity in the whole country. Hence, national (state-level) sta-tistics can conceal as much as they reveal.

Development in Spatial Perspective

Various schemes to group the world's states into economic-geographic categories have come and gone, and others will probably arise in the future. For our purposes, the classifi-cation scheme used by the World Bank (one of the agen-cies that monitor economic conditions across the globe) is the most effective. It sorts countries into four categories based on the success of their economies: (1) high-income, (2) upper-middle-income, (3) lower-middle-income, and (4) low-income. These categories, when mapped, display interesting regional clustering (see Fig. G-11). Compare this map to our global framework (Fig. G-3), and you can see the role of economic geography in the layout of the world's geographic realms. Also evident are regional boundaries within realms—for instance, between Brazil and its western neighbors, between South Africa and most of the rest of Subsaharan Africa, and between west and east in Europe.

In the year 2000, the United Nations introduced its so-called *Millennium Development Goals*: by 2015, a number of basic goals had to be reached to improve the lot of the world's poor. For example, the proportion of people forced to live on less than U.S. $1.25 per day is to be halved; children everywhere should complete primary education; and girls and boys should be educated equally. Other goals involve mortality rates, access to medicine, safe environmental conditions, and the like. Some progress has been made, but here too it is very difficult to assess well-being and development at the aggregate global scale when there is such enormous spatial variation. For instance, much of the very recent improvement in global statistics is the result of China's rapid development. More generally, whereas many countries as a whole have bettered their positions, it is increasingly the case that even "middle-income" countries contain significant percentages of very poor people. In other words, economic development becomes geographically more complex and more uneven at certain scales.

Economic geography is not the entire story, but along with factors of physical geography (such as climate), cultural geography (including resistance or receptivity to change and innovation), and political geography (history of colonialism, growth of democracy), it plays a powerful role in shaping our variable world.

As suggested above, the economic prowess of states is hugely important in their overall political influence in the world. Here, size does matter. Now have a look at impressionistic Figure G-12 which shows the countries of the world, their areal size proportionate to the size of their (estimated) national economies in 2015. This type of pseudo-map is called a **cartogram [29]**; it serves not to reflect actual political space but to blow up and shrink down countries relative to their international prominence. It makes for a powerful image of who rules in the world today. Note the new dominance of China (which is expected to surpass the United States within the next few years), India, and much of East and Southeast Asia. Observe, too, how Russia, western Asia, and most of Africa are squeezed into the background.

Uneven Development

It has been obvious for a very long time that human success on the Earth's surface has focused on certain areas and bypassed others. The earliest cities and states of the Fertile Crescent, the empires of the Incas and the Aztecs, the dominance of ancient Rome, and many other hubs of activity tell the story of development and decay, of growth and collapse. In their heyday, such centers of authority, innovation, production, and expansion were the earliest **core areas [30]**, places of dominance whose inhabitants exerted their power over their surroundings near and far. Such core areas grew rich and, in many cases, endured for long periods because their occupants skillfully exploited these surroundings— controlling and taxing the local population, forcing workers to farm the land and mine the resources at their command. This created a **periphery [31]** that sustained the core for as long as the system endured, so that core-periphery interactions, one-sided though they were, created wealth for the former and enforced stability in the latter.

Uneven development, therefore, exists at a range of scales, from the urban to the global. Indeed, most functional regions, in essence, are *spatial networks* with nodes of various centrality and importance, and this often translates into different levels of economic development. Except for a few special cases, all countries contain core areas. These national core areas are often anchored by the capital

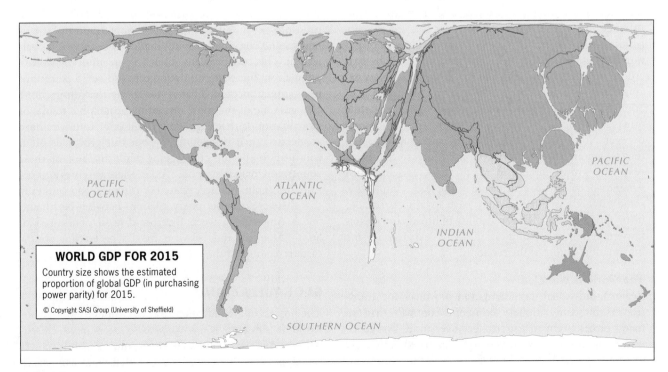

WORLD GDP FOR 2015

Country size shows the estimated proportion of global GDP (in purchasing power parity) for 2015.

© Copyright SASI Group (University of Sheffield)

FIGURE G-12

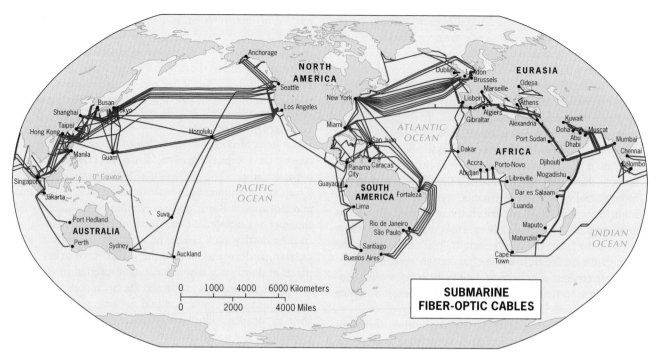

FIGURE G-13

© H. J. de Blij, P. O. Muller, J. Nijman, and John Wiley & Sons, Inc.

and/or largest city in the country: Paris (France), Tokyo (Japan), Buenos Aires (Argentina), and Bangkok (Thailand) are just a few notable examples. Larger countries may have more than one core area, such as Australia with its east and west coast cores and intervening periphery.

One useful illustration of core-periphery relations and of a global spatial network is found in Figure G-13, which shows submarine fiber-optic cables connecting different parts of the world. These cables are used for digital communications; they are much faster and more reliable than communication via satellites, and they are vital to Internet and telephone users. Consider that an ever-expanding part of the modern economy involves information and data processing and that broadband availability is crucial for many companies and public agencies. Now examine the map and notice how uneven this network is. Some realms and regions are very well connected while others are virtually excluded. Most of the connections are between North America, Europe, and parts of East Asia— that is, among core countries. Connections to Southeast Asia have grown rapidly in recent years, and in South America Brazil and Argentina are in the lead. Until 2010, some parts of East Africa did not have any direct access to fiber-optic networks. The availability of broadband creates important new opportunities and triggers expansion of the information sectors of the economy.

The World Is Not Flat!

Regions, individual states, and even provinces and counties exhibit core-periphery economic contrasts that, in many places, seem to get stronger over time. But advocates of *globalization* (a concept to be discussed in the next section) say that the old rules about exploiting cores and exploited peripheries are being reversed on a worldwide scale, and that a "flatter" and fairer world is in the offing. In his widely read 2005 book, *The World Is Flat*, Thomas Friedman argued that the world is becoming so mobile, so interconnected, and so integrated that historic barriers are falling, including those separating cores from peripheries.

But is it? The world continues to exhibit major differences in productivity and well-being. The *global core*, anchored by North America and flanked by Europe to the east and Japan and Australia to the west, not only constitutes an assemblage of the most affluent states and the most prosperous cities, but is also home to most of the financial and corporate empires that drive economic globalization. This core contains about 15 percent of the world's population, but that population earns some 75 percent of total annual income. That is not to say that things don't change. One of the most intriguing geographic results of globalization is that a growing number of countries have accelerated their development—but this growth is often confined to specific city-regions while the rest of these countries remains very poor. Today, there are cores and peripheries at almost every scale. For instance, you could say that there are core-periphery relations between North and South America, between Brazil and most of the rest of South America, and between Brazil's pacesetting southeastern region and the rest of that country.

GLOBALIZATION

The process of **globalization [32]** continues to change our world. It is essentially a *geographical* process in which spatial relations—economic, cultural, political—shift to ever broader scales. What that means is that what happens in one place has repercussions in places ever more distant, thereby integrating the entire world into an ever "smaller" global village. Globalization comes into our homes via television: news travels faster than ever before, and sometimes even government leaders turn to cable news stations to get the latest reports on international events. The process is in part driven by rapid advances in communication and transportation technologies.

Globalization is not something entirely new. The mid- and late nineteenth century, for instance, also witnessed major advances in the intensification of global interdependence. It was particularly affected by new technologies such as the steamship, the railway, and the telegraph, which subsequently were followed by the first motor vehicles and airplanes. Today's key technologies are satellite television, mobile phones, and, of course, the Internet. If the world is getting ever more interconnected, geography and our knowledge of the world's realms and regions become ever more important—because what happens elsewhere will have consequences for where *you* are.

Global Challenges, Shared Interests

Globalization plays out in various spheres, from the environmental to the cultural to the economic. Today's most pressing environmental issue, no doubt, is global warming, a threat to the world at large. It is clear that we must confront this problem together, but it is far from easy to agree on strategies: some countries are bigger polluters than others, some have more resources than others, and some are more developed than others. How to divide the burdens? At the Copenhagen Conference on Climate Change in 2009, four countries (India, China, Brazil, and South Africa) threatened to walk out of the proceedings to protest efforts by global-core countries to impose emission reductions that these four considered too damaging to their ongoing industrial development. Why should they have to sacrifice so much when the rich countries (that are now primarily *post*industrial) never had to face this kind of interference in their economic development?

Culturally, too, the world is coming ever closer together, and this is most apparent in global migration flows. Migration used to be uncommon (most people were rooted in their home environment, and that is where they lived their entire lives), and when it did occur it used to be one-way: people moved from one place to another and stayed there. In the global era, migration has picked up in terms of intensity, in part because people now possess much more knowledge about opportunities elsewhere. In addition, many more opportunities arise to travel back and forth, allowing people to maintain close ties with their original home countries. *Transnational migrants* are highly mobile and are frequently instrumental in the spreading of cultures around the world. Examples include Algerians in Paris, Haitians in Montreal, Cubans in Miami, Mexicans in Los Angeles, Indians in Singapore, and Indonesians in Sydney.

But it is important to keep in mind that, while the world is highly uneven and some parts are much better off than others, people's mobility is often restrained. Rich countries are a magnet for migrants, but very often they do not get access. Millions of workers hope to leave the periphery, which contains the world's poorest regions, to seek a better life somewhere in the core. Many of them die every year in the waters of the Mediterranean, the Caribbean, and the Atlantic. Others risk their lives at barriers and barricades that encircle the global core, from the "security fence" between Mexico and the United States to the walls that guard Israel's safety to the razor wire that encircles Spain's outposts on North Africa's shores.

When the world economy experienced a sharp downturn in 2008, unemployment skyrocketed and opportunities for migrants declined accordingly. In Europe, undocumented migration plunged 33 percent from 2008 to 2009, and along the U.S.-Mexican border the number of apprehensions fell by 23 percent during the same 12-month period. But when the economy recovers, legal and illegal migration rates are expected to increase concomitantly.

Winners and Losers

If a geographic concept can arouse strong passions, globalization is it. To leading economists, politicians, and businesspeople, this is the best of all possible worlds—the march of international capitalism, open markets, and free trade. In theory, globalization breaks down barriers to international trade, stimulates commerce, brings jobs to remote places, and promotes social, cultural, political, and other kinds of exchanges. High-tech workers in India are employed by computer firms based in California. Japanese cars are assembled in Thailand. American shoes are made in China. Fast-food restaurant chains spread standards of service and hygiene as well as familiar (and standard) menus from Tokyo to Tel Aviv to Tijuana. If wages and standards of employment are lower in peripheral countries than in the global core, production will shift there and the gap will shrink. Everybody wins. But that's only theory.

To millions of poor farmers and powerless citizens in debt-mired African and Asian countries, it represents a system that will keep them in poverty and subservience forever. Economic geographers can prove that global economic integration allows the overall economies of poorer countries to grow faster: compare their international trade to their national income, and you will find that the **gross national income (GNI)*** of those that engage in more international trade (and thus are more "globalized") rises,

---

*Gross national income (GNI) is the total income earned from all goods and services produced by the citizens of a country, within or outside of its borders, during a calendar year. **Per capita GNI** is a widely used indicator of the variation of spendable income around the globe and is reported for each country in the world in the farthest-right column of the Data Table in Appendix B.

This edition of *Regions* appears at a time when the global economy remains in recession and debate rages over the merits (and demerits) of globalization, American-style capitalism, European-style socialism, and other systems of economics and governance. As the crisis deepened, leaders of the so-called Group of 20 Countries, comprising the world's largest and most influential economies, met in London—one of the "capitals of globalization"—in April 2009 to coordinate their responses. But thousands of protestors, including anticapitalist, anti-war, and environmental activists, took to the streets of London, blocking traffic, and kindling fierce violence in the city's financial district. Barricades around bank buildings and angry slogans on banners made London the hub of global discord.

© ZHANG YUWEI/Xinhua/Landov

while the GNI of those with less international trade actually declines. The problem is that those poorest countries contain more than 2 billion people, and their prospects in this globalizing world are worsening, not rising. In countries such as the Philippines, Kenya, and Nicaragua, income per person has shrunk, and there globalization is seen as a culprit, not a cure.

And there is another, more complicated issue. Although many countries, even lesser-developed ones that were able to latch onto globalization, have witnessed accelerated economic growth and rising per capita incomes, inequality within these countries has often increased just as fast. In other words, uneven development within countries has become more pronounced. This is quite obvious in China, the fastest-growing economy in the world over the past two decades: much of this growth took place in its Pacific coastal zone, not in the interior of the country, and income differentials became ever wider. And the same is true in India and most other *emerging markets*. This is why a *regional* approach is so important to understanding what is going on in the world economy.

Globalization in the economic sphere is proceeding under the auspices of the World Trade Organization (WTO), of which the United States is the leading architect. To join, countries must agree to open their economies to foreign trade and investment. As of mid-2011, the WTO had 153 member-states, all expecting benefits from their participation. But the leading global-core countries themselves do not always oblige when it comes to creating a "level playing field." The case of the Philippines is often cited: Filipino farmers found themselves competing against North American and European producers who receive subsidies to support the production as well as the export of their products—and losing out. Meanwhile, low-priced, subsidized U.S. corn appeared on Filipino markets. As a result, the Philippine economy lost several hundred thousand farm jobs, wages went down, and WTO membership had the effect of severely damaging its agricultural sector. Not surprisingly, the notion of globalization is not popular among rural Filipinos. Nor is high-income-country protectionism confined to agriculture. When cheap foreign steel began to threaten what remained of the American steel industry in 2001, the U.S. government erected tariffs to guard domestic producers against unwanted competition.

The Future

As with all major transformations, the overall consequences of globalization are uncertain. Critics underscore that one of its outcomes is a growing gap between rich and poor, a polarization of wealth that will destabilize the world. Proponents argue that, as with the Industrial Revolution, it will take time for the benefits to spread—but that globalization's ultimate effects will be advantageous to all. Indeed, the world is functionally shrinking, and we will find evidence for this throughout this book. But the "global village" still retains its distinctive neighborhoods, and globalization did not erase their particular properties—in some cases it even sharpened the contrasts. In the chapters that follow, we use the vehicle of geography to visit and investigate them.

REALMS AND REGIONS: THE STRUCTURE OF THIS BOOK

At the beginning of this chapter, we outlined a map of the great geographic realms of the world (Fig. G-3). We then addressed the task of dividing these realms into regions, and we used criteria ranging from physical geography to economic geography. The result is Figure G-14.

From the Field Notes . . .

© H. J. de Blij

"Approaching Instanbul's European shore of the Bosphorus produced this instructive moment full of symbolism. In the far distance you can see the domed basilica of the Hagia Sophia, the Church of Holy Wisdom, built in the sixth century, when Constantinople was the center of the Byzantine Empire. On the right, the seventh-century Venetian Tower rises over the cityscape on the north side of the Golden Horn. But most of what you see here dates from modern times, when Turkey rose from the ashes of the Ottoman Empire, Constantinople was renamed Istanbul, the Asian side of the city grew exponentially, Ankara became Turkey's capital, and the trappings of globalization made this a megacity of 10.8 million whose economic gains more than compensated for its political downgrading."

www.conceptcaching.com

On this map, note that we display not only the world geographic realms but also the regions into which they subdivide. The numbers in the legend reveal the order in which the realms and regions are discussed, starting with Europe (1) and ending with the Pacific Realm (12).

Before we begin our survey, here is a summary of the 12 geographic realms and their regional components.

Europe (1)

Territorially small, politically still fragmented but economically united, Europe remains disproportionately influential in global affairs. Generally speaking, Europe's core is to the west, and it has a wide periphery curving across the realm's southern, eastern, and far northern domains.

Russia (2)

Territorially enormous and politically unified, Russia was the dominant force in the former Soviet Union that disintegrated in 1991. Undergoing a difficult transition from dictatorship to democracy and from communism to capitalism, Russia is geographically complex and changing. We define five regions: the Russian Core in the west, the Southeastern Frontier, Siberia, the Far East, and Transcaucasia.

North America (3)

Another realm in the global core, North America consists of the United States and Canada. We identify nine regions: the North American Core, the Maritime Northeast, French Canada, the South, the Southwest, the Pacific Hinge, the Western Frontier, the Continental Interior, and the Northern Frontier. Five of these regions extend across the U.S.-Canada border.

Middle America (4)

Nowhere in the world is the contrast between the global core and periphery as sharply demarcated as it is between North and Middle America. This small, fragmented realm divides into four regions: Mexico, Central America, and the Caribbean Basin's Greater and Lesser Antilles.

South America (5)

The continent of South America also defines a geographic realm in which Iberian (Spanish and Portuguese) influences dominate the cultural geography but Amerindian imprints survive. We recognize four regions: the Caribbean North, composed of Caribbean-facing states; the Andean West, with its strong Amerindian influences; the Southern Cone; and Brazil, the realm's giant.

Subsaharan Africa (6)

Between the African Transition Zone in the north and the southernmost Cape of South Africa lies Subsaharan Africa. The realm consists of five regions: Southern Africa, East Africa, Equatorial Africa, West Africa, and the African Transition Zone.

North Africa/Southwest Asia (7)

This vast geographic realm has several names, extending as it does from North Africa into Southwest and, indeed, Central Asia. It is a very complex and volatile realm, and much of this is related to its particular regional geographies. There are

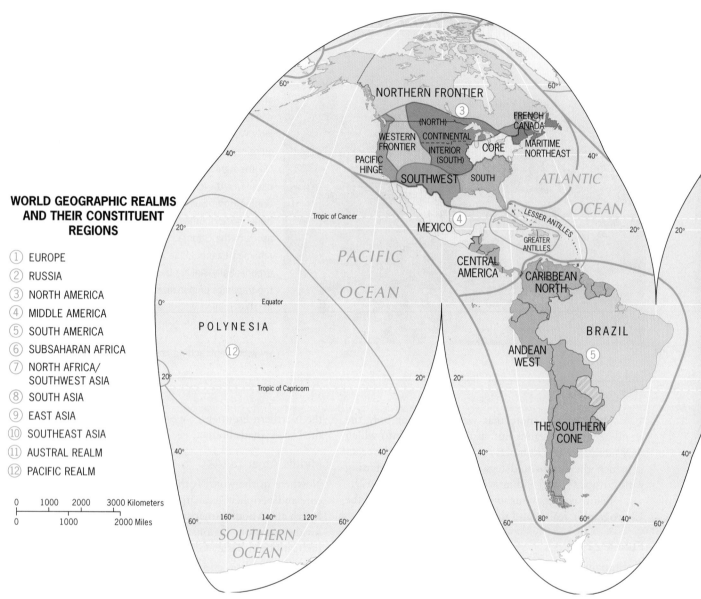

WORLD GEOGRAPHIC REALMS AND THEIR CONSTITUENT REGIONS

1. EUROPE
2. RUSSIA
3. NORTH AMERICA
4. MIDDLE AMERICA
5. SOUTH AMERICA
6. SUBSAHARAN AFRICA
7. NORTH AFRICA/ SOUTHWEST ASIA
8. SOUTH ASIA
9. EAST ASIA
10. SOUTHEAST ASIA
11. AUSTRAL REALM
12. PACIFIC REALM

FIGURE G-14

© H. J. de Blij, P. O. Muller, and John Wiley & Sons, Inc.

six regions: Egypt and the Lower Nile Basin, the Maghreb in North Africa, the Middle East, the Arabian Peninsula, and toward the east, the Empire States and Turkestan.

South Asia (8)

Physically, South Asia is one of the most clearly defined geographic realms, but has a complex cultural geography. It consists of five regions: India at the center, Pakistan to the west, Bangladesh to the east, the Mountainous North, and the Southern Islands that include Sri Lanka and the Maldives.

East Asia (9)

The vast East Asian geographic realm extends from the deserts of Central Asia to the tropical coasts of the South China Sea and from Japan to the Himalayan border with India. We identify seven regions: China's Coastal Core,

Interior, and Western Periphery; Mongolia; the Korean Peninsula; Japan; and Taiwan.

Southeast Asia (10)

Southeast Asia is a varied and intriguing mosaic of natural landscapes, cultures, and economies. Influenced by India, China, Europe, and the United States, it includes dozens of religions and hundreds of languages plus economies representing both the global core and periphery. Physically, Southeast Asia consists of a broad peninsular mainland and an offshore arc consisting of thousands of islands. The two regions (Mainland and Insular) are based on this distinction.

Austral Realm (11)

Australia and its neighbor New Zealand form the Austral geographic realm by virtue of continental dimensions,

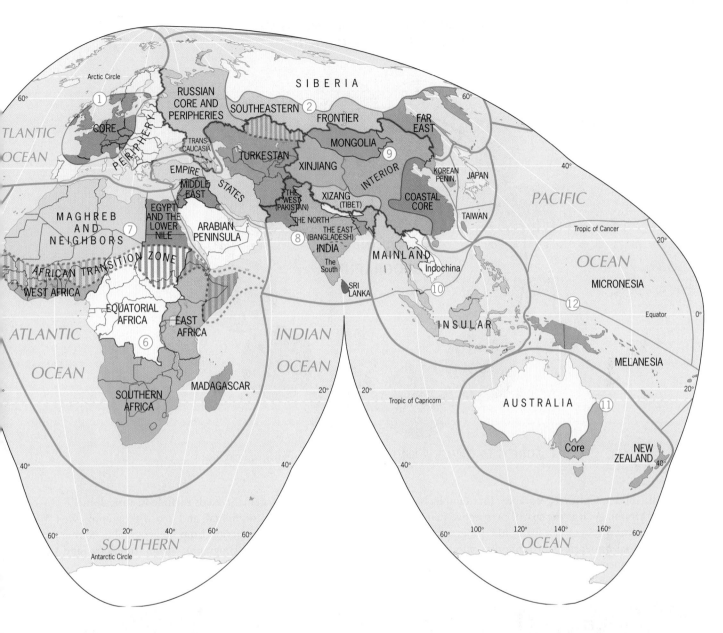

insular separation, and predominantly Western cultural heritage. The regions of this realm are defined by physical as well as cultural geography: in Australia, a highly urbanized, two-part core and a vast, desert-dominated interior; and in New Zealand, two main islands that exhibit considerable geographic contrast.

Pacific Realm (12)

The vast Pacific Ocean, larger than all the landmasses combined, contains tens of thousands of islands large and small. Dominant cultural criteria warrant three regions: Melanesia, Micronesia, and Polynesia.

As this introductory chapter demonstrates, our world regional survey is no mere description of places and areas. We have combined the study of realms and regions with a look at geography's ideas and concepts—the notions, generalizations, and basic theories that make the

discipline what it is. We continue this method in the chapters ahead so that we will become better acquainted with the world and with geography. By now you are aware that geography is a wide-ranging, multifaceted discipline. It is often described as a social science, but that is only half the story: in fact, geography straddles the divide between the social and the physical (natural) sciences. Many of the ideas and concepts you will encounter have to do with the multiple interactions between human societies and natural environments.

Regional geography allows us to view the world in an all-encompassing way. As we have seen, regional geography borrows information from many sources to create an overall image of our divided world. Those sources are not random. They represent topical or ***systematic geography.*** Research in the systematic fields of geography makes our world-scale generalizations possible. As Figure G-15 shows, these systematic fields relate closely to those of

THE RELATIONSHIP BETWEEN REGIONAL AND SYSTEMATIC GEOGRAPHY

FIGURE G-15 © H. J. de Blij, P. O. Muller, and John Wiley & Sons, Inc.

other disciplines. Cultural geography, for example, is allied with anthropology; it is the spatial perspective that distinguishes cultural geography. Economic geography focuses on the spatial dimensions of economic activity; political geography concentrates on the spatial imprints of political behavior. Other systematic fields include historical, medical, behavioral, environmental, and urban geography. We will also draw on information from biogeography, marine geography, population geography, geomorphology, and climatology (as we did earlier in this chapter).

What Do Geographers Do?

A SYSTEMATIC SPATIAL perspective and an interest in regional study are the unifying themes and enthusiasms of geography. Geography's practitioners include physical geographers, whose principal interests are the study of geomorphology (land surfaces), research on climate and weather, vegetation and soils, and the management of water and other natural resources. There also are geographers whose research and teaching concentrate on the ecological interrelationships between the physical and human worlds. They study the impact of humankind on our globe's natural environments and the influences of the environment (including such artificial contents as air and water pollution) on human individuals and societies.

Other geographers are regional specialists, who often concentrate their work for governments, planning agencies, and multinational corporations on a particular region of the world. Still other geographers—who now constitute the largest group of practitioners—are devoted to topical or systematic subfields such as urban geography, economic geography, and cultural geography (see Fig. G-15 above). They perform numerous tasks associated with the identification and resolution (through policy-making and planning) of spatial problems in their specialized areas. And, as in the past, there remain many geographers who combine their fascination for spatial questions with technical expertise.

Computerized cartography, geographic information systems (GIS), remote sensing, and even environmental engineering are among the specializations listed by the 10,000-plus professional geographers of North America. On the book's website, you will find much information on the discipline, how one trains to become a geographer, and the many exciting career options that are open to the young professional.

using GEOGRAPHY

Geographic Connections

1. If you were asked by someone without a geographic background to define the geographic characteristics of your home region, how would you respond? Where do you perceive its boundaries to lie, on what criteria do you base these, and are they sharp or transitional? (You will be tempted to draw a sketch map to support this part of your answer.) Where is your home with respect to the urban framework in your region, and do you consider yourself a resident of a city, suburbia, exurbia, or a rural hinterland? What elements of the formal and the functional can you discern in your regional experience? How would you summarize your region's natural environment(s), and, in your lifetime, have you noticed changes that might be considered trends? If so, how have these affected you?

2. Imagine that you are working in your university's study-abroad office. Students will be planning to study for a semester in numerous countries, and they know that you have some geographic experience. How would you answer their questions about the climate and weather they are likely to experience in those countries, using Figure G-7?

What should a student from Vancouver in western Canada's British Columbia expect in Auckland, New Zealand? Would a student from Charleston, South Carolina feel at home (weather-wise) in Shanghai, China? What can you tell a student from San Francisco about the weather in Cape Town, South Africa? How would you prepare a student from Phoenix, Arizona for the weather in Paris, France?

3. A well-known economist once wrote the following: "Virtually all of the rich countries of the world are outside the tropics, and virtually all the poor countries are in them . . . climate, then, accounts for quite a significant proportion of the cross-national and cross-regional disparities of world income." Comment on this assertion, using Figures G-7 and G-11 as well as text pages 30–32. How might a geographer's perspective on such an issue differ from an economist's?

Geographic Literature: A comprehensive listing of noteworthy books and recent articles on global geography, as well as some introductory works on the discipline itself, can be found in the *References and Further Readings* section of this book's website at www.wiley.com/college/deblij.

These systematic fields of geography are so named because their approach is global, not regional. Take the geographic study of cities, urban geography. Urbanization is a worldwide process, and urban geographers can identify certain human activities that all cities in the world exhibit in one form or another. But cities also display regional properties. The typical Japanese city is quite distinct from, say, the African city. Regional geography, therefore, borrows from the systematic field of urban geography, but it injects this regional perspective.

In the following chapters we call upon these systematic fields to give us a better understanding of the world's realms and regions. As a result, you will gain insights into the discipline of geography as well as the regions we investigate. This will prove that geography is a relevant and practical discipline when it comes to interpreting, comprehending, and coping with our rapidly transforming world (see box titled "What Do Geographers Do?" and the book's website for a discussion of career opportunities).

POINTS TO PONDER

- Within the next 10 years or so, China will have surpassed the United States as the biggest national economy in the world. How will it affect the world?

- By the time you read this, the global human population will have passed the 7 billion mark. Is there a limit to the Earth's carrying capacity?

- At the present time, almost 1 billion people survive on less than one U.S. dollar a day.

- Global warming is expected to cause a significant sea-level rise by the end of the twenty-first century, with estimates varying widely.

- Globalization causes the world to "shrink," but there remain marked differences among realms and regions.

IN THIS CHAPTER

- Why Europe is one of the world's most complex and dynamic realms
- Unification in the face of diversity
- The perennial question of Europe's eastern boundary
- Is Islam settling in western Europe?
- Aging Europe
- Provinces and regions rile national governments

CONCEPTS, IDEAS, AND TERMS

Physiography	1
Land hemisphere	2
City-state	3
Local functional specialization	4
Industrial Revolution	5
Nation-state	6
Nation	7
Indo-European language family	8
Complementarity	9
Transferability	10
Metropolis	11
Central business district (CBD)	12
Centrifugal forces	13
Centripetal forces	14
Supranationalism	15
Four Motors of Europe	16
Devolution	17

FIGURE 1A-1   © H. J. de Blij, P. O. Muller, and John Wiley & Sons, Inc.

It is appropriate to begin our investigation of the world's geographic realms in Europe because over the past five centuries Europe and Europeans have influenced and changed the rest of the world more than any other realm or people has done. For good or bad, much of the world would look very different today if it had not been for Europe. The realm's empires spanned the globe and transformed societies far and near. European colonialism propelled an early wave of globalization. Millions of Europeans migrated from their homelands to the Old World as well as the New, changing (and sometimes nearly obliterating) traditional communities and creating new societies from Australia to North America. Colonial power and economic incentive combined to impel the movement of millions of imperial subjects from their ancestral homes to distant lands: Africans to the Americas, Indians to Africa, Chinese to Southeast Asia, Malays to South Africa's Cape, Native Americans from east to west. In agriculture, industry, politics, and other spheres, Europe generated revolutions—and then exported those revolutions across the world, thereby consolidating the European advantage.

But throughout much of that 500-year period of European hegemony, Europe also was a cauldron of conflict. Religious, territorial, and political disputes precipitated bitter wars that even spilled over into the colonies. And during the twentieth century, Europe twice plunged the world into war. The terrible, unprecedented toll of World War I (1914–1918) was not enough to stave off World War II (1939–1945), which again drew in the United States and ended with the first-ever use of nuclear weapons in Japan. In the aftermath of that war, Europe's weakened powers lost most of their colonial possessions and a new rivalry emerged: an ideological Cold War between the communist Soviet Union and the capitalist United States. This Cold War lowered an Iron Curtain across the heart of Europe, leaving most of the east under Soviet control and most of the west in the American camp. Western Europe proved resilient, overcoming the destruction of war and the loss of colonial power to regain economic strength. Meanwhile, the Soviet communist experiment failed at home and abroad, and in 1990 the last vestiges of the Iron Curtain were lifted. Since then, a massive effort has been underway to reintegrate and reunify Europe from the Atlantic coast to the Russian border, the key geographic story of this chapter.

GEOGRAPHICAL FEATURES

As Figure 1A-1 shows, Europe is a realm of peninsulas and islands on the western margin of the world's largest landmass, Eurasia. It is a realm of just under 600 million people and 40 countries, but it is territorially quite small. Yet despite its modest proportions it has had—and continues to have—a major impact on world affairs. For many centuries Europe has been a hearth of achievement, innovation, invention, and domination.

Europe's Eastern Boundary

The European realm is bounded on the west, north, and south by Atlantic, Arctic, and Mediterranean waters, respectively. But where is Europe's eastern limit? Each episode in the historical geography of eastern Europe has left its particular legacy in the cultural landscape.

Twenty centuries ago, the Roman Empire ruled much of it (Romania is a cartographic reminder of that period); for most of the second half of the twentieth century, the Soviet Empire controlled nearly all of it. In the intervening two millennia, Christian Orthodox church doctrines spread from the southeast, and Roman Catholicism advanced from the northwest. Turkish (Ottoman) Muslims invaded and created an empire that reached the environs of Vienna. By the time the Austro-Hungarian Empire ousted the Turks, millions of eastern Europeans had been converted to Islam. Albania and Kosovo today remain predominantly Muslim countries. Meanwhile, it is often said that western Europe's civilization had its cradle in ancient Greece, but that lies farther still to the southeast, beyond the former Yugoslavia.

Eastern Europe's tumultuous history, itself an expression of the absence of clear natural boundaries, played out on a physical stage of immense diversity, its landscapes ranging from open plains and wide river basins to strategic mountains and crucial corridors. Epic battles fought centuries ago remain fresh in the minds of many people

major geographic qualities of EUROPE

1. The European geographic realm lies on the western flank of the Eurasian landmass.

2. Though territorially small, Europe is heavily populated and fragmented into 40 states.

3. Europe's enduring world influence results mainly from advantages accrued over centuries of colonial and imperial domination.

4. European natural environments are highly varied, and Europe's resource base is rich and diverse.

5. Europe's geographic diversity, cultural as well as physical, created strong local identities, specializations, and opportunities for trade and commerce.

6. European nation-states, based in durable and powerful core areas, survived the loss of colonies and evolved into modern democratic states.

7. Europe's states are engaged in a historic effort to achieve multinational economic integration and, to a lesser degree, political coordination.

8. Europe's relatively prosperous population is highly urbanized, rapidly aging, and in demographic decline partly offset by significant immigration.

9. Local demands for greater autonomy as well as cultural challenges posed by immigration, are straining the European social fabric.

10. Despite Europe's momentous unification efforts, east-west contrasts still mark the realm's regional geography.

11. Relations between Europe and neighboring Russia have often been problematic and remain so today.

living here today; pivotal past migrations are celebrated as though they happened yesterday. Nowhere in Europe is the cultural geography as complex as in the southeastern part of the realm. Illyrians, Slavs, Turks, Magyars, and other peoples converged here from near and far. Ethnic and cultural differences kept them in chronic conflict.

As we shall see, the geographic extent of Europe has always been debatable, and today it is a particularly contentious issue in terms of European Union expansion and with respect to relations with Russia. Europe's eastern boundary is a dynamic one, and it has changed with history. Some would say that there really is no clear boundary, that Europe's atmosphere, so to speak, just gets thinner toward the east. Our definition, at the present time, places Europe's eastern boundary between Russia and its numerous European neighbors to the west. This definition is based on several geographic factors including European-Russian contrasts in territorial dimensions, geopolitical developments, cultural properties, and history.

Climate and Resources

From the balmy shores of the Mediterranean Sea to the icy peaks of the Alps, and from the moist woodlands and moors of the Atlantic fringe to the semiarid prairies north of the Black Sea, Europe presents an almost infinite range of natural environments (Fig. 1A-2).

Europe's peoples have benefited from a large and varied store of raw materials. Whenever the opportunity or need arose, the realm proved to contain what was required.

Early on, these requirements included cultivable soils, rich fishing waters, and wild animals that could be domesticated; in addition, extensive forests provided wood for houses and boats. Later, coal and mineral ores propelled industrialization. More recently, Europe proved to contain substantial deposits of oil and natural gas.

Landforms and Opportunities

In the Introduction, we noted the importance of physical geography in the definition of geographic realms. The natural landscape with its array of landforms (such as mountains and plateaus) is a key element in the total physical geography—or **physiography** [1]—of any part of the terrestrial world. Other physiographic components include climate and the physical features that mark the natural landscape, such as vegetation, soils, and water bodies.

Europe's area may be small, but its physical landscapes are varied and complex. Regionally, we identify four broad units: the Central Uplands, the southern Alpine Mountains, the Western Uplands, and the North European Lowland (Fig. 1A-3).

The ***Central Uplands*** form the heart of Europe. It is a region of hills and low plateaus loaded with raw materials whose farm villages grew into towns and cities when the Industrial Revolution transformed this realm.

The ***Alpine Mountains***, a highland region named after the Alps, extend from the Pyrenees on the French-Spanish border to the Balkan Mountains near the Black

EUROPE'S CLIMATES

B DRY CLIMATE
- BS Semiarid
- BW Arid

h=hot
k=cold

C HUMID TEMPERATE CLIMATE
- Cfa
- Cfb } No dry season
- Cfc

- Csa
- Csb } Dry summer (Mediterranean)

a=hot summer
b=cool summer
c=short, cool summer
d- very cold winter

D HUMID COLD CLIMATE
- Dfa
- Dfb } No dry season
- Dfc

E COLD POLAR CLIMATE
- E Tundra and ice

H HIGHLAND CLIMATE
- H Undifferentiated highlands

0 400 800 1200 1600 Kilometers
0 200 400 600 800 1000 Miles

FIGURE 1A-2

© H. J. de Blij, P. O. Muller, and John Wiley & Sons, Inc.

Sea, and include Italy's Appennines and the Carpathians of eastern Europe.

The **Western Uplands**, geologically older, lower, and more stable than the Alpine Mountains, extend from Scandinavia through western Britain and Ireland to the heart of the Iberian Peninsula in Spain.

The **North European Lowland** extends in a lengthy arc from southeastern Britain and central France across Germany and Denmark into Poland and Ukraine, from where it continues well into Russia. Also known as the Great European Plain, this has been an avenue for human migration time after time, so that complex cultural and economic mosaics developed here together with a jigsaw-like political map. As Figure 1A-3 shows, many of Europe's major rivers and connecting waterways serve this populous region, where a number of Europe's leading cities (London, Paris, Amsterdam, Copenhagen, Berlin, Warsaw) are located.

Locational Advantages

Europe is endowed with some exceptional locational advantages. Its **relative location**, at the crossroads of the **land hemisphere [2]**, creates maximum efficiency for contact with much of the rest of the world (Fig. 1A-4). A "peninsula of peninsulas," Europe is nowhere far from the ocean and

its avenues of seaborne trade and conquest. Hundreds of kilometers of navigable rivers, augmented by an unmatched system of canals, open the interior of Europe to its neighboring seas and to the shipping lanes of the world. The Mediterranean and Baltic seas, in particular, were critical in the development of trade in early modern times, and in the emergence of Europe's early trading cities such as Venice (Italy) in the south and Lübeck (Germany) in the north.

Also consider the scale of the maps of Europe in this chapter. Europe is a realm of moderate distances and close proximities. Short distances and large cultural differences make for intense interaction, the constant circulation of goods and ideas. That has been the hallmark of Europe's geography for more than a millennium.

ANCIENT EUROPE

Modern Europe was peopled in the wake of the Pleistocene's most recent glacial retreat and global warming—a gradual warming that caused tundra to give way to deciduous forest and ice-filled valleys to turn into grassy vales. On Mediterranean shores, Europe witnessed the rise of its first great civilizations, on the islands and peninsulas of Greece and later in what is today Italy.

FIGURE 1A-3

© H. J. de Blij, P. O. Muller, and John Wiley & Sons, Inc.

Ancient Greece and Imperial Rome

Ancient Greece lay exposed to influences radiating from the advanced civilizations of Mesopotamia and the Nile Valley, and in their fragmented habitat the Greeks laid the foundations of European civilization. Their achievements in political science, philosophy, the arts, and other spheres have endured for 25 centuries. One of the essential legacies of ancient Greece involved the formation of **city-states [3]** such as Athens and Sparta: relatively small territories comprised of cities and their hinterlands that were ruled by elected governments. This is where Western democracy had its origins. But the ancient Greeks never managed to unify their domain, and their persistent conflicts proved fatal when the Romans challenged them from the west. By 147 BC, the last of the sovereign Greek intercity leagues (alliances) had fallen to the Roman conquerors.

The center of civilization and power now shifted to Rome in present-day Italy. Borrowing from Greek culture,

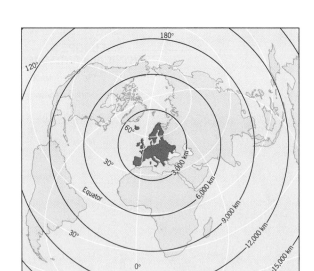

3000 kilometers=ca. 1900 miles

FIGURE 1A-4 © H. J. de Blij, P. O. Muller, and John Wiley & Sons, Inc.

the Romans created an empire that stretched from Britain to the Persian Gulf and from the Black Sea to Egypt; they made the Mediterranean Sea a Roman lake carrying armies to distant shores and goods to imperial Rome. With an urban population that probably exceeded 1 million, Rome was the first metropolitan-scale urban center in Europe.

The Romans founded numerous other cities throughout their empire and linked them to the capital through a vast system of overland and water routes, facilitating political control and enabling economic growth in their provinces. It was an unparalleled *infrastructure*, much of which long outlasted the empire itself.

Triumph and Collapse

Roman rule brought disparate, isolated peoples into the imperial political and economic sphere. By guiding (and often forcing) these groups to produce particular goods or materials, the Romans launched Europe down a road for which it would become famous: local functional specialization [4]. The workers on Elba, a Mediterranean island, mined iron ore. Those near Cartagena in Spain produced silver and lead. Certain farmers were taught irrigation to produce specialty crops. Others raised livestock for meat or wool. The *production of particular goods by particular people in particular places* became and remained a hallmark of the realm.

The Romans also spread their language across the empire, setting the stage for the emergence of the *Romance languages*; they disseminated Christianity; and they established durable systems of education, administration, and commerce. But when their empire collapsed in the fifth century AD, disorder ensued and massive migrations soon brought Germanic and Slavic peoples to their present positions on the European stage. Capitalizing on Europe's weakness, the Arab-Berber Moors from North Africa,

energized by Islam, conquered most of Iberia and penetrated France. Later the Ottoman Turks invaded eastern Europe and reached the outskirts of Vienna.

EARLY MODERN EUROPE

Europe's revival—its *Renaissance*—did not begin until the fifteenth century. After a thousand years of feudal turmoil marking the "Dark" and "Middle" Ages, powerful monarchies began to lay the foundations of modern states. The discovery of continents and riches across the oceans opened a new era of *mercantilism*, the competitive accumulation of wealth chiefly in the form of gold and silver. Best placed for this competition were the kingdoms of western Europe. Europe was on its way to colonial expansion and world domination.

Even as Europe's rising powers reached for world domination overseas, they fought with each other in Europe itself. Powerful monarchies and land-owning ("landed") aristocracies had their status and privilege challenged by ever-wealthier merchants and businesspeople. Demands for political recognition grew; cities mushroomed with the development of industries; the markets for farm products burgeoned; and Europe's population, more or less stable at about 100 million since the sixteenth century, began to increase.

The Agrarian Revolution

We know Europe as the focus of the Industrial Revolution, but before this momentous development occurred another revolution was already in progress: the *agrarian revolution*. Port cities and capital cities thrived and expanded, and their growing markets created economic opportunities for farmers. This led to revolutionary changes in land ownership and agricultural methods. Improved farm practices, better equipment, superior storage facilities, and more efficient transport to the urban markets marked a revolution in the countryside. The colonial merchants brought back new crops (the American potato soon became a European staple), and market prices rose, drawing more and more farmers into the economy.

The City-States of Early Modern Europe

From the fourteenth to the seventeenth centuries, the growing surplus production in agriculture combined with expanding interregional trade fostered the emergence of a fairly large number of city-states. These were cities that were dominated by merchant classes with powerful economic interests. The most important concentrations of city-states were on or near the Mediterranean and Baltic seas, locations that allowed for long-distance commerce. Famous examples were Venice (for some time the richest place on the face of the Earth), Genoa, and Florence, all in present-day Italy; Lübeck and Hamburg in what is now northern Germany;

and Bruges in today's Belgium. These city-states existed in a Europe in which the modern nation-state had not yet appeared and whose political geography was fragmented at the finest of scales. But in economic terms the realm was being integrated at what were then unprecedented levels; along the Baltic Sea; along the Mediterranean; along the North Sea; along the Rhine and other major rivers; and overland.

MODERN HISTORICAL GEOGRAPHY

The Industrial Revolution

The term **Industrial Revolution** [5] suggests that an agrarian Europe was suddenly swept up in wholesale industrialization that changed the realm in a few decades. In reality, seventeenth- and eighteenth-century Europe had been industrializing in many spheres, long before the chain of events known as the Industrial Revolution began. From the textiles of England and Flanders to the iron farm implements of Saxony (in present-day Germany), from Scandinavian furniture to French linens, Europe had already entered a new era of *local functional specialization*. Nonetheless, by the end of the eighteenth century, industrial Europe took off as never before.

British Primacy

Britain was at the epicenter of this revolution. In the 1780s, the Scotsman James Watt and others devised a steam-driven engine, which was soon adapted to numerous industrial applications. At about the same time, coal (converted into carbon-rich coke) was recognized as a vastly superior substitute for charcoal in smelting iron. These momentous innovations had a rapid effect. The power loom revolutionized the weaving industry. Iron smelters, long dependent on Europe's dwindling forests for fuel, could now be concentrated near coalfields. Engines could move locomotives as well as power looms. Ocean shipping entered a new age.

Britain had an enormous advantage, for the Industrial Revolution occurred when British influence reigned worldwide and the significant innovations were achieved in Britain itself. The British controlled the flow of raw materials, they held a monopoly over products that were in global demand, and they alone possessed the skills necessary to make the machines that manufactured the products. Soon the fruits of the Industrial Revolution were being exported, and the modern industrial spatial organization of Europe began to take shape. In Britain, manufacturing regions developed near coalfields in the English Midlands, at Newcastle to the northeast, in southern Wales, and along Scotland's Clyde River around Glasgow.

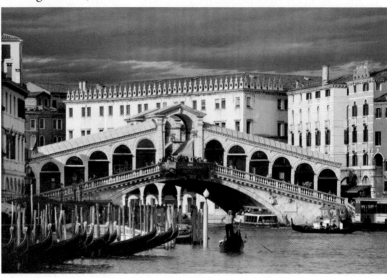

The Rialto Bridge, one of Venice's famous landmarks, dates from the time that Venice was one of Europe's most powerful and richest city-states. The first bridge over the Grand Canal at this point dates back to the twelfth century; the present span was completed in 1591.
© Visions of Our Land/The Image Bank/GettyImages, Inc.

Diffusion Onto the Continent

The Industrial Revolution diffused (spread) eastward from Britain onto the European mainland throughout the middle and late nineteenth century (Fig. 1A-5). Population skyrocketed, emigration mushroomed, and industrializing cities burst at the seams. European states already had acquired colonial empires before this revolution started; now colonialism gave Europe an unprecedented advantage in its dominance over the rest of the world.

In mainland Europe, a belt of major coalfields extends from west to east, roughly along the southern margins of the North European Lowland, due eastward from southern England across northern France and Belgium, Germany (the Ruhr), western Bohemia in the Czech Republic, Silesia in southern Poland, and the Donets Basin (Donbas) in eastern Ukraine. Iron ore is found in a broadly similar belt and together with coal provides the key raw material for the manufacturing of steel.

As in Britain, this cornerstone industry now spawned new concentrations of economic activity, growing steadily as millions migrated from the countryside to fill expanding employment opportunities. Densely populated and heavily urbanized, these emerging agglomerations became the backbone of Europe's world-scale population cluster (as shown in Fig. G-8).

Two centuries later, this east-west axis along the coalfield belt remains a major feature of Europe's population distribution map (Fig. 1A-6). It should also be noted that while industrialization produced new cities, another set of manufacturing zones arose in and near many existing urban centers. London—already Europe's leading urban focus and Britain's richest domestic market—was

FIGURE 1A-5

© H. J. de Blij, P. O. Muller, and John Wiley & Sons, Inc.

typical of these developments. Many local industries were established here, taking advantage of the large supply of labor, the ready availability of capital, and the proximity of so great a number of potential buyers. Although the Industrial Revolution thrust other places into prominence, London did not lose its primacy: industries in and around the British capital multiplied.

Political Revolutions

The Industrial Revolution unfolded against the backdrop of the ongoing formation of national states, a process that had been under way since the 1600s. Europe's political revolutions took many different forms and affected diverse peoples and countries, but in general it involved the ongoing centralization of power by royal courts and the usurpation of the former city-states in larger national territories. This was often a particularly violent reconfiguration of the realm's political geography. Historians often point to the Peace (Treaty) of Westphalia in 1648 as a key step in the evolution of Europe's state system, ending decades of war and recognizing territories, boundaries, and the sovereignty of countries. But this was only the beginning. The

so-called *absolutist states*, in which monarchs held all the power and the people had few if any rights, would not last.

Competing Ideologies

One of the most dramatic episodes to follow was the French Revolution (1789–1795), which ended the era of absolutist states ruled by all-powerful monarchs. More gradual political transformations occurred in the Netherlands, Britain, and the Scandinavian countries. Other parts of Europe would for much longer remain under the control of authoritarian (dictatorial) regimes headed by monarchs or despots. By the late nineteenth century, Europe became the arena for the competing ideologies of **liberalism**, **socialism**, and **nationalism** (national spirit, pride, patriotism). The rise of extreme nationalism (***fascism***) in the twentieth century threw the realm into the most violent wars the world had ever seen.

A Fractured Map

Europe's political map has always been one of intriguing complexity. As a geographic realm, Europe occupies only about 5 percent of the Earth's land area; but that tiny area is fragmented into some 40 countries—

**EUROPE
POPULATION DISTRIBUTION: 2012**
One dot represents 50,000 persons

0   500   1000 Kilometers
0   250   500 Miles

Arctic Circle

Longitude East of Greenwich

FIGURE 1A-6

© H. J. de Blij, P. O. Muller, and John Wiley & Sons, Inc.

more than one-fifth of all the states in the world today. Therefore, when you look at Europe's political map, the question that arises is how did so small a geographic realm come to be divided into so many political entities? Europe's map is a legacy of its feudal and royal periods, when powerful kings, barons, dukes, and other rulers, rich enough to fund armies and powerful enough to extract taxes and tribute from their domains, created bounded territories in which they reigned supreme. Royal marriages, alliances, and conquests actually simplified Europe's political map. In the early nineteenth century, there still were 39 German States; a unified Germany as we know it today would not emerge until the 1870s.

State and Nation

Europe's political revolution produced a form of political-territorial organization known as the nation-state, a state embodied by its culturally distinctive population. But what is a **nation-state [6]**? The term **nation [7]** has multiple meanings. In one sense it refers to a people with a single language, a common history, a similar ethnic background. In the sense of **nationality**, it relates to legal membership in the state, that is, citizenship. Very few states today are so homogeneous culturally that the culture is conterminous with the state. Europe's prominent nation-states of a century ago—France, Spain, the United Kingdom, Italy—have become multicultural societies, their nations defined more by an intangible "national spirit" and emotional commitment than by cultural or ethnic homogeneity.

CONTEMPORARY EUROPE: A DYNAMIC REALM

Cultural Diversity

The European realm is home to peoples of numerous cultural-linguistic stocks, including not only Latins, Germanics, and Slavs but also minorities such as Finns, Magyars (Hungarians), Basques, and Celts. This diversity of ancestries continues to be an asset as well as a liability. It has generated not only interaction and exchange, but also conflict and war.

It is worth remembering that Europe's territory is just over 60 percent the size of the United States, but that the population of Europe's 40 countries is about twice as large as America's. This population of nearly 600 million speaks numerous languages, almost all of which belong to the **Indo-European language family [8]** (Figs. 1A-7, G-9). But most of those languages are not mutually understandable; some, such as Finnish and Hungarian, are not even members of the Indo-European family. When Europe's unification efforts began after World War II, one major problem was to determine which languages to recognize as "official." That problem still prevails, although English has become the realm's unofficial **lingua franca** (common language). During a visit to Europe, though, you would find that English is more commonly usable in the big cities than in the countryside, and more in western Europe than farther east. Europe's multilingualism remains a rich cultural legacy, but is also a barrier to integration.

Another divisive force confronting Europeans involves religion. Europe's cultural heritage is steeped in Christian traditions, but sectarian strife between Catholics and Protestants, that plunged parts of the realm into bitter and widespread conflict, still divides communities and, as until recently in Northern Ireland, can still arouse violence. Some political parties still carry the name "Christian," for example, Germany's Christian Democrats.

More generally, Christianity has gradually lost adherents since secularization gathered momentum in the late 1960s, especially in western Europe. The Roman Catholic church, long very powerful in much of Europe to the point that it could be considered a "state within the state," has been losing its grip on society, and many church institutions—schools, universities, unions, political parties, charities, clubs—have been hollowed out. Moreover, churches have closed down or are now used for such other purposes as art exhibitions, public meetings, and even corporate offices.

Today, a new factor roils the religious landscape: the rise of Islam. In southeastern Europe, this takes the form of new Islamic assertiveness in an old Muslim bastion: the (Turkish) Ottoman Empire left behind millions of converts from Bosnia to Bulgaria among whom many are demanding greater political representation and power. In the west, this Islamic resurgence results from the relatively recent infusion of millions of Muslim immigrants from North Africa and other parts of the Islamic world. Here, as mosques overflow with the faithful, churches stand nearly empty as secularism among Europeans, as noted above, is on the rise. Some scholars have even gone so far as to describe Europe as having entered a "post-Christian" era.

For so small a realm, Europe's cultural geography is sharply varied. The popular image of Europe tends to be formed by British pageantry, the French wine country, or historic cities such as Venice or Amsterdam—but go beyond this core area, and you will find isolated Slavic communities in the mountains facing the Adriatic, Muslim towns in poverty-mired Albania, Roma (Gypsy) villages in the interior of Romania, farmers using traditional methods unchanged for centuries in rural Poland. That map of 40 countries does not begin to reflect the diversity of European cultures.

Spatial Interaction

If not culture, then what does unify Europe? The answer lies in the realm's outstanding opportunities for productive interaction. The European realm is best understood as an enormous **functional region**, an interdependent realm that is held together through highly developed spatial economic and political networks. Modern Europe has seized on the realm's abundant geographic opportunities to create a huge, intensively-used network of spatial interaction linking places, communities, and countries in countless ways. This interaction operates on the basis of two key principles.

First, regional **complementarity [9]** means that one area produces a surplus of a commodity required by another area. The mere existence of a particular resource or product is no guarantee of trade: it must be needed elsewhere. When two areas each require the other's products, we speak of double complementarity. Europe exhibits countless examples of this complementarity, from local communities to entire countries. Industrial Italy needs coal from western Europe; western Europe needs Italy's farm products.

Second, the ease with which a commodity can be transported by producer to consumer defines its **transferability [10]**. Distance and physical obstacles can raise

LANGUAGES OF EUROPE

MAJOR INDO-EUROPEAN BRANCHES	OTHER INDO-EUROPEAN BRANCHES	URALIC LANGUAGE FAMILY	

GERMANIC GROUP

WESTERN GERMANIC / NORTHERN GERMANIC
1 Dutch 5 Danish 8 Icelandic
2 German 6 Swedish 9 Faeroese
3 Frysian 7 Norwegian
4 English

ROMANCE GROUP
10 Portuguese 16 Rhaeto-Romansch
11 Spanish 17 Romanian
12 Catalan 18 Corsican-Italian
13 Provençal 19 Sardinian-Italian
14 French 20 Walloon
15 Italian

SLAVIC GROUP

WEST SLAVONIC EAST SLAVONIC SOUTH SLAVONIC
21 Polish 25 Russian 28 Slovene
22 Slovak 26 Ukrainian 29 Serbo-Croatian
23 Czech 27 Belarussian 30 Macedonian
24 Lusatian 31 Bulgarian

CELTIC GROUP
BRITANNIC GAELISH
32 Breton 34 Irish Gaelic
33 Welsh 35 Scots Gaelic

BALTIC GROUP
36 Latvian 37 Lithuanian

HELLENIC
38 Greek

THRACIAN/ILLYRIAN GROUP
39 Albanian

INDO-IRANIAN GROUP
40 Romani (dispersed)

FINNO-UGRIC GROUP
41 Finnish 44 Estonian
42 Karelian 45 Hungarian
43 Saami 46 Komi

SAMOYEDIC GROUP
47 Samoyedic

ALTAIC LANGUAGE FAMILY
TURKIC GROUP
48 Turkish

OTHER LANGUAGES
BASQUE
49 Basque

Areas with significant concentrations of other languages (usually adjacent national languages)

Boundary between languages

After Murphy, 1998.

FIGURE 1A-7 © H. J. de Blij, P. O. Muller, and John Wiley & Sons, Inc.

METROPOLITAN LONDON

Built-up area — Road
Greenbelt — Railroad
Central Business District

0   10   20   30   40 Kilometers
0   10   20 Miles

FIGURE 1A-8
© H. J. de Blij, P. O. Muller, and
John Wiley & Sons, Inc.

the cost of a product to the point of unprofitability. But Europe is small, distances are short, and Europeans have built the world's most efficient transport system of roads, railroads, and canals linking navigable rivers. Taken together, Europe's enormously diverse economic regions and its particularly efficient transportation infrastructure make for a highly interdependent economic realm.

A Highly Urbanized Realm

About three of every four Europeans live in towns and cities, an average that is far exceeded in the west (see the Data Table in Appendix B) but not yet attained in much of the east. Large cities are production centers as well as marketplaces, and they also form the crucibles of their nations' cultures. Europe's major cities tend to have long histories and are compact, and in general the European cityscape looks quite different from the American.

Seemingly haphazard inner-city street systems impede traffic; central cities may be picturesque, but they are also cramped. The urban layout of the London region (Fig. 1A-8) reveals much about the internal spatial structure of the European **metropolis** [11] (the central city plus its suburban ring). Such a *metropolitan area* remains focused on the large city at its center, especially the downtown **central business district (CBD)** [12], which is the oldest commercial part

of the urban agglomeration. European city centers tend to be more vibrant than those in the United States. They offer a mix of businesses, government functions, shopping facilities, educational and art institutions, and entertainment as well as housing for upper-income groups.

Wide residential sectors radiate outward from the CBD across the rest of the central city, often inhabited by particular income groups. Beyond the central city lies a sizeable suburban ring, but even here residential densities are much higher than those in the United States because the European tradition is one of setting aside recreational spaces (in "greenbelts") and living in apartments rather than in detached single-family houses. There also is a greater reliance on public transportation, which further concentrates the suburban development pattern. That has allowed many nonresidential activities to suburbanize as well, and today ultramodern outlying business centers increasingly compete with the CBD in many parts of urban Europe (see photo below).

If you were to be asked what city might be shown here, would Paris spring to mind? Most images of France's capital show venerable landmarks such as the Eiffel Tower or the Arc de Triomphe or other landmarks of the Old City. But this is another Paris, the district called La Défense, where Paris shows its ultramodern face. This is where Paris escapes the height restrictions and architectural limitations of the historic center, where glass-boxed highrises reflect the vibrant global city this is, and where the landmark is the "Cube," a huge open structure admired as well as reviled (as was the Eiffel Tower in its time). Set just slightly off the axis created by the Champs Elysées beyond, the Cube is the focus for the high-tech, financial, and service hub La Défense has become. Look in the distance and you will see the Arc de Triomphe appropriately (as some would view it) diminished, and the Eiffel Tower to its right, just a needle rising from the vast urban tapestry that is Paris today.

© Pascal Crapet/Stone/Getty Images, Inc.

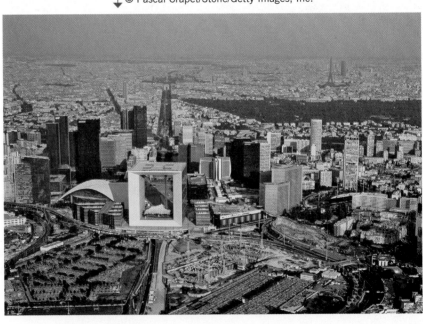

A CHANGING POPULATION

Negative Natural Population Growth

There was a time when Europe's population was (in the terminology of population geographers) exploding, sending millions to the New World and the colonies and still growing at home. But today Europe's native population, unlike most of the rest of the world's, is actually shrinking. To keep a population from declining, the (statistically) average woman must bear 2.1 children. For Europe as a whole that figure was 1.6 in 2010. Several countries recorded a number as low as 1.3, such as Germany, Portugal, Romania, and Hungary. And one eastern European country (Bosnia) even recorded 1.2—the lowest ever seen in any human population. Such negative population growth poses serious challenges for any nation. When the population pyramid becomes top-heavy, the number of workers whose taxes pay for the social services of the aged goes down, leading to reduced pensions and dwindling funds for health care.

Immigration

Meanwhile, immigration is partially offsetting Europe's population deficit. But it could be a mixed blessing, wherein demographic stability comes at the cost of social disruption. Millions of Turks, Turkish Kurds, Algerians, Moroccans, West Africans, Pakistanis, and West Indians are changing the social fabric of what once were homogeneous nation-states, especially France, Germany, the United Kingdom, Spain, Belgium, and the Netherlands. As noted earlier, one key dimension of this change is the spread of Islam in Europe. Muslim populations in eastern Europe (such as Albania's, Kosovo's, and Bosnia's) are indigenous communities converted during the period of Ottoman rule. The Muslim sectors of western European countries, on the other hand, represent more recent immigrations.

The majority of these immigrants are generally more religious than the Christian natives. They arrived in a Europe where native populations are stagnant or declining, where religious institutions are weakened, and where certain cultural norms are incompatible with Islamic traditions. Integration and assimilation of Muslim communities into the national mainstream has been slow, their education and income levels considerably lower than average. In western European countries particularly, Islamic immigrants are highly concentrated in metropolitan areas. Thus in cities like Amsterdam, London, and Brussels, the proportion of Muslim populations is considerably higher than the national average.

The Growing Multicultural Challenge

The social and political implications of Europe's cultural transformation are numerous and far-reaching. Long

From the Field Notes . . .

"After taking the train from Schiphol Airport to Rotterdam and checking into my downtown hotel, I took a train from the "Central Station" (which was in the middle of one huge construction site) to what the Dutch call "Hoek van Holland" ("Corner of Holland") on the North Sea coast, where huge ferries disembark thousands of travelers from England and elsewhere daily. A colleague had advised me to get off at Maassluis and walk the streets for a sense of the history as well as the modern changes going on here. Certainly the growing Islamic presence is transforming cultural landscapes in Europe, but not just in the cities. Even in small towns like this, the minarets of mosques rise between the spires of churches and, as is the case here, the local mosque is the centerpiece for a multipurpose Muslim cultural center that serves as a social (and political) center in ways that churches do not. A couple of generations ago, the people living in those apartments might have expected to overlook a church, but today their vista is quite different."

© H. J. de Blij

www.conceptcaching.com

The 2009 referendum in Switzerland on new minaret construction produced some ugly symbolism (note the rocket-like depiction of minarets on the Swiss flag).
© AP/Wide World Photos

EUROPEAN UNIFICATION

Realms, regions, and countries can all be subject to dividing and unifying forces that cause them to become more or less cohesive and stable political units. Where the region or country is home to diverse populations with different political agendas, it may prove difficult to avoid divergence and territorial fragmentation. Political geographers use the term **centrifugal forces [13]** to identify and measure the strength of such division, which may result from religious, racial, linguistic, political, economic, or other regional factors.

Centrifugal forces are measured against **centripetal forces [14]**, the binding, unifying glue of the state or region. General satisfaction with the system of government and administration, legal institutions, and other functions of the state (notably including its treatment of minorities) can ensure stability at the state level. In the case of the former Yugoslavia, the centrifugal forces unleashed after the end of the Cold War exceeded the weak centripetal forces in that relatively young state, and it disintegrated with dreadful consequences.

Since the Second World War, Europe has witnessed a steady process of integration and unification at a much broader geographic scale. A growing majority of European states and their leaders recognize that closer association and regional coordination form the key to a more stable, prosperous, and secure future. A realmwide union has been in the making now for more than a half-century. Despite tenacious challenges and occasional setbacks, centripetal forces have prevailed.

Background

At the end of World War II, much of Europe lay shattered, its cities and towns devastated, its infrastructure wrecked, its economies ravaged. If this was one of the world's most developed economic realms at the beginning of the twentieth century, its economic prowess had been almost completely destroyed by 1945. One of the primary motives for integration and collaboration among western European countries, therefore, was rapid economic recovery.

The United States played a leading role, initially, in spurring on such cooperation. In 1947, U.S. Secretary of State George C. Marshall proposed a European Recovery Program designed to help counter all this dislocation and to create stable political conditions in which democracy would survive. Over the next four years, the United States provided about $13 billion in assistance to Europe (well over $100 billion in today's money). The United States was driven by economic and political motives. It had become the largest producer in the world of manufactured goods and was eager to restore European markets. Politically, the ending of World War II witnessed accelerating tensions with the Soviet Union, which had taken control over the bulk of eastern Europe. In addition, communist parties seemed poised to dominate the

known for tolerance and openness, European societies are now trying to restrict immigration in various ways, and political parties with anti-immigrant platforms are gaining ground. The new Islamic presence has been a cause of social tension in certain countries, sometimes seriously but occasionally giving way to populist political theatrics. In a controversial referendum in 2009, the Swiss electorate approved a law that prohibits the construction of minarets. A majority of the Swiss seemed to feel that their culture was under threat from Islam and that this type of mosque architecture gave expression to an overly assertive minority—even though there were only four minarets in the entire country. Others were strongly opposed to the law, arguing that the petition organizers were playing on ungrounded fears and xenophobia, and that such a law would violate basic religious freedoms. No doubt, Switzerland's 400,000 Muslims (5 percent of the population, mainly Turks, and not known to be particularly zealous) must have felt uneasy about the vote. The posters suspended above were distributed by organizations in favor of the law and plastered on walls across Switzerland in the weeks prior to the referendum.

SUPRANATIONALISM IN EUROPE

1944 Benelux Agreement signed.

1947 Marshall Plan created (effective 1948–1952).

1948 Organization for European Economic Cooperation (OEEC) established.

1949 Council of Europe created.

1951 European Coal and Steel Community (ECSC) Agreement signed (effective 1952).

1957 Treaty of Rome signed, establishing European Economic Community (EEC) (effective 1958), also known as the Common Market and "The Six." European Atomic Energy Community (EURATOM) Treaty signed (effective 1958).

1959 European Free Trade Association (EFTA) Treaty signed (effective 1960).

1961 United Kingdom, Ireland, Denmark, and Norway apply for EEC membership.

1963 France vetoes United Kingdom EEC membership; Ireland, Denmark, and Norway withdraw applications.

1965 EEC–ECSC–EURATOM Merger Treaty signed (effective 1967).

1967 European Community (EC) inaugurated.

1968 All customs duties removed for intra-EC trade; common external tariff established.

1973 United Kingdom, Denmark, and Ireland admitted as members of EC, creating "The Nine." Norway rejects membership in the EC by referendum.

1979 First general elections for a European Parliament held; new 410-member legislature convenes in Strasbourg. European Monetary System established.

1981 Greece admitted as member of EC, creating "The Ten."

1985 Greenland, acting independently of Denmark, withdraws from EC.

1986 Spain and Portugal admitted as members of EC, creating "The Twelve." Single European Act ratified, targeting a functioning European Union in the 1990s.

1987 Turkey and Morocco make first application to join EC. Morocco is rejected; Turkey is told that discussions will continue.

1990 Charter of Paris signed by 34 members of the Conference on Security and Cooperation in Europe (CSCE). Former East Germany, as part of newly reunified Germany, incorporated into EC.

1991 Maastricht meeting charts European Union (EU) course for the 1990s.

1993 Single European Market goes into effect. Modified European Union Treaty ratified, transforming EC into EU.

1995 Austria, Finland, and Sweden admitted into EU, creating "The Fifteen."

1999 European Monetary Union (EMU) goes into effect.

2002 The euro is introduced as historic national currencies disappear in 12 countries.

2003 First draft of a European Constitution is published to mixed reviews from member-states.

2004 Historic expansion of EU from 15 to 25 countries with the admission of Cyprus, the Czech Republic, Estonia, Hungary, Latvia, Lithuania, Malta, Poland, Slovakia, and Slovenia.

2005 Proposed EU Constitution is rejected by voters in France and the Netherlands.

2007 Romania and Bulgaria are admitted, bringing total EU membership to 27 countries. Slovenia adopts the euro.

2008 Cyprus and Malta adopt the euro.

2009 Slovakia adopts the euro.

2010 Financial crisis strikes heavily indebted Greece, requiring massive EU bailout. Later in the year, Ireland follows raising fears that the same may happen in Portugal, Spain, and Italy. The future of the EMU is clouded; the value of the euro declines after a long rise against the dollar.

2011 Estonia adopts the euro.

---

political life of major western European countries. The United States was also intent, through the Marshall Plan, to have a firm hand in western Europe and keep communist influences at bay. The Marshall Plan applied solely to 16 European countries, including defeated (West) Germany and Turkey.

Northwestern European countries themselves were also driven by political considerations. The two world wars had clearly shown the dangers of excessive nationalism and had laid bare the devastating problems that can arise from a lack of political cooperation and collaborative efforts (if the Allies had acted as one sooner, Hitler might have been stopped before things escalated into war in 1939). From the perspective of countries like France, the Netherlands, and Belgium, one of the key issues was to control

Germany in the postwar years, and this could only be done through close political cooperation.

As the economies recovered and Germany became firmly embedded in a pan-European structure, the motives for European unification received a different emphasis. Increasingly, the process has been driven by the need to facilitate an ever larger and more efficient open market that can compete globally with the United States, Japan, and China. Politically, the goals today are more and more about stabilizing a much larger and diverse Europe that now approaches the Russian frontier. It is not just about Germany anymore but rather about a much bigger and more complicated geopolitical zone that has to accommodate a large number of (still) sovereign states and at the same time maintain good relations with the Russian giant next door.

The Unification Process

The Marshall Plan did far more than stimulate European economies. It confirmed European leaders' conclusion that their countries needed a joint economic-administrative structure. Such a structure was needed not only to coordinate the financial assistance, but also to ease the flow of resources and products across Europe's mosaic of boundaries, to lower restrictive trade tariffs, and to seek ways to improve political cooperation.

For all these needs Europe's governments had some guidelines. While in exile in Britain, the leaders of three small countries—Belgium, the Netherlands, and Luxembourg—had been discussing an association of this kind even before the end of the war. There, in 1944, they formulated and signed the Benelux Agreement, intended to achieve total economic integration. When the Marshall Plan was launched, the Benelux precedent helped speed the creation of the Organization for European Economic Cooperation (OEEC), which was established to coordinate the investment of America's aid (see the box entitled "Supranationalism in Europe" on p. 55).

Soon the economic steps led to greater political cooperation as well. In 1949, the participating governments created the Council of Europe, the beginnings of what was to become a European Parliament meeting in Strasbourg, France. Europe was embarked on still another political revolution, the formation of a multinational union involving a growing number of European states. This is a classic example of supranationalism [15], which geographers define as a voluntary association in economic, political, or cultural spheres of three or more independent states willing to yield some measure of sovereignty for their mutual benefit.

Under the Treaty of Rome, six countries joined to become the European Economic Community (EEC) in 1958, also called the "Common Market." In 1973 the United Kingdom, Ireland, and Denmark joined, and the renamed European Community (EC) now encompassed nine members. As Figure 1A-9 shows, membership reached 15 countries in 1995, after the organization had been renamed yet one more time to become the *European Union (EU)*. Since then, the number of member-states has climbed to 27.

The EU's administrative, economic, and even political framework has become so advanced that the organization now has a headquarters with many of the trappings of a capital city. Early on, EU planners chose Brussels (already the national capital of Belgium, a member of Benelux) as the organization's center of governance. And in order to avoid giving Brussels too much prominence, they chose Strasbourg, in the northeastern corner of France, as the seat of the European Parliament, whose elected membership represents all EU countries.

CONSEQUENCES OF UNIFICATION

The European Union is not just a paper organization: it has a major impact on national economies, on the role of individual states, and on the daily lives of its member-countries' citizens.

One Market

EU directives are aimed at the creation of a single market for producers and consumers, businesses and workers. Corporations should be able to produce and sell

From the Field Notes . . .

"The French like flying their flag. You see the tricolor everywhere, on public as well as private buildings and facilities. And here's another: I can't remember a country where the European Union flag is more commonplace, often flying beside (not below) the French flag as seen here at the regional headquarters in Rouen. This reflects the French government's enthusiasm for the "European Project" as they like to call it here: with the German government, these are the two key pillars of the EU. But surveys suggest that the general public in both countries is rather more ambivalent."

© H. J. de Blij

www.conceptcaching.com

FIGURE 1A-9

© H. J. de Blij, P. O. Muller, and John Wiley & Sons, Inc.

anywhere in the Union without legal impediments, whereas workers should be able to move anywhere in the EU and find employment without legal restrictions. In order to make this happen and to keep things manageable, member-states have had to harmonize a wide range of national laws from taxation to the protection of the environment to educational standards. One major step forward came with the introduction of a single central bank (with considerable power over, for instance, interest rates) and a single currency, the *euro*. The single currency also symbolizes Europe's strengthening unity and establishes a joint counterweight to the once mighty American dollar. In 2002, twelve of the (then) 15 EU

countries withdrew their currencies and began using the euro, with only the United Kingdom, Denmark, and Sweden staying out (Fig. 1A-9). Most recently, Slovenia adopted the euro in 2007, Cyprus and Malta in 2008, Slovakia in 2009, and Estonia became the seventeenth adopter in 2011.

A New Economic Geography

The establishment and continuing growth of the European Union have generated a new economic landscape that today not only transcends the old but is fundamentally reshaping the realm's regional geography. By investing heavily in new

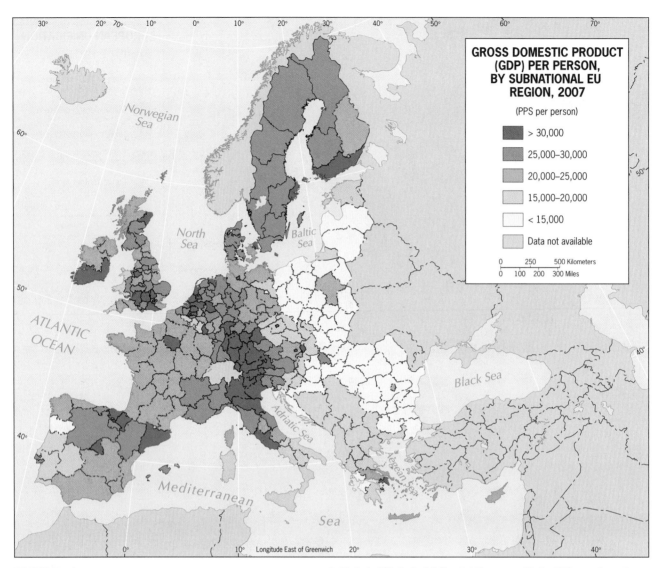

FIGURE 1A-10   © H. J. de Blij, P. O. Muller, J. Nijman, and John Wiley & Sons, Inc.

infrastructure and by smoothing the flows of money, labor, and products, European planners have dramatically reduced the divisive effects of their national boundaries. And by acknowledging demands for greater freedom of action by their provinces, States, departments, and other administrative units of their countries, European leaders unleashed a wave of economic energy that transformed some of these subnational units into powerful engines of growth (Fig. 1A-10). Four of these growth centers are especially noteworthy, to the point that geographers refer to them as the **Four Motors of Europe [16]**: (1) France's southeastern **Rhône-Alpes Region**, centered on the country's second-largest city, Lyon; (2) **Lombardy** in north-central Italy, focused on the industrial city of Milan; (3) **Catalonia** in northeastern Spain, anchored by the cultural and manufacturing center of Barcelona; and (4) **Baden-Württemberg** in southwestern Germany, headquartered by the high-tech city of Stuttgart.

Another change in regional economic geography involves agricultural activities and is closely related to deliberate policies designed in Brussels. Taxes tend to be high in Europe, and those collected in the richer member-states are used to subsidize growth and development in the less prosperous ones. This is one of the burdens of membership that is not universally popular in the EU, to say the least. But it has strengthened the economies of Portugal, Spain, and other national and regional economies to the betterment of the entire organization. Some countries also object to the terms and rules of the Common Agricultural Policy (CAP), which, according to some critics, supports farmers far too much and, according to others, far too little. (France in particular obstructs efforts to move the CAP closer to consensus, subsidizing its agricultural industry relentlessly while arguing that this protects its rural cultural heritage as well as its farmers.)

Despite economic integration and the harmonization of policies, Figure 1A-10 shows that major differences persist among Europe's regions and within EU member-states. This is an important reminder that, as noted in the introductory chapter, uneven development

Regional ISSUE — Europe: How Desirable Is Economic and Political Union?

IN SUPPORT OF UNION

"As a public servant in the French government I have been personally affected by the events of the past half-century. I am in favor of European unification because Europe is a realm of comparatively small countries, all of which will benefit from economic and political union. Europe's history is bedeviled by its fragmentation—its borders and barriers and favored cores and disadvantaged peripheries, its different economies and diverse currencies, its numerous languages and varied cultures. Now we have the opportunity and determination to overcome these divisions and to create an entity whose sum is greater than its parts, a supranational Europe that will benefit all of its members.

"In the process we will not only establish a common currency, eliminate trade tariffs and customs barriers among our members, and enable citizens of one member-state to live and work in any of the others. We will also make European laws that will supersede the old "national" laws, and create regulations that will be adhered to by all members on matters ranging from subsidies to help small farms survive to rules regarding environmental protection. In order to assist the poorer members of this European Union, the richer members will send billions of euros to the needy ones, paid for from the taxes levied on their citizens. This has already lifted Spain and Ireland from poverty to prosperity. But to ensure that 'national' economies do not get out of line, there will be rules to govern their fiscal policies, including a debt limit of 3 percent of GDP.

"The EU will require sacrifices from people and governments, but the rewards will make these worth it. Our European Parliament, with representatives from all member-countries, is the harbinger of a truly united Europe. The European Commission deals with the practical problems as we move toward unification, including the momentous EU enlargement of 2004. A half-century after the first significant steps were taken, the EU has 27 member-states including former wartime enemies, reformed communist economies, and newly independent entities. Rather than struggling alone, these member-states will form part of an increasingly prosperous and economically powerful whole.

"I hope that administrative coordination and economic union will be followed by political unification. The ultimate goal of this great movement should be the creation of a federal United States of Europe, a worthy competitor and countervailing force in a world dominated by another federation, the United States of America."

IN OPPOSITION TO THE EUROPEAN UNION

"I am a taxi driver in Vienna, Austria, and here's something to consider: the bureaucrats and big business, not the workers, are the great proponents of this European Union plan. Let me ask you this: how many member-countries' voters have had the chance to tell their governments whether they want to join the EU and submit to all those confounded regulations? It all happened before we realized the implications. At least the Norwegians had common sense: they stayed out of it, because they didn't want some international body to tell them what to do with their oil and gas and their fishing industry. But I can tell you this: if the workers of Europe had had the chance to vote on each step in this bureaucratic plot, we'd have ended it long ago.

"So what are we getting for it? Higher prices for everything, higher taxes to pay for those poor people in southern Italy and needy Portugal, costly environmental regulations, and—here's the worst of it—a flood of cheap labor and an uncontrolled influx of immigrants. Notice that the recent 2004 expansion wasn't voted on by the Germans, and do you know why? Because the German government knew well and good that the people wouldn't support this ridiculous scheme. This whole European Union program is concocted by the elite, and we have to accept it whether we like it or not.

"Does make me wonder, though. Do these EU planners have any notion of the real Europe they're dealing with? They talk about how those 'young democracies' joining in 2004 will have their democratic systems and their market reforms locked into place by being part of the Union, but I have friends who try to do business in Poland and Slovakia, and they tell me that there may be democracy, but all they experience is corruption, from top to bottom. I'll bet that those subsidies we'll be paying for to help these 'democracies' get on their feet will wind up in Swiss bank accounts. By the way, the Swiss, you'll notice, aren't EU members either. Wonder why?

"A United States of Europe? Not as far as I am concerned. I am Austrian, and I am and will always be a nationalist as far as my country is concerned. I speak German, but already English is becoming the major EU language, though the Brits were smart enough not to join the euro. I'm afraid that these European Commission bureaucrats have their heads in the clouds, which is why they can't see the real Europe they're putting in a straitjacket."

Vote your opinion at www.wiley.com/college/deblij

From the Field Notes . . .

"The Grasshopper, one of Amsterdam's many 'coffee shops.' For about four decades now, the Netherlands has effectively legalized sales and consumption of small amounts of marijuana. The general consensus is that marijuana is not as harmful as hard drugs and that legalization helps to decriminalize its commerce and use. This policy is at odds with legislation in most of the rest of the EU, especially with bordering Germany, Belgium, and nearby France. It used to be that coffee shops were found only in Amsterdam and other major cities, but in recent years they have opened in small towns along the border to attract a foreign clientele (another byproduct of heightened cross-border interaction). As a result, once quiet and conservative rural villages (where lifestyles differ enormously from those in Amsterdam) started to experience heavy traffic of buyers and users from abroad—not something the town folk were likely to appreciate. This has been one factor in what may be a major imminent change: the national government is entertaining a proposal to turn all coffee shops into membership-only establishments whereby only Dutch citizens and residents are eligible to join. As you might expect, reactions to this proposal vary geographically: it is supported in the rural border areas, but most people in Amsterdam view it as an infringement on traditional liberties and a bad business decision to boot. Said one Amsterdammer and steady customer of The Grasshopper with a tone of sarcasm: 'I never thought I'd experience this: some provincials deciding what happens in Amsterdam. But I have to see it before I believe it. We'll find a way around it.'"

© Jan Nijman

www.conceptcaching.com

---

is a tenacious phenomenon that plays out at various geographic scales.

Diminished State Power and New Regionalism

As states relinquish some of their power to Brussels and express agreement with the ideal of realmwide integration, some of the provinces and regions *within* states have seized the opportunity to assert their cultural identity and particular economic interests. Often, the local governments in these subregions simply bypass the governments in their national capitals, dealing not only with each other but even with foreign governments as their business networks span the globe. In this they are imitated by other provinces, all seeking to foster their local economies and, in the process, strengthen their political position relative to the state.

Provinces, States, and other subnational political units on opposite sides of international boundaries can now cooperate in pursuit of shared economic goals. Such cross-border cooperation creates a new economic map that seems to ignore the older political one, creating economically powerful regions that are more or less independent of surrounding national states. Note how European unification seems to simultaneously erode the power of states from above and from below: in order to retain membership, states are more or less compelled to concede major decision-making power to Brussels while at the same time the demands from subnational (and cross-national) regional authorities, from Catalonia to Lombardy, eat away at their territorial control at home.

Thus even as Europe's states have been working to join forces in the EU, many of those same states are confronting severe centrifugal stresses. The term **devolution [17]** has come into use to describe the powerful centrifugal forces whereby regions or peoples within a state, through negotiation or active rebellion, demand and gain political strength and sometimes autonomy at the expense of the center. Most states exhibit some level of internal regionalism, but the process of devolution is set into motion when a key centripetal binding force—the nationally accepted idea of what a country stands for—erodes to the point that a regional drive for autonomy, or for outright

FIGURE 1A-11

© H. J. de Blij, P. O. Muller, and John Wiley & Sons, Inc.

secession, is launched. As Figure 1A-11 shows, numerous European countries are affected by devolution.

States respond to devolutionary pressures in various ways, ranging from accommodation to suppression. One way to deal with these centrifugal forces is to give historic regions (such as Scotland or Catalonia) certain rights and privileges formerly held exclusively by the national government. Another answer is for EU member-states to create new administrative divisions that will allow the state to meet regional demands, often in consultation with, or under pressure of, the ***European Commission*** (the name of the EU's central administration based in Brussels).

DEFINING EUROPE: DIFFICULT CHOICES

Widening or Deepening?

Expansion has always been an EU objective, and the subject has always aroused passionate debate. Will the incorporation of weaker economies undermine the strength of the whole? Should the ties and cooperation between existing members be deepened and solidified before other, less prepared countries are invited to join? Remember that member-states must adhere to strict economic policies and harmonize their political systems. This is much easier for prosperous countries with longstanding democratic traditions than for poorer nations with a volatile political past.

Despite such misgivings, negotiations to expand the EU have long been in progress and still continue. In 2004, a momentous milestone was reached: ten new members were added, creating a greater European Union with 25 member-states. Geographically, these ten came in three groups: three Baltic states (Estonia, Latvia, and Lithuania); five contiguous states in eastern Europe, extending from Poland and the Czech Republic through Slovakia, Hungary, and Slovenia; and two Mediterranean island-states, the mini-state of Malta and the still-divided state of Cyprus. And in 2007, both Romania and Bulgaria were incorporated, raising the number of members to the current 27 and extending the EU to the shores of the Black Sea (Fig. 1A–9).

Numerous structural implications arise from this expansion, affecting all EU countries. A common agricultural policy is now even more difficult to achieve, given the poor condition of farming in most of the new member-states. Also, some of the former EU's poorer countries, which were on the receiving end of the subsidy program that aided their development, now will have to pay up to support the much poorer new eastern members. And disputes over representation at EU's Brussels headquarters arose quickly. Even before the newest members' accession, Poland was demanding that the representative system favor medium-sized members (such as Poland and Spain) over larger ones (such as Germany and France).

The Remaining Outsiders

This momentous expansion is having major geographic consequences for all of Europe, and not just the EU. As Figure 1A-9 shows, following the admission of Romania and Bulgaria in 2007, a number of countries and territories still remain outside the Union, and their prospect of joining seems a mixed bag. The first group includes the states emerging from the former Yugoslavia plus Albania. In this cluster of western Balkan states, most of them are troubled politically and economically; only Slovenia has achieved full membership. The remainder rank among Europe's poorest and most ethnically fractured states, but they contain more than 20 million people whose circumstances could worsen outside "the club." Better, EU leaders reason, to move them toward membership by demanding political, social, and economic reforms, although the process will take many years. Discussions with Croatia began in 2005 and with newly independent Montenegro in 2006, but the aftermath of war and other problems are holding up Serbia's invitation. Elsewhere, ethnically and politically divided Bosnia seemed far from ready for negotiations, Albania was even less prepared, and unresolved issues continue to roil Macedonia as well as newly independent Kosovo. Note that this area of least-prepared countries also

is the most Islamic corner of Europe, which could be another impediment for accession.

The second group of outsiders is comprised of Ukraine and its neighbors. Four former Soviet republics in Europe's "far east" could some day join the EU (even Belarus, until recently the most disinterested, may now be leaning in that direction). The government of Ukraine has shown interest, although its electorate is strongly (and regionally) divided between a pro-EU west and a pro-Russian east. Moldova views the EU as a potential supporter in a struggle against devolution along its eastern flank, and in 2006 a country not even on the map—Georgia, across the Black Sea from Ukraine—proclaimed its interest in EU membership for similar reasons: to find an ally against Russian intervention in its internal affairs.

Finally, we should take note of still another important candidate for EU membership: Turkey. Some EU leaders would like to include a mainly "Muslim country" in what Islamic states sometimes call the "Christian Club," but Turkey needs progress on social standards, human rights, and economic policies before its accession can be contemplated—and it would have to overcome the entrenched conviction of many that Turkey is just not European enough.

Once Again: In Search of Europe's Eastern Boundary

In the past, "Eastern Europe" incorporated all of Europe east of Germany, Austria, and Italy; north of Greece; and south of Finland. The Soviet communist domination of Eastern Europe (1945–1990) behind an ideological and strategic "Iron Curtain" served to reinforce the division between "West" and "East." The collapse of the Soviet Empire in the early 1990s freed the European countries that had been under Soviet rule, and they swiftly turned their gaze from Moscow to the West. Meanwhile, the European Union had been expanding eastward, and membership in the EU became an overriding goal for the majority of the liberated eastern states. Today, with the Soviet occupation mostly a distant memory and the European Union extending from the British Isles to the Black Sea, the boundary between Western and Eastern Europe has disappeared, at least in a political sense. Economically, there remain significant contrasts between west and east (see the Data Table in Appendix B and compare, for instance, Bulgaria and France); but in this respect, too, things are evening out.

During the Cold War, Europeans (especially in the west) were preoccupied with the "Iron Curtain" and much less so with the question of an eastern boundary of Europe as a whole. After all, eastern Europe was drawn so tightly into the Soviet orbit politically, militarily, economically, and even culturally (the main foreign language taught in schools was Russian) that it seemed a moot point. But deep down, most of those eastern components remained quite European in spirit and tradition, and possessed historical ties that could not be easily erased. So now we are once again left to contemplate where, between Moscow and Vienna, Europe's eastern boundary lies or should be placed. More than anything else, it is the eastward expansion of the EU that is testing this question.

EU versus NATO

The emphasis placed nowadays on developments in the EU tends to overshadow another very important pan-European construct, the *North Atlantic Treaty Organization (NATO)*. This alliance partially overlaps with the EU and includes many European countries (see Fig. 1A-12). But its relationship to the EU is tenuous and uneven, and NATO is in some respects an entirely different creature. First, NATO is primarily concerned with politics and security and not with economics; NATO's purpose is to provide military security to its members. Second, NATO is led by the United States, which provides the bulk of the alliance's military muscle, and also includes Canada. Third, NATO includes non-EU member Turkey, a country that is, in fact, hugely important because it is strategically located on Europe's southeastern edge and close to the Middle East (one can almost hear the Turks think "if

we are good enough for NATO, why aren't we welcome in the EU?").

NATO is and always has been concerned with military matters. It was created in 1949 to provide stability in postwar Europe and to cement the involvement there of the United States. Interestingly, the main target at the time was Germany with the goal of preventing another round of German aggression. But West Germany itself became a member in 1955, and by that time the perception of threat had shifted to the Soviet Union. Europe had become divided into Western and Eastern halves with NATO providing a security umbrella for the Western side (in response, the Soviets and their allies created the Warsaw Pact).

When the Cold War ended in 1991 with the disintegration of the Soviet Union, NATO lapsed into something of an identity crisis. But its existence was soon relegitimized as a more general security organization that would provide stability across Europe—and that opened its doors to membership for eastern European countries. This eastward expansion, not unlike that of the EU, continues to this day (Fig. 1A-12).

At a summit in Lisbon, Portugal in November 2010, the alliance officially adopted a new strategic direction. First and foremost, NATO is now more concerned with

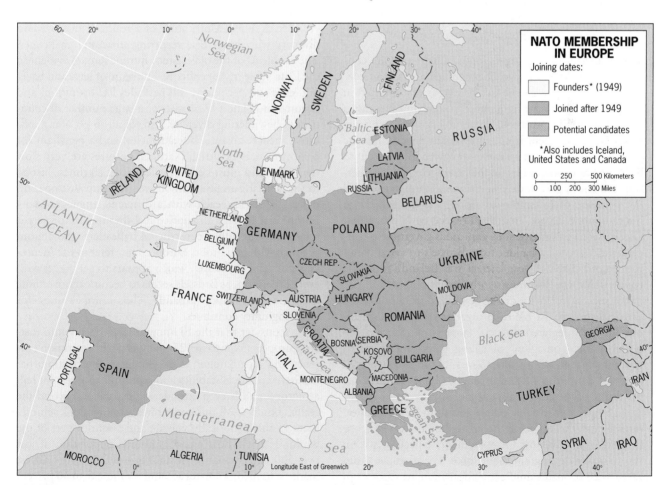

FIGURE 1A-12

© H. J. de Blij, P. O. Muller, J. Nijman, and John Wiley & Sons, Inc.

the possibility of twenty-first-century types of threats to its member-states: terrorism, attacks by "rogue" states with weapons of mass destruction, and cyber war (a very real and growing danger with critical global implications). This also means that the alliance, while still mainly focused on Europe, is increasingly active on the global scene. For instance, one of the items atop its agenda is a possible nuclear missile threat from Iran. In the past few years, the largest military operation of the alliance has been far away from Europe, in Afghanistan. Most troops in Afghanistan are American, but nearly one-third are European, all operating under the banner of NATO that has been on the ground in that country for more than a decade (also see Chapters 7A and 7B).

Of course, both economic issues (EU) and security issues (NATO) are deeply political and are therefore often intertwined. Having two separate organizations to deal with these two spheres adds another layer of complexity to the European realm. Economically, the United States and Europe are often perceived as competitors, and the Europeans expressly go their own way. But where security is concerned the United States still functions as a guarantor of peace in the realm, with most Europeans happy to defer to the Americans who pick up the lion's share of the bill. The United States, in turn, views strategic involvement in Europe as fundamental to its own security interests.

Impact of the Global Economic Crisis

The challenges to ongoing European integration are considerable, and they seem to be magnified at times of economic distress. The global financial crisis that erupted in 2008 is still reverberating around Europe and has cast a shadow over EU prospects. Economic growth rates are down; unemployment is up; government deficits in some cases have skyrocketed; politics in many countries seem to be increasingly polarized, and the benefits of EU membership are once again widely questioned; immigration and the presence of (unemployed) foreigners are subject to intense public debate. In late 2009, the news broke that Greece had accumulated such enormous government deficits that it needed a bailout from Brussels to avoid bankruptcy and chaos. A year later, it was Ireland's turn: the country that was not long ago hailed as one of Europe's major success stories needed to be rescued from imminent financial collapse. In 2011, Portugal needed a similar bailout. In the integrated financial system that the EU has become, the "contagion" of failing banks and faltering governments is a major concern—as if Europe constituted a row of dominoes. Not surprisingly, observers feared for the future of the euro as a stable global currency.

Some countries had experienced high immigration rates before the onset of the crisis, but then were confronted with rampant unemployment, especially within the unskilled immigrant labor force. For example, from 2000 to 2007 Spain's immigrant population increased from 2 to 10.4 percent of the population. The country now has

some 5 million immigrants who came mainly from northern Africa. Over the same period, Ireland's immigrant population, the majority coming from eastern Europe, grew from 3.3 to 10.5 percent. This did not seem to be much of a problem until the economy ground to a halt in the late 2000s. From January 2008 to January 2010, unemployment rose from 8.8 to 19 percent in Spain, and from 4.8 to 13 percent in Ireland. In the EU area as a whole, these numbers resembled those of the United States, with unemployment rates rising to about 10 percent.

In many European countries, the economic downturn was accompanied by a negative political reaction to immigration, and in some cases there were signs that foreigners have become scapegoats for the economic malaise. More generally, at times of economic hardship, individual countries and government are less likely to be agreeable to any type of sacrifice for the common European good. These are not the best of times for the EU to advance.

Future Prospects of the EU

But let us focus on the big picture. Assuming that these problems will dissipate as the recession makes way for a new period of economic growth, the history of unification as outlined above underscores how far-reaching and seemingly irreversible Europe's current transformation is—even if Europe's ultimate supranational goals have not yet been attained. Unification of a realm of this size and diversity will always face recurring structural challenges. The problems of the old Eastern Europe remain evident in the Data Table in Appendix B. The sense of have and have-not, core and periphery, still pervades EU operations and negotiations. The EU consists of a large number of countries of bewildering diversity, and for all to get along infinite patience, gigantic bureaucracies, and a penchant for compromise is essential. National cultures are very slow to change and must be acknowledged at all times. Just go to the official European Union website (http://europa.eu): it has 23 portals, each with a different language. Many Europeans, perhaps a majority, feel that the EU project is a bureaucratic edifice constructed by leaders who are out of touch with grass-roots Europe. They refer to the "democratic deficit" of the EU, which appears to get bigger with every expansion, as Brussels seems to become more remote every time the Union expands. Thus Europe's quest for enduring unity continues.

Consider that the EU now reaches deeply into eastern Europe; encompasses 27 members; has a common currency and a parliament; is developing a constitution; has facilitated political stability and economic progress for half a century; and is even considering expansion beyond the realm's borders. All this constitutes a tremendous achievement in this historically volatile and fractious part of the world. The EU now has a combined population of just over 500 million that constitutes one of the world's richest markets; its member-states account for more than 40 percent of the world's exports.

It is remarkable that this has been accomplished in little more than five decades. Some of the EU's leaders want more than economic union: they envisage a United States of Europe, a political as well as an economic competitor for the United States. To others, such a "federalist" notion is an abomination not even to be mentioned (the British in general are particularly wary of such an idea). Whatever happens, Europe is going through still another of its revolutionary transformations, and when you study its evolving map you are looking at history in the making.

● POINTS TO PONDER

- Switzerland, situated at the heart of Europe but still outside the EU by choice, may be reconsidering its position.

- Some western European countries accepted relatively large numbers of religious Muslim immigrants at a time when secularization among native populations reached a peak.

- NATO's main purpose is to provide security to its member-states in Europe plus the United States and Canada, yet its most important military operation is in Afghanistan.

- EU foreign ministers in 2009 approved a measure that would allow Croatia to join the EU in 2012.

- In 2010, Greece and Ireland descended into a financial morass and had to be bailed out by other member-states; Portugal followed suit in 2011. The EU feared for the contagious effects throughout the realm's financial system and for the value of the euro.

EUROPE: MODERN CO
AND HISTORIC REGIO

- - - European Core Bound

Western Europe

British Isles

Northern (Nordic) Eur

Mediterranean Europe

Eastern Europe

0   250   500 Kilo
0   100   200   300 Miles

ICELAND

Arctic Circle

Norwegian Sea

Faeroe Islands (Den.)

60°

North Sea

NORWAY

SWEDEN

FINLAND

ESTONIA

LATVIA   *Western Dvina R.*

LITHUANIA

RUSSIA

50°

IRELAND

UNITED KINGDOM

Irish Sea

DENMARK

Baltic Sea

RUSSIA

BELARUS

ATLANTIC OCEAN

NETHERLANDS

GERMANY

POLAND

Vistula R.

Dnieper R.

RUSSIA

Thames R.

BELGIUM

Rhine R.

Elbe R.

CZECH REP.

Dniester R.

UKRAINE

LUXEMBOURG

Seine R.

Danube R.

SLOVAKIA

MOLDOVA

Loire R.

LIECHTENSTEIN

FRANCE

AUSTRIA

HUNGARY

ROMANIA

SWITZERLAND

Rhône R.

SLOVENIA

Po R.

CROATIA

40°

Ebro R.

Adriatic Sea

BOSNIA   SERBIA   *Danube R.*

Black Sea

PORTUGAL

SPAIN

Tagus R.

ITALY

MONTENEGRO

KOSOVO   BULGARIA

MACE-DONIA

ALBANIA

TURKEY

GREECE

Aegean Sea

Mediterranean

MOROCCO

ALGERIA

TUNISIA

MALTA

Sea

CYPRUS

0°

10°   Longitude East of Greenwich   20°   30°

IN THIS CHAPTER

- ◆ Financial crises and the threat to the euro
- ◆ Why Britain is different
- ◆ The rebirth of East-Central Europe
- ◆ Germany's leadership: Needed but feared
- ◆ Ethnicity and politics in the Balkans
- ◆ Is Europe advancing toward Russia's doorstep?

CONCEPTS, IDEAS, AND TERMS

FIGURE 1B-1   © H. J. de Blij, P. O. Muller, and John Wiley & Sons, Inc.

© Jan Nijman
Where were these photos taken?
Find out at www.wiley.com/college/deblij

Our objective in this and later regional (**B**) chapters is to become familiar with the regional and national frameworks of the realms under investigation, and to discover how various regions and individual states fit into the larger geographic mosaic. Even in this era of globalization, and even in integrating Europe, the state still plays a key role in that process. The geography of these various regions and states, how they interact, and what their prospects may be in this changing world are among the key questions we will address.

EUROPE'S REGIONAL COMPLEXITY

Europe presents us with a particular challenge because of its numerous countries and territories, and because of the continuing changes affecting its geography. Readers of this book who reside in North America tend to be accustomed to a familiar and relatively stable map of two countries, the United States and Canada. In Europe, on the other hand, just since 1990 no less than 15 new national names have appeared on the political map, some of them restorations of old entities such as Estonia and Lithuania, others new and less familiar (Montenegro, Moldova). And there may be more to come: Kosovo became nominally independent in 2008, and Belgium may yet fracture the way Czechoslovakia did in 1993.

An additional contrast to North America lies in the long and tumultuous history of Europe. Many of Europe's cities were first built long before settlers reached the shores of the New World; a stroll through Venice, Bruges, or Prague takes us back many centuries. Because Europe's twenty-first-century cultures and political boundaries are the result of long and deeply rooted historical processes that were conditioned by geography, we should keep that in mind when we examine the realm's current spatial framework.

● MAJOR CITIES OF THE REALM

Metropolitan Area	Population* (in millions)
Paris, France	10.6
London, UK	8.7
Madrid, Spain	6.0
Frankfurt, Germany	5.9
Barcelona, Spain	5.2
Berlin, Germany	5.1
Milan, Italy	4.3
Stuttgart, Germany	3.7
Rome, Italy	3.4
Athens, Greece	3.3
Kiev, Ukraine	2.9
Brussels, Belgium	1.9
Hamburg, Germany	1.8
Vienna, Austria	1.7
Budapest, Hungary	1.7
Warsaw, Poland	1.7
Lyon, France	1.5
Stockholm, Sweden	1.3
Prague, Czech Republic	1.2
Amsterdam, Netherlands	1.1
Dublin, Ireland	1.1

*Based on 2012 estimates.

And Europe displays something else North America does not: microstates [1] that do not have the attributes of "complete" states but are on the map as tiny yet separate entities nonetheless, such as Monaco, San Marino, Andorra, and Liechtenstein. Add to these the Russian exclave (outlier) of Kaliningrad on the Baltic Sea and the British dependency of Gibraltar at the outlet of the Mediterranean Sea, and Europe seems to display a bewildering political mosaic indeed.

Traditional Formal Regions, Modern Spatial Network

In the old days, Europe divided rather easily into Western, Northern (Nordic), Mediterranean (Southern), and Eastern regions, and countries were grouped accordingly. Some of this traditional historical geography lingers in cultural landscapes and still has relevance. But in the era of European unification, a more dominant *core-periphery* framework has been superimposed on that regional scheme, defined in terms of how central—or peripheral—countries are to the workings of the EU. Put differently, Europe can be understood as a set of formal, more or less homogeneous, cultural regions; but at the same time, the entire realm functions as a tightly interdependent economic and political regional system, a spatial network that revolves around the EU.

The original Common Market of 1957 still anchors what has become a *core area* for all of the European realm. The British Isles form part of this Core, but the British—whose EU membership was delayed by a French veto and who never adopted the euro—still are not the full-fledged members the Germans, the French, and the Irish are. It is therefore appropriate to consider the United Kingdom as a geographically distinct component of the European Core.

Take a close look at the map in Figure 1B-1 and note that today the core area of the EU does not coincide with national borders—the core is primarily defined on the basis of regional economic performance, not political geography. For example, Northern Italy (but not the South) or Southern Sweden (but not the North) are part of this core. You see how complicated and challenging the work of geographers can be when trying to understand the world's spatial organization. And the European realm illustrates that challenge like no other. But this chapter will bring clarity to what at first glance may appear to be a chaotic regional landscape.

In the discussion that follows, we will focus first on countries such as Germany and France that lie entirely within the Core region. Next, we discuss countries with significant regions inside as well as outside the Core such as Spain, Italy, and Sweden. Lastly, our attention focuses on Europe's Periphery.

▪▪ THE MAINLAND CORE

Not counting the microstate of Liechtenstein, eight states form the Mainland Core of Europe: dominant Germany and France, the three Benelux countries, the two landlocked mountain states of Switzerland and Austria, and the Czech Republic (Fig. 1B-2). This is the European region sometimes still referred to as "Western Europe," and its most populous country (Germany) also has Europe's largest economy.

Germany

Twice during the twentieth century Germany plunged Europe and the world into war, until, in 1945, the defeated and devastated German state was divided into two parts, West and East (see the delimitation in red in Fig. 1B-3), the latter of which had to surrender some important industrial areas to Poland.

West and East Germany then set forth on widely different economic and political trajectories. Soviet rule in East Germany was established on the Soviet-communist model, and given the extreme hardships the USSR had suffered at German hands during the war, it was harshly punitive. The American-led authority in West Germany was less strict and aimed more at rehabilitation. When the Marshall Plan was instituted, West Germany was included, and its economy recovered rapidly. Meanwhile, West Germany was reorganized politically into a modern federal state based on democratic foundations.

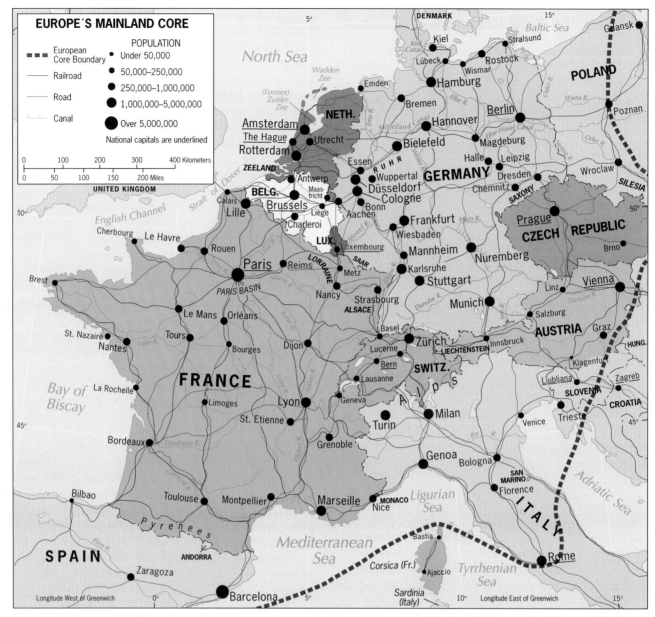

FIGURE 1B-2

© H. J. de Blij, P. O. Muller, and John Wiley & Sons, Inc.

West Germany's economy soon thrived. Between 1949 and 1964 its gross national income (GNI) tripled while industrial output rose 60 percent. Simultaneously, West Germany's political leaders participated enthusiastically in the negotiations that led to the six-member Common Market in 1957. Geography worked in West Germany's favor: it had common borders with all but one of the initial member-states. Its transport infrastructure, rapidly rebuilt, was second to none in the realm. More than compensating for its loss of Saxony and Silesia were the expanding Ruhr (in the hinterland of the Dutch port of Rotterdam) and the newly emerging industrial complexes centered on Hamburg in the north, Frankfurt (the leading financial hub as well) in the center, and Stuttgart in the south. West Germany exported huge quantities of iron, steel, motor vehicles, machinery, textiles, and farm products. Until just a

few years ago, Germany had the largest export economy by value in the world; today it is surpassed by China, but Germany still outranks the United States in second place.

In 1990, West Germany had a population of about 62 million and East Germany 17 million. Communist misrule in the East had yielded outdated factories, crumbling infrastructures, polluted environments, drab cities, inefficient farming, and inadequate legal and other institutions. Reunification that year was more a rescue than a merger, and the cost to West Germany was gargantuan. When the West German government imposed sales-tax increases and an income-tax surcharge on its citizens, many Westerners began to doubt the wisdom of reunification. It was projected that it would take decades to reconstruct Virginia-sized East Germany: 10 years later, exports from the former East still contributed only about 7 percent of

STATES (*LÄNDER*) OF REUNIFIED GERMANY

GDP PER CAPITA,
IN EUROS, 2005

Over 26,000

22,000–26,000

Below 22,000

Railroads

Roads

City population

● Under 50,000

● 50,000–250,000

● 250,000–1,000,000

● 1,000,000–5,000,000

● Over 5,000,000

0   50   100 Kilometers

0   25   50 Miles

National capitals are underlined
Data source: Eurostat, 2008

FIGURE 1B-3

© H. J. de Blij, P. O. Muller, and John Wiley & Sons, Inc.

the national total. And 20 years later, in 2010, unemployment rates in the East were twice as high as in the West. Regional disparity will continue to afflict Germany for some time to come.

Upon reunification, East Germany was reorganized into six new States to fit the federal system of West Germany. Today's reunified Federal Republic of Germany consists of 16 States (or *Länder*). Figure 1B-3 makes a key point: regional disparity in terms of gross domestic product (GDP) per person remains a serious problem between the former East and West. Note that five of former East Germany's six States (Berlin being the sole exception) are in the lowest income category, while most of the ten for-

mer West German States rank in the two higher income categories. In the first decade of this century, Germany's economy was stagnant, raising unemployment and slowing former East Germany's recovery.

Another challenge facing Germany, along with other western European countries, is the integration of sizeable immigrant minorities. Germany has nearly 10 million immigrants, about 12 percent of the total population of 81.3 million. They are culturally different and economically worse off. About 40 percent of immigrant households live below the official poverty line, and immigrants are far more likely to be unemployed. They crowd together in the big cities, and immigrant children go to schools that have difficulty teaching a curriculum in the German language. Only about 8 percent of the immigrant teenagers pursue vocational training (one of the traditional strengths of the German educational system) compared to 37 percent of native teenagers. Currently, as elsewhere in western Europe, there is something of a backlash against immigration. It will hopefully dissipate as the economy turns the corner again in the near future, because the western European economies, Germany's included, need immigrant labor.

Germany has been pulling through the global recession better than any other country in the realm, and it is again exerting its dominance over a continental Europe in which it has no peer. But it is sometimes said that the "German question" never dies, it just mutates. The EU needs Germany's leadership, but fears its dominance. If Germany is to openly assume that mantle of leadership, it will have to do so in the most diplomatic of ways.

France

German dominance in the European Union is a constant concern in the other leading Mainland Core country. The French and the Germans have been rivals in Europe for centuries. France (population: 63.5 million) is an old state, by most measures the oldest in this region. Germany is a young country, created in 1871 after a loose association of German-speaking states had fought a successful war against . . . the French.

Territorially, France is larger than Germany, and the map suggests that France has a superior relative location, with coastlines on the Mediterranean, the Atlantic Ocean, and, at Calais, even a window on the North Sea. But France does not have any good natural harbors, and oceangoing ships cannot navigate its rivers and other waterways very far inland.

The map of the Mainland Core region (Fig. 1B-2) reveals a significant contrast between France and Germany in terms of population distribution. France has one dominant city, Paris, and no other city comes close either in size or centrality: Paris has 10.6 million residents, whereas its closest rival, Lyon, contains only 1.5 million. Germany has no city to match Paris, but it does have a range of medium-sized cities and it is more highly urbanized overall (89 percent) than France (77 percent).

Why should Paris, without major raw materials nearby, have grown so large? Whenever geographers investigate the evolution of a city, they focus on two important locational qualities: its **site [2]** (the physical attributes of the place it occupies) and its **situation [3]** (its location relative to surrounding areas). The site of the original settlement at Paris lay on an island in the Seine River, a defensible place where the river was often crossed. This island, the Île de la Cité, (photo p. 68), a Roman outpost 2000 years ago; for centuries its security ensured continuity. Eventually the island became overcrowded, and the city expanded along the banks of the river (Fig. 1B-4A).

SITE AND SITUATION OF PARIS, FRANCE

▪▪▪▪ 12th Century Wall   —— Roads   ▨ Forests, Parks

══ Wall of 1840s   —— Railroads   ▧ Built-up area

FIGURE 1B-4   © H. J. de Blij, P. O. Muller, and John Wiley & Sons, Inc.

● AMONG THE REALM'S GREAT CITIES . . .

PARIS

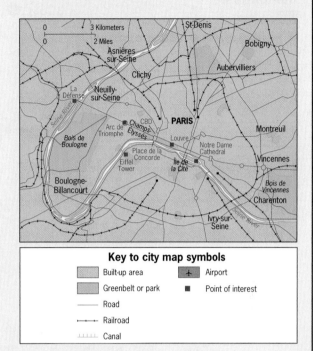

© H. J. de Blij, P. O. Muller, and John Wiley & Sons, Inc.

IF THE GREATNESS of a city were to be measured solely by its number of inhabitants, Paris (10.6 million) would not even rank in the world's top 20. But if greatness is measured by a city's historic heritage, cultural content, and international influence, Paris has no peer. Old Paris, near the Île de la Cité that housed the original village where Paris began (Fig. 1B-4A) and carries the eight-century-old Notre Dame Cathedral, contains an unparalleled assemblage of architectural and artistic landmarks old and new. The Arc de Triomphe, erected by Napoleon in 1806 (though not completed until 1836), commemorates the emperor's victories and stands as a monument to French neoclassical architecture, overlooking one of the world's most famous streets, the Champs Elysées, which leads to the grandest of city squares, the Place de la Concorde, and on to the magnificent palace-turned-museum, the Louvre.

Even the Eiffel Tower, built for the 1889 International Exposition over the objections of Parisians who regarded it as ugly and unsafe, became a treasure. From its beautiful Seine River bridges to its palaces and parks, Paris embodies French culture and tradition. It is perhaps the ultimate primate city in the world.

As the capital of a globe-girdling empire, Paris was the hearth from which radiated the cultural forces of Francophone assimilation, transforming much of North, West, and Equatorial Africa, Madagascar, Indochina, and many smaller colonies into societies on the French model. Distant cities such as Dakar, Abidjan, Brazzaville, and Saigon acquired a Parisian atmosphere. France, meanwhile, spent heavily to keep Paris, especially Old Paris, well maintained—not just as a relic of history, but as a functioning, vibrant center, an example to which other cities can aspire.

Today, Old Paris is ringed by a new and different Paris. Stand on top of the Arc de Triomphe and turn around from the Champs Elysées, and the tree-lined avenue gives way to the vista of *La Défense*, an ultramodern high-rise complex that is one of Europe's leading business districts (see photo, p. 52). But from atop the Eiffel Tower you can see as far as 80 kilometers (50 mi) and discern a Paris visitors rarely experience: grimy, aging industrial quarters, and poor, crowded neighborhoods where discontent and unemployment fester—and where immigrants cluster in a world apart from the splendor of the old city.

Soon the settlement's advantageous situation stimulated its growth and prosperity. Its fertile agricultural hinterland thrived, and, as an enlarging market, Paris's focality increased steadily. The Seine River is joined near Paris by several navigable tributaries (the Oise, Marne, and Yonne). When canals extended these waterways even farther, Paris was linked to the Loire Valley, the Rhône-Saône Basin, Lorraine (an industrial area in the northeast), and the northern border with Belgium. When Napoleon reorganized France and built a radial system of roads—followed later by railroads—that focused on Paris from all parts of the country, the city's *primacy* was assured (Fig. 1B-4B). A **primate city** [4] is one that is disproportionately large compared to all others in the urban system and exceptionally expressive of the nation's culture. Paris is the quintessential example, not only personifying France but also serving as its capital.

The difference in urban systems between Germany and France must also be seen in historical context. France is one of Europe's oldest countries where an absolutist state formed more than 500 years ago, one that was always ruled from Paris. In few other countries were the centralizing powers and territorial designs of the monarchy as successful as in France. Germany, in contrast, remained a highly politically fragmented region until as late as 1871 when it was first united as a modern state. It was ruled from Berlin—a powerful place—but it had to compete with other major cities that had formed in earlier decades and centuries, from Hamburg in the north to Munich in the south and from Köln (Cologne) in the west to Leipzig in the east. After World War II, the division into East and West Germany implied that Berlin was cut off from the West altogether, and a number of West German cities

REGIONS OF FRANCE

GDP PER CAPITA,
IN EUROS, 2005

▨ Over 26,000

▨ 22,000–26,000

☐ Under 22,000

— Railroad

Data Source: Eurostat, 2008

City population

• Under 50,000

• 50,000–250,000

● 250,000–1,000,000

● 1,000,000–5,000,000

⬤ Over 5,000,000

National capitals are underlined

DÉPARTEMENTS

FIGURE 1B-5

© H. J. de Blij, P. O. Muller, and John Wiley & Sons, Inc.

continued to evolve into important centers. Here we can very clearly see the interesting ways in which history and geography are entwined.

France developed early on the strong tradition of a highly centralized state, where the central government in Paris maintained tight control over the provinces near and far. With Napoleon's rule in the early 1800s, the country was subdivided into a great number of small *départements* (eventually 96), each of which had representation in Paris, but with power concentrated in the capital. It remained so for nearly two centuries, with the entire country tightly integrated and without notable regional opposition (see Fig. 1B-5 inset map). Only the island *département* of Corsica produced a rebel movement, whose violent opposition to

French rule continued for decades and even touched the mainland. In 2003, voters in Corsica rejected an offer of special status for their island, including limited autonomy. They want more, and trouble lies ahead.

Today, France is decentralizing. A new layer of governance, consisting of 26 larger provinces named *régions*, has been inserted at the level between Paris and the *départements* to accommodate devolutionary forces as well as seize the opportunities of regional and local growth throughout the country (Fig. 1B-5). These regions (one of which is Corsica and four of which are located overseas) have substantial autonomy in such spheres as taxation, borrowing, and development spending. The cities that anchor them benefit because they are the seats of governing

From the Field Notes . . .

"No country has a more appropriate name than the Netherlands: most of its land, and the majority of its population, lies below sea level. Over centuries of increasingly sophisticated engineering, the Dutch created a system of artificial levees, dikes, pumps (originally windmill-driven), drainage canals, and locks to keep the water out. The southwestern part of the Netherlands is marked by major waterways, low-lying islands, and brackish ecosystems teeming with wildlife. Because the area lies largely below sea level it is protected by the so-called Delta Works, an engineering marvel created to keep the land from flooding during North Sea storms, and to simultaneously maintain the natural habitat. This is the Haringvliet Sluizen, one of a series of adjustable storm-surge barriers. Structures of this type, installed across inlets, provide safety and avoid the need to build protective barriers along all the shorelines further inland. The last major flood, which killed almost 2000, occurred more than a half-century ago, before the Delta Works were built."

© Jan Nijman

www.conceptcaching.com

regional councils that can attract investment, not only within France but also from abroad.

France has one of the world's most productive and diversified economies, based in one of humanity's richest cultures and vigorously promoted and protected (notably its heavily subsidized agricultural sector). Northern French agriculture remained Europe's most productive and varied, exploiting the country's wide range of soils and microclimates and enjoying state subsidies and protections.

Lyon, France's second city and headquarters of the region named Rhône-Alpes, has become a focus for growth industries and multinational firms. This region is evolving into a self-standing economic powerhouse that is becoming a driving force in the European economy. Indeed, it is one of the Four Motors of Europe (see p. 58) with its own international business connections to countries as far away as China and Chile.

Today France's economic geography is marked by new high-tech industries. It is a leading producer of high-speed trains, aircraft, fiber-optic communications systems, and space-related technologies. It also is the world leader in nuclear power, which currently supplies more than 80 percent of its electricity and thereby reduces dependence on foreign oil imports.

France is one of those European states with a rapidly aging population, posing some major challenges for the years ahead. Natural population growth is negative and must be compensated for through immigration. But most immigrants hail from poor Islamic countries with very differ- ent cultures, and their integration and acceptance in French society has been a painful process. The graying of the population means that in the future more and more elderly people must be cared for with the incomes earned by a shrinking productive workforce. Back in 1983, when things still looked quite different, the retirement age was lowered from 65 to 60—but now the government may be forced to move it back to 65 because the country cannot afford to support that many retirements of such long duration. But in the dire economic climate of the early 2010s, the French electorate is not taking kindly to this and other austerity measures: in September 2010, a record two million workers joined a massive one-day nationwide strike that was followed by others throughout the fall. By the spring of 2011, President Sarkozy's popularity had dropped to an all-time low.

Benelux

Three small countries are crowded into the northwestern corner of the Mainland Core—the Netherlands, Belgium, and Luxembourg—and are collectively referred to by their first syllables (**Be-Ne-Lux**). Their total population, 28 million, reminds us that this is one of the most densely peopled corners of our planet. Not coincidentally, the Low Countries, as they are also sometimes called, are situated at the **estuary [5]** of the great Rhine and Scheldt rivers and have access to the seas that has, throughout history, been the envy of the Germans and the French. They are a highly productive trio: both the Netherlands and Belgium rank among

VOICE FROM THE

Region

© Alice Driesen
Alice Driesen, 3rd year student in Earth Sciences, University of Amsterdam

AMSTERDAM'S CITY CENTER DECLARED A UNESCO WORLD HERITAGE SITE, 2010

"I feel honored that Amsterdam's Canal District is now included in the UNESCO World Heritage list, and I think it is rightly so. It is quite amazing that this entire area that was mostly built in the seventeenth century still exists and is so well maintained. It is a major attraction for tourists from all over the world, but it's also a real part of the city where we live and work, a kind of a living outdoor museum. Maybe it is unique in that sense. Some people complain that there will be lots of rules and regulations from UNESCO about what we can and cannot do here and that it may get in the way of economic development. But our own government put lots of rules like that in place a long time ago—that is why it still looks so beautiful. This will just add another layer of protection to make sure that 100 years from now, and beyond, the historic center will remain intact and that is just wonderful. It will probably attract ever more tourists, which is also good for business. I do hope the place will not get as crowded as, for example, Venice, and that it won't get in the way of us just living our lives here. We love tourists, but most of all we love our own city!"

the top 20 economies of the world, and tiny Luxembourg has the world's highest per capita gross national income. The seafaring Dutch had a thriving agricultural economy and a rich colonial empire; the Belgians forged ahead during the Industrial Revolution; and Luxembourg came into its own after the Second World War as a financial center for the evolving European Union and, indeed, the world.

The Netherlands, one of Europe's oldest democracies and a constitutional monarchy today, has for centuries been expanding its territory—not by warring with its neighbors but by wresting land from the sea. Its greatest project so far, the draining and reclaiming of almost the entire Zuider Zee (Southern Sea), began in 1932 and continues. In the southwestern province of Zeeland, islands are being connected by dikes and the water is being pumped out, adding still more *polders* (as the Dutch call inhabited land claimed from the sea lying behind dikes and below sea level) to the national territory. Among the technologies in which the Dutch lead the world, not surprisingly, is the engineering of coastal systems to control the sea and protect against storms (see photo, p. 75).

The regional geography of this highly urbanized country (16.7 million) is noted for the **Randstad**, a roughly triangular urban core area anchored by Amsterdam, the constitutional capital; Rotterdam, Europe's largest port; and The Hague, the seat of government. This **conurbation [6]**, as geographers call large urban areas when two or more cities merge spatially, now forms a ring-shaped complex that surrounds a still-rural center (in Dutch, *rand* means edge or margin; *stad* means city).

The Netherlands' economic geography, like that of the other states in this subregion, is heavily dominated by services, finance, and trade; manufacturing contributes about 15 percent of the value of the GDP annually, and farming, once a mainstay, less than 3 percent. Amsterdam's airport, Schiphol, is regularly recognized as Europe's best; Rotterdam, one of the world's busiest ports, is rated as the most efficient as well.

Belgium also has a thriving economy and a major port in the city of Antwerp. But Belgium is a much younger state than the Netherlands, becoming independent within its present borders as recently as 1830. With 10.8 million people, Belgium's regional geography is dominated by a cultural fault line that cuts diagonally across the country, separating a Flemish-speaking majority (54 percent) centered on Flanders in the northwest from a French-speaking minority in southeastern Wallonia (36 percent). Brussels, the mainly French-speaking capital, lies like a cultural island in the Flemish-speaking sector; but the city also is one of Belgium's greatest assets because it serves as the headquarters, and in many ways as the functional capital, of the European Union (though it is important to note that the EU does not have an official capital city). Still, devolution is a looming problem for Belgium, with political parties espousing Flemish separatism roiling the social landscape.

Luxembourg, one of Europe's many ministates, lies between Germany, Belgium, and France, with a Grand Duke as head of state (its official name is the Grand Duchy of Luxembourg), a territory of only 2600 square kilometers (1000 sq mi), and a population of just half a million. Luxembourg has translated sovereignty, relative location, and stability into a haven for financial, service, and information-technology industries. In 2007, there were more than 160 banks in this tiny country and nearly 14,000 holding companies (corporations that hold controlling stock in other businesses in Europe and worldwide). With its unmatched per capita income (U.S. $64,320 in 2010), Luxembourg is in some ways the greatest beneficiary of the advent of the European Union. And in no country in Europe is support for the EU stronger than it is here.

The Alpine States

Switzerland, Austria, and the *microstate* of Liechtenstein on their border share an absence of coasts and the mountainous terrain of the Alps—and little else (Fig. 1B-2). Austria speaks one language; the Swiss speak German in their north, French in the west, Italian in the southeast, and even a bit of Rhaeto-Romansch in the remote central highlands (Fig. 1A-7). Austria has a large primate city; multicultural Switzerland does not. Austria has a substantial range of domestic raw materials; Switzerland does not. Austria is twice the size of Switzerland and has a larger population, but far more trade crosses the Swiss Alps between western and Mediterranean Europe than crosses Austria.

Switzerland, not Austria, is in most ways the leading state in the Alpine subregion of the Mainland Core. Mountainous terrain and **landlocked location [7]** can constitute crucial barriers to economic development but, as is so often the case, geography poses constraints in one way while offering opportunities in another. That is why Switzerland is such an important lesson in the complexities of human geography. Through the skillful maximization of their limited opportunities (including the transfer needs of their neighbors), the Swiss have transformed their seemingly restrictive environment into a prosperous state. They deftly utilized the waters cascading from their mountains to generate hydroelectric power to develop highly specialized industries. Swiss farmers perfected ways to optimize the productivity of mountain pastures and valley soils. Swiss leaders converted their country's isolation into stability, security, and neutrality, making it a world banking giant, a global magnet for money. Zürich, in the German-speaking sector, is the financial center; Geneva, in the French-speaking sector, is one of the world's most internationalized cities. The Swiss feel that they do not need to join the EU—and they have not done so to this day.

Austria, which joined the EU in 1995, is a remnant of the Austro-Hungarian Empire and has a historical geography that is far more reminiscent of unstable eastern Europe than that of Switzerland. Even Austria's physical geography seems to demand that the country look eastward: it is at its widest, lowest, and most productive in the east, where the Danube links it to Hungary, its old ally in the anti-Muslim wars of the past.

Vienna, by far the Alpine subregion's largest city, also lies on the country's eastern perimeter. One of the world's most expressive primate cities with magnificent architecture and monumental art, Vienna today is the Mainland Core's easternmost city, but Vienna's relative location changed dramatically with EU enlargement in 2004 and 2007. Peripheral to the EU and a vanguard of the Core until 2004, Vienna found itself in a far more centralized position when the EU border shifted eastward. But many Austrians had their doubts about their neighbors to the east becoming EU members of potentially equal standing, and in Austria public support for the European Union has fallen to the lowest level of any member-state.

The Czech Republic

The Czech Republic, product of the 1993 Czech-Slovak "velvet divorce," centers on the historic province of Bohemia, the mountain-encircled national core area that focuses on the capital, Prague. This is a classic primate city as well, its cultural landscape faithful to Czech traditions, and it is also an important industrial center. The surrounding mountains contain many valleys with small towns that specialize, Swiss-style, in fabricating high-quality goods. In the old Eastern Europe, even during the communist period, the Czechs always were leaders in technology and engineering; their products could be found on markets in foreign countries near and far.

Bohemia always was cosmopolitan and Western in its exposure, outlook, development, and linkages; Prague lies in the basin of the Elbe River, its traditional outlet through northern Germany to the North Sea. Today, the Czech Republic (10.5 million) is reclaiming its position at the center of European action.

▪▪ THE CORE OFFSHORE: THE BRITISH ISLES

As Figure 1B-1 shows, two countries form the maritime portion of the European Core Area: the United Kingdom and Ireland. These countries lie on two major islands, surrounded by a constellation of tiny ones. The larger island, a mere 34 kilometers (21 mi) off the mainland at the closest point, is *Britain*; its smaller neighbor to the west is *Ireland* (Fig. 1B-6).

The names attached to these islands and the countries they encompass are the source of some confusion. They still are called the British Isles, even though British dominance over most of Ireland ended in 1921. The state that occupies Britain and the northeastern corner of Ireland is officially called the United Kingdom—UK by abbreviation. But this country often is referred to simply as Britain, and its people are known as the British. The state of Ireland officially is known as the Republic of Ireland (Eire in Irish Gaelic), but it does not include the entire island of Ireland.

During the long British occupation of Ireland, which is overwhelmingly Catholic, many Protestants from northern Britain settled in northeastern Ireland. In 1921, when British domination ended, the Irish were set free—except in that corner in the north, where London kept control to protect the area's Protestant settlers. That is why the country to this day is officially known as the United Kingdom of Great Britain and Northern Ireland.

Northern Ireland was home not only to Protestants from Britain, but also to a substantial population of Irish Catholics who found themselves on the wrong side of the border when Ireland was liberated. Ever since, conflict has intermittently engulfed Northern Ireland and spilled over into Britain.

Although all of Britain lies in the United Kingdom, political divisions exist here as well. England is the largest of these units, the center of power from which the rest of the region was originally brought under unified control. The English conquered Wales in the Middle Ages, and Scotland's link to England, cemented when a Scottish

THE BRITISH ISLES

FIGURE 1B-6

© H. J. de Blij, P. O. Muller, and John Wiley & Sons, Inc.

king ascended the English throne in 1603, was ratified by the Act of Union of 1707. Thus England, Wales, Scotland, and Northern Ireland became the United Kingdom.

Having united the Welsh, Scots, and Irish, the British set out to forge what would become the world's largest colonial empire. An era of mercantilism (competitive accumulation of wealth among countries) and domestic manufacturing (the latter based on water power from streams flowing off the Pennines, Britain's highland backbone) foreshadowed the momentous Industrial Revolution that transformed Britain—and much of the world.

But British **hegemony** [8] (political dominance) came to an end in the early twentieth century, challenged by the rise of other powers such as Russia, Germany, Japan, and the United States. World War II marked the end of the age of colonial empires and the beginning of U.S. global dominance—now with Britain as a "junior partner." For Britain, change after 1945 was profound: not only was it forced to let go of most of its overseas territories (though many remained part of the largely symbolic British Commonwealth), but it also had to deal with the rapid resurgence of the European mainland. Always ambivalent about

"The Troubles" between Protestants and Catholics, pro- and anti-British factions that have torn Northern Ireland apart for decades at a cost of more than 3000 lives, are etched in the cultural landscape. The so-called 'Peace Wall' across West Belfast, shown here separating Catholic and Protestant neighborhoods, is a tragic monument to the failure of accommodation and compromise, a physical manifestation of the emotional divide that still runs deep. As recently as the summer of 2010, there was a spate of attacks across Belfast by extremist 'Republicans' using booby traps and car bombs. Political negotiations and agreements have in a way become even more difficult as the original Irish Republican Army (IRA) has split into numerous factions with different perspectives and strategies. What an analyst said to the author back in 2005 still is the view across the cityscape you see here: "Good fences make good neighbors . . . the Peace Wall will have to stay up for a few more decades yet."

© Barry Chin/The Boston Globe/Landov LLC

the EC and EU, and with its first membership application vetoed by the French in 1963, Britain (admitted in 1973) has worked to restrain moves toward tighter integration. When most member-states adopted the new euro in favor of their national currencies, the British kept their pound sterling and delayed their participation in the European Monetary Union (EMU). As for a federalized Europe, to Britain that is out of the question. In this as in other respects, Britain's historic, insular standoffishness continues.

The United Kingdom

The UK, with an area about the size of Oregon and a population of 62.7 million, is by European standards quite a large country. Based on a combination of physiographic, historical, cultural, economic, and political criteria, the United Kingdom can be divided into four subregions (numbered in Fig. 1B-6):

1. **England.** So dominant is this subregion of the United Kingdom that the entire country is sometimes referred to by this name. Small wonder: England is anchored by the huge London metropolitan area, which by itself contains more than one-seventh of the UK's total population. Indeed, along with New York, London is regarded as one of the most prominent leading **world cities [9]** of the current era of globalization, with financial, high-technology, communications, engineering, and related industries reflecting the momentum of its long-term growth and agglomeration. To the north and west, England was the hearth of the Industrial Revolution, its cities synonymous with manufactures of matchless variety.

2. **Wales.** This nearly rectangular, rugged territory was a refuge for ancient Celtic peoples, and in its western counties more than half the inhabitants still speak Welsh. Because of the high-quality coal reserves in its southern tier, Wales too was engulfed by the Industrial Revolution, and Cardiff, the capital, was once the world's leading coal exporter. But the fortunes of Wales also declined, and many Welsh emigrated. Among the 3 million who remained, however, the flame of Welsh nationalism survived, and in 1997 the voters approved the establishment of a Welsh Assembly to administer public services in Wales, a first devolutionary step.

3. **Scotland.** Nearly twice as large as the Netherlands and with a population of just 5.3 million, Scotland is a major component of the United Kingdom. Most of Scotland's people live in the Scottish Lowlands anchored by Edinburgh, the capital, in the east and Glasgow in the west. Attracted there by the labor demands of the Industrial Revolution (coal and iron reserves lay in the area), the Scots developed a world-class shipbuilding industry. Decline and obsolescence were followed by high-tech development, notably in the hinterland of Glasgow, and Scottish participation in the exploitation of oil and gas reserves under the North Sea, which transformed the eastern ports of Aberdeen and Leith (Edinburgh). But many Scots feel that they are disadvantaged within the UK and should play a major role in the EU. Therefore, when the British government put the option of a Scottish Parliament before the voters in 1997, 74 percent approved. Many Scots still hope that sovereignty lies in their future, and in local elections in 2007 the independence-minded Scottish National Party won more seats in the Parliament than any of the other parties.

4. **Northern Ireland.** A declining majority of the overall population of 1.8 million, now barely 50 percent, are Protestants and trace their ancestry to Scotland

LONDON

© H. J. de Blij, P. O. Muller, and John Wiley & Sons, Inc.

SAIL WESTWARD UP the meandering Thames River toward the heart of London, and be prepared to be disappointed. London does not overpower or overwhelm with spectacular skylines or beckoning beauty. It is, rather, an amalgam of towns—Chelsea, Chiswick, Dulwich, Hampstead, Islington—each with its own social character and urban landscape. Some of these towns come into view from the same river the Romans sailed 2000 years ago: Silvertown and its waterfront urban renewal; Greenwich with its famed Observatory; Thamesmead, the model modern commuter community. Others somehow retain their identity in the vast metropolis (see Fig. 1A-8) that remains, in many ways, Europe's most civilized and cosmopolitan city. And each contributes to the whole in its own way: every part of London, it seems, has its memories of empire, its memorials to heroes, its monuments to wartime courage.

Along the banks of the Thames, London displays the heritage of state and empire: the Tower and the Tower Bridge, the Houses of Parliament (officially known as the Palace of Westminster), the Royal Festival Hall. Step ashore, and you find London to be a memorable mix of the historic and (often architecturally ugly) modern, of the obsolete and the efficient, of the poor and the prosperous. Public transportation, by world standards, is excellent, even though facilities are aging rapidly; traffic, however, often is chaotic, gridlocked by narrow streets. Recreational and cultural amenities are second to none. London seems to stand with one foot in the twenty-first century and the other in the nineteenth. Its airports are ultramodern. But the Eurostar TGV train from Paris had to wait for decades until upgraded tracks were finally opened in 2007 to allow it to operate at its high-speed norm between the Channel Tunnel and London's St. Pancras Station; the extension of such a rail network into other parts of Britain remained only a dream in 2011.

In the summer of 2012, London will be the first city in the world to host the Olympic Games for the third time (the last occasion was in 1948). Why do countries and cities compete so fanatically to host the Games? It is a matter of prestige and visibility, resulting in an increase of tourism and investment. But it is often also a way to redevelop parts of a city in ways that require massive funding that would otherwise be unavailable. So it is in London. The former industrial wasteland in the eastern part of the city, for more than a century London's poorest area, will be transformed into an Olympic Park with plenty of new facilities and housing that should be fit for everyday usage after the Games.

London remains one of the world's most livable metropolises for its size (8.7 million) largely because of far-sighted urban planning that created and maintained, around the central city, a so-called greenbelt set aside for recreation, farming, and other nonresidential, noncommercial uses (Fig. 1A-8). Although London's growth eroded this greenbelt, in places leaving only "green wedges," the design preserved critically needed open space in and around the city, channeling suburbanization outward into the zone beyond 40 kilometers (25 mi) from the center. Today's map reveals the design's continuing impact.

or England; a growing minority, currently around 46 percent, are Roman Catholics and share their religion with the Irish Republic on the other side of the border. Although Figure 1B-6 suggests that there are majority areas of Protestants and Catholics in Northern Ireland, no clear separation exists; mostly they live in clusters throughout the territory, including walled-off neighborhoods in the major cities of Belfast and Londonderry. Partition is no solution to a conflict that has raged for more than four decades at a cost of thousands of lives; Catholics accuse London as well as the local Protestant-dominated administration of discrimination, whereas Protestants accuse Catholics of seeking union with the Republic of Ireland. The Northern Ireland Assembly, to which powers are supposed to be devolved from London, had to be suspended from 2002 to 2007 before the parties could be brought together at the table again. The prospects for greater autonomy are clouded.

Republic of Ireland

What Northern Ireland has been missing because of its conflicts is evident to the south, in the Irish Republic itself. Here participation in the EU, adoption of the euro, business-friendly tax policies, comparatively low wages, an

English-speaking workforce, and an advantageous relative location combined to produce (until recently) the highest rate of economic growth in the entire European Union. This booming, service-based economy, accompanied by burgeoning cities and towns, fast-rising real estate prices, mushrooming industrial parks, and bustling traffic, transformed a country long known for emigration into a magnet for industrial workers, producing new social challenges for a closely knit, long-isolated society. Among these immigrants were thousands of people of Irish descent returning from foreign places to take jobs at home, workers from elsewhere in the European Union (including large numbers of Poles), and job-seekers from African and Caribbean countries.

The Republic of Ireland fought itself free from British colonial rule just three generations ago. Its cool, moist climate had earlier led to the adoption of the American potato as the staple crop, but excessive rain and a blight in the late 1840s, coupled with colonial mismanagement, caused famine and cost over a million lives. Another 2 million Irish left for North America and other shores.

Hard-won independence in 1921 did not bring real economic prosperity until the last few decades of the twentieth century, when Ireland became known as the ***Celtic Tiger*** (likening it to the miraculous economic development of the so-called *Asian Tigers*). For some time, Ireland had

the highest economic growth rates of the EU, and Dublin became an international business node almost overnight.

By 2007, however, Ireland's economic boom started to fade, and a year later things proceeded to fall apart: the real estate market declined, service industries found more favorable conditions in eastern Europe, and unemployment rose. By the end of 2010, it became painfully clear that the country's financial troubles were far greater than initially thought: many banks were failing after overextending credit (part of the global financial crisis that unfolded in the United States in 2008), and the government itself had amassed huge deficits and had to be bailed out by the EU. By the beginning of 2011, the economy had been shrinking for two full years; unemployment was nearly 14 percent, and hit the recent foreign immigrants (over 10 percent of the population) especially hard. The Irish economy should be able to stage a comeback, but not without first having inflicted substantial pain on investors, corporations, and the Irish people.

■■ THE DISCONTINUOUS SOUTH

As Figures 1B-1 and 1B-7 show, the northern sectors of two major southern states (Italy and Spain) form parts of the European Core. Portugal, on the western flank of the Iberian Peninsula, remains outside the European Core Area,

FIGURE 1B-7

© H. J. de Blij, P. O. Muller, and John Wiley & Sons, Inc.

FIGURE 1B-8   © H. J. de Blij, P. O. Muller, and John Wiley & Sons, Inc.

far less urbanized, much more agrarian, and not strongly integrated into it. The fourth national entity in this southern domain is the island ministate of Malta, south of Sicily, an historically important crossroads with a population of about 400,000 and a booming tourist industry.

Italy

Centrally located in Mediterranean Europe, most populous of the realm's southern states, best connected to the European Core, and economically most advanced is Italy (population: 60.5 million), a charter member of Europe's Common Market.

Administratively, Italy is organized into 20 internal regions, many with historic roots dating back centuries (Fig. 1B-8). Several of these regions have become powerful economic entities centered on major cities, such as Lombardy (Milan) and Piedmont (Turin); others are historic hearths of Italian culture, including Tuscany (Florence) and Veneto (Venice). These regions in the northern half of Italy stand in strong social, economic, and political contrast to such southern regions as Calabria (the "toe" of the Italian "boot") and Italy's two major Mediterranean islands, Sicily and Sardinia. Not surprisingly, Italy is often described as two countries—a progressive north and a stagnant south (known as the ***Mezzogiorno***). The urbanized,

ROME

FROM A HIGH vantage point, Rome seems to consist of an endless sea of tiled roofs, above which rise numerous white, ochre, and gray domes of various sizes; in the distance, the urban perimeter is marked by high-rises fading in the urban haze. This historic city lives amid its past as perhaps no other as busy traffic encircles the Colosseum, the Forum, the Pantheon, and other legacies of Europe's greatest empire.

Founded about 3000 years ago at an island crossing point on the Tiber River about 25 kilometers (15 mi) from the sea, Rome had a high, defensible site. A millennium later, with a population some scholars estimate as high as 1 million, it was the capital of a Roman domain that extended from Britain to the head of the Persian Gulf and from the shores of the Black Sea to North Africa. Rome's emperors endowed the city with magnificent, marble-faced, columned public buildings, baths, stadiums, obelisks, arches, and statuary; when Rome became a Christian city, the domes of churches and chapels added to its luster.

It is almost inconceivable that such a city could collapse, but that is what happened after the center of Roman power shifted eastward to Constantinople (now Istanbul). By the end of the sixth century, Rome probably had fewer than 50,000 inhabitants, and in the thirteenth, a mere 30,000. Papal rule and a Renaissance revival lay ahead, but in 1870, when Rome became the capital of newly united Italy, it still had a population of only 200,000.

Now began a growth cycle that eclipsed all previous records. As Italy's political, religious, and cultural focus (though not an industrial center to match), Rome grew to 1 million by 1930, to 2 million by 1960, and subsequently to 3.4 million where it has leveled off today. The religious enclave of Vatican City, Roman Catholicism's headquarters, makes Rome a twin capital; the Vatican functions as an independent entity and has a global influence that Italy cannot equal.

© H. J. de Blij, P. O. Muller, and John Wiley & Sons, Inc.

Rome today remains a city whose economy is dominated by service industries: national and local government, finance and banking, insurance, retailing, and tourism employ three-quarters of the labor force. The new city sprawls far beyond the old, walled, traffic-choked center where the Roman past and the Italian future come face to face.

industrialized north is part of Europe's Core; the low-income south typifies the Periphery.

North and south are bound by the ancient headquarters, Rome, which lies astride the narrow transition zone between Italy's contrasting halves. This clear manifestation of Europe's Core-Periphery contrast is referred to in Italy as the *Ancona Line*, named after the city on the Adriatic coast where it reaches the other side of the peninsula (Fig. 1B-8, blue line). Whereas Rome remains Italy's capital and cultural focus, the functional core area of Italy has shifted northward into Lombardy in the basin of the Po River. Here lies southern Europe's leading manufacturing complex. The Milan–Turin–Genoa triangle exports appliances, instruments, automobiles, ships, and many specialized products. Meanwhile, the Po Basin, lying on the margins of southern Europe's dominant Mediterranean climatic regime (with its hot, dry summers), enjoys a

more even pattern of rainfall distribution throughout the year, making it a productive agricultural zone as well.

Metropolitan Milan embodies the new, modern Italy. Not only is Milan (4.3 million) Italy's largest city and leading manufacturing center—making Lombardy one of Europe's Four Motors—but it also is the country's financial and service-industry headquarters. Today the Milan area, a cornerstone of the European Core, contains just 7 percent of Italy's population but accounts for fully one-third of the entire country's national income.

Spain, Portugal, and Malta

At the western end of southern Europe lies the Iberian Peninsula, separated from France and western Europe by the rugged Pyrenees and from North Africa by the narrow Strait of Gibraltar. Spain (population: 47.4 million) occupies most of this compact Mediterranean landmass, and

FIGURE 1B-9

© H. J. de Blij, P. O. Muller, and John Wiley & Sons, Inc.

peripheral Portugal lies in its southwestern corner. Both countries benefited enormously from their admission to the European Union in 1986.

Spain followed the leads of Germany and France and decentralized its administrative structure in response to devolutionary pressures. These pressures were especially strong in Catalonia, the Basque Country, and Galicia, and in response the Madrid government created so-called *Autonomous Communities (ACs)* for all 17 of its regions (Fig. 1B-9). Every AC has its own parliament and administration that control planning, public works, cultural affairs, education, environmental policy, and even, to some extent, international commerce. Each AC can negotiate its own degree of autonomy with the central government in the capital. Some Spanish observers feel that devolution has gone too far and that a federal system (such as Germany's) would have been preferable, but now there is no turning back.

Even so, the AC framework has not totally defused the most violent of the secessionist-minded movements, propelled by a small minority of Basques. On the other hand, relations between Madrid and Catalonia have improved. Centered on the prosperous, productive coastal city of Barcelona, the AC named Catalonia is Spain's leading industrial area and has become one of Europe's Four Motors. Catalonia

is endowed with its own distinctive language and culture that find vivid expression in Barcelona's urban landscape.

As Catalans like to remind visitors to their corner of Spain, most of the country's industrial raw materials are found in the northwest, but most of its major industrial development has occurred in the northeast, where innovations and skills drive a high-technology regional economy. In recent years, Catalonia—with 6 percent of Spain's territory and 15 percent of its population—has annually produced 25 percent of all Spanish exports and nearly 40 percent of its industrial exports. Such economic strength translates into political power, and in Spain the issue of Catalonian separatism is never far from the surface.

Relations between Catalonia and Castile (where Madrid is located) are also a product of history. During the Spanish Civil War (1936–1939), the military dictator Francisco Franco brought all of Spain under his command by violently crushing the resistance in several corners of the country, especially in independent-minded Catalonia. From 1939 to 1975, Franco's dictatorial rule was bitterly resented in Barcelona. Even though things have greatly improved since Spain converted to democracy in 1975, there has always remained a sense of animosity that is nowadays perhaps best expressed (and more happily so) in "*el clasico,*"

From the Field Notes . . .

"Catalonian nationalism is visible both obviously and subtly in Barcelona's urban landscape. Walking toward the Catalonian Parliament, I noticed that the flags of Spain (left) and Catalonia (right) flew from slightly diverging flagpoles above the entrance to the historic building. Is there a message here?"

www.conceptcaching.com

© H. J. de Blij

the highly anticipated annual soccer match between the country's leading teams, Barcelona and Real Madrid.

As Figure 1B-9 shows, Spain's capital and largest city, Madrid, lies near the geographic center of the state. It also lies within an economic-geographic transition zone. In terms of personal annual income, the Spanish north is far more affluent than the south, a direct result of the country's distribution of resources, climate (the south suffers from drought and inferior soils), and overall development opportunities. The most prosperous ACs are Barcelona-centered Catalonia and Madrid; between them, the contiguous group of ACs extending from the Basque Country to Aragon rank next. In the ACs of northern Spain, economies tend to do better in the east, where tourism and wine production (especially in La Rioja, a famous name in Spanish wine) are among the mainstays, than in the west, where industrial obsolescence, dwindling raw-material sources, the decline of the fishing industry, and emigration

Gibraltar, Ceuta, and Melilla

ALTHOUGH SPAIN AND the United Kingdom are both EU members committed to cooperative resolution of territorial problems, the two countries are embroiled in a dispute over a sliver of land at the southern tip of Iberia: legendary *Gibraltar* (see lower inset map, Fig. 1B-9). "The Rock" was ceded by Spain to the British in perpetuity in 1713 (though not the neck of the peninsula linking it to the mainland, which the British later occupied) and has been a British colony ever since. Its 30,000-odd residents are used to British institutions, legal rules, and schools. Successive Spanish governments have demanded that Gibraltar be returned to Madrid's rule, but the colony's residents are against it. And under their 1969 constitution, Gibraltarians have the right to vote on any transfer of sovereignty.

Now the British and Spanish governments are trying to reach an agreement under which they would share the administration of Gibraltar for an indefinite period; since both are EU members, little would have to change. The advantages would be many: economic development is being slowed by the ongoing dispute, border checks would be lifted, EU benefits would flow. But Gibraltarians have their doubts and want to put the issue to a referendum. Spain will not accept the idea of a referendum, and the matter is far from resolved.

Spain's refusal to allow, and abide by, a referendum in Gibraltar stands in marked contrast to its demand for a referendum in its own outposts, two small *exclaves* on the coast of Morocco, *Ceuta* and *Melilla* (Fig. 1B-9). Morocco has been demanding the return of these two small cities, but Spain has refused on the grounds that the local residents do not want this. The matter has long simmered quietly but came to international attention in 2002 when a small detachment of Moroccan soldiers seized the island of Perejil, an uninhabited, Spanish possession off the Moroccan coast also wanted by Morocco. In the diplomatic tensions that followed, the entire question of Spain's holdings in North Africa (and its anti-Moroccan stance in the larger matter of Western Sahara [an issue discussed in Chapter 7B]) exposed some contradictions Madrid had preferred to conceal.

plague local economies. Spain's southernmost ACs have long been the least developed economically and are more emblematic of Europe's Periphery. Here, too, three additional tiny entities mark the political map in the far south (see the box titled "Gibraltar, Ceuta, and Melilla").

Economically, Spain is going through a rough patch. Economic growth declined in both 2009 and 2010, and there were fears that the country's banking and financial sectors might follow those of Greece, Ireland, and Portugal. Unemployment in early 2011 approached 20 percent, and government debt ranked among the highest in the EU. One of the fastest-growing immigration countries since 2000, job-seeking foreigners are now widely viewed as a social problem. Since Spain's economy represents a much bigger part of the EU than Greece's or Ireland's, a Spanish default could be far more consequential. Whether Spain's financial sector will hold, and enable the country to turn the economic corner once this stubborn recession gives way to renewed growth, remains to be seen.

Much the same can be said of **Portugal** (10.7 million), which occupies the southwestern corner of the Iberian Peninsula. Because EU agricultural policies entail transfers from rich to poor regions in the Union and because the EU funds major infrastructural projects in hitherto isolated areas, Portugal has benefited enormously from its admission to the EU—but it remains a far cry from Europe's leading economies. Unlike Spain, which has major population clusters on its interior plateau as well as its coastal lowlands, the Portuguese are concentrated along and near the Atlantic coast. Lisbon and the second city, Porto, are coastal cities. The best farmlands lie in the moister western and northern zones of the country; but the farms are small and inefficient, and even though Portugal remains dominantly rural, it must import as much as half of its food. Portugal's poor finances and its struggle to get beyond the recession required an EU bailout in 2011, raising additional concern for the stability of the Union.

Southernmost Europe also contains the ministate of **Malta**, located in the central Mediterranean Sea just south of Sicily. Malta is a small archipelago of three inhabited and two uninhabited islands with a population of just over 400,000 (Fig. 1B-7, inset map). An ancient crossroads and culturally rich with Arab, Phoenician, Italian, and British infusions, Malta became a British dependency and served British shipping and its military. It suffered terribly during World War II bombings, but despite limited natural resources recovered strongly during the postwar era. Today Malta has a booming tourist industry and a relatively high standard of living, and was one of the ten new member-states to join the European Union as part of the historic expansion of 2004.

Greece and Cyprus

If the label "discontinuous geography" applies anywhere, it is the area between Austria and the Aegean Sea, covering the Balkans (former Yugoslavia) and Greece. This is an area encompassing various (and conflicting) religions and linguistic groups; prosperous and very poor nations; lands with volatile and violent histories; EU members and non-members (some without any prospect of gaining entry anytime soon); and, furthest removed from the Core is Greece, the birthplace of European civilization and a longtime EU member and Western ally. We turn first to Greece and the Greek-connected island-state of Cyprus, and then to the Balkans.

Seemingly dangling from the southern end of eastern Europe, **Greece** was the wellspring of one of the ancient world's greatest civilizations, its scientists and philosophers still cited to this day, its famous tragedies still staged in the amphitheaters built more than 2000 years ago. Its familiar peninsulas and islands are bounded by Turkey to the east and by Bulgaria, Macedonia, and Albania to the north (Fig. 1B-10). Note that some of Greece's islands in the Aegean Sea lie on Turkey's very doorstep; in addition, Greeks represent the majority of the population of Cyprus in the distant northeast corner of the Mediterranean Sea.

A member of the EU since 1981, Greece has had a turbulent modern history, with alternating leftist and fascist dictatorships giving way to more democratic government in the 1980s and economic upheavals marking the EU period. Economic stagnation in the 1980s was followed by such progress that Greece came to be called the "locomotive of the Balkans," an EU beacon in its remote corner of the Union. Metropolitan Athens burgeoned, benefiting from EU assistance, the hosting of the 2004 Olympics, and membership in the EMU. Arriving by air, you find yourself at one of the world's most modern airports followed by a ride on world-class superhighways or subways. The metropolitan area of Athens and its busy port, Piraeus, contain about one-third of the country's population of 11.3 million.

But EU membership had other, less favorable consequences. When the Union expanded to include poorer states such as Bulgaria and Romania, Greece lost much of its EU subsidy as Brussels had to divert its equalization funds to needier members. This meant that Greece had to tighten its financial belt, but most citizens had become used to early retirement, loose tax enforcement, lifetime government employment, and protection against competition (for example, the notion of private universities elicits angry street protests). And so, as the first decade of our new century came to a close, Greece was at the center of an unfolding tragedy as its inefficient, hamstrung government was mired in debt, deadly street riots greeted efforts at painful reform, and a proud historic nation found itself forced to conform to the bailout terms imposed by the EU powers providing the necessary loans.

This southeastern corner of the European Periphery also contains the island country of **Cyprus** (an EU member since 2004), whose political geography merits special attention because of the complications it created, and continues to create, for the EU. It is useful to elaborate on this issue because it is such a good illustration of the political intricacies and challenges of the unification process. As Figure 1B-1 shows, Cyprus lies closer to Turkey than it does to any part

FIGURE 1B-10

© H. J. de Blij, P. O. Muller, J. Nijman, and John Wiley & Sons, Inc.

of Europe, but it is predominantly peopled by Greeks who have been there the longest. In 1571, the Turkish Ottoman Empire conquered Cyprus, until then ruled by Venice, and controlled it until 1878 when the British took over. Most of the island's Turks arrived during the Ottoman era.

When the British were ready to give Cyprus independence after World War II, the 80-percent-Greek majority mostly preferred union with Greece. Ethnic

conflict followed, but in 1960 the British granted Cyprus independence under a constitution that prescribed majority rule but guaranteed minority rights. This fragile order broke down in 1974, and civil war engulfed the island. Turkey sent in troops and massive dislocation followed, resulting in the partition of Cyprus into northern Turkish and southern Greek sectors (Fig. 1B-10, inset maps). In 1983, the 40 percent of Cyprus under Turkish control,

ATHENS

TAKE A MAP of Greece. Draw a line to circumscribe its land area as well as all the islands. Now find the geographic center of this terrestrial and maritime territory, and you will find a major city nearby. That city is Athens, capital of modern Greece, culture hearth of ancient Greece, source of Western civilization.

Athens today is only part of a greater metropolis that sprawls across, and extends beyond, a mountain-encircled, arid basin that opens southward to the Bay of Phaleron, an inlet of the Aegean Sea. Here lies Piraeus, Greece's largest port, linked by rail and road to the adjacent capital. Other towns, from Keratsinion in the west to Agios Dimitrios in the east, form part of an urban area that not only lies at, but is, the heart of modern Greece. With a population of 3.3 million, most of the country's major industries, its densest transport network, and a multitude of service functions ranging from government to tourism, Greater Athens is a well-defined national core area.

Arrive at the new Eleftherios Venizelos Airport east of Athens and drive into the city, and you can navigate on the famed Acropolis, the 150-meter (500-ft) high hill that was ancient Athens' sanctuary and citadel, crowned by one of humanity's greatest historic treasures, the Parthenon, temple to the goddess Athena. Only one among many relics of the grandeur of Athens 25 centuries ago, it is now engulfed by mainly low-rise urbanization and all too often obscured by smog generated by factories and vehicles and trapped in the basin. Conservationists decry the damage air pollution has inflicted on the city's remaining historic structures, but in truth the greatest destruction was wrought by the Ottoman Turks, who plundered the country during their occupation of

© H. J. de Blij, P. O. Muller, and John Wiley & Sons, Inc.

it, and the British and Germans, who carted away priceless treasures to museums and markets in western Europe.

Travel westward from the Middle East or Turkey, and Athens will present itself as the first European city on your route. Travel eastward from the heart of Europe, however, and your impression of its cultural landscape may differ: behind the modern avenues and shopping streets are bazaar-like alleys and markets where the atmosphere is unlike that of any other in this geographic realm. And beyond the margins of Greater Athens lies a Greece that lags far behind this bustling, productive city.

---

with about 100,000 inhabitants (plus some 30,000 Turkish soldiers), declared itself the independent **Turkish Republic of Northern Cyprus**. Only Turkey recognizes this ministate (which today contains a population of about 400,000); meanwhile, the international community recognizes the government on the Greek side as legitimate.

Things got even more complicated when only the Greek side of the island was admitted to the EU in 2004. Resentment was high on the Turkish side and in Turkey itself as well—just as discussions on Turkey's own admission to the EU were getting under way. It was—and remains—a reminder that Cyprus's "Green Line" separating the Greek and Turkish communities constitutes not just a regional border but a boundary between geographic realms.

The Balkans

We now turn to a component of Europe that has undergone wrenching changes since 1990, a process that started with the violent dismantling of communist Yugoslavia during the

final decade of the twentieth century and continues—fortunately less violently—today. Here the European Union has made only slight progress, and just one of the countries that emerged from Yugoslavia's disintegration, Slovenia, has joined the EU. Discussions with Croatia, Slovenia's neighbor, may yield results in 2012. But no other state in this area is a serious candidate for admission at present.

This part of Europe has had a particularly volatile history, and it is a classic example of what geographers call a **shatter belt [10]**, a zone of persistent splintering and fracturing. Geographic terminology uses several expressions to describe the breakup of established order, and these tend to have their roots in this part of the world. One of those expressions is **balkanization [11]**. The southern half of eastern Europe is referred to as the Balkans or Balkan Peninsula (named after a mountain range in Bulgaria). Balkanization denotes the recurrent division and fragmentation of a region.

The key state on the new map is **Serbia**, the name of what is left of a much larger domain once ruled by the

Serbs—who were dominant in the former Yugoslavia. Centered on the historic capital of Belgrade on the Danube River, Serbia (7.2 million people) is the largest and potentially most important new country in the area. But the Serbs are having to accommodate some major changes. First, more than 1 million of them live in neighboring Bosnia, where they have an uneasy relationship with the local Muslims and Croats (Fig. 1B-11). Second, the coastal province named *Montenegro* broke away in 2006, when voters there decided to form an independent state. Third, its Muslim-majority province of *Kosovo* declared independence in 2008, its sovereignty immediately recognized by the United States and a majority of (but not all) European governments. And fourth, Serbia still incorporates a Hungarian minority of some 400,000 in its northern province of *Vojvodina* on the northern side of the Danube at a time when Hungary has already joined the EU. Clearly, the Balkans represent a virtual kaleidoscope of ethnic identities that severely complicates any kind of territorial arrangement (see Fig. 1B-11).

If EU expansion occurs in this southeastern sector, it may involve **Croatia**, the crescent-shaped country with prongs along the Hungarian border as well as the Adriatic coast. But here, too, there are problems. About 90 percent of Croatia's 4.4 million citizens are Croats, but the country's Serb minority has faced discrimination and has

declined from 12 percent of the population to under 4. The EU took a dim view of human rights issues in Croatia, but these improved when EU membership beckoned. Meanwhile, about 800,000 Croats live outside Croatia in Bosnia, where their relationships with Muslims and Serbs are not always cooperative.

When the former Yugoslavia collapsed, **Bosnia** was the cauldron of calamity. No ethnic group was overwhelmingly dominant here, and this multicultural, effectively landlocked triangle of territory, lying between the Serbian stronghold to the east and the Croatian republic to the west and north, fell victim to disastrous conflict among Serbs, Croats, and *Bosniaks* (now the official name for Bosnia's Muslims, who constitute about 50 percent of the population of 3.8 million). As many as 250,000 people perished in concentration camps associated with *ethnic cleansing* practices; in 1995, a U.S. diplomatic effort resulted in a truce that partitioned the country as shown in Figure 1B-11. This is one rough corner of Europe.

The southernmost "republic" of former Yugoslavia was **Macedonia**, which emerged from the collapse as a state with a mere 2 million inhabitants, of whom two-thirds are Macedonian Slavs. As the map shows, Macedonia adjoins Muslim Albania and Kosovo, and its northwestern corner is home to the 25 percent of the Macedonians who are nominally

FIGURE 1B-11

© H. J. de Blij, P. O. Muller, and John Wiley & Sons, Inc.

Muslims. The remainder of this culturally diverse population are Turks, Serbs, and Roma (see box on page 96). Macedonia is one of Europe's poorest countries, landlocked and powerless. Even its very name caused it problems: Macedonia's Greek neighbors argued that this name was Greek property and so would not recognize it. Next, Macedonia faced an autonomy movement among its Albanian citizens, requiring allocation of scarce resources in the effort to hold the fledgling state together. Macedonians cling to the hope that eventual EU admission will bring it subsidies and better economic times.

The ministate of **Montenegro** has a mere 630,000 inhabitants, one-third of them Serbs, a small capital (Podgorica), some scenic mountains, and a short but spectacular Adriatic coastline—and very little else to justify its rank as one of Europe's 40 countries. Some tourism, a black market, trickles of Russian investment, and a scattering of farms sum up the assets of this country.

In 1999, amid the chaos of disintegrating Yugoslavia, NATO took charge in **Kosovo**, then still a formal Serbian province, after a brief but damaging military campaign. With an overwhelmingly Albanian-Muslim population of more than 2 million and a Serb minority of about 150,000 mostly tucked away in its northern corner, landlocked Kosovo has few of the attributes of a full-fledged state. NATO eventually turned over Kosovo's administration to the United Nations, and in 2008 the capital of Pristina witnessed independence celebrations as the UN administration yielded to a newly elected national government.

The only other dominantly Muslim country in Europe is **Albania**, where over 75 percent of the population of 3.2 million adhere to Islam. Albania shares with one other country—Moldova—the status of being Europe's poorest state. Albania has one of Europe's higher rates of natural population growth, and many Albanians try to emigrate to the EU by crossing the Adriatic Sea to Italy. Most Albanians subsist on livestock herding and farming, but the poverty-stricken Gegs in the north lag behind the somewhat better-off Tosks in the south, with the capital of Tirane lying close to this cultural divide. For all of Europe's globalization, Albania would represent the symptoms of the global periphery anywhere in the world.

▪▪ THE DISCONTINUOUS NORTH

It is a long way from the Balkans to the Baltic region—in terms of distance, physiography, and culture. The north's remoteness, isolation, and environmental severity, especially at higher latitudes, have had positive binding effects for part of this domain and seem to have fostered similar cultures. The Scandinavian Peninsula lay removed from most of the wars of mainland Europe and developed in relatively tranquil circumstances (although Norway was overrun by Nazi Germany during World War II). The three major languages—Swedish, Norwegian, and Danish—are mutually intelligible, and in terms of religion there is overwhelming adherence to the same Lutheran church in the three Scandinavian countries as well as Iceland, Finland, Estonia, and Latvia. Lithuania and Latvia have distinct languages, and in Lithuania the Roman Catholic religion prevails—but all the Baltic states have a shared history on the shores of the Baltic Sea and in proximity to Russia. In Scandinavia, democratic and representative governments emerged early, and individual rights and social welfare have long been carefully protected. Women participate more fully in government and politics here than in any other part of the world.

But consider the implications of Figures 1B-1 and 1B-12, and it is obvious why we refer to the *Discontinuous North*: the southern, coastal, and urban areas of the region form part of the Core, but the rest does not. In the aggregate, economic indicators for most of the region are very strong, but almost all development is concentrated within that smaller Core zone. Territorially, most of northern Norway, Sweden, and Finland form part of the European Periphery.

The countries of this northern domain of Europe have a combined population of only 32 million, about half of Italy's. Nothing here compares to Italy's north or Spain's Catalonia, but the core areas of the three Scandinavian countries (Sweden, Norway, and Denmark) make up in prosperity and external linkages what they lack in dimensions.

Sweden

Sweden is the Discontinuous North's largest country in terms of both population (9.4 million) and territory. Most Swedes live south of 60° North latitude (which passes through Uppsala just north of the capital, Stockholm), in what is climatically the most moderate part of the country (Fig. 1B-12). Here lie the primate city, core area, and the main industrial districts; here, too, are the main agricultural areas that benefit from the lower relief, better soils, and milder climate.

Sweden long exported raw or semifinished materials to industrial countries, but today the Swedes make finished products themselves, including motor vehicles, electronics, stainless steel, furniture, and glassware. Much of this production is based on local resources, including a major iron ore reserve at Kiruna in the far north (there is a steel mill at Luleå). Swedish manufacturing, in contrast to that of several western European countries, is based in dozens of small and medium-sized towns specializing in particular products. Energy-poor Sweden was a pioneer in the development of nuclear power, but a national debate over the risks involved has reversed that course.

Norway

Norway does not need a nuclear power industry to supply its energy needs. It has found its economic opportunities on, in, and beneath the sea. Norway's fishing industry, now augmented by highly efficient fish farms, long has been a cornerstone of the economy, and its merchant marine spans the world. But since the 1970s, Norway's economic

FIGURE 1B-12

© H. J. de Blij, P. O. Muller, and John Wiley & Sons, Inc.

life has been transformed by the bounty of oil and natural gas discovered in its sector of the North Sea.

With its limited patches of cultivable soil, high relief, extensive forests, frigid north, and spectacular *fjords*, Norway has nothing to compare to Sweden's agricultural or industrial development. Its cities, from the capital Oslo and the North Sea port of Bergen to the historic national focus of Trondheim as well as Arctic Hammerfest, lie on the coast and have difficult overland connections. The isolated

northern province of Finnmark has even become the scene of an autonomy movement among the reindeer-herding indigenous Saami (Fig. 1A-11). The distribution of Norway's population of 4.9 million has been described as a necklace, its beads linked by the thinnest of strands. But this has not constrained national development. Norway in the late 2000s had the second-lowest unemployment rate in Europe (after tiny Luxembourg). In terms of income per capita, Norway is one of the richest countries in the world.

Norwegians have a strong national consciousness and a spirit of independence. In the mid-1990s, when Sweden and Finland voted to join the European Union, the Norwegians again said no. They did not want to trade their economic independence for the regulations of a larger, even possibly safer, Europe.

Denmark

Territorially small by Scandinavian standards, Denmark has a population of 5.5 million, the North's second-largest country after Sweden. It consists of the Jutland Peninsula and several islands to the east at the gateway to the Baltic Sea; it is on one of these islands, Sjaelland, that the capital of Copenhagen is located. Copenhagen, the "Singapore of the Baltic," has long been a port that collects, stores, and transships large quantities of goods. This **break-of-bulk** [12] function exists because many oceangoing vessels cannot enter the shallow Baltic Sea, making the city an **entrepôt** [13] where transfer facilities and activities prevail.

The completion of the Øresund bridge-tunnel link to southern Sweden in 2000 enhanced Copenhagen's situation (Fig. 1B-12).

Denmark remains a kingdom, and in centuries past Danish influence spread far beyond its present confines. Remnants of that period now challenge Denmark's governance. Greenland came under Danish rule after union with Norway (1380) and continued as a Danish possession when that union ended (1814). In 1953, Greenland's status changed from colony to province, and in 1979 the 60,000 inhabitants were given home rule with an Inuit name: **Kalaallit Nunaat**. They promptly exercised their rights by withdrawing from the European Union, of which they had become a part when Denmark joined.

Greenland is still highly dependent on financial support from Denmark but this may change in the near future. After decades of exploration, oil was discovered in 2010 off Greenland's west coast, north of the Arctic Circle. Oil extraction in the Arctic carries substantial environmental risk, and it may be a while before operations

From the Field Notes . . .

"The most cheerful corner of Nuuk, the capital of Greenland, is enlivened by the bright colors of the homes built on what was the original site of settlement here on the Davis Strait. Nuuk offers a few surprises: a supermarket featuring vegetables grown in Greenland, a nine-hole golf course being expanded into 18 holes, and a vigorous debate over the prospect of independence. With a population of only about 60,000, considerable autonomy from former colonial power Denmark, and substantial investment from Copenhagen, the citizens of Kalaallit Nunaat are nonetheless divided on the question of their future. Climate change, the prospect of oil reserves to be found offshore, and freedom to fish (including whales) like the Norwegians do, are factors seen by many indigenous Greenlanders as potential rewards of independence. Danish residents and those of Danish (but local) ancestry tend to see it differently. In the June 2009 election, the party representing indigenous interests did better than expected, and 'KN' seemed to be on course toward sovereignty."

www.conceptcaching.com

© H. J. de Blij

are in place and revenues start flowing in; but when they do, they are likely to bring profound change to Greenland.

Another restive Danish dependency is the Faroe Islands, located between Scotland and Iceland. These 17 small islands and their 45,000 inhabitants were awarded self-government in 1948, complete with their own flag and currency, but even this was not enough to defuse demands for total independence. A referendum in 2001 confirmed that even Denmark is not immune from Europe's devolutionary forces.

Finland

Finland, territorially almost as large as Germany, has only 5.4 million residents, most of them concentrated in the southern triangle formed by the capital, Helsinki, the textile-producing center, Tampere, and the shipbuilding center, Turku (Fig. 1B-12). A land of evergreen forests and glacial lakes, Finland's national income has long been sustained by wood and wood-product exports. But the Finns have now developed a more diversified economy with staple agricultural crops and the manufacture of precision machinery and telecommunications equipment (including Nokia mobile phones).

Finland's environmental challenges and relative location have created cultural landscapes similar to those of Sweden and Norway, but the Finns are not a Scandinavian people; their linguistic and historic links are instead with the Estonians to the south across the Gulf of Finland. Ethnic groups speaking these Finno-Ugric languages are widely dispersed across what is today western Russia.

Estonia

The northernmost of the three "Baltic states," Estonia has longstanding ethnic and linguistic ties to Finland. But during the period of Soviet control from 1940 to 1991, Estonia's demographic structure changed drastically: today about 25 percent of its 1.3 million inhabitants are Russians, most of whom came there as colonizers.

After a difficult period of adjustment, Estonia today is forging well ahead of its Baltic neighbors, Latvia and Lithuania (see Appendix B). Busy traffic links Tallinn, the capital, with its counterpart in Finland, Helsinki, and a free-trade zone at Muuga Harbor facilitates commerce with Russia. But more important for this country's future was its entry into the European Union in 2004. Estonia has gained attention in recent years due to its economic advances and especially its creative entrepreneurship in the high-tech and software industries. Skype, for instance, was originally an Estonian company. Having entered the euro zone in 2011, Estonia is well on its way toward soon joining the ranks of the European Core.

Latvia

As Figure 1B-12 reminds us, the boundary of the European Core that traverses Europe's North includes southern Norway and Sweden but excludes Estonia and the other two Baltic states. The latter have experienced economic improvement in recent years, but are not yet on a par with Estonia, let alone Scandinavia. Latvia, the middle Baltic state centered on the port and capital city of Riga, was tightly integrated into the Soviet system during Moscow's long domination. Still today, only 59 percent of the population of 2.2 million is Latvian, and about 28 percent is Russian.

After independence ethnic tensions arose between these Baltic and Slavic sectors, but the prospect of EU membership required the end of discriminatory practices. Latvians concentrated instead on the economy, which had been left in dreadful shape, with the result that it qualified for 2004 admission. Consider this: 25 years ago, virtually all of Latvia's trade was with the Soviet Union. Today its principal trading partners are Germany, the United Kingdom, and Sweden. Russia figures in only one category: Latvia's import of oil and gas.

Lithuania

This southernmost Baltic state of 3.3 million has a residual Russian minority of only about 5 percent, but relations with its giant neighbor are worse than Latvia's—this despite Lithuania's greater dependence on Russia as a trading partner. One reason for this bad relationship has to do with neighboring Kaliningrad, Russia's **exclave [14]** on the Baltic Sea (Fig. 1B-12). When Kaliningrad became a Russian territory after World War II (Russia views the exclave as a potentially important outpost at a time when the EU and NATO are advancing toward its doorstep), Lithuania was left with only about 80 kilometers (50 mi) of Baltic coastline and a small port that was not even connected by rail to the interior capital of Vilnius. Lithuania in 2005 called for the demilitarization of Kaliningrad as a matter of national security.

Despite these problems, Lithuania's economy during 2003 had the highest growth rate in Europe, spurred by foreign investment and by profits from its oil refinery at Mazeikiai, facilitating EU admission in 2004.

Iceland

Iceland, the volcanic, glacier-studded island in the frigid waters of the North Atlantic just south of the Arctic Circle, is the eighth country of the Discontinuous North. Inhabited by people with Scandinavian ancestries (population: 340,000), Iceland and its small neighboring archipelago, the Westermann Islands, are of special scientific interest because they lie on the Mid-Atlantic Ridge. Here, the Eurasian and North American tectonic plates of the Earth's crust are diverging (see Fig. G-4), new land can be seen forming, and spectacular volcanic eruptions are periodically on display (as occurred memorably in April 2010 when the Eyjafjallajökull volcano brought transatlantic and domestic air traffic across northern Europe to a halt for several days).

Iceland's population is almost totally urban, and the capital, Reykjavik, contains about half the country's inhabitants. The nation's economic geography is traditionally oriented to the surrounding waters, whose seafood harvests have given Iceland a high standard of living. In the 1990s, Iceland embarked on economic liberalization policies and its financial industries and banks grew rapidly. For a while, Iceland was referred to as the *Nordic Tiger*— but not unlike Ireland, which went through a similar boom, the economy went into a nosedive with the onset of global recession in 2008. The Icelandic government, closely involved with some of the troubled banks, had to be bailed out by the International Monetary Fund.

▪▪ THE EASTERN PERIPHERY

Europe's Eastern Periphery is a fascinating region, not least because it is so difficult to know exactly where its boundaries are. It is a region that in some ways becomes less European going east and less Slavic going west; one that throughout history has found itself in the way of territorial ambitions of major powers on either side; and with a changeable political geography to match. In a recent history book, the area extending from the Baltic to the Black Sea was morbidly dubbed as "Bloodlands," the region where in the second quarter of the twentieth century Hitler's Germany and Stalin's Soviet Union did their most murderous work. More people died here than in any other part of Europe. After World War II, eastern Europe was, forcibly for the most part, incorporated into the Soviet orbit, its various national identities suppressed along with basic freedoms and its economies subordinated to Moscow's interests.

Europe has changed fundamentally in the two decades since the Cold War ended in 1991. The old Eastern Europe, sharply defined along the Iron Curtain, no longer exists, and in some ways we have seen the rebirth of the even older cultural region known as Central Europe or, in its prevailing German language, *Mitteleuropa*. Many of these changes were consolidated as the EU expanded eastward. As we saw, the Czech Republic has already become integrated into the Core region, and Slovenia is likely to be the first former State of Yugoslavia to merit similar inclusion. By 2007, a relatively large number of eastern countries had already joined the Union, and others are knocking on the door. Undoubtedly, the transformation of this sector of the European realm will continue to advance steadily in the 2010s.

East-Central Europe

As Figures 1B-1 and 1B-13 show, the four states in this group are all immediately adjacent to or already partially integrated into the European Core. The most important of these four countries is **Poland**, which was also the largest and most populous country among the ten that joined the EU in 2004. With 38.3 million people and lying between two historic enemies, Poland has borders that have shifted time and again; but its current status may be more durable than in the past. As the map shows, the historic and once-central capital of Warsaw now lies closer to Russia than to Germany, but the country now looks west, not east.

From the Field Notes . . .

"Turbulent history and prosperous past are etched on the cultural landscape of the old Hanseatic city of Gdansk, Poland. Despite major wartime destruction, much of the old architecture survives, and restoration has revived not only Gdansk but also its twin (port) city, Gdynia. Gdansk was the initial stage for the rebellion in the 1980s of the Solidarity labor union against Poland's communist regime, and after that regime was toppled, the country went through difficult economic times. Today Poland, a full member of the European Union, is making rapid progress; its historic cities, from Gdansk in the north to Krakow in the south, attract throngs of foreign visitors and the tourist industry is booming. Note: don't let that Dutch tricolor mislead you. Attracting foreign visitors is one way to raise revenues, and the historic old city on the Baltic is becoming a tourist draw."

© Sigmund Malinowski

www.conceptcaching.com

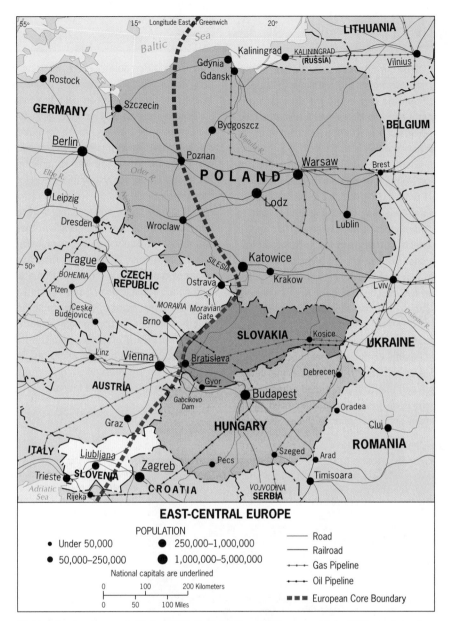

FIGURE 1B-13   © H. J. de Blij, P. O. Muller, J. Nijman, and John Wiley & Sons, Inc.

During the Soviet-communist period, Silesia in the southwest was its industrial heartland, and Katowice, Wroclaw, and Krakow grew into major manufacturing centers amid some of the world's worst environmental degradation. The Soviets invested far less in agriculture, collectivizing farms without modernizing technologies and leaving post-Soviet farming in abysmal condition. All this made governing Poland difficult, but the prospect of EU membership (with the promise of EU subsidies) motivated the government to get its house in order.

After entering the EU, Poland saw hundreds of thousands of workers depart for jobs in the European Core, but many have returned and the economy is growing. Today, Poland is in the best shape it has been in for a very long time. The economy, despite the global recession, has never been more robust, the country's geopolitical position has

never been more secure (it joined NATO in 1999), and its government is stable and highly regarded. Poland's primary goals are to stimulate further economic growth and stay firmly embedded in European economic and political organizations.

Government also was the problem in neighboring, landlocked **Slovakia**, where during the communist period the people were far more pro-Soviet than the Czechs next door. Many observers of the 2004 EU expansion wondered whether misgoverned Slovakia should be admitted because the capital, Bratislava, had become synonymous with corruption and inefficiency. Slovakia's Hungarian minority, comprising about 10 percent of the population of 5.4 million and concentrated in the south along the Danube River, was at odds with the Slovak regime. Moreover, additional concerns were raised over

Europe's Stateless Nation

EVERY ETHNIC GROUP in Europe's intricate social mosaic, it seems, has its own homeland, either in a state or a subnational unit. From the Frisians to the Corsicans and from the Norwegians to the Greeks, everyone has a historic base, whether it is a State or a province or some other political entity, no matter how small.

But there is one significant exception: the **Roma**. Perhaps as many as 11 million Roma, formerly (and derisively) called *gypsies*, constitute Europe's troubled, stateless nation. Although their origins are uncertain, the Roma are believed to have originated in India and migrated westward along several routes, one of which carried them via present-day Iran and Turkey into Europe. They never established a domain; they have retained their mobile, nomadic lifestyle to the present day; and they live, mostly in poverty, in a discontinuous arc across Bulgaria, Romania, Hungary, Slovakia, and the Czech Republic. Wherever they have gone, they have found themselves facing discrimination, resentment, unemployment, and poor health conditions (see photo below). Their mobile lifestyle contributes to low educational levels and to the perception and reality of associated crime.

The Roma were an issue during the debate preceding the EU's 2004 enlargement and again prior to the accession of Bulgaria and Romania in 2007, when many Europeans talked of a "Roma deluge" as a reason to oppose expansion. Even before 2004, some Roma had been entering the European Core as asylum-seekers, arriving in English towns and setting up their wagons and encampments in public parks and village squares. When several Core countries strengthened their asylum rules, their objective was not only to stem the flow from Islamic countries but also within the EU itself, with the Roma the primary target.

But they came anyway: there are now close to a million Roma in Spain and perhaps half that number in France, and their presence has been cause for highly charged public debate. In August 2010, the French government expelled a group of more than a thousand Roma after violent clashes with police in the town of Saint Aignan and the city of Grenoble, both sites of illegal Roma encampments. The problem is that the situation for the Roma in eastern European countries is even worse. EU efforts to help the Roma in their source countries through subsidies and other assistance have been hampered by the high levels of corruption in those states where Roma minorities are largest.

Europe's largest minority, the stateless Roma, also are the realm's poorest. Slovakia is one of several European countries with substantial Roma populations, and its government has been criticized by the EU for its treatment of Roma citizens. This depressing photograph of a Roma settlement in Hermanovce shows a cluster of makeshift dwellings virtually encircled by a moat bridged only by a walkway. The village of Hermanovce may itself not be very prosperous, but that moat represents a social chasm between comparative comfort and inescapable deprivation.

© Tomasz Tomaszewski/ngs/GettyImages, Inc.

reports of mistreatment of the smaller Roma (Gypsy) minority (see the box titled "Europe's Stateless Nation"). But again the promise of EU membership led to some reforms, and after 2004 the economy perked up. Following another downturn in 2007, a more enlightened administration took over, and in 2008 Slovakia surprised many by meeting the terms of admission to the EMU, adopting the euro in 2009 before the Poles or the Hungarians could.

Many economic geographers anticipated a bright future for also-landlocked **Hungary** following the collapse of the Soviet Empire and the dismantling of the Iron Curtain. The Hungarians (Magyars) moved into the middle Danube Basin more than a thousand years ago from an Asian source; they have neither Slavic nor Germanic roots. They converted their fertile lowland into a thriving nation-state and created an imperial power that held sway over an area far larger than present-day Hungary. Ethnic Magyar remnants

FIGURE 1B-14

© H. J. de Blij, P. O. Muller, and John Wiley & Sons, Inc.

of this Greater Hungary can still be found in parts of Romania, Serbia, and Slovakia (Fig. 1B-14), and the government in the twin-cities capital astride the Danube River, Budapest, has a history of irredentism toward these external minorities.

The concept of **irredentism [15]**—a government's support for ethnic or cultural cohorts in neighboring or more distant countries—derives from a nineteenth-century campaign by Italy to incorporate the territory inhabited by an Italian-speaking minority of Austria, calling it *Italia Irredenta* or "Unredeemed Italy." One advantage of joining the EU, obviously, was that much of the reason for Hungarian irredentism disappeared. If and when Serbia joins the EU, it may disappear altogether.

With a population of 9.9 million, a distinctive culture, and a considerable and varied resource base, Hungary should have good prospects, and its economic potential was a strong factor in its 2004 admission into the EU. Most recently, this country's newfound status was confirmed with its assumption of the presidency of the EU in 2011.

The success story among the four countries in this regional grouping remains modest, with most of it accounted for by progressive **Slovenia**, which lies wedged against Austria and Italy in the hilly terrain near the head of the Adriatic Sea. With 2.1 million people, a nearly homogeneous ethnic complexion, and a productive economy, Slovenia was the first "republic" of the seven that emerged from the collapse of Yugoslavia to be invited to join the EU. Shortly thereafter, Slovenia also became part of the euro zone.

The Southeast: Romania and Bulgaria

Just how Romania managed to persuade EU leaders to endorse its 2007 accession remains a question for many Europeans. As the Data Table in Appendix B indicates, Romania has some of Europe's worst social indicators: its economy is weak, incomes are low, its governmental system has not been sufficiently upgraded from communist times (a number of "apparatchiks" have acquired state assets under the guise of "privatization" and control the political process), and political infighting and corruption are endemic.

But Romania is an important country, located in the lower basin of the Danube River and occupying much of the heart of eastern Europe. With 21.4 million inhabitants and a pivotal situation on the Black Sea, Romania is a bridge between central Europe and the realm's southeastern corner, where EU member Greece and would-be member Turkey face each other across land and water (Fig. 1B-15).

Romania's drab and decaying capital of Bucharest (once known as the "Paris of the Balkans") and its surrounding core area lie in the interior, linked by rail to the Black Sea port of Constanta. The country's once-productive oilfields have now been fully depleted, about a third of the labor force works in agriculture, poverty is widespread in the countryside as well as the towns, and unemployment is high. Many talented Romanians continue to leave the country in search of opportunities elsewhere.

Across the Danube lies Romania's southern neighbor, Bulgaria. The rugged Balkan Mountains form Bulgaria's physiographic backbone, separating the Danube and Maritsa basins. As the map shows, Bulgaria has five neighbors, several of which are in political turmoil.

The Bulgarian state appeared in 1878, when the Russian czar's armies drove the Turks out of this area. The Slavic Bulgars, who form 85 percent of the population of 7.5 million, were (unlike the Romanians) loyal allies of Moscow during the Soviet period. But they did not treat their Turkish minority, about 10 percent of the population, very kindly, closing mosques, prohibiting use of the Turkish language, and forcing Turkish families to adopt Slavic names. Conditions for the remaining Turks improved somewhat after the demise of the Soviet Union. They improved further as Bulgaria became a candidate and then in 2007 a member of the European Union, which required judicial and other social reforms.

Bulgaria has a Black Sea coast and an outlet, the port of Varna, but the main advantage it derives from its coastal location is the tourist trade its beaches generate. The capital, Sofia, lies not on the coast but near the opposite border with Serbia, and in its core-area hinterland some foreign investment is changing the economic landscape—but slowly (Fig. 1B-15). Bulgaria's GNI is on a par with Romania's, although the telltale agricultural sector is much smaller here.

Bulgarians, too, are emigrating in droves—and this worries the countries of western Europe, where an uncontrolled influx of immigrants with newly-won access to their job markets would create serious problems. In 2006, the United Kingdom announced that it would place severe restrictions on workers from Romania and Bulgaria—a surprising reversal for a country that has long championed openness in the EU job market. It was a signal that admitting these two states had been a stretch, the wisdom of which remains to be proven.

Europe's Eastern Edge

In Europe's far east, and on Russia's doorstep, are three countries of which one, the key state of Ukraine, is territorially the largest in all of Europe. Demographically, with a population of 45.5 million, Ukraine ranks in the second tier of European states, along with Poland and Spain. As Figure 1B-15 shows, Ukraine's relative location is crucial. Not only does it link the core of Russia to the periphery of the European Union: it forms the northern shore of the Black Sea from Russia to Romania, including the strategically important Crimea Peninsula. And most importantly, oil and gas pipelines connect Russian fields to European markets across Ukrainian territory.

Ukraine's capital, Kiev (Kyyiv), is a major historic, cultural, and political nexus. Briefly independent before the communist takeover of Russia in 1917, Ukraine regained its sovereignty as a much-changed country in 1991. Once a land of farmers tilling its famously fertile soils, Ukraine emerged from the Soviet period with a huge industrial complex in its east—and with a substantial (17 percent) Russian minority. Ukraine's boundaries also changed during the Soviet era. In 1954, a Soviet dictator capriciously transferred the entire Crimea Peninsula, including its Russian inhabitants, to Ukraine as a reward for its productivity.

The Dnieper River forms a useful geographic reference to comprehend Ukraine's spatial division (Fig. 1B-15). To its west lies agrarian, rural, mainly Roman Catholic Ukraine; in its great southern bend and eastward lies industrial, urban, Russified (and Russian Orthodox) Ukraine. Soviet planners made eastern Ukraine a communist Ruhr based on abundant local coal and iron ore, making the Donets Basin (*Donbas* for short) a key industrial complex. Meanwhile, the Russian Soviet Republic supplied Ukraine with oil and gas.

As the map shows, Ukraine, with the exception of its eastern and urban-concentrated Russian minority (still 17 percent today), has an ethnically homogeneous population by eastern European standards. Ukraine is a critically important country for Europe's future as it has access to international shipping lanes, a large resource base, massive farm production, educated and skilled labor, and a large domestic market. But it suffers from numerous problems

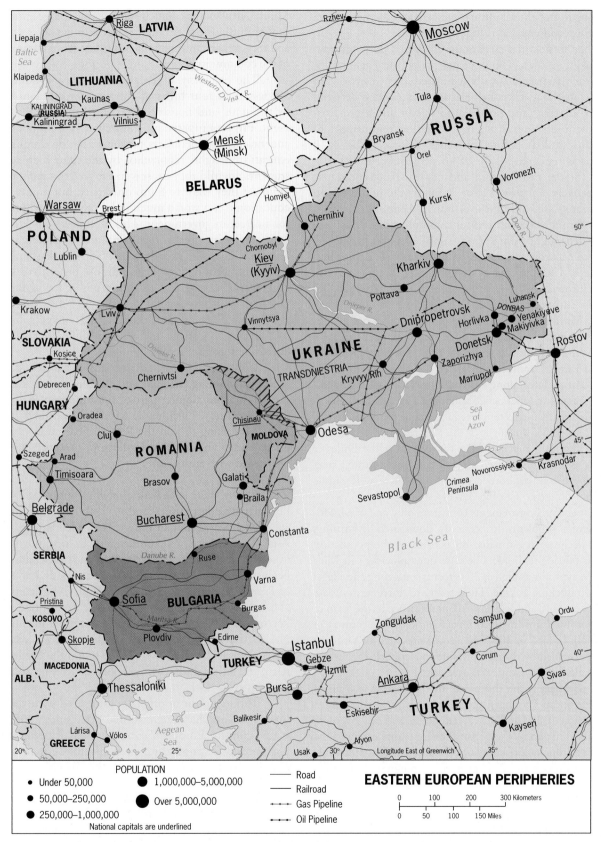

© H. J. de Blij, P. O. Muller, J. Nijman, and John Wiley & Sons, Inc.

FIGURE 1B-15

ranging from political mismanagement and corruption to a faltering economy and rising crime.

A presidential election in 2004 first drew international attention to Ukraine's unique political geography, and the pattern was neatly replicated in the elections of 2010. In effect, Ukraine's electorate is divided between a pro-Western (and pro-EU) northwest and a pro-Russian southeast, resulting in a situation reminiscent of Czechoslovakia's before its "velvet divorce." The 2010 elections rendered a split of 46 percent versus 49 percent for the pro-Western and pro-Russian candidates, respectively, putting Victor Yanukovych in office. In Ukraine's case, the transit of much needed energy supplies through the east creates a complication. Ukraine's own dependence on Russian supplies makes it vulnerable to Moscow's decisions on prices as well as supplies. Some of Ukraine's leaders urge speedy integration into the European sphere; others prefer a middle road between Europe and Russia.

Moldova, Ukraine's small and impoverished neighbor (by many measures Europe's poorest country), was a Romanian province seized by the Soviets in 1940 and made into a landlocked "Soviet Socialist Republic." A half-century later, along with other such "republics," Moldova gained independence when the Soviet Union disintegrated. Romanians remain in the majority among its 4.1 million people, but most of the Russians and Ukrainians (each about 13 percent) have moved across the Dniester River to an industrialized strip of land between that river and the Ukrainian border, proclaiming there a "Republic of Transdniestria" (see Fig. 1A-11).

Separatist movements constitute only one of Moldova's several problems. Its economy, dominated by farming, is in decline; an estimated 40 percent of the population works outside its borders because unemployment in Moldova is typically as high as 30 percent; smuggling and illegal arms trafficking are rife. These many misfortunes translate into weakness, and Russia's malign influence in the form of support for Transdniestria's separatists keeps the country in turmoil. In fact, in 2007 the elected president of Moldova was persuaded by Russian pressure to recognize the regime of breakaway Transdniestria as legitimate.

Belarus is the third state in this eastern margin of the European Periphery and in many ways it is the most peripheral of all. Landlocked, autocratic, and misgoverned, sustained in part by the transit of energy supplies from Russia to Europe's Core countries, Belarus has few functional links to Europe and little prospect of progress. More than 80 percent of Belarus's 9.5 million inhabitants are Belarussians ("White" Russians), a West Slavic people; only some 11 percent are (East Slavic) Russians. A small Polish minority occasionally complains of mistreatment and discrimination. Belarus was always a borderland of sorts, of Lithuania during medieval times and of Russia since the eighteenth century. Devastated during World War II, Belarus became one of Moscow's most loyal satellites, and the Soviets built

Minsk (Mensk), the country's capital, into a large industrial center. But in the post-Soviet era Belarus has lagged badly, and its government functions in ways that are reminiscent of the communist period.

The economy of Belarus is from a bygone era. No less than four-fifths of the workforce is employed by the state, and housing is provided through employment. Private property, for most citizens, does not exist, and neither do chambers of commerce or unions. All education is tightly controlled by the state: high school students are told they must join the government's Youth Union if they want to go to university, but a university education is not worth much unless you want to prepare for a career in government. This is a society stuck in old Soviet ways in a realm that continues to undergo profound change. The president, Alexander Lukashenko, is a dictator seemingly with no purpose other than keeping Belarus as it is and retaining power for as long as possible. In his speeches, Lukashenko refers to Russia as "mommy." In effect, Belarus trades energy subsidies for serving as a stable and willing buffer to European encroachment on Russia's doorstep.

Unlike Ukrainians, Belarussians so far express little interest in joining the EU, and Belarus cannot even be viewed as a functioning part of the European Periphery. On the contrary, Lukashenko has made overtures toward Moscow, seeking to join the Russian Federation in some formal way. The Russian leadership, however, has not responded with enthusiasm.

Finally, there is **Turkey**, a mighty country of 75 million that straddles Europe and Asia. Turkey is the successor of the great Ottoman Empire that ruled large parts of Asia, North Africa, and southeastern Europe, but it also experienced notable secularization and Westernization beginning in the 1920s under the charismatic leadership of Kemal Atatürk. It is a Muslim country with considerable influence in the Arab and Islamic worlds, but also a key military ally of the West and a longtime member of NATO. Bordering no less than eight countries, controlling the straits between the Black Sea and the Mediterranean, and facing three seas, Turkey holds a key geostrategic position (Fig. 1B-1).

Turkey has been officially associated with the EU for decades and has sought formal entry since 2005. But is this country European enough? Or is it perhaps too big (second in population size only to Germany) and potentially destabilizing? Or is the Turkish case less urgent than others because it is already firmly embedded in Western security structures (NATO) led by the United States? At any rate, Turkish membership is on hold. The main reasons, it is said, are the poor treatment of Kurds in eastern Turkey and the apparent inability of the Turkish government to conform to EU economic standards—issues that are not likely to be resolved in the immediate future.

With nearly 600 million inhabitants in 40 countries, including some of the world's highest-income economies, a politically stable and economically integrated Europe would be a superpower in any new world order. However, Europe's political geography is anything but stable as devolutionary forces and cultural stresses continue to trouble the realm even as EU expansion proceeds. The global economic crisis of the past few years has underscored the enormous difficulty of managing such a large and diverse realm as a single economic system. Europeans have not yet found a way to give voice to collective viewpoints in the world, or to generate collective action in times of crisis. Importantly, even after the latest expansion of 2007, the EU incorporates only two-thirds of the realm's national economies, and additional enlargement will become increasingly difficult for economic, political, and cultural reasons. Europe has always been a realm of revolutionary change, and it remains so today.

POINTS TO PONDER

- In 2009, the Swiss voted to ban the further construction of minarets on mosques. In 2010, France passed legislation banning the wearing of the *burqa* in public places.

- In northern European countries, the south tends to be more developed than the north, whereas in southern European countries the north tends to be more developed than the south.

- In 2010, the Autonomous Community of Catalonia outlawed bullfighting, a pointed rebuke to the Spain of which it is a part.

- Compared to the United States, cities in Europe tend to be much older, more compact, and display higher population densities.

using GEOGRAPHY

1. You are a member of the staff of a multinational corporation doing business in several European countries. Your management has been used to dealing with governments in the capitals of a number of countries including the Netherlands (The Hague), Poland (Warsaw), and Hungary (Budapest). Now the company has announced plans to expand operations in Spain at Barcelona, in Belgium at Antwerp, in Ukraine at Donetsk, and in the United Kingdom at Edinburgh. You have been asked to prepare a "hurdles and pitfalls" report as part of the preparation for this initiative. What advice will you provide based on your knowledge of Europe's cultural geography and devolutionary pressures?

2. The European Union expanded from 15 to 27 member-states during the middle years of the last decade, and voters in every one of the joining countries endorsed their governments' desire for membership. But even as other countries hope to enter the EU in the foreseeable future, certain states have actually declined membership. What countries are these, and why would they stay out of the EU? Under what circumstances might a member-state decide to leave the EU?

3. The largest and most populous country in Europe that is not a member of the EU is predominantly Christian Ukraine, and in 2010 Ukraine was not even in discussions aimed at EU membership. Yet Turkey, not in Europe and almost entirely Muslim, is in the early stages of negotiations toward eventual membership. What cultural-geographic and geopolitical factors are at play in this situation?

Elevation (m)

3000
1500
600
300
150
0

below sea level
0
-150
-1500
-3000

Roads
Railroads

0 km   200   400   600   800   1000   1200

0 miles   200   400   600   800

Lambert Azimuthal Equal-Area Projection
Scale 1:20,000,000

UNITED
KINGDOM

Amsterdam
NETHERLANDS

North
Sea

HAMBURG

GERMANY

BERLIN

Elbe

Oder

POLAND

WARSAW

Vistula

COPENHAGEN
DENMARK

Kaliningrad
(RUSSIA)

LITHUANIA
VILNIUS

Baltic Sea

Gulf of Riga

Riga
LATVIA

ESTONIA
Tallinn

Gulf of Finland

Helsinki

Stockholm

SWEDEN

SCANDINAVIA

NORWAY

Gulf of Bothnia

FINLAND

Lake
Ladoga

Lake
Onega

Lake
Peipus

Pskov

ST. PETERSBURG

Novgorod

Murmansk

KOLA
PENINSULA

Arctic Circle

Lovozero

Barents Sea

Svalbard
(NORWAY)

Alexandra Land
George Land

Franz Josef
Land

Wilczek Land

Graham
Bell I.

Ushakova

October R

ARCT

Kolguyev I.

Severodvinsk

Arkhangelsk

White Sea

Petrozavodsk

Onega

Cherepovets

Rybinsk

Tver

BELARUS

MINSK

DNIEPER LOWLAND

KYYIV/KIEV

UKRAINE

Dnieper

MOLDOVA
Chisinau

ODESA

CRIMEAN
PEN.

Sevastopol

Black
Sea

Novorossiysk

Sochi

Krasnodar

Maykop

Stavropol

Armavir

Cherkessk

Nalchik

Nazran

Grozny

Mt. Elbrus
(5,633 m,
18,481 ft.)

CAUCASUS MOUNTAINS

Vladikavkaz

TBILISI
GEORGIA

YEREVAN
ARMENIA

AZER.

AZERBAIJAN

BAKU

TURKEY

IRAQ

TABRIZ

IRAN

ZAGROS M

ELBURZ MOUNTAINS

TEHRAN

KARAJ

QOM

Makhachkala

Caspian
Sea

Astrakhan

CASPIAN DEPRESSION

Elista

Volga

Ural

Garabogaz
Bay

USTYURT
PLATEAU

Aral
Sea

UZBEKISTAN

TURKMENISTAN

Ashgabat

TASHKENT

QIZILQUM

KAZAKHSTAN

KAZAKH

UPLANDS

MOYYNQUM

SARYESIK-ATYRAU
DESERT

Lake
Balqash

Bishkek

ALM

JUNGGAR
BASIN

Smolensk

Kaluga

Bryansk

Tula

Orel

Kursk

Staryy
Oskol

Belgorod

KHARKIV

DONETSK

Taganrog

Sea of
Azov

Shakhty

ROSTOV

VOLGOGRAD

Volzhsky

Volga

Don

Tambov

Lipetsk

Voronezh

Ryazan

MOSCOW

Vladimir

Dzerzhinsk

NIZHNIY NOVGOROD

Cheboksary

Yoshkar-Ola

KAZAN

Naberezhnye Chelny

Nizhnekamsk

Kama

Izhevsk

Vyatka

Kudymkar

Syktyvkar

Northern Dvina

Vologda

Kostroma

Yaroslavl

Ivanovo

Zelenograd

Pueva

Volga

EUROPEAN LOWLAND

Penza

Saransk

Ulyanovsk

Tolyatti

SAMARA

Syzran

Engels

Saratov

Balakovo

Orenburg

Orsk

Magnitogorsk

Sterlitamak

Ufa

PERM

YEKATERINBURG

Nizhny-Tagil

Mt. Narodnaya
(1,895 m, 6,214 ft.)

Ob

PECHORA
BASIN

Pechora

Naryan Mar

Nanyan Mar

Vorkuta

Gornyatskiy

Salekhard

Novyy Port

Ob

YAMAL PENINSULA

Gulf of Ob

Belyy I.

Dickson

Dudinka

Nori

GYDA
PENINSULA

Kara Sea

Novaya Zemlya

Vaygach I.

Urengoy

Sergino

WEST

SIBERIAN

RUSSIA

PLAIN

Khanty-
Mansiysk

Nizhnevartovsk

Surgut

Tyumen

Kurgan

CHELYABINSK

Petropavl

OMSK

Irtysh

Kolpashevo

Lesosibirs

Tomsk

Krasnoyars

Kemerovo

NOVOSIBIRSK

Prokopyevsk

Novokuznetsk

Barnaul

Biysk

Semey

Rubtsovsk

Gorno-
Altaysk

Mt. Belukha
(4,506 m, 14,783 ft.)

ALTAI

SAY

Astana

URAL MOUNTAINS

OCEAN

Bering Strait

Chukchi
Sea

CHUKCHI
PENINSULA

Gulf
of
Anadyr

St. Lawrence I.

UNITED
STATES

St. Matthew I.

Bering
Sea

Aleutian Islands

Wrangel I.

CHUKCHI
RANGE

Pevek

Ambarchik

Anadyr

KORYAK RANGE

Mt. Ledyanaya
(2,562 m, 8,406 ft.)

East
Siberian
Sea

New Siberian Islands

Novaya
Sibir I.

Kotelnyy I.

Malyy Lyakhovskiy I.

Bolshoy Lyakhovskiy I.

KOLYMA RANGE

Karaginskiy I.

Karaga

Commander
Islands
(RUSSIA)

rth Land
vernaya Zemlya)
omolets I.

Laptev
Sea

Gulf of
Yana

Kolyma

Srednekolymsk

Shelikhov
Gulf

KAMCHATKA
PENINSULA

Klyuchevskaya
(4,750 m, 15,584 ft.)

UNITED STATES

evik I.

Begichev I.

Nordvik

Tiksi

Omsukchan

Petropavlovsk-
Kamchatskiy

NORTH SIBERIAN LOWLAND

CHERSKIY RANGE

Ust Nera

Magadan

Okhotsk

Bolsheretsk

Cape Lopatka

PENINSULA

S I B E R I A

Mt. Mus Khaya
(2,959 m, 9,708 ft.)

Sea
of
Okhotsk

Paramushir I.

CENTRAL

VERKHOYANSK RANGE

YAKUTSK
BASIN

Lena

Namtsy

Yakutsk

Ust Maya

Aldan

B

E

SIBERIAN

Vilyuy

Suntar

Kurile Islands

Tura

Mirnyy

Lena

Urup

S

PLATEAU

I

A

Lensk

DZHUGDZHUR RANGE

Uda Bay

Shantar
Islands

Noglikl

Sakhalin
Island

Etorofu

Angara

STANOVOY RANGE

Lazarev

Tatar Strait

Yuzhno Sakhalinsk

Novikovo

Kunashiri

Shikotan

Habomai

Ust Ilimsk

Mt. Skalistyy
(2,467 m, 8,094 ft.)

Skovorodino

Komsomolsk

Vanino

La Perouse Strait

Wakkanai

Ust Kut

Amur

SIKHOTE-ALIN RANGE

Tayshet

Bratsk

YABLONOVYY RANGE

Heihe

Blagoveshchensk

Birobidzhan

Mt. Tardoki-Yani
(2,090 m,
6,857 ft.)

Khabarovsk

SAPPORO

HOKKAIDO

Kushiro

Lake
Baykal

Chita

Amur

Dalnegorsk

JAPAN

PACIFIC
OCEAN

Ust Ordynskiy

Aginskoye

GREATER KHINGAN RANGE

Songhua

Lake
Khanka

Vladivostok

Nakhodka

Komori

UNTAINS

Angarsk

Irkutsk

Ulan-Ude

HARBIN

JILIN

East Sea
(Sea of Japan)

SENDAI

Munku Sardyk
m. (11,453 ft.)

Ulaanbaatar

NORTHEAST
CHINA
PLAIN

CHANGCHUN

SHENYANG

NORTH
KOREA

Mt. Fuji
(3,776 m,
12,388 ft.)

TOKYO

KAWASAKI

YOKOHAMA

NAGOYA

KYOTO

OSAKA

GAYN MOUNTAINS

MONGOLIA

MONGOLIAN
PLATEAU

Xiliao

Liao

Liao

CHINA

PYONGYANG

FIGURE 2A-1   © H. J. de Blij, P. O. Muller, and John Wiley & Sons, Inc.

Every geographic realm has a dominant, distinguishing feature: Europe's jigsaw-puzzle map of more than 40 countries, South America's familiar triangle, Subsaharan Africa's straddling of the equator, Southeast Asia's thousands of islands and peninsulas. But the Russian geographic realm has what is perhaps the most distinctive hallmark of all—its gigantic territorial size. Almost (but not quite) all of this realm is constituted by the state that dominates it, which is why we call this the Russian realm. By itself, Russia is nearly twice as large as Canada, the world's second-largest country. Russia is three times as big as neighboring Europe. Stretching east-west from the Pacific Ocean to the Baltic Sea, Russia has *nine* time zones. When Russians are getting ready for bed in Moscow, they're having breakfast in Vladivostok.

As Figure 2A-1 reveals and a globe shows even better, Russia's entire northern coast, from the border with Finland in the west to the Bering Strait in the east, faces the Arctic Ocean, with all the climatic consequences you would expect. And for all its huge dimensions, most of the Russian realm lies at high latitudes, shut out by mountains and deserts from the warmer and moister parts of Eurasia. That has been a geographic problem for Russian rulers and governments for as long as Russia has existed as a state. Russia's ports on the Pacific Ocean in the east lie about as far from where most Russians live as they could be. And in the west, where Russia's core area is situated, every maritime outlet is hamstrung in some way. The Arctic ports operate on seasonal access. The Baltic Sea ports lie far from the open ocean. The Black Sea ports (some leased from Ukraine) require navigation through narrow straits across Turkish territory just to reach the Mediterranean Sea.

The Russian realm is vast, but its human numbers are modest. As the Data Table in Appendix B reports, this entire realm's population of less than 160 million is about

major geographic qualities of
RUSSIA

1. Russia is the largest territorial state in the world. Its area is nearly twice as large as that of the next-ranking country (Canada).

2. Russia is the northernmost large and populous country in the world; much of it is cold and/or dry. Extensive rugged mountain zones separate Russia from warmer subtropical air, and the country lies open to Arctic air masses.

3. Russia is by far the "widest" realm on Earth, stretching from west to east for some 10,000 kilometers (over 6000 miles) and covering nine time zones.

4. Russia was one of the world's major colonial powers. Under the czars, the Russians forged the world's largest contiguous empire; the Soviet rulers who succeeded the czars took over and expanded that empire.

5. For so large a territory, Russia's shrinking population of just over 140 million is comparatively small. The population remains heavily concentrated in the westernmost one-fifth of the country.

6. Development in Russia is concentrated west of the Ural Mountains; here are the major cities, leading industrial regions, densest transport networks, and most productive farming areas. National integration and economic development east of the Urals extend mainly along a narrow corridor that stretches from the southern Urals region to the southern Far East around Vladivostok.

7. Russia is a multicultural state with a complex domestic political geography and administrative structure.

8. Its large territorial size notwithstanding, Russia faces land encirclement within Eurasia; it has few good and suitably located ports.

9. Regions long part of the Russian and Soviet empires are still realigning themselves in the postcommunist era. Eastern Europe and the heavily Muslim Southwest Asia realm are encroaching on Russia's imperial borders.

10. The failure of the Soviet communist system left Russia in economic disarray. It is often considered to be one the world's major emerging economies, but its growth is fragmentary and precarious. Russia is highly dependent on exports of oil and gas.

Winter in the heart of southern Siberia.
TassPhoto/WireImage/Getty Images, Inc.

political, economic, and social conditions that continue to produce negative demographic effects today. This issue has come to the forefront among Russia's internal concerns at a time when Russia's leaders want to restore Russia's power and prestige in the wider world. To accomplish this restoration, Russia will need to strengthen its social institutions, diversify its economy, and restore the confidence lost when the Soviet system disintegrated. The Russian geographic realm as defined by our framework consists of four political entities: the giant Russian state itself and three small countries in Transcaucasia, the mountainous area between the Black Sea and the Caspian Sea. Georgia, Armenia, and Azerbaijan are small in size and their combined population is just over 10 percent of the realm's total. They are located in a historically turbulent zone of conflict between Russian and non-Russian peoples.

Take a look at the Russian realm in Figure G-3, and you will note that many of Russia's neighbors, even those countries that formed part of the former Soviet Union, are now absorbed into other realms. The Baltic states, for

equal to those of such individual countries as Nigeria and Bangladesh—and what is more, the population has been shrinking rapidly. For a combination of reasons we will examine later, the collapse of the Soviet Union created example, have become an integral part of Europe and the European Union. The Central Asian countries once under Moscow's sway, such as Kazakhstan and Uzbekistan, now form a discrete region in an adjoining realm in which Islam is resurgent. But the three Transcaucasian states are not collectively European (although Georgia has European aspirations), or Islamic (only in Azerbaijan does Islam have a strong foothold), or Russian. What these three states have in common is Russia's still-powerful influence in one form or another. Georgia endures Russian political, economic, and even military intervention; Azerbaijan seeks ways to avoid having to depend on Russian transit for its energy exports; and Armenia continues to view Russia as its most dependable ally. For these reasons, elaborated in the discussion below, we map these states as part of the Russian geographic realm.

Even though our first task is to look in some detail at the giant physical stage on which the Russian geographic realm is built, we should note that the human geography of this realm is not neatly defined by sharp boundaries. Look again at Figure G-3, and you will see that the margins of the Russian realm in two prominent places (and elsewhere not shown at that map's small scale) are marked by *transition zones*. There, geographic features of this realm spill over into neighboring realms, sometimes creating social and political problems for the adjacent countries affected. But first, let us examine the physical landscapes, climate, and ecologies of the Russian realm.

PHYSICAL GEOGRAPHY OF THE RUSSIAN REALM

Physiographic Regions

The first feature you notice when looking at the physiographic map of the Russian realm is a prominent north-south trending mountain range that extends from the Arctic Ocean southward to Kazakhstan (Figs. 2A-1 and 2A-2), dividing Russia into two parts: the Russian Plain to the west and Siberia to the east. This range, the Ural Mountains, is sometimes designated as the "real" eastern boundary of Europe, but as we noted in Chapters 1A and 1B, Russia is not Europe. Cultural life to the east of the Urals is pretty much the same as it is to the west, and there is no geographic justification for putting the city of Samara (see Fig. 2A-1) in Europe but Chelyabinsk, on the other side of the Urals, outside of it. Neither is European; both are Russian.

The Russian Plain

The Russian Plain ①, which lies west of the Ural Mountains, is the eastward continuation of the North European Lowland, and here lies Russia's *core area*. Travel northward from the centrally located capital of Moscow, and the countryside soon changes to coniferous (needleleaf) forests like those of Canada. Centuries ago, those forests afforded protection to the founders of the Russian state when horse-riding invaders from eastern Asia swept into this region. The Volga River, Russia's Mississippi, flows southward in a wide arc across the plain from the forested

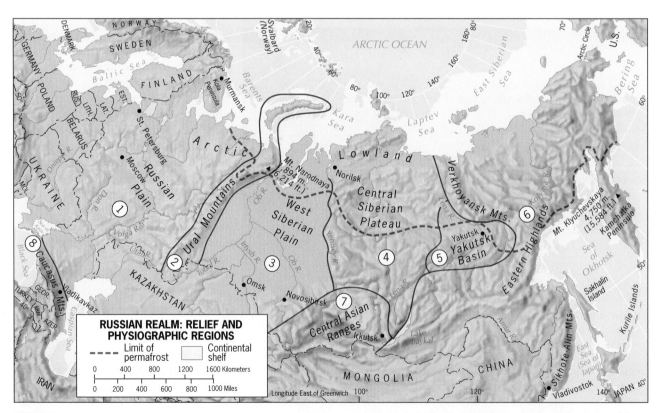

FIGURE 2A-2

© H. J. de Blij, P. O. Muller, and John Wiley & Sons, Inc.

north—but unlike the Mississippi, the Volga drains into a landlocked lake, the Caspian "Sea." Travel southward from Moscow today, and the land is draped in grain fields and pastures. Eastward, the Ural Mountains ② form a range not tall enough to create a major barrier to transportation, but it is prominent because it separates two vast expanses of low *relief* (elevation range).

In geopolitical texts, the western Russian Plain has also been dubbed the Eurasian heartland [1] since it lies deep within the world's greatest landmass, and because it has had a major influence throughout history on the shaping of societies to the west (Europe), southeast (Central Asia), and far east (China, Japan, Korea, and even across the Bering Strait into North America). But this exceptional geographic position not only provides opportunities for expansion; it also implies potential vulnerability and a sense of encirclement. This apparent contradiction has been a feature of the realm for centuries.

Siberia

The West Siberian Plain ③, on the far side of the Urals, is often described as the world's largest unbroken lowland, and here the rivers flow north, not south. Over the last 1600 kilometers (1000 mi) of its course to the Arctic Ocean, the Ob River falls less than 90 meters (300 ft) in huge meanders flanked by forests until the trees give way in the near-Arctic cold to mosses and lichens. Follow an imaginary trail eastward, and in just about the middle of Russia the West Siberian Plain gives way to the higher relief of the Central Siberian Plateau ④, one of the most sparsely populated areas in the habitable world. Where the West Siberian Plain meets the Central Siberian Plateau, the Yenisey River, also flowing northward into the Arctic Ocean, marks the change of landscape.

Continuing our eastward transect, we next encounter the Yakutsk Basin ⑤, drained by the Lena River and forming the last vestige of moderate *topography* (surface configuration). Now we approach the realm's mountainous eastern perimeter (look again at Fig. 2A-1), where the few roads and railroads must wend their way through tunnels and hairpin valley turns. We map this jumble of ranges collectively as the Eastern Highlands ⑥, but this is one of the most spectacular, diverse, and still-remote regions on Earth.

Kamchatka and Sakhalin

In its farthest eastern reaches, the Russian realm makes contact with the Pacific Ring of Fire (see Fig. G-5). There are no active volcanoes in Siberia and earthquakes are rare, but the Kamchatka Peninsula has plenty of both. This is one of the most volatile segments of the Pacific Ring of Fire, with nearly 70 volcanoes, many of them active or dormant, forming the spine of the peninsula. Nothing here resembles Siberia (the climate is moderated by the Pacific waters offshore; the vegetation is mixed, not coniferous). It is one of the most difficult places on Earth to live, virtually disconnected from the rest of Russia (and entirely so overland).

Also on the Pacific fringe of the Russian realm is an island named Sakhalin, where earthquakes rather than volcanic activity are the leading environmental hazard. Long a battleground between Russians and Japanese, Sakhalin finally fell to the Russians at the end of World War II. In the postwar period the island and its maritime environs proved to contain major reserves of oil and natural gas, making it a valuable asset in Russia's energy-based export economy.

The Southern Perimeter

A final look at Figures 2A-1 and 2A-2 reminds us that mountainous topography encircles much more than just the eastern edge of the Russian realm. Some of the highest relief in Eurasia prevails in the southern interior, where the Eastern Highlands meet the Central Asian Ranges ⑦; here lies Lake Baykal in a tectonic trough that is more than 1500 meters (5000 ft) deep—the deepest lake of its kind in the world. And between the Caspian Sea and the Black Sea, the Caucasus Mountains ⑧, geologically an extension of Europe's Alps, rise like a wall between Russia

The village of Tara on the Irtysh River in the Omsk Region in the Siberian Federal District displays little or no evidence of modernization. Czarist-era wooden houses and wobbly sheds are seen here on a sunny summer day. But although this is one of Russia's southernmost administrative units (see Fig. 2B-2), it lies exposed to the harsh Siberian winter; when the river freezes over you can walk to the inhabited meander knoll seen in the near distance. The short growing season allows cultivation of spring wheat and some other early-ripening crops, and small farm settlements like this dot the flat landscape.

© Jon Arnold Images Ltd/Alamy.

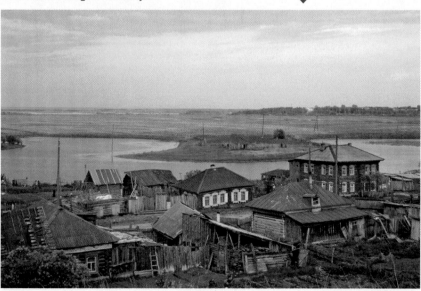

and the lands beyond—a multi-tiered barrier over which Russians have fought with their neighbors for centuries.

When you approach the forbidding Caucasus ranges from the low relief of the Russian Plain to the north, they seem impenetrable, and you realize why so few routes cross them, even today. It is also clear why Russia's armies had such difficulty pushing across the Caucasus into the areas beyond. At present, these mountains shelter insurgents defying Russia's authority in such territories as Chechnya and Ingushetiya. And the zone beyond is no easy target: to the south of the Caucasus, high relief dominates all the way into neighboring Turkey and Iran.

Harsh Environments

The historical geography of Russia is the story of Slavic expansion from its populous western heartland across interior Eurasia to the east, and into the mountains and deserts of the south. This eastward march was hampered not only by vast distances but also by harsh natural conditions. As the northernmost populous country on Earth, Russia has virtually no natural barriers against the onslaught of Arctic air. Moscow lies farther north than Edmonton, Canada, and St. Petersburg lies at latitude 60° North—the latitude of the southern tip of Greenland. Winters are long, dark, and bitterly cold in most of Russia; summers are short and growing seasons limited. Many a Siberian frontier outpost was doomed by cold, snow, and hunger.

It is therefore useful to view Russia's past, present, and future in the context of its climate. Even today, in this age of global warming, Russia still suffers from severe cold and associated drought. Precipitation, even in western Russia, ranges from modest to minimal because the warm, moist air carried across Europe from the North Atlantic Ocean loses much of its warmth and moisture by the time it reaches Russia. Figure 2A-3 reveals the consequences. Russia's climatic **continentality [2]** (inland climatic environment remote from moderating and moistening maritime influences) is expressed by its prevailing *Dfb* and *Dfc* conditions. Compare the Russian map to that of North America (Fig. G-7), and you note that, except for a small corner off the Black Sea, Russia's climatic conditions resemble those of the Upper Midwest of the United States and interior Canada. Along its entire northern edge, Russia has a zone of *E* climates, the most frigid on the planet. In these Arctic latitudes originate the polar air masses that dominate its environments.

The Russian realm's harsh northern climates affect people, animals, plants—and even the soil. In Figure 2A-2 you can find a direct consequence of what you see in Figure 2A-3: a dashed blue line that starts at the shore of the Barents Sea, crosses Siberia eastward, and reaches the coast of the Pacific Ocean north of the neck of the Kamchatka Peninsula. North of this line, water in the ground is permanently frozen, creating an even more formidable obstacle to settlement and infrastructure than the severe

CLIMATES OF THE RUSSIAN REALM

B DRY CLIMATE	C HUMID TEMPERATE CLIMATE	D HUMID COLD CLIMATE	E COLD POLAR CLIMATE	H HIGHLAND CLIMATE
BS Semiarid	Cf No dry season	Df No dry season	E Tundra and ice	H Unclassified highlands
BW Arid	Cw Dry winter	Dw Dry winter		
h=hot k=cold	Cs Dry summer	a=hot summer b=cool summer / c=short, cool summer d=very cold winter	Taiga (snowforest)	

FIGURE 2A-3

© H. J. de Blij, P. O. Muller, and John Wiley & Sons, Inc.

As Figure 2A-3 shows, Siberia is cold and, in Russia's far northeast, dry as well. In the north lies the tundra, treeless and windswept, and beyond is the ice of the Arctic. But where somewhat more moderate conditions prevail (moderate being a relative concept in northwestern Russia and Siberia), coniferous forests known as taiga cover the countryside. Also called boreal (cold-temperate) forests, these evergreen, needleleaf pines and firs create a dense and vast high-latitude girdle of vegetation across northern Eurasia as well as northern North America. This view from the air shows how tightly packed the trees are; they are slow-growing, but long-lived. Most of the world's taiga forest, among the largest surviving stands of primary forest on the planet, remains protected by distance from the threat of exploitation—but lumbering is nevertheless taking its toll. Recent studies, however, indicate that climate change in high latitudes is enabling the forest to expand northward at a faster rate than it diminishes due to lumbering activities elsewhere, a rare case of good news relating to global warming.

© Arcticphoto/Alamy

forest of coniferous and deciduous trees. Although the taiga prevails from northern Scandinavia to the Russian Far East, this is the vegetative landscape most often associated with Siberia: endless expanses of rolling countryside draped in dense stands of evergreen pine trees (see photo at left).

Climates and Peoples

Climate and weather (there is a distinction: *climate* refers to a long-term average, whereas *weather* describes a set of existing atmospheric conditions at a given place and time) have always challenged the peoples of this realm. The high winds that drive the bitter Arctic cold southward deep into the landmass, the blizzards of Siberia, the temperature extremes, rainfall variability, and short and undependable growing seasons even in the more moderate western parts of this realm have always made farming difficult. That was true during the time of the ruling czars, when the threat of famine never receded. It was also a problem during the Soviet-communist period, when Moscow was the capital of an empire far larger than this realm and the non-Russian sectors of that empire (such as Ukraine and parts of Central Asia) produced food staples Russia needed. Despite a massive effort to restructure agriculture throughout the Soviet Empire by means of collectivization and irrigation projects, the Russians often had to import grain.

As we noted in the Introduction, and as we will observe time and again throughout this book, humanity's long-term dependence on agriculture remains etched on the population map, even as our planet becomes ever more urbanized and its economy more globalized. By studying the climates of the Russian realm, we can begin to understand what the map of population distribution shows (Fig. 2A-4). The overwhelming majority of the realm's nearly 160 million people remain concentrated in the west and southwest, where environmental conditions were least difficult at a time when farming was the mainstay of most of the people. To the east, the population is sparser and tends to cluster along the southern margin of the realm, becoming even more thinly distributed east of Lake Baykal. If you consider that almost three-quarters of this realm's population today lives in cities and towns, it is not difficult to imagine just how empty vast stretches of countryside must be—especially in frigid northern latitudes.

Climate Change and Arctic Prospects

As every map of this high-latitude realm shows, its northern coast lies entirely on the polar side of the Arctic Circle. Such a lengthy coastline on the Arctic Ocean does not exactly constitute an advantage: most of the Arctic Ocean is frozen much of the year, and only some warmth from the

weather alone. It is referred to as **permafrost [3]**, and it affects other high-latitude environments as well (Alaska, for example). Where the permafrost ends, seasonal temperature changes cause alternate thawing and freezing, with destructive impacts on buildings, roads, railroad tracks, and pipelines.

Examine Figure 2A-3 carefully, and you will see two ecological terms: *tundra* and *taiga*. **Tundra [4]**, as the map suggests, refers to both climate and vegetation. The blue area mapped as *E* marks the coldest and ice-affected environmental zone in Russia and elsewhere in the high Arctic: this is frigid, treeless, windswept, low-elevation terrain where bare ground and rock prevail and mosses, lichens, bits of low grass, and a few hardy shrubs are all that grows. **Taiga [5]**, the stippled area on the map, a Russian word meaning "snowforest" (also called boreal forest), extends across vast reaches of Eurasia as well as northern North America and is dominated by coniferous (as in pine cones) trees. As the map shows, the taiga extends southward into relatively moderate environs, and there it becomes a mixed

FIGURE 2A-4

© H. J. de Blij, P. O. Muller, and John Wiley & Sons, Inc.

North Atlantic Drift ocean current keeps the ports of Murmansk and Arkhangelsk open a bit longer. Ports like this (and even St. Petersburg on the Gulf of Finland, an arm of the Baltic Sea) would never have developed the way they did if Russia had better access to the world ocean, but as we noted, this has been one of the country's historic impediments.

Now it looks as though nature may give Russia a helping hand. If, as most climatologists anticipate, global warming causes long-term melting of large sections of the Arctic Ocean's ice cover, that body of water may come to play a new and different role in this realm's future. Milder atmospheric conditions may shrink the area of permafrost mapped in Figure 2A-2; moister air masses may improve agriculture on the Russian Plain; warmer water may keep Arctic ports open year-round; and a so-called Russian Maritime Passage may even open up between the Bering Strait and the North Sea, shortening some international sea routes by thousands of kilometers and ushering in a new era of trade.

That, at least, is how many Russians envision global warming—as a potential advantage nature has long denied them. And indeed, the Arctic provides much evidence for such warming: the melting of parts of the Greenland Ice Sheet, the shrinking of the average extent of permanent Arctic Ocean ice, and the reduced incidence of icebergs. Russian economic planners look at the shallow waters off-shore along the Arctic Ocean coastline (see the white- and light-blue areas in Fig. 2A-1) and hope that oil and gas

reserves beneath those waters will come within reach of exploitation.

These developments may even have the effect of ex-panding the Russian geographic realm far into the Arctic. As we note when we discuss the polar areas in more detail in our final chapter, states with coastlines in the Arctic are likely to demand certain exclusive rights not only over wa-ters offshore but also the resources on and beneath the ocean floor. It is therefore in their interest to extend these rights as far from shore as possible, even hundreds of (nau-tical) miles outward. In 2007, the Russian government even placed a metal Russian flag at the North Pole on the seafloor, under the permanent ice of the Arctic Ocean, symbolizing its intentions (see photo, p. 603). The map of the northern perimeter of the Russian realm is changing—but not always for the better.

Ecologies at Risk

If, as computer models predict, global warming continues and perhaps intensifies in the polar latitudes, these eco-logically sensitive environments will be severely affected. Animals as well as humans have long adapted to prevailing climatic conditions, and such adaptations would be sig-nificantly disrupted by environmental change. The intri-cate web of relationships among species and their environ-ments on the one hand, and among species themselves on the other, can be quickly damaged by temperature change. The polar bear is just one prominent example: it depends on ample floating sea ice to hunt as well as raise cubs.

When sea ice contracts and wider stretches of open water force polar bears to swim greater distances during the cub-rearing season, fewer infants make it to adulthood. If ice-free Arctic summers are indeed in the offing, the polar bear could become extinct in your lifetime. Rapid ecosystem change would also endanger seal, bird, fish, and other Arctic populations.

Such changes affect human populations as well. For example, Inuit (formerly called Eskimo) communities still pursuing traditional lives in parts of the Arctic domain (though not in the Russian realm) are similarly adapted to the harsher environments of the past. Their traditions, already under pressure from political and economic forces resulting from their incorporation into modern states, would be further threatened by environmental changes that are likely to alter, or even destroy, the ways of life they developed over thousands of years.

And if a new era of oil and natural gas exploration and exploitation is indeed about to open, offshore environments face even greater peril. As technologies of recovery put ever more of these reserves within reach, oil platforms, drills, pumps, and pipelines will make their appearance—along with risks of oil spills, pollution, disturbance, and damage of the kind we have seen all over the lower-latitude world. Clearly, the forces of globalization are finally penetrating a part of the world long protected from it by distance and nature.

RUSSIA'S NATURAL RICHES

Given the enormous territorial dimensions of the Russian realm, it is not surprising that its natural resources are vast and varied. The Russians today are excited about the energy reserves that are to be found and exploited beneath Arctic waters, but this realm already is a major producer and exporter of oil and natural gas from a growing number of fields dispersed across Russia from the North Caucasus in the west to the island of Sakhalin in the east and from western Siberia in the north to the Caspian Basin in the south (Fig. 2A-5).

And petroleum and natural gas are not the only underground riches of this realm. Deposits of so many minerals lie within Russian territory that virtually all of the raw materials required by modern industry are present. Nor are oil and natural gas the only energy reserves. Major coalfields are found east as well as west of the Urals, and in Siberia as well as in more southerly latitudes (the string of mines producing high-quality coal along the Trans-Siberian Railroad corridor were vital in Russia's industrial development and its successful war against Nazi Germany). Large deposits of iron ore, too, lie widely scattered across this enormous realm, from the so-called Kursk Magnetic Anomaly near the border with Ukraine to the remotest corners of Siberia's Arctic north. And when it comes to other metals, this realm has it all—or almost all—from gold to lead and from platinum to zinc. Moreover, one of

FIGURE 2A-5   © H. J. de Blij, P. O. Muller, and John Wiley & Sons, Inc.

the world's largest concentrations of nonferrous (non-iron) metals lies in and around the Ural Mountains, where Russia's metallurgical industries emerged.

And there is likely to be much more. Vast reaches of this far-flung realm have not yet been fully explored, especially in the far northeast. Given such material assets, you would imagine that this realm has always been in the forefront of industrial production and economic diversification. But as we will also see, the availability of raw materials is only one part of such development. You are likely to buy Chinese and Japanese products on an almost daily basis—but you are likely to own very few if any items of Russian manufacture. That is part of the story we relate in Chapter 2B.

Environment—in the form of climate, relief, vegetation, and other elements—has much to do with the way Russia evolved as the dominant power over this vast realm. To better understand the map, let us briefly trace the historical geography of Russia, its defenders, and rulers who eventually pushed Russian influence far beyond the Russian Plain—even into western North America. It may come as a surprise to many Americans that Russians not only colonized Alaska, but even reached California and built a fort not very far north of present-day San Francisco. So let us focus first on place and time in Russia's evolution.

RUSSIAN ROOTS

A thousand years ago, Eurasian peoples of many ethnic sources and cultural backgrounds were migrating across the plains south of the taiga (coniferous snowforest) in search of new and secure homelands; Scythians, Sarmatians, Goths, Huns, and others came, fought, settled, survived, absorbed neighbors, or were driven off. Eventually the Slavs, comparative latecomers to this turbulent stage, established settlements in the area of present-day Ukraine, north of the Black Sea and in the southwestern corner of the physiographic region we have defined as the Russian Plain (Fig. 2A-2).

Here the Slavic peoples found fertile soils, a relatively moderate climate, and a physical landscape with many advantages and opportunities: the Dnieper River and its tributaries, stands of forest, relatively low relief, and easy contact among settlements. The Slavs used the name *Rus* to designate such settlements, and the largest and most successful of these early Russes was the one located where the capital of modern Ukraine, Kiev (Kyyiv), lies today. This is one reason why many "Russians" today cannot imagine an independent, Europe-oriented Ukraine: this, after all, is their historic heartland. And from this southern base, the Slavs expanded their domain across the Russian Plain, first establishing their northern headquarters at Novgorod on Lake Ilmen (see Fig. 2A-1). This northern Rus was well positioned to benefit from the trade between the Hanseatic ports on the Baltic Sea and the trading centers on the Black and Mediterranean seas. During the eleventh and twelfth centuries, the Kievan Rus and the Novgorod Rus combined to form a large and prosperous state astride both the northern forest and southern **steppe** [6] (semiarid grassland).

The Mongol Invasion

Prosperity attracts attention, and knowledge about the Russes spread far and wide. In the distant east, north of China, another successful state had been building: the empire of the Mongol peoples under Genghis Khan. Also under the sway of this legendary ruler was a group of nomadic, Turkic-speaking peoples known as the Tatars. Together, Mongol-Tatar armies rode westward on horseback into the domain of the Russes to challenge the power of the Slavs.

Geography had much to do with their early success: on the open steppes of the southern Russian Plain, the Russes lay exposed to the fast-charging Mongol forces, and by the middle of the thirteenth century, the Kievan Rus had fallen. Slavic refugees were fleeing into the northern forests, where they reorganized to face their enemies in newly built Russes. The forest environment was their ally: Mongol tactics were effective for the open plain, but not in the woods. What ensued was not a victory for either side, but a standoff: the Tatars threatened and tried to lay siege, but they could not win outright. So the leaders of the forest-based Russes paid tribute to the Mongol-Tatar invaders, in exchange for which they were left alone.

One of these Russes was Moscow, deep in the forest on the Moscow River and destined to become the capital of a vast empire. On a defensible site and in a remote situation, Moscow's leaders began to establish trade links with even safer Novgorod, and the city grew and thrived. When the Mongols, worried about Moscow's growing power and influence, attacked and were successfully repulsed, Moscow emerged as the leader among Russes, its destiny assured. The Mongol-Tatar campaign had essentially failed.

Before we go on, we should take note of a crucial geographic development. Whereas the Mongols and Tatars fought together, their failure to conquer the Slavic-Russian heartland had both cultural and geographic consequences that have endured into the present. Although most of the Mongols withdrew following their unsuccessful fourteenth-century assault on Moscow, many Tatars remained on what, at the time, was the periphery around the Slavic/Russian core in such places as the Volga River Basin, the Crimean Peninsula, and other smaller pockets. There they were converted in the wave of Islam that was rolling into the Black Sea Basin and beyond, creating a new kind of division between Christian Slavs and their old-time adversaries.

Grand Duchy of Muscovy

During the fourteenth century the Grand Duchy of Muscovy rose to preeminence under the rule of princes or dukes. They extended Moscow's trade links from the Baltic to Black Sea shores and forged valuable religious ties with the leadership of the Eastern Orthodox Church in Constantinople. Then, from around 1450, there began a period lasting more than three centuries during

FIGURE 2A-6

© H. J. de Blij, P. O. Muller, and John Wiley & Sons, Inc.

which powerful Russian rulers etched their imprints on the map of Russia, of Eurasia, and, by extension, the world (Fig. 2A-6).

By the sixteenth century, during the reign of Ivan IV, better known as the infamous *Ivan the Terrible*, the Grand Duchy of Muscovy had been transformed into a major military power and an imperial state. He expanded Moscow's empire by conquering the Islamic regions to the south and gaining control over present-day Estonia and Latvia on the Baltic Sea. It was a time of almost continuous warfare, and Ivan the Terrible acquired his reputation from the reign of terror he unleashed in his pursuit of military discipline, centralized administrative control, and retaliation against many members of the nobility.

BUILDING THE RUSSIAN EMPIRE

Czarist Russia

When Peter the Great became czar (he ruled from 1682 to 1725), Moscow already lay at the center of a great empire—great, at least, in terms of the territories it controlled. The Islamic threat had been ended with the defeat of the Tatars. The influence of the Russian Orthodox Church was represented by its distinctive religious architecture and powerful bishops. Peter consolidated Russia's gains and endeavored

to make a modern, European-style state out of his loosely knit country. He built St. Petersburg as a **forward capital** **[7]** on the doorstep of Swedish-held Finland, fortified it with major military installations, and made it Russia's leading port.

Peter the Great, an extraordinary leader, was in many ways the founder of modern Russia. In his desire to remake Russia—to pull it from the forests of the interior to the seas of the west, to open it to outside influences, and to relocate its population—he left nothing to chance. Prominent merchant families were forced to move from other cities to St. Petersburg. Ships and wagons entering the city had to bring building stones as an entry toll. The czar himself, aware that to become a major power Russia had to be strong at sea as well as on land, went to the Netherlands to work as a laborer in the famed Dutch shipyards to learn the most efficient method for building ships. Meanwhile, the czar's forces continued to conquer people and territory: Estonia was incorporated in 1721, widening Russia's window to the west, and in Siberia major expansion soon occurred south of the city of Tomsk (Fig. 2A-6).

Under Czarina Catherine the Great, who ruled from 1760 to 1796, Russia's empire in the Black Sea Basin grew at the expense of the Ottoman Turks. The Crimea Peninsula, the port city of Odesa (Odessa), and the entire

From the Field Notes . . .

"Not only the city of St. Petersburg itself, but also its surrounding suburbs display the architectural and artistic splendor of czarist Russia. The czars built opulent palaces in these outlying districts (then some distance from the built-up center), among which the Catherine Palace, begun in 1717 and completed in 1723 followed by several expansions, was especially majestic. During my first visit in 1994, the palace, parts of which had been deliberately destroyed by the Germans during World War II, was still being restored; large black and white photographs in the hallways showed what the Nazis had done and chronicled the progress of the repairs during communist and postcommunist years. A return visit in 2000 revealed the wealth of sculptural decoration on the magnificent exterior (top) and the interior detail of a set of rooms called the 'golden suite,' of which the ballroom (bottom) exemplifies eighteenth-century Russian Baroque at its height."

www.conceptcaching.com

© H. J. de Blij

northern coastal zone of the Black Sea fell under Russian control. Also during this period, the Russians made a fateful move: they penetrated the corridor between the Black and Caspian seas, whose spine is the mountainous Caucasus with its dozens of ethnic and cultural groups, many of which were Islamized. The cities of Tbilisi (now in Georgia), Baki (Baku) in Azerbaijan, and Yerevan (Armenia) were captured. Eventually, the Russian push toward an Indian Ocean outlet was halted by the British, who held sway in Persia (modern Iran), and also by the Turks.

Meanwhile, Russian colonists (descendants of Cossacks who originally hailed from Ukraine and southwestern Russia) had advanced far to the east, settling along the southeastern frontier, crossing the Bering Strait, and entering Alaska. Catherine had made Russia a huge colonial power, yet within a century the Russians gave up on their North American outposts (see box titled "Russians in North America").

Nineteenth-Century Expansion

Although Russia had withdrawn from North America, Russian expansionism during the nineteenth century continued in Eurasia. While extending their empire southward, the Russians also took on the Poles, old enemies to the west, and succeeded in taking most of what is today the Polish state, including the capital of Warsaw. To the northwest, Russia took over Finland from the Swedes in 1809.

Russians in North America

THE FIRST WHITE settlers in Alaska were Russians, not western Europeans, and they came across Siberia and the Bering Strait, not across the Atlantic and North America. Russian hunters of the sea otter, valued for its high-priced pelt, established their first Alaskan settlement at Kodiak Island in 1784. Moving south along the North American coast, the Russians founded a chain of additional villages and forts to protect their tenuous holdings until they reached as far as the area just north of San Francisco Bay, where they built Fort Ross in 1812.

But these Russian settlements were isolated and vulnerable. European fur traders began to put pressure on their Russian competitors, and St. Petersburg soon found the distant settlements a burden and a risk. In any case, American, British, and Canadian hunters were decimating the sea otter population, and profits declined. When U.S. Secretary of State William Seward offered to purchase Russia's North American holdings in 1867, St. Petersburg quickly agreed—for $7.2 million. Thus Alaska, including its lengthy southward coastal extension, became U.S. territory and, in 1959, the forty-ninth State. Although Seward was ridiculed for his decision—Alaska was called "Seward's Folly" and "Seward's Icebox"—he was vindicated when gold was discovered there in the 1890s. The twentieth century further underscored the wisdom of Seward's action, strategically as well as economically. At Prudhoe Bay off Alaska's northern Arctic slope, large oil reserves were tapped in the 1970s and are still being exploited. And like Siberia, Alaska probably contains yet unknown riches. From the point of view of the United States, besides the 24 dollars allegedly paid to acquire Manhattan Island, this $7.2 million transaction is often regarded as the most lucrative real estate deal in world history.

During most of the nineteenth century, however, the Russians were preoccupied with Central Asia—the region between the Caspian Sea and western China—where Tashkent and Samarqand (Samarkand) came under St. Petersburg's control (Fig. 2A-6). The Russians here were still bothered by raids of nomadic horsemen, and they sought to establish their authority over the Central Asian drylands as far as the edges of the high mountains that lay to the south. Thus Russia gained many Muslim subjects, because this was Islamic Asia they were penetrating. Under czarist rule, though, these people retained some autonomy.

Much farther to the east, a combination of Japanese expansionism and a decline of Chinese influence led Russia to annex from China a number of provinces beyond the Amur River. Soon thereafter, in 1860, the Russians founded the port of Vladivostok on the Pacific. Now began the series of events that were to lead to the first involuntary halt in the Russian drive for territory. As Figure 2A-1 shows, the most direct route from western Russia to the port of Vladivostok lay across northeastern China, the territory then still called Manchuria. The Russians had begun construction of the Trans-Siberian Railroad in 1892, and they wanted China to permit the track to cross Manchuria. But the Chinese resisted. Then, taking advantage of the Boxer Rebellion in China in 1900 (see Chapter 9A), Russian forces occupied Manchuria so that railway construction might proceed.

That move, however, threatened Japanese interests in this area, and Japan confronted Russia in the Russo-Japanese War of 1904–1905. Not only were the Russians beaten and forced out of Manchuria: Japan even took possession of the southern part of Sakhalin Island, which they named Karafuto and retained until 1945.

Thus Russia, like Britain, France, and other European powers, expanded through *imperialism*. Whereas the other European powers expanded overseas, Russian influence traveled overland into Central Asia, Siberia, China, and the Pacific coastlands of the Far East. What emerged was not the greatest empire but the largest territorially contiguous empire in the world.

The czars embarked on their imperial conquests in part because of Russia's relative location: Russia always lacked warm-water ports. Had the Revolution not intervened, their southward push might have reached the Persian Gulf or even the Mediterranean Sea. Czar Peter the Great envisaged a Russia open to trading with the entire world; he developed St. Petersburg on the Baltic Sea into Russia's leading port. But in truth, Russia's historical geography is one of remoteness from the mainstreams of change and progress, as well as one of self-imposed isolation.

A Multinational Empire

Centuries of Russian expansionism did not confine itself to empty land or unclaimed frontiers. The Russian state became an imperial power that annexed and incorporated many nationalities and cultures. This was done by employing force of arms, by overthrowing uncooperative rulers, by annexing territory, and by stoking the fires of ethnic conflict. By the time the ruthless Russian regime had begun to confront revolution among its own citizens, czarist Russia controlled peoples representing more than 100 nationalities. The winners in the ensuing revolutionary struggle—the communists who forged the Soviet Union—did not liberate these subjugated peoples. Rather, they changed the empire's framework, binding the peoples colonized by the czars into a new system that would in theory give them autonomy and identity. In practice, it doomed those peoples to bondage and, in some cases, extinction. The Soviet Union, created in the wake of the communist revolution of 1917, constituted another effort to incorporate all these peoples and regions into this vast multinational state, notwithstanding the standard socialist condemnations of imperialism.

THE SOVIET UNION

Communism found fertile ground in the Russia of the 1910s and 1920s. In those days Russia was infamous for the wretched serfdom of its peasants, the cruel exploitation of its workers, the excesses of its nobility, and the ostentatious palaces and riches of the czars. Ripples from the European Industrial Revolution introduced a new age of misery for those laboring in factories. There were workers' strikes and ugly retributions, but when the czars finally tried to better the lot of the poor, it was too little too late.

There was no democracy, and the people had no way to express or channel their grievances. Europe's democratic revolution had passed Russia by, and its economic revolution touched the czars' domain only slightly. Most Russians, as well as tens of millions of non-Russians under the czars' control, faced exploitation, corruption, starvation, and harsh subjugation. When the people began to rebel in 1905, there was no hint of what lay in store; even after the full-scale Revolution of 1917, Russia's political future hung in the balance.

The Political Framework

Russia's great expansion had brought many nationalities under czarist control; now the revolutionary government sought to organize this heterogeneous ethnic mosaic into a smoothly functioning state. The czars had conquered, but they had done little to bring Russian culture to the peoples they ruled. In 1917, when the old order was overthrown, the Russians themselves constituted only about one-half of the population of the entire empire.

In the wake of the Revolution, the first response of many of Russia's subject peoples was to proclaim independent republics, as occurred in Ukraine, Georgia, Armenia, Azerbaijan, and even Central Asia. But Vladimir Lenin, the communist leader and chief architect of the new political system, had no intention of permitting the Soviet state to break up. In 1923, when his blueprint for the new Soviet Union went into effect, the last of these briefly independent units was fully absorbed into the sphere of the Moscow regime.

The political framework for the Soviet Union was based on the ethnic identities of its many incorporated peoples. Given the size and cultural complexity of the empire, it was impossible to allocate territory of equal political standing to all the nationalities (the communists controlled the destinies of well over 100 peoples, both large nations and small isolated groups). It was decided to divide the vast realm into 15 *Soviet Socialist Republics* (*SSRs*), each of which was delimited to correspond broadly to one of the major nationalities (Figure 2A-7). As was noted, Russians constituted only about half of the developing Soviet Union's population, but they also were (and still are) the most widely dispersed ethnic

FIGURE 2A-7

© H. J. de Blij, P. O. Muller, and John Wiley & Sons, Inc.

group in the realm (see Fig. 2A-8). The Russian Republic, therefore, was by far the largest designated SSR, encompassing just over three-quarters of the total Soviet territory.

Within the SSRs, smaller minorities were assigned political units of lesser rank. These were called Autonomous Soviet Socialist Republics (ASSRs), which in effect were republics within republics; other areas were designated Autonomous Regions or other nationality-based units. It was a complicated, cumbersome, often poorly designed framework, but in 1924 it was launched officially under the banner of the *Union of Soviet Socialist Republics* (*USSR*).

A Phantom Federation

The Soviet planners called their system a **federation [8]**. Federalism involves the sharing of power between a country's central government and its political subdivisions (provinces, States, or, in the Soviet case, "Socialist Republics"). Study the map of the former Soviet Union (Fig. 2A-7) and an interesting geographic corollary emerges: every one of the 15 SSRs had a boundary with a non-Soviet neighbor. This seemed to give geographic substance to the notion that any Republic was free to leave the USSR if it so desired. Reality, of course, was different, and Moscow's absolute control over the SSRs made the Soviet Union a federation in theory only.

But it was very difficult to accommodate the shifting multinational mosaic of the Soviet realm. The republics quarreled among themselves over boundaries and territory. Demographic change, forced migrations, war, and economic factors soon made much of the layout of the 1920s obsolete. Moreover, the communist planners made it Soviet policy to relocate entire peoples from their homelands in order to better fit the grand design, and to reward or punish—sometimes capriciously. The overall effect, however, was to move minority peoples eastward and to replace them with Russians. This **Russification [9]** of the Soviet Empire produced substantial ethnic Russian minorities in all the non-Russian republics.

The centerpiece of the tightly controlled Soviet "federation" was the Russian Republic. With half of the vast state's population, the capital city, the realm's core area, and 76 percent of the Soviet Union's territory, Russia was the empire's nucleus. In other republics, "Soviet" often was simply equated with "Russian"—it was the reality with which the lesser republics lived. Russians came to the other republics to teach (Russian was taught in the colonial schools), to organize (and frequently dominate) the local Communist Party, and to implement Moscow's economic decisions. This was colonialism, but somehow the communist disguise—how could socialists, as the communists called themselves, be colonialists?—and the contiguous spatial nature of the empire made it appear to the rest of the world as something else. Indeed, on the world stage the Soviet Union became a champion of oppressed peoples, a force in the decolonization process. It was an astonishing contradiction that would, in time, be fully exposed.

The Soviet Economic Framework

The geopolitical changes that resulted from the establishment of the Soviet Union were accompanied by a gigantic economic experiment: the conversion of the empire from a czarist autocracy with a capitalist veneer to communism. From the early 1920s onward, the country's economy would be *centrally planned*—the communist leadership in Moscow would make all decisions regarding economic planning and development. Soviet planners had two principal objectives: (1) to accelerate industrialization and (2) to **collectivize** agriculture. For the first time ever on such a scale, an entire country was organized to work toward national goals prescribed by a central government.

The Soviet planners believed that agriculture could be made more productive by organizing it into huge state-run enterprises. The holdings of large landowners were expropriated, private farms were taken away from their farmers, and the land was consolidated into collective farms. Initially, all such land was meant to be part of a *sovkhoz*, literally a grain-and-meat factory in which agricultural efficiency, through maximum mechanization and minimum labor requirements, would be at its peak.

Soviet agriculture never attained such productivity and those who obstructed the communists' grand design suffered a dreadful fate. It is estimated that between 30 and 60 million people lost their lives from imposed starvation, constant political purges, Siberian exile, and forced relocation. The Soviet experiment was an incalculable human tragedy, but the secretive character of Soviet officialdom made it possible to hide everything from the world.

The USSR practiced a **command economy [10]**, in which state planners assigned the production of particular manufactures to particular places, often disregarding the rules of economic geography. For example, the manufacture of railroad cars might be assigned (as indeed it was) to a factory in Latvia. No other factory anywhere else would be permitted to produce this equipment—even if supplies of raw materials would make it cheaper to build them near, say, Volgograd 2000 kilometers (1250 mi) away. Yet, despite an expanded and improved transport network (see Fig. 2A-9), such practices made manufacturing in the USSR extremely expensive, and the absence of competition made managers complacent and workers far less productive than they might have been.

Of course, the Soviet planners never imagined that their experiment would fail and that a market-driven economy would replace their command economy. When that happened, the transition was predictably difficult; indeed, it is far from over and continues to severely stress the now more democratic, postcommunist state.

THE NEW RUSSIA

On December 25, 1991, the inevitable occurred: the Soviet Union imploded. The centrally planned economy was a structural failure, the arms race with the United States during the Cold War had drained resources, and efforts at

FIGURE 2A-8

© H. J. de Blij, P. O. Muller, and John Wiley & Sons, Inc.

Russification had only fueled the drive for independence among so many of the peoples that had been forcibly incorporated into the Union. The last Soviet president, Mikhail Gorbachev, resigned, and the Soviet hammer-and-sickle flag flying atop the Kremlin was lowered for the last time and immediately replaced by the white, red, and blue Russian tricolor. A new and turbulent era began, but Soviet structures and systems will long cast their shadows over transforming Russia.

When the Soviet system failed and the Soviet Socialist Republics became independent states in 1991, Russia was left without the empire that had taken centuries to build and consolidate—and that contained crucial agricultural and mineral resources. No longer did Moscow control the farms of Ukraine and the oil and natural gas reserves of Central Asia. But look at Figures 2A-1 and (especially) 2B-2 and you will see that, even without its European and Central Asian colonies, Russia remains an empire. Russia lost the "republics" on its periphery, but Moscow still rules over a domain that extends from the borders of Finland to North Korea. Inside that domain Russians are in the overwhelming majority, but many subjugated nationalities, from Tatars to Yakuts, still inhabit ancestral homelands. Accommodating these many indigenous peoples is one of the challenges facing the Russian Federation today.

A Complex Cultural Mosaic

Although Russia's dominance of this geographic realm justifies our naming it as such, this is a culturally and ethnically diverse part of the world whose traditions and customs spill over into neighboring realms even as neighbors have come to live here. Russians still form the great major-

ity, but sizeable parts of this realm are home to non-Russian peoples—and not just along the borders. As Figure 2A-8 shows, the Russian realm contains Finnish, Turkic, Armenian, and dozens of other "nationalities," and the scale of this map cannot begin to reflect the complexity of the overall ethnic mosaic. Here is an interesting example: in the southwest, facing the Caspian Sea, there is a small "republic" that is part of modern Russia, named Dagestan. It is about half the size of Maine, has a population of 3 million, and contains almost 30 ethnic "nationalities" speaking their own languages, most of which are variants of languages spoken in the Caucasus, Turkey, and Iran.

As the map shows, the Slavic peoples known collectively as the Russians not only form the majority of the population but also are the most widely dispersed. Although the Russian Plain is the core area of the Russian state and was its historic hearth, Russian settlement extends from the shores of the Arctic Sea to the Black Sea coast and from St. Petersburg on the Gulf of Finland to Vladivostok on the Sea of Japan. Discontinuous nuclei and ribbons of Russian settlement are scattered across Siberia, but as we noted in Figure 2A-4, the population, Russian and non-Russian, tends to concentrate in the southern sector of the realm.

Russians are not the only Slavic peoples. More than a thousand years ago, when many ethnic groups struggled to establish themselves in Europe, the early Slavs, whose original home may have been on the North European Plain in the area north of the Carpathians, achieved stability and security in a homeland that expanded steadily, not only eastward into what is today Russia, but also westward and southwestward into the valley of the Danube and beyond.

As we noted in Chapters 1A and 1B, the Serbs, Croats, Slovaks, and Czechs are Slavic peoples, as are the Poles, Ukrainians, and Belarussians, among others (see Fig. 1B-14). Today you can discern evidence of this distant past in the Slavic languages these descendants speak (see Fig. 1A-7), which still have much in common despite centuries of divergence.

Even as the early Russian state consolidated, non-Russian peoples incubated in and beyond the Caucasus Mountains in the corridor between the Black and Caspian seas. Georgians, Armenians, Azeris, and many others created a jigsaw puzzle of nationalities in this region (the aforementioned Dagestan is only one small part of it). Then, after about AD 1400, Islam propagated by the Ottoman Empire pushed northward here, and by 1500 Islam's vanguards were challenging Slavic peoples along the north shore of the Black Sea. Remnants of such advances and invasions still remain on the cultural map (Fig. 2A-8): look at the crescent of Turkic peoples (shaded red) extending from near the city of Nizhniy Novgorod to the border of Kazakhstan, and you see part of the evidence. Today, Russia has a far higher percentage of Muslims in its population than western European countries do; although data are not precise, the consensus is that about 12 percent of Russia's inhabitants are Muslims.

Long before Islam appeared on the scene, Slavic peoples from one end of their domain to the other had accepted the teachings of the Eastern Orthodox Church, and in the Russian realm the Russian church became Eastern Orthodoxy's dominant institution. It stayed that way until the triumph of the communist revolutionaries, who in 1917 put an end to rule by the Russian czars and began the disestablishment of the church. For seven decades, atheism was official policy—until the communist system disintegrated. Since the early 1990s, the Russian Orthodox Church has made a vigorous comeback, attended by unrestrained nationalist and ethnic propaganda. The postcommunist-era Patriarch Alexy II in his sermons often asserted that Orthodox beliefs and "the Slavic soul" are one and the same, so that the former is the spiritual and cultural foundation of the latter. In an increasingly heterogeneous society with Islamic, Buddhist, and other non-Christian constituencies, such doctrinaire assertions may have worrisome implications for the future.

Cities Near and Far

When the czars ruled Russia, people living in villages and on homesteads in the countryside far outnumbered those residing in towns and cities. It is not that the czars wanted to keep it that way, but their policies—social, economic, and otherwise—obstructed change and impeded opportunities that draw people to urban areas. Even under Peter the Great, who admired urbanizing western Europe and wanted to make St. Petersburg a glittering Russian window on the wider world, Russia's urbanization lagged far behind Europe's. Compare Russia's level of urbanization today (73 percent) to that of western Europe (84 percent), and this contrast endures. Figure 1A-5 reminds us that the Industrial Revolution came late to this realm, and that its impact was slowed until the revolutionary communist organizers who succeeded the czars in the 1920s undertook a massive effort to catch up.

The three countries that form the Transcaucasus region are even less urbanized than Russia: in Azerbaijan and Georgia barely 50 percent of the people live in urban areas, and even in Armenia urbanization lags (see the Data Table in Appendix B to compare these numbers to, say, East Asia or South America). Here the twentieth-century period of Russian influence did little to dislodge rural traditions or stimulate change.

Nonetheless, the cities, when mapped by size (Fig. 2A-9), reveal much about this realm's expansive regional geography. Dominant Moscow and northerly but coastal

The landscape of Tula, a city of 530,000 about 150 kilometers (100 mi) south of Moscow, is typical of urban centers in western Russia. Tula's townscape today is a mixture of pre-Soviet historic buildings, drab Soviet-era tenements, and scattered post-Soviet high-rises and single-family houses. The urban cultivators in the foreground are a common sight in a country with limited agricultural land and short growing seasons. Fresh produce is particularly coveted and is widely used for bartering among friends and neighbors in exchange for other hard-to-get commodities. Informal farming on fertile plots of urban land has increased since 1991 as the redistribution of population has channeled rural migrants into cities, where their slow absorption into urban society often requires them to fall back on their food production and preservation skills to supplement their diets and make meager incomes stretch further.

© MauroGalligani/Contrasto/ReduxPictures.

FIGURE 2A-9

© H. J. de Blij, P. O. Muller, and John Wiley & Sons, Inc.

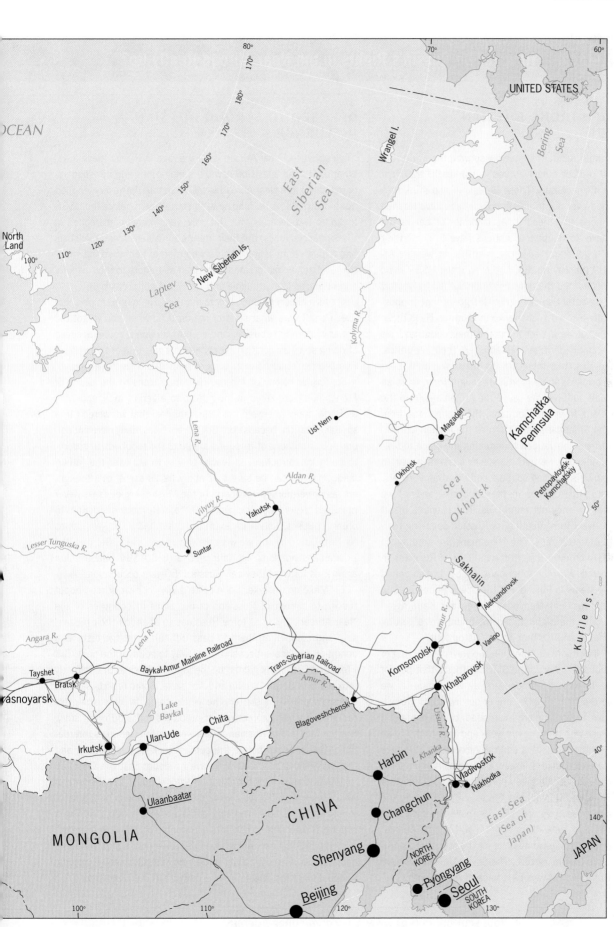

Regional ISSUE   How Far Do Russia's Rights in the Near Abroad Really Go?

IN FAVOR OF STRONG RUSSIAN INFLUENCE

"I am delighted that President Medvedev showed the world that we Russians will protect ourselves and our allies in the Near Abroad wherever it's necessary. Those Georgians who think they can ride roughshod over our friends the Ossetians have learned a lesson, and I hope they heed it. As a former officer in the Soviet Army, I have no doubt that nations have to show their strength or they'll get trampled on. I'm still sorry we left Afghanistan in 1989 the way we did. And even if the USSR is no more, we Russians earned the right not to be bullied by upstart leaders like that dreadful Saakashvili in Georgia or that supposedly democratically-elected Yushchenko in Ukraine. Have those people forgotten what we did for them and their countries? We created their national identities, established their republics, built their economies, put them on the map. We built their roads, railroads, bridges, and airports; we taught them Russian and nourished their cultures. We saved the Armenians from the Turks, the Estonians from the Germans, the Mongolians from the Chinese. If the Americans hadn't armed the Afghan extremists, we wouldn't have the Taliban threatening the place today.

"So what thanks do we get? Saakashvili starts a war in South Ossetia and kills Russian soldiers and Ossetian villagers. Chechen terrorists kill Russians in the Caucasus, and all we hear about is that we mistreat minorities. The Latvians insult and offend us at every turn, treating its Russian-speaking residents as second-class citizens. The Ukrainians are cozying up to the European Union even though the millions of Russians in the country oppose it. Now the Americans want to put so-called missile defense systems in Poland and the Czech Republic, and I'm pleased that President Medvedev played our Kaliningrad card. I've had it with the high-handed treatment we're getting around our perimeter, and it's time to put a stop to it.

"You can look at the Near Abroad in two ways: as the ring of countries that encircles us or as the Slavic outposts that persist from Kazakhstan to Kosovo. Whichever way you define it, we Russians reserve the right to take care of our own, and we will not stand by as our kinspeople are mistreated, or our homeland threatened, by those who think we are weak and impotent. I hope the Ukrainians and the Moldovans and the Albanians and others who might miscalculate take a lesson from what happened in Georgia. If we have to, we will take appropriate action, and that includes military action. Our military forces suffered during the Yeltsin years, but those days are over. We have the money to rebuild our forces and we're doing so. We will not be pushed around, and it would be a grave mistake for the rest of the world to see us as weak or lacking resolve. The Soviet Union may have collapsed, but Russia will always be a force to reckon with."

OPPOSED TO STRONG RUSSIAN INFLUENCE

"Talk about the Near Abroad is all the rage in Russia these days. Whether it's the television news, the magazines and newspapers, or talk radio, the Near Abroad is the topic that gets everyone riled up. In truth, this is nothing new—it started even before the USSR broke up two decades ago. Not only were Russians marooned in the former republics, but their friends and allies among the locals had a tough time as well. Those who tried to help save the situation Gorbachev had created were seen as traitors to their people, and in places like Lithuania and Georgia it got pretty ugly.

"As a history teacher, I try to put things in long-term perspective. I don't care what state or nation it is, when you've got the power you tend to abuse it, and even if you don't, you're accused of doing so by association. That certainly was true of the western European colonial powers. Look at France and its 'Near Abroad' in its former colony of Algeria. How many tens of thousands of Frenchmen were killed in the run-up to Algeria's independence, and how many afterward? We Russians feel that we weren't in a similar situation, because our Soviet republics really were not colonies. But millions of Russians did go to the republics to govern, to build infrastructure, to teach, or just to work for the fatherland, and you can be sure that many locals saw these 'foreigners' as power-hungry outsiders. In the Muslim societies we never managed to persuade the people of the irrelevance of religion. In other republics, Latvia for example, there had been sympathies for Nazi Germany and we were viewed as occupiers. Get this: the Lithuanians not long ago launched a (U.S.) $34 billion claim against us for what they call Russia's '50-year occupation.'

"Make no mistake: I don't like it when I hear or read about the mistreatment of Russians or our allies in countries of the Near Abroad. Trying to force Russians to abandon their citizenship and compelling them to pass difficult language tests as a condition for local citizenship is no way to get over the past. But I feel that nothing is gained by the kind of violence that recently took place in South Ossetia and Abkhazia. Some reports say that Georgia's president started the whole thing by unleashing attacks on Russian protective forces; others tell a different story. But the fact is that Russian forces were within Georgia's international borders even before the conflict took its deadly turn, and that was asking for trouble. I have no doubt that the Georgian regime was guilty of discrimination against South Ossetians (and Abkhazians as well), but we're talking here about only tens of thousands of people in a tiny enclave—was it worth risking a wider war over what is essentially a minority-treatment issue?

"I don't think so, but let me tell you—I'm in a small minority here. The small shooting war in Georgia has galvanized Russians. I hope it's not the beginning of a new era in the Near Abroad."

Vote your opinion at www.wiley.com/college/deblij

St. Petersburg anchor the Russian core area. Historic urban centers like Novgorod, Kazan, and Yekaterinburg mark crucial episodes in Russia's historical geography. The cities on the Volga River spearheaded the country's post-czarist transformation. As the ribbon of cities and towns thins out eastward beyond the Urals, we observe the names associated with the communist Soviet Union's industrial might and heroic resistance against the Nazi incursion: Omsk, Krasnoyarsk, and chill-inducing Novosibirsk. In the Russian Far East, Vladivostok symbolizes Soviet naval power and more recent Russian indifference. And on the Kamchatka Peninsula, people by the tens of thousands are leaving Petropavlovsk, once an outpost favored by the Soviets, now abandoned by Moscow.

As for Transcaucasia, the three leading cities, also the capitals of their respective republics, all signify a singular and dominant issue: oil and its export routes in Baki (Baku), Azerbaijan; conflict with Russia in Tbilisi, Georgia; and landlocked weakness in Yerevan, Armenia.

The Near Abroad

The ending of the Cold War implied an abrupt termination of Soviet geopolitical strategies that had been aimed at confronting U.S.-led alliances in Europe, the Middle East, and South as well as Southeast Asia. Indeed, Soviet foreign policy and power was projected beyond its own frontiers to other continents, from Africa to Middle America. The collapse of the Soviet Union meant that the new Russia, shorn of **satellite states [11]** in eastern Europe and abandoned by all of the other former Soviet republics, had to completely rethink its position in the world and most immediately in Eurasia. And it had to do so in the knowledge that a total makeover of its economy was the top priority.

The Russian government introduced the term **Near Abroad [12]** for the newly-formed countries that surround Russia today. These are for the most part the other former Soviet republics, from the Baltic states to Kazakhstan. These territories may now be "abroad" but seen from Moscow they are so "near"—and together form such an encirclement of Russia that it is entitled to take a special interest in them (compare Figs. 2A-7 and 2A-9). What this really means is that Russia will not allow any of these bordering countries to develop a threat of any sort and that it reserves the right to intervene if necessary. It is not that hard to imagine this feeling of encirclement in Russia, particularly with the eastward expansion of the EU (now practically on Russia's western doorstep), Islamic fundamentalism in the south, and the swift rise and growing assertiveness of China in the east.

And then there is the issue of Russian minorities who now find themselves living abroad. Although many Russian nationals returned from the former SSRs, millions of others stayed on. They lived in areas where Russians were in the majority (for example, in eastern Ukraine and in northern Kazakhstan) and where post-Soviet life was not all that different from what it was during Moscow's rule. But elsewhere, Russians who remained in the former SSRs found themselves mistreated in various ways, and they often appealed to Moscow for help. Thus the concept of a Russian "Near Abroad" found its way into Russian discourse. Beyond Russia's borders but of concern to Russia's leaders and people, the "Near Abroad" became geographic shorthand for a Russian sphere of influence, whereby Moscow stood ready to help its fellow ethnics in case of trouble. On the map, the most vivid example of this is the so-called Russian Transition Zone in northern Kazakhstan (Fig. 2A-9), which was tightly integrated into the adjacent Russian sphere during Soviet times. (A larger-scale map would show similar but smaller Russian clusters in Estonia, Latvia, Moldova, and elsewhere.)

A REALM IN TRANSITION

Twenty years ago, the Soviet Empire collapsed and disintegrated. What remains today is the vast centerpiece of that empire, the Russian Federation. Two decades may seem like a long time, but the present situation still bears the marks of the Soviet era and this is still a realm in transition. Demographically, economically, administratively, politically, and geostrategically, Russia is likely to continue to adjust in the years ahead. Ideally, relations with the Near Abroad remain cordial and peaceful, the Russian Federation itself maintains cohesion, and economic well-being improves. Clearly, this does not only depend on decision-making in Moscow. The geography of the Russian realm is such that it finds itself at the heart of the Eurasian continent, the Russian Federation itself bordering no less than 14 countries and four other realms. Perhaps, in such circumstances, overall long-term stability is an illusion. To be sure, it is a challenge to manage a country of this massive size and diversity. In Chapter 2B, we will discuss the evolving administrative (spatial) structure of the new Russia and take a closer look at the different regions that constitute this unique realm.

POINTS TO PONDER

- High-speed train travel, badly needed in this enormous realm, has been slow to develop. In 2010 the first such route opened between the capital, Moscow, and the leading port and second city, St. Petersburg.

- Global warming produces winners as well as losers, and the Russians regard themselves as potential winners as Siberia's harsh environments moderate and the Arctic Ocean's ice recedes, opening up sea lanes and access to submerged energy reserves.

- The Russian Federation borders 14 countries as well as four other realms. Compare that to the conterminous United States, which borders only two countries and two other realms.

REGIONS OF THE RUSSIAN REALM

Peripheries in transition

| 0 | 500 | 1000 Kilometers |

| 0 | 200 | 400 | 600 Miles |

Longitude East
of Greenwich

FIGURE 2B-1

© H. J. de Blij, P. O. Muller, and John Wiley & Sons, Inc.

REGIONS

IN THIS CHAPTER

- Running a country with nine time zones
- Moscow: from Soviet capital to global city
- Where Russia meets China, Japan, and the Korean Peninsula
- The cold beauty of Siberia
- Trouble in Transcaucasia
- Terrorism comes to Moscow

CONCEPTS, IDEAS, AND TERMS

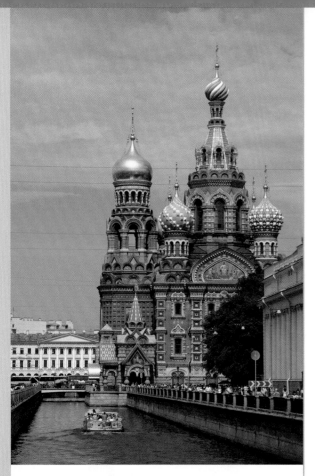

© Connie Coleman/GettyImages, Inc.
Where were these photos taken?
Find out at www.wiley.com/college/deblij

No other single state so thoroughly dominates its geographic realm as does Russia. It is important to remember the many changes that have taken place in recent decades along the edges of this realm. After the USSR imploded in 1991 and all of the other former Soviet Republics had seceded from the Union, Russia was left all by itself—a gigantic country of unmatched proportions, but one that had lost control of major bordering regions in Europe and Southwest Asia, from Estonia to Kazakhstan. Today we refer to the Russian realm as comprising the country of Russia as well as the three comparatively tiny Transcaucasian countries of Georgia, Azerbaijan, and Armenia. To be sure, Russia's historic influences continue to spill over into now-independent countries all around its borders from the Baltic to the Far East, but most of the latter are now drawn more firmly into the orbits of other realms. However, such is not the case with Transcaucasia, which remains very tightly embedded within the Russian sphere.

Russia itself is huge and diverse—physiographically, ethnically, and economically—and could be subdivided in a variety of ways and at different scales. Let us try to keep things simple and delimit what we might call first-order regions, recognizing that each of this realm's enormous regions contains numerous subdivisions. Four of the five regions we discuss lie entirely inside Russia (Fig. 2B-1):

1. The Russian Core

2. The Southeastern Frontier

3. Siberia

4. The Russian Far East

and only one lies outside the Russian state:

5. Transcaucasia

© Jochem D. Wijnands/GettyImages, Inc.

POST-SOVIET RUSSIA

When the USSR dissolved in 1991, Russia's former empire devolved into 14 independent states, and Russia itself was a changed nation. Russians now made up about 83 percent of the population of just under 150 million, a far higher proportion than in the days of the Soviet Union. But numerous minority peoples remained under Moscow's new flag, and millions of Russians found themselves under new governments in the former Republics.

Russia's Changing Political Geography

The new Russian government also faced an old Soviet problem: the sheer size of the country, its vast distances, and the remoteness of many of its Regions. Geographers refer to the principle of **distance decay [1]** in order to explain how increasing distances between places tend to reduce interactions among them. Because Russia is the world's largest country, distance is a significant factor in the relationships between the capital and outlying areas. Furthermore, Moscow lies in the far west of this gigantic country, half a world away from the shores of the Pacific (see box titled "Russia Permanently on Daylight Saving Time"). Not surprisingly, one of the most obstreperous Regions has been remote Primorskiy, the Region of Vladivostok (Fig. 2B-1). Still another problem is the stupendous size variation among the country's administrative units, as can be seen in Figure 2B-2. Whereas the (territorially) smallest divisions are concentrated in the Russian core area, the largest lie far to the east, where Sakha is nearly a thousand times as large as Ingushetiya. On the other hand, the populations of the enormous eastern divisions are tiny compared to those of the smaller ones in the Russian Core. Such diversity spells administrative difficulty.

But perhaps the most serious current problem for Russia is the intensifying social disharmony between Moscow and the subnational political entities. Almost wherever one goes in Russia today, Moscow is disliked and often berated by angry locals. The capital is seen as the privileged playground for those who have benefited most from the post-Soviet transition—bureaucrats and hangers-on whose economic policies have driven down standards of living, whose greed and corruption have hurt the economy, whose actions in Chechnya have been a disaster, who have allowed foreigners (prominently the United States) to erode Russian power and prestige, who fail to pay the wages of workers toiling in industries still owned by the state, and who do not represent the Russian people. Complaints about the capital are not uncommon in free countries, but in Russia the mistrust between capital and subordinate areas has become so serious that it constitutes a key challenge for the government.

The Soviet Legacy

Soviet planners had created a highly complex administrative structure for their "Russian Soviet Federative Socialist Republic," and Russia's postcommunist leaders have been forced to use this framework to make their country function. Figure 2B-2 reflects this complex Russian administrative system that dates back to Soviet times. There are now 83 entities in all: 2 Autonomous Federal Cities (Moscow and St. Petersburg), 21 Republics, 46 Provinces (Oblasts), 9 Territories (Krays), 4 Autonomous Regions (Okrugs), and 1 Autonomous Oblast. These administrative units operate with varying degrees of power and autonomy from the federal government. The 21 Republics, established and recognized to accommodate substantial ethnic minorities, include a cluster in the far south, another to the east of Moscow, and a third which lies in the Mongolia border zone.

Russia Permanently on Daylight Saving Time

IN EARLY 2011, President Medvedev announced that Russia will remain on daylight saving time starting in the fall of that year, arguing that switching clocks twice a year is detrimental to people's health.

Switching to and from daylight saving time in the spring and the fall, respectively, is common in more than 100 countries around the world, and it is especially significant in higher latitudes where winter and summer exhibit the greatest differential in hours of sunlight. Moscow sees less than 8 hours of daylight in January and more than 17 in June—and the swing becomes even more extreme in Russia's northern reaches.

Switching the time back and forth in an environment where daylight is so highly valued (along with agreeable temperatures) is thought to cause considerable mental and even physical stress. "Every fall and every spring we are swearing at this system," Medvedev said. "Our biorhythms are damaged. We are all angry. We either oversleep and turn up late for work or wake up too early and don't know what to do with this free time. Let alone poor cows and other animals that can't understand why they should have their meals or be milked earlier or later." An extensive study by the Russian Academy of Medical Sciences showed that after time changes the number of heart attacks increased by 150 percent, the suicide rate jumped by 66 percent, and far more people called ambulance services. The effects were greatest on children and the elderly, who experienced heightened stress and sleep disorders as well as cardiovascular, immune-system, and metabolic problems.

From now on, all of Russia will remain permanently on daylight saving time. For Moscow, that is GMT + 4 (4 hours ahead of Greenwich Mean Time, in London). Midwinter sunrise in Moscow is not until around 10 am, and it turns dark again before 5:00 pm. According to polls, 60 percent of all Russians support this decision.

RUSSIA'S ADMINISTRATIVE STRUCTURE

Ethnic Republics and Autonomous Regions

Russian Regions

National capital is underlined

0   200   400   600   800 Kilometers
0 100 200 300 400 500 Miles

37.1 Population estimate for 2009, in millions

−9% Percent decline in population, since 1991

● Federal District capital

━━ Federal District boundary

RUSSIAN ADMINISTRATIVE UNITS
(named after their capitals)

1. Astrakhan	10. Kurgan	19. Penza	28. Tambov
2. Belgorod	11. Kursk	20. Pskov	29. Tula
3. Bryansk	12. Lipetsk	21. Rostov	30. Tver
4. Chelyabinsk	13. Mariy-el	22. Ryazan	31. Ulyanovsk
5. Ivanovo	14. Moscow	23. St. Petersburg	32. Vladimir
6. Kaluga	15. Novosibirsk	24. Samara	33. Volgograd
7. Kemerovo	16. Omsk	25. Saratov	34. Voronezh
8. Kostroma	17. Orenburg	26. Smolensk	35. Yaroslavl
9. Krasnodar	18. Orel	27. Stavropol	

FIGURE 2B-2

© H. J. de Blij, P. O. Muller, J. Nijman, and John Wiley & Sons, Inc.

OCEAN

North
Land

Laptev
Sea

East
Siberian
Sea

Wrangel I.

New Siberian Is.

Kolyma R.

Bering
Sea

CHUKOTSKIY
(CHUCKCHI)

UNITED STATES

Lena R.

SAKHA
(YAKUTIYA)

FAR EASTERN

6.5
-17%

Vilyuy R.

Aldan R.

MAGADAN

KAMCHATKA

Sea
of
Okhotsk

KHABAROVSK

SAKHALIN

Kurile Is.

RASNOYARSK

Lesser Tunguska R.

BERIAN

19.6
-5%

Angara R.

Lena R.

IRKUTSK

BURYATIYA

L. Baykal

CHITA

Amur R.

AMUR
OBLAST

Amur R.

YEVREVSKAYA

Khabarovsk

Ussuri R.

PRIMORSKIY
KRAY

L. Khanka

TYVA

MONGOLIA

CHINA

NORTH
KOREA

SOUTH
KOREA

East Sea
(Sea of
Japan)

JAPAN

When Russia's post–1991 government took over, it faced complicated administrative problems. Not only were some entities reluctant to sign the Russian Federation Treaty, but despite the ranking implied by the list above, certain Regions were more equal than others—that is, some were used to privilege under the old system, and their local leaders expected that to continue. Meanwhile, a multinational, multicultural state that had been accustomed to authoritarian rule and government control over virtually everything—from factory production to everyday life—now had to be governed in a new way. Democratization of the political system, transition to a market economy, the sale of state-owned industries (privatization), and other far-reaching changes had to come quickly or the country risked chaos.

Unitary versus Federal Options

Russia's leaders recognized that their options were limited. They could continue to hold as much power as possible at the center, making decisions in Moscow that would apply to all the Republics, Regions, and other subdivisions of the state. Such a **unitary state system [2]**, with its centralized government and administration, marked authoritarian kingdoms of the past and serves totalitarian dictatorships of the present. Or they could share power with the Republics and Regions, allowing elected regional leaders to come to Moscow to represent the interests of their people. That is the federal system Russia chose as the only way to accommodate the country's economic and cultural diversity.

In a *federal system*, the national government usually is responsible for matters such as defense, foreign policy, and foreign trade. The Regions (or provinces, States, or other subdivisions) retain authority over affairs ranging from education to transportation. A federal system does not create unity out of diversity, but it does allow diverse components of the state to coexist, their common interests represented by the national government and their regional interests by their local administrations. Some countries owe their survival as coherent states to their federal frameworks; India and Australia are cases in point.

But to maintain a generally acceptable balance of power between the center and Regions (or States) is difficult. Disputes concerning "States' rights" even continue to roil the U.S. political scene more than two centuries after the Constitution was adopted. Early on, the Russian government decided to end, for all practical purposes, the hierarchical regional system the Soviets had established. The Republics retained their special status, but all the others—okrugs, krays, and so forth—were redesignated as Regions. This was intended to address the favoritism of Soviet times and to streamline the system in general.

Russia's New Federal Structure

In 1992, most of Russia's internal "republics," autonomous regions, oblasts, and krays (all of them components of the administrative hierarchy) signed a document known as the Russian Federation Treaty, which committed them to cooperate in the new federal system. At first a few units refused to sign, including Tatarstan, scene of Ivan the Terrible's brutal conquest more than four centuries earlier, and a republic in the remote Caucasus periphery, Chechnya-Ingushetiya, where Muslim rebels waged a campaign for independence (Fig. 2B-2).

Chechnya-Ingushetiya next split into two separate republics, whose names at the time were spelled Chechenya and Ingushetia. Eventually, only Chechnya refused to sign the Russian Federation Treaty, and subsequent Russian military intervention led to a prolonged and violent conflict, with disastrous consequences for Chechnya's people and infrastructure (the capital, Groznyy, was completely destroyed). The conflict in Chechnya continues today and is a disaster for Russia's government as well.

In 2000, the new Putin administration moved to diminish the influence of the Regions by creating a new geographic framework that combined the 83 Regions, Republics, and other entities into eight new administrative units—not to enhance their influence in Moscow, but to increase Moscow's authority over them. As Figure 2B-2 shows, each of these new Federal Districts has a capital, elevating cities such as Rostov and Novosibirsk to a status secondary only to Moscow. In a related move, the Russian president proposed, and the parliament approved, that Regional governors would be appointed rather than elected—thereby concentrating still more power in Moscow. Russia's federal system now appears to be moving in a unitary direction.

A Shrinking Population

Some of these administrative problems are compounded by an alarming trend: the Russian population has been shrinking steadily for the past two decades. And, as Figure 2B-2 shows, certain parts of the country have witnessed a veritable **population implosion [3]**. When the Soviet Union disintegrated in 1991, Russia's population totaled about 149 million. Today, two decades later, this number is down to less than 142 million—despite the in-migration of several million ethnic Russians from the former Soviet SSRs beyond Russia's borders. Since the end of communist rule, Russia has witnessed over 10 million more deaths than births, with severe social dislocation the primary cause.

Undoubtedly, the transition away from Soviet rule is one reason: uncertainty tends to cause families to have fewer children, and abortion is widespread in Russia. But the birth rate has stabilized at around 12 per thousand; it is Russia's death rate that has skyrocketed, now exceeding more than 14 per thousand. This produces an annual loss of population that exceeds 0.2 percent, or more than a quarter-million per year.

Russian males are the ones most affected. Male *life expectancy* dropped from 71 in 1991 to 62 in 2010 (female life expectancy has also declined, though markedly less, to 74). Males are more likely to be afflicted by alcoholism and related diseases, by AIDS (significantly underreported in Russia, according to international agencies), by heavy smoking, and by suicide, accidents, and murder. On average, a Russian male is nine times more likely to die a violent or an accidental death than his European counterpart. Shockingly, fewer than half of today's male Russian teenagers will survive to age 60.

From 2009 to 2010, the latest years for which statistics were available at press time, Russia seemed to be bucking the trend: for the first time in 20 years the overall population expanded again, albeit only by about 25,000. Still, things look grim at the regional scale. Look again at the percentages of population decline on the map showing the Districts (Fig. 2B-2): clearly, out-migration plays an important role too. People are no longer forced to stay put in these challenging environs by a dictatorial regime, and they are drawn to the far more prosperous and opportunity-rich western parts of Russia. But the emptying out of villages has been occurring all across Russia: as many 15,000 former villages have turned into ghost towns—no less than 10 percent of all villages in the country two decades ago.

What is the answer? Russia's leaders hope that improvements in the social circumstances of the average Russian will reduce the overall rate of population loss. Campaigns against alcoholism and careless lifestyles are under way. Immigration is another option, and should Russia relax its rules, hundreds of thousands of Koreans and Chinese would move into the Russian Far East to counter the outflow of Russians there. But Moscow is not eager to see its Pacific Rim transformed into an extension of East Asia. Thus most of the ghost towns are unlikely to be reborn.

A Volatile Economy

During the past 15 years or so, Russia's economy has turned 180 degrees. Private property, upstart companies, trade, foreign investment, and stock exchanges, all the things we consider vital components of the free market, have found their way into Russia. The city of Moscow has changed beyond recognition with a new high-rise skyline and unprecedented linkages to the global economy. After some major (and for many people painful) adjustments were made, Russia's GDP grew vigorously and the Moscow stock market skyrocketed. In common parlance, Russia is included as one of the **BRICs [4]**, the world's biggest emerging markets (Brazil, Russia, India, China). Russia's *nouveaux riche* are known for conspicuous shopping trips to Paris, Milan, or New York, for sending their children to boarding schools in England, and even for buying up leading soccer clubs in Europe (as some billionaires have done).

But Russia's rise seems precarious: much of it is based exclusively on exports of oil and gas; incoming foreign investment is quite limited compared to other emerging markets; the get-rich-quick schemes of a small number of highly successful businessmen (women are rarely part of this group) do not benefit the masses; and the enormous state apparatus left by the Soviets is responsible for widespread corruption and close involvement with organized crime. The value of the "bribes market" is estimated to be about 20 percent of the national economy. Unlike the other BRICs, the Russian economy was sharply impacted by the global financial crisis of 2008–2010 when economic growth turned negative. Government spending at the outset of this decade was around 40 percent of GDP.

Seventy years of Soviet-style socialism have left their mark. If statistics on the Russian economy suggest robust growth, there is also ample evidence that most of the people aren't seeing the fruits. And that only makes it harder to preserve this federation, which is not helped by its enormously complicated administrative spatial structure that is permeated with corruption and intransigence.

It is now time to turn our attention to the realm's regional geography. As was noted at the beginning of this chapter, so vast is Russia, so varied its physiography, and so diverse its cultural landscape that regionalization requires a small-scale perspective and a high level of generalization. We will proceed to outline the five-region framework that is mapped in Figure 2B-1: (1) the *Russian Core*, everything lying west of the Urals; (2) the *Southeastern Frontier*, the eastward extension of the Core into the heart of Eurasia; (3) *Siberia*, the frigid territory sprawling across the country's huge northeast quadrant; (4) the *Far East*, Russia's elongated window on the Pacific Ocean; and (5) *Transcaucasia*, whose three small republics lie just outside Russia in the corridor between the Black and Caspian seas. As we shall see, each of these first-order regions contains major subregions.

● MAJOR CITIES OF THE REALM

City	Population* (in millions)
Moscow, Russia	10.6
St. Petersburg, Russia	4.6
Baki (Baku), Azerbaijan	2.0
Novosibirsk, Russia	1.4
Yekaterinburg, Russia	1.4
Nizhniy Novgorod, Russia	1.3
Kazan, Russia	1.2
Chelyabinsk, Russia	1.1
Tbilisi, Georgia	1.1
Yerevan, Armenia	1.1
Volgograd, Russia	1.0
Vladivostok, Russia	0.6

*Based on 2012 estimates.

● AMONG THE REALM'S GREAT CITIES . . .

MOSCOW

© H. J. de Blij, P. O. Muller, and John Wiley & Sons, Inc.

IN THE VASTNESS of Russia's territory, Moscow, capital of the Federation, seems to lie far from the center, close to its western margin. But in terms of Russia's population distribution, Moscow's centrality is second to none among the country's cities (see Fig. 2A-4). Moscow lies at the heart of Russia's primary core area and at the focus of its circulatory systems.

On the banks of the meandering Moscow River, Moscow's skyline of onion-shaped church domes and modern buildings rises from the forested flat Russian Plain like a giant oasis in a verdant setting. Archeological evidence points to ancient settlement of the site, but Moscow enters recorded history only in the middle of the twelfth century. Forest and river provided defenses against Tatar raids, and when a Muscovy force defeated a Tatar army in the late fourteenth century, Moscow's primacy was assured. A huge brick Kremlin (citadel; fortress), with walls 2 kilometers (1.2 mi) in length and with 18 towers, was built to ensure the city's security. From this base Ivan the Terrible expanded Muscovy's domain and laid the foundations for Russia's vast empire.

The triangular Kremlin and the enormous square in front of it (Red Square of Soviet times), flanked by St. Basil's Cathedral and overlooking the Moscow River (photo at right), is the center of old Moscow and is still the heart of this metropolitan complex of 10.6 million. From here, avenues and streets radiate in all directions to the Garden Ring and beyond. Until the 1970s, the Moscow Circular Motorway encircled most of the built-up area, but today the metropolis sprawls far beyond this beltway. For all its size, Moscow never developed a world-class downtown skyline. Communist policy was to create a "socialist city" in which neighborhoods—consisting of apartment buildings, workplaces, schools, hospitals, shops, and other amenities—would be clustered so as to obviate long commutes into a high-rise city center.

Moscow may not be known for its architectural appeal, but the city contains noteworthy historic as well as modern structures, including the Cathedral of the Archangel and the towers of Moscow State University.

In the postcommunist era, the core of Moscow has changed tremendously as Soviet-dictated urbanism made way for the emergence of a fast-paced international business center—one that increasingly resembles global cities elsewhere. Once the free market, however imperfectly, shaped up, Moscow's center became the preferred location for foreign investors, multinational companies, upscale shopping, and elite housing. It quickly became one of the most expensive residential areas in the world, its exclusiveness even more striking in the heart of a metropolis that is still mostly dominated by dull, basic Soviet architecture. So for the vast majority of Muscovites, life just became a lot more expensive.

■■ THE RUSSIAN CORE

The heartland of a state is its *core area*. Here much of the population is concentrated, and here lie its biggest cities, leading industries, densest transportation networks, most intensively cultivated lands, and other key components of the country. Core areas of long standing strongly reflect the imprints of culture and history. The Russian Core, broadly defined, extends from the western border of the Russian realm to the Ural Mountains in the east (Fig. 2B-1). This is the Russia of Moscow and St. Petersburg, of the Volga River and its industrial cities.

Moscow is the megacity hub of a wider area comprising some 50 million inhabitants (more than one-third of the country's entire population), most of them concentrated in cities that, during the Soviet period, specialized in assigned industries such as Nizhniy Novgorod (the "Soviet Detroit" where automobiles were made) and Ivanovo (known for its textiles). But today, Russia is less a diversified manufacturing country than a commodity producer and exporter (mainly oil and natural gas). The tall steel-and-glass towers you see rising above the central business districts, such as the one now being built in the heart of

St. Petersburg's historic district, for the most part belong to state-controlled energy companies such as Gazprom. And the cars being built are likely to display foreign names such as Fiat or Volkswagen.

St. Petersburg (the former Leningrad) remains Russia's second city, with a population of 4.6 million. Under the czars, St. Petersburg was the focus of Russian political and cultural life, and Moscow was a distant second city. Today, however, St. Petersburg possesses none of Moscow's locational advantages, at least not with respect to the domestic market. It lies well outside the Central Industrial Region (Fig. 2B-3) near the northwestern corner of the country, 650 kilometers (400 mi) from the capital. Neither is it better off than Moscow in terms of resources: fuels, metals, and foodstuffs must all be brought in, mostly from far away. The former Soviet emphasis on self-sufficiency even reduced St. Petersburg's asset of being on the Baltic seacoast because some raw materials could have been imported much more cheaply across the Baltic from foreign sources than from domestic sites in distant Central Asia (only bauxite deposits lie nearby, at Tikhvin).

Moscow's heart, with the triangular Kremlin (right of center) fronting on adjacent Red Square.
© Travelwide/Alamy.

Central Industrial Region

At the heart of the Russian Core lies the Central Industrial Region (Fig. 2B-3). The precise definition of this subregion varies, because all regional definitions are subject to

FIGURE 2B-3

© H. J. de Blij, P. O. Muller, and John Wiley & Sons, Inc.

ST. PETERSBURG

© H. J. de Blij, P. O. Muller, and John Wiley & Sons, Inc.

CZAR PETER THE Great and his architects transformed the islands of the Neva Delta, at the head of the Baltic Sea's Gulf of Finland, into the Venice of the North, its palaces, churches, waterfront façades, bridges, and monuments giving St. Petersburg a European look unlike that of any other city in Russia. Having driven out the Swedes, Peter laid the foundations of the Peter and Paul Fortress on the bank of the wide Neva River in 1703, and the city was proclaimed the capital of Russia in 1712.

Peter's "window on Europe," St. Petersburg (named after the saint, not the czar) was to become a capital to match Paris, Rome, and London. During the eighteenth century, St. Petersburg acquired a magnificent skyline dominated by tall, thin spires and ornate cupolas, and graced by baroque and classical architecture. The Imperial Winter Palace and the adjoining Hermitage (one of the biggest art museums in the world) at the heart of the city are among a host of surviving architectural treasures (see photo, p. 114).

Revolution and war lay in St. Petersburg's future. In 1917, the Russian Revolution began in the city (then named Petrograd), and following the communist victory it lost its functions as a capital to Moscow and its name to Lenin. Renamed Leningrad, it suffered through the 872-day Nazi siege during World War II, holding out heroically through endless bombardment and starvation that took nearly a million lives and severely damaged many of its buildings.

The communist period witnessed the neglect and destruction of some of Leningrad's most beautiful churches, the intrusion and addition of crude monuments and bleak apartment complexes, and the rapid industrialization and growth of the city (which now totals 4.6 million). Immediately after the collapse of the Soviet Union, the city regained its original name and a new era opened. The Russian Orthodox Church revived, building restoration went forward, and tourism boomed. But the transformation from communist to capitalist ways has been accompanied by social problems that include crime and poverty.

Today, St. Petersburg's ports on the Gulf of Finland are still very important to Russia's trade with the rest of the world, and the city is a major port of call for many cruise ships. If Moscow is nowadays primarily defined in terms of global finance and producer services, St. Petersburg's niche is trade, culture, and tourism.

---

debate. Some geographers prefer to call this the Moscow Region, thereby emphasizing that for over 400 kilometers (250 mi) in all directions from the capital, everything is oriented toward this historic focus of the state. As Figure 2B-3 (and even better, Fig. 2A-9) shows, Moscow has maintained its decisive **centrality** [5]: roads and railroads converge in all directions from Ukraine in the south; from Minsk (Belarus) and the rest of eastern Europe in the west; from St. Petersburg and the Baltic coast in the northwest; from Nizhniy Novgorod and the Urals in the east; from the cities and waterways of the Volga Basin in the southeast (a canal links Moscow to the Volga, Russia's single most important navigable river); and even to the subarctic northern periphery that faces the Barents Sea, where the strategic naval port of Murmansk and the lumber-exporting outpost of Arkhangelsk lie.

Povolzhye: The Volga Region

Another major subregion within the Russian Core is the **Povolzhye**, the Russian name for an area that extends along the middle and lower valley of the Volga River as well as its tributary, the Kama, above the city of Kazan. It would be appropriate to call this the Volga Region, for this greatest of Russia's rivers is its lifeline and most of the Povolzhye's cities lie on its banks (Fig. 2B-3). In the 1950s, a canal was completed to link the lower Volga with the lower Don River (and thereby directly access the Black Sea).

VOICE FROM THE

Region

CITY LIFE IN MOSCOW

© Mitia Kolossov
Mitia Kolossov, Moscow
City life in the capital

"I was born and raised in Moscow, not far from the campus of Moscow State University. The main university building is 236 meters [775 ft] high; I could see the tower from my room in the house where I grew up. My parents and grandparents are also from Moscow. That is pretty rare because the capital has always been a city of migrants. Moscow has changed incredibly in the last 15 years. For me, the most important thing is that it is a city with so many opportunities: for instance, Moscow now has more than 200 theaters, dozens of art galleries, and its restaurants represent all possible world cuisines. Unlike the Soviet past, the city is now brightly illuminated and looks nicer, more romantic at night than during the day. But it has also become much busier and hectic everywhere, with endless and often unpredictable traffic jams and very limited parking spaces. Some years ago I graduated from Moscow State University and am presently in the advertising business. For a large number of my clients, both international and local companies, this city became the main center of control and decision making. About 95 percent of Russian advertising activity is located in Moscow, and you cannot enter the market from anywhere else. Sometimes I dream about living in a smaller town, being tired of the monster-city tempo. But Moscow is where the action is. And it is home."

© Mitia Kolossov
Moscow State University,
Main Building

The Volga corridor was an important historic route in old Russia, but for a long time neighboring regions overshadowed it. The Moscow area and Ukraine were far ahead in industry and agriculture. The Industrial Revolution that came late in the nineteenth century to the Central Industrial Region did not have much effect in the Povolzhye. Its major function remained the transit of foodstuffs and raw materials to and from other regions.

The elongated Povolzhye (the *vol* refers to the Volga) forms the eastern flank of the Central Industrial Region, its cities and farms sustained by Russia's greatest river. For a long time this subregion was bypassed in the Soviet industrialization drive, but during World War II it suddenly took on much greater importance because its cities lay relatively remote from the German armies invading from the west. Accelerated industrial development put the Volga Valley's cities on the map as never before. Then, in the 1950s, the Volga-Don Canal was opened, creating new linkages and opportunities. Next, the full dimensions of the oil- and gasfields in and near the Povolzhye became clear (see Fig. 2A-5), and for some time these reserves were the largest in the entire Soviet Union. And finally, the Volga Region's northwestward linkages, via the Moscow and Mariinsk canals to the Baltic Sea, were improved and supplemented by rail and road connections. Again, the industrial composition of the Volga's riverside cities continues to change, but if the Central Industrial Region is Russia's heart, the Povolzhye is its key artery.

The Urals Region

The Ural Mountains form the eastern edge of the Russian Core. They are not particularly high; in the north they consist of a single range, but southward they broaden into a hilly zone. Nowhere are they an obstacle to east-west surface transportation. An enormous storehouse of metallic mineral resources located in and near the Urals has made this area a natural place for industrial development. Today, the Urals Region, well connected to the nearby Volga and Central Industrial Regions, extends from Serov in the north to Orsk in the south (Fig. 2B-3).

As Figure 2B-3 reveals, the Soviet legacy survives on Russia's economic landscape, despite the country's conversion from a domestic manufacturing system to what some observers now call a "petro-ruble state." Not only the Central Industrial Region but also the great Volga corridor, the Urals manufacturing complex, the industrial nodes of the Southeastern Frontier, and even the outliers of the Far East still mark the manufacturing map, even if many factories have fallen silent and people are leaving. Urban dwellers tend to find new ways to make their living, and such cities as Volgograd, Chelyabinsk, Novosibirsk, Irkutsk, or Vladivostok will never be empty.

▪▪ THE SOUTHEASTERN FRONTIER

From the southeastern flanks of the Ural Mountains to the headwaters of the Amur River, and from the latitude of Tyumen to the northern zone of neighboring Kazakhstan, lies Russia's vast Southeastern Frontier region, product of a gigantic experiment in the eastward extension of the Russian Core (Fig. 2B-1). As the maps of cities and transportation (Fig. 2A-9) as well as population distribution (Fig. 2A-4) suggest, this region is more densely peopled and more fully developed in the west than in the east. To the east of the Yenisey River, linear settlement prevails, marked by ribbons and clusters along the east-west railroads. Two subregions dominate the human geography: the Kuznetsk Basin in the west and the Lake Baykal area in the east.

The container port at Krasnoyarsk, where the south-to-north-flowing Yenisey River meets the east-west Trans-Siberian Railroad near the center of the Southeastern Frontier region (see Fig. 2B-1). Supplies brought by train from the west are shipped northward to settlements in the Siberian interior; raw materials from Siberia arrive by boat and barge for dispatch to Russia's factories and markets. Large storage facilities stand across the river.

© Wolfgang Kaehler/Alamy.

The Kuznetsk Basin (*Kuzbas*)

Some 1500 kilometers (950 mi) east of the Urals lies another of Russia's primary regions of heavy manufacturing resulting from the communist period's national planning: the Kuznetsk Basin, or ***Kuzbas*** (Fig. 2B-3). In the 1930s, it was opened up as a supplier of raw materials (especially coal) to the Urals, but that function became less important as local industrialization accelerated. The original plan was to move coal from the Kuzbas west to the Urals and allow the returning trains to carry iron ore east to the coalfields—a classic case of **double complementarity [6]**. However, good-quality iron ore deposits were subsequently discovered near the Kuznetsk Basin itself. As the new resource-based Kuzbas industries expanded, so did its urban centers. The leading city, located just outside the Kuzbas, is Novosibirsk, which stands at the intersection of the Trans-Siberian Railroad and the Ob River as the symbol of Russian enterprise in the vast eastern interior. To the northeast lies Tomsk, one of the oldest Russian towns east of the Urals, founded in the seventeenth century and now caught up in the modern development of the Kuzbas. Southeast of Novosibirsk lies Novokuznetsk, a city that produces steel for this subregion's machine and metal-working plants as well as aluminum products from Urals bauxite.

The Lake Baykal Area (*Baykaliya*)

East of the Kuzbas, development becomes more insular, and distance becomes a stronger adversary. North of the central segment of the Mongolian border zone and eastward around Lake Baykal, larger and smaller settlements cluster along the two railroads to the Pacific coast (Fig. 2B-3). West of the lake, these rail corridors lie in the headwater zone of the Yenisey River and its tributaries. A number of dams and hydroelectric projects serve the valley of the Angara River, particularly the city of Bratsk. Mining, lumbering, and some farming sustain life here, but isolation dominates it. The city of Irkutsk, near the southern end of Lake Baykal, is the principal service center for the enormous Siberian region to the north as well as for a lengthy east-west stretch of southeastern Russia.

Beyond Lake Baykal, the Southeastern Frontier really lives up to its name: this is southern Russia's most rugged, remote, and forbidding country. Settlements are rare, with many being mere camps. The Buryat Republic (see Fig. 2B-2) is part of this zone; the territory bordering it to the east was taken from China by the czars and may well become an issue in the future. Where the Russian-Chinese boundary turns southward, along the Amur River, the region called the Southeastern Frontier ends and Russia's Far East begins.

▪▪ SIBERIA

Before we assess the potential of Russia's Pacific Rim, we should remember that the ribbons of settlement just discussed hug the southern perimeter of this gigantic country, avoiding the huge Siberian region to the north (Fig. 2B-1). Siberia extends from the Ural Mountains to the Kamchatka Peninsula—a vast, frigid, forbidding land. Larger than the conterminous United States but inhabited by only an estimated 15 million people, Siberia quintessentially symbolizes the Russian environmental plight: vast distances, cold temperatures worsened by strong Arctic winds, difficult terrain, poor soils, and limited options for survival.

But Siberia also possesses resources. From the days of the first Russian explorers and Cossack adventurers, Siberia's riches have beckoned. Gold, diamonds, and other precious minerals were found. Later, metallic ores including iron and bauxite were discovered. Still more recently, the Siberian interior proved to contain massive quantities of oil and natural gas (Fig. 2A-5) that now contribute significantly to Russia's energy supply and exports.

As the physiographic map (Fig. 2A-2) shows, major rivers—the Ob, Yenisey, and Lena—flow gently northward across Siberia and the Arctic Lowland into the Arctic Ocean. Hydroelectric power development in the basins of

Oymyakon is known as the coldest town on Earth. It is located north of the city of Yakutsk in eastern Siberia. The average daily temperature during winter hovers around minus 46° Celsius (minus 51° Fahrenheit) and the record low recorded here in 1933 was −68°C (−90°F). A few hundred people live here, and there is even a school—which does not open if the temperature drops below −52°C (−62°F). Cell phones would not work here, even if there was satellite coverage, and visitors are advised not to wear glasses—they freeze to your face. Not surprisingly, the town has no hotel.

Dean Conger/©Corbis

these rivers has generated electricity used to extract and refine local ores, and run the lumber mills that have been set up to exploit the vast Siberian forests.

The human geography of Siberia is fragmented, and most of the region is virtually uninhabited (Fig. 2A-8). Ribbons of Russian settlement have developed; the Yenisey River, for instance, can be traced on this map of Soviet peoples (a series of small settlements north of Krasnoyarsk), and the upper Lena Valley is similarly fringed by ethnic Russian settlement. Yet hundreds of kilometers of empty territory separate these thin ribbons and other islands of habitation.

Siberia, Russia's freezer, is stocked with goods that may become mainstays of future national development. Already, precious metals and mineral fuels are bolstering the Russian economy. In time, we may expect Siberian resources to play a growing role in the economic development of the Southeastern Frontier and the Russian Far East as well. One step in that process was already taken during Soviet times: the completion of the BAM (Baykal-Amur Mainline) Railroad in the 1980s. This route, lying north of and parallel to the old Trans-Siberian Railroad, extends 3500 kilometers (2200 mi) eastward from near the important center of Krasnoyarsk directly to the Far East city of Komsomolsk (Fig. 2B-3). In the post-Soviet era, the BAM Railroad has been beset by equipment breakdowns and workers' strikes. Nonetheless, it is a key element of the

infrastructure that will serve the economic growth of easternmost Russia in the twenty-first century.

▪▪ THE RUSSIAN FAR EAST

When the government in Moscow in 2000 superimposed its framework of Federal Districts over Russia's matrix of Soviet-era regions (Fig. 2B-2), it gave the name *Far East* to the largest one (in terms of territory) of all. In one stroke, the Far East incorporated four residual ethnic republics and six Russian Regions. But many Russian citizens have a different view of what constitutes the geographic region they call the Far East. Justifiably, they see most of the northern zone of the official Far East as a continuation of Siberia. To them, the real Far East is constituted by the area beyond the Southeastern Frontier to the Pacific coast, the island of Sakhalin, the Kamchatka Peninsula, and a narrow stretch of land along the northern shore of the Sea of Okhotsk (Figs. 2B-1 and 2B-4).

In Soviet times, citizens who agreed to move to this distant region to work in factories or for the administration were rewarded with special privileges to compensate for their hardship. Make no mistake: the Pacific Ocean's proximity tends to temper natural environments here, but life in the Russian Far East is tough. Moreover, in the postcommunist era, Moscow's favors were withdrawn, so that hundreds of thousands of inhabitants under the new

The Siberian tiger, the largest member of the cat family, is native to this environment. It needs a vast and undisturbed habitat, its coat protects against the cold, and its partially white fur serves as camouflage. But as a result of human encroachment and poaching, even in these remote parts, there are only a few hundred left and these animals have become a protected species. The Siberian tiger can now only be found in the forests and mountain ranges of Russia's Far East, and sightings are rare.

Konrad Wothe/Minden Pictures, Inc.

FIGURE 2B-4

© H. J. de Blij, P. O. Muller, and John Wiley & Sons, Inc.

democracy voted with their feet and departed. Take the largest city (Petropavlovsk) on the Kamchatka Peninsula. In 1991, just before the Soviet Union collapsed, it had a population of 248,000; by 2002, it was down to 198,000; today, it contains barely more than 180,000 residents. And Petropavlovsk also has the distinction of being the largest place on Earth without an overland link to the outside world: there simply is no road off the peninsula.

Almost everywhere in the Far East, people now feel abandoned by Moscow. The once-great naval base of Vladivostok (the city's name means "We Own the East") lies dormant. Factory closures are widespread. And when the locals tried something new—clandestinely importing

used Japanese cars—the government sent its federal police to close down the business. Meanwhile, the nearby container port at Nakhodka suffers repeated breakdowns. Cross-border trade with China is minimal. Trade with Japan is inconsequential. The two railroads that were to stimulate cross-country traffic and trade cannot operate efficiently without huge subsidies—available during Soviet times but not in the twenty-first century's market economy.

The Far East does have significant reserves of what underpins the Russian economy—oil and natural gas in and around the island of Sakhalin. And, as Figure 2B-4 reminds us, the southern sector of the Russian Far East lies between other reserves and potential markets: China next

From the Field Notes . . .

© H. J. de Blij

"The Russian Far East is losing people at the fastest rate among all Russian Federal Districts, and my field trip to Petropavlovsk on the Kamchatka Peninsula only reinforced what I had learned in Vladivostok some years ago. During Soviet times, Russians willing to move to the Far East were given special privileges (an older man told me that 'being as far away from Moscow as possible was reward enough'), but today locals here feel abandoned, even repressed. In 1991, Petropavlovsk had about 250,000 residents; by 2002 the population was down to under 200,000, and today it is estimated to be only about 180,000. It is not just the bleak environment—the ubiquitous ash and soot from dozens of active volcanoes on Kamchatka, covering recent snowfalls and creating a pervasive black mud—it is also loss of purpose people feel here. There was a time when the Soviet windows on the Pacific were valuable and worthy; now the people scramble to make a living and when they find a way, for example by selling used cars bought cheaply in Japan and brought to the city by boat, Moscow sends law enforcement to halt the illegal trade. Petropavlovsk has no surface links to the rest of the world, and a springtime (April 2009) view of the city at the foot of the volcano Koryakskaya (top) and a street scene near the World War II museum (bottom) suggest why the Far East is depopulating."

www.conceptcaching.com

door and Japan across a narrow sea. Already, the Japanese and the Chinese are both bidding for Russia's energy supplies, and undoubtedly there will be construction of pipelines and other energy-related infrastructure. But what the region needs is a diversified economy of the kind that is reviving China's Northeast across the Amur and Ussuri rivers. Such development, however, is currently not in sight.

THE SOUTHERN PERIPHERY

Having followed Russia's eastward expansion and organization, let us take another look at the realm's regional map (Fig. 2B-1) to focus on the most complicated region of all—the area where Russia's leaders, communists as well as czars, pushed the expanding state's influence across and beyond the Caucasus Mountains. On this map, it looks as though the Russian Core extends to the international boundary (which coincides, more or less, with the

crestline of this massive range). But it is not that simple. When you travel southward from Volgograd or Rostov, the landscape—physical as well as cultural—does not change; the Russian Plain's gently rolling countryside with its villages and farms is very much like that south of Moscow. Taking the train south from Astrakhan in the Volga River Delta, you soon see the Caucasus looming in the distance, but you may not realize that you have entered a different world until you get off at Groznyy, the capital of Chechnya, one of Russia's ethnic "republics" (Fig. 2B-5). Just observing the mix of people makes you aware that you have left the heartland and reached Russia's periphery.

Figure 2B-5 is important enough to spend some serious time studying it. Here the communist leaders who succeeded the czars faced multiple challenges. First, inside the Russian state itself, there were many minority peoples (such as the Chechens of Chechnya) who had fought the czars' armies and who now faced new rulers. Second, there

FIGURE 2B-5

© H. J. de Blij, P. O. Muller, and John Wiley & Sons, Inc.

were peoples who had sided with the czars and who wanted to cooperate with the Soviet regime, such as the Ossetians. To recognize the aspirations of these peoples, some (but not all) of whom were Muslim, the communist administration designated ethnic "republics" for them. On our map, these are colored pink to distinguish them from the rest of Russia, which is shown in tan.

Russia's Internal Periphery

From Buddhist-infused Kalmykia on the Caspian perimeter westward to mainly Orthodox Christian Adygeya near the Black Sea, the Internal Periphery of southern Russia is defined by eight ethnic republics, each with its own identity (Fig. 2B-5). In the east, Kalmykiya is Buddhist-infused, and, as noted earlier, Dagestan is a hodgepodge of small, culturally discrete communities.

Physiographically as well as culturally divided Chechnya, as the inset map suggests, lies in a transition zone between plain and mountains, the mountains serving as refuge for those opposed to Moscow's rule. When the post-Soviet Russian government asked all Regions and Republics within Russia to sign its proposed Russian Federation Treaty, Chechnya's leaders refused, believing that they now had the opportunity to finally end Russian control. Its western neighbor, Ingushetiya, also got caught up in anti-Russian activism, and the ensuing conflict took a terrible toll in adjacent North Ossetia where, in 2004, more than 350 students, teachers, and parents died in a terrorist attack on a school. Still farther west, things remained calmer, and no problems at all arose in generally pro-Moscow Adygeya.

This tier of ethnic republics along Russia's southern border display all the properties of a disadvantaged periphery: not only do they share a history of subjugation by the powerful core to the north, but they also lag behind the rest of Russia by almost any measure of social progress. In terms of health, education, income, opportunity, and other such indices, the peoples of the Internal Periphery, whether they are pro-Russian or not, have a long way to go to achieve Russian living standards. And there is much simmering resentment of Russian rule, past and present, in such republics as Chechnya and Ingushetiya.

In the case of Chechnya, such anger has its most recent roots in the Second World War, when Soviet leaders accused the Chechens of sympathizing, and even collaborating, with the Nazi invaders. Josef Stalin, then the Soviet dictator, ordered the entire Chechen population loaded on trains and exiled to the desert of Kazakhstan across the Caspian Sea. Tens of thousands died along the way or after arriving there, and even though Stalin's successor Nikita Khrushchev in 1957 allowed the survivors to return to their homeland, the Chechens never forgave their tormentors. Then in 1991, when the Soviet Union disbanded, they saw their opportunity, refusing to sign the Russian Federation Treaty that would make their republic an integral part of the new Russian state and mounting a campaign for independence that led to a bitter and costly war

that destabilized a number of neighboring republics as well. Chechen terrorists carried this campaign to the very heart of Moscow, killing hundreds of civilians there and changing Russia's political landscape. When presidential candidate Vladimir Putin promised victory, the Russians elected him overwhelmingly in 2000. Massive military intervention did overpower the Chechens, but the Internal Periphery remains restive and unstable.

▪▪ TRANSCAUCASIA: RUSSIA'S EXTERNAL PERIPHERY

Russian (and later Soviet) manipulation of the ethnic map did not end at the boundary we see in Figure 2B-5, but it did take a different turn. Beyond the Russian border, the Soviet Empire extended its power over three adjoining Transcaucasian entities where the Russians had for centuries exerted influence and waged war. These three units were **Georgia** (mapped in yellow in Fig. 2B-5), facing the Black Sea; **Azerbaijan** (green), bordering on the Caspian Sea; and **Armenia** (lavender), landlocked in the middle. During Soviet times, Georgia was a loyal partner (Josef Stalin was a Georgian by birth). Christian Armenia appreciated membership in the Soviet Union because of the security it provided next door to Islamic Turkey. And Islamic Azerbaijan, whose ethnic majority is a people called the *Azeris*—who have close kinship with Iranian Azeris directly across their southern boundary with Iran—was a valued member of the USSR because of its rich oil reserves.

As if this were not enough, note in Figure 2B-5 that Azerbaijan has a large exclave on the Iranian border (Naxcivan), separated from the main part of the country by Armenian territory—and that the Armenians hold another exclave called Nagorno-Karabakh inside what would seem to be Azerbaijan's territory (which is home to 150,000 Christian Armenian citizens who were placed under Islamic Azerbaijan's jurisdiction by Soviet mandate).

Soviet rule kept the lid on potential conflict in this historically turbulent part of Transcaucasia, and after the collapse of the USSR the Russians sought to retain their influence and maintain stability in this sector of their Near Abroad. The Russians wanted to keep Azerbaijan's oil flowing to and through Russia (see the pipeline that leads from Baki [Baku] via Chechnya's Groznyy to the Black Sea terminal at Novorossiysk in Fig. 2B-5); they also wanted to keep Georgia in the Slavic fold, and they sought ways to mitigate the still-ongoing conflict between Azerbaijan and Armenia over the enclave of Nagorno-Karabakh.

But things have not gone well here for Moscow. Spurred by Western investments, Azerbaijan in 2006 began exporting oil via a new pipeline across Georgia and Turkey to the Mediterranean coast, providing it with an alternative to Russian transit. Armenia and Azerbaijan persisted in their sometimes-violent quarrel over Nagorno-Karabakh. And things really fell apart in Georgia, where the government pursued pro-European policies, allegedly mistreated its pro-Russian Ossetian minority, and where

From the Field Notes . . .

"Upon first entering Sukhum, the capital of the 'de facto' state of Abkhazia on the Black Sea coast (see Fig. 2B-5), I visited the former government house—now an eerie and unintended war memorial. It was the scene of an intense battle in September 1993 when the Georgian goverment unsuccessfully tried to suppress Abkhazian separatists. The building bears the marks of bullets, bombs, and fires, and is deliberately kept in this state as a reminder of the war. The Abkhazians repelled the Georgians and insist on their own spelling of the city's name (Georgians use *Sokhumi*). The legacies of war in the Caucasus are dramatically evident in the landscape with the omnipresence of ruined buildings, war memorials, empty houses of displaced persons, refugee camps, military checkpoints, minefields, abandoned farms, ruined roads, and other damaged infrastructure. Like many other buildings (over half by some estimates) in Sukhum, this one remains as it was when the fighting ended, but its vast size and high-rise prominence in a city of small, low buildings makes it visible from everywhere in town and a daily reminder of the war's brutality."

© John O'Loughlin.

www.conceptcaching.com

Russia openly supported devolutionary initiatives in the northwestern corner of the country, a small ethnic entity known as Abkhazia (Fig. 2B-5).

Soon after the USSR's disintegration in 1991, the argument between Russia and Georgia escalated into serious discord. The Russians closed the Georgian border, shutting Georgian farmers out of their crucial market. Then the Russians started issuing Russian passports to Abkhazian citizens of Georgia. Next, the Russians signaled even stronger support for their Ossetian allies, and in a still-disputed sequence of events during 2008 Russian forces entered South Ossetia and engaged the Georgian military in a brief but costly war. After a cease-fire was brokered by international action, Moscow took the extraordinary step of recognizing the two pro-Russian autonomous regions of Georgia—Abkhazia and South Ossetia—as independent states.

This action sent shudders through Europe and much of the world. If Russia could intervene militarily in this part of its Near Abroad, where else might Russian forces be sent to enforce Russia's will? Georgia, as the map shows, has a lengthy border with Russia, but so does Ukraine, where Russia could claim to have even stronger interests including leased military bases. Unlike Azerbaijan, mostly agricultural Georgia has no energy card to play—although it did agree to allow the new pipeline from Azerbaijan to Turkey to traverse its territory, thereby angering the Russians even further. And the Georgian government has courted the European Union and even NATO, seeking the security Georgians have lacked for centuries. The Azeris of Azerbaijan, whose dominant religion is Shi'ite Islam like that of Iran, are caught in the global web (and political geography) of energy demand and supply, but most ordinary people have benefited little from their country's oil wealth. And the landlocked 3 million Armenians live in a world of unsettled scores: with the Turks over their decimation by Ottoman Muslims, with the Azeris over their exclave of Nagorno-Karabakh, and with the international community over the recognition of their plight. Transcaucasia may be the smallest region of the Russian realm, but it is also the second-most populous and undoubtedly the most volatile.

An Uncertain Future

It seems that Russia is paying an increasingly high price for its efforts to control the Southern Periphery. Terrorism has not only come to the Federation's capital: it is becoming ever more lethal as well. In the spring of 2010, suicide bombers killed 39 passengers on two subway trains in Moscow; a few months earlier a similar number had been killed when terrorists bombed the country's main railroad line connecting the capital and St. Petersburg. And in January 2011, suicide bombers killed 35 bystanders and wounded dozens more at Moscow's newest and busiest airport. The perpetrators are generally assumed to be extremists from the northern Caucasus.

In many ways, this hurts Moscow's 2 million Muslims the most (about 20 percent of the metropolitan population), because they tend to be associated with the terror unleashed by other Russians. Moscow has a growing number of immigrants and as many as a million may be there illegally (a large number of whom are also Muslim). They often have low-paying jobs at the bottom of the social ladder and are poorly integrated. Some radical and reactionary Russian groups are taking their frustration out on these immigrants. In 2009 alone, local "vigilantes" and xenophobic groups like *Russky Obraz* (Russian Way) killed 50 immigrants from Central Asia and the North Caucasus. Not surprisingly, Moscow is becoming increasingly edgy and ever more deeply polarized.

Even after the dismantling of the Soviet Empire, keeping a country together the size of Russia is a colossal challenge. Russia has not yet found a balance, not in dealing with free markets, not in a partnership with the West, not in accommodating minorities or forging stable relations with the ex-Soviet republics that surround it. At the same time, Russia is both losing people and facing mounting challenges. However, this is still a very powerful country that will continue to throw its weight around on the Eurasian continent.

▶ POINTS TO PONDER

- Is Russia too large to be ruled effectively as one state? Might there be some "ideal" size for the proper functioning of a democracy?
- One in ten Russian villages have been abandoned and are now ghost towns.
- Terrorism originating in the remote Caucasus area threatens Russia at its very center: Moscow.
- Since the end of communism, economic and mental well-being have declined for the majority of Russians.
- Russia is far and away the most unpredictable of the four BRICs in terms of economic development.

using GEOGRAPHY

1. The American Geographical Society in New York, the country's oldest geographical society, presents travelers the opportunity not only to visit remote parts of the world, but to do so accompanied by a professional geographer who specializes in the region or area being visited. Other organizations also appoint geographers as lecturers and guides on ships, planes, and even trains and buses. Imagine that you have been asked to accompany a group of your fellow students taking a riverboat down the Volga and Don rivers in July from Moscow to Rostov near the Sea of Azov, an arm of the Black Sea. In your first briefing, what will you tell them about the weather they should expect, about the minority "republics" of which they will get a glimpse, the crops and land uses they will encounter, the cities and towns they will see, and how they will get from the Volga to the Don? What might be the most interesting side trips along this itinerary?

2. Is it important for Russia to stop the emptying out of Siberia and the Far East? How could that be accomplished?

3. Make a list of the advantages and disadvantages of Russia's relative location in Eurasia's heartland.

FIGURE 3A-1

© H. J. de Blij, P. O. Muller, and John Wiley & Sons, Inc.

DEFINING the Realm

IN THIS CHAPTER

- ◆ Spectacular scenery, natural wealth
- ◆ The American creed
- ◆ Bilingualism in Canada
- ◆ The geography of home foreclosures
- ◆ Immigration and the future of multiculturalism

CONCEPTS, IDEAS, AND TERMS

Cultural pluralism	**1**
Physiographic regions	**2**
Rain shadow effect	**3**
Fossil fuel	**4**
Urban system	**5**
American Manufacturing Belt	**6**
Outer city	**7**
Deindustrialization	**8**
Gentrification	**9**
Sunbelt	**10**
Migration	**11**
Melting pot	**12**

We turn next to the Western Hemisphere, the two great interconnected continents that separate the Atlantic and Pacific oceans and extend, very nearly, from pole to pole, flanked by numerous islands large and small, indented by gulfs and bays of historic and economic import, and endowed with an enormous range of natural resources. Two continents—North and South—form the Americas, but three geographic realms blanket them (North, Middle, and South). In the context of physical (natural) geography, North America from Canada's Ellesmere Island in the far north to Panama in the south is a continent. In terms of modern human geography, the northern continent is divided into North American and Middle American realms along a transition zone marked by a political as well as physical boundary between the United States and Mexico (Fig. 3A-1). From the Gulf of Mexico to El Paso/Ciudad Juarez, the Rio Grande forms this border. From El Paso westward to the Pacific Ocean, straight-line boundaries, reinforced by fences and walls, separate the North from the Middle. Here, global core and global periphery meet, sometimes contentiously. We begin in North America.

North America is constituted by two of the world's most highly advanced countries by virtually every measure of social and economic development. Blessed by an almost unlimited range of natural resources and bonded by trade as well as culture, Canada and the United States are locked in a mutually productive embrace that is reflected by the economic statistics: in an average year of the recent past, more than 80 percent of Canadian exports went to the United States, and about two-thirds of Canada's imports came from its southern neighbor. For the United States, Canada is not quite as dominant, but it is still its leading export—and import—market.

The realm is also defined on the basis of a range of broad cultural traits, from urban landscapes and a penchant for mobility to religious beliefs, language, and political persuasions. Both countries rank among the world's most highly urbanized: nothing symbolizes the North American city as strongly as the skyscrapered panoramas of New York, Chicago, and Toronto—or the vast, beltway-connected suburban expanses of Los Angeles, Washington, and Houston. North Americans are also the most mobile people in the world. Each year about one out of every eight individuals changes his or her residence, a proportion that has recently declined but still leads the world.

Despite notable multilingualism, in most areas English is both countries' *lingua franca*. And notwithstanding all the religions followed by minorities, the great majority of both Canadians and Americans are Christians. Both states are stable democracies, and both have federal systems of government.

Canadians and Americans also share a passion for the same sports. Baseball and a modified version of American football are popular sports in Canada; in 1992, the Toronto Blue Jays became the first team outside the United States to win the World Series. Canada's national sport, ice hockey, now has more NHL teams in America than in Canada, though curling, a slower-paced Canadian pastime, has not exactly captured the American imagination.

Driving from Michigan through southern Ontario to upstate New York does not involve a sharp contrast in cultural landscapes. Americans visiting Canadian cities

San Francisco from the north: Fisherman's Wharf (foreground), Telegraph Hill crowned by Coit Tower (center), and the downtown skyline beyond.

© H. J. de Blij

major geographic qualities of
NORTH AMERICA

1. North America encompasses two of the world's biggest states territorially (Canada is the second-largest in size; the United States is third).

2. Both Canada and the United States are federal states, but their systems differ. Canada's is adapted from the British parliamentary system and is divided into ten provinces and three territories. The United States separates its executive and legislative branches of government, and it consists of 50 States, the Commonwealth of Puerto Rico, and a number of island territories under U.S. jurisdiction in the Caribbean Sea and the Pacific Ocean.

3. Both Canada and the United States are plural societies. Canada's pluralism is most strongly expressed in regional bilingualism. In the United States, divisions occur largely along ethnic, racial, and income lines.

4. A substantial number of Quebec's French-speaking citizens support a movement that seeks independence for the province. The movement's high-water mark was reached in the 1995 referendum in which (minority) non-French-speakers were the difference in the narrow defeat of separation. Prospects for a breakup of the Canadian state have diminished since 2000 but have not disappeared.

5. North America's population, not large by international standards, is one of the world's most highly urbanized and mobile. Largely propelled by a continuing wave of immigration, the realm's population total is expected to grow by more than 40 percent over the next half-century.

6. By world standards, this is a rich realm where high incomes and high rates of consumption prevail. North America possesses a highly diversified resource base, but nonrenewable fuel and mineral deposits are consumed prodigiously.

7. North America is home to one of the world's great manufacturing complexes. The realm's industrialization generated its unparalleled urban growth, but today both of its countries are experiencing the emergence of a new postindustrial society and economy.

8. The two countries heavily depend on each other for supplies of critical raw materials (e.g., Canada is a leading source of U.S. energy imports) and have long been each other's chief trading partners. Today, the North American Free Trade Agreement (NAFTA), which also includes Mexico, is linking all three economies ever more tightly as the remaining barriers to international trade and investment flow are dismantled.

9. Continued immigration and high transnational mobility make for an exceptionally diverse multicultural realm.

(and vice versa) find themselves in mostly familiar settings. Canadian cities have fewer impoverished neighborhoods, no ethnic ghettos (although low-income ethnic districts do exist), lower crime rates, and, in general, better public transportation; but rush hour in Toronto very much resembles rush hour in Chicago. The boundary between the two states is porous, even after the events of 9/11. Americans living near the border shop for more affordable medicines in Canada; Canadians who can afford it seek medical treatment in the United States.

None of this is to suggest that Canadians and Americans don't have their differences—they do. Most of these differences may seem subtle to people from outside the realm, but they are important nonetheless, particularly so to Canadians who at times feel dominated by Americans whose political or cultural values they don't always share. You can even argue over names: if this entire hemisphere is called the Americas, aren't Canadians, Mexicans, and Brazilians as well as others Americans too?

POPULATION CLUSTERS

Although Canada and the United States share many historical, cultural, and economic qualities, they also differ in significant ways, diversifying this realm. The United States, somewhat smaller territorially than Canada, occupies the heart of the North American continent and, as a result, encompasses a greater environmental range. The U.S. population is dispersed across most of the country, forming major concentrations along both the (north-south trending) Atlantic and Pacific coasts (Fig. 3A-2). The overwhelming majority of Canadians, on the other hand, live in an interrupted east-west corridor that extends across southern Canada, mostly within 300 kilometers (200 mi) of the U.S. border. The United States, again unlike Canada, is a fragmented country in that the peninsula of Alaska is part of it (offshore Hawai'i, however, belongs in the Pacific Realm).

Figure 3A-2 reveals that the great majority of both the U.S. and Canadian population resides to the east of

The central business districts (CBDs) of North America's large cities have distinctive skylines, often featuring architectural icons like the Empire State Building. Indeed, you should have little trouble identifying this CBD: the black profile of Hancock Tower on the shoreline of Lake Michigan just to the right of center is one of the signature structures of downtown Chicago, America's third-largest city. In fact, the only skyscraper more famous than Hancock Tower is Willis Tower, the Western Hemisphere's tallest building, from whose 110-story-high Skydeck this photo was taken. Willis Tower??? Absolutely—although it is still far more widely known by its former name of Sears Tower (the original 1973 tenant that relocated its corporate headquarters to suburban Hoffman Estates well over a decade ago). Today, according to its website, Willis Tower is reinventing itself as one of the world's greenest buildings, in the vanguard of a movement that marks Chicago as the U.S. leader in working toward a more sustainable urban future.

© Vito Palmisano/Photographer's Choice/Getty Images, Inc.

of ancestral and traditional backgrounds). Indeed, Canada recognizes two official languages, English and French (the United States does not even designate English as such); and by virtue of its membership in the British Commonwealth, East and South Asians form a larger sector of Canada's population than Asians and Pacific Islanders do in the United States. On the other hand, cultural pluralism in the United States reflects large Hispanic (16 percent) and African-American (13 percent) minorities and a wide range of other ethnic backgrounds.

Robust urbanization, substantial immigration, and cultural pluralism are defining properties of the human geography of the North American realm. But first we need to become familiar with the physical stage on which the human drama is unfolding.

NORTH AMERICA'S PHYSICAL GEOGRAPHY

Physiographic Regions

One of the distinguishing properties of the North American landmass, all of which lies on the North American tectonic plate, is its remarkable variegation into regional physical landscapes. So well defined are many of these landscape regions that we use their names in everyday language—for example, when we say that we flew over the Rocky Mountains or drove across the Great Plains or hiked in the Appalachians. These landscapes are called physiographic regions [2], and in the entire hemisphere their diversity is greatest to the north of the Rio Grande (Fig. 3A-3).

In the Far West, the Pacific Mountains extend all the way from Southern California through coastal Canada to Alaska. In the western interior, the Rocky Mountains form a continental backbone from central Alaska to New Mexico. Around the Great Lakes, the low-relief landscapes of the Interior Lowlands and the Great Plains to the west are shared by Canada and the United States, and the international boundary even divides the Great Lakes. In the east, the Appalachian Mountains (as the cross-sectional inset shows, no match for the Rockies) form a corridor of ridges, valleys, and plateaus that represent a familiar topography from Alabama and Georgia to Nova Scotia and Newfoundland. If there is a major physiographic province that belongs to only one of the realm's two countries, it is the Canadian Shield, scoured bare by the Pleistocene glaciers that deposited their pulverized rocks as fertile soil in the U.S. Midwest, that is, in the Interior Lowlands.

a line drawn down the middle of the realm, reflecting the historic core-area development in the east and the later and still-continuing shift to the west, and, in the United States, to the south. Certainly this map shows the urban concentration of North America's population: you can easily identify cities such as Toronto, Chicago, Denver, Dallas-Fort Worth, San Francisco, and Vancouver. Just under 80 percent of the realm's population is concentrated in cities and towns, a higher proportion even than Europe's.

As the Data Table in Appendix B indicates, the population of the United States, which in 2010 passed the 310-million mark, is growing at a rate 50 percent higher than Canada's, and the 400-million mark may be reached as soon as 2040. This is an unusually high rate of growth for a high-income economy, resulting from a combination of natural increase (about 60 percent) and substantial immigration.

Although Canada's overall growth rate is significantly lower than that of the United States, immigration contributes proportionately even more to this increase than in the United States. With just under 35 million residents, Canada, like the United States, has been relatively open to legal immigration, and as a result both societies exhibit a high degree of cultural pluralism [1] (a diversity

FIGURE 3A-2

© H. J. de Blij, P. O. Muller, and John Wiley & Sons, Inc.

Climate

This diversity of landscape is matched by a variety of climates. Take another look at Figure G-7, and you can see the lineaments of landscape mirrored in the contrasts of climate. North America may not have it all—there are no areas of true tropical environment in the North American realm to speak of except at the southern tip of Florida—but does exhibit a great deal of variation. North America contains moist coastal zones and arid interiors, well-watered

plains, and even bone-dry deserts. On the world map, *Cf* and *Df* climates are especially good for commercial farming; note how large North America's share of these environments is.

The map leaves no doubt: the farther north you go, the colder it gets, and even though coastal areas derive moderation from warmer offshore waters, which is why there is a Vancouver and a Halifax, the rigors of continentality set in not far from the coast. Hot summers, frigid

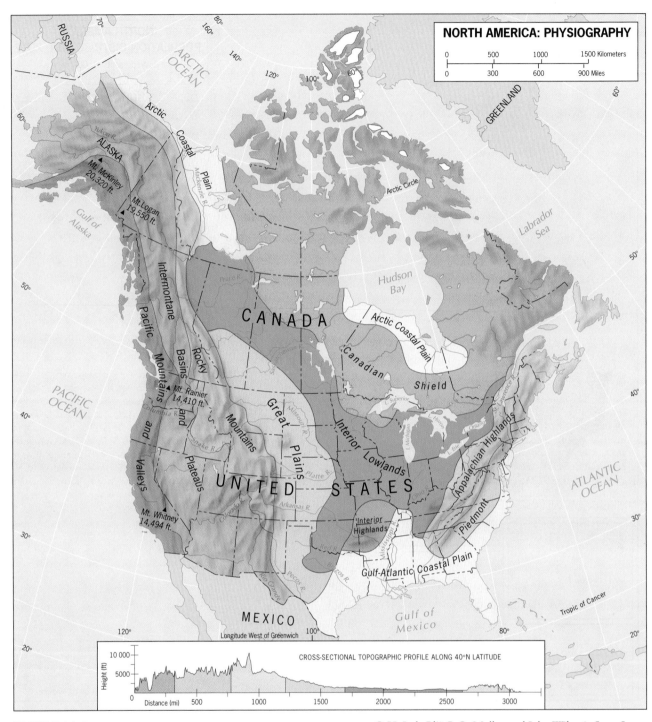

FIGURE 3A-3

© H. J. de Blij, P. O. Muller, and John Wiley & Sons, Inc.

winters, and limited precipitation make high-latitude continental interiors difficult places to make it on the land. Figure G-7 has much to do with the southerly concentration of Canadian population and with the lower densities in the interior throughout the realm.

In the west, especially in the United States, you can see what the Pacific Mountains do to areas inland. Moisture-laden air arrives from the ocean, the mountain wall forces the air upward, cools it, condenses the moisture in it, and

produces rain—the rain for which Seattle and Portland and other cities of the Pacific Northwest are (in)famous. By the time the air crosses the mountains and descends on the landward side, most of the moisture has been drawn from it, and the forests of the ocean side give way to scrub and brush. This **rain shadow effect [3]** extends all the way across the Great Plains; North America does not turn moist again until the Gulf of Mexico sends humid tropical air northward into the eastern interior via the Mississippi-Missouri River Basin.

On August 29, 2005, Category-4 Hurricane Katrina cut a 230,000-square-kilometer (90,000-sq-mi) path of destruction along the Louisiana and Mississippi Gulf Coast after making landfall just west of the mouth of the Mississippi River. Most devastated was the city of New Orleans, 80 percent of which lay underwater in the immediate aftermath of the storm. The pair of aerial photos above shows central New Orleans under normal conditions in the spring of 2004 (left) and two days after the tropical cyclone struck (right). Katrina was by far the most costly natural disaster in U.S. history, and its toll was truly staggering: more than 1800 lives lost, at least 200,000 homes destroyed, nearly a million people displaced, and more than $25 billion in insured property damage. Despite $180 billion in federal aid during the years following the disaster, rebuilding has been patchy and painfully slow. More than five years after the storm, the impact of Katrina continued to weigh heavily on the Crescent City. More than one-fourth of the pre-2005 population (ca. 100,000) has never returned. An astonishing 50,000 residential properties remain unoccupied, most of them still damaged. The proportion of the population living in poverty is now more than 20 percent, and the number of homeless people exceeds 12,000. Although still fragile, the tourist economy of New Orleans has fared better, but the Big Easy remains highly vulnerable to hurricanes as it continues its struggle toward an acceptable level of restoration.

© Digital Globe/Eurimage/Photo Researchers, Inc.

In a very general way, therefore, and not including the coastal strips along the Pacific, nature divides North America into an arid west and a humid east. Again the population map reveals more than a hint of this: draw a line approximately from Lake Winnipeg to the mouth of the Rio Grande, and look at the contrast between the (comparatively humid) east and drier west when it comes to population density. Water is a large part of this story.

Between the Rocky Mountains and the Appalachians, North America lies open to air masses from the frigid north and tropical south. In winter, southward-plunging polar fronts send frosty, bone-dry air masses deep into the heart of the realm, turning even places like Memphis and Atlanta into iceboxes; in summer, hot and humid tropical air surges northward from the Gulf of Mexico, giving Chicago and Toronto a taste of the tropics. Such air masses clash in low-pressure systems along weather fronts loaded with lightning, thunder, and, frequently, dangerously destructive tornadoes. And the summer heat brings an additional threat to the Gulf-Atlantic Coastal Plain (Fig. 3A-3): hurricanes capable of inflicting catastrophic devastation on low-lying areas. Such hurricanes also prune natural vegetation, replenish underground water reservoirs, fill natural lakes, and flush coastal channels.

Great Lakes and Great Rivers

Two great drainage systems lie between the Rockies and the Appalachians: (1) the five Great Lakes that drain into the St. Lawrence River and the Atlantic Ocean, and (2) the mighty Mississippi-Missouri system that carries water from a vast interior watershed to the Gulf of Mexico, where the Mississippi forms one of the world's major *deltas*. Both natural systems have been modified by human engineering. In the case of the St. Lawrence Seaway, a series of locks and canals has created a direct shipping route from the Midwest to the Atlantic. The Mississippi and Missouri rivers have been fortified by artificial levees that, while failing to contain the worst of flooding (such as in May 2011), have enabled farmers to cultivate the most fertile of American soils.

EUROPEAN SETTLEMENT AND EXPANSION

When the first Europeans set foot on North American soil, the continent was inhabited by millions of people whose ancestors had reached the Americas from Asia via Alaska, and possibly also across the Pacific, more than 14,000 years before (and perhaps as long as 30,000 years ago). In search of Asia, the Europeans misnamed them Indians. These Native Americans or *First Nations*—as they are now called, respectively, in the United States and Canada—were organized into hundreds of nations with a rich mosaic of languages and a great diversity of cultures (Fig. 3A-4). Canada's First Nations also include Métis (of mixed native and European descent) and the Inuit (formerly Eskimos) of the far north.

The eastern nations were the first to bear the brunt of the European invasion. By the end of the eighteenth century, ruthless and land-hungry settlers had driven most of the native peoples living along the Atlantic and Gulf

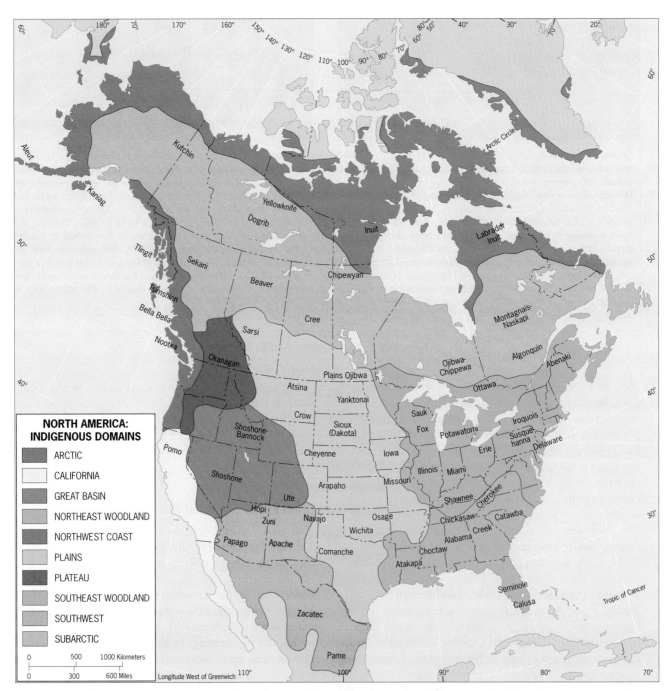

FIGURE 3A-4   © H. J. de Blij, P. O. Muller, and John Wiley & Sons, Inc.

coasts from their homes and lands, initiating a westward push that was to devastate indigenous society. Today, the remaining Native American nations are left with only about 4 percent of U.S. territory in the form of mostly impoverished reservations.

The current population geography of North America is rooted in the colonial era of the seventeenth and eighteenth centuries that was dominated by Britain and France. The French sought mainly to organize a lucrative fur-trading network, while the British established settlements along the coast of what is now the eastern seaboard of the United States (the oldest, Jamestown, was founded in Virginia just over four centuries ago).

The British colonies soon became differentiated in their local economies, a diversity that was to endure and later shape much of North America's cultural geography. The northern colony of New England (Massachusetts Bay and environs) specialized in commerce; the southern Chesapeake Bay colony (Virginia and Maryland) emphasized the plantation farming of tobacco; the Middle Atlantic area lying in between (southeastern New York, New Jersey, eastern Pennsylvania) formed the base for a number of smaller, independent-farmer colonies.

These neighboring colonies thrived and yearned to expand, but the British government responded by closing the inland frontier and tightening economic controls. This policy provoked general opposition in the colonies, followed by the revolutionary challenge that was to lead to independence and the formation of the United States of America by the late 1780s. The western frontier of the fledgling nation now swung open, and the area north of the Ohio River was occupied soon after it was discovered that soils and climate in these Interior Lowlands (Fig. 3A-3) were more favorable to farming than those of the Atlantic Coastal Plain and the Appalachian foothills of the Piedmont to its west. Deforestation cleared huge tracts for farming and livestock, and new cultural landscapes spread apace. The acquisition of more distant western territories followed in short order (Fig. 3A-5).

CULTURAL FOUNDATIONS

The modern American creed, if one can be identified, has been characterized by urban geographer Brian Berry as exhibiting an adventurous drive, a liking for things new, an ability to move, a sense of individualism, an aggressive pursuit of goals and ambition, a need for societal acceptance, and a firm sense of destiny. These qualities are not unique to modern U.S. culture, of course, but in combination they create a particular and pervasive mindset that is reflected in many ways in this geographic realm. Generally speaking, these tendencies tend to translate into an intense pursuit of educational and other goals and high aspirations for upward socioeconomic mobility.

Facilitating these aspirations is *language*. None of these goals would be within reach for so many were it not for the use of English throughout most of the realm. In Europe, by contrast, language inhibits the mobility so routinely practiced by Americans: workers moving from Poland to Ireland found themselves at a disadvantage in competition with immigrants from Ghana or Sri Lanka. In North America, a worker from Arkansas would not even consider language to be an issue when applying for a job in Calgary.

As we shall see, the near-universality of English in North America is diminishing, but English continues to be the dominant medium of interaction. Interestingly, English in the U.S. is undergoing a change that is affecting it worldwide: in areas where it is the second language it is blending with the local tongue, producing hybrids just as is happening in Nigeria ("Yorlish"), Singapore ("Singlish"), and the Philippines ("Taglish"). Thus English may become the Latin of the twenty-first century: Latin, too, produced blends that eventually consolidated into Italian, French, Spanish, and the other Romance languages.

Also reflecting the cultural values cited above is the role of *religion*, which sets American (more so than Canadian) culture apart from much of the rest of the Western world. The overwhelming majority of North Americans express a belief in God, and a large majority regularly attend church, in contrast to much of Europe. Eight out of ten people in the United States say they belong to a religion, and four out of ten say they attend a religious service once a week (more than in Iran!). Religious observance is a virtual litmus test for political leaders; no other developed country prints "In God We Trust" on its currency.

Nonetheless, most people who say they are religious tend to be fairly moderate in their convictions and tolerant of other faiths or denominations. This is most likely a product of centuries of immigration of (and coexistence with) different peoples with different persuasions. A study in 2010 showed that about half of all U.S. marriages involve partners who grew up in different religious environments, and that most individuals will maintain social networks encompassing people from a variety of religious backgrounds. The unfortunate exception may be Islam: the same study indicated that 45 percent of U.S. residents believe that the values of Islam are "at odds" with the American way of life.

Figure 3A-6 shows the mosaic of Christian faiths that blankets the realm. A map at this scale can only suggest the broadest outlines of a much more intricate pattern: Protestant denominations are estimated to number in the tens of thousands in North America, and no map could show them all. Southern (and other) Baptists form the majority across the U.S. Southeast from Texas to Virginia, Lutherans in the Upper Midwest and northern Great Plains, Methodists in a belt across the Lower Midwest, and Mormons in the interior West centered on Utah. Roman Catholicism prevails in most of Canada as well as the U.S. Northeast and Southwest, where ethnic Irish and Italian adherents form majorities in the Northeast and Hispanics in the Southwest. Behind these patterns lie histories of proselytism, migration, conflict, and competition, but tolerance of

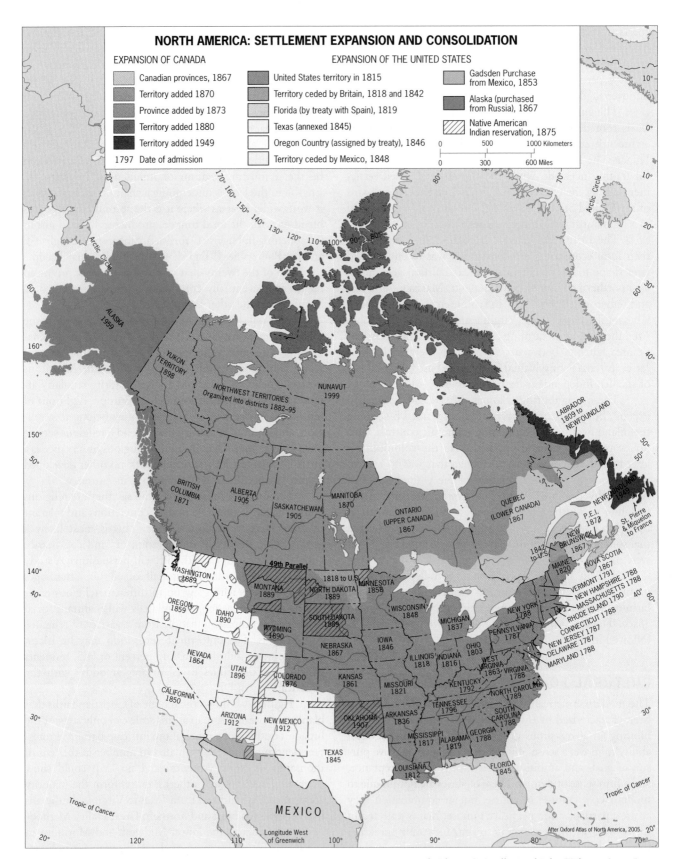

NORTH AMERICA: SETTLEMENT EXPANSION AND CONSOLIDATION

EXPANSION OF CANADA

- Canadian provinces, 1867
- Territory added 1870
- Province added by 1873
- Territory added 1880
- Territory added 1949
- 1797 Date of admission

EXPANSION OF THE UNITED STATES

- United States territory in 1815
- Territory ceded by Britain, 1818 and 1842
- Florida (by treaty with Spain), 1819
- Texas (annexed 1845)
- Oregon Country (assigned by treaty), 1846
- Territory ceded by Mexico, 1848
- Gadsden Purchase from Mexico, 1853
- Alaska (purchased from Russia), 1867
- Native American Indian reservation, 1875

After Oxford Atlas of North America, 2005.

FIGURE 3A-5

© H. J. de Blij, P. O. Muller, and John Wiley & Sons, Inc.

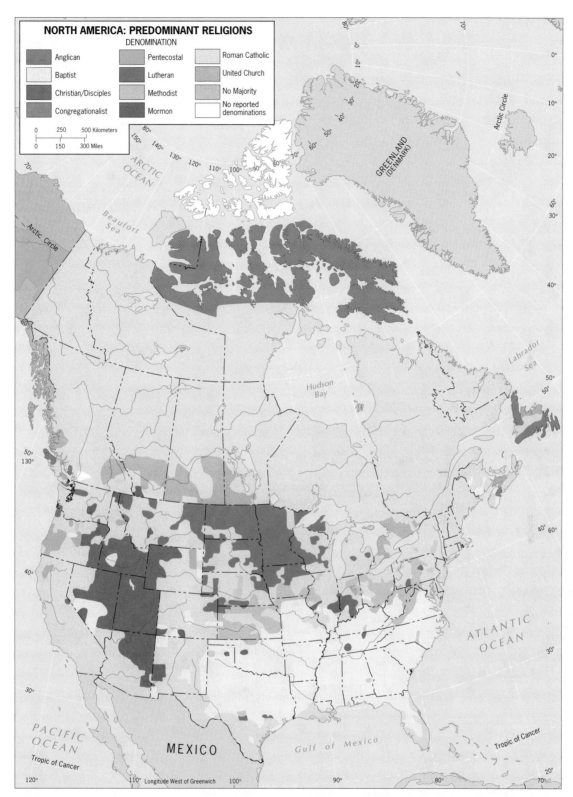

NORTH AMERICA: PREDOMINANT RELIGIONS
DENOMINATION

- Anglican
- Baptist
- Christian/Disciples
- Congregationalist
- Pentecostal
- Lutheran
- Methodist
- Mormon
- Roman Catholic
- United Church
- No Majority
- No reported denominations

0   250   500 Kilometers
0   150   300 Miles

FIGURE 3A-6

© H. J. de Blij, P. O. Muller, and John Wiley & Sons, Inc.

diverse religious (even nonreligious) views and practices is a hallmark of this realm.

On the other hand, U.S. culture does have its complexities, and one example will suffice. This is in many respects the freest country in the world, and it certainly prides itself as such. Yet the United States also has the largest prison population of the developed Western world: about 2.4 million or 1 for every 100 adults. That is 8 times as many as in Canada, 9 times as many as in Germany, and 12 times as many as in Japan. It is also 3 times the rate in Iran and even

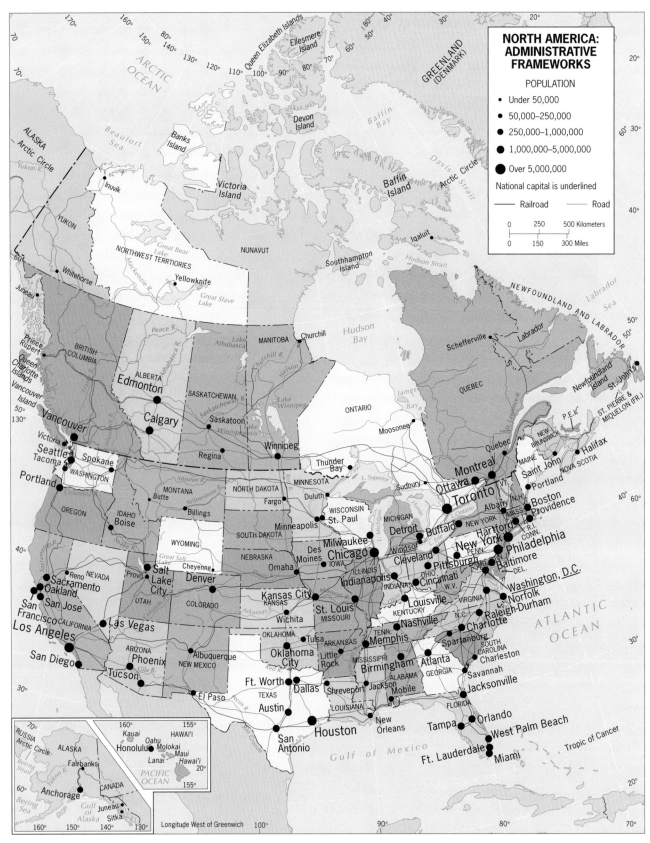

FIGURE 3A-7

© H. J. de Blij, P. O. Muller, and John Wiley & Sons, Inc.

exceeds the number in Russia by a wide margin. It wasn't always like this: the proportion of the population behind bars has quadrupled since 1970, mainly as a result of (overly?) harsh laws and sentencing. This may have produced a safer country, but the cost is enormous—financially (at around $25,000 per year per inmate) as well as socially, because incarceration affects so many people (not just the inmates but also their relatives, dependents, and social networks).

THE FEDERAL MAP OF NORTH AMERICA

The two states that constitute North America may have arrived at their administrative frameworks with different motives and at different rates of speed, but the result is unmistakably similar: their internal political geographies are dominated by straight-line boundaries of administrative convenience (Fig. 3A-7). In comparatively few places, physical features such as rivers or the crests of mountain ranges mark internal boundaries, but by far the greater length of boundaries is ruler-straight. Even most of the international boundary between Canada and the United States west of the Great Lakes coincides with a parallel of latitude, 49°N.

The reasons are all too obvious: the framework was laid out before, in some areas long before, significant white settlement occurred. There was something arbitrary about these delineations, but by delimiting the internal administrative units early and clearly, governments precluded later disputes over territory and resources. In any case, neither Canada nor the United States was to become a unitary state: both countries are *federal states*. The Canadians call their primary subdivisions provinces, whereas in the United States they are, as the country's name implies, States.

It is no matter of trivial pursuit to take a careful look at some of the provinces and States that compartmentalize this realm, because some are of greater consequence than others. We have already noted the special significance of mainly French-cultured Quebec, which still occupies a special position in federal Canada. In many ways the key Canadian province is Ontario, containing the country's largest and most globalized city (Toronto) as well as the capital (Ottawa) on its eastern river border with Quebec. In the United States, key States in the political and economic geography of the federation (indeed the four most populated, in descending order) are California facing the Pacific, Texas bordering Mexico, New York with its window on the Atlantic, and Florida pointing toward the Caribbean and South America beyond.

The rectangular layout of the realm seems to symbolize North America's modernity and its stability, whereas the federal structure itself (with considerable autonomy at the State or provincial level) reflects the culture of freedom and independence. If federalism works within these high-income economies, shouldn't the rest of the world emulate it? Federalism does have its assets, but it also carries some significant liabilities.

THE DISTRIBUTION OF NATURAL RESOURCES

One of these liabilities has to do with the variable allotment of natural resources among the States and provinces. In Canada, Alberta is favored with large energy reserves (and more may be in the offing), but its provincial government is not always eager to see the federal administration take that wealth to assist poorer provinces. In the United States, when oil prices are high Texas and Oklahoma boom while California's budget suffers. So it is important to compare the map of natural resources (Fig. 3A-8) with the one of States and provinces.

Water certainly is a natural resource, and North America as a realm is comparatively well supplied with it despite concerns over long-term prospects in the American Southwest and the Great Plains, where some States depend on sources in other States. Another concern focuses on lowering water tables in some of North America's most crucial aquifers (underground water reserves) in which a combination of overuse and decreasing replenishment suggests that supply problems may arise.

North America is endowed with abundant reserves of *minerals* that are mainly found in three zones: the Canadian Shield north of the Great Lakes, the Appalachian Mountains in the east, and the mountain ranges of the west. The Canadian Shield contains substantial iron ore, nickel, copper, gold, uranium, and diamonds. The Appalachians yield lead, zinc, and iron ore. And the western mountain

Wind "farms" are arising in many more places in North America as the clean-energy campaign to generate electricity from sources other than fossil fuels accelerates. This cluster of modern wind turbines on a cattle ranch in southern Wyoming (the livestock provide a sense of scale) exploits a flat-floored windstream corridor between areas of higher relief.

© Jim Harvey/Alamy

FIGURE 3A-8

© H. J. de Blij, P. O. Muller, and John Wiley & Sons, Inc.

zone has significant deposits of copper, lead, zinc, molybdenum, uranium, silver, and gold.

In terms of **fossil fuel [4]** energy resources (oil, natural gas, coal), North America is also quite well endowed, though the realm's voracious demand, the highest in the world, cannot be met by domestic supplies alone and necessitates an enormous need for imports. Figure 3A-8 displays the distribution of oil, natural gas, and coal.

The leading *oil*-production areas lie (1) along and offshore from the Gulf Coast, where the floor of the Gulf of Mexico is yielding a growing share of the output; (2) in the Midcontinent District, from western Texas to eastern Kansas; and (3) along Alaska's North Slope facing the Arctic Ocean. An important development is taking place in

Canada's northeastern Alberta, where oil is being drawn from deposits of *tar sands* in the vicinity of the boomtown of Fort McMurray. The process is expensive and can reward investors only when the price of oil is comparatively high, but the reserves of oil estimated to be contained in the tar sands may exceed those of Saudi Arabia. In 2008, when the price of oil skyrocketed, activity in this area soared. When the price dropped later that year, production slowed—only to rebound again in 2009. It is a pattern that is likely to continue for years to come (we will have more to say about the impact of Alberta's tar sands in Chapter 3B).

The distribution of **natural gas** reserves resembles that of oilfields because petroleum and natural gas tend to be found in similar geological formations (that is, the floors of ancient, shallow seas). What the map cannot reveal is the volume of production, in which this realm leads the world (Russia and Iran lead in terms of proven reserves). Natural gas has risen rapidly as the fuel of choice for electricity generation in North America, raising fears that the supply will not keep up with growing demand.

The *coal* reserves of North America, perhaps the largest on the planet, are found in Appalachia, beneath the Great Plains of the United States as well as Canada, and in the southern Midwest among other places. These reserves guarantee an adequate supply for centuries to come, although coal has become a less desirable fuel due its polluting emissions that contribute to global warming.

Given the relative scarcity of oil and the geopolitical disadvantages of having to import more than half of what is needed in the United States (51 percent in 2009), you might expect nuclear energy to be more ubiquitous, especially since it is relatively cheap to produce. There are more than 100 nuclear power plants in the United States providing about one-fifth of all electrical energy (Canada has 18 plants, producing 15 percent of the country's electricity). But ever since the notorious leak and near-calamity at Three Mile Island (Pennsylvania) in 1979, the fear of nuclear accidents is pervasive. Since then, not a single new nuclear plant has been approved by the government. As of 2010, two plants were under construction, but neither had yet obtained full approval.

URBANIZATION AND THE SPATIAL ECONOMY

Industrial Cities

When the Industrial Revolution crossed the Atlantic and touched down in America in the 1870s, it took hold so successfully and advanced so robustly that only 50 years later North America was surpassing Europe as the world's mightiest industrial complex. Industrialization progressed in tandem with urbanization, as manufacturing plants in need of large labor supplies were built in and near cities. This, in turn, propelled rural-urban migration (including the migration of African Americans from the South to northern cities during the first half of the twentieth century) and the growth of individual cities. At a larger scale, a *system* of new cities emerged that specialized in the collection and processing of raw materials and the distribution of manufactured products. This **urban system [5]** was interconnected through a steadily expanding network of railroads, themselves a product of the Industrial Revolution.

As new technologies and innovations emerged, and specializations such as Detroit's automobile industry strengthened, an **American Manufacturing Belt [6]** (that extended into southern Ontario) evolved into the foundation of a North American Core (Fig. 3A-9). This

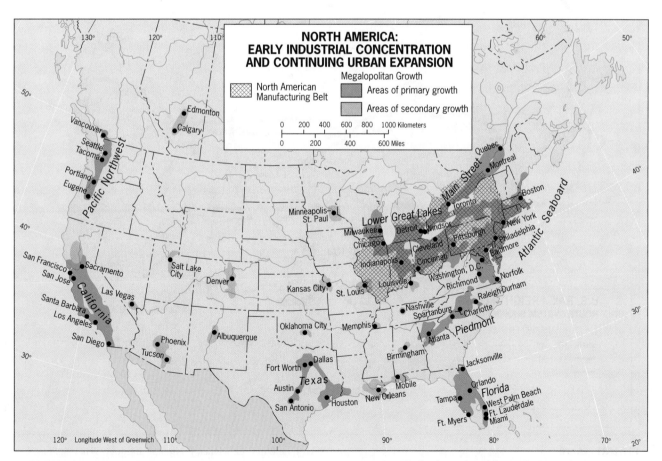

FIGURE 3A-9

© H. J. de Blij, P. O. Muller, and John Wiley & Sons, Inc.

core area, on the way to becoming the world's most productive and important, contained the majority of the realm's industrial activity and leading cities, including New York, Chicago, Toronto, and Pittsburgh (the "steel city").

In the course of intertwined urbanization and industrialization, the spatial economy of North America underwent profound changes. The *primary* sector, involving the extraction of raw materials from nature (agriculture, mining, fishing), was rapidly mechanized, and the workforce in this sector shrank considerably. Employment in the *secondary* sector, using the input of raw materials and manufacturing them into finished products, grew rapidly. The *tertiary* sector, entailing all kinds of services to support production and consumption (banking, retail, transport) expanded as well. Both the secondary and tertiary sectors were heavily concentrated in cities.

Realm of Railroads

The United States first became a single, integrated, continental-scale economy when long-distance railroads were built during the middle decades of the nineteenth century. This rail network was launched in the Northeast and soon expanded into and across the Midwest, the realm's industrial heartland. With the completion of the transcontinental railway in 1869, the west coast was incorporated as well, especially San Francisco, a great natural harbor and vital U.S. port and naval base. In the 1880s, the port of Los Angeles began its rise once its harbor was improved to handle large cargo vessels, and to this day rail traffic to and from that city remains among the busiest in all of North America. The realm's network grew denser in the first half of the twentieth century, and the two national rail systems were tightly integrated throughout the lengthy U.S.-Canada border zone (Fig. 3A-10). But then, after 1950,

FIGURE 3A-10

© H. J. de Blij, P. O. Muller, J. Nijman, and John Wiley & Sons, Inc.

the network entered a period of stagnation as the construction of the interstate highway system increasingly facilitated competition by long-haul trucks.

Since around 1980, rail freight has experienced a revival that is generally attributed to the deregulation of rail transport services and a decline in shipping costs. Between 1981 and 2011, the latter dropped by more than 50 percent, making the movement of goods by rail in the United States and Canada among the cheapest in the world—half of what it costs in Europe or in Japan.

Interestingly, usage of this transportation mode in North America is the opposite of Europe's: here the rail-

roads primarily carry cargo rather than passengers, while in the European realm it is the other way around. Rail freight in North America accounts for about 43 percent of the entire freight market (that also includes trucking, shipping, and air transport), which ranks it among the highest in the world's wealthy countries. Although the largest single cargo share is claimed by coal (45 percent), the fastest growing category of cargo is what is called "intermodal"—truck-trailer containers that are stacked onto railroad flatcars at ports and other break-of-bulk points.

This enormous rail infrastructure requires constant maintenance and updating. One example is the recently

---

AMONG THE REALM'S GREAT CITIES . . .

NEW YORK

NEW YORK IS much more than the largest city of the North American realm. It is one of the most famous places on Earth; it is the hemisphere's gateway to Europe and the rest of the Old World; it is a tourist mecca; it is the seat of the United Nations; it is one of the globe's most important financial centers—a true "world city" in every sense of that overused term. Unfortunately, the terrorists who planned and executed the attacks of 9/11 were only too well aware of New York's symbolic importance on the international stage.

New York City consists of five boroughs, centered by the island of Manhattan, which contains the CBD south of 59th Street (the southern border of Central Park). Here is a skyscrapered landscape unequaled anywhere, studded with more fabled landmarks, streets, squares, and commercial facilities than any other cities except London and Paris. This is also the cultural and media capital of the United States, which means that the city's influence constantly radiates across the planet thanks to New York-based television networks, newspapers and magazines, book publishers, fashion and design leaders, and artistic and new media trendsetters.

At the metropolitan scale, New York forms the center of a vast urban region, 250 kilometers (150 mi) square (population: 22.3 million), which sprawls across parts of three States in the heart of Megalopolis. That outer city has become a giant in its own right with its population of 14-plus million, massive business complexes, and flourishing suburban downtowns. Thus, despite its global connections, New York is caught in the currents that affect U.S. central cities today. Its ghettos of disadvantaged populations continue to expand; its aging port facilities and industrial base are deteriorating; its corporate community is increasingly stressed by the ever higher cost of doing business in ultra-expensive Manhattan—and its Wall Street banking and finance community may never regain its international prominence following the economic crisis that began in 2008.

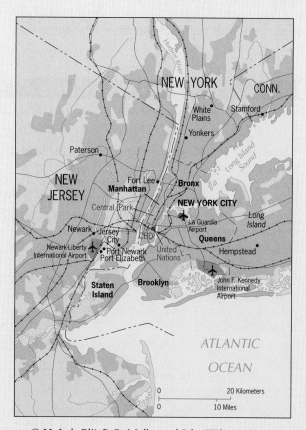

© H. J. de Blij, P. O. Muller, and John Wiley & Sons, Inc.

These challenges notwithstanding, New York today is seeking to reinforce its position as a great, sustainable, twenty-first-century city that will continue to lure the hypermobile, super-affluent global elite. Its bold new plans are geared to meet its biggest problems head-on: a major increase in affordable housing for a growing population, across-the-board upgrading of a disintegrating infrastructure, and a sizeable expansion of parks and other outdoor recreational spaces.

completed *Alameda Corridor*, a huge project involving a 20-mile-long rail cargo express line that links the ports of Long Beach and Los Angeles to the Southern California mainlines near downtown Los Angeles (which from there connect to the national rail system). It consists of a series of bridges and underpasses that allow fast and uninterrupted transport, separated from all other traffic. And in this decade, even more rail capacity will be needed. For instance, in 2014 the Panama Canal will open its enlarged, second lane (see diagram, p. 225). That is bound to translate into a major increase of trans-Pacific cargo headed for U.S. ports on the Gulf of Mexico and the Atlantic seaboard; this, in turn, will require a commensurate increase in overland transfer by rail to get these goods to their destinations.

Deindustrialization and Suburbanization

Cities continued to evolve, along with the nature of the economy and with the development of new transportation and communication technologies. The mass introduction of automobiles and the accompanying construction of a large-scale highway system, especially after World War II, had two major effects on cities. First, it resulted in a much higher mobility and interconnectedness between cities. Second, it propelled a process of suburbanization: the change from compact city to widely dispersed metropolis, the evolution of residential suburbia into a complete **outer city [7]** with its own businesses and industries, sports and entertainment complexes, and other amenities. As the newly urbanized suburbs increasingly captured major economic activities, many central cities saw their comparative status diminish.

Importantly, this development coincided with the **deindustrialization [8]** that hit American cities in the 1960s: the loss of manufacturing (and jobs) due to automation and relocation of production to countries with lower wages. While the suburban outer cities became the destination of intrametropolitan migrants and of

---

© H. J. de Blij, P. O. Muller, and John Wiley & Sons, Inc.

AMONG THE REALM'S GREAT CITIES . . .

TORONTO

TORONTO, CAPITAL OF Ontario and Canada's largest metropolis (5.6 million), is the historic heart of English-speaking Canada. The landscape of much of its center is dominated by exquisite Victorian-era architecture and surrounds a healthy downtown that is one of Canada's leading economic centers. Landmarks abound in this CBD, including the Sky Dome stadium (Rogers Centre), the famous City Hall with its facing pair of curved high-rises, and mast-like CN Tower, which at 553 meters (1815 ft) is the world's fourth-tallest freestanding structure. Toronto also is a major port and industrial complex, whose facilities line the shore of Lake Ontario.

Livability is one of the first labels Torontonians apply to their city, which has retained more of its middle class than central cities of its size in the United States. *Diversity* is another leading characteristic because this is North America's richest urban ethnic mosaic. Among the largest of more than a dozen thriving ethnic communities are those dominated by Italians, Portuguese, Chinese, Greeks, and Ukrainians; overall, Toronto now includes residents from about 170 countries who speak more than 100 languages; and the immigrant inflow continues strongly, with those born outside Canada constituting nearly half of the city's population (in all the world only two other million-plus cities exhibit a higher percentage). *Vibrancy* is yet another hallmark of this remarkable city, whose resiliency was tested but not broken by the SARS-virus outbreak of 2003.

Toronto has worked well in recent decades, thanks to a metropolitan government structure that fostered central city–suburban cooperation. But that relationship is increasingly stressed as the outer city gains a critical mass of population, economic activity, and political clout. Toronto has responded with plans to become a "city-state," but first it must overcome problems of municipal finance and aging infrastructure.

From the Field Notes . . .

"Monitoring the urbanization of U.S. suburbs for the past four decades has brought us to Tyson's Corner, Virginia on many a field trip and data-gathering foray. It is now hard to recall from this 2008 view that less than 50 years ago this place was merely a near-rural crossroads. But as nearby Washington, D.C. steadily decentralized, 'Tyson's' capitalized on its unparalleled regional accessibility (its Capital Beltway location at the intersection with the radial Dulles Airport Toll Road) to attract a seemingly endless parade of high-level retail facilities, office complexes, and a plethora of supporting commercial services." [Today, this suburban downtown ranks among the largest business districts in all of North America. But it also exemplifies urban sprawl, and in 2009 its developers, tenants, and county government formed a coalition to transform Tyson's into a true city characterized by smarter, greener development. The elements of the new plan include much higher residential densities, mass transit in the form of four stations on a new Metro line connecting to Washington and Dulles Airport, and major new facilities to accommodate pedestrians and bicycles. Details can be found in a prominent article in the June 22, 2009 issue of *Time*.]

© Andy Ryan Photography

www.conceptcaching.com

economic activity and employment in the service sector, many once-thriving and dominant CBDs were now all but reduced to serving the less affluent populations that increasingly dominated the central city's close-in neighborhoods. The fate of Detroit is often cited as exemplifying this sequence of events, but to a certain degree all major U.S. cities suffered.

The Information Economy and City Regions

Since the early 1980s, many North American cities (but not all) have bounced back from the destructive effects of deindustrialization and unemployment. The ***information economy***—embodied in the *quaternary* sector—had arrived. Many northern cities had lost population and had been "hollowed out" for nearly two decades, but the tide turned again. Employment in the tertiary sector started to grow rapidly, especially in high-technology, producer services (such as consulting, advertising, and accounting), finance, research and development, and the like. Indeed, with the advent of computerization, most of these information-rich services evolved into quaternary activities. The geography of the information economy is certainly very different from that of the old manufacturing industries and is still not fully understood. Some of it

concentrates in established CBDs like Midtown Manhattan; some of it locates around important hubs of the Internet infrastructure to benefit from maximum bandwidth (e.g., Dallas); and still other activities cluster in suburban areas containing a high density of highly skilled workers such as Silicon Valley or North Carolina's Research Triangle.

Northern California's ***Silicon Valley***—the world's leading center for computer research and development and the headquarters of the United States' microprocessor industry—illustrates the locational dynamics of this newest sector of the spatial economy. Proximity to Stanford, a world-class research university; not far from cosmopolitan San Francisco; the availability of a large pool of highly educated and skilled workers; a strong business culture; ample local investment capital; good housing; and a scenic area with good weather—all combining to make Silicon Valley a prototype for similar developments elsewhere, and not just in North America. Under different names (*technopolis* is used in Brazil, France, and Japan; *science park* in China, Taiwan, and South Korea), such ultramodern, campus-like complexes symbolize the information-dominated era just as the smoke-belching factory did the industrial age of the past.

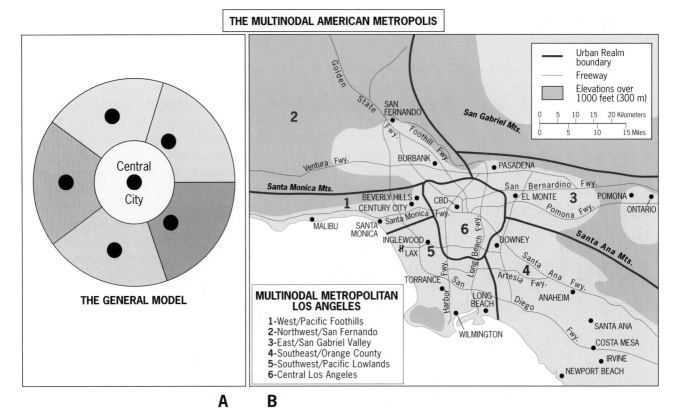

THE MULTINODAL AMERICAN METROPOLIS

THE GENERAL MODEL

A   B

FIGURE 3A-11

© H. J. de Blij, P. O. Muller, and John Wiley & Sons, Inc.

Polycentric Cities

Fly into a contemporary U.S. or Canadian city, and you will see the high-rises of suburban downtowns encircling the old central-city CBD, some of them boasting their own impressive skylines (see photo, p. 163). In Canada, where such deconcentration has had more to do with lower land values outside the city centers, this spatial pattern has become common as well, although most Canadian cities remain more compact than the far-flung U.S. metropolis.

Thus the overall structure of the modern North American metropolis is polycentric and resembles a pepperoni pizza (Fig. 3A-11) in its typical form. The traditional CBD still tends to be situated at the center, much of its former cross-traffic diverted by beltways; but the outer city's CBD-scale nodes are ultramodern and thriving. Efforts to attract businesses and more affluent residents back to the old CBD sometimes involve the construction of multiple-use high-rises that often displace low-income residents, resulting in conflicts and lawsuits. The revitalization and upgrading—or **gentrification [9]**—of crumbling downtown-area neighborhoods raises real estate values as well as taxes, which tends to drive lower-income, long-time residents from their homes. Restoring the old CBDs and their immediate surroundings may be attainable in some central cities, but the growth and development of large interconnected metropolitan regions marked by multiple urban centers is certain to continue.

Effects of the Great Recession

The financial crisis that erupted in the United States in mid-2008, and soon assumed global proportions, was concentrated in the banking sector; it had much to do with excessive mortgage lending and borrowing, and with the increasingly nontransparent trade in mortgages. While many banks got into serious trouble for having accumulated what were clinically labeled "toxic assets," millions of American households fell victim as their mortgages became ever more difficult to afford.

The resulting major economic downturn—quickly dubbed the Great Recession—took on unusually clear geographical dimensions because it affected some places far more than others. The impact was most devastating where housing values had risen unrealistically in the years just prior to 2008. In those real estate markets, there had seemed no limit to the escalation of home prices. Buyers were betting on continued appreciation, and many bought for speculative purposes with the intention of selling in the near future for a tidy profit. The banks propelled everything along with easy-to-obtain home loans, sometimes with variable interest rates they knew would prove fatal to buyers at a later stage. When the bubble finally burst, housing prices nosedived and myriad homeowners went under water as their home now became worth less than the value of their mortgage. This situation is referred to as ***negative equity*** (equity being the difference between market value and mortgage

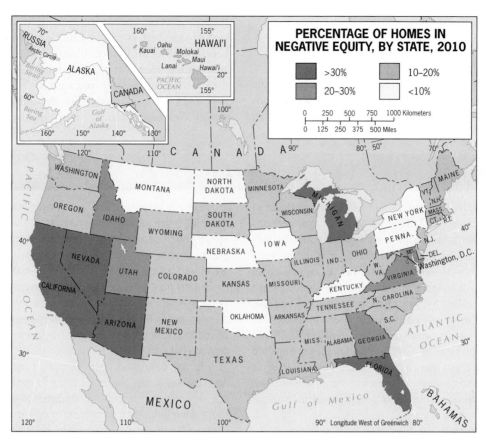

FIGURE 3A-12   © H. J. de Blij, P. O. Muller, J. Nijman, and John Wiley & Sons, Inc.

balance), a problem because it renders homeowners less affluent. In these circumstances, they cannot afford to sell because to pay off the bank they would end up with a significant loss.

The scale of this crisis was enormous. By late 2010, a staggering 23 percent of all mortgaged residential properties in the United States—almost 11 million homes—were affected. Figure 3A-12 shows that these numbers varied considerably by State. Hardest hit were California, Nevada, Arizona, and Florida, where the housing bubble had been at its greatest earlier in the 2000s. This is also a result of the large influx of migrants to those States, which further heightened demand in the housing market. Note that in Nevada relatively few homeowners have accumulated any equity, whereas at the other end of the spectrum almost half of New York's homeowners have built more than 50 percent equity.

It should also be noted that the crisis is most severe in what were until recently fast-growth urban areas. In Nevada, worst off were Las Vegas and its suburbs as well as Reno; in Florida, cities such as Miami, Orlando, and Tampa; in Arizona, Phoenix and especially its suburb of Glendale; and in California, the Central Valley's Stockton, Modesto, and Bakersfield. Michigan ranked high as well, but here the situation is different because this State's economic problems are deeply rooted in the expanding unemployment triggered by the long-term

decline of the automobile industry that began more than four decades ago. This is yet another reminder that deindustrialization in the American Manufacturing Belt has forged a widening economic gap between cities like Detroit and such Sunbelt counterparts as Phoenix, Atlanta, and Dallas.

As the U.S. economy spiraled downward into the Great Recession in 2008 and 2009, unemployment rose, stock exchanges experienced major jolts, and more and more people had difficulty maintaining homes financed by negative-equity-plagued mortgages. This swiftly led to an alarming rise in the number of foreclosures. As you would expect, foreclosure rates are highest where negative equity is at its highest; in 2010, Nevada topped all other States as an astonishing 1 in 79 of its mortgaged properties were foreclosed

Tellingly, the effects of the Great Recession in Canada have been decidedly less severe. Canadian banks were far more conservative than their U.S. counterparts in extending credit, and the housing market remained generally healthy. In fact, the volume and value of home sales in 2010 reached record highs! In other sectors as well, the Canadian economy maintained its relative strength: unemployment rates were considerably lower than in the United States and Europe, and the federal government's budget deficit was one of the lowest in the Western world.

THE MAKING OF A MULTICULTURAL REALM

The Virtues of Mobility and Immigration

Given the mobility of North America's population, the immigrant streams that continue to diversify it, the economic changes affecting it, and the forces acting upon it, the population distribution map (Fig. 3A-2) should be viewed as the latest still in a motion picture, one that has been unreeling for four centuries. Slowly at first, then with accelerating speed after 1800, North Americans pushed their settlement frontier westward to the Pacific. Even today, such shifts continue. Not only does the "center of gravity" of population continue to move westward, but

The Migration Process

BOTH CANADA AND the United States are products of international **migration [11]** (residential relocation intended to be permanent). Europeans first crossed the Atlantic with the intent of establishing permanent colonies in the early 1600s, and from these colonies evolved the two countries of North America. The Europeanization of North America doomed the realm's indigenous societies, but this was only one of many areas around the world where local cultures and foreign invaders came face to face. Between 1835 and 1935, perhaps as many as 75 million Europeans departed for distant shores—most of them bound for the Americas (Fig. 3A-13). Some sought religious freedom, others escaped poverty and famine, still others simply hoped for a better life. A comparative few were transported against their will to penal colonies.

Studies of the **migration decision** show that migration flows vary in size with (1) the perceived difference between home (source) and destination; (2) the effectiveness of information flow, that is, the news about the destination that emigrants send back to those left behind waiting to decide; and (3) the distance between source and destination (shorter moves attract many more migrants than longer ones).

Every migration stream produces a counter-stream of returning migrants who cannot adjust, are unsuccessful, or are otherwise persuaded or compelled to return home. Migration studies also conclude that several discrete factors are at work in the process. **Push factors** motivate people to move away from an undesirable locale that may be afflicted by famine, armed conflict, religious persecution, or some other adversity. **Pull factors** attract them to destinations perceived to hold a promise of security, opportunity, or another desired goal.

To the early (and later) European immigrants, North America was a new frontier, a place to escape persecution and acquire a piece of land. Opportunities were reported to be unlimited. That perception of opportunity has never changed. Immigration continues to significantly shape the human-geographic complexion of the United States as well as Canada. Today's immigrants account for at least 40 percent of the annual population growth of the United States. Never in its history, however, has the United States received so large an undocumented (illegal) immigration flow in addition to the legitimate stream, and the ongoing influx from Middle to North America is one of the largest population shifts in human history.

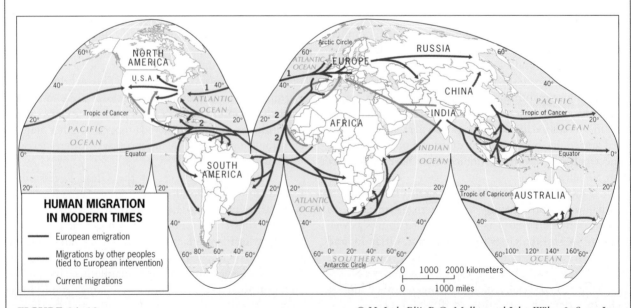

HUMAN MIGRATION IN MODERN TIMES

— European emigration

— Migrations by other peoples (tied to European intervention)

— Current migrations

FIGURE 3A-13

© H. J. de Blij, P. O. Muller, and John Wiley & Sons, Inc.

within the United States it is also moving southward—the latter a drift that gained momentum in the 1970s as the advent of universal air conditioning made the so-called Sunbelt [10] States of the U.S. southern tier ever more attractive to internal migrants.

The current population map is the still-changing product of numerous forces. For centuries, North America has attracted a pulsating influx of immigrants who, in the faster-growing United States, were rapidly assimilated into the societal mainstream. Throughout the realm, people have sorted themselves out to maximize their proximity to evolving economic opportunities, and they have shown little resistance to relocating as these opportunities successively favored new locales.

During the past century, such transforming forces have generated a number of major migrations, of which the still-continuing shift to the west and south is only the latest. Five others were: (1) the persistent growth of metropolitan areas, first triggered by the late-nineteenth-century Industrial Revolution's impact in North America; (2) the large-scale movement of African Americans from the rural South to the urban North during the height of the industrial era; (3) the shift of tens of millions of urban residents from central cities to suburbs and subsequently to exurbs even farther away from the urban core; (4) the return migration of millions of African Americans from the deindustrializing North back to the growing opportunities in the South (in metropolises such as Atlanta and Charlotte); and (5) the strong and steady influx of immigrants from outside North America including, in recent times, Cubans, Mexicans, and other Latinos; South Asians from India and Pakistan; and East and Southeast Asians from Hong Kong, Vietnam, and the Philippines.

Several of the migration streams affecting North America rank among the largest in world history, creating the pair of plural societies that characterize this realm (see the box titled "The Migration Process"). North America functions, in some ways, as a global magnet for human resources, and its persistent prosperity hinges at least in part on wave after wave of ambitious immigrants, be they engineers from China, physicians from India, nurses from Jamaica, or farm workers from Mexico.

The Challenge of Multiculturalism

But growing pluralism, especially at a time of globalization and *transnationalism*, comes with challenges. The sheer dimensions of immigration into the United States, from virtually all corners of the world, creates an increasingly complex ethnic and cultural mosaic that tests melting pot [12] assertions at every geographic scale. Today there are more people of African descent living in the United States than in Kenya. The number of Hispanic residents in America is nearing half the population of Mexico. Miami is the second-biggest Cuban city after Havana, and Montreal the largest French-speaking city in the world after Paris. In short, there are sufficient immigrant numbers in

the United States for them to create durable societies within the national society. The challenge is to ensure that the great majority of the immigrants become full participants in that larger society.

The issue is compounded as well as politicized by the fact that never in its history has the realm received so large an infusion of undocumented (illegal) immigration in addition to the legitimate stream. The ongoing influx from Middle to North America, principally from or via Mexico, is one of the biggest population shifts in human history. A considerable part of it is illegal, producing a sometimes acrimonious debate over border security and law enforcement. In 2010, an estimated 11 million illegal immigrants from Mexico were in the United States at a time when the Hispanic minority had already become the country's largest (about 50 million or 16 percent of the total population)—transforming neighborhoods, cities, and even entire regions.

Such are the numbers, and so diverse are the cultural traditions, that the melting pot that America once was is changing into something different. The United States is becoming a *mosaic culture*, an increasingly heterogeneous complex of separate, more or less uniform "tiles" that spend less time interacting and "melting." This applies not only to new immigrant groups but also to existing communities, underscored by the proliferation of "gated communities" in metropolitan areas across the realm. There is a serious downside to this, as Bill Bishop writes in his book, *The Big Sort: Why the Clustering of Like-Minded America Is Tearing Us Apart*. As is the case in the world at large, balkanization fueled by people wanting to interact only with those closely resembling themselves leads to misunderstanding of others, miscalculations when it comes to decisions and policies, and widening fissures between communities. This could threaten the very survival of the democratic values that have underpinned the evolution of American society.

In Canada, the melting-pot notion was put to the test during the 1990s when East Asian immigrants from Hong Kong arrived during and following the 1997 takeover of the British crown colony by Beijing's communist government. Affluent Chinese families arriving in Vancouver proceeded to purchase and renovate (or replace) traditional homes in long-stable Vancouver neighborhoods, arousing the ire of numerous locals. But in general, Canadian views of immigration differ from those in the United States. Canada has long faced critical labor shortages, especially in its western provinces including energy-booming Alberta and Asian-trade-burgeoning British Columbia. In need not only of professionals but also truck drivers and dock workers, Canada tries to balance its legal immigration process to keep it compatible with the country's employment as well as demographic needs. Across the realm, multicultural tendencies now vary on a regional basis depending on the subtle interplay of economics, existing cultural patterns, immigration, and local politics.

Regional ISSUE   Immigrants: How Many Can North America Accommodate?

IMMIGRATION BRINGS BENEFITS—THE MORE THE MERRIER!

"The United States and Canada are nations of immigrants. What would have happened if our forebears had closed the door to America after they arrived and stopped the Irish, the Italians, the eastern Europeans, and so many other nationalities from entering this country? Now we're arguing over Latinos, Asians, Russians, Muslims, you name it. Fact is, newcomers have always been viewed negatively by most of those who came before them. When Irish Catholics began arriving in the 1830s, the Protestants already here accused them of assigning their loyalty to some Italian pope rather than to their new country, but Irish Catholics soon proved to be pretty good Americans. Sound familiar? Muslims can be very good Americans too. It just takes time, longer for some immigrant groups than others. But don't you see that America's immigrants have always been the engine of growth? They become part of the world's most dynamic economy and make it more dynamic still.

"My ancestors came from Holland in the 1800s, and the head of the family was an architect from Rotterdam. I work here in western Michigan as an urban planner. People who want to limit immigration seem to think that only the least educated workers flood into the United States and Canada, depriving the less-skilled among us of jobs and causing hardship for citizens. But in fact America attracts skilled and highly educated as well as unskilled immigrants, and they all make contributions. The highly educated foreigners, including doctors and technologically skilled workers, are quickly absorbed into the workforce; you're very likely to have been treated by a physician from India or a dentist from South Africa. The unskilled workers take jobs we're not willing to perform at the wages offered. Things have changed! A few decades ago, American youngsters on summer break flooded the job market in search of temporary employment in hotels, department stores, and restaurants. Now they're vacationing in Europe or trekking in Costa Rica, and the managers of those establishments bring in temporary workers from Jamaica and Romania.

"And our own population is aging, which is why we need the infusion of younger people immigration brings with it. We don't want to become like Japan or some European countries, where they won't have the younger working people to pay the taxes needed to support the social security system. I agree with opponents of immigration on only one point: what we need is legal immigration, so that the new arrivals will get housed and schooled, and illegal immigration must be curbed. Otherwise, we need more, not fewer, immigrants."

LIMIT IMMIGRATION NOW

"The percentage of recent immigrants in the U.S. population is the highest it has been in 70 years, and in Canada in 60 years. America is adding the population of San Diego every year, over and above the natural increase, and not counting illegal immigration. This can't go on. By 2020, more than one-sixth of the U.S. population will consist of recent immigrants. At least one-third of them will not have a high school diploma. They will need housing, education, medical treatment, and other social services that put a huge strain on the budgets of the States they enter. The jobs they're looking for often aren't there, and then they start displacing working Americans by accepting lower wages. It's easy for the elite to pontificate about how great immigration is for the American melting pot, but they're not the ones affected on a daily basis. Immigration is a problem for the working people. We see company jobs disappearing across the border to Mexico, and at the same time we have Mexicans arriving here by the hundreds of thousands, legally and illegally, and more jobs are taken away.

"And don't talk to me about how immigration now will pay social security bills later. I know a thing or two about this because I'm an accountant here in Los Angeles, and I can calculate as well as the next guy in Washington. Those fiscal planners seem to forget that immigrants grow older and will need social security too. And as for that supposed slowdown in the aging of our population because immigrants are so young and have so many children, over the past 20 years the average age in the United States has dropped by four months. So much for that nonsense. What's needed is a revamping of the tax structure, so those fat cats who rob corporations and then let them go under will at least have paid their fair share into the national kitty. There'll be plenty of money to fund social services for the aged. We don't need unskilled immigrants to pay those bills.

"And I'm against this notion of amnesty for illegal immigrants being talked about these days. All that would do is to attract more people to try to make it across our borders. I heard the president of Mexico propose opening the U.S.-Mexican border the way they're opening borders in the European Union. Can you imagine what would happen? What our two countries really need is a policy that deters illegal movement across that border, which will save lives as well as jobs, and a system that will confine immigration to legal channels. These days, that's not just a social or economic issue; it's a security matter as well."

Vote your opinion at www.wiley.com/college/deblij

POINTS TO PONDER

- The North Atlantic Ocean for generations was the most important external trade route for North America, by volume and value. Now it is the Pacific Ocean.

- There are only two states in the world that share the same country code for international telephone traffic: the United States and Canada.

- The United States does not have an official language. Should it?

- Whereas more than 70 percent of Canada's population lives within 200 kilometers (125 mi) of the northern U.S. border, only about 12 percent of the Mexican population lives that close to the southern U.S. border.

- The U.S. population (today totaling 313 million) passed the 300-million milestone in 2006. Get ready for 400 million in less than 30 years.

FIGURE 3B-1

© H. J. de Blij, P. O. Muller, and John Wiley & Sons, Inc.

REGIONS

IN THIS CHAPTER

- U.S.-Canada cross-border linkages
- The Hispanicization of the Southwest
- The natural riches of the Northern Frontier
- Miami: South of the South
- China's impact on the Pacific Hinge
- The Alaskan frontier and U.S. geopolitics

CONCEPTS, IDEAS, AND TERMS

© Jan Nijman

Where were these photos taken?
Find out at www.wiley.com/college/deblij

Geology, climate, culture, and history have combined to create an interesting matrix of regions in North America (Fig. 3B-1). Mountain ranges, coastal zones, deserts, low-lying plains, and a range of climatic environments all demand different kinds of human adaptation. Successive waves of immigrants brought with them a variety of cultures; sometimes the differences were subtle, but sometimes they were quite stark. The emergence of two large countries, the United States and Canada, provided a pair of major political and institutional territorial structures within which regional differences have been accommodated. As Figure 3B-1 shows, a majority of North America's regions traverse both countries regardless of the international boundary. But some other regional developments are best understood within the national context of each country. As much as Canada and the United States have in common, there is ample reason to focus the first part of this chapter on each individually. In the discussion that follows, therefore, we begin by highlighting the regional geographies of two concordant neighbors—in the knowledge that Canadians tend to know more about the United States than Americans know about Canada. After that, we will examine the nine evolving regions of the realm as a whole.

REGIONALISM IN CANADA: DIVISIVE FORCES

Canada's Spatial Structure

Territorially the second-largest state in the world after Russia, Canada is administratively divided into only 13 subnational entities—10 provinces and 3 territories (the United States has 50). The provinces—where 99.7 percent of all Canadians live—range in dimensions from tiny,

© Jan Nijman

MAJOR CITIES OF THE REALM

City	Population* (in millions)
New York	22.3
Los Angeles	18.2
Chicago	9.9
San Francisco	7.8
Dallas-Ft. Worth	7.0
Philadelphia	6.7
Houston	6.2
Washington, DC	5.9
Miami-Ft. Lauderdale	5.7
Toronto, Canada	5.6
Atlanta	5.5
Detroit	5.1
Boston	4.6
Phoenix	4.5
Seattle	4.4
Montreal, Canada	3.8
Denver	3.2
San Diego	3.2
Vancouver, Canada	2.3
Ottawa, Canada	1.2

*Based on 2012 estimates.

Delaware-sized Prince Edward Island to Quebec, nearly twice the size of Texas (Fig. 3B-2).

As in the United States, the smallest provinces lie in the northeast. Canadians call these four the Atlantic Provinces (or simply Atlantic Canada): Prince Edward Island, Nova Scotia, New Brunswick, and the mainland-island province named Newfoundland and Labrador. Gigantic, mainly Francophone (French-speaking) Quebec and populous, heavily urbanized Ontario, both flanking Hudson Bay, form the heart of the country. Most of western Canada (which is what Canadians call everything west of Ontario) is organized into the three Prairie Provinces: Manitoba, Saskatchewan, and Alberta. In the far west, beyond the Canadian Rockies and extending to the Pacific Ocean, lies the tenth province, British Columbia.

Of the three territories in Canada's Arctic North, the Yukon is the smallest—but nothing is small here in Arctic Canada. With the Northwest Territories and recently created Nunavut, the three Territories cover almost 40 percent of Canada's total area. The population, however, is extremely small: only about 100,000 people inhabit this vast, frigid frontier zone.

The case of Nunavut merits special mention. Created in 1999, this newest Territory is the outcome of a major aboriginal land claim agreement between the *Inuit* people (formerly called Eskimos) and the federal government, and encompasses all of Canada's eastern Arctic as far north as Ellesmere Island (Fig. 3B-2), an area far larger than any

other province or territory. With around one-fifth of Canada's total area, Nunavut—which means "our land"—has about 34,000 residents, of whom 80 percent are Inuit.

Canada's population is clustered in a discontinuous ribbon, 300 kilometers (ca. 200 mi) wide, nestled against the U.S. border and along ocean shores, whereas most of the country's energy reserves lie far to the north of this ribbon. As such, the map of Canada's human settlement (Figs. 3A-2 and 3B-2) bears out the generalization and emphasizes the significance of environment. People choose to live in the more agreeable areas to the south, but territorial control of abundant northern resources is highly important. The population pattern creates cross-border affinities with major American cities in several places (Toronto-Buffalo, Windsor-Detroit, Vancouver-Seattle), although very little of this occurs in the three Prairie Provinces of the Canadian West, which for the most part adjoin sparsely settled North Dakota and Montana.

Cultural Contrasts

Canada's capital, Ottawa, is symbolically located astride the Ottawa River, which marks the boundary between English-speaking Ontario and French-speaking Quebec. Though comparatively small, Canada's population is markedly divided by culture and tradition, and this division has a pronounced regional expression. Just under 60 percent of Canada's citizens speak English as their mother tongue, 23 percent speak French, and the remaining 17 percent other languages. Only about 18 percent of the population is thoroughly bilingual in English and French. The spatial concentration of more than 85 percent of the country's Francophones in Quebec accentuates this division (Fig. 3B-2). Quebec's population is nearly 80 percent French-Canadian, and Quebec remains the historic, traditional, and emotional focus of French culture in Canada.

Over the past half-century a strong nationalist movement has emerged in Quebec, and at times it has demanded outright separation from the rest of Canada. This ethnolinguistic division continues to be the litmus test for Canada's federal system, and the issue continues to pose a latent threat to the country's future national unity. The historical roots of the matter are related to French-British competition since the seventeenth century in which Britain gained the upper hand. To this day, the British monarch remains the official head of state (as Canada remains part of the British Commonwealth), represented by the governor general. It is one of the issues that riles Quebec nationalists.

There does seem to be an overall decline in support for Quebec's independence. This is in part due to a number of laws that reassure the primacy of Québécois culture in the province (a 2006 federal parliamentary resolution effectively recognizes the Québécois as a *distinct nation*). In addition, there appears to be a growing realization among the French-speaking people that continued integration in Canada combined with a high degree of autonomy is the best possible option. But that struggle for maximum

FIGURE 3B-2

© H. J. de Blij, P. O. Muller, and John Wiley & Sons, Inc.

autonomy continues, along with the accompanying political and ethnic tensions.

The Ascendancy of Indigenous Peoples

On the wider Canadian scene, the culture-based Quebec struggle has stirred ethnic consciousness among the country's 1.4 million native peoples—dominated by **First Nations [1]**, but also including Métis and Inuit. They, too, have received a sympathetic hearing in Ottawa, and in recent years breakthroughs have been achieved with the creation of Nunavut as well as local treaties for limited self-government in northern British Columbia.

A foremost concern among these peoples is that their aboriginal rights be protected by the federal government against the provinces of which they are a part. This is especially true for the First Nations of Quebec, the Cree, whose historic domain covers the northern half of the province. As Figure 3B-2 shows, the territory of the Cree is no unproductive wilderness: it contains the James Bay Hydroelectric Project, a massive scheme of dikes and dams that has transformed much of northwestern Quebec

and generates hydroelectric power for a huge market within and outside the province. In 2002, after the federal courts supported its attempts to block construction of the project, the Cree negotiated a groundbreaking treaty with the Quebec government whereby this First Nation dropped its opposition in return for a portion of the income earned from electricity sales. The Cree also secured the right to control their own economic and community development.

Centrifugal Forces

The events of the past four decades have had a profound impact on Canada's national political geography, and not only in Quebec. Leaders of the western provinces objected to federal concessions to Quebec, insisting that the equal treatment of all ten provinces is a basic principle that precludes the designation of special status for any single one. Recent federal elections and opinion surveys have clearly revealed the emerging fault lines that surround Quebec and set off the western provinces and British Columbia from the rest of the country. Thus, with national unity

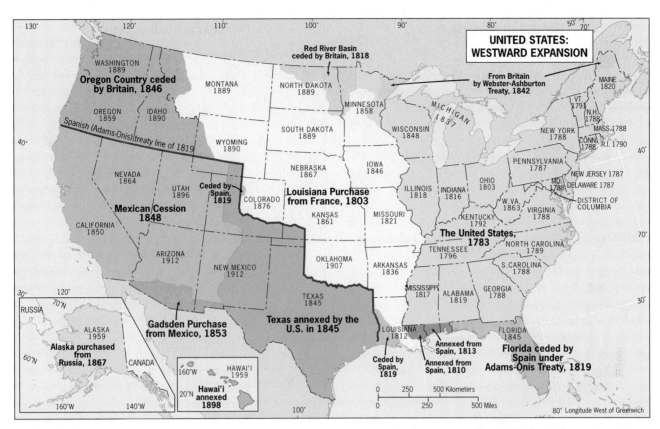

FIGURE 3B-3

© H. J. de Blij, P. O. Muller, and John Wiley & Sons, Inc.

always a challenge in this comparatively young federation, Canada confronts forces of **devolution [2]** that threaten to weaken the state.

In addition, the continuing economic integration of Canada and the United States (galvanized by the North American Free Trade Agreement of 1994) tends to pull the southern parts of Canada's most populous provinces into an ever closer relationship with its U.S. neighbor. Local **cross-border linkages [3]** are already strongly developed and are likely to intensify in the future: the Atlantic Provinces with neighboring New England; Quebec with New York State; Ontario with Michigan and adjacent Midwestern States; the Prairie Provinces with the Upper Midwest; and British Columbia with the (U.S.) Pacific Northwest. Such functional reorientations constitute yet another set of potentially powerful devolutionary challenges confronting the Ottawa government.

REGIONALISM AND ETHNICITY IN THE UNITED STATES

Nothing in the United States compares with the devolutionary challenges Canada has faced throughout its evolution as a modern state. The United States has about four times as many subnational political units as Canada, but none since the Civil War (1861–1865) has mounted a serious campaign for **secession [4]** of the kind Quebec has. And in only one (Hawai'i) have indigenous people demanded recognition of a part of the State as sovereign territory.

In Canada, indigenous peoples were able to hold on to sparsely populated regions in the north. In the United States, on the other hand, the relentless push westward by European settlers and the U.S. government (Fig. 3B-3) in effect dispossessed—and decimated—the original inhabitants, in the end confining them to some 300 reservations located mainly in the western half of the country (Fig. 3B-4, upper-left map). There are only about 5.2 million Native Americans left today, with the Navajo and Cherokee being the largest surviving nations. Whereas in Canada the First Nations hold significant territorial power, in the United States they are in a far weaker position to challenge the federal government.

Within the United States, much more so than in Canada, the social fabric and ethnic relations are determined by past and present waves of immigration. In certain cases, immigration has helped forge a particular regional geography. This was especially true in the case of the hundreds of thousands of Africans who were forcibly brought to the United States in bondage from the seventeenth to the mid-nineteenth centuries and put to work as slaves on plantations in the American South.

Slavery was abolished in 1865, but racial segregation continued in various legal forms until the civil rights movement peaked in the 1960s. In violation of the Constitution, a wide range of State and local Jim Crow laws were enacted to codify the repression of African Americans. This historical legacy is still visible today in the geography of race in the United States, with blacks regionally concentrated in the Southeast, from eastern Texas to Maryland (Fig. 3B-4,

FIGURE 3B-4

© H. J. de Blij, P. O. Muller, and John Wiley & Sons, Inc.

upper-right map). We noted earlier that substantial numbers of blacks migrated north to work in the industrial centers of the American Manufacturing Belt during the second quarter of the twentieth century; major African-American population clusters still exist in those northern cities, but they are not clearly apparent at the scale of the national map.

Race continues to be a sensitive issue in the United States, more so than in Canada, because of this terrible history and because until recently many African Americans were effectively denied equal opportunities. The 2008 election of the first black president, Barack Obama, may serve as a catalyst of progressive change.

Another highly visible regional expression of ethnic migration concerns the presence of Hispanics in the southwestern United States (Fig. 3B-4, lower-left map). This migration, too, should be understood in historical and geographical contexts. In these parts of the country, many Hispanic place names reflect Mexico's ouster from a vast area acquired in 1848 by the United States at the end of the Mexican War through the Treaty of Guadalupe Hidalgo. This treaty established the Mexican-American border at the Rio Grande and assigned to the United States no less than 1.3 million square kilometers (525,000 sq mi) of Mexican territory, including present-day California, Arizona, Nevada, New Mexico, Texas (which had already declared independence from Mexico), southern Colorado, and Utah (Fig. 3B-3). In Mexico, this

disastrous event (which precipitated that country's civil war) is much more vivid in the national consciousness and memory than it is in the United States. Certain Mexican nationalists even proclaim that much of the current migration across the U.S.-Mexican border, illegal though it may be, is a matter of resettlement of unjustly lost lands.

Since 1980, the number of Hispanics in the United States has more than tripled and now exceeds 50 million, outnumbering the African-American minority and changing the ethnic-cultural map of America. Although Hispanics are now dispersing into many parts of the country—including rural areas in the Southeast and Midwest as well as urban communities from Massachusetts to Washington State—they remain strongly clustered in the Southwest, California, and Florida (see Fig. 3B-4, lower-left map). In fact, along virtually the entire multi-State tier north of the Mexican border, America has become a cultural transition zone (think of the expression Tex-Mex), where immigration, legal as well as illegal, has become a contentious issue. Language, not a point of controversy for the African-American community, is an emotional matter for some Anglophone Americans who fear erosion of the status of English in their communities and in the country as a whole.

Finally, the Asian immigrant population, whose arrival rate has also risen in recent decades, is geographically the most agglomerated of the main ethnic minorities

People have walled and fenced themselves off since the earliest days, as the Great Wall of China and Hadrian's Wall (built by the Romans in northern Britain) remind us. But no wall—whether the Berlin Wall during the Cold War or the Israeli "Security Barrier" of today—has ever halted all migration. Neither will migration be stopped by a fortified fence along the Mexican border from the Pacific coast to the Gulf of Mexico. However, long stretches of the U.S.–Mexico boundary already look like this. This photo shows a ground-level view of the improvised wall that slashes westward, between Tijuana, Mexico and the southernmost suburbs of San Diego, California, as it makes its way toward the Pacific shore on the horizon. The poverty-stricken landscape of this part of Tijuana extends almost to the rusty barrier itself, while development on the U.S. side has kept its distance from this well-patrolled segment of the border.

© Diane Cook and Ken Jenshel/Getty Images, Inc.

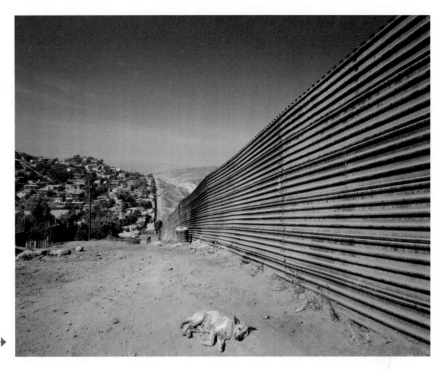

(Fig. 3B-4, lower-right map). Chinese remain the most numerous among the 17-plus million Americans who identify themselves as Asian, but this diverse minority also includes Japanese who have been in the United States for generations, Filipinos who began arriving after World War II, Koreans who have migrated since the 1950s, Vietnamese who were resettled after the Indochina War, more gradual arrivals of Pacific Islanders including Hawaiians, and considerable numbers of South Asian Indians in the past few decades. Whatever their origins, the great majority have

From the Field Notes . . .

© Jan Nijman

"On this street corner in central Los Angeles, four different cultures converge. Los Angeles is one of the most ethnically diverse metropolitan areas in the United States and in the world. More than a third of the population was born abroad, and more than half speak another language than (or in addition to) English. The largest non-native ethnic group are Hispanics (mainly Mexicans and Central Americans), but there are also many Asians, especially Chinese and Koreans. Cesar E. Chavez, after whom this avenue is named, was a charismatic Mexican-American farm worker turned union organizer and became nationally known as a leading civil rights activist. His birthday, March 31, is a State holiday in both California and Texas—and an optional holiday in Arizona (!). Chavez Avenue is a major east-west thoroughfare through downtown Los Angeles, and even a short stroll along it offers a sampling of global cultures."

www.conceptcaching.com

remained in California, where many have achieved notable economic success and upward mobility.

This regional geography of ethnicity within the national context of the United States was largely shaped by migration histories as well as the processes of voluntary or involuntary inclusion and exclusion. Now let us broaden the scope and consider the North American realm as a whole by making a number of more comprehensive regional distinctions on the basis of physical, cultural, and historical geography.

REGIONS OF NORTH AMERICA

Ample justification exists for recognizing the two countries of the North American realm as discrete regions, based on several contrasts we have discussed. There is,

however, another way to conceptualize North America's regional geography: by combining functional and formal principles. The result, Figure 3B-1 (p. 170), is by no means the only solution because the drawing of a regional boundary, though well-informed, usually remains arbitrary to some degree. As we note in the following regional survey based on our map, there is more to the framework of North America than ethnicity and politics.

▪▪ THE NORTH AMERICAN CORE (1)

The historic Core region of the North American realm integrates what have long been the most prominent parts of the United States and Canada (Fig. 3B-1). They contain the largest cities and federal capitals of both countries, the leading financial markets, the largest number of corporate

AMONG THE REALM'S GREAT CITIES . . .

CHICAGO

CHICAGO LIES NEAR the southern end of Lake Michigan not far from the geographic center of the conterminous United States. Its centrality and crossroads location were evident to its earliest indigenous settlers, who developed the site as a portage where canoes could be hauled from the lake to the headwaters of a nearby stream that led to the Mississippi River. Centuries later, when the modern U.S. spatial economy emerged, Chicago became the leading hub on the continental transport network. As a freight-rail node, it still reigns supreme. And even though most long-distance passengers switched to jet planes a half-century ago, Chicago remains the quintessential North American hub city, with O'Hare International consistently ranking among the world's busiest airports.

Poet Carl Sandburg described Chicago as "the city of big shoulders," underscoring its personality and prowess as the leading U.S. manufacturing center during the century following the Industrial Revolution of the 1870s and 1880s. Among the myriad products that emanated from Chicago's industrial crucible were the first steel-frame skyscraper, elevated railway, refrigerated boxcar, cooking range, electric iron, nuclear reactor, and window envelope. Chicago also became a major commercial center, spawning a skyscraper-dominated central business district second only to New York's in size and influence (see photo, p. 148).

Chicago today (metro population: 9.9 million) is trying hard to make the transition to the postindustrial age. Its diversified economy has proven to be a precious asset in this quest, and the city is determined to reinvent itself as a first-order service center. The new forces it must contend with, however, are symbolized in the fate of the emptying Sears (now Willis) Tower, the city's—and the nation's—tallest building. Sears & Roebuck itself abandoned the structure in 1992 for a new headquarters campus along Interstate-90 in the far northwestern suburb of Hoffman Estates. This shift

© H. J. de Blij, P. O. Muller, and John Wiley & Sons, Inc.

acknowledged the rise of Chicagoland's thriving outer city, one of the country's largest and most successful. Around O'Hare, a huge and still-expanding suburban office/industrial-park complex has truly become Chicago's "Second City," underscoring that major airports themselves will increasingly function as urban growth magnets in the ever more globalized spatial economy of the twenty-first century.

headquarters, dominant media centers, prestigious universities, cutting-edge research complexes, and the busiest airports and inter-city expressways. Moreover, both the U.S. and Canadian portions of the North American Core still contain more than one-third of their respective national populations. Political and business decisions, investments, and many other commitments made here affect not just North America, but the world at large.

As the manufacturing cities thrived, they grew larger, vertically as well as horizontally. City centers acquired characteristic and symbolic skylines (with Manhattan the synonym for the most spectacular of all), and urban peripheries expanded to the degree that cities near each other coalesced. The geographer Jean Gottmann in the 1950s combined the Greek words "megalo" (great) and "polis" (city) to coin the term megalopolis [5] to describe such coalescing metropolitan areas. The most prominent one of all extended along the Atlantic seaboard from north of Boston to south of Washington, D.C., and has become known simply as Megalopolis (with a capital "m"). Sometimes also referred to as "Bosnywash," this was the economic anchor of the North American core area: the seat of the U.S. government, the nucleus of business and finance, the hearth of culture, and the trans-Atlantic trading interface between much of the realm and Europe. In Figure 3A-9, note that Canada's predominant megalopolis is its most highly urbanized zone extending from Windsor (adjacent to Detroit) through Toronto to Montreal and Quebec City; urban geographers call this Windsor-Quebec axis **Main Street**, and it also forms part of the realm's core area.

The North American Core is also one of the world's foremost centers of globalization, with innovations radiating outward in countless fields of endeavor. And in the new era of the information economy, the Core's high-technology centers—located in the Virginia suburbs of Washington (Internet Alley lining the expressway to Dulles Airport) and the beltway corridor (circumferential Route 128) that girdles Boston—rival Silicon Valley.

Nonetheless, the dominance of the historic Core has been declining. As was true of the American Manufacturing Belt [6]—the region with which the Core coincides—it once had unmatched supremacy. But deindustrialization and the subsequent rise of the information economy have eroded that dominance as competitors to the south and west continue to siphon away some of its key functions. As already noted, the State of Michigan is experiencing long-term economic decline as jobs in the automobile industry have steadily disappeared over the past few decades. In fact, household incomes in Michigan fell by more than 21

Small fishing ports still dot the New England coast: Chatham, Massachusetts, at the elbow of Cape Cod.
© H. J. de Blij

percent just between 2000 and 2009, the most precipitous drop in the nation. Second and third on this unenviable list were two other States in the old Manufacturing Belt: Indiana and Ohio.

■■ THE MARITIME NORTHEAST (2)

When you travel north from Boston into New Hampshire, Vermont, and Maine, you move across one of North America's historic culture hearths whose identity has remained strong for four centuries. Even though Massachusetts, Connecticut, and Rhode Island—traditional New England States—share these qualities, they have become part of the Core, so that on a functional as well as formal basis, the Maritime Northeast extends from the northern border of Massachusetts to Newfoundland, thus incorporating all four Canadian Atlantic Provinces (Fig. 3B-1).

Difficult environments, a maritime orientation, limited resources, and a persistent rural character have combined to slow economic development here. Primary industries—fishing, logging, some farming—are mainstays, although recreation and tourism have boosted the regional economy in recent decades. Upper New England experiences the spillover effect of the Boston area's prosperity to some degree, but not enough to ensure steady and sustained economic growth. Atlantic Canada has also endured economic hard times in the recent past. For example, overfishing has depleted fish stocks. Alternate opportunities focus on this subregion's spectacular scenery and the tourism it attracts. Economic prospects have also been boosted in Canada's poorest province, Newfoundland and Labrador, by the discovery of significant offshore oil reserves.

■■ FRENCH CANADA (3)

Francophone Canada constitutes the inhabited, southern portion of Quebec from near Montreal to the mouth of the St. Lawrence River (Fig. 3B-1). It also includes the French-speaking Acadians who reside in neighboring New Brunswick (Fig. 3B-2). The French cultural imprint on the region's cities and towns is matched in the rural areas by the narrow, rectangular *long lots* perpendicular to the river, also of French origin.

The economy of French Canada was historically less industrial and more rural than the adjacent Core region (and Toronto far outranks Montreal here). Nevertheless, Montreal's economic role has advanced in recent years, with information technology, telecommunications, and biopharmaceuticals all growing in importance. But Quebec's economic prospects tend to be shadowed by the region's political uncertainties, and bursts of nationalism (such as laws prohibiting shops from advertising in English) have at times eroded business confidence. When the provincial government sought to address the declining birth rate resulting from accelerating urbanization by encouraging immigration, tolerance for multiculturalism in Quebec proved to be so low that a special commission held hearings to investigate the matter. Parochialism has cost Quebec dearly and sets this region off from the rest of the Canada that virtually encircles it.

The Acadians in the neighboring province of New Brunswick, Canada's largest cluster of French-speakers outside Quebec, take a different view. Not only do they reject the notion of independence for themselves, but they also actively promote all efforts to keep Quebec within the Canadian federation. Accommodation with the Anglophone majority and acceptance of multiculturalism in New Brunswick set an example for the rest of French Canada.

Yet in some ways, French Canada is a very rich part of North America. Northern Quebec produces so much hydroelectricity that it exports the surplus to Ontario and to the northeastern United States. Most of the power is generated through the James Bay Project, a series of dams and power stations on the eastern side of Hudson Bay along the La Grande River (Fig. 3B-2). Quebec is so well off in this regard that a recent discovery of natural gas in the St. Lawrence River Basin was met with tempered enthusiasm. Many Québécois feel that they have no need for gas; that unlike hydroelectric power it is nonrenewable energy; that unlike hydropower it will heighten carbon emissions; and that gas exploitation carries risks of water pollution. Not surprisingly, these views are disputed by energy-industry interests pushing for extraction of this gas. It remains to be seen if drilling can begin in the foreseeable future.

An unmistakably French cultural landscape in North America: the Rue Saint-Louis in Quebec City, the capital of Quebec Province. Not an English sign in sight in this center of French/Québécois culture. The green spires in the background belong to Le Chateau Frontenac, a grand hotel built on the site of the old Fort St. Louis (the Count of Frontenac was a prominent governor of what was then known as New France). French-speaking, Roman Catholic Quebec lies at the heart of French Canada.

© Andre Jenny/Alamy Limited

■■ THE SOUTHEAST (4)

For more than a century following the Civil War, the Southeast (Figs. 3B-1, 3B-5) remained in economic stagnation and cultural isolation from the rest of the country. In the 1970s, however, things changed so fast that a New South arose almost overnight. The Sunbelt migration drove people and enterprises into the long-dormant cities. Core-region companies looking for headquarters or subsidiary offices found Atlanta, Charlotte, Miami-Fort Lauderdale, Tampa, and other urban areas economical and attractive, turning them into boomtowns virtually overnight. Racial segregation had been dismantled in the wake of the civil rights movement. A new social order was matched by new facilities ranging from airports (Atlanta's quickly became one of the world's busiest) to theme parks open to all. Soon the nation was watching Atlanta-based CNN on television and following space flights that originated at Cape Canaveral.

The geography of development in the South, however, is very uneven. While many cities and some agricultural areas have benefited, others have not; the South contains some of the realm's poorest rural areas, and the gap between rich and poor here is wider than in any other region in the realm. Virginia's Washington suburbs, North Carolina's Research Triangle, Tennessee's Oak Ridge complex, and Atlanta's corporate campuses are locales that typify the New South; but Appalachia and

AMONG THE REALM'S GREAT CITIES . . .

MONTREAL

© H. J. de Blij, P. O. Muller, and John Wiley & Sons, Inc.

MONTREAL LIES ON a large triangular island in the St. Lawrence River, its historic core laid out along the base of what Montrealers call "the mountain"—flat-topped Mount Royal, from which the city takes its name. Montreal is the cultural capital of French Canada, which with 3.8 million people is the largest French-speaking city in the world after Paris, and therein lie its opportunities and challenges. Without a doubt, this is still one of the realm's most cosmopolitan cities, a strong European flavor pervading its stately CBD, its lively neighborhoods, and its bustling street life. To escape the long, harsh winters here, much of downtown Montreal has become an underground city with miles of handsome passageways lined with shops, restaurants, and cinemas, all of it served by one of the world's most modern and attractive subway systems.

Yet despite the sparkle and continental ambience, Montreal has not been a happy place in recent times. With the escalation of tensions between Francophone and Anglophone Canada over the past four decades, Quebec's largest metropolis became a crucible of confrontation. For most of that period, the ethnolinguistic division was most visible along the Boulevard St. Laurent, the longstanding boundary between the French-speakers of Montreal's East End (who constitute about 67 percent of the city's population) and the English-speakers who clustered in the West End (the remnant of a larger Anglophone community that was reduced by 40 percent when the French-supremacy language laws were implemented). These days, however, things are improving as linguistically-mixed communities proliferate throughout Montreal, driven by an influx of immigrants from outside North America who settle in both English- and French-speaking neighborhoods. At the same time, polarization continues to abate as stores, newspapers, and universities cater to increasingly bilingual clienteles, and intermarriage between Anglophones and Francophones is on the rise.

Montreal's economy has also been affected. During the 1990s, hundreds of Anglophone firms (and thousands of their workers) fled to Ontario and points west, leaving in their wake a commercial real estate tailspin, rising unemployment, and a noticeable decline in the city's prosperity. Today, with the secession issue no longer on the front burner in Quebec, Montreal is steadily progressing in its comeback. Tourists have returned in droves; new foreign-trading ties, especially with the United States, are reviving the business community; and vibrant suburbs are leading the metropolis to reinvent itself as a hub for information technology, telecommunications, and biopharmaceutical companies.

rural Mississippi still represent the Old South, where depressed farming areas and stagnant small industries restrict both incomes and change.

Climatically, the South is warmer than the Core region to the north and more humid than the Southwest. It is not for nothing that this is where the tobacco, cotton, and sugar plantations thrived. But the summer heat brings a serious threat to the Gulf-Atlantic Coastal Plain of the South: hurricanes capable of inflicting catastrophic devastation on low-lying areas. These tropical cyclones are fueled by the rapidly rising air over the warm waters of the Gulf of Mexico and the Caribbean that reach peak temperatures in late summer. Coastal Florida and southern Louisiana have been particularly hard hit in recent decades.

A major disaster of another kind hit the Gulf Coast in 2010: an explosion on April 20 at a British Petroleum rig about 100 kilometers (60 mi) southeast of the Mississippi Delta caused the worst oil spill in U.S. history. It took BP five months to seal the resulting seafloor leak, but not before an estimated 4.9 million barrels of crude oil spewed into the Gulf, with much of it washing ashore from western Louisiana all the way east to Florida's Panhandle (Fig. 3B-5; photo, p. 182). Investigations into the damage to wildlife, fisheries, and local industries (including tourism) were still in progress at press time. This environmental disaster exposed the enormous risks of deep-sea oil extraction, and the prospects for expanding offshore drilling elsewhere in U.S. waters were dealt a major setback.

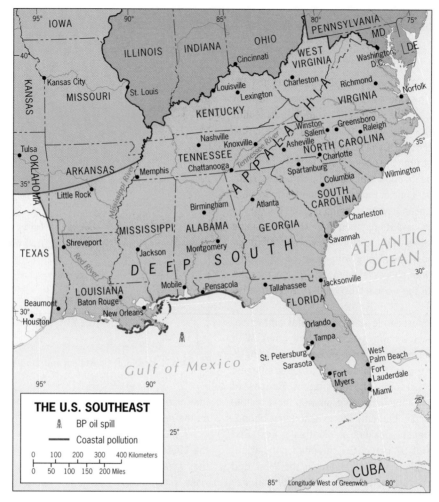

FIGURE 3B-5 © H. J. de Blij, P. O. Muller, J. Nijman, and John Wiley & Sons, Inc.

There is another, relatively small part of this region that warrants special attention. South Florida used to be a highly remote and nearly unpopulated corner of the United States. Before the twentieth century, it was too far from the northeastern core of the country (and even from closer urban centers like Atlanta) to warrant development; the climate and mosquitoes posed a formidable challenge to prospective settlers; and there were no natural resources to exploit. Indeed, it was not until the 1960s that this subregion of the Southeast became fully integrated into the national economy and urban system—thanks to the widespread introduction of air conditioning and affordable air travel, not to mention the mass immigration of Cubans who fled Fidel Castro's communist regime. In short order, now-booming South Florida became part of an international region that encompassed the Caribbean Basin and much of Middle and South America. Located south of the American "Deep South," today it continues to evolve into a transition zone between North America and the rest of the hemisphere.

South Florida's largest urban center, Miami, emerged in the 1980s as an important *world city*

Oil-drilling always poses ecological risks, and the massive blowout of BP's notorious Deepwater Horizon well in the Gulf of Mexico just south of the Mississippi Delta on April 20, 2010 quickly became one of the worst environmental disasters in American history. This appalling scene is the shoreline of Louisiana's Grand Isle, the barrier island on the west side of the delta, taking the brunt of the first onslaught of oil about a month into the spill.

© Julie Dermansky/©Corbis

From the Field Notes . . .

"Domino Park lies at the center of Little Havana, perhaps the most iconic neighborhood in Greater Miami. When Fidel Castro came to power in 1959, thousands of Cubans fled the

© Jan Nijman

island and resettled here. Today there are more than 800,000 Cubans in the metropolitan area, the biggest concentration of any single ethnic group in the United States. In fact, Miami would qualify as the second largest city in Cuba. Domino Park is situated on Southwest 8th Street, better known locally as 'Calle Ocho.' It is only a few blocks from the Bay of Pigs Monument that commemorates the fallen soldiers of the failed invasion of Cuba by exiles in 1961. Taking a stroll along Calle Ocho, you can stop at a cafe for a *cortadito*, watch cigars being made by hand, or get fitted for a guayabera shirt. Domino Park recently opened to women (!), but it is mostly older men who come here to play games, talk politics, and meet with friends—just as they would in Cuba."

www.conceptcaching.com

and is often viewed as a sort of safety valve between the hemisphere's north and south. This new role is certainly a result of the growth of its large, well-educated Hispanic population, but that is not the entire story. Take another look at the world map in Figure G-1 and note Miami's longitude: it is not only located at the southern end of North America's Atlantic seaboard, but also lies on the same meridian as Guayaquil, Ecuador, the westernmost major city in South America. Now compare Miami's relative location to that of Los Angeles (three time zones to the west) and it becomes clear why Miami is far better connected to countries such as Brazil, Colombia, and Venezuela—while L.A.'s southern linkages are almost entirely with Mexico. In fact, Los Angeles is mainly a part of the Pacific Rim, whereas Miami is the realm's most important world city focused on the Americas.

■■ THE SOUTHWEST (5)

The Southwest has taken on its own regional identity in recent years, and today it is firmly established in the North American consciousness for several reasons. One is environmental: here, steppes and deserts dominate. Another is cultural: Anglo, Hispanic, and Native American cultures coexist here, sometimes amicably, at other times uneasily.

As Figure 3B-1 shows, by our definition the Southwest begins in eastern Texas and extends all the way to the eastern edge of Southern California. Texas leads this region in most respects. Its economy, once dependent on oil and natural gas, has been restructured and diversified so that today the Dallas-Forth Worth–Houston–San Antonio triangle has become one of the world's most productive **technopoles [7]** (state-of-the-art, high-technology industrial complexes) including Austin at the heart of it. This is also a hub of international trade and the northern anchor

VOICE FROM THE

Region

© Cinthia Carvajal
Cinthia Carvajal,
Fourth-year student
in Anthropology and
Sustainability, Arizona
State University

ILLEGAL IMMIGRATION IN ARIZONA

Senate Bill 1070 was signed into law in Arizona in 2010. This law was created to reduce illegal immigration in the State, where crossing the international border without documents has occurred for years. As a Hispanic immigrant living in Arizona, I can see how the law has affected friends, family, and other people around me. Basically, it has caused polarization in Arizona at a scale not seen before. Conversations about immigration and the new law are not pleasant. The law has divided the State, has divided its classrooms, and has divided people who once coexisted in peaceful and respectful ways. Those who support the law are convinced it is the best solution and has no effects on documented and law-abiding immigrants—that it just targets those who are undocumented. Others oppose the law; they are people of many races and ethnicities, but most are documented immigrants, or children of immigrants whose background can be seen in their faces or heard in their accents. People like me. The ambiguity of the law has caused many to leave the State. Those of us who stay behind live every day with that little doubt in our minds: who is going to be the next person to doubt my 'legality'?

of a NAFTA-generated transnational growth corridor that extends into Mexico as far as Monterrey.

New Mexico, in the middle of the Southwest region, is the least developed economically and ranks low on many U.S. social indices, but its environmental and cultural attractions benefit Albuquerque and Santa Fe, with Los Alamos a famous name in research and development. In the west, Arizona's technologically transformed desert now harbors two large coalescing metropolises, Phoenix and Tucson. Growth continues, but concerns about climate change add to the uncertainties of water supplies already stretched to the limit.

The huge State of Texas, larger than France, is surpassed in areal size only by Alaska and is also the second most populous U.S. State after California (26 and 37 million, respectively). It is a most affluent State, with much of its wealth still emanating from its energy sector: Texas claims about a quarter of all U.S. oil reserves and nearly a third of its natural gas deposits. Today, the Lone Star State has become a leader in the generation of wind energy as well, with wind turbines dominating much of the sparsely settled landscape of West Texas.

Texas and the other southwestern States border Mexico, and all are home to substantial Mexican populations, both legal and illegal. In Arizona, the Hispanic population has nearly tripled since 1990 as the majority of non-Hispanic whites dwindled from 72 to 57 percent. Among the elderly 83 percent are non-Hispanic whites, but the State's 18-and-younger age cohort is now 44 percent Hispanic. In 2010, Arizona passed a widely debated (and criticized) law making it mandatory for people to carry proof of legal residency, failure of which results in detention and presumably deportation. Much of the debate centered on the apparent inevitability of **racial profiling [8]** that this would involve (who will be asked to produce such documentation?). Legal challenges and appeals have been filed, but the mood among Arizonans seemed clear: 81 percent of registered voters were in favor of the law. By early 2011, an estimated 100,000 Latinos had already departed for other States or had returned to Mexico.

▪▪ THE PACIFIC HINGE (6)

The Southwest meets the Far West near metropolitan San Diego on California's border with Mexico, and from there the region we call the Pacific Hinge extends all the way north into Canada where Vancouver forms the northern anchor in southwestern British Columbia (Fig. 3B-1). We include almost all of California in this region, but (for environmental as well as economic reasons) only the western portions of Oregon and Washington State. This is a particularly important part of North America and increasingly a counterweight to the historic Core in the U.S. Northeast.

In many parts of the world, this region would constitute a top-ranked core area, with such leading cities as San Diego, Los Angeles, San Francisco, Portland, and

Seattle plus America's most populous State: in fact, California's economy in recent times has ranked among the world's ten largest—*by country*. The Pacific Hinge also includes one of the realm's most productive agricultural areas in California's Central Valley, magnificent scenery, and a culturally diverse population drawn by its long-term economic growth and pleasant living conditions.

Our regional definition is mainly based on economic considerations and intensifying trade connections across the Pacific Ocean. This part of North America is deeply involved in the development of countries in eastern Asia. In the postwar era, Japan's success had a salutary impact here, but what has since taken place in China, South Korea, Taiwan, Vietnam, Singapore, and other Asian-Pacific economies has created unprecedented opportunities. The term **Pacific Rim [9]** has come into use to describe the discontinuous regions surrounding the great Pacific Ocean that have experienced miraculous economic growth and progress over the past three decades: coastal China and various parts of East and Southeast Asia but also Australia, South America's Chile, and the western shores of Canada and the United States. That is why in this chapter we speak of the *Pacific Hinge* of North America, now the key interface between the North American realm and the Pacific Rim.

The Pacific Rim is therefore a good example of a functional region, with economic activity in the form of capital flows, raw-material movements, and trading linkages that generate urbanization, industrialization, and labor migration. As part of this process, human landscapes from Sydney to Santiago are being transformed within a 32,000-kilometer (20,000-mi) corridor that lines the globe's largest body of water.

Greater Los Angeles, the region's largest and most dynamic activity complex, is profiled in the box to the right. Farther north, the San Francisco Bay Area has seen rather more orderly expansion, with the Silicon Valley technopole the key component of its high-technology success story. In what Americans refer to as the Pacific Northwest, the Seattle-Tacoma area originally benefited from cheap hydroelectricity generated by Columbia River dam projects, attracting aluminum producers and aircraft manufacturers (especially Boeing). Meanwhile, the technopole centered on suburban Redmond (Microsoft's headquarters) heralds the Seattle area's transformation into a prototype metropolis of the new information economy.

Nonetheless, at the time of this writing, the west-coast States were struggling through a tough economic recession, particularly California. Unemployment there was just under 12 percent during the spring of 2011 (second highest in the United States), and the Golden State was one of the hardest hit by the still-ongoing housing crisis (see the discussion of foreclosures on pages 164–165). Moreover, California's government was running a deficit in excess of $9 billion (nearly 10 percent of the budget), making extremely painful expenditure and services cuts unavoidable.

LOS ANGELES

IT IS HARD to think of another city that has been in more headlines in the past several years than Los Angeles. Much of this publicity has been negative. But despite its earthquakes, mudslides, wildfires, pollution episodes showcase trials, and riots, L.A.'s glamorous image has hardly been dented. This is compelling testimony to the city's resilience and vibrancy, and to its unchallenged position as the Western world's entertainment capital.

The plane-window view during the descent into Los Angeles International Airport, almost always across the heart of the metropolis, gives a good feel for the immensity of this urban landscape. It not only fills the huge natural amphitheater known as the Los Angeles Basin, but it also oozes into adjoining coastal strips, mountain-fringed valleys and foothills, and even the margins of the Mojave Desert more than 90 kilometers (60 mi) inland. This quintessential spread city, of course, could only have materialized in the automobile age. In fact, most of it was built rapidly since 1920, propelled by the swift postwar expansion of a high-speed freeway network unsurpassed anywhere in metropolitan America. In the process, Greater Los Angeles became so widely dispersed that today it has reorganized within a sprawling, multinodal geographic framework (see Fig. 3A-11 on p. 164).

The metropolis as a whole is home to 18.2 million Southern Californians, constitutes North America's second largest urban agglomeration, and forms the southern anchor of the huge California megalopolis that parallels the Pacific coast from San Francisco southward to the Mexican border.

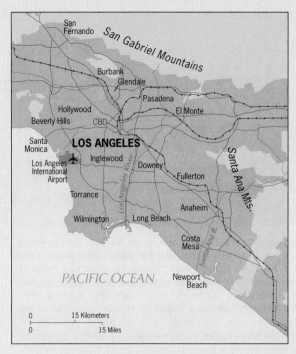

© H. J. de Blij, P. O. Muller, and John Wiley & Sons, Inc.

It also is the Pacific Hinge's leading trade, manufacturing, and financial center. In the global arena, Los Angeles is the eastern Pacific Rim's dominant city and the origin of the greatest number of transoceanic flights and sailings to Asia.

Vancouver, located at the northern edge of the Pacific Hinge, has the locational advantage of being closer, on air and sea routes, to East Asia than any other major North American city. Ethnic Chinese here constitute almost 20 percent of the metropolitan population, and the total number of resident Asians now approaches 50 percent.

▪▪ THE WESTERN FRONTIER (7)

Where the forests of the Pacific Hinge yield to the scrub of the rain shadow on the inland slopes of the mountain wall that parallels the west coast, and the terrain turns into intermontane ("between the mountains") basins and plateaus, lies a region stretching from the Sierra Nevada and Cascades to the Rockies, encompassing parts of southern Alberta and British Columbia, eastern Washington State and Oregon, all of Nevada, Utah, and Idaho plus the western segments of Montana, Wyoming, and Colorado. As Figure 3B-1 shows, Edmonton, Calgary, and Denver are situated along the inland edge of this aptly named Western Frontier region, and

Salt Lake City, anchoring the Software Valley technopole and symbolizing the new high-tech era, lies at the heart of it.

Remoteness, dryness, and sparse population typified this region for a very long time. It became the redoubt of the Mormon faith and a place known for boom-and-bust cycles of mining, logging, and, where possible, livestock raising. However, in recent decades advances in communications and transportation technologies, sunny climates, wide open spaces, lower costs of living, and growing job opportunities have combined to form effective pull factors for myriad outsiders. Every time an earthquake struck in California, eastward migration got a boost. And during the most recent ten-year period, millions of Californians emigrated from the overcrowded, overpriced Pacific coast to the high desert.

Development in the Western Frontier centers on its widely dispersed urban areas, which are making this the realm's fastest growing region (albeit from a low base). Here thousands of new high-technology manufacturing and specialized service jobs propelled a two-decade-long

For years the hottest growth area in the Western Frontier region was its southernmost corner—the Las Vegas Valley. Nevada was one of America's fastest-growing States, attracting new residents by the hundreds of thousands and visitors to its burgeoning recreation industry by the tens of millions annually. Between 1990 and 2000, this Las Vegas suburb of Henderson ranked as the fastest-growing municipality in the United States. Sprawling development on a massive scale converted this patch of high desert into a low-density urban checkerboard whose expansion seemed without end. But by 2009 the growth cycle had ground to a halt as thousands of homeowners were forced into foreclosure, unemployment was rising, and national and global economic crises were clouding the future of Boomtown USA.

© Ethan Miller/Getty Images, Inc.

influx, slowed but not reversed by the economic crisis that began in 2008. Yet this growth has not come without problems as locals see land values rise (see photo above), conservative ideas challenged, and traditional lifestyles endangered.

One of the fastest-growing metropolitan areas in North America, Las Vegas, lies on the border between the Western Frontier and Southwest regions. Far more than a gambling and entertainment center that welcomes some 40 million visitors annually, Las Vegas attracts immigrants because of its history of job creation, relatively cheap land and low-cost housing, moderate taxes, and sunny if seasonally hot weather. No doubt, proximity to Southern California has had much to do with its explosive growth, but Las Vegas is also known far and wide as the ultimate frontier city—where (almost) anything goes.

▪▪ THE CONTINENTAL INTERIOR (8)

The Western Frontier meets the Continental Interior region along a line marked by cities that influence both regions, from Denver in the south to Edmonton in the north. This vast North American heartland extends from interior Canada to the borders of the Southeast and Southwest

(Fig. 3B-1). This region contains many noteworthy cities, including Kansas City, Omaha, Minneapolis, and Winnipeg, but agriculture is the dominant story here. As the map of farm resource regions (Fig. 3B-6) shows, this is North America's breadbasket. This is the land of the beef- and pork-producing Corn Belt (*Meat Belt* would be a more accurate name), which extends into the western half of the Core region; of the mighty soybean, America's and the world's most rapidly expanding crop of the past half-century; and of spring wheat in the Dakotas and Canada's Prairie Provinces as well as winter wheat in Kansas. Indeed, the cities and towns of this region share histories of food processing, packing, and marketing; flour milling; and soybean, sunflower, and canola oil production. And this is also the scene of the struggle for survival of the family farm as the unrelenting incursion of large-scale corporate farming threatens a longstanding way of life.

This may be farmland, but it is not immune from what might seem to be the nonagricultural realities of the contemporary world, not even the energy issue. Corn continues to be grown mainly as feed for meat-producing hogs and beef cattle, but a sizeable proportion of the crop is now dedicated to making ethanol, the gasoline substitute (or additive). Whereas farmers hope that rising prices will revitalize the Corn Belt, others worry that rising corn prices will increasingly strain meat-loving consumers. In any case, the amount of corn needed to make a real dent in U.S. gasoline consumption is many times greater than the entire annual crop. In 2007, when almost half of the corn grown in the Midwest was converted into biofuels, ethanol accounted for only a paltry 4 percent of national gasoline consumption while raising alarms about possible food shortages.

Even a cursory look at demographic statistics makes it clear why many States in the Continental Interior are losing population or gaining far fewer than the national average. The Great Plains, the western component of this region, is especially hard hit: younger people and those better off are leaving, whereas older and less affluent residents stay behind. Villages and small towns continue to die, and the notion of abandoning certain areas to return them to their natural state is being seriously discussed.

But things are not the same everywhere. The Continental Interior may not have as prominent a presence on the national scene as New York or California, and the region is not as wealthy, but Wyoming and North Dakota had the highest increases in household income in the United States during the first decade of this century (at 8

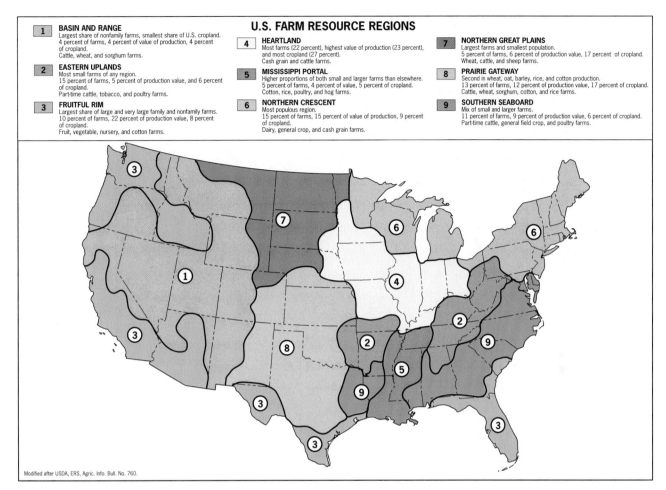

U.S. FARM RESOURCE REGIONS

1 BASIN AND RANGE
Largest share of nonfamily farms, smallest share of U.S. cropland.
4 percent of farms, 4 percent of value of production, 4 percent of cropland.
Cattle, wheat, and sorghum farms.

2 EASTERN UPLANDS
Most small farms of any region.
15 percent of farms, 5 percent of production value, and 6 percent of cropland.
Part-time cattle, tobacco, and poultry farms.

3 FRUITFUL RIM
Largest share of large and very large family and nonfamily farms.
10 percent of farms, 22 percent of production value, 8 percent of cropland.
Fruit, vegetable, nursery, and cotton farms.

4 HEARTLAND
Most farms (22 percent), highest value of production (23 percent), and most cropland (27 percent).
Cash grain and cattle farms.

5 MISSISSIPPI PORTAL
Higher proportions of both small and larger farms than elsewhere.
5 percent of farms, 4 percent of value, 5 percent of cropland.
Cotton, rice, poultry, and hog farms.

6 NORTHERN CRESCENT
Most populous region.
15 percent of farms, 15 percent of value of production, 9 percent of cropland.
Dairy, general crop, and cash grain farms.

7 NORTHERN GREAT PLAINS
Largest farms and smallest population.
5 percent of farms, 6 percent of production value, 17 percent of cropland.
Wheat, cattle, and sheep farms.

8 PRAIRIE GATEWAY
Second in wheat, oat, barley, rice, and cotton production.
13 percent of farms, 12 percent of production value, 17 percent of cropland.
Cattle, wheat, sorghum, cotton, and rice farms.

9 SOUTHERN SEABOARD
Mix of small and larger farms.
11 percent of farms, 9 percent of production value, 6 percent of cropland.
Part-time cattle, general field crop, and poultry farms.

Modified after USDA, ERS, Agric. Info. Bull. No. 760.

FIGURE 3B-6

© H. J. de Blij, P. O. Muller, and John Wiley & Sons, Inc.

and 7.4 percent, respectively—while the national figure was *negative* 7.1 percent). Moreover, unemployment rates here rank among the lowest in the country. With an economy largely driven by agriculture and mining, and with oil and gas exploration a leading growth industry, the effects of the Great Recession and its accompanying housing crisis were minimal in this region.

▪▪ THE NORTHERN FRONTIER (9)

Figure 3B-1 leaves no doubt as to the dimensions of this final North American region: it is by far the largest of the realm, covering nearly 90 percent of Canada and all of the biggest U.S. State, Alaska. Not only does this region include the northern parts of seven of Canada's provinces, but it also comprises the Yukon and Northwest Territories as well as recently established Nunavut. The sparsely peopled Northern Frontier remains a land of isolated settlement based on the exploitation of newly discovered resources. The Canadian Shield, underlying most of the eastern half of the region, is a rich storehouse of mineral resources, including metallic ores such as nickel, uranium, copper, gold, silver, lead, and zinc. The Yukon and Northwest Territories have proven especially bountiful, with

gold- and diamond-mining (Canada now ranks among the world's top five producers); at the opposite, eastern edge of the Shield, near Voisey's Bay on the central coast of Labrador, the largest body of high-grade nickel ore ever discovered is now open for business.

The province of Alberta is experiencing once of the most spectacular economic booms in the realm's history. Recent estimates suggest that Alberta's oil reserves, including the **tar sands [10]** around Fort McMurray (see Fig. 3B-7 and the photo on page 189), are second only to those of Saudi Arabia, and possibly even larger. Alberta has become Canada's fastest-growing province, and Calgary its fastest-expanding major city. The boom makes Calgary a magnet for business and highly-skilled workers; the city's skyline is in spectacular transformation, comparable to what has recently occurred on the Arabian Peninsula. The big question is how long these good times will last.

Productive locations in the Northern Frontier comprise a far-flung network of mines, oil and gas installations, pulp mills, and hydropower stations that have spawned hundreds of small settlements as well as thousands of kilometers of interconnecting transportation and communications lines. Inevitably, these activities have infringed on the lands of indigenous peoples without the preparation of

FIGURE 3B-7   © H. J. de Blij, P. O. Muller, J. Nijman, and John Wiley & Sons, Inc.

treaties or agreements, which has led to recent negotiations between the government and leaders of First Nations over resource development in the Northern Frontier. But essentially this vast region remains a frontier in the truest sense of that term.

Alaska's regional geography differs from the rest of the Northern Frontier in that the State contains several urban settlements and an incipient core area in the coastal zone around Anchorage. Alaska's population, which is now nearing 750,000, accounts for more than one-third of the region's total. Here also lies the only metropolis of any size, Anchorage (300,000). The State's internal communications (air and surface) are also better developed.

Another difference lies in Alaska's North Slope oil exploitation, one of the hemisphere's key energy sources since the 1300-kilometer (800-mi) Trans-Alaska Pipeline began operating in 1977. The North Slope is the area north of the Brooks Range that lies along the Arctic Ocean. Dwindling supplies at the main reserve at Prudhoe Bay are now compelling producers to turn to huge additional reserves in Alaska's northland, but opposition from preservationist organizations has slowed their plans. The issue remains in contention—as is another project to construct a natural gas pipeline to parallel the oil pipeline to send gas, a byproduct of oil drilling, to the growing market in the Lower-48.

Climate change is likely to affect the Northern Frontier in as-yet unforeseeable ways. As the map shows, this region extends all the way north to Ellesmere Island and thus includes waters that, in the event of longer-term sustained global warming and associated recession of Arctic ice, will open, possibly year-round, to create natural waterways that would dramatically alter shipping routes and intercontinental distances (a topic discussed in Chapter 12). The question of ownership of these waters—and the submerged seafloor beneath them, coming within reach of exploitation—will be a matter of significant international contention in the decades ahead.

In 2010, Canada solidified its reputation as an environmentally conscious nation through the signing of an unprecedented agreement. The Canadian Boreal Forest Agreement involves 21 logging firms and 9 environmental organizations as well as the Canadian government. As indicated in Figure 3B-7, it imposes a moratorium on logging and any other development in a vast zone of 290,000 square kilometers (112,000 sq mi). Greenpeace, which originated in Canada, called it the all-time biggest agreement to preserve nature. The word "boreal" is derived from the Latin word for north; **boreal forests [11]**—identical to the Russian *taiga* (snowforest) discussed on page 109—are dominated by dense stands

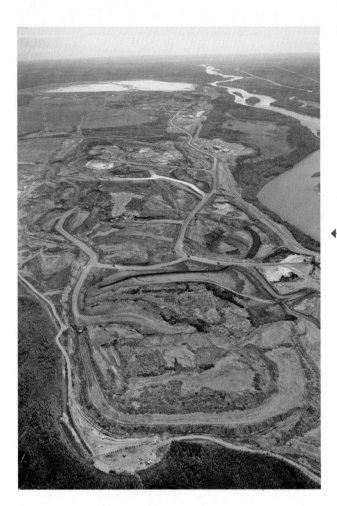

As the price of a barrel of oil (and a gallon of gasoline) rises, it becomes profitable to derive oil from sources other than liquid reserves. The Canadian province of Alberta contains vast deposits of "oil sands" in which the petroleum is mixed with sand, requiring a relatively expensive and complicated process to extract it. The quantity of oil locked in these Athabasca Tar Sands is estimated to constitute one of the world's largest reserves, and this huge open-pit mining project is under way around the town of Fort McMurray to recover it. Given the enormous size of the area to be mined, exploitation is not without its environmental critics. Those who assume that Canada will sell this oil to the United States should be aware that the Chinese are offering to fund construction of a pipeline across the mountains to Canada's Pacific coast; the rising cost of oil has political as well as economic ramifications.

© Jim Wark

POINTS TO PONDER

- Arizona's tough new laws on illegal immigration have caused an estimated 100,000 Hispanics (illegal and legal) to leave the State through mid-2011, with thus far unknown economic consequences.

- Nearly 700,000 foreign students attend American colleges and universities annually, and one-third of them in academic year 2009–10 came from just two countries—China and India.

- In 2010, 72 percent of all mortgaged homes in Las Vegas had market values that were less than the balance of the mortgage.

- Counting all sources—onshore as well as offshore, proven oil and gas reserves, and particularly Alberta's tar-sands deposits—Canada now ranks first among the world's states in actual and potential fossil-fuel energy resources.

of coniferous (cone-bearing) needleleaf trees such as spruce, fir, and pine. The boreal forests of the Northern Frontier also constitute the habitat of such endangered wildlife species as the caribou, lynx, American black bear, and wolverine.

using GEOGRAPHY

1. Imagine that you are the chief executive of a successful young company that manufactures software for the newest high-speed computers. Where would you locate your company and why? Establish a "short list" of three possible sites, and choose one by a process of elimination in which your decision is based upon the most important locational variables for your plant and its highly-skilled workforce.

2. If you could pick a job somewhere today, how would housing issues affect your choice as to where to settle? Given the choice, would you buy or rent?

3. Does immigration affect your daily life, and if so, how? How might it affect you when you are ready to enter the job market following completion of your education?

4. We have discussed urban regions. Use Google Earth to zoom in and out of large metropolitan areas such as Los Angeles or Miami-Fort Lauderdale and notice how various cities and towns have coalesced into megalopolises. Now do the same for the Bosnywash Megalopolis corridor, particularly where it crosses New Jersey and Connecticut. Do you see how difficult it has become to delineate cities and individual urban regions?

Channel Is.
120°
Salton Sea
Colorado
110°
SONORAN DESERT
MOUNTAINS
Amarillo
100°
Oklahoma City
Arkansas

SAN DIEGO
TIJUANA
Mexicali
PHOENIX
ROCKY
Rio Grande
Lubbock
Little Rock
Mississippi

Ensenada
Tucson
U N I T E
Fort Worth
DALLAS
Jacks

30°
Cape San Quintin
Nogales
Las Cruces
El Paso
CIUDAD JUÁREZ
S T A T
Austin
Baton Rouge
Beaumont

Guadalupe (MEXICO)
BAJA CALIFORNIA
Tiburón Island
Hermosillo
Chihuahua
EDWARDS PLATEAU
Del Rio
Ciudad Acuña
Rio Grande
SAN ANTONIO
HOUSTON
Galveston
Ne Or

Archangel Island
Gulf of California (Sea of Cortés)
C O A S T A
MISS RIVER

Cedros Island
Eugenia Point
Los Mochis
Culiacán
Nuevo Laredo
Laredo
Corpus Christi
Gul Mex

Tropic of Cancer
La Paz
Durango
Torreón
Saltillo
MONTERREY
Guadalupe
Reynosa
McAllen
Brownsville
Matamoros

False Cape
Cabo San Lucas
Mazatlán
Zacatecas
SIERRA MADRE ORIENTAL
Ciudad Victoria

20°
Marias Islands
Tepic
San Luis Potosí
Aguascalientes
M E X I C O
Tampico
Bay of Campeche
Campeche
YU

Revillagigedo Islands (MEXICO)
Cape Corrientes
GUADALAJARA
Puerto Vallarta
LEÓN
Guanajuato
Querétaro
MEXICO CITY
Pachuca
Poza Rica
Xalapa

Colima
Morelia
Toluca
Popocatépetl (Vol.) (5,465 m, 17,930 ft.)
PUEBLA
Pico de Orizaba (Vol.) (5,747 m, 18,855 ft.)
Veracruz
Ciudad del Carmen
Ch

SIERRA MADRE DEL SUR
Chilpancingo
Tuxtepec
Coatzacoalcos
Villahermosa
Ciudad Pemex
Monclova
PETÉN

Acapulco
Oaxaca
ISTHMUS OF TEHUANTEPEC
Tuxtla Gutiérrez
Be

Salina Cruz
Gulf of Tehuantepec
GUATEMA

Tapachula
Quetzaltenange
Guate City

10°
Santa Ana
San Salvador
EL SALVAD

P A C I F I C

O C E A N

Elevation (m)

3000
1500
600
300
150
0

below sea level
0
-150
-1500
-3000
-6000

Roads
Railroads

0 km   200   400   600   800
0 miles 100   200   300   400   500
Albers Equal-Area Projection
Scale 1:15,000,000

Archipiélago de (Galápagos Islan
(ECUADOR)
Pinta I.
Marchena

FIGURE 4A-1   © H. J. de Blij, P. O. Muller, and John Wiley & Sons, Inc.

L ook at a world map, and it is obvious that the Americas comprise two landmasses: North America extending from Alaska to Panama and South America from Colombia to Argentina. But here we are reminded that continents and geographic realms do not necessarily coincide. In Chapters 3A and 3B we discussed a North American realm whose southern boundary is the U.S.-Mexican border and the Gulf of Mexico. Between North America and South America lies the small but important geographic realm known as Middle America. Consisting of a mainland corridor and numerous Caribbean islands, Middle America is a highly fragmented realm.

From Figure 4A-1 it is clear that Middle America is much wider than it is long. The distance from Baja California to Barbados is about 6000 kilometers (3800 mi), but from the latitude of Tijuana to Panama City is only half that distance. In terms of total area, Middle America is the second-smallest of the world's geographic realms. As the map shows, the dominant state of this realm is Mexico, larger than all its other countries and territories combined.

Middle America may be a small geographic realm by global standards, but comparatively it is densely peopled. Its population in 2012 was exactly 200 million, more than half of it in Mexico alone, and the rate of natural increase (at 1.6 percent) remains above the world average. Figure 4A-2 shows Mexico's populous interior core area quite clearly, but note that the population in Guatemala and Nicaragua tends to cluster toward the Pacific rather than the Caribbean coast. Among the islands, this map reveals how crowded Hispaniola (containing Haiti and the Dominican Republic) is, but at this scale we cannot clearly discern the pattern on the smaller islands of

major geographic qualities of
MIDDLE AMERICA

1. Middle America is a relatively small realm consisting of the mainland countries from Mexico to Panama and all the islands of the Caribbean Basin to the east.

2. Middle America's mainland constitutes a crucial barrier between Atlantic and Pacific waters. In physiographic terms, this is a land bridge connecting the continental landmasses of North and South America.

3. Middle America is a realm of intense cultural and political fragmentation. The presence of many small, insular, and remote countries poses major challenges to economic development.

4. Middle America's cultural geography is complex. Various African and European influences dominate the Caribbean, whereas Spanish and Amerindian traditions survive on the mainland.

5. The realm contains the Americas' least-developed territories. New economic opportunities may help alleviate Middle America's endemic poverty.

6. In terms of area, population, and economic potential, Mexico leads the realm.

7. Mexico and Panama are uniquely connected beyond the realm, the former because of its border with the United States and the latter because of the Panama Canal; the rest of the realm is relatively isolated.

the eastern Caribbean and the Bahamas, some of which are also densely populated.

What Middle America lacks in size it makes up in physiographic and cultural diversity. This is a realm of soaring volcanoes and spectacular shorelines, of tropical forests and barren deserts, of windswept plateaus and scenic islands.

It holds the architectural and technological legacies of ancient indigenous civilizations. Today it is a mosaic of immigrant cultures from Africa, Europe, and elsewhere, richly reflected in music and the visual arts. Material poverty, however, is endemic: island Haiti is the poorest country in the Americas; Nicaragua, on the mainland, is almost as badly off. As we will discover, a combination of factors has produced a distinctive but challenged realm between North and South America.

GEOGRAPHICAL FEATURES

Sometimes you will see Middle and South America referred to in combination as "Latin" America, alluding to their prevailing Spanish-Portuguese heritage. This is an imperfect regional name, just as "Anglo-America," a term once commonly used for North America, was also improper. Such culturally based terminologies reflect historic power and dominance, and they tend to make outsiders out of those people they do not represent.

In North America, the term *Anglo* (as a geographic appellation) was offensive to many Native Americans, African Americans, Hispanics, Quebecers, and others. In Middle (and South) America, millions of people of indigenous-American, African, Asian, and European ancestries do not fit under the "Latin" rubric. You will not find the cultural landscape very "Latin" in the Bahamas, Barbados, Jamaica, Belize, or large parts of Guatemala and Mexico. So let us adopt the geographic neutrality of North, Middle, and South.

Automobiles wait to enter the United States at the Tijuana-San Ysidro border (San Diego). According to most accounts, this is the busiest border crossing in the world: here, every day some 50,000 vehicles and 25,000 pedestrians enter the United States and every year about 300 million people cross in either direction. In 2011 construction began to expand the number of inspection booths from 24 to 46 to reduce the waiting time for vehicles, usually 2-3 hours.

© David R. Frazier Photolibrary, Inc./Alamy

FIGURE 4A-2

© H. J. de Blij, P. O. Muller, and John Wiley & Sons, Inc.

But is Middle America sufficiently different from either North or South America to merit distinction as a realm? Certainly, in this age of globalization and migration, many border areas are becoming transition zones, as is happening along the U.S.-Mexican boundary. Still, consider this: North America encompasses only a pair of states, and the entire continent of South America only 12 (plus France's dependency on the northeast coast). But far smaller Middle America, as we define it, incorporates more than three-dozen political entities, including several dependencies (or quasi-dependencies) of the Netherlands, the United Kingdom, and France, but not counting constituent territories of the United States. Therefore, unlike South America, Middle America is a multilingual patchwork of independent states, territories in political transition, and residual colonial dependencies, with strong continuing ties to the United States and non-Iberian Europe. Middle America is defined in large measure by its vivid cultural-geographic pluralism.

The Realm's Northern Land Boundary

The 3169-kilometer (1969-mi) land border between North America and Middle America is the longest in the world separating a rich realm from a poor one. The U.S.-Mexican boundary crosses the continent from the Pacific to the Gulf, but Mexican cultural influences penetrate deeply into the southwestern States and American impacts reach far into Mexico. To Mexicans the border is a reminder of territory lost to the United States in historic conflicts; to Americans it is a symbol of economic contrasts and illegal immigration.

Along the Mexican side of this border the effects of **NAFTA** [1] (the North American Free Trade Agreement between Canada, the United States, and Mexico that went into effect in 1994) are transforming Mexico's economic geography.

During the early and mid-1990s, the implementation of NAFTA led to an economic boom as Mexico became part of a free-trade zone and market encompassing more than 400 million people. A major beneficiary was the strip of land that ran along the boundary with the United States: every day thousands of trucks and close to a million people crossed the international border in either direction, as did some $8 billion worth of legal commercial traffic. The resulting boom changed urban landscapes all along the emerging border zone—but it could not, of course, close the massive economic gap between the two sides.

Under NAFTA, factories based in Mexico could assemble imported, duty-free raw materials and components into finished products, which were then exported back to the U.S. market. Logically, most of these factories, called **maquiladoras** [2], located as close to the U.S. border as possible (see upper photo on next page). Thus manufacturing employment in the cities and towns along that border, from Tijuana on the Pacific to Matamoros at the mouth of the Rio Grande, expanded rapidly. By 2001, after only seven years of NAFTA's existence, there were some 4000 factories with more than 1.2 million workers in the border zone and in northern Yucatán (where Mérida was part of the process).

Mexico's economic linkages with its mighty northern neighbor have had a positive overall effect on the national economy, but they have increased regional inequalities and also made Mexico more vulnerable to economic

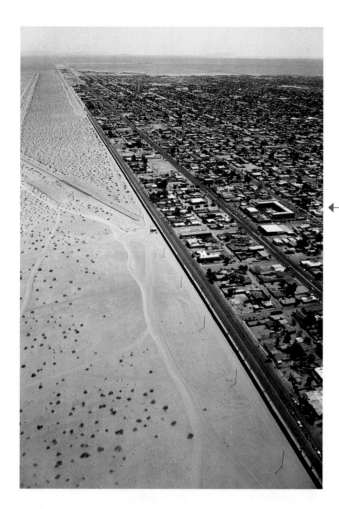

The boundary between Mexico and the United States displays some stunning cross-border contrasts. In this scene, the crowded Mexican town of San Luis Colorado ends at a tall metal fence, where the Arizona desert takes over. This photo, therefore, looks eastward; south is to the right. Like other Maxican towns pressed against the border, San Luis Rio Colorado (locals simply call it San Luis) grew rapidly and chaotically in response to the opening of several maquiladora assembly plants. Those factories make products for export to the United States, so that twinned towns along the border tend to be much larger on the Mexican than on the American side. San Luis, for example, has 160,000 inhabitants; San Luis Arizona, not visible in this picture, has fewer than 30,000 residents.

© Jim Wark/AirPhoto

States are countless. Over the past decade, American authorities have apprehended an average of about 1 million would-be, illegal migrants per year, many of them repeat offenders. Some were so desperate to make it across that they risked their lives: the number of deaths in the border zone since 2000 has varied between 200 and 500 annually. It is a painful reminder of the enormous gap in living standards and life opportunities between the two realms.

The cocaine trade, and to a lesser degree the smuggling of marijuana, is another major concern. Most cocaine is produced in South America, and the bulk of what is destined for the U.S. market passes through Mexico (see the map on page 217). Much of the marijuana consumed in the United States is grown in western Mexico and finds its way overland across the border as well. The contraband is carried in SUVs, trucks, or cars, often disguised as regular cargo in myriad ways: stuffed in teddy bears, inside hollowed out furniture, or simply in boxes labeled anything from candy to

crises in the United States. In 2008, Mexico entered a deep slump as exports to the Great Recession-plagued United States declined sharply; only at the end of 2010 did signs of recovery finally begin to appear. Some in Mexico measure their economic success in comparison with China: both countries rely on exports of labor-intensive products to the United States. Mexico now provides roughly 12 percent of all U.S. imports annually, the third-largest source after China (17 percent) and Canada (15 percent). But lately China has been increasing its share more rapidly, even though it faces tariffs that NAFTA-member Mexico does not. This puts pressure on Mexico to improve its competitiveness in order to build an even closer economic relationship with the United States.

A Troubled Border Zone

Despite the recent economic development of the borderlands, the boundary itself has become increasingly troubled since 2005, with the United States seeking to reinforce it to deter illegal immigration and the smuggling of drugs. Illegal crossings from Mexico into the United

On February 6th, 2010, groups of students in Mexico City staged a protest against the drug trafficking violence in Ciudad Juárez, one of Mexico's most violent places. More than 2600 people were murdered in Juárez in 2009, and in 2010 the number climbed to 3075—the most violent city in the hemisphere.

© ALEX CRUZ/epa/Corbis

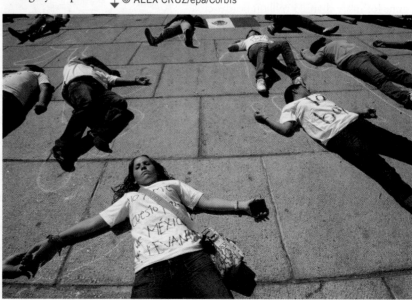

is the designated highway (indeed, the only expressway) east or west in the United States. As the power of Mexican organized crime syndicates has grown enormously over the past few years, the narcotics trade has experienced a commensurate, frightening increase of violence in northern Mexico.

Juárez itself recorded an astounding 3075 murders in 2010, the highest total of any city in the Americas; by early 2011, the level of violence had already driven away at least 15 percent of the civilian population and Juárez was increasingly referred to as a dying city. (We return to the topic of Mexico's drug wars at the national scale in Chapter 4B.)

In 2006, the U.S. government, responding to both homeland security considerations (potential terrorists are thought to enter the country via Mexico) and drug (mainly cocaine) smuggling, initiated the construction of a fortified fence along the entire length of the Mexican border. The very idea seems at odds with the ideals of closer economic and political cooperation, and many security experts question its efficacy. At any rate, in early 2011, with about one-third of the border (ca. 1000 kilometers [650 mi])

FIGURE 4A-3   © H. J. de Blij, P. O. Muller, J. Nijman, and John Wiley & Sons, Inc.

engine parts. It is the sheer volume of this cross-border traffic that makes it virtually impossible for authorities to conduct frequent inspections; and it is the infrastructure of the cities and industrial areas on the Mexican side of the border that facilitates the operation of the narcotics trade.

The city of Ciudad Juárez, probably the largest node in the hemispheric drug trade, is ideally located to handle this clandestine illicit activity. It is situated directly on the international boundary and forms a single built-up urban area with El Paso, Texas on the U.S. side, though the two cities are separated by a continuous artificial barrier running northwest-southeast across the entire metropolis (Figure 4A-3). There are four official ports of entry that together account for more than 23 million border crossings a year (see photo, page 193). Juárez has grown quite rapidly in the post–1993 NAFTA era, today exhibiting long stretches of manufacturing districts as well as a sprawling mosaic of neighborhoods, ranging from clusters of upper-income *colonias* to the teeming slums and shacktowns that house the poorest of the new migrants from the south. With a plentiful supply of poor people willing to take the risk of driving or carrying drugs across the border, Juarez also has dozens of maquiladoras that can be bribed to store, conceal, and transport the contraband north. With Mexican authorities having little or no control over the area, it is hardly surprising that local law enforcement personnel are often implicated in the drug business. Once the illegal goods reach El Paso across the border, Interstate-10

Hundreds of thousands of North American retirees, and many affluent purchasers of second homes, have sought the sun and low-cost residential opportunities of Middle America, converting some areas into virtual exclaves of the North. From Mexico to Panama, waterfront real estate is among the attractions for these permanent and seasonal migrants, as here in Cabo San Lucas at the southern tip of Maxico's Baja California peninsula.

© Liane Cary/Age Fotostock America, Inc.

hardened by walls and fences, the project to build a "virtual fence" to fill the remaining gaps had become such a boondoggle that it was cancelled. With the physical barriers concentrated around all the main crossing points, it is unclear how the economic health of the border zone they divide will be affected.

The Regions

As Figure 4A-1 suggests, Middle America can be divided into four distinct regions. Dominant **Mexico** occupies the largest part of the collective territory, with much of its northern border defined by the Rio Bravo (Rio Grande) from El Paso to the Gulf of Mexico. To the southeast, Mexico yields to the region called **Central America**, consisting of the seven republics of Guatemala, Belize, Honduras, El Salvador, Nicaragua, Costa Rica, and Panama. Sometimes the entire mainland portion of the realm is incorrectly referred to as Central America, but only these seven countries constitute this region. The **Greater Antilles** is the regional name that refers to the four large islands in the northern sector of the Caribbean Sea: Cuba, Jamaica, Hispaniola, and Puerto Rico; two countries, Haiti and the Dominican Republic, share the island of Hispaniola. And the **Lesser Antilles** form an extensive arc of smaller islands from the Virgin Islands off Puerto Rico to Aruba near the northwestern coast of Venezuela; they also include the Bahamas island chain north of the Greater Antilles. In the Caribbean islands large and small, moist winds sweep in from the east, watering windward (wind-facing) coasts while leaving leeward (wind-protected) areas dry. On the map, the "Leeward" and "Windward" Islands actually are not "dry" and "wet," these terms having navigational rather than environmental implications.

PHYSICAL GEOGRAPHY

A Land Bridge

The funnel-shaped mainland, a 4800-kilometer (3000-mi) connection between North and South America, is wide enough in the north to contain two major mountain chains and a vast interior plateau, but narrows to a slim 65-kilometer (40-mi) ribbon of land in Panama. Here this strip of land—or **isthmus**—bends eastward so that Panama's orientation is east-west. Thus mainland Middle America is what physical geographers call a land bridge [3], an isthmian link between continents.

If you examine a globe, you can see some other present and former land bridges: Egypt's Sinai Peninsula between Asia and Africa, the (now-broken) Bering land bridge between northeasternmost Asia and Alaska, and the shallow waters between New Guinea and Australia. Such land bridges, though temporary features in geologic time, have played crucial roles in the dispersal of animals and humans across the planet. But even though mainland Middle America forms a land bridge, its internal fragmentation has always inhibited movement. Mountain ranges, swampy coastlands, and dense rainforests make contact and interaction difficult.

Island Chains

As shown in Figure 4A-1, the approximately 7000 islands of the Caribbean Sea stretch in a lengthy arc from Cuba and the Bahamas eastward and then southward to Trinidad, with numerous outliers outside (such as Barbados) and inside (e.g., the Cayman Islands) the main chain. As we noted above, the four large islands—Cuba, Hispaniola (containing Haiti and the Dominican Republic), Puerto Rico, and Jamaica—are called the Greater Antilles, and all the remaining smaller islands constitute the Lesser Antilles. The entire Antillean archipelago [4] (island chain) consists of the crests and tops of mountain chains that rise from the floor of the Caribbean, the result of collisions between the Caribbean Plate and its neighbors (Fig. G-4). Some of these crests are relatively stable, but elsewhere they contain active volcanoes, and almost everywhere in this realm

This shocking street scene in central Port-au-Prince is typical of the widespread devastation resulting from the January 2010 earthquake that struck the Haitian capital. The human casualty toll was enormous, thousands of buildings were made uninhabitable, debris clogged the streets, and the economy was wrecked. Survivors depended almost exclusively on assistance from foreign governments and international relief agencies. But life somehow found a way to go on, as these resourceful people prove. All too soon, though, Haiti's plight disappeared from the headlines even as the human drama continued. The hemisphere's poorest country and weakest economy faces a colossal struggle to rebuild, and that struggle will remain one of the Caribbean's most riveting geographical stories for a long time to come.

© Craig Ruttle/Alamy

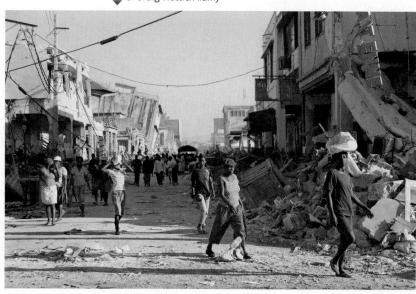

earthquakes are an ever-present danger—in the islands as well as on the mainland (Fig. G-5). Add to this the realm's seasonal exposure to Atlantic/Caribbean hurricanes, and it amounts to some of the highest-risk real estate on Earth.

Dangerous Landscapes

The danger from below is dramatically illustrated in the Haitian, western half of Hispaniola, which is laced with geologic fault lines associated with the nearby boundary that separates the North American and Caribbean plates (see Fig. 4B-11, inset map). On January 12, 2010, Haiti was hit by a massive earthquake; the epicenter of that 7.0 temblor was located just outside Port-au-Prince and virtually destroyed this teeming, impoverished capital city of 2.2 million (see photo, p. 197). At least 300,000 people died, and within a week more than a million had fled to the countryside. It was the worst natural disaster in Haiti's history, but certainly not the first. In 2008, it was struck by three hurricanes and a tropical storm, repeatedly flooding cities and towns, destroying homes, and dislocating hundreds of thousands of people.

The environmental hazard from above comes in the form of hurricanes, powerful tropical cyclones that annually threaten the Caribbean Basin and its surrounding coastlines. Middle America is one of the two most hurricane-prone areas in the world (the other is the zone of seaboards and islands along the southeastern coast of Asia, whose even more frequent tropical cyclones are called typhoons). One of the key conditions for the formation of hurricanes is very warm ocean water, because it further heats the hot moist air rising above it to "fuel" the evolving storm. The prolonged Atlantic/Caribbean hurricane season lasts from June 1 to December 1, with the greatest number of tropical cyclones occurring in August and September when seawater reaches its highest temperatures.

Most storms travel in a westerly direction from their low-latitude spawning ground off the coast of West Africa, steered across the Atlantic by the trade winds to reach the Caribbean Sea. Once there, many of these cyclones follow similar routes within **"Hurricane Alley" [5]**—whose wide axis lies along all of the Greater Antilles and then broadens to include southern Florida, Mexico's Yucatán Peninsula, and the Gulf of Mexico. On average, every season sees the development of four to eight major hurricanes—and rarely does a year go by (2010 was one) without a destructive landfall on at least one of these densely populated areas.

No part of Middle America is safe from environmental onslaught, but let's keep in mind that certain places are riskier than others—and whenever disaster strikes, the poorest inhabitants, often living in the least desirable and most hazardous locales, suffer disproportionately. A big earthquake in Haiti is a far greater tragedy than a similar event in California, and a hurricane in Cuba is going to be deadlier than an identical hit in Florida—because in these poorer countries the people and their weaker structures are more vulnerable. Being safe from the vagaries of nature is a matter of geography, but so too of the ability to cope.

Altitudinal Zonation of Environments

Continental Middle America and the western margin of South America are areas of high relief and strong environmental contrasts. Even though settlers have always favored temperate intermontane basins and valleys, people also cluster in hot tropical lowlands as well as high plateaus just below the snow line in South America's Andes Mountains. In each of these zones, distinct local climates, soils, vegetation, crops, domestic animals, and modes of life prevail. Such **altitudinal zones [6]** (diagrammed in Fig. 4A-4) are known by specific names as if they were regions with distinguishing properties—as, in reality, they are.

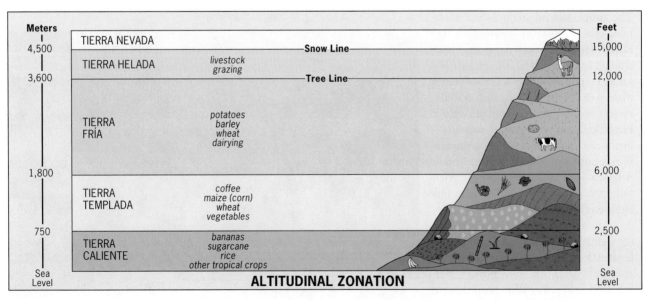

FIGURE 4A-4

© H. J. de Blij, P. O. Muller, and John Wiley & Sons, Inc.

The lowest of these vertical zones, from sea level to 750 meters (2500 ft), is known as the *tierra caliente* [7], the "hot land" of the coastal plains and low-lying interior basins where tropical agriculture predominates. Above this zone lie the tropical highlands containing Middle and South America's largest population clusters, the *tierra templada* [8] of temperate land reaching up to about 1800 meters (6000 ft). Temperatures here are cooler; prominent among the commercial crops is coffee, while corn (maize) and wheat are the staple grains. Still higher, from about 1800 to 3600 meters (6000 to nearly 12,000 ft), is the *tierra fría* [9], the cold country of the higher Andes where hardy crops such as potatoes and barley are mainstays. Above the tree line, which marks the upper limit of the *tierra fría*, lies the *tierra helada* [10]; this fourth altitudinal zone, extending from about 3600 to 4500 meters (12,000 to 15,000 ft), is so cold and barren that it can support only the grazing of sheep and other hardy livestock. The highest zone of all is the *tierra nevada* [11], a zone of permanent snow and ice associated with the loftiest Andean peaks. As we will see, the varied human geography of mainland Middle and western South America closely reflects these diverse environments.

Despite scattered attempts to reverse this landscape scourge, deforestation continues to afflict Central America. Its worst effects plague the steeper slopes of interior highlands, as here in western Panama. In the wake of recent deforestation, the land near the top of this hill has already begun to erode, because in the absence of binding tree roots the copious tropical rains are making short work of the unprotected topsoil.

© Humberto Olarte Cupas/Alamy

Tropical Deforestation

Before the Europeans arrived, two-thirds of continental Middle America (at lower altitudes) was covered by tropical rainforests. It is estimated that at present only about 10 percent of this vegetation remains. Over the past decade alone, about 1.2 million hectares (3 million acres) of Central American and Mexican woodland disappeared annually. El Salvador has already lost nearly all of its forests, and most of the six other republics in its region will soon approach that stage.

The causes of **tropical deforestation** [12] are related to the persistent economic and demographic problems of disadvantaged countries. In Central America, the leading cause has been the need to clear rural lands for cattle pasture as many countries, especially Costa Rica, became meat producers and exporters. Because tropical soils are so nutrient-poor, newly deforested areas are able to function as pastures for only a few years at most. These fields are then abandoned for other freshly cut lands and quickly become a ravaged landscape (see photo above). Without the protection of trees, local soil erosion and flooding immediately become problems, affecting still-productive areas nearby. A second cause of deforestation is the rapid logging of tropical woodlands as the timber industry increasingly turns from the exhausted forests of the midlatitudes to harvest the rich tree resources of the equatorial zones, responding to accelerating global demands for housing, paper, and furniture. The third major contributing factor is related to the region's population explosion: as more and more peasants are required to extract a subsistence from inferior lands, they have no choice but to cut down the remaining forest for both firewood and additional crop-raising space, and their intrusion prevents the trees from regenerating.

CULTURAL GEOGRAPHY

Mesoamerican Legacy

Mainland Middle America was the scene of the emergence of a major ancient civilization. Here lay one of the world's true **culture hearths** [13], a source area from which new ideas radiated and whose population could expand and make significant material and intellectual progress. Agricultural specialization, urbanization, and transportation networks developed, and writing, science, art, and other spheres of achievement saw major advances. Anthropologists refer to the Middle American culture hearth as ***Mesoamerica***, which extended southeast from the vicinity of present-day Mexico City to central Nicaragua. Its development is especially remarkable because it occurred in highly different geographic environments, each presenting obstacles that had to be overcome in order to unify and integrate large areas. First, in the low-lying tropical plains of what is now northern Guatemala, Belize, and Mexico's Yucatán Peninsula, and perhaps simultaneously in Guatemala's highlands to the south, the Maya civilization arose more than 3000 years ago. Later, far to the northwest on the high plateau in central Mexico, the Aztecs founded a major civilization centered on the largest city ever to exist in pre-Columbian times.

The Lowland Maya

The Maya civilization is the only major culture hearth in the world that arose in the lowland tropics. Its great cities, with their stone pyramids and massive temples, still yield archeological information today. Maya culture reached its zenith from the third to the tenth centuries AD. The Maya civilization, anchored by a series of city-states, unified an area larger than any of the present-day Middle American countries except Mexico. Its population probably totaled between 2 and 3 million; certain Maya languages are still used in the area to this day. The Maya city-states were marked by dynastic rule that functioned alongside a powerful religious hierarchy, and the great cities that now lie in ruins were primarily ceremonial centers. We also know that this culture produced skilled artists and scientists, and the Maya achieved a great deal in agriculture and trade as well. They grew cotton, created a rudimentary textile industry, and exported cotton cloth by seagoing canoes to other parts of Middle America in return for valuable raw materials. They domesticated the turkey and grew cacao, developed writing systems, and studied astronomy.

The Highland Aztecs

In what is today the intermontane highland zone of Mexico, significant cultural developments were also taking place. Here, just north of present-day Mexico City, lay Teotihuacán, the first true urban center in the Western Hemisphere, which prospered for nearly seven centuries after its founding around the beginning of the Christian era.

The Aztec state, the pinnacle of organization and power in pre-Columbian Middle America, is thought to have originated in the early fourteenth century with the founding of a settlement on an island in a lake that lay in the *Valley of Mexico* (the area surrounding what is now Mexico City). This urban complex, a functioning city as well as a ceremonial center, named Tenochtitlán, was soon to become the greatest city in the Americas and the capital of a large powerful state.

The Aztec state soon began to conquer territories to the east and south. The Aztecs' expansion of their empire was driven by their desire to subjugate peoples and towns in order to extract taxes and tribute. As Aztec influence spread throughout Middle America, the state grew ever richer, its population mushroomed, and its cities thrived and expanded.

The Aztecs produced a wide range of impressive accomplishments, although they were better borrowers and refiners than they were innovators. They developed irrigation systems, and they constructed elaborate walls to terrace slopes where soil erosion threatened. Indeed, the greatest contributions of Mesoamerica's Amerindians surely came

From the Field Notes . . .

"We spent Monday and Tuesday upriver at Lamanai, a huge, still mostly overgrown Maya site deep in the forest of Belize. On Wednesday we drove from Belize City to Altun Ha, which represents a very different picture. Settled around 200 BC, Altun Ha flourished as a Classic Period center between AD 300 and 900, when it was a thriving trade and redistribution center for the Caribbean merchant canoe traffic and served as an entrepôt for the interior land trails, some of them leading all the way to Teotihuacán. Altun Ha has an area of about 6.5 square kilometers (2.5 sq mi), with the main structures, one of which is shown here, arranged around two plazas at its core. I climbed to the top of this one to get a perspective, and sat down to have my sandwich lunch, imagining what this place must have looked like as a bustling trade and ceremonial center when the Roman Empire still thrived, but a more urgent matter intruded. A five-inch tarantula emerged from a wide crack in the sun-baked platform, and I noticed it only when it was about two feet away, apparently attracted by the crumbs and a small piece of salami. A somewhat hurried departure put an end to my historical-geographical ruminations."

© H. J. de Blij

www.conceptcaching.com

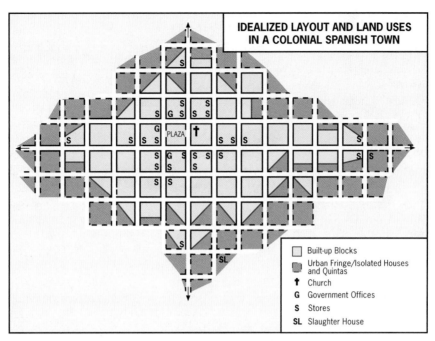

FIGURE 4A-5   © H. J. de Blij, P. O. Muller, and John Wiley & Sons, Inc.

from the agricultural sphere. Corn (maize), the sweet potato, various kinds of beans, the tomato, squash, cacao
beans (the raw material of chocolate), and tobacco are just
a few of the crops that grew in Mesoamerica when the
Europeans first made contact.

Spanish Conquest

Spain's defeat of the Aztecs in the early sixteenth century
opened the door to Spanish penetration and supremacy.
The Spaniards were ruthless colonizers but not more so
than other European powers that subjugated other cultures. True, the Spaniards first enslaved the Amerindians
and were determined to destroy the strength of indigenous
society. But biology accomplished what ruthlessness could
not have achieved in so short a time: diseases introduced
by the Spaniards and slaves imported from Africa killed
millions of Amerindians.

Middle America's cultural landscape was drastically
modified. Unlike the Amerindians, who had used stone
as their main building material, the Spaniards employed
great quantities of wood and used charcoal for heating,
cooking, and smelting metal. The onslaught on the forests
was immediate, and rings of deforestation swiftly expanded around the colonizers' towns. The Spaniards also
introduced large numbers of cattle and sheep, and people
and livestock now had to compete for available food (requiring the opening of vast areas of marginal land that further disrupted the region's food-production balance).
Moreover, the Spaniards introduced their own crops (notably wheat) and farming equipment, and soon large
wheatfields began to encroach upon the small plots of corn
that the Amerindians cultivated.

The Spaniards' most far-reaching cultural changes
derived from their traditions as town-dwellers. Amerindians were moved off their land into nucleated villages and
towns that the Spaniards established and laid out. In these
settlements, the Spaniards could exercise the kind of rule
and administration to which they were accustomed (Fig.
4A-5). The internal focus of each Spanish town was the
central *plaza* or market square, around which both the
local church and government buildings were located. The
surrounding street pattern was deliberately laid out in
gridiron form, so that any insurrections could be contained by having a small military force seal off the affected
blocks and then root out the troublemakers. Each town
was located near what was thought to be good agricultural
land (which was often not so good), so that the Amerindians could go out each day and work in the fields. Packed
tightly into these towns and villages, they came face to face
with Spanish culture. Here they (forcibly) learned the Europeans' Roman Catholic religion and Spanish language,
and they paid their taxes and tribute to a new master.
Many of Middle America's leading cities still bear this
Spanish imprint.

Collision of Cultures

But Middle America is not Spain. The cultural fabric of
Middle America reflects the collision of Amerindian,
Spanish, and other European influences. Indeed, in more
remote areas in southeastern Mexico and inner Guatemala, the nucleated indigenous village survived and to this
day native languages prevail over Spanish (see Fig. 4B-5).

In Middle America outside Mexico, only Panama,
with its twin attractions of interoceanic transit and gold

FIGURE 4A-6

© H. J. de Blij, P. O. Muller, and John Wiley & Sons, Inc.

deposits, became an early focus of Spanish activity. From there, following the Pacific side of the isthmus, Spanish influence radiated northwestward through Central America and into Mexico. The major arena of international competition in Middle America, however, lay not on the Pacific side but on the islands and coasts of the Caribbean Sea. Here the British gained a foothold on the mainland, controlling a narrow coastal strip that extended southeast from Yucatán to what is now Costa Rica. As the colonial-era map (Fig. 4A-6) shows, in the Caribbean the Spaniards faced not only the British but also the French and the Dutch, all interested in the lucrative sugar trade, all searching for instant wealth, and all seeking to expand their empires.

Much later, after centuries of European colonial rivalry in the Caribbean Basin, the United States entered the picture and made its influence felt in the coastal areas of the mainland, not through colonial conquest but through the introduction of widespread, large-scale, banana plantation agriculture. The effects of these plantations were as far-reaching as the impact of colonialism on the Caribbean islands. Because the diseases the Europeans

had introduced were most rampant in these hot, humid lowlands (as well as the Caribbean islands to the east), the Amerindian population that survived was too small to provide a sufficient workforce. This labor shortage was quickly remedied through the trans-Atlantic slave trade from Africa that transformed the Caribbean Basin's demography.

The cultural variety of the Caribbean Basin is especially striking, and it is hardly an arena of exclusive Hispanic cultural heritage. For example, Cuba's southern neighbor, Jamaica (population 2.8 million, mostly of African ancestry), has a legacy of British involvement, while to the east in Haiti (10.2 million, overwhelmingly of African ancestry) the strongest imprints have been African and French. The Lesser Antilles also exhibit great cultural diversity. There are the (once Danish) U.S. Virgin Islands; French Guadeloupe and Martinique; a group of British-influenced islands, including Barbados, St. Lucia, and Trinidad and Tobago; and the Dutch St. Maarten (shared with the French) as well as the now-autonomous islands of the former Netherlands Antilles—Aruba, Curaçao, and Bonaire—off the northwestern Venezuelan coast.

From the Field Notes . . .

"Driving around the small Caribbean island of Bonaire and coming face to face with this surreal landscape, I first thought I had experienced a mirage. These glistening white, perfectly cone-shaped hills are actually salt piles. This small, Dutch-controlled island off the coast of Venezuela possesses the perfect geographic conditions for salt production: it has a series of salt water inlets, it is very hot and dry, and lies in the zone of persistent trade winds. Remember that salt, nowadays taken for granted in every household, has long been one of the world's most precious spices, especially when used for the preservation of meat and fish. The Dutch began large-scale production of salt in the 1620s; today this local industry is in the hands of the Antilles International Salt Company and continues to be an important source of Bonaire's foreign revenues."

www.conceptcaching.com

© Jan Nijman

Concept Caching

POLITICAL AND ECONOMIC FRAGMENTATION

Independence

Independence movements stirred Middle America at an early stage. On the mainland, revolts against Spanish authority (beginning in 1810) achieved independence for Mexico by 1821 and for the Central American republics by the end of the 1820s, resulting in the creation of eight different countries. The United States, concerned over European designs in the realm, proclaimed the Monroe Doctrine in 1823 to deter any European power from reasserting its authority in the newly independent republics or from further expanding its existing domains.

By the end of the nineteenth century, the United States itself had become a major force in Middle America. The Spanish-American War of 1898 made Cuba independent and put Puerto Rico under the U.S. flag; soon afterward, the Americans were in Panama constructing the Panama Canal. Meanwhile, with U.S. corporations driving a boom based on huge banana plantations, the Central American republics had become colonies of the United States in all but name.

Independence came to the Caribbean Basin in fits and starts. Afro-Caribbean Jamaica as well as Trinidad and Tobago, where the British had brought in a large South Asian population, attained full sovereignty from the United Kingdom in 1962; other British islands (among them Barbados, St. Vincent, and Dominica) became independent later on. France, however, retains Martinique and Guadeloupe as overseas *départements* of the French Republic, and the Dutch islands are at various stages of autonomy. No less than 33 states are found on the political map of the Caribbean Basin today.

Regional Contrasts

There are some striking contrasts, socially and economically, between the Middle American uplands on the one hand and the Caribbean coasts and islands on the other (Fig. 4A-7). These were conceptualized by cultural geographer John Augelli into the **Mainland-Rimland framework**. The Euro-Amerindian *Mainland*, from Mexico southeast to Panama, is dominated by European (Spanish) and Amerindian influences and also includes **mestizo [14]** sectors where the two ancestries mixed. As Figure 4A-7 shows, the Mainland is subdivided into several areas based on the strength of the Amerindian legacy. The Mainland's Caribbean coastal strip and all of the Basin's islands to the east constitute the *Rimland*, which for the most part has a very different cultural heritage, based on a fusion of European and African influences.

Supplementing these contrasts are regional differences in outlook and orientation. The Caribbean Rimland was an area of sugar and banana plantations, of high accessibility, of seaward exposure, and of maximum cultural contact and mixture. The Middle American Mainland, being farther removed from these contacts, was an area of greater isolation. The Rimland was the domain of the great **plantation [15]**, and its commercial economy was

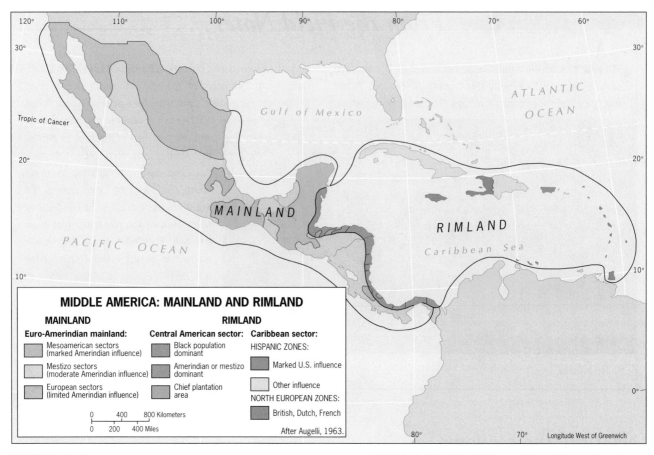

FIGURE 4A-7

© H. J. de Blij, P. O. Muller, and John Wiley & Sons, Inc.

therefore susceptible to fluctuating world markets and tied to overseas investment capital. The Mainland was dominated by the hacienda [16], which was far more self-sufficient and less dependent on external markets.

This contrast between plantation and hacienda land tenure in itself constitutes strong evidence for the Rimland-

Mainland dichotomy. The hacienda was a Spanish institution, whereas the modern plantation was the concept of northwestern Europeans. On their haciendas, Spanish landowners possessed a domain whose productivity they might never push to its limits: the very possession of such a vast estate brought with it social prestige and a comfortable

Contrasting land uses in the Middle American Rimland and Mainland give rise to some very different rural cultural landscapes. Huge stretches of the realm's best land continue to be controlled by (often absentee) landowners whose haciendas yield export or luxury crops, or foreign corporations that raise fruits for transport and sale on their home markets. The banana plantation shown here (left) lies near the Caribbean coast of Costa Rica. The vast fields of banana plants stand in strong contrast to the lone peasant who ekes out a bare subsistence from small cultivable plots of land, often in high-relief countryside where grazing some goats or other livestock is the only way to use most of the land.

(Left): © John Coletti/Getty Images, Inc. (Right): © Jean-Gerard Sidaner/Photo Researchers lifestyle. Native workers lived on the land—which may once have been their land—and had plots where they could grow their own subsistence crops. All this is written as though it is mostly in the past, but the legacy of the hacienda system, with its inefficient use of land and labor, still exists throughout mainland Middle America.

The plantation, in contrast, is all about efficiency and profit. Foreign ownership and investment is the norm, as is production for export. Most plantations grow only a single crop, be it sugar, bananas, or coffee. Much of the labor is seasonal—needed in large numbers mainly during the harvest period—and such labor has been imported because of the scarcity of Amerindian workers. With its "factory-in-the-field" operations (left photo, p. 204), the plantation is more efficient in its use of land and labor than the hacienda. Profit and wealth, rather than social prestige, are the dominant motives for the plantation's establishment and operation. The Rimland is also subdivided, with the most obvious partitioning the one between the mainland-coastal plantation zone and the islands of the Caribbean. Notice, too, that the islands themselves are classified according to their cultural heritage (Figs. 4A-6 and 4A-7).

If in the past the plantations brought their owners considerable wealth, little of that ever reached their workers, and inequality seemed a structural part of life. In recent decades, the near monopoly of some of the Caribbean crops (particularly sugar) came to an end with competition from elsewhere (for example, the United States). Revenues declined and unemployment increased. Millions of workers were pushed into a life of subsistence, and many took refuge in the cities, where conditions and opportunities are hardly better.

Connections Matter

The unusual layout of the Middle American realm offers an opportunity to speculate about the role of spatial factors in (economic) development. Geographic fragmentation and size undoubtedly play a part, and so does accessibility versus remoteness in the form of **connectivity [17]**. This is true for both the Caribbean and mainland components of the realm. As pointed out earlier, Mexico's northern margin is strongly connected to the United States, and this provides an impulse for the entire Mexican economy even if it tends to accentuate regional differences (discussed in Chapter 4B). Among Central American countries, there is no question that Panama is by far the best connected to the world economy; indeed, the Panama Canal provides an enormous range of global linkages in trade, finance, producer services, and other sectors of international business.

Now take a look at Figure 4A-8 and note how connectivity correlates with economic development. Of the eight mainland countries, Mexico and Panama have the highest GDP per capita; moreover, as we move away from the realm boundaries with North and South America toward the increasingly remote central interior of the

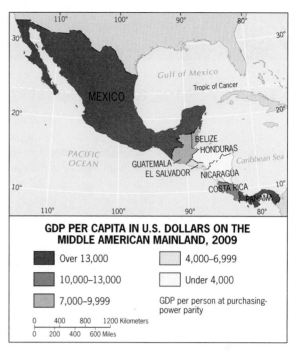

GDP PER CAPITA IN U.S. DOLLARS ON THE MIDDLE AMERICAN MAINLAND, 2009

- Over 13,000
- 10,000–13,000
- 7,000–9,999
- 4,000–6,999
- Under 4,000

GDP per person at purchasing-power parity

FIGURE 4A-8   © H. J. de Blij, P. O. Muller, J. Nijman, and John Wiley & Sons, Inc.

Maxico-Panama land corridor, GDP per capita declines substantially. Nicaragua, Honduras, and El Salvador, tucked away in the middle of that narrowing land bridge, are the most distant from the bigger and more vibrant economies of the neighboring realms and clearly lack the expanding external connections of Panama—or even the historical connections of Belize (to the UK, its former colonial ruler) or Costa Rica (to the United States, sparked well over a century ago by the banana trade).

Is Small Beautiful?

As noted, the Middle American realm is exceptional in its modest size, in terms of both territory and population. But the number of countries is considerable, and they tend to be quite small. Caribbean islands often invoke images of beautiful scenery, white sandy beaches, palm trees, tropical cocktails, and shiny blue waters. But the economic realities are invariably harsh. The limited land areas of the Lesser Antilles each average well below half a million people, and that, combined with insularity, remoteness, and low connectivity, poses formidable challenges that are common to these **small-island developing economies [18]**. First, natural resources are often limited, which requires a heavy reliance on imports made more expensive by added transport costs. Second, the cost of government is relatively high on a per capita basis: even the smallest population will require services such as schools, hospitals, waste disposal, military capabilities, and general administration. Third, specialized services must often be brought in from elsewhere. And fourth, local production cannot really benefit from economies of scale; this means that local producers

Regional ISSUE   The Role of the Tourist Industry in Middle American Economies

IN SUPPORT OF THE TOURIST INDUSTRY

"As the general manager of a small hotel in St. Maarten, on the Dutch side of the island, I can tell you that without tourists, we would be in deep trouble economically. Mass tourism, plain and simple, has come to the rescue in the Caribbean and in other countries of Middle America. Look at the numbers. Here it's the only industry that is growing, and it already is the largest worldwide industry. For some of the smaller countries of the Caribbean, this is not just the leading industry but the only one producing external revenues. Whether it's hotel patrons or cruise-ship passengers, tourists spend money, create jobs, fill airplanes that give us a link to the outside world, require infrastructure that's good not just for them but for locals too. We've got better roads, better telephone service, more items in our stores. All this comes from tourism. I employ 24 people, most of whom would be looking for nonexistent jobs if it weren't for the tourist industry.

"And it isn't just us here in the Caribbean. Look at Belize. I just read that tourism there, based on their coral reefs, Mayan ruins, and inland waterways, brought in U.S. $100 million last year, in a country with a total population of only about 300,000! That sure beats sugar and bananas. In Jamaica, they tell me, one in every four workers has a job in tourism. And the truth is, there's still plenty of room for the tourist industry to expand in Middle America. Those Americans and Europeans can close off their markets against our products, but they can't stop their citizens from getting away from their awful weather by coming to this tropical paradise.

"Here's another good thing about tourism. It's a clean industry. It digs no mine shafts, doesn't pollute the atmosphere, doesn't cause diseases, doesn't poison villagers, isn't subject to graft and corruption the way some other industries are.

"Last but not least, tourism is educational. Travel heightens knowledge and awareness. There's always a minority of tourists who just come to lie on the beach or spend all their time in some cruise-ship bar, but most of the travelers we see in my hotel are interested in the place they're visiting. They want to know why this island is divided between the Dutch and the French, they ask about coral reefs and volcanoes, and some even want to practice their French on the other side of the border (don't worry, no formalities, just drive across and start talking). Tourism's the best thing that happened to this part of the world, and other parts too, and I hope we'll never see a slowdown."

CRITICAL OF THE TOURIST INDUSTRY

"You won't get much support for tourism from some of us teaching at this college in Puerto Rico, no matter how important some economists say tourism is for the Caribbean. Yes, tourism is an important source of income for some countries, like Kenya with its wildlife and Nepal with its mountains, but for many countries that income from tourism does not constitute a real and fundamental benefit to the local economies. Much of it may in fact result from the diversion to tourist consumption of scarce commodities such as food, clean water, and electricity. More of it has to be reinvested in the construction of airport, cruise-port, overland transport, and other tourist-serving amenities. And as for items in demand by tourists, have you noticed that places with many tourists are also places where prices are high?

"Sure, our government people like tourism. Some of them have a stake in those gleaming hotels where they can share the pleasures of the wealthy. But what those glass-enclosed towers represent is globalization, powerful multinational corporations colluding with the government to limit the opportunities of local entrepreneurs. Planeloads and busloads of tourists come through on prearranged (and prepaid) tour promotions that isolate those visitors from local society.

"You geographers talk about cultural landscapes. Well, picture this: luxury liners sailing past poverty-stricken villages, luxury hotels towering over muddy slums, restaurants serving caviar when, down the street, children suffer from malnutrition. If the tourist industry offered real prospects for economic progress in poorer countries, such circumstances might be viewed as the temporary, unfortunate byproducts of the upward struggle. Unfortunately, the evidence indicates otherwise. Name me a tourism-dependent economy where the gap between the rich and poor has narrowed.

"As for the educational effect of tourism, spare me the argument. Have you sat through any of those 'culture' shows staged by the big hotels? What you see there is the debasing of local culture as it adapts to visitors' tastes. Ask hotel workers how they really feel about their jobs, and you'll hear many say that they find their work dehumanizing because expatriate managers demand displays of friendliness and servitude that locals find insulting to sustain.

"I've heard it said that tourism doesn't pollute. Well, the Alaskans certainly don't agree—they sued a major cruise line on that issue and won. Not very long ago, cruise-ship crews routinely threw garbage-filled plastic bags overboard. That seems to have stopped, but I'm sure you've heard of the trash left by mountain-climbers in Nepal, the damage done by off-road vehicles in the wildlife parks of Kenya, the coral reefs injured by divers off Florida. Tourism is here to stay, but it is no panacea."

Vote your opinion at www.wiley.com/college/deblij

can be put out of business by cheaper imports, which in turn causes unemployment.

Given the Caribbean Basin's limited economic options, does the tourist industry offer better opportunities? Opinions on this question are divided (see box at left). The resort areas, scenic treasures, and historic locales of Caribbean America attract between 15 and 20 million visitors annually, with about half of them traveling on Florida-based cruise ships. Certainly, Caribbean tourism is a prospective money-maker for many islands. In Jamaica alone, this industry accounts for about one-fifth of the gross domestic product and employs more than one-fourth of the labor force.

But Caribbean tourism also has serious drawbacks. The invasion of poor communities by affluent tourists contributes to rising local resentment, which is further fueled by the glaring contrasts of shiny new hotels towering over substandard housing and luxury liners gliding past poverty-stricken villages. At the same time, tourism can have the effect of debasing local culture, which often is adapted to suit the visitors' tastes at hotel-staged "culture" shows. In addition, the cruise industry tends to monopolize revenues (accommodations, meals, drinks, entertainment) with relatively few dollars flowing into the local economy. Finally, even though tourism does generate income in the Caribbean, the intervention of island governments and multinational corporations removes opportunities from local entrepreneurs in favor of large operators and major resorts.

The Push for Regional Integration

Another challenge for the Middle American realm is to foster greater economic integration. Many of the countries on the mainland as well as in the Caribbean are poorly connected within the realm and are heavily dependent on big outside countries, the United States in particular. For the large majority of these countries, the United States is the primary trading partner. Consider this: of all the trade involving Middle American countries, less than 10 percent takes place inside the realm. And less than 1 percent takes place between the Caribbean Basin and the Middle American mainland.

Over the years, efforts have been made to advance economic integration and to turn this realm into more of a *functional* region. The mainland saw the creation of the Central American Common Market in 1960, but it became moribund within a decade because of the 1969 war between El Salvador and Honduras that was followed by further intraregional conflict throughout the 1970s and 1980s (some with U.S. involvement). The organization was resuscitated in the 1990s but since 2005 it has been overshadowed by **CAFTA**, the Central American Free Trade Agreement with the United States. CAFTA seems a mixed blessing: it may increase access to U.S. markets and lead to cheaper imports, but it may also galvanize the singular domination of the United States at the cost of greater intraregional integration.

Within the Basin, **CARICOM** (Caribbean Community) was established in 1989 and now consists of 15 full members, including nearby Guyana and Suriname in South America. CARICOM follows the example, in some respects, of the European Union, and in 2009 even introduced a common passport. But economic change has been slow. The geography of the Middle American realm poses many significant challenges, and these are not easily overcome even with the best political and economic intentions.

> **◖ POINTS TO PONDER ◗**
>
> - The 3169-kilometer (1969-mi) land border between North America and Middle America is the longest in the world separating a rich realm from a poor one.
>
> - The United States is the single most important economic partner of just about every country and territory in this realm—is that a positive or a negative for Middle America?
>
> - "Poor Mexico . . . so far from God and so close to the United States"—former Mexican President Porfirio Diaz.
>
> - Most of the small Caribbean island-nations count less than half a million people—is that enough to sustain a viable economy?

© H. J. de Blij, P. O. Muller, and John Wiley & Sons, Inc.

FIGURE 4B-1

REGIONS OF MIDDLE AMERICA

- ☐ Mexico
- ☐ Central America
- ☐ Greater Antilles
- ☐ Lesser Antilles } Caribbean

POPULATION
Under 50,000
50,000–250,000
250,000–1,000,000
1,000,000–5,000,000
Over 5,000,000

—— Road
—— Railroad

National capitals are underlined

0   200   400   600 Kilometers
0   100   200   300 Miles

■: REGIONS

Mexico

Central America

Caribbean Basin

(Greater Antilles)

(Lesser Antilles)

IN THIS CHAPTER

- ◆ Mexico's drug wars
- ◆ Indigenous peoples demand recognition and rights
- ◆ Projected Panama Canal expansion fuels boom in Panama City
- ◆ The 2010 killer earthquake in Haiti
- ◆ The debate over Puerto Rico's status

CONCEPTS, IDEAS, AND TERMS

© Jan Nijman
Where were these photos taken?
Find out at www.wiley.com/college/deblij

he Middle American realm stands out in terms of its overall small population, limited land cover, territorial fragmentation, and the insulated position of many its countries. Mexico is the obvious exception when it comes to size, accounting for 57 percent of the realm's population and 72 percent of its area; all other countries vary from small to tiny. Fragmentation and isolation tend to go hand in hand in this realm. All territories are either islands or have long coastlines (most of them on both sides of the narrow mainland land bridge). Only one country, Guatemala, shares borders with as many as four other countries; Mexico and Honduras have three neighbors; the five remaining Central American republics have only two. All other territories in the realm have either one or none at all as they are completely surrounded by water. Not surprisingly, regional integration, a condition for political stability and economic progress, has been one of the foremost challenges in this part of the world.

Middle America consists of four distinct geographic regions: (1) *Mexico*, the giant of the realm in every respect; (2) *Central America*, the string of seven small republics occupying the land bridge from Mexico to South America; (3) the four islands of the *Greater Antilles*, Cuba, Jamaica, Hispaniola (containing Haiti and the Dominican Republic), and Puerto Rico; and (4) the numerous small islands of the *Lesser Antilles* (Fig. 4B-1).

▪▪ MEXICO

Physiography

The physiography of Mexico is reminiscent of that of the conterminous western United States, although its

© Jan Nijman

environments are more tropical. Figure 4B-2 shows several prominent features: the elongated peninsula of Baja (Lower) California in the northwest, the far eastern Yucatán Peninsula, and the Isthmus of Tehuantepec in the southeast where the Mexican landmass tapers to its narrowest extent. Here in the southeast, Mexico most resembles Central America physiographically; a mountain backbone forms the isthmus, curves southeastward into Guatemala, and extends northwestward toward Mexico City. Shortly before reaching the capital, this mountain range divides into two chains, the Sierra Madre Occidental in the west and the Sierra Madre Oriental in the east (Figs. 4B-1, 4B-2). These diverging ranges frame the funnel-shaped Mexican heartland, the center of which consists of the rugged, extensive Plateau of Mexico (the important Valley of Mexico lies near its southeastern end). As Figure G-7 reveals, Mexico's climates are characterized by dryness, particularly in the broad, mountain-flanked north. Most of the better-watered areas lie in the southern half of the country where a number of major population concentrations have developed (see Fig. 4A-2).

Regions of Mexico

Physiographic, demographic, economic, historical, and cultural criteria combine to reveal a regionally diverse Mexico extending from the lengthy ridge of Baja California to

FIGURE 4B-2

© H. J. de Blij, P. O. Muller, and John Wiley & Sons, Inc.

● MAJOR CITIES OF THE REALM

City	Population* (in millions)
Mexico City, Mexico	29.7
Guadalajara, Mexico	4.5
Monterrey, Mexico	4.0
San Juan, Puerto Rico	2.8
Puebla, Mexico	2.4
Santo Domingo, Dominican Rep.	2.3
Port-au-Prince, Haiti	2.2
Havana, Cuba	2.1
Tijuana, Mexico	1.7
San José, Costa Rica	1.6
San Salvador, El Salvador	1.6
Panama City, Panama	1.5
Ciudad Juárez, Mexico	1.3
Guatemala City, Guatemala	1.2
Tegucigalpa, Honduras	1.1
Managua, Nicaragua	1.0

*Based on 2012 estimates.

the tropical lowlands of the Yucatán Peninsula, and from the economic dynamism of the NAFTA North to the Amerindian traditionalism of the Chiapas southeast (Fig. 4B-3). In the Core Area, anchored by Mexico City, and the West, centered on Guadalajara, lies the transition zone from the more Hispanic-mestizo north to the heavily Amerindian-infused mestizo south. East of the Core Area lies the Gulf Coast, once dominated by major irrigation projects and sprawling livestock-raising schemes but now the mainland center of Mexico's petroleum industry. The semiarid, scrub-vegetated Balsas Lowland separates the Core Area from the rugged, Pacific-fronting Southern Highlands, where Acapulco's luxurious resorts stand in stark contrast to the Amerindian villages and communal-farm settlements of the interior.

The dry, vast north stands in the sharpest contrast to these southern regions. The NAFTA North region is still formative and discontinuous, but, as we will see, it is changing northern Mexico significantly. This is true even in Yucatán, where the city of Mérida and environs are also substantially affected by NAFTA development. To go from comparatively well-off northern Yucatán to poverty-stricken southern Chiapas is to observe the entire range of Mexico's regional geography.

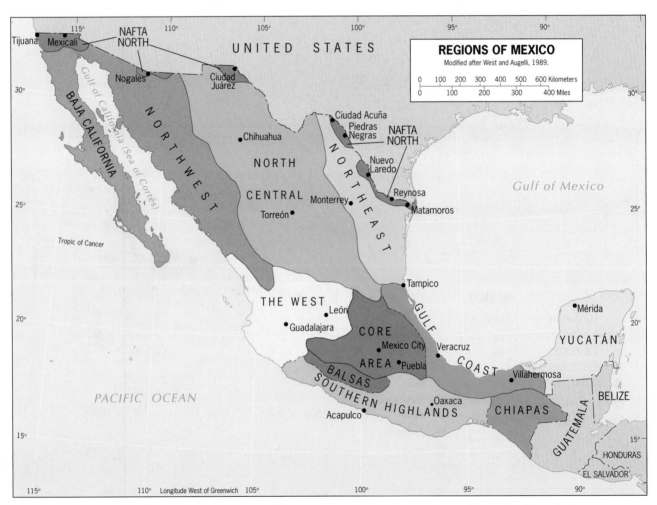

FIGURE 4B-3

© H. J. de Blij, P. O. Muller, and John Wiley & Sons, Inc.

Population Patterns

Mexico's population expanded rapidly during the last three decades of the twentieth century, doubling in just 28 years; but demographers have recently noted a sharp drop in fertility, and they are predicting that Mexico's population (currently at 113.7 million) will stop growing altogether by about 2050. This will have enormous implications for the country's economy and for the United States as well since it will reduce cross-border migration in the future.

The distribution of Mexico's population and its association with the country's 31 internal States is shown in Figure 4B-4 (and even more precisely in Fig. 4A-2). The largest concentration, containing more than half the Mexican people, extends across the densely populated "waist" of the country from Veracruz State on the eastern Gulf Coast to Jalisco State on the Pacific. The center of this corridor is dominated by the most populous State, Mexico (**3** on the map), at whose heart lies the Federal District of Mexico City (**9**). In the dry and rugged terrain to the north of this central corridor lie Mexico's least-populated States. Southern Mexico also exhibits a sparsely peopled periphery in the hot and humid lowlands of the Yucatán, but here most of the highlands of the continental spine contain sizeable populations.

Another major feature of Mexico's population map is urbanization, driven by the pull of the cities (with their perceived opportunities for upward mobility) in tandem with the push of the economically stagnant countryside. Today, 77 percent of the Mexican people reside in towns and cities, a surprisingly high proportion for a less-developed country. Undoubtedly, these numbers are affected by the explosive recent growth of Mexico City, which now totals almost 30 million (making it the largest urban concentration on Earth) and is home to an astonishing 26 percent of the national population. Urbanization rates at the other end of Mexico, however, are at their lowest in those remote uplands where Amerindian society has been least touched by modernization (Fig. 4B-4).

A Mix of Cultures

Nationally, the Amerindian imprint on Mexican culture remains strong. Today, 62 percent of all Mexicans are mestizos, 21 percent are predominantly Amerindian, and about 7 percent are full-blooded Amerindians; most of the remaining 10 percent are Europeans. The Spanish influence in Mexico has been profound, but it has been met with an equally powerful thrust of Amerindian culture. It has

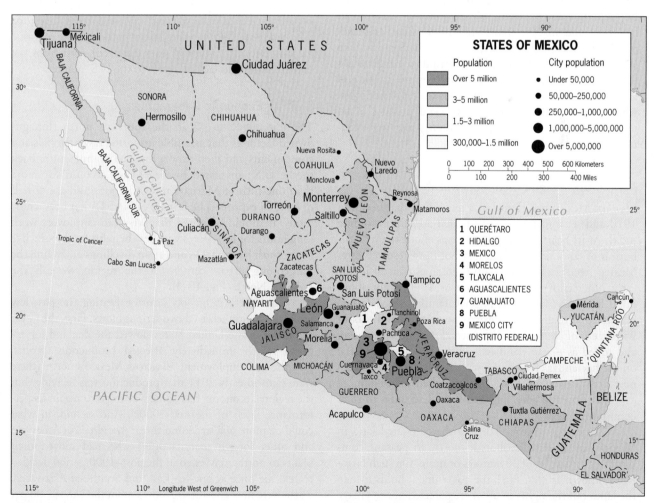

FIGURE 4B-4

© H. J. de Blij, P. O. Muller, and John Wiley & Sons, Inc.

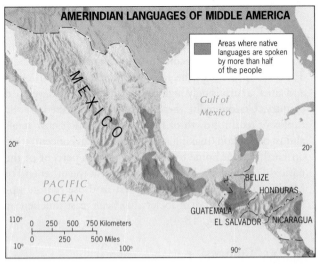

FIGURE 4B-5   © H. J. de Blij, P. O. Muller, and John
Wiley & Sons, Inc.

therefore not been a case of one-way European-dominated **acculturation [1]** but rather **transculturation [2]**—the two-way exchange of culture traits between societies in close contact. In the southeastern periphery (Fig. 4B-5), several hundred thousand Mexicans still speak only an Amerindian language, and millions more still utilize these languages daily even though they also speak Mexican Spanish. The latter has been strongly shaped by Amerindian influences, as have Mexican modes of dress, foods and cuisine, artistic and architectural styles, and folkways. This fusion of heritages, which makes Mexico unique, is the end product of an upheaval that began to reshape the country just over a century ago.

Agriculture: Fragmented Modernization

Modern Mexico was forged in a revolution that began in 1910 and set into motion events that are still unfolding today. At its heart, this revolution was about the redistribution of land, an issue that had not been resolved after Mexico freed itself from Spanish colonial control in the early nineteenth century. As late as 1900, only about 8000 haciendas (large, traditional, family-owned farms) blanketed virtually all of Mexico's better farmland. About 95 percent of all rural families owned no land whatsoever and toiled as *peones* (landless, constantly indebted serfs) on the haciendas. The triumphant revolution produced a new constitution in 1917 that launched a program of expropriation and parceling out of the haciendas to rural communities.

Since 1917, more than half the cultivated land in Mexico has been redistributed, mostly to peasant communities consisting of 20 families or more. On such farmlands, known as *ejidos* **[3]**, the government holds title to the land, but the rights to use it are parceled out to village communities and then to individuals for cultivation. This

system of land management is an Amerindian legacy, and not surprisingly most *ejidos* lie in central and southern Mexico where Amerindian social and agricultural traditions are strongest. About half of Mexico's land continues to be held in such "social landholdings." But the reforms did not lead to increased production. Some of the land is excessively fragmented, causing both low crop yields and widespread rural poverty. In the 1990s, the Mexican government attempted to privatize *ejidos*, hoping to promote consolidation and increase productivity, but that effort did not work. To date, less than 10 percent of the *ejidos* have been privatized.

Although traditional subsistence agriculture and the output of the inefficient *ejidos* have not changed a great deal in the poorer areas of rural Mexico, larger-scale commercial agriculture has diversified during the past three decades and made major gains with respect to both domestic and export markets. The country's arid northern tier has led the way as major irrigation projects have been built on streams flowing down from the interior highlands. Along the booming northwestern coast of the mainland, which lies within a day's drive of Southern California, mechanized large-scale cotton production now supplies an increasingly profitable export trade. Here, too, wheat and winter vegetables are grown, with fruit and vegetable cultivation attracting myriad foreign investors. Still, many small farmers are having a tough time of it due to competition of cheap (and subsidized) corn from the United States and lack of access to the U.S. market.

Shifting Economic Geographies

The **maquiladoras [4]** in the country's north (foreign-owned factories that assemble imported raw materials and components and then export finished manufactures back to the U.S.; discussed on page 194) account for more than one-fifth of Mexico's industrial jobs and about half of its total exports. The impact of NAFTA has been enormous. It should be noted that Mexican employees work long hours for low wages with few benefits, and many live in the most basic shacks and slum dwellings encircling the burgeoning urban centers of the region we call the NAFTA North (Fig. 4B-3).

And there is no job security. During the past few years, this region received two serious blows. First, the Great Recession in the United States slowed the demand for exports, which immediately affected production, profits, growth, and employment all across Mexico's north (there were signs in early 2011 that production was reviving, but the regional economy had some way to go to regain its performance levels of the mid-2000s). And second, in what could be more of a structural threat, hundreds of American and other foreign manufacturers who had moved their plants to northern Mexico in the post–1990 period decided to relocate once again—to East and Southeast Asia where wages were now considerably lower than the U.S. $2.00 per hour average paid by the maquiladoras. Although factories

AMONG THE REALM'S GREAT CITIES . . .

MEXICO CITY

MIDDLE AMERICA HAS only one great metropolis: Mexico City. With 29.7 million inhabitants, Mexico City is home to just over one-fourth of Mexico's population and grows by about 200,000 each year. Even more significantly, Mexico City has surpassed Tokyo, Japan (current population: 26.7 million) to become the world's largest urban agglomeration.

Lakes and canals marked this site when the Aztecs built their city of Tenochtitlán here seven centuries ago. The conquering Spaniards made it their headquarters, and following independence the Mexicans made it their capital. Centrally positioned and well connected to the rest of the country, Mexico City, hub of the national core area, became the quintessential primate city.

Vivid social contrasts mark the cityscape. Historic plazas, magnificent palaces, churches, villas, superb museums, ultramodern skyscrapers, and luxury shops fill the city center. Beyond lies a zone of comfortable middle-class and struggling, but stable, working-class neighborhoods. Outside this belt, however, lies a ring of more than 500 slums and countless, even poorer *ciudades perdidas*—the "lost cities" where newly arrived peasants live in miserable poverty and squalor. (These squatter settlements contain no less than one-third of the metropolitan area's population.) Mexico City's more affluent residents have also been plagued by problems in recent times as the country's social and political order came close to unraveling. Rampant crime remains a serious concern, much of it associated with corrupt police.

Environmental crises parallel the social problems. Local surface waters have long since dried up, and groundwater supplies are approaching depletion; to meet demand, the metropolis must now import a third of its water by pipeline from across the mountains (with about 40 percent lost through leakages in the city's crumbling water pipe network). Air pollution here is among the world's worst as more than 4 million motor vehicles and 40,000 factories churn out smog that in Mexico City's thin, high-altitude air sometimes reaches 100 times the acceptably safe level. And

© H. J. de Blij, P. O. Muller, and John Wiley & Sons, Inc.

add to all this a set of geologic hazards: severe land subsidence as underground water supplies are overdrawn; the ever-present threat of earthquakes that can wreak havoc on the city's unstable surface (the last big one occurred in 1985); and even the risk of volcanic activity as nearby Mount Popocatépetl occasionally shows signs of ending centuries of dormancy.

In spite of it all, the great city continues to beckon, and every year more than 100,000 of the desperate and the dislocated arrive with hope—and little else.

---

assembling bulky products such as motor vehicles and major home appliances were still better off situated right across the U.S. border, many others producing lighter and smaller goods such as electronic equipment and cameras were quick to relocate to China, Vietnam, and other Asia-Pacific countries. As a result, thousands of Mexican workers became unemployed, with many crossing into the United States.

Mexico's only way to reverse this trend is through the growth of higher-paying jobs in more advanced sectors—particularly electronics—that are less likely to be lost to other parts of the developing world. Expansion of education and training in high-tech and management

fields is a critical first step. The northeastern city of Monterrey in the relatively high-income State of Nuevo Léon has become the most successful model in this effort, nurturing both a substantial international business community and a critical mass of modern industrial facilities that have attracted leading multinational corporations. Here indeed lies hope for Mexico's future.

States of Contrast

Countries with strong regional disparities face serious challenges that can be difficult to overcome and may worsen over time. Mexico's southernmost States—Chiapas, Oaxaca, and Guerrero, all bordering the Pacific Ocean—are by far the poorest. States bordering the United States in the north, including Nuevo Léon, Chihuahua, and Baja California, have the highest incomes. Using rural poverty as a measure, only about 10 percent of people in the countryside in the north are in the poorest category, but nearly 50 percent in the south.

Mexico's north-south divide is especially evident in the economic data mapped in Figure 4B-6. In general terms, annual per capita income in the northern States exceeds U.S. $10,000; in the southern States it falls below

$5000. Economic growth in the northern States has averaged more than 4 percent in recent years; in the southern States it is less than 2 percent. Mexico's infrastructure, already inadequate, serves the south far less well than the north. Whatever the index—literacy, electricity use, water availability—the south lags.

This is a serious problem for Mexico because the south is not only culturally the most Amerindian part of the country, but also the least well educated, the most isolated and remote, the least productive agriculturally, and lagging badly in terms of infrastructure investment and

FIGURE 4B-6

© H. J. de Blij, P. O. Muller, and John Wiley & Sons, Inc.

overall development. Since 1994, a radical group of Maya peasant farmers in Chiapas State calling themselves the Zapatista National Liberation Army (ZNLA) have engaged in guerrilla warfare, demanding better treatment for Mexico's 32 million Amerindian citizens (who mostly live in the badly impoverished south). Despite substantial public support, their struggle has not yielded significant results. In fact, NAFTA has only widened the gap between north and south. To upwardly mobile Mexicans in the north, the cause of the Zapatistas is a distant one indeed.

These still-widening contrasts were thrown into sharp relief in 2006, when Mexico's most recent presidential election was contested by three candidates, of whom the two leading ones were a conservative and a populist. When the ballots were counted there was a near-tie, a result so close that the populist at first refused to concede. But on the map, it was clear that the north was won by the conservative candidate and the south by the populist (see inset map in Fig. 4B-6). The country's regional economic disparities were producing serious social and political consequences. It will be most interesting to see if the 2012 presidential election will produce any change in this pattern.

The Drug Wars

Since the late 2000s, Mexico has been bedeviled by yet another obstacle in its struggle to achieve sustained economic progress. This problem has received worldwide attention as the cocaine-producing drug cartels centered in Colombia established new bases in U.S.-border cities in northern Mexico and launched a vicious war for supremacy. They responded in part to the success of the antidrug campaign in Colombia, but the cartels also saw new opportunities beckoning in Mexican territory adjacent to their main market in the United States. So serious was the situation by 2010 that the drug barons and their allies were killing law-enforcement officers by the hundreds as well as competitors (and often innocent bystanders) by the thousands, intimidating government and threatening all who stood in their way. Fueling this rampage was the steady stream of guns, purchased easily in the United States with the flood of drug money generated by that illicit trade, which flowed across the border and further empowered the cartels.

Figure 4B-7 provides an overview of the geography of this drug war. Cocaine is produced in Colombia, Bolivia, and Peru (see Chapter 5B), and most of it these days enters

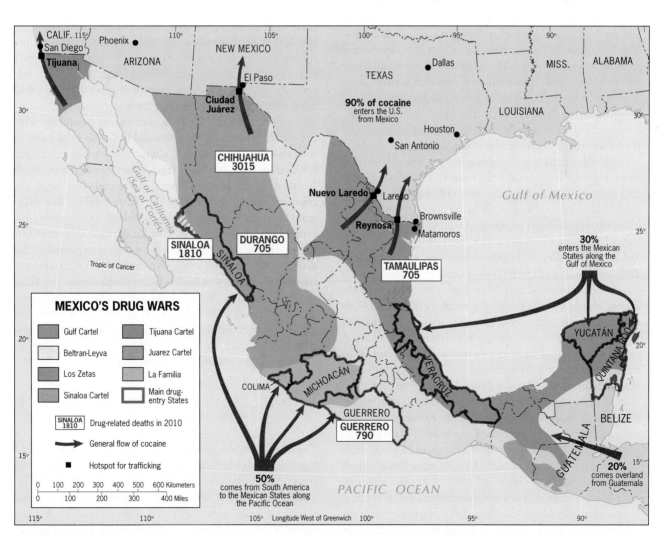

FIGURE 4B-7

© H. J. de Blij, P. O. Muller, J. Nijman, and John Wiley & Sons, Inc.

the United States through Mexico. Bolivian and Peruvian cocaine, constituting about half of Mexico's "imports," is shipped by sea and illegally enters the country in the Pacific-Coast States of Guerrero, Michoacán, Colima, and Sinaloa. The other half originates in Colombia and enters either overland via the Middle American land bridge through Guatemala, or by boat across the Caribbean and Gulf of Mexico to be smuggled into the eastern States of Yucatán, Quintana Róo, and Veracruz. The illicit powder is then repackaged (and/or crystallized to form crack) in smaller quantities for "retailing" purposes before being transferred into the United States, mainly through the border cities of Reynosa, Nuevo Laredo, Ciudad Juárez, and Tijuana (Fig. 4B-7).

Seven major cartels are involved in this drug trafficking operation, and competition today is literally murderous. Territorial control is key in this "business"; each cartel dominates its own turf, with various pieces being heavily contested from time to time. The cartels themselves are not particularly stable, with countless mergers and splits occurring since 2007. The two leading rivals are *Los Zetas* and the *Sinaloa Cartel*, which control huge swaths of eastern and western Mexico, respectively. The *Sinaloa* recently formed an alliance with two smaller cartels, *La Familia* and the *Gulf*, creating what is referred to as *The Federation*—which in early 2011 was engaged in a violent conflict for supremacy with *Los Zetas*. The other major organizations are *Beltran-Leyva*, the *Tijuana Cartel*, and the *Juárez Cartel*. The first controls the States of Sonora and a small portion of neighboring Sinaloa, while the latter two are mainly concentrated in the metropolitan areas of Tijuana and Juárez where they control two of the key "export valves."

Much of the worst violence occurs where control is contested, especially around the most crucial ports of entry and primary transport routes. In 2010, nearly half of the total 15,273 killings took place in the States of Sinaloa (cocaine entry; marijuana production) and Chihuahua (which contains Juárez, the biggest U.S. port of entry). By early 2011, the number of murders since the war erupted in 2007 went well beyond 35,000. Even though the Mexican government has called in more and more troops to fight the cartels, they have only achieved limited success. In 2011, as noted in Chapter 4A, the heightening level of violence was rapidly driving away the civilian population, particularly in Juárez which has already lost more than 15 percent of its residents since 2008. In fact, in May 2011 the city officially changed its name to Heroica Ciudad Juárez (Heroic City of Juárez), coincidentally to mark the hundredth anniversary of a decisive local battle in the Mexican Revolution.

As the conflict intensified, some U.S. observers began to regard Mexico not only as a threat to national security but also as a **failed state [5]**, because the Mexican government had lost control of parts of its own territory. This is more than a little ironic since the United States is the main market for the northward-flowing drugs, with the profits earned on American streets funneled clandestinely back into Mexico. At the same time, the guns used in the drug war overwhelmingly originate in the United States, making it difficult to view the conflict as an exclusively Mexican problem.

Even as the war raged, the cartels have worked to diversify their activities. One example will suffice. During 2008 and 2009, the *Los Zetas* cartel, which was founded by former military commandos, siphoned away more than $1 billion worth of oil from state-owned pipelines in the Veracruz sector of the Gulf Coast. The *Zetas* do not tap the oil themselves, but are said to effectively "own" lengthy stretches of pipeline and "tax" anyone who has the technology to siphon it off. Most of the stolen oil is then sold across the border to U.S. companies (who deny knowing anything about its source), not only bleeding the national treasury but also underscoring the state's impotence.

Mexico's Future

The future of the country depends on the ability of the government to meet an array of formidable challenges. In the short term, it must end the violence that has destabilized much of the country. In the medium to long term, it must close the gap between rich and poor to reduce vast—and widening—regional inequalities. And from the standpoint of economic geography, the government must spearhead the effort to spread the positive effects of NAFTA from north to south, especially through infrastructural-upgrade projects, increased investment in education, and more effective antipoverty programs.

The future of Mexico is inextricably bound up with the United States, in good times and bad. The good refers to the two countries' economic interaction: Mexico's economy has undoubtedly benefited from closer economic ties with North America, and the United States continues to be the country's most important trading partner. Given this foundation, Mexico's longer-term problems may seem less acute—but it would be a mistake for the federal government to lose sight of them, no matter how urgent its pursuit of the drug cartels. Clearly, Mexico is in the midst of a tumultuous transformation, and its leaders face multiple opportunities as well as challenges. When times are prosperous, Mexicans dream of possibilities such as the so-called **dry canal [6]** across the narrowest part of the country that would compete with the Panama Canal (see Fig. 4B-2). When times are bad, the very integrity of the state seems at risk.

▪▪ THE CENTRAL AMERICAN REPUBLICS

A Land Bridge

Crowded onto the narrow segment of the Middle American land bridge between Mexico and the South American continent are the seven countries of Central America (Fig. 4B-8). Territorially, they are all quite small; their population sizes range from Guatemala's more than 15 million down to Belize's 300,000. The land bridge here consists of

FIGURE 4B-8   © H. J. de Blij, P. O. Muller, and John Wiley & Sons, Inc.

a highland belt flanked by coastal lowlands on both the Caribbean and Pacific sides (Fig. 4B-9). These highlands are studded with volcanoes, and local areas of fertile volcanic soils are scattered throughout them.

The land bridge has a fascinating geologic and evolutionary history, and one famous study refers to it as the Monkey Bridge. For some 50 million years, North and South America were separated; the land bridge was formed "only" 3 million years ago, becoming a highway of sorts for evolutionary exchange. The region contains only 1 percent of the Earth's land but 7 percent of all the world's natural species. The southern part (Costa Rica and Panama) is known as a global **biodiversity hot spot [7]**, even though deforestation has been a major problem (see photo, p. 199). Human inhabitants have always been concentrated in the upland zone, where tropical temperatures are moderated by elevation and rainfall is sufficient to support a variety of crops.

Central America is not a large region, but because of its physiography it contains many isolated, comparatively inaccessible locales. The population tends to concentrate in the much cooler uplands—*tierra templada* [8]—and population densities are generally greater toward the Pacific than toward the Caribbean side (see Fig. 4A-2). The most significant exception is El Salvador, whose political boundaries confine its inhabitants mostly to its tropical coastal lowlands—*tierra caliente* [9]—tempered here by the somewhat cooler Pacific offshore.

Central America, as noted earlier, actually begins within Mexico, in Chiapas and in Yucatán, and the region's republics face many of the same problems as the least-developed parts of Mexico. Population pressure is one of them. A population explosion began in the mid-twentieth century, increasing the region's inhabitants from 9 million to almost 45 million by 2012.

Today, the region also continues its slow emergence from a grim period of turmoil that lasted through the 1980s well into the 1990s. Devastating inequities, repressive governments, external interference, and the frequent unleashing of armed forces have destabilized Central America for much of its modern history.

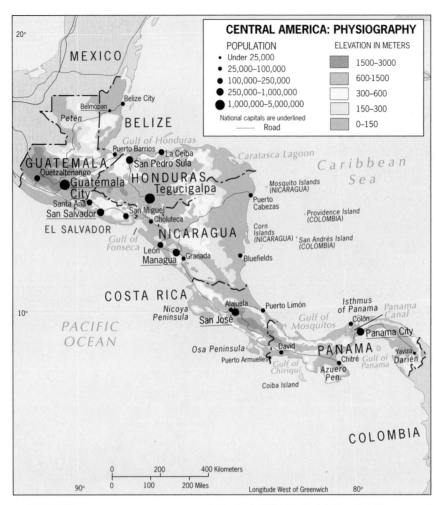

FIGURE 4B-9

© H. J. de Blij, P. O. Muller, J. Nijman, and John Wiley & Sons, Inc.

Guatemala

The westernmost of Central America's republics, Guatemala has more land neighbors than any other. Straight-line boundaries across the tropical forest mark much of the border with Mexico, creating the box-like region of Petén between Chiapas State on the west and Belize on the east; also to the east lie Honduras and El Salvador (Fig. 4B-8). This heart of the ancient Maya Empire, which remains strongly infused by Amerindian culture and traditions, has just a small window on the Caribbean but a longer Pacific coastline. Guatemala was still part of Mexico when the Mexicans threw off the Spanish yoke, and though independent from Spain after 1821, it did not become a separate republic until 1838. Mestizos, not the Amerindian majority, secured the country's independence.

Most populous of the seven republics with 15.2 million inhabitants (mestizos are in the majority with 59 percent, Amerindians 41 percent), Guatemala has seen much conflict. Repressive regimes brokered deals with U.S. and other foreign economic interests that stimulated development, but at a high social cost. Over the past half-

century, military regimes have dominated political life. The deepening split between the wretchedly poor Amerindians and the better-off mestizos, who here call themselves *ladinos*, generated a civil war that started in 1960 and claimed more than 200,000 lives as well as 50,000 "disappearances" before it ended in 1996. An overwhelming number of the victims were of Maya descent; the mestizos continue to control the government, army, and land-tenure system.

Guatemala's tragedy is that its economic geography has considerable potential, but has long been shackled by unceasing internal conflicts and the widening of one of the hemisphere's largest gaps between rich and poor. In tandem, they keep the income of at least 60 percent of the population below the poverty line (some estimates range as high as 75 percent). The country's mineral wealth includes nickel in the highlands and oil in the lower-lying north. Agriculturally, soils are fertile and moisture is ample over highland areas large enough to produce a wide range of crops, including excellent coffee.

In 2011, a dangerous new problem threatened the country as the drug wars began to spill across the Mexican border. Figure 4B-7 shows that both *Los Zetas* and the *Sinaloa Cartel* have staked out turfs in southwestern Guatemala, the staging area for the 20 percent of Colombian cocaine that enters Mexico overland. In May 2011, assailants presumably affiliated with *Los Zetas* massacred 27 people at a ranch in the as-yet-unclaimed Petén region—very likely to open up a new route into Mexico through northern Guatemala for smuggling both narcotics and (increasingly profitable) illegal migrants headed for the U.S. If this activity intensifies as local authorities expect, it raises a huge new challenge for the Mexican government: already stretched thin fighting in the drug-related conflicts in its northern border zone, the country's security forces will increasingly need to also police the opposite, 929-kilometer (577-mi) border with Guatemala. And should that fail to contain the invasion of Mexico's drug cartels, the stability of the entire land bridge would be endangered. In the words of President Alvaro Colom of Guatemala, ". . . either the countries of Central America join together to fight them or they will defeat us and finish off our democracies."

Belize

Strictly speaking, Belize is not a Central American republic in the same tradition as the other six. Until 1981, this country, a wedge of land between northern Guatemala, Mexico's Yucatán Peninsula, and the Caribbean Sea (Fig. 4B-8), was a dependency of the United Kingdom known as British Honduras. Slightly larger than Massachusetts and with a minuscule population of just over 300,000 (many of African descent), Belize is much more reminiscent of a Caribbean island than of a continental Middle American state. Today, all that is changing as the demographic complexion of Belize is being reshaped. Thousands of residents of African descent have recently emigrated (many went to the United States) and were replaced by tens of thousands of Spanish-speaking immigrants. Most of the latter are refugees from strife in nearby Guatemala, El Salvador, and Honduras, and their proportion of the Belizean population has risen from 33 to just under 50 percent since 1980. Within the next few years the newcomers will be in the majority, Spanish will emerge as the *lingua franca*, and Belize's cultural geography will become increasingly Hispanicized.

The Belizean transformation extends to the economic sphere as well. No longer just an exporter of sugar and bananas, Belize is producing new commercial crops, and its seafood-processing and clothing industries have become major revenue earners. Also important is tourism, which annually lures more than 250,000 vacationers to the country's Mayan ruins, resorts, and newly legalized casinos; a growing specialty is *ecotourism*, based on the natural attractions of the country's near-pristine environment, including its magnificent offshore coral reefs. Belize is also known as a center for **offshore banking [10]**—a financial haven for foreign companies and individuals who want to avoid paying taxes in their home countries.

Honduras

Honduras is a country still hampered by the struggle to rebuild its battered infrastructure and economy more than a decade after taking a direct hit from Category-5 Hurricane Mitch in 1998. With 8 million inhabitants, about 90 percent mestizo, bedeviled Honduras still has years to go even to restore what was already the third-poorest economy in the Americas (after Haiti and Nicaragua). Agriculture,

From the Field Notes . . .

© Jan Nijman

"Invited by the Honduran government to survey the country's ecotourism potential, I toured the beautiful countryside with a National Geographic Society delegation in 2006. There is no doubt that Honduras is exceptionally endowed by nature and that there is considerable potential for a vibrant tourist industry. But this is a poor country, and the government has great difficulty making the required investments—in infrastructure, for example. We flew in these small propeller planes to get from the capital, Tegucigalpa, to the Mayan ruins of Copán. There was no airport nearby and overland travel is arduous and time-consuming. Ecotourism by its very nature is small-scale and the revenues tend to be relatively modest, so it is a major challenge to raise the substantial funds needed for costly infrastructural improvements."

www.conceptcaching.com

livestock, forestry, and limited mining formed the mainstays of the pre-1998 economy, with the familiar Central American products—bananas, coffee, shellfish, and apparel—earning most of the external income.

Honduras, in direct contrast to Guatemala, has a lengthy Caribbean coastline and a small window on the Pacific (Fig. 4B-8). The country also occupies a critical place in the political geography of Central America, flanked as it is by Nicaragua, El Salvador, and Guatemala—all continuing to grapple with the aftermath of years of internal conflict and, more recently, natural disaster. Comparable in natural beauty and biodiversity to Costa Rica (see photo on previous page), Honduras is seeking to exploit its potential for ecotourism but is constrained by a poor infrastructure and lack of funds for new facilities.

El Salvador

This is Central America's smallest country territorially, smaller even than Belize, but with a population 21 times as large (6.4 million) it is the most densely peopled. With Belize, it is one of only two continental republics that lack coastlines on both the Caribbean and Pacific sides (Fig. 4B-8). El Salvador adjoins the Pacific in a narrow coastal plain backed by a chain of volcanic mountains, behind which lies the country's heartland. Unlike neighboring Guatemala, El Salvador has a quite homogeneous population (85 percent mestizo and just 1 percent Amerindian). Yet ethnic homogeneity has not translated into social or economic equality or even opportunity. Whereas other Central American countries were called banana republics, El Salvador was a coffee republic, and the coffee was produced on the huge landholdings of a few landowners and on the backs of a subjugated peasant labor force. The military supported this system and repeatedly suppressed violent and desperate peasant uprisings.

From 1980 to 1992, El Salvador was torn by a devastating civil war that was worsened by outside arms supplies from the United States (supporting the government) and Nicaragua (aiding the Marxist rebel forces). But ever since the negotiated end to that war, efforts have been under way to prevent a recurrence because El Salvador is having difficulty overcoming its legacy of searing inequality. The civil war did have one positive result: affluent citizens who left the country and did well in the United States and elsewhere send substantial funds back home; these *remittances* now provide the largest single source of foreign revenues. This has helped stimulate such industries as apparel and footwear manufacturing as well as food processing. But a major stumbling block to revitalization of the agricultural sector has again been land reform, and El Salvador's future still hangs in the balance.

Nicaragua

This country is best approached by reexamining the map (Fig. 4B-8), which underscores Nicaragua's position tucked away in the heart of Central America. The Pacific coast follows a southeasterly direction here, but the Caribbean coast is oriented north-south so that Nicaragua forms a triangle of land, with its lakeside capital, Managua, located in a valley on the mountainous, earthquake-prone, Pacific side (the country's core area has always been located here). The Caribbean side, where the uplands yield to a coastal plain of tropical rainforest, savanna, and swampland, has for centuries been home to Amerindian peoples such as the Miskito, who have been remote from the focus of national life.

Until the 1990s, Nicaragua had a checkered history of political instability, conflict, and economic backwardness. The strife ended in 1990 when the first democratic government in decades was voted into office. But economic progress has been scant. For the past three decades, Nicaragua's economy has ranked as continental Middle America's poorest. In 1998, Hurricane Mitch struck here too, devastating the country's farms and driving tens of thousands into the impoverished towns just at a time when the agricultural sector was recovering and land reform promised a better life for some 200,000 peasant families.

The country's options are limited: agriculture dominates the economy and manufacturing is weak. There is a growing reliance on remittances from Nicaraguans who have emigrated (now 15 percent of the economy) as well as on foreign aid. For years there has been talk of a *dry canal* transit role for this country, but other potential overland routes are more promising. Meanwhile, the accelerated growth of Nicaragua's population continues (2012 total: 6.2 million), dooming hopes for a rise in standards of living.

Costa Rica

If there is one country that underscores Middle America's endless variety and diversity it is Costa Rica—because it differs significantly from its neighbors and from the norms of Central America as well. Bordered by two volatile countries (Nicaragua to the north and Panama to the east), Costa Rica is a nation with an old democratic tradition and, in this cauldron, no standing army for the past six decades! Although the country's Hispanic imprint is similar to that found elsewhere on the mainland, its early independence, its good fortune to lie remote from regional strife, and its leisurely pace of settlement allowed Costa Rica the luxury of concentrating on its economic development. Perhaps most important, internal political stability has prevailed over much of the past 175 years.

Like its neighbors, Costa Rica is divided into environmental zones that parallel its coastlines. The most densely settled is the central highland zone, lying in the cooler *tierra templada*, whose heartland is the *Valle Central* (Central Valley), a fertile basin that contains Costa Rica's main coffee-growing area and the leading population cluster focused on the capital, San José (Fig. 4B-8)—the most cosmopolitan urban center between Mexico City and the primate cities of northern South America.

The long-term development of Costa Rica's economy has given it the region's highest standard of living, literacy rate, and life expectancy (though even here, one-fifth of the

From the Field Notes . . .

"The Panama Canal remains an engineering marvel 90 years after it opened in August 1914. The parallel lock chambers each are 1000 feet long and 110 feet wide, permitting vessels as large as the *Queen Elizabeth II* to cross the isthmus. Ships are raised by a series of locks to Gatún Lake, 85 feet above sea level. We watched as tugs helped guide the *QEII* into the Gatún Locks, a series of three locks leading to Gatún Lake, on the Atlantic side. A container ship behind the *QEII* is sailing up the dredged channel leading from the Limón Bay entrance. The lock gates are 65 feet wide and 7 feet thick, and range in height from 47 to 82 feet. The motors that move them are recessed in the walls of the lock chambers. Once inside the locks, the ships are pulled by powerful locomotives called mules that ride on rails that ascend and descend the system. It was still early morning, and a major fire, probably a forest fire, was burning near the city of Colón, where land clearing was in progress. This was the beginning of one of the most fascinating days ever."

© H. J. de Blij

www.conceptcaching.com

Concept Caching

population is trapped in poverty). Agriculture continues to dominate (with bananas, coffee, tropical fruits, and seafood the leading exports), and tourism has expanded steadily. Despite the deforestation of more than 80 percent of its original woodland cover, Costa Rica is well known for its superb scenery and for its efforts to protect what is left of its diverse tropical flora and fauna.

Panama

Panama owes its existence to the idea of a canal connecting the Atlantic and Pacific oceans to avoid the lengthy circumnavigation of South America. The Panama Canal (see the inset map in Fig. 4B-8) was opened in 1914, a symbol of U.S. power and influence in Middle America. The Canal Zone was held by the United States under a treaty that granted it "all the rights, powers, and authority" in the area "as if it were the sovereign of the territory." Such language might suggest that the United States held rights over the Canal Zone in perpetuity, but the treaty nowhere stated specifically that Panama permanently yielded its own sovereignty in that transit corridor. In the 1970s, as the canal was transferring more than 14,000 ships per year (that number today is about the same, but the cargo tonnage is up significantly) and generating hundreds of millions of dollars in tolls, Panama sought to terminate U.S. control in the Canal Zone. Delicate negotiations began. In 1977, an agreement was reached on a staged withdrawal by the United States from the territory, first from the Canal Zone and then from the Panama Canal itself (a process completed on December 31, 1999).

Panama today reflects some of the usual geographic features of the Central American republics. Its population of 3.6 million is about 70 percent mestizo and also contains substantial Amerindian, European, and black minorities. Spanish is the official language, but English is also widely used. Ribbon-like and oriented east-west, Panama's topography is mountainous and hilly. Eastern Panama, especially Darien Province adjoining Colombia, is densely covered by tropical rainforest, and here is the only remaining gap in the intercontinental Pan American Highway (Fig. 4B-8). Most of the rural population lives in the uplands west of the canal; there, Panama produces bananas, coffee, sugarcane, rice, and, along its narrow coastal strips, shrimps and other seafood. Much of the urban population is concentrated in the vicinity of the waterway, anchored by the cities at each end of the canal.

Near the northern, Caribbean end of the Panama Canal lies the city of Colón, site of the Colón Free Zone, a huge trading *entrepôt* designed to transfer and distribute goods bound for South America. It is augmented by the Manzanillo International Terminal, an ultramodern port facility capable of transshipping more than 1000 containers a day. Near the southern, Pacific end lies Panama City, often likened to Miami because of its waterfront location and high-rise-dominated skyline. The capital is the financial center that handles the revenues generated by the Canal, but its high-rise profile also reflects the proximity of Colombia's illicit drug industry and associated money-laundering and corruption. The city's modern image presents a stark contrast to the (rural) poverty that afflicts one-third of Panama's total population.

The expansion of the Panama Canal will surely boost interoceanic traffic and increase business opportunities in Panama. U.S. ports on the Gulf of Mexico and the

NEW FEATURES

1 Retaining basins
fill lock chambers
and retain 60
percent of the
water for re-use.

2 Rolling lock gates
slide in recesses,
saving space and
making maintenance
work much easier.

3 Tugboats
maneuver vessels
into position,
replacing costly
towing locomotives.

Locomotives

**Current
locks ▼**

**Proposed
new lane ▶**

*Locks transfer vessels
from one elevation to another
by filling closed chambers with
water, or emptying them. When
the level is even with that of the
next chamber, the gates open.*

Tugboat

POST-PANAMAX SHIP

PANAMAX SHIP
Current
maximum size

Adapting to larger ships

The most recent container vessels and
tankers are known as post-Panamax.
They can hold up to 12,000 containers,
2.5 times the cargo capacity of ships
able to fit in the existing locks.

160 ft
60 ft
180 ft

106 ft   42 ft
110 ft

**Diagram showing the planned expansion and upgrading of the
Panama Canal's lock system in order to accommodate ships that
currently exceed the waterway's ("Panamax") capacity. The target
date for completing the project is August 15, 2014, the one-
hundredth anniversary of the first interocean transit of the Canal
by the passenger-cargo vessel, S.S. Ancon.**

© NG Image Collection

Atlantic seaboard will also get busier as they handle this heightened volume of trade. The ports of the New York–New Jersey urban region have already begun to expand their capacity, and leading retailers such as Walmart (a voracious importer of Chinese goods) are building additional storage facilities at the port of Houston and other key coastal distribution points.

Nevertheless, China will be the biggest beneficiary because its rapid recent development has made it the world's number-one exporter. Even during the post–2007 global recession, which primarily affected the United States and western Europe, the Chinese economy kept moving forward. And China's imports coming through an expanded Panama Canal are expected to be considerably cheaper, further enhancing its economic strength and standing on the international scene. Most of all, this is yet another example of just how tightly interconnected the world economy has become—and how the relative location of one small country in Middle America is of the utmost global significance.

THE CARIBBEAN BASIN

Fragmentation and Insularity

As Figure 4B-1 reveals, the Caribbean Basin, Middle America's island region, consists of a broad arc of numerous islands extending from the western tip of Cuba to the southern coast of Trinidad. The four larger islands or **Greater Antilles** (Cuba, Hispaniola, Jamaica, and Puerto Rico) are clustered in the western segment of this arc. The smaller islands or **Lesser Antilles** of the eastern segment extend southward within a crescent-shaped chain from the Virgin Islands to Trinidad and Tobago. (Breaking this tectonic-plate-related regularity are the Bahamas and the Turks and Caicos, located north of the Greater Antilles, and numerous other islands too small to appear on a map at the scale of Fig. 4B-1.)

On these innumerable islands, whose combined land area accounts for only 9 percent of Middle America, lie 33 states and many other political entities. (Europe's colonial flags have not totally disappeared from this region, and the U.S. flag flies over Puerto Rico and the Virgin Islands.) The populations of these states and territories, however, constitute 21 percent of the entire geographic realm, making this the most densely peopled part of the Americas.

The island-nations of the Caribbean Basin are generally (very) small, and their territories are separated and often at considerable distances from other islands. These fragmented geographic conditions create a set of circumstances that has proved to be challenging. Economic opportunities are few, most things are relatively expensive, and due to limited interaction with the outside world the island societies tend to be static.

Ethnicity and Class

For one thing, **social stratification [11]** in most Caribbean islands is rigid and social mobility is limited. Class structures tend to be closely associated with ethnicity, and as such the region still carries imprints of colonial times. The historical geography of Cuba, the Dominican Republic, and Puerto Rico is suffused with Hispanic culture; Haiti and Jamaica carry stronger African legacies. But the reality of this ethnic diversity is that European lineages still hold the advantage. Hispanics tend to be in the best positions in the Greater Antilles; people who have mixed European-African ancestries, and who are described as **mulatto [12]**, rank next. The largest part of this social pyramid is also the least advantaged: the Afro-Caribbean majority. In virtually all Caribbean societies, the minorities hold disproportionate power and exert overriding influence. In Haiti, the mulatto minority accounts for less than 5 percent of the population but has long held most of the power. In the adjacent Dominican Republic, the pyramid of power puts Hispanics (16 percent) at the top, the mixed sector (73 percent) in the middle, and the Afro-Caribbean minority (11 percent) at the bottom. In the Caribbean social mosaic, historic advantage has a way of perpetuating itself.

The composition of the population of the islands is further complicated by the presence of Asians from both China and India. During the nineteenth century, the emancipation of slaves and subsequent local labor shortages brought some far-reaching solutions. Some 100,000 Chinese emigrated to Cuba as indentured laborers; and Jamaica, Guadeloupe, and especially Trinidad saw nearly 250,000 South Asians arrive for similar purposes. To the African-modified forms of English and French heard in the Caribbean Basin, therefore, can be added several Asian languages. The ethnic and cultural variety of the plural societies of Caribbean Middle America is indeed endless.

■■ THE GREATER ANTILLES

The four islands of the Greater Antilles (whose populations constitute 90 percent of the Caribbean Basin's total) contain five political entities: Cuba, Haiti, the Dominican Republic, Jamaica, and Puerto Rico (Fig. 4B-1). Haiti and the Dominican Republic share the island of Hispaniola.

Cuba

The largest Caribbean island-state in terms of both territory (111,000 square kilometers/43,000 sq mi) and population (11.3 million), Cuba lies only 145 kilometers (90 mi) from the southernmost island of Florida (Fig. 4B-10). Havana, the now-dilapidated capital, lies almost directly south of the outermost Florida Keys on the northwest coast of the elongated island. Cuba was a Spanish possession up until the late 1890s when, with U.S. help in the Spanish-American War, it achieved independence. Fifty years later, an American-backed dictator was in control, and by the 1950s Havana had become an American playground. The island was ripe for revolution, and in 1959 Fidel Castro's insurgents gained control, thereby converting Cuba into a communist dictatorship and a Soviet client. Nearly a million Cubans fled the island for the United States, and Miami became the second-largest Cuban city after Havana.

In the fall of 1962, the world was on the brink of nuclear war when the United States called on the Soviet Union to remove its newly installed nuclear missiles from Cuba (aimed at the United States) or face reprisals. In the end, the Soviets conceded and in return Washington promised not to attack Cuba. The United States also continued to retain the Guantánamo Bay Naval Base located near the southeastern tip of the island (Fig. 4B-10), which was perpetually leased to the United States by the Cuban government in 1903. Since 2002, the Guantánamo base has acquired notoriety as a detention camp for 9/11-related terrorists as well as prisoners of war from Iraq and Afghanistan.

Castro's rule survived the 1991 collapse of the Soviet Empire despite the loss of subsidies and sugar markets

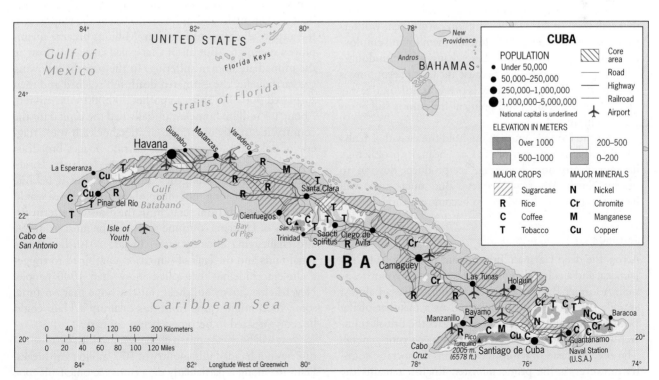

FIGURE 4B-10

© H. J. de Blij, P. O. Muller, and John Wiley & Sons, Inc.

on which it had long relied. As the map shows, sugar was Cuba's economic mainstay for many years; the plantations, once the property of rich landowners, extend all across the island. But sugarcane is losing its position as the leading Cuban foreign exchange earner. Mills have been closing down, and cane fields are being cleared for other crops as well as for pastures. Cuba has additional economic opportunities, however, especially in its highlands. There are three mountainous areas, of which the southeastern *Sierra Maestra* is the highest and most extensive. These highlands create considerable environmental diversity as reflected by extensive timber-producing tropical forests and varied soils on which crops ranging from tobacco to coffee to rice to subtropical and tropical fruits are grown. Rice and beans are the staples, but Cuba is unable to meet its dietary needs and so must import food. The savannas of the center and west support livestock. Even though Cuba has only limited mineral reserves, its nickel deposits are extensive and have been mined for more than a century.

Early in the twenty-first century, Cuba found a crucial new supporter in Venezuela's leader, Hugo Chávez. Cuba has no domestic petroleum reserves, but Venezuela is oil-rich and, since 2003, has been providing all of the fuel that Cuba needs. In return, Castro sent 30,000 health workers and other professionals to Venezuela, where they aid the poor.

Poverty, crumbling infrastructure, crowded slums, and unemployment mark the Cuban cultural landscape, but the regime still seems to have support among the general population. In 2006, Fidel Castro became too ill to lead the country and turned the reins over to his brother Raúl. Things stayed the same for the most part, with a few modest liberalization measures slowly taking effect in the years that followed. In 2011, half a million government workers were scheduled to be laid off, forcing them to find employment in the private sector. These changes have undoubtedly been born of necessity because the Cuban economy is in such woeful shape, but there is precious little in the way of constructive reform and political freedom is still completely suppressed.

In the United States, many hold the view that Cuba could be the shining star of the Caribbean, its people free, its tourist economy booming, its products flowing to nearby North American markets. But much more will have to change for that vision to become a reality.

Jamaica

Across the deep Cayman Trench from southern Cuba lies Jamaica, and a cultural gulf separates these two countries as well. A former British dependency, Jamaica has an almost entirely Afro-Caribbean population. As a member of the British Commonwealth, it still recognizes the British monarch as the chief of state, represented by a governor-general. The effective head of government in this democratic country, however, is the prime minister. English remains the official language here, and British customs still linger.

Smaller than Connecticut and with 2.8 million people, Jamaica has experienced a steadily declining national income over the past few decades despite its relatively slow population growth. Tourism has become the largest source of income, but the markets for bauxite (aluminum ore), of which Jamaica is a major exporter, have dwindled. And like other Caribbean countries, Jamaica has trouble making money from its sugar exports. Jamaican farmers also produce crops ranging from bananas to tobacco, but the country faces the disadvantages on world markets common to those in the global periphery. Meanwhile, Jamaica must import all of its oil and much of its food because the densely populated coastal flatlands suffer from overuse and shrinking harvests.

The capital, Kingston, lies on the southeast coast and reflects Jamaica's economic struggle. Almost none of the hundreds of thousands of tourists who visit the country's beaches, explore its Cockpit Country of (karst) limestone towers and caverns, or populate the cruise ships calling at Montego Bay or other points along the north coast get even a glimpse of what life is like for the ordinary Jamaican.

Haiti

Already the poorest state in the Western Hemisphere for several decades, Haiti in the fall of 2008 was staggered by no less than four tropical cyclones that struck within a few weeks of each other. The storms and the floods they unleashed killed more than 800 people and dislocated some 800,000. Then, barely 15 months later, a killer earthquake on January 12, 2010 caused devastation on an unimaginable scale (see photo, page 197). The capital, Port-au-Prince, lay in ruins, the infrastructure collapsed, and the Haitian government became effectively invisible. Schools, hospitals, and railroads as well as the seaport and the airport all ceased to function. Millions became unemployed overnight. But all of that paled in comparison to the immediate human suffering. In the months following the earthquake, the estimated death toll reached and then surpassed 300,000; at least another 300,000 were injured. About 1.5 million homeless initially fled the capital for the countryside, partly out of fear of aftershocks. It was a tragedy of colossal proportions, its horrific images broadcast around the world on television and the Internet. Large-scale international emergency aid arrived, but it became clear that, in the long run, much of the affected area would have to be rebuilt from scratch. In the disaster's aftermath, millions of Haitians continue to live in the most primitive and challenging circumstances. In late 2010, heavy seasonal rains and the lack of sanitation combined to trigger a cholera epidemic that killed more than 1000 people. Nevertheless, many still held out the hope that, in time, some good might come from this calamity if Haiti could be reconstructed to become a better country.

It will be a gargantuan challenge, not least because nature has not handed this country of 10.2 million any breaks. Haiti lies along a particularly dangerous tectonic fault zone where the North American and Caribbean plates meet.

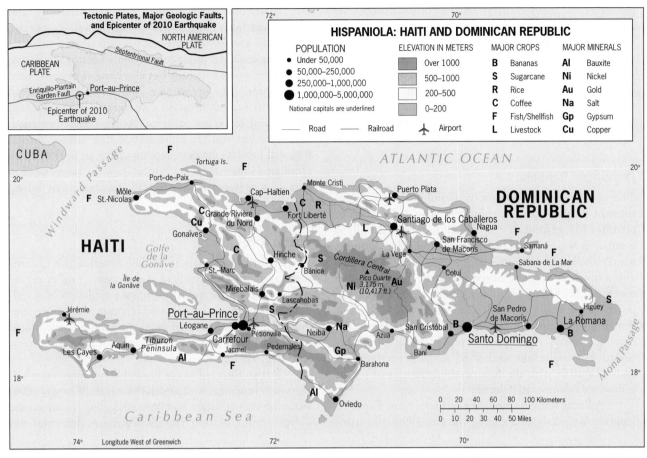

FIGURE 4B-11

© H. J. de Blij, P. O. Muller, and John Wiley & Sons, Inc.

The quake of 2010, geologists say, had been in the making for centuries, and more could easily follow in the future (Fig. 4B-11, inset map). The country also lies in the center of Hurricane Alley, and even a single year (fortunately including 2010) without a major storm is something to be grateful for. Finally, Haiti has few natural resources and little to offer in the international trading arena. Few countries in the world have a more checkered history than Haiti, where political instability, repression, and material deprivation have been constants for well over a century. In 2009, Haiti's GNI

From the Field Notes . . .

"Fly along the political boundary between Haiti and the Dominican Republic, and you see long stretches of the border marked by a stark contrast in vegetation: denudation prevails to the west in Haiti while the forest survives on the Dominican (eastern) side. Overpopulation, lack of governmental control, and mismanagement on the Haitian side combine to create one of the region's starkest spatial contrasts."

www.conceptcaching.com

© H. J. de Blij

per capita had fallen to barely one-fifth of Jamaica's, a level lower than that of many poor African countries. Compared to other countries in Middle and South America, its infant mortality rate was nearly three times as high. Only 50 percent of elementary school-age girls actually received an education. Most of the country's limited public expenditures were made possible through foreign aid, and private consumption relied heavily on remittances from the Haitian Diaspora in places like Miami, Paris, New York, and Montreal.

Dominican Republic

The mountainous Dominican Republic has a larger share of Hispaniola than Haiti (Fig. 4B-11) in terms of territory (64 percent of the island), but at 10.2 million is equal in size to the Haitian population. The "Dominican," too, has a wide range of natural environments but also a much stronger resource base than its neighbor to the west. Nickel, gold, and silver have long been exported along with sugar, tobacco, coffee, and cocoa, but tourism—the great opportunity lost to Haiti—is the leading industry. A long period of dictatorial rule punctuated by revolutions and U.S. military intervention ended in 1978 with the first peaceful transfer of power following a democratic election.

Political stability brought the Dominican Republic handsome rewards, and during the late 1990s the economy, based on manufacturing, high-tech industries, and remittances from Dominicans abroad as well as tourism, grew at an average of 7 percent per year. But in the early 2000s the economy imploded, not only because of a downturn in the world economy but also because of bank fraud and corruption in government. Suddenly the Dominican currency collapsed, inflation skyrocketed, unemployment surged, and blackouts prevailed. As the people protested, lives were lost and the self-enriched elite blamed foreign financial institutions that were unwilling to lend the government additional money. Once again the hopes of ordinary citizens were dashed by greed and corruption among those in power.

Puerto Rico

The largest U.S. domain in Middle America, this easternmost and smallest island of the Greater Antilles region covers 3500 square miles and has a population of 4 million. Puerto Rico is larger than Delaware and more populous than Oregon. Most of the population is concentrated in the urbanized northeastern sector of this rectangular island (Fig. 4B-12).

Puerto Rico fell to the United States more than 110 years ago during the Spanish-American War of 1898. Since the Puerto Ricans had been struggling for some time to free themselves from Spanish hegemony, this transfer of

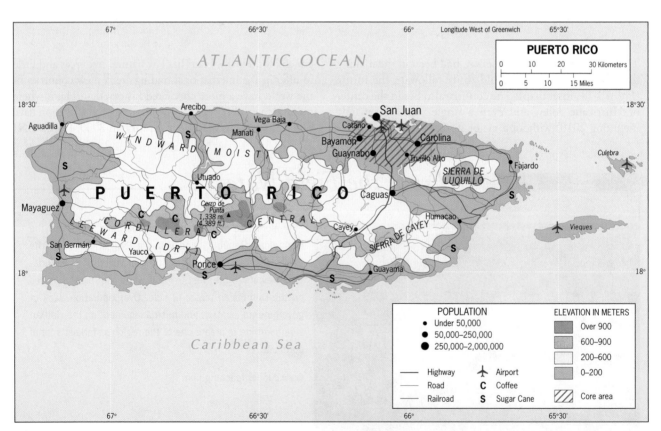

FIGURE 4B-12

© H. J. de Blij, P. O. Muller, and John Wiley & Sons, Inc.

power was, in their view, only a change from one colonial master to another. As a result, the first half-century of U.S. administration was difficult, and it was not until 1948 that Puerto Ricans were permitted to elect their own governor. When the island's voters approved the creation of a Commonwealth in a 1952 referendum, Washington, D.C. and San Juan, the two seats of government, entered into a complicated arrangement. Puerto Ricans are U.S. citizens but pay no federal taxes on their local incomes. The Puerto Rican Federal Relations Act governs the island under the terms of its own constitution and awards it considerable autonomy. Puerto Rico also receives a sizeable annual subsidy from Washington, totaling more than U.S. $4 billion in recent years.

Despite these apparent advantages in the poverty-mired Caribbean, Puerto Rico has not thrived under U.S. administration. Long dependent on a single-crop economy (sugar), the island based its industrialization during the 1950s and 1960s on its comparatively cheap labor, tax breaks for corporations, political stability, and special access to the U.S. market. Consequently, pharmaceuticals, electronic equipment, and apparel top today's list of exports, not sugar or bananas. But this industrialization failed to stem a tide of emigration that carried more than 1 million Puerto Ricans to the New York City area alone. The same wages that favored corporations kept many Puerto Ricans poor or unemployed. By some estimates, unemployment on the island stands at more than 25 percent today. Most receive federal support. Another 30 percent work in the public sector, that is, in government. The private sector is severely underdeveloped.

Although Puerto Ricans are vocal in demanding political change, it is not surprising that referendums in both 1993 and 1998 resulted in the retention of the status quo: continuation of Commonwealth status rather than either Statehood or independence.

▪▪ THE LESSER ANTILLES

As Figure 4B-1 shows, the Greater Antilles are flanked by two clusters of islands: the extensive Bahamas/Turks and Caicos archipelago to the north and the Lesser Antilles to the southeast. **The Bahamas**, the former British colony that is now the closest Caribbean neighbor to the United States, alone consists of nearly 3000 coral islands—most of them rocky, barren, and uninhabited—but about 700 carrying vegetation, of which some 30 are inhabited. Centrally positioned New Providence Island houses most of the country's 320,000 inhabitants and also contains the capital, Nassau, a leading tourist attraction.

The Lesser Antilles are grouped geographically into the Leeward Islands and the Windward Islands, a (climatologically incorrect) reference to the prevailing airflows in this tropical area. The Leeward Islands extend from the U.S. Virgin Islands to the French dependencies of Guadeloupe and Martinique, and the Windward Islands from St. Lucia to the Dutch-affiliated islands of Aruba and Curaçao off the northwestern Venezuelan coast.

It would be impractical to detail the individual geographic characteristics of each of the Lesser Antilles, but we should note that most share the insularity and environmental risks of this region—earthquakes, volcanic eruptions, and hurricanes; that they confront, to varying degrees, similar socioeconomic problems in the form of limited domestic resources, overpopulation, soil deterioration, land fragmentation, and market limitations; and that tourism has become the leading industry for many.

Under these circumstances one looks for certain hopeful signs, and such an indication comes from **Trinidad and Tobago**, the two-island republic at the southern end of the Lesser Antilles (Fig. 4B-1). This country (population: 1.3 million) has embarked on a natural-gas-driven industrialization boom that could well turn it into an economic tiger. Trinidad has long been an oil producer, but lower world prices and dwindling supplies in the 1990s forced a reexamination of its natural gas deposits to help counter the downturn. That quickly resulted in the discovery

The capital city of Trinidad and Tobago may have an historic colonial name (Port of Spain), but after nearly three centuries of Spanish rule, the British took control here. English became the *lingua franca*, democratic government followed independence in 1962, and natural-gas reserves propelled a thriving economy. As can be seen here, the harbor of Port of Spain is bursting at the seams as a car carrier delivers automobiles from Japan, and containers are stacked high on the docks.

© Boutin/Sipa Press

VOICE FROM THE
Region

© Megan Elizabeth Chong
Megan Elizabeth Chong,
Curaçao

THE PULL OF CURAÇAO'S ETHNIC MOSAIC

"Growing up on the island of Curaçao, I have had the opportunity to interact with people of various races, ethnicities, religions, and nationalities. As is true for most Caribbean islands, immigrants have come from many different parts of the world. This ethnic mix makes for the unique culture of Curaçao. Personally, I have an American mother of German origin and an Antillean father of Chinese, Amerindian, and Spanish descent. It would be difficult for me to find someone with identical origins, but that goes for many people living here. Since the island is so small, only 60 kilometers (38 mi) long and 15 kilometers (9 mi) wide, most people know each other even if they have very different backgrounds. I think this has helped me to develop a deeper understanding and appreciation for others. On the other hand, the opportunities for work and studies are quite limited on small islands like Curaçao. Each year, many graduating high school students move abroad to attend college. The majority end up in the Netherlands, with which we have historical ties and where students from Curaçao get the same subsidies as Dutch nationals. Some choose to study in the United States, but a good number stay on the island to take classes at the local university or look for work. This year, I will be attending Maastricht University in the Netherlands to major in psychology. Many of the students who leave Curaçao to study abroad do not return after their studies—or at least not right away—as they normally begin a career elsewhere. I will probably do the same, having connections in Europe as well as the United States; but I know I will always be drawn back to Curaçao, the island I am proud to call home."

of major new supplies, and this cheap and abundant fuel has sparked a local gas-production boom as well as an influx of energy, chemical, and steel companies from Europe, Canada, and even India. (Meanwhile, Trinidad has become the largest supplier of liquefied natural gas for the United States.) Many of the new industrial facilities have agglomerated at the ultramodern Point Lisas Industrial Estate outside the capital, Port of Spain, and they have propelled Trinidad to become the world's leading ex-

porter of ammonia and methanol. Natural gas is also an efficient fuel for the manufacturing of metals, and steelmakers as well as aluminum refiners have been attracted to locate here. With Trinidad lying only a few kilometers from the Venezuelan coast of South America, it is also a sea-lane crossroads that is well connected to the vast, near-coastal supplies of iron ore and bauxite that are mined in nearby countries, particularly the Brazilian Amazon.

But Trinidad is an exception. It is a reflection of the predicament of **small-island developing economies [13]** that, after the initial wave of decolonization in the 1950s and 60s, several territories decided that they were probably better off staying with the European country that had ruled them. Thus, for example, Guadeloupe remained with France and the Cayman Islands are still British. The former Netherlands Antilles forsook full independence to continue their affiliation with the Dutch but did so separately and with considerable autonomy: **Curaçao**, **Aruba**, **and St. Maarten** are now "countries" within the Kingdom of the Netherlands, whereas the other Caribbean islands of the former Dutch Empire have acquired the status of overseas municipalities. Although economic logic seems to dictate continued affiliation with the former colonial powers, local politics and pride demand a minimal degree of autonomy and avoidance of paternalistic European interference. The combined result is often a complicated legal framework that shapes the political status of the islands.

POINTS TO PONDER

- Northern Mexico highlights the volatility of globalization: the inception of NAFTA led to foreign investment and the proliferation of maquiladoras, but a decade later many companies were leaving for East Asian destinations that offered even lower wages.

- Costa Rica has known political stability for more than 175 years, and it has been a continuous democracy for over half a century. What could other Central American countries learn from this example?

- The geography of a small Middle American country, Panama, significantly impacts the global economy and especially its two biggest players, the United States and China.

- Critical parts of Haiti will have to be rebuilt almost from the ground up. Will its economic and geographic conditions ever allow self-sustained development?

using GEOGRAPHY

1. When we studied the expansion of the European Union (EU) in Chapter 1A, the issue of Turkey's application for membership was raised. Turkish leaders desired admission and worked to meet the conditions of membership, but Turkey still fails to meet these criteria in important respects. The views of EU members range across the spectrum, from enthusiastic to ambivalent to negative. Now imagine a North American Union of which Canada and the United States are members, and which Mexico wishes to join (NAFTA is a far less comprehensive union than the EU, in effect only being a first step toward one). Use your geographic insights to analyze the prospects of such a merger, and predict reactions in Canada and the United States. In what ways would Mexico qualify for membership, and in what respects would it not? Do you see any similarities between current Turkey–Europe and Mexico–North America relationships? In what ways are the two situations crucially different?

2. If you could retire and buy a home in Middle America (as quite a few U.S. residents have done), where would you choose to buy and for what reasons? Think of how far your dollars would go but also about natural amenities and hazards, accessibility, and political conditions.

EL SALVADOR
BELIZE
NICARAGUA
Managua
Lake Nicaragua
COSTA RICA
San José
ISTHMUS OF PANAMA
PANAMA
Panama City
Gulf of Panama
Turbo

Caribbean Sea
Curaçao
Aruba
Bonaire (NETH.)
Lesser Antilles

ST. VINCENT & THE GRENADINES
GRENADA
BARBADOS
TRINIDAD & TOBAGO

Santa Marta
BARRANQUILLA
CARTAGENA
MARACAIBO
Maracay
Barquisimeto
VALENCIA
CARACAS
Barcelona
Maturín
Boca Grande

Cúcuta
Bucaramanga
San Cristóbal
V E N E Z U E L A

MEDELLÍN
Pereira
Obando
Boa Vista
BOGOTÁ
COLOMBIA
San José del Guaviare
Mitú

Cocos Island (COSTA RICA)
Malpelo Island (COLOMBIA)
Buenaventura
CALI

Tumaco
Pasto
Esmeraldas
Equator
QUITO
Cotopaxi (5,897 m, 19,347 ft.)
Chimborazo (6,267 m, 20,561 ft.)
ECUADOR
GUAYAQUIL
Gulf of Guayaquil
Machala
Cuenca
Talara
Piura

Archipiélago de Colón (Galápagos Islands) (ECUADOR)

La Pedrera
Iquitos
Marañón
Cruzeiro do Sul
Putumayo
Tefé
Amazon

GUIANA
Angel Falls
GUYANA
Georgetown
Paramaribo
SURINAME
FRENCH GUIANA
Kourou
Cayenne

HIGHLANDS

A M A Z O N
Negro
Branco
Japurá
MANAUS
Balbina Reservoir
Alenquer
Santarém
Itaituba

B A S I N

ATLANTIC OCEAN

Equator

Mouths of the Amazon
Macapá
Marajó Bay
Marajó Island
Ananindeua
BELÉM

Porteira
Amazon

Parnaíba
São Luís
Sobral
FORTA

Aguja Point
Chiclayo
Cajamarca
Pucallpa
Trujillo
Chimbote
Huascarán (6,768 m, 22,205 ft.)
Huánuco
Callao
LIMA
Huancayo
Cerro de Pasco
Cuzco
Chincha Alta
Ayacucho
Ica
Juliaca
La Paz
Illimani (6,462 m, 21,201 ft.)
Arequipa
Tacna
Arica
Iquique
Calama
Chuquicamata
Antofagasta
Copiapó
Mt. Ojos del Salado (6,880 m, 22,572 ft.)

Porto Velho
Rio Branco
Riberalta
Ji-Paraná (Rondônia)
Reyes
Trinidad
MATO GROSSO PLATEAU
Cuiabá
Rondonópolis
Rio Verde
Corumbá

B R A Z I L
B R A Z I L I A N
Araguaína
Palmas
BRASÍLIA
GOIÂNIA
Uberlândia

Madeira
Tapajós
Marabá
Imperatriz
Teresina
Floriano
Mossoró
Na
João Pes
Cape Bra
REC
Maceió
Feira de Santana
SALVAD

SERTÃO
Barreiras
Petrolina
Sobradinho Reservoir
São Francisco
Montes Claros
Vitória da Conquist

H I G H L A N D S
BELO HORIZONTE
Juiz de Fora
Vila Velha (Espírito Santo)
RIO DE JANEIRO
Cape Frio

MONTAÑA
P E R U
A N D E S
ALTIPLANO
BOLIVIA
Cochabamba
SANTA CRUZ
Sucre
Potosí
Tarija
Fortín Madrejón
Filadelfia
P A R A G U A Y
Campo Grande
Dourados
Londrina
Ponta Grossa
Foz do Iguaçu
Iguaçu Falls
CURITIBA
Joinville
Florianópolis
Passo Fundo
Caxais do Sul
PORTO ALEGRE
Pelotas

SÃO PAULO
São José dos Campos
Santos
CAMPINAS

Jujuy
Salta
Tucumán
GRAN CHACO
Asunción
Ciudad del Este
Corrientes
Resistencia
Posadas
Santiago del Estero
Reconquista
Santa María
Uruguaiana
Catamarca
La Rioja
ENTRE RÍOS
URUGUAY
Salto
Melo
Mercedes

ATACAMA DESERT
PACIFIC OCEAN
San Félix I. San Ambrosio I. (CHILE)

Coquimbo
CÓRDOBA
Santa Fe
Mendoza
San Luis
ROSARIO
Paraná
BUENOS AIRES
La Plata
Punta del Este
MONTEVIDEO

Viña del Mar
Valparaíso
SANTIAGO
Juan Fernández Islands (CHILE)

A R G E N T I N A
Mt. Aconcagua (6,960 m, 22,834 ft.)
P A M P A S
Mar del Plata

ATLANTIC OCEAN

Talca
Talcahuano
Concepción
Los Ángeles
Temuco
Chillán
Santa Rosa
Neuquén
Bahía Blanca

Osorno
Puerto Montt
Chiloé Island
Corcovado Gulf
Chonos Archipelago
Coihaique
Gulf of Penas
Wellington Island

San Carlos de Bariloche
Esquel
Trelew
Puerto Madryn
Viedma
San Matías Gulf
Valdés Peninsula (-40 m, -131 ft.)
Comodoro Rivadavia
Gulf of San Jorge
Jaramillo
Gran Bajo de San Julián (-105 m, -344 ft.)

A N D E S
M O U N T A I N S
P A T A G O N I A

Coihaique

Puerto Natales
Puerto Santa Cruz
Río Gallegos
TIERRA DEL FUEGO
Río Grande
Punta Arenas
Strait of Magellan
Cape Horn

Falkland Islands (U.K.)
West Falkland
East Falkland
Stanley

Tropic of Capricorn
Tropic of Ca

Elevation (m)

3000	
1500	
600	
300	
150	
0	below sea level
0	-150
	-1500
	-3000
	-6000

— Roads
— Railroads

0 km 400 800 1200 1600
0 miles 200 400 600 800 1000

Lambert Azimuthal Equal-Area Projection
Scale 1:30,000,000

South Geo

IN THIS CHAPTER

- The South American commodity boom
- The growing power of indigenous peoples
- Inequality and violence: A hallmark of South America?
- Brazil on the move
- The Chinese are coming . . .

CONCEPTS, IDEAS, AND TERMS

FIGURE 5A-1 © H. J. de Blij, P. O. Muller, and John Wiley & Sons, Inc.

Of all the continents, South America has the most familiar shape—a gigantic triangle connected by mainland Middle America's tenuous land bridge to its neighbor in the north. South America also lies not only south but mostly east of its northern counterpart. Lima, the capital of Peru—one of the continent's westernmost cities—lies farther east than Miami, Florida. Thus South America juts out much more prominently into the Atlantic Ocean toward southern Europe and Africa than does North America. But lying so far eastward means that South America's western flank faces a much wider Pacific Ocean, with the distance from Peru to Australia nearly twice that from California to Japan.

As if to reaffirm South America's northward and eastward orientation, the western margins of the continent are rimmed by one of the world's longest and highest mountain ranges, the Andes, a giant wall that extends unbroken from Tierra del Fuego near the continent's southern tip in Chile to northeastern Venezuela in the far north (Fig. 5A-1). The other major physiographic feature of South America dominates its central north—the Amazon Basin; this vast humid-tropical amphitheater is drained by the mighty Amazon River, which is fed by several major tributaries. Much of the remainder of the continent can be classified as plateau, with the most important components being the Brazilian Highlands that cover most of Brazil southeast of the Amazon Basin, the Guiana Highlands located north of the lower Amazon Basin, and the cold Patagonian Plateau

The old city in historic Salvador, Brazil, former capital and port of entry for tens of thousands of Africans.

© H. J. de Blij

that blankets the southern third of Argentina. Figure 5A-1 also reveals two other noteworthy river basins beyond Amazonia: the Paraná-Paraguay Basin of south-central South America, and the Orinoco Basin in the far north that drains interior Colombia and Venezuela.

PHYSIOGRAPHY

Explorers' Continent

It was here in northern South America that the great German explorer and scientist Alexander von Humboldt, one of the founders of the modern discipline of geography, embarked on his legendary expeditions in the early nineteenth century. After landing on the coast of Venezuela and trekking across the continent's northern interior, the 30-year-old von Humboldt was struck by the area's biodiversity, its majestic natural beauty, and the adaptive abilities of the human populations. He discovered and named many species of flora and fauna, traversed tropical grasslands and jungles, met with indigenous peoples, crossed dangerous rivers, and reached the summit of the highest mountain in the Americas climbed by Europeans at the time (Ecuador's Chimborazo). He compiled large numbers of maps and was one of the first scientists to note how the coastlines of eastern South America and western Africa fitted together like pieces of a jigsaw puzzle, speculating about the continents' geologic movements.

Von Humboldt was most important in the rise of the geographic discipline because of his many discoveries as well as his views on the **unity of place [1]**: that in a particular locale or region intricate connections exist between climate, geology,

major geographic qualities of
SOUTH AMERICA

1. South America's physiography is dominated by the Andes Mountains in the west and the Amazon Basin in the central north. Much of the remainder is plateau country.

2. Almost half of the realm's area and just under half of its total population are concentrated in one country—Brazil.

3. South America's population remains concentrated along the continent's periphery. Most of the interior is sparsely peopled, but sections of it are now undergoing significant development.

4. Interconnections among the states of the realm are improving rapidly. Economic integration has become a major force, particularly in southern South America.

5. Regional economic contrasts and disparities, both in the realm as a whole and within individual countries, remain strong.

6. Cultural pluralism exists in almost all of the realm's countries and is often expressed regionally.

7. Rapid urban growth continues to mark much of the South American realm, and urbanization overall is today on a par with the levels of the United States and western Europe.

8. This realm has plenty of natural riches, and it has benefited in recent years from increased global demand for raw materials.

biology, and human cultures. As such, he laid the foundation for modern geography as an *integrative discipline* with a spatial perspective. It took him about four decades to produce his magnum opus that articulated this holistic perspective, the highly ambitious and appropriately titled *Cosmos*, published in five volumes in 1845.

Almost three centuries earlier, the opposite end of South America was the scene of a crucial stage in the first circumnavigation of the globe and expedition led by Ferdinand Magellan. Once it became clear that Columbus had stumbled upon America, not India, Spanish and Portuguese explorers continued their efforts to discover a westward passage from the Atlantic to the Pacific. This took them to the far south along what is now the Argentinean coast. Magellan and his crew spent some five months in Patagonia, named by Magellan for "big-foot people" (official, but never-verified, reports of the voyage told of the crew's mysterious encounters with 8-foot giants). With the onset of spring in 1520, Magellan's five ships set sail for the treacherous waters of what he named the *Estrecho de Todos los Santos*—now called the Strait of Magellan (Fig. 5A-1). Of the five ships in Magellan's fleet, only three survived the daring, 600-kilometer (375-mile)-long passage; the other two crashed on the rocks in the icy waters of the Southern Ocean.

Myriad Climates and Habitats

If the Russian realm is the widest in east-west extent, the South American realm is the longest measured from north to south. Within South America, no other country is more emblematic of this elongated geography than Chile, averaging only about 150 kilometers (90 mi) in width but 4000 kilometers (2500 mi) in length. As a consequence of this latitudinal span, fully one-tenth of our planet's circumference, the realm contains an enormous range of climates and vegetation. Combine this with substantial variation in relief from west to east, and it is clear why South America has such an impressive range of natural habitats.

Take another look at the map showing the global distribution of climates (Fig. G-7) and note the variation in climatic types, particularly in the realm's northwest and southern half. Travel northeast from Lima, Peru for about 600 kilometers (375 mi) and you encounter no less than four different climate zones: arid, highland, and two varieties of humid tropical. A transect of similar length from Santiago, Chile eastward across the continent to Buenos Aires, Argentina will take you through five climate zones: interior highland, arid, and semiarid environments bracketed by a different humid temperate climate along each coast. Vegetation in South America varies accordingly, from lush tropical rainforests to rocky and barren snow-covered mountain tops to grasslands, fertile as well as parched. This natural diversity also makes for considerable cultural differences, as we shall soon see.

STATES ANCIENT AND MODERN

Thousands of years before the first European invaders appeared on the shores of South America, peoples now referred to as **Amerindians** had migrated into the continent via North and Middle America and founded societies in coastal valleys, in river basins, on plateaus, and in mountainous locales. These societies achieved different and remarkable adaptations to their diverse natural environments, and by about 1000 years ago, a number of regional cultures thrived in the elongated valleys between mountain ranges of the Andes from present-day Colombia

From the Field Notes . . .

"Traveling on Peru's Cuzco *altiplano* in 2009, it was easy to get a sense of the advantages of this highland environment to the Incas of the past. The soils are fertile, temperatures are moderate, and the surrounding snowcapped ranges of the Andes feed the Urubamba River and many other streams. The Urubamba Range in the background, partially hidden behind the clouds, rises up to nearly 6000 meters (19,000 ft) and the city of Cuzco itself sits at 3350 meters (11,000 ft). For centuries, this area has produced abundant corn (seen in the foreground), potatoes, and other *tierra fría* crops. But the retreating glaciers in the higher elevations are now a cause of concern. Global warming may jeopardize what has been nature's gift to this *altiplano* for many centuries: a reliable water supply."

© Jan Nijman

www.conceptcaching.com

southward to Bolivia and Chile. These high-altitude valleys, called **altiplanos** [2], provided fertile soils, reliable water supplies, building materials, and natural protection to their inhabitants.

The Inca State

One of these *altiplanos*, at Cuzco in what is now Peru, became the core area of South America's greatest indigenous empire, that of the **Inca**. The Inca were expert builders whose stone structures (among which Machu Picchu near Cuzco is the most famous), roads, and bridges helped unify their vast empire; they also proved themselves to be efficient administrators, successful farmers and herders, and skilled manufacturers; their scholars studied the heavens, and physicians even experimented with brain surgery. Great military strategists, the Inca integrated the peoples they vanquished into a stable and well-functioning state, an amazing accomplishment given the high-relief terrain they had to contend with.

As a minority ruling elite in their far-flung empire, the Inca were at the pinnacle in their rigidly class-structured, highly centralized society. So centralized and authoritarian was their state that a takeover at the top was enough to gain immediate power over all of it—as a small army of Spanish invaders discovered in the 1530s. The European invasion brought a quick end to thousands of years of Amerindian cultural development and changed the map forever.

The Iberian Invaders

The modern map of South America started to take shape when the Iberian colonists began to understand the location and economies of the Amerindian societies. The Inca, like Mexico's Maya and Aztec peoples, had accumulated gold and silver at their headquarters, possessed productive farmlands, and constituted a ready labor force. Not long after the defeat of the Aztecs in 1521, Francisco Pizarro sailed southward along the continent's northwestern coast, learned of the existence of the Inca Empire, and withdrew to Spain to organize its overthrow. He returned to the Peruvian coast in 1531 with 183 men and two dozen horses, and the events that followed are well known. In 1533, his party rode victorious into Cuzco.

At first, the Spaniards kept the Incan imperial structure intact by permitting the crowning of an emperor who was under their control. But soon the breakdown of the old order began. The new order that gradually emerged in western South America placed the indigenous peoples in serfdom to the Spaniards. Great haciendas were formed by **land alienation** [3] (the takeover of indigenously held land by foreigners), taxes were instituted, and a forced-labor system was introduced to maximize the profits of exploitation.

Lima, the west-coast headquarters of the Spanish conquerors, soon became one of the richest cities in the world, its wealth based on the exploitation of vast Andean silver deposits. The city also served as the capital of the Viceroyalty of Peru, as the Spanish authorities quickly integrated the new possession into their colonial empire (Fig. 5A-2). Subsequently, when Colombia and Venezuela came under Spanish control and, later, when Spanish settlement expanded into what is now Argentina and Uruguay, two additional viceroyalties were added to the map: New Granada and La Plata.

Meanwhile, another vanguard of the Iberian invasion was penetrating the east-central part of the continent, the coastlands of present-day Brazil. This area had become

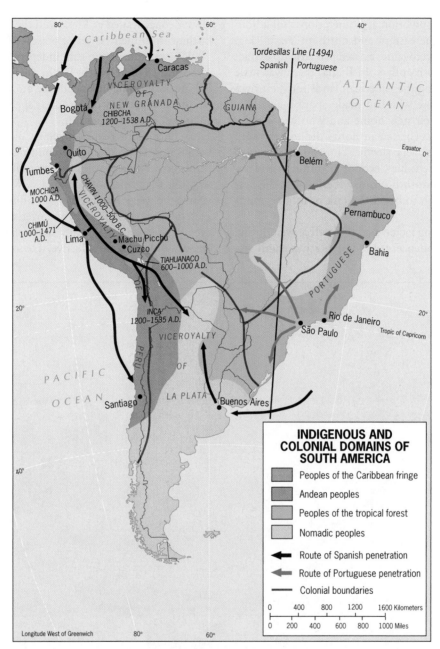

FIGURE 5A-2 © H. J. de Blij, P. O. Muller, and John Wiley & Sons, Inc.

a Portuguese sphere of influence because Spain and Portugal had signed a treaty in 1494 to recognize a north-south line 370 leagues west of the Cape Verde Islands as the boundary between their New World spheres of influence. This border ran approximately along the meridian of 50°W longitude, thereby cutting off a sizeable triangle of eastern South America for Portugal's exploitation (Fig. 5A-2). But a brief look at the political boundaries of South America (Fig. 5A-1) shows that this treaty did not limit Portuguese colonial territory to the east of the 50th meridian. Instead, Brazil's boundaries were bent far inland to include almost the entire Amazon Basin, and the country came to be only slightly smaller in territorial size than all the other South American countries combined.

This westward thrust was the result of Portuguese and Brazilian penetration, particularly by the *Paulistas*, the settlers of São Paulo who needed Amerindian slave labor to run their plantations.

Independence and Isolation

Despite their adjacent location on the same continent, their common language and cultural heritage, and their shared national problems, the countries that arose out of South America's Spanish viceroyalties (together with Brazil) until quite recently existed in a considerable degree of isolation from one another. Distance and physiographic barriers reinforced this separation, and the realm's major population agglomerations still adjoin the coast, mainly the eastern and northern coasts (Fig. 5A-3 below). The viceroyalties existed primarily to extract riches and fill Spanish coffers. In Iberia, there was little interest in developing the American lands for their own sake. Only after those who had made Spanish and Portuguese America their new home and who had a stake there rebelled against Iberian authority did things begin to change, and then very slowly. Thus South America was saddled with the values, economic outlook, and social attitudes of eighteenth-century Iberia—not the best tradition from which to begin the task of forging modern nation-states.

Certain isolating factors had their effect even during the wars for independence. Spanish military strength was always concentrated at Lima, and those territories that lay farthest from their center of power—Argentina as well as Chile—were the first to gain independence from Spain (in 1816 and 1818, respectively). In the north, Simón Bolívar led the burgeoning independence movement, and in 1824 two decisive military defeats there spelled the end of Spanish power in South America.

This joint struggle, however, did not produce unity because no fewer than nine countries emerged from the three former viceroyalties. It is not difficult to understand why this fragmentation took place. With the Andes intervening between Argentina and Chile and the Atacama Desert between Chile and Peru, overland distances seemed even greater than they really were, and these obstacles to contact proved quite effective. Hence, from their outset

SOUTH AMERICA: POPULATION DISTRIBUTION, 2012
One dot represents 50,000 persons

0   600   1200 Kilometers
0   300   600 Miles

FIGURE 5A-3   © H. J. de Blij, P. O. Muller, and John Wiley & Sons, Inc.

the new countries of South America began to grow apart amid friction and even wars. Only within the past two decades have the countries of this realm finally begun to recognize the mutual advantages of increasing cooperation and to make lasting efforts to steer their relationships in this direction.

THE CULTURAL MOSAIC

When we speak of the interaction of South American countries, it is important to keep in mind just who does the interacting. The fragmentation of colonial South America into ten individual republics, and the subsequent postures of each of these states, was the work of a small minority that constituted the landholding, upper-class elite. Therefore, in every country a vast majority—be they Amerindians in Peru or people of African descent in Brazil—could only watch as their European masters struggled with one another for supremacy.

The Population Map—Then and Now

If we were able to reconstruct a map of South America's population before the arrival of the Europeans (a "pre-Columbian" map, as it would be called), it would look quite different from the current one in Figure 5A-3. Indigenous Amerindian societies inhabited not only the Andes and adjacent lowlands but also riverbanks in the Amazon Basin, where settlements numbering in the thousands subsisted on fishing and farming. They did not shy away from harsh environments such as those of the island of Tierra del Fuego in the far south, where the fires they kept going against the bitter cold led the Europeans to name the place "land of fire."

Today the map looks quite different. Many of the indigenous societies succumbed to the European invaders, not just through warfare but also because of the diseases the Hispanic conquerors brought with them. Geographers estimate that 90 percent of native Amazonians died within a few years of contact, and the peoples of Tierra del Fuego also are no longer there to build their fires. From one end of South America to the other, the European arrival spelled disaster.

Spanish and Portuguese colonists penetrated the interior of South America, but the great majority of the settlers stayed on or near the coast, a pattern still visible today. Almost all of the realm's major cities have coastal or near-coastal locations, and the current population distribution map gives you the impression of a continent yet to be penetrated and inhabited. But look carefully at Figure 5A-3, and you will see a swath of population located well inland from the settlements along the northwest coast, most clearly in Peru but also extending northward into Ecuador and southward into Bolivia. That is the legacy of the Inca Empire and its incorporated peoples, surviving in their mountainous redoubt and still numbering in the millions.

Amerindian Reawakening

Today, South America's long-downtrodden Amerindians are staging a social, political, and economic reawakening. In several South American countries where their numbers are large enough to translate into political strength, they have begun to realize their potential. Amerindians in Peru constitute 47 percent of the national population, and in Bolivia they are in the majority at 57 percent.

Newly empowered Amerindian political leaders are emerging and bringing the plight of the realm's indigenous peoples to both local and international attention. Amerindians were conquered, decimated by foreign diseases, robbed of their best lands, subjected to involuntary labor, denied the right to grow their traditional crops, socially discriminated against, and swindled out of their fair share of the revenues from resources in their traditional domains. They may still be the poorest of the realm's poor, but they are increasingly asserting themselves. For some South American states, the consequences of this movement will be far-reaching.

The Amerindian reawakening is in part related to changing religious practices in South America. Officially, more than 90 percent of the population is Roman Catholic, and traditionally South Americans tend to be viewed as devout followers of the Vatican. But recent surveys show that many do not attend church regularly and that more than half of all Catholics describe themselves as believers of the doctrine, rather than adherents of the Church. Since the late nineteenth century, the Catholic Church has often been criticized for its conservative position on social issues and for siding with the establishment. During the 1950s, a powerful movement known as Liberation theology [4] emerged in South America and subsequently gained followers around the world. The movement was a blend of Christian religion and socialist thinking that interpreted the teachings of Christ as a quest to liberate the impoverished masses from oppression. Although the Church has lately been trying to make amends, it could not avoid losing popular support, especially among Amerindians. In recent years, Protestant-evangelical faiths have also gained adherents, mainly at the expense of Catholicism.

African Descendants

As Figure 5A-2 shows, the Spaniards initially got very much the better of the territorial partitioning of South America—not just in land quality but also in the size of the aboriginal labor force. When the Portuguese began to develop their territory, they turned to the same lucrative activity that their Spanish rivals had pursued in the Caribbean—the plantation cultivation of sugar for the European market. And they, too, found their workforce in the same source region, as millions of Africans (nearly half of all who came to the Americas) were brought in bondage to the tropical Brazilian coast north of Rio de Janeiro (see Fig. 6A-6). Not surprisingly, Brazil now has South America's largest black population, which is still heavily concentrated

From the Field Notes . . .

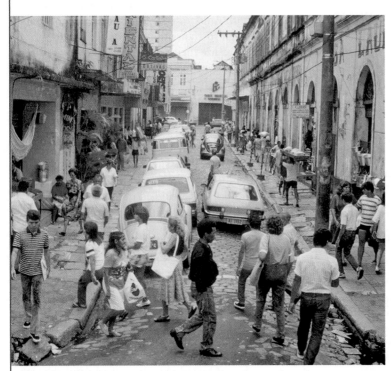

© H. J. de Blij

"Near the waterfront in Belém lie the now-deteriorating, once-elegant streets of the colonial city built at the mouth of the Amazon. Narrow, cobblestoned, flanked by tiled frontages and arched entrances, this area evinces the time of Dutch and Portuguese hegemony here. Mapping the functions and services here, we recorded the enormous diversity of activities ranging from carpentry shops to storefront restaurants and from bakeries to clothing stores. Dilapidated sidewalks were crowded with shoppers, workers, and people looking for jobs (some newly arrived, attracted by perceived employment opportunities in this growing city of 2.3 million). The diversity of population in this and other tropical South American cities reflects the varied background of the region's peoples and the wide hinterland from which these urban magnets have drawn their inhabitants."

www.conceptcaching.com

in the country's poverty-mired northeastern States. With Brazilians of direct or mixed African ancestry today accounting for just over half of the population of 198 million, the Africans decidedly constitute a major immigration of foreign peoples into South America.

Ethnic Landscapes

The cultural landscape of South America, similar to that of Middle America, is a layered one. Amerindians cultivated and crafted diverse landscapes throughout the continent, some producing greater impacts than others. When the Europeans arrived, the cultural transformation that resulted from depopulation severely impacted the environment. Native peoples now became minorities in their own lands, and Europeans introduced crops, animals, and ideas about land ownership and land use that irreversibly changed South America. They also brought in Africans from various parts of Subsaharan Africa. Europeans from non-Iberian Europe also started immigrating to South America, especially during the first half of the twentieth century. Japanese settlers arrived in Brazil and Peru during the same era. All of these elements have contributed to the present-day ethnic complexion of this realm. Figure 5A-4 shows the distinct concentrations of Amerindian and African cultural dominance, as well as areas where these

groups are largely absent and people of European ancestry predominate.

Of course, ethnic origins are not always so straightforward, and patterns can change as a result of internal migrations and ethnic mixing. In recent years, research on individual DNA and genetics has taught us a lot about the regional and group origins of people. On the map, Argentina is indicated as having predominantly European ancestry. More specifically, recent research shows that, for the average Argentine, nearly 80 percent of his or her genetic structure is European, 18 percent Amerindian, and 2 percent African. If nobody had any mixed ancestors, these percentages would translate perfectly into European, Amerindian, and African shares of the population—but of course that is not the case. Many people do not have perfect knowledge about their ethnic ancestry, and while some may be strictly of one single origin, many others will have some degree of mixed ancestry. In the aggregate, however, there is no doubt that the population of cone-shaped southern South America is predominantly of European origin. South America, therefore, is a realm of **plural societies [5]**, where Amerindians of different cultures, Europeans from Iberia and elsewhere, Africans mainly from western tropical Africa, and even some Asians from India, Japan, and Indonesia cluster in adjacent areas but generally do not mix. The bottom line is a cultural mosaic of almost endless variety.

FIGURE 5A-4   © H. J. de Blij, P. O. Muller, and John Wiley
& Sons, Inc.

ECONOMIC GEOGRAPHY

Agricultural Land Use and Deforestation

The internal divisions of this cultural kaleidoscope are further reflected in the realm's economic landscape. In South
America, larger-scale **commercial [6]** or market (for-
profit) and smaller-scale **subsistence [7]** (primarily for
household use) **agriculture** exist side by side to a greater
degree than anywhere else in the world (Fig. 5A-5). The
geography of plantations and other commercial agricultural systems was initially tied to the distribution of landholders of European background, whereas subsistence
farming (such as highland mixed subsistence-market,
agroforestry, and shifting cultivation) is historically associated with the spatial patterns of indigenous peoples as
well as populations of African and Asian descent (see map
above).

At a broader level of generalization, agricultural systems and land use in South America vary in close relationship to physiography, as one would expect. Forestry and
agroforestry prevail in the Amazon Basin; ranching dominates in the grasslands of the Southern Cone; and different
forms of agriculture, with or without irrigation, are found
in an extensive zone from northeastern Brazil to northern
Argentina, as well as scattered among pockets of moderate

elevation in the Andean highlands. Land uses throughout
South America are changing rapidly today, mainly due to
ongoing deforestation and the introduction or expansion
of new crops. The fastest-growing crop is soybeans, the
cultivation of which now dominates much of east-central
Brazil and spills over into adjacent areas of Paraguay, Uruguay, and Argentina (Fig. 5A-5).

Deforestation is a particularly acute problem in
northern Brazil. In the past, deforestation was attributed
mainly to small-scale landholders, colonists who had made
their way into the Amazon rainforest to eke out a new living. The usual pattern of settlement often went like this.
As main and branch highways were cut through the forest,
settlers followed (often in response to government-sponsored land occupation schemes), moving out laterally to
clear land for farming (see photo, p. 280). Subsistence
crops were planted, but within a year or two weed infesta-
tion and declining soil fertility made these plots unproductive. As soil fertility continued to decline, the settlers
planted grasses and then sold their land to cattle ranchers.
The peasant farmers then moved on to newly opened
nearby areas, cleared more land for planting, and the cycle
repeated itself.

Unfortunately, this activity not only entrenches
low-grade land uses across widespread areas, but also

SOUTH AMERICA: AGRICULTURAL SYSTEMS

Oases
Scattered Pastoralism
Non-Soybean Grain Crops
Dairy Products and Flowers for Export
Mixed Dry Farming
Soybeans
Cattle

Plantation
Agroforestry & Shifting Cultivation
Highland Mixed Subsistence-Market
Mixed
Nonagricultural areas

0   500   1000   1500 Kilometers
0   500   1000 Miles

FIGURE 5A-5   © H. J. de Blij, P. O. Muller, and John Wiley & Sons, Inc.

contribution to global trade in grains, soybeans, coffee, orange juice, sugar, and many other crops is significant and growing. And in addition to all these successful agricultural endeavors, illegal farming thrives in the form of narcotics production, particularly cocaine (see box titled "The Geography of Cocaine").

Industrial Development

Manufacturing is growing rapidly in South America, ranging from chemicals to electronics and from textiles to biofuels. But the geography of industrial production varies markedly. Brazil, Chile, and Argentina have traditionally been ahead, while countries like Peru, Ecuador, and Bolivia have long struggled to modernize their economies and improve standards of living. Within countries, there is considerable **uneven development [8]** as well. Most manufacturing is concentrated in and around major urban centers, leaving vast empty spaces in the interior of the realm.

Brazil, in particular, is increasingly drawing attention for its momentous growth and the rapid sophistication of a number of economic sectors. In the past, typical Brazilian exports were foods and footwear. Today, the leading exports include oil (thanks to the fortuitous discovery of huge offshore reserves in the past few years), steel, and state-of-the-art Embraer aircraft. For almost a decade now, Brazil has been included as one of the four biggest emerging markets in the world, along with Russia, India, and China—the so-called **BRICs**.

Despite recent impressive economic growth across much of the realm, it is clear that the longstanding isolation of its individual countries remains a serious obstacle. Nonetheless, most governments agree that tighter integration, political as well as in terms of transportation infrastructure, would greatly enhance economic opportunities.

Economic Integration

The separatism that has so long characterized international relations in this realm is giving way as South American countries discover the benefits of forging new partnerships with one another. With mutually advantageous trade the catalyst, a new continentwide spirit of cooperation is blossoming at every level. Periodic flare-ups of boundary disputes now rarely escalate into open conflict. Cross-border rail, road, and pipeline projects, stalled for years, are multiplying steadily. In southern South America, five formerly contentious nations are developing the *hidróvia* (water highway), a system of river locks that is opening most of the Paraná-Paraguay Basin to barge transport; similar proposals have been advanced to connect this basin to the Amazon River system. And importantly, investments today flow more freely than ever from one country to another, particularly in the agricultural sector.

entails the burning and clearing of vast stands of tropical woodland (in fact, since the 1980s, an area of rainforest about the size of Ohio has been lost *annually* in northern Brazil). To make matters even worse, the deforestation crisis is now intensifying as *agro-industrial* operations engaged in large-scale production for export markets increasingly penetrate Amazonia and transform its land cover.

As South American economies modernize, agriculture remains highly important. More than one-fifth of this realm's workforce is still employed in the primary sector that includes farming, cattle raising, and fishing—a much greater share than in North America. The South American

The Geography of Cocaine

ANY GEOGRAPHICAL DISCUSSION of northwestern South America today must take note of one of its most wide-spread activities: the production of illegal narcotics. Of the enormous flow of illicit drugs that enter the United States each year, the most widely used substance is cocaine—all of which comes from South America, mainly Colombia, Peru, and Bolivia. Within these three countries, cocaine annually brings in billions of (U.S.) dollars and "employs" tens of thousands of workers, constituting an industry that functions as a powerful economic force. Those who operate the industry have accumulated considerable power through bribery of politicians, kidnappings and other forms of intimidation, threats of terrorism, and alliances with guerrilla groups in outlying zones beyond governmental control. The cocaine industry itself is structured within a tightly organized network of territories that encompass the various stages of this drug's production.

The first stage of cocaine production is the extraction of coca paste from the coca plant, a raw-material-oriented activity that is located near the areas where the plant is grown. The coca plant was domesticated in the Andes by the Incas centuries ago; millions of their descendants still chew coca leaves for stimulation and brew them into coca tea, the leading beverage of high-altitude South America. The main zone of coca-plant cultivation is along the eastern slopes of the Andes and in adjacent tropical lowlands in Bolivia, Peru, and Colombia.

Today, five areas dominate in the growing of coca leaves for narcotic production: Bolivia's Chaparé district in the marginal Amazon Lowlands northeast of the city of Cochabamba; the Yungas Highlands north of the Bolivian capital, La Paz; north-central Peru's Huallaga and neighboring valleys; south-central Peru's Apurimac Valley, southeast of Huancayo; and the areas around the guerrilla-controlled territories of southern Colombia (Fig. 5A-6).

These areas, which can produce as many as nine crops per year, are especially conducive to high leaf yields thanks to favorable local climatic and soil environments that also allow plants to develop immunity to many diseases and insect ravages. Despite government efforts to aerially spray herbicides and manually eradicate plants in the field, all of these areas continue to thrive as coca cultivation has become a full-fledged cash crop.

Operations in guerrilla-controlled portions of Colombia have reached the scale of plantations, inducing thousands of local subsistence farmers (sometimes at gunpoint) to join the more lucrative ranks of the field workforce. In Peru and Bolivia, the specialized coca-cultivation zones have lured an even larger peasant-farmer population to abandon the less profitable production of food crops (coca's per-acre income is now seven times greater than cocoa's), thereby further reducing the capability of these nutrition-poor countries to feed themselves. As (U.S.-supported) government efforts to eliminate coca raising have intensified in recent years, the geography of production has made some adjustments. The most aggressive pursuit occurred within Colombia, and the net effect was to displace coca cultivation to nearby Andean countries. But this shift actually appears to have strengthened rather than weakened the industry: recent press reports indicate that not only are coca-leaf prices—and overall productivity—rising steadily, but also that street prices for cocaine are falling worldwide.

Coca leaves harvested in the source areas of the Andes and nearby interior lowlands make their way to local collection centers, located at the convergence of rivers and trails, where coca paste is extracted and prepared. This begins the second stage of production, which involves the refining of that coca paste (ca. 40 percent pure cocaine) into cocaine hydrochloride (more than 90 percent pure), a lethal concentrate that is diluted with substances such as sugar or flour before being sold on the streets to consumers. Cocaine refining requires sophisticated chemicals, carefully controlled processes, and a labor force skilled in their supervision, and here Colombia has predominated. Most of this activity takes place in modern processing centers located in the rebel-held territory of the lowland central-south and east, beyond the reach of the Bogotá government. Interior Colombia also possesses the geographic advantage of intermediate location, lying between the source areas to the south (as well as locally) and the U.S. market to the north.

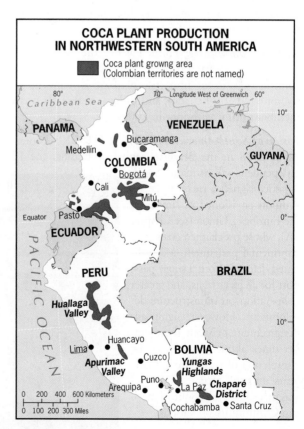

FIGURE 5A-6 © H. J. de Blij, P. O. Muller, J. Nijman, and John Wiley & Sons, Inc.

The final stage of production entails the distribution of cocaine to the marketplace, which depends on an efficient, clandestine transportation network that leads into the United States. Private planes operating out of remote airstrips were the preferred "exporting" method until about a decade ago, but aggressive U.S. measures along the coasts of Florida as well as the Gulf of Mexico have effectively closed down trans-Caribbean flight paths. Most cocaine now travels overland to be smuggled by sea through northwestern South America's Pacific and Caribbean seaports. As noted in Chapter 4A, its main destination is Mexico, which today supplies at least 90 percent of all the cocaine that enters the United States. This rerouting through Mexico has been aided by the increased northward flow of goods into the United States under NAFTA—providing ever greater opportunities for smuggling via the millions of trucks that cross the border each year. It has also triggered the rapid rise of organized crime (see pages 217–218) as several Mexican drug gangs evolved into sophisticated international cartels that control both the cocaine inflow and retail distribution within the United States, and increasingly dictate the terms of operation to their South American suppliers.

Recognizing that free trade may well solve many of the realm's economic-geographic problems, governments are now pursuing multiple avenues of economic supranationalism. In 2011, South America's republics were affiliating with the following major trading blocs:

- *Mercosur/l* (*Mercosur* in Spanish; *Mercosul* in Portuguese): Launched in 1995 by countries of the Southern Cone and Brazil, this Common Market established a free-trade zone and customs union linking Brazil, Argentina, Uruguay, and Paraguay. Colombia, Chile, Ecuador, Peru, and Bolivia are associate members. Venezuela's admission has been blocked for years, first by Brazil and now by Paraguay. This organization is becoming the dominant free-trade organization for South America.

- *Andean Community:* Formed as first the Andean Pact in 1969 but restarted in 1995 as a customs union with common tariffs for imports, this bloc is made up of Colombia, Peru, Ecuador, and Bolivia. Venezuela was a member until it withdrew in 2006.

- *Union of South American Nations (UNASUR):* Founded in 2008 in the Brazilian capital of Brasília, the 12 independent countries of South America signed a treaty to create a union envisioned as similar to the European Union (see Chapter 1A), whose proclaimed goals are a continental parliament, a coordinated defense effort, a single passport for all its citizens, and greater cooperation on infrastructure development. However, significant disagreement between the member-states about details puts the completion of these efforts many years into the future. UNASUR was preceded by the South American Community of Nations.

- *Free Trade Area of the Americas (FTAA):* The United States and other NAFTA proponents have tried to move this hemispheric free-trade idea forward, but it has been resisted by peasants and workers in South America, and formally opposed by Mercosur/l. As long as the terms of trade remain set by the North, the Southern partners will be reluctant to participate in this initiative.

The Commodity Boom

From 2005 through 2010, the South American economy in the aggregate grew robustly by about 5 percent annually.

A mid-2009 aerial image of one of South America's most spectacular raw-material sites: northern Brazil's Carajás Iron Ore Mine, located where the southeastern rim of the Amazon Basin meets the Brazilian Highlands in the center of Pará State. This may be the world's largest single deposit of iron ore, whose open-pit mines steadily deepen and sprawl outward as minerals are extracted from the surface—one stripped-away, environmentally-degrading layer at a time. In a typical year, 300 million metric tons are hauled away via a dedicated, 850-kilometer (535-mi) railroad to the Atlantic port of São Luis (see map, p. 274).

NASA image created by Jesse Allen, using EO-1 ALI data provided courtesy of the NASA EO-1 Team

Much of this growth occurred in Brazil, Chile, Peru, and Argentina, but most of the rest of the continent did quite well too. A leading reason has been an increased demand for the realm's abundant raw materials in the global marketplace—demand in no small part driven by the explosive growth (and voracious appetite) of China and India. These two Asian giants have been gobbling up Brazilian soybeans, Chilean copper, Peruvian silver, Venezuelan oil, and more. In the years leading up to 2012, commodity prices have increased substantially and South America has reaped the benefits. As a result, much of the realm was virtually unaffected by the global financial crisis that peaked in 2008–2010.

Heavy reliance on the production of raw materials (52 percent of the realm's exports today) is not always a good thing. As the saying goes, what goes up must come down: world commodity prices can be notoriously unstable and, along with them, these countries' national incomes. A second cautionary note is that high demand for commodities can drive up a nation's currency value, which can shrink sales of other exports because they get more expensive for foreign trading partners. Third, natural resource management, because of complex ownership issues and exploitation rights, invites corruption and complacency: note that discoveries of major oil reserves do not always result in a society's betterment. Finally, and most pertinently, some raw materials (gold, oil, silver, minerals) are finite, and it is important for producing countries to plan for a future without them. And that's a major challenge to politicians who tend to think no further than the next election. But if governments wisely manage their resources and economy, the commodity boom can only contribute to their development; it is a comparative advantage that several other parts of the world view with envy.

URBANIZATION

Rural-Urban Migration

As in most other realms, South Americans are leaving the land and migrating to towns and cities. South America's urbanization process got an early start and has been intensifying since 1950 as the growth of urban settlements averaged 5 percent a year. Meanwhile, rural areas grew by less than 2 percent annually over those same six decades. As a result, the realm's urban population has climbed to its current level of 82 percent, ranking it with those of western Europe and the United States. These numbers underscore not only the dimensions but also the durability of the **rural-to-urban migration [9]** from the countryside to the cities.

In South America, as in Middle America, Africa, and Asia, people are attracted to the cities and driven from the poverty of the rural areas. Both ***pull*** and ***push*** factors are at work. Rural land reform has been slow in coming, and for this and other reasons every year tens of thousands of farmers simply give up and leave, seeing little or no possibility for economic advancement. The urban centers lure them because they are perceived to provide opportunity— the chance to earn a regular wage. Visions of education for their children, better medical care, upward social mobility, and the excitement of life in a big city draw hordes to places such as São Paulo and Lima.

From the Field Notes . . .

"Two unusual perspectives of Rio de Janeiro form a reminder that here the wealthy live near the water in luxury high-rises, such as these overlooking Ipanema Beach—while the poor have million-dollar views from their hillslope *favelas*, such as Rocinho."

www.conceptcaching.com

© H. J. de Blij

RIO DE JANEIRO

© H. J. de Blij, P. O. Muller, J. Nijman, and John Wiley & Sons, Inc.

SAY "SOUTH AMERICA" and the first image most people conjure up is Sugar Loaf Mountain, Rio de Janeiro's landmark sentinel that guards the entrance to beautiful Guanabara Bay. Nicknamed the "magnificent city" because of its breathtaking natural setting, Rio replaced Salvador as Brazil's capital in 1763 and held that position for almost two centuries until the federal government shifted its headquarters to interior Brasília in 1960. Rio de Janeiro's primacy suffered yet another blow in the late 1950s: São Paulo, its urban rival 400 kilometers (250 mi) to the southwest, surpassed Rio to become Brazil's largest city—a gap that has been widening ever since. Although these events triggered economic decline, Rio (12.2 million) remains a major *entrepôt*, air-travel and tourist hub, and center of international business. As a focus of cultural life, Rio de Janeiro is unquestionably Brazil's leader, and the *cariocas* (as Rio's residents call themselves) continue to set the national pace with their entertainment industries, universities, museums, and libraries.

On the darker side, this city's reputation is increasingly being tarnished by the widening abyss between Rio's affluent and poor populations—symbolizing inequities that rank among the world's most extreme (see photos, p. 245). All great cities experience problems, and Rio de Janeiro is currently bedeviled by the drug use and crime waves emanating from its most desperate hillside *favelas* (slums) that continue to grow explosively. In November of 2010, after days of relentless gang violence in which at least 25 died and nearly 100 cars and buses were set on fire, the police (backed by an army of 800 federal troops) moved into the notorious *favela* known as *Complexo do Alemão* and searched the homes of leading suspects. Some 190 gang members were detained. It is a hopeful sign that the Brazilian authorities are getting serious, finally, in dealing with this problem.

Rio's planners recently launched a major project (known as "Rio-City") to improve urban life for all residents. This ambitious scheme is designed to reshape nearly two dozen of the aging city's neighborhoods, introduce an ultramodern crosstown expressway to relieve nightmarish traffic congestion, and—most importantly—bring electrical power, paved streets, and a sewage-disposal network to the beleaguered *favelas*.

In 2014, Rio will host the Olympics *and* soccer's World Cup tournament, and local and federal authorities will do all they can to make their city fit the "happy-go-lucky" image of Brazilian soccer. At a time of robust economic growth in Brazil, this may be a rare opportunity for the city's poor to improve their lot.

But the actual move can be traumatic. Cities in developing countries are surrounded and often invaded by squalid slums, and this is where the urban immigrant most often finds a first—and frequently permanent—abode in a makeshift shack without even the most basic amenities and sanitary facilities. Unemployment remains persistently high, often exceeding 25 percent of the available labor force. But still the people come, hopeful for a better life, the overcrowding in the shantytowns worsens, and the threat of regional-scale disease (and other disasters) rises.

Regional Patterns

The generalized spatial pattern of South America's urban transformation is displayed in Figure 5A-7, which shows a **cartogram** of the continent's population. Here we see not only the realm's countries in population space relative to each other, but also the proportionate sizes of individual large cities within their total national populations.

Regionally, the Southern Cone (countries colored green) is the most highly urbanized. Today in Argentina, Chile, and Uruguay, almost all of the population resides in

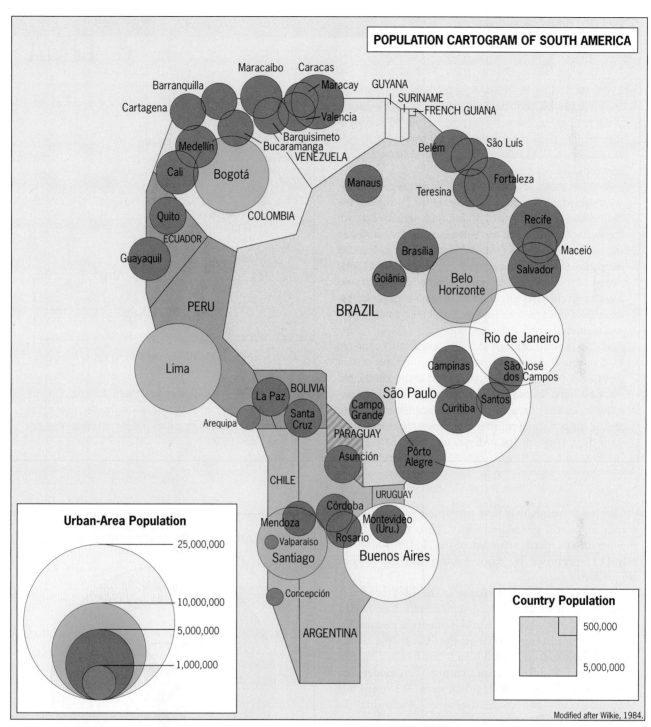

POPULATION CARTOGRAM OF SOUTH AMERICA

Urban-Area Population

— 25,000,000

— 10,000,000

— 5,000,000

— 1,000,000

Country Population

500,000

5,000,000

Modified after Wilkie, 1984.

FIGURE 5A-7   © H. J. de Blij, P. O. Muller, and John Wiley & Sons, Inc.

cities. Ranking next in urbanization is Brazil. The next highest group of countries (beige) borders the Caribbean Sea in the north. Not surprisingly, the Andean countries (brown) constitute the realm's least urbanized zone. Figure 5A-7 tells us a great deal about the relative positions of major metropolises in their countries. Three of them—Brazil's São Paulo and Rio de Janeiro, and Argentina's Buenos Aires—rank among the world's **megacities [10]** (cities whose pop-

ulations exceed 10 million). But even the population of the Amazon Basin is now more than 70 percent urban.

The "Latin" American City Model

The urban experience in the South and Middle American realms varies because of diverse historical, cultural, and economic influences. Nevertheless, there are a number

LIMA

EVEN IN A realm marked by an abundance of primate cities, Lima (9.4 million) stands out. Here reside 29 percent of the Peruvian population—who produce over 70 percent of the country's gross national income, 90 percent of its collected taxes, and 98 percent of its private investments. Economically, at least, Lima is Peru.

Lima began as a modest oasis in a narrow coastal desert squeezed between the cold waters of the Pacific and the soaring heights of the nearby Andes. Near here the Spanish conquistadors discovered one of the best natural harbors on the western shoreline of South America, where they founded the port of Callao; but they built their city on a site 11 kilometers (7 mi) inland where soils and water supplies proved more beneficial. This city was named Lima and quickly became the Spaniards' headquarters for all their South American territories.

Since independence in 1821, Lima has continued to dominate Peruvian national life. The city's population growth was manageable through the end of the 1970s, but the past three decades have witnessed a disastrous doubling in its size. As usual, the worst problems are localized in the peripheral squatter shantytowns that now house close to one-half of the metropolitan population. Lima's

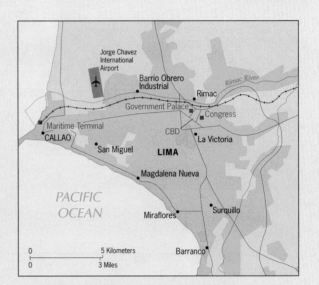

© H. J. de Blij, P. O. Muller, and John Wiley & Sons, Inc.

CBD, however, is worlds removed from this squalor, its historic landscape dotted with five-star hotels, new office towers, and prestigious shops as foreign investments have poured in since 1990.

of common threads that have prompted geographers to search for useful generalizations. One is the model of the intraurban spatial structure of the **"Latin" American city [11]** proposed by Ernst Griffin and Larry Ford (Fig. 5A-8).

As noted in earlier chapters, the idea behind a **model** is to create an idealized representation of reality, displaying as many key real-world elements as possible. In the case of South America's cities, the basic spatial framework of city structure, which blends traditional elements of South and Middle American culture with modernization forces now reshaping the urban scene, is a composite of radial sectors and concentric zones. Anchoring the model is the **CBD**, which is the primary business, employment, and entertainment focus of the surrounding metropolis. The CBD contains many modern high-rise buildings but also mirrors its colonial beginnings. As shown in Figure 4A-5, by colonial law Spanish colonizers laid out their cities around a central square, or plaza, dominated by a church and government buildings. Santiago's (and Cuzco's—see photo next page) *Plaza de Armas*, Bogotá's *Plaza Bolívar*, and Buenos Aires's *Plaza de Mayo* are classic examples. The plaza was the hub of the city, which later outgrew its old core as new commercial districts formed nearby; but to this day the plaza remains a ceremonial center and an important link with the past.

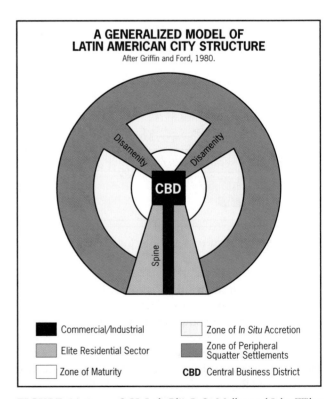

FIGURE 5A-8 © H. J. de Blij, P. O. Muller, and John Wiley & Sons, Inc.

From the Field Notes . . .

"Cuzco is the historic capital of the Inca Empire, located in the Andes of southeastern Peru. The city center is a UNESCO-designated World Heritage Site and attracts well over a million tourists each year. I drove out of the valley up the hillside to get a good view of the layout of the center, so typical of Spanish colonial cities. The *Plaza de Armas* with its parks and benches is surrounded by the Cathedral of Santo Domingo and other religious and municipal buildings, and adjacent areas around them are used for commerce and retailing. Located farther away from the Plaza are more recent residential areas. This historic Inca capital was sacked in the early sixteenth century by Pizarro, and only some Inca ruins remain in the center; outside the city lies the impressive fortress Sacsayhuaman with most of its walls still intact. This is not a very large city—it is too remote for that—but the population did triple over the past two decades to about 350,000."

© Jan Nijman

www.conceptcaching.com

Radiating outward from the urban core along the city's most prestigious axis is the commercial *spine*, which is adjoined by the *elite residential sector* (shown in green in Fig. 5A-8). This widening corridor is essentially an extension of the CBD, featuring offices, retail facilities, and housing for the upper and upper-middle classes.

The three remaining concentric zones are home to the less fortunate residents of the city, with income level and housing quality decreasing as distance from the CBD increases. The *zone of maturity* in the inner city contains housing for the middle class, who invest sufficiently to keep their aging dwellings from deteriorating. The adjacent *zone of* **in situ** *accretion* is one of much more modest housing interspersed with unkempt areas, representing a transition from inner-ring affluence to outer-ring poverty and taking on slum characteristics.

The outermost *zone of peripheral squatter settlements* is home to the millions of comparatively poor and unskilled workers who have recently migrated to the city. Here many newcomers earn their first cash income by becoming part of the informal sector [12], in which workers are undocumented and money transactions are beyond the control of government. The settlements consist mostly of self-help housing, vast shantytowns known as *barrios* [13] in Spanish-speaking South America and *favelas* in Brazil. Some of their entrepreneurial inhabitants succeed more than others, transforming parts of these shantytowns into beehives of activity that can propel resourceful workers toward a middle-class existence.

A final structural element of many South American cities forms an inward, narrowing sectoral extension of the zone of peripheral squatter settlements and is known as the *zone of disamenity*. It consists of undesirable land along highways, rail corridors, riverbanks, and other low-lying areas; people here are so poor that they are forced to live in the open. Thus the realm's cities present enormous contrasts between poverty and wealth, squalor and comfort—harsh contrasts all too frequently observed in the cityscape.

FUTURE PROSPECTS

The Need for Stability

In 2010, a number of South American countries were celebrating two centuries of independence. What mattered most, for nationalistic purposes, was the date when independence was declared, even if it took additional years for Spain to recognize that new political status. The wars with Spain lasted from 1808 (not coincidentally, soon after Spain had been weakened by Napoleon's invasion of Iberia) to 1838. Brazil's independence from Portugal was achieved during that same period, in 1822.

But two centuries of freedom have not seen the evolution of the kind of stable, mature political climates and institutions that one might expect. South American countries have been in frequent political turmoil. Dictatorial regimes ruled from one end of the realm to the other; unstable governments fell with damaging frequency. Widespread poverty, harsh regional disparities, poor internal surface connections, limited international contact, and economic stagnation prevailed.

Today, South America is a realm in dramatic transition. Entering the second decade of the twenty-first century, democracy seems to be taking hold almost everywhere. Long-isolated countries are becoming more interconnected through new transport routes and trade agreements. New settlement frontiers are being opened. Energy resources, some long exploited and others newly discovered, are boosting national economies as world prices have risen. Foreign states and corporations have appeared on the scene to buy commodities and invest in infrastructure. The pace of globalization has increased from Bogotá to Buenos Aires.

But these exciting developments must be seen against a backdrop of persistent problems. The realm's giant, Brazil, is embarked on a program of land reform, a campaign against poverty, and an effort to maintain financial rigor that have all run up against endemic corruption in government. The economy of Argentina in early 2011 was recovering after an implosion that shook the country to its core. Whereas Chile has been the realm's success story, emerging from the horrors of the Pinochet years as a stable and vibrant democracy with a thriving economy, neighboring Bolivia remains in the grip of a social revolution arising in part from the realization of its energy riches. And on the north coast lies Venezuela, its oil reserves among the largest in the world (and by far the largest in the realm) and its political life dominated by a one-time coup leader whose closest ideological ally is Cuba's communist ruler and whose major adversary is the U.S. government.

Problems of Inequality and Violence

Even if economic growth continues, several South American governments will have to address, more effectively than they have done in the past, the enormous inequality among their social classes. By some measures, the disparity between rich and poor is still wider in this realm than in any other, and wealth is disproportionately concentrated in the hands of a small minority (the richest 20 percent of the people control 70 percent of all wealth, while the poorest 20 percent own 2 percent). South American leaders are forced to walk a fine line between fueling growth while improving the lives of the masses, and some (such as Luiz Inácio da Silva, the former president of Brazil) have seemed to master that balancing act better than others (Hugo Chávez, in Venezuela). Moreover, in certain parts of the realm internal divisions are complicated by the resurgence of Amerindian identity among native peoples, and this,

too, requires delicate and even-handed maneuvering by national governments.

Inequality and socioeconomic polarization carry a price in terms of their harmful effects on civil society and social cohesion. This can be seen across the South America realm. In Colombia, civil strife has been the status quo for nearly half a century, claiming almost one million lives. It has divided the country, drawn poor peasants into the production of cocaine, and deeply politicized the population. In Argentina, thousands of homeless occupied a park in central Buenos Aires in late 2010 demanding housing, and two of them were killed by police bullets when they were evicted. Venezuela has a murder rate nearly ten times higher than that of the United States. In São Paulo and other big cities, gated communities for the middle and upper classes have become the norm, armed with state-of-the-art security systems to stave off crimes committed by the poor. The slums of Rio de Janeiro are not just poor places; they are also dangerous and deadly. According to a survey in 2007, 13 percent of slum dwellers had experienced or witnessed a shooting and 8 percent had a friend, neighbor, or family member who had met a violent death.

This type of dislocation, so characteristic of South (and Middle) America, can only be cured through more economic opportunities for the poor, more inclusive development, and better government. In this regard, the realm seems to offer a mixed bag of future directions.

The Shadow of the United States

The United States has long played a key role in this realm, for better or for worse. Beginning with the Monroe Doctrine in 1823 (which stated that European colonial powers had no rights to South America), the United States has consistently claimed a special interest in this part of the hemisphere. Over time, and especially during the Cold War in the second half of the twentieth century, the United States became politically involved in a number of countries. Preoccupation with efforts to keep Soviet influence out of the Western Hemisphere sometimes led the American government to choose the wrong side in local political struggles (as in Chile during the early 1970s); such things are not easily forgotten by the people of these countries. Indeed, anti-Americanism never seems far from the surface in this realm and is mainly based on past U.S. support for dictatorships or perceived imperialistic behavior.

To be sure, relations with the United States have never been smooth or always asymmetrical. For instance, South America today represents only about 4 percent of all U.S. foreign trade, and it is far less important to the United States than, say, Canada or western Europe. But the United States remains the biggest trading partner for South America as a whole, accounting for almost one-fifth of the exports and imports of all South American countries.

It was hardly a coincidence that *dependencia* theory [14] originated here in South America during the 1960s. It was a new way of thinking about economic development and underdevelopment that explained the persistent poverty of some countries (or of the masses inside those countries) in terms of the unequal relations with other (wealthy) countries. In this view, South American nations may have become officially independent of Spain in the early nineteenth century, but effectively they wound up becoming subservient to the United States (and other rich industrialized countries). In the present political and economic climate, however, most would consider *dependencia* theory as dated or at least unable to provide a way forward.

Nevertheless, the asymmetry remains, and U.S. foreign policy, as seen from South America, is not always credible. Interestingly, U.S. special interests in the realm are now under increasing pressure from another major power that would have been furthest from President Monroe's mind back in 1823: China.

China Calling

By 2010, China had displaced the United States as the leading trading partner of Brazil and Chile, and had become the second-largest for both Colombia and Peru. This remarkable development has yet to fully register in the United States—China is on the way to becoming the most important great power in the South American realm.

Nearly invisible only ten years ago, China is now making its presence felt across South America in a variety of ways that include establishing new embassies and consulates, buying up companies, partnering joint ventures, financing infrastructure projects and development assistance, and sending and inviting high-level trade delegations. Their motives are clear: the Chinese need raw materials such as oil, copper, cement, and minerals to fuel the enormous economic growth of their country (see photo above). They also are seeking to expand markets for Chinese exports, and the growing middle classes of Brazil and other countries make enticing targets.

China has none of the historical baggage that handicaps America's efforts to maintain a decisive influence in the realm. It has never had any political designs on South America, and its objectives are seen as strictly economic. From the point of view of Brazil, at least, it also represents a far more equal relationship: as BRICs, both rank among the biggest emerging economies in the world.

Whether it is the United States or China, South Americans will need to carefully consider how and where to focus their economic attention. Within the realm itself, there is still much to be done in terms of political and economic integration. Perhaps those Chinese investments in the infrastructural projects mentioned above will help to build the bridges so badly needed to bring this realm together.

Nothing symbolizes China's new economic drive in South America more than its growing interest in its burgeoning fellow BRIC. In 2011, the two countries concluded a (U.S.) $7 billion deal in which Chinese investors will finance a major expansion of Brazil's already-booming soybean production. And in 2012, billions of additional dollars of Chinese investment will come to fruition as the new *Superporto do Acu* opens for business (this photo shows a Chinese delegation visiting the construction site in 2010). Acu is located about 400 kilometers (250 mi) northeast of Rio de Janeiro, and its massive three-kilometer (2-mi)-long concrete pier jutting out into the Atlantic will contain 10 berths capable of accommodating the largest ships afloat (which are also Chinese). Brazilians are already calling it the "Highway to China", which upon completion immediately becomes the world's largest port handling bulk cargoes.

© Fabio Rossi/Globo/Getty Images Inc.

● POINTS TO PONDER

- The Amazonian rainforest is mainly located inside Brazil, but sometimes is referred to as "the lungs of the Earth." Should its preservation be a global responsibility?

- Is it a coincidence that both *dependencia* theory and Liberation Theology originated in this realm?

- The map of South America's religions is changing. Brazil is still nominally the world's largest Roman Catholic country, but the number of adherents is declining and now accounts for less than 75 percent of the national population (down from 90 percent three decades ago).

- Should China's growing involvement in South America be a cause of concern for the United States?

SOUTH AMERICA: POLITICAL UNITS AND MODERN REGIONS

POPULATION
- · Under 50,000
- · 50,000–250,000
- ● 250,000–1,000,000
- ● 1,000,000–5,000,000
- ● Over 5,000,000

National capitals are underlined

REGIONS
- The North
- The West
- The South
- Brazil

| 0 | 400 | 800 | 1200 Kilometers |

| 0 | 200 | 400 | 600 | 800 Miles |

FIGURE 5B-1

© H. J. de Blij, P. O. Muller, and John Wiley & Sons, Inc.

REGIONS

The Caribbean North

The Andean West

The Southern Cone

Brazil

IN THIS CHAPTER

◆ The cocaine curse

◆ Oil and politics in Venezuela

◆ Amerindian resurgence

◆ Brazil's unmatched natural resources

CONCEPTS, IDEAS, AND TERMS

© Jan Nijman
Where were these photos taken?
Find out at www.wiley.com/college/deblij

On the basis of physiographic, cultural, and political criteria, South America can be divided into four rather clearly defined regions (Fig. 5B-1):

1. **The Caribbean North**, located almost entirely north of the equator, consists of five entities that display a combination of Caribbean and South American features: Colombia, Venezuela, and those that represent three historic colonial footholds by Britain (Guyana), the Netherlands (Suriname), and France (French Guiana).

2. **The Andean West** is formed by four republics that share a strong Amerindian cultural heritage as well as powerful influences resulting from their Andean physiography: Ecuador, Peru, Bolivia, and, transitionally, Paraguay.

3. **The Southern Cone**, for the most part located south of the Tropic of Capricorn (latitude 23½°S), includes three countries: Argentina, Chile, and Uruguay (all with strong European imprints and little remaining Amerindian influence) plus aspects of Paraguay. The far northern part of this region that lies east of the Andes is occupied by the *Gran Chaco*, a sparsely populated, hot, semiarid, grassy lowland centered on western Paraguay.

4. **Brazil** occupies an enormous part of interior and eastern South America. In the Amazonian north, its own interior overlaps with the continent's "green heart"— sometimes called the world's "lungs"—the largest tropical rainforest on Earth. As South America's giant, accounting for just about half the realm's territory as well as population, it is swiftly developing into the Western Hemisphere's second superpower. In Brazil the dominant Iberian influence is Portuguese, not Spanish,

© Jan Nijman

MAJOR CITIES OF THE REALM

City	Population* (in millions)
São Paulo, Brazil	26.2
Buenos Aires, Argentina	12.8
Rio de Janeiro, Brazil	12.2
Lima, Peru	9.4
Bogotá, Colombia	8.9
Santiago, Chile	6.1
Belo Horizonte, Brazil	6.1
Brasília, Brazil	4.1
Caracas, Venezuela	3.2
Guayaquil, Ecuador	2.8
Asunción, Paraguay	2.1
Manaus, Brazil	1.9
Quito, Ecuador	1.9
La Paz, Bolivia	1.7
Montevideo, Uruguay	1.6

*Based on 2012 estimates.

and here Africans, not Amerindians, form a significant component of demography and culture. And Brazil is in some ways a bridge between the Americas and Africa, which is why we discuss it last in this chapter, just before we turn to Subsaharan Africa in Chapter 6A.

Each of these four regions shares some important commonalities in terms of physical environment, ethnic origins, cultural milieu, and international outlook. But there also are important differences, especially involving economic performance, democratic functioning, and social stability.

■■ THE CARIBBEAN NORTH

The countries of South America's northern tier have something in common besides their coastal location: each has a coastal tropical-plantation zone based on the Caribbean colonial model. Especially in the three Guianas, early European plantation development encompassed the forced immigration of African laborers and eventually the absorption of this element into the population matrix. Far fewer Africans were brought to South America's northern shores than to Brazil's Atlantic coasts, and tens of thousands of South Asians also arrived as contract laborers and stayed as settlers, so the overall situation here is not comparable to Brazil's. And it is also distinctly different from that of the rest of South America.

Today, Guyana, Suriname, and French Guiana still display the coastal orientation and plantation dependency with which the colonial period endowed them, although the logging of their tropical forests is penetrating and ravaging the interior. In Colombia and Venezuela, however, farming, ranching, and mining drew the population inland,

overtaking the coastal-plantation economy and creating diversified economies (Fig. 5B-2).

Colombia

Imagine a country more than twice the size of France, not at all burdened by overpopulation, with an environmental geography so varied that it can produce crops ranging from the temperate to the tropical, and possessing world-class oil reserves as well as other natural resources. This country is situated in the crucial northwest corner of South America, with 3200 kilometers (2000 mi) of coastline on both Atlantic (Caribbean) and Pacific waters, closer than any of its neighbors to the markets of the north and sharing a border with giant Brazil to the southeast. Its nation uses a single language and adheres to one dominant religion. Wouldn't such a country thrive?

The answer, unfortunately, is no. Colombia has a history of strife and violence, its politics unstable, its economy damaged, and its future clouded. Colombia's cultural uniformity did not produce social cohesion. Its spectacular, scenic physical geography also divides its population of 46.8 million into clusters not sufficiently interconnected to foster integration; even today, this huge country has less than 800 kilometers (500 mi) of four-lane highways. Its proximity to U.S. markets is a curse as well as a blessing: at the root of Colombia's latest surge of internal conflict lies its role as one of the world's largest producers of illicit drugs.

History of Conflict

Colombia's recent disorder is not its first. In the past, civil wars between conservatives and liberals (based on Roman Catholic religious issues) developed into conflicts pitting rich against poor, elites against workers. In Colombia today, people still refer to the last of these wars as *La Violencia*, a decade of strife beginning in 1948 during which as many as 200,000 people died. In the 1970s, disaster struck again. In remote parts of the country, groups opposed to the political power structure began a campaign of terrorism, damaging the developing infrastructure and destroying confidence in the future. Simultaneously, the U.S. market for narcotics expanded rapidly, and many Colombians got involved in the drug trade. Powerful and wealthy drug cartels formed in major cities such as Medellín and Cali, with networks that influenced all facets of Colombian life from the peasantry to the politicians. In all, in the past half-century, it is estimated that one million people died violently. The fabric of Colombian society unraveled.

People and Resources

As Figure 5B-2 reveals, Colombia's physiography is mountainous in its Andean west and north and comparatively flat in its interior. Colombia's scattered population (see Fig. 5A-3) tends to cluster in the west and north, where the resources and the agricultural opportunities (including the coffee for which Colombia is famous) lie.

Colombia's population clusters are not well interconnected. Several lie on the Caribbean coast, centered on

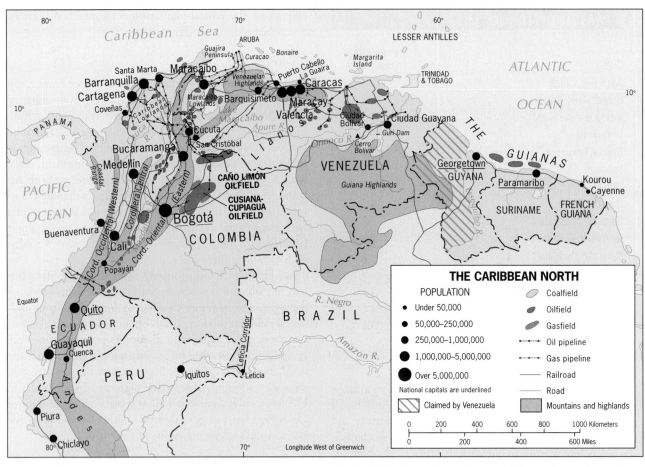

FIGURE 5B-2

© H. J. de Blij, P. O. Muller, and John Wiley & Sons, Inc.

Barranquilla, Cartagena, and Santa Marta, old colonial entry points. Others are anchored by major cities such as Medellín and Cali. What is especially interesting about Figure 5B-2 in this context is how little development has taken place on the country's lengthy Pacific coast, where the port city of Buenaventura, across the mountains from Cali, is the only place of any size. The map suggests that Colombia's Pacific Rim era, already in full force in Chile, Peru, and Ecuador, has yet to arrive.

The map also shows that Colombia and neighboring Venezuela share the oil and gas reserves in and around the "Lake" Maracaibo area ("Lake" because this is really a gulf with a narrow opening to the sea), but Venezuela has the bigger portion. Recent discoveries, however, have boosted Colombia's production along the base of the easternmost Andean cordillera, and our map shows a growing system of pipelines from the interior to the coast. Meanwhile, the vast and remote interior proved fertile ground for that other big money-maker in Colombia: drugs.

Cocaine's Curse

The rise of the narcotics industry, fueled by outside (especially U.S. but also Brazilian and European) demand, and coupled with its legacy of violence, crippled the state for decades and threatened its very survival. Drug cartels based in the cities controlled vast networks of producers

and exporters; they infiltrated the political system, corrupted the army and police, and waged wars with each other that cost tens of thousands of lives and destroyed Colombia's social order. The drug cartels also organized their own armed forces to combat attempts by the Colombian government to control the illegal narcotics economy. Meanwhile, the owners of large haciendas in the countryside hired private security guards to protect their properties, banding together to expand these units into what became, in effect, private armies. Colombia was in chaos as narcoterrorists committed appalling acts of violence in the cities, rebel forces and drug-financed armies of the political "left" fought paramilitaries of the political "right" in the countryside, and Colombia's legitimate economy, from coffee-growing to tourism, suffered fatally. To further weaken the national government, the rebels even took to bombing oil pipelines.

Threats of the Insurgent State

What happened in Colombia beginning in the 1970s and escalating in the four decades following was not unique in the world—states have succumbed to chaos in the past—but this was an especially clear-cut case of a process long studied and modeled by political scientists. By the turn of this century, certain parts of Colombia were beyond the control of its government and armed forces. There,

FIGURE 5B-3   © H. J. de Blij, P. O. Muller, and John Wiley & Sons, Inc.

insurgents of various stripes created their own domains, successfully resisting interference and demanding to be left alone to pursue their goals, illegitimate though they might be. Leaders of some of these insurgent domains even sent emissaries to Bogotá to negotiate their terms for "independence."

Viewing this process institutionally, political scientists recognized three evolutionary stages. During the first, *contention*, a rebellion erupts, sustains itself, and becomes based in some part of a country. During the second stage, *equilibrium*, the rebels gain effective control over a territorially-defined sector of the state, establish a capital, impose rules of conduct, and have sufficient strength to bring the national government to the negotiating table. If they succeed, they may achieve secession from the state or even bring the government down and take control of all of it. But equilibrium may also be followed by a third stage, *counteroffensive*, in which the state, perhaps with the support of outside forces, resumes the conflict and ultimately defeats the insurgents. We will encounter instances of all three stages in this model in later chapters.

We use the concept of an **insurgent state [1]** to represent the second stage, that of equilibrium: it is the moment of truth when the rebel domain takes on the elements of a nascent state complete with boundaries, a core

area as well as a capital, a local government, and schools and other social services that substitute for those the formal state may have provided previously.

Colombia in the 1990s and early 2000s contained several entities that were taking on the properties of insurgent states, and one of these, the red-striped zone shown in Figure 5B-3, even acquired a name—"Farclandia"—after the initials of the Revolutionary Armed Forces of Colombia (FARC), one of the most brutal and by far the most powerful among Colombia's insurgent groups. At the end of 1999, FARC had the Colombian government on the ropes, forcing it to demilitarize its area south of Bogotá (about the size of Switzerland) and announcing plans for a second insurgent state centered on the remote southeastern town of Mitú. It appeared to be just a matter of time before the national government in Bogotá would lose control, equilibrium would turn into disintegration, and Colombia would devolve into a **failed state [2]** (a country whose institutions have collapsed and in which anarchy prevails).

But for the past decade, Colombia has mounted a twin campaign to defeat the rebels militarily and to persuade them through legal means to give up their arms. The government succeeded in transforming Colombia from a nearly failed state to a more stable, safer, and prosperous country. Securing assistance from the United States, it was able to increase the pressure on armed rebels as well as coca growers, invaded "Farclandia," and scored a number of noteworthy successes in killing and arresting leaders as well as freeing hostages held by the rebels. Colombia's nascent insurgent states weakened, and disarray marked the once-formidable rebel armies.

A New Economic Future?

The government matched its domestic counteroffensive with an international campaign to help revive the economy, promoting market-oriented, business-friendly policies. Under the political circumstances, it is remarkable that economic growth before the global downturn of 2008 was as high as 8 percent annually, led by the revenues from energy, metals, and agricultural products; the comparative lull in violence boosted the exporting of coffee and cut flowers (Colombia ranks second in the world in this business) as well as a rise in tourism.

But Colombia remains a nation in crisis, not all of it of its own making. About 70 percent of the cocaine consumed in the United States comes from Colombia; most of the remainder goes to Brazil or Europe. American support for the government's antidrug campaign (*Plan Colombia*), logistical as well as financial, raises political issues in Bogotá, where extraditions of Colombian drug lords to the United States are not universally popular. Conversely, the U.S. Congress refused to approve a free-trade agreement with Colombia because of the Colombian government's tilt toward right-wing paramilitaries in conjunction with the amnesty program. Colombians argue that the cocaine keeps flowing because Americans continue to buy it and that Americans should look in the mirror before they accuse others of bad behavior.

And cocaine is not Colombia's only problem. Economic development is hampered by extreme inequality that has only gotten worse over time, especially in the countryside. In 1954, 3 percent of landowners controlled 55 percent of the land; in 2005, less than half of 1 percent of the landowners held a whopping 63 percent of all agricultural lands. If the government wants to restore belief in the political system among the rural poor (many of whom have turned to employment in the cocaine industry and/or have been engaged by the rebels), land reform is a must. In 2011, Colombia's newly elected government faced a decidedly clouded future.

Venezuela

A long and tortuous boundary separates Colombia from Venezuela, its neighbor to the east. Much of what is important in Venezuela is concentrated in the northern and western parts of the country, where the Venezuelan Highlands form the eastern spur of the north end of the Andes system. Most of Venezuela's 29.7 million people are concentrated in these uplands, which include the capital of Caracas, its early rival Valencia, and the commercial/industrial centers of Barquisimeto and Maracay.

The Venezuelan Highlands are flanked by the Maracaibo Lowlands and "Lake" Maracaibo to the northwest and by an immense plainland of savanna country, known as the *Llanos*, in the Orinoco Basin to the south and east (Fig. 5B-2). The Maracaibo Lowlands, at one time a disease-infested, sparsely peopled coastland, today constitute one of the world's leading oil-producing areas; much of the oil is drawn from reserves that lie beneath the shallow waters of the lake itself. The country's third-largest city, Maracaibo, is the focus of the petroleum industry that transformed the Venezuelan economy in the 1970s; however, as we shall see, since then oil has been more of a curse than a blessing.

The *Llanos* on the southern side of the Venezuelan Highlands and the Guiana Highlands in the country's southeast, like much of Brazil's interior, are in a nascent stage of development. The 300- to 650-kilometer (180- to 400-mi)-long *Llanos* slope gently from the base of the northernmost Andean spur to the Orinoco's floodplain. Their mixture of savanna grasses and scrub woodland supports cattle grazing on higher ground, but widespread wet-season flooding of the more fertile lower-lying areas has thus far inhibited the plainland's commercial farming potential (much of the development of the *Llanos* to date has been limited to the exploitation of its substantial oil reserves). Crop-raising conditions are more favorable in the **tierra templada** areas at more moderate altitudes in the Guiana Highlands. Economic integration of this more remote interior zone with the heartland of Venezuela has been spearheaded by the discovery of rich iron ores on the northern side of the Guiana Highlands southwest of Ciudad Guayana. Local railroads now connect with the Orinoco, and from there ores are shipped directly to foreign markets.

Oil and Politics

Despite these opportunities, Venezuela since the late 1990s has been in upheaval as longstanding economic and social problems finally intensified to the point where the electorate decided to push the country into a new era of radical political change. For nearly two decades since the euphoric 1970s, oil had not bettered the lives of most Venezuelans. A primary reason was that the government unwisely acquired the habit of living off oil profits, forcing the country to suffer the consequences of the prolonged global oil depression that began in the early 1980s. Venezuela found itself heavily burdened by a huge foreign debt it had incurred through borrowing against future oil revenues that were not materializing fast enough. By the mid-1990s, the government was required to sharply devalue the currency, and a political crisis ensued that resulted in a severe recession and widespread social unrest.

With more and more Venezuelans enraged at the way their oil-rich country was approaching bankruptcy without making progress toward the more equitable distribution of the national wealth, voters resoundingly turned in an extreme direction in the 1998 presidential election.

Expressing their disgust with Venezuela's ruling elite of both political parties, they elected Hugo Chávez, a former colonel who in 1992 had led a failed military coup. Reaffirming their decision in 2000, Venezuelans gave nearly 60 percent of their votes to Chávez, apparently providing him with a mandate to act as strongman on behalf of the urban poor and the increasingly penurious middle class.

Venezuela's Autocratic Turn

Chávez, clinging to power tenaciously, has indeed pursued this course since entering office in 1999, sweeping aside Congress and the Supreme Court, supervising the rewriting of the Venezuelan constitution in his own image, and proclaiming himself the leader of a "peaceful leftist revolution" that will transform the country. Although he professes that social equality ranks highest on his agenda, Chávez has stirred up racial divisions by actively promoting mestizos (about 70 percent of the population) over people of European background (19 percent).

In the international arena, Chávez sparks controversy at every turn: infuriating the government of neighboring Colombia by expressing his "neutrality" in its confrontation with cocaine-producing insurgent forces; unsettling another neighbor, Guyana, by aggressively reviving a century-old territorial claim to the western zone of that country (the striped zone in Fig. 5B-2); embracing the Cuban regime; proclaiming his solidarity with Iran's president; and, in January 2010, openly declaring himself a Marxist. More generally, Chávez has obstructed the globalization plans of the United States throughout South America, using the windfall income from high-priced oil to reward allies in this effort with substantial subsidies. For example, considerable financial support has been provided to Bolivia's President Evo Morales, the realm's first national leader of Amerindian ancestry, who faced powerful opposition from the country's affluent Hispanic minority.

Chávez has positioned himself as the champion of Venezuela's (and South America's) poor, trumpeting his "Bolivarian Revolution" as a local alternative to what he describes as the insidious encroachment of U.S. imperialism. Most of all, his leftist ideological rhetoric serves to appeal to the poor and to disguise his increasingly autocratic regime.

In 2012, Chávez's second presidential term will come to an end, but in 2009 he won a referendum that abolished term limits. It was a huge victory, paving the way for his continued reign in years to come. Vilified by much of the Western world and by many of Venezuela's entrepreneurial citizens (a good number of whom have fled the country), Chávez has a message that seems to resonate at least to some extent among the poor South American masses, in itself testimony to the continuing inequality that marks this realm.

The "Three Guianas"

Three small entities form the eastern flank of the realm's northern region: Guyana, Suriname, and French Guiana. They are good reminders of why the name "Latin" America

is inappropriate: the first is a legacy of British colonialism and employs English as its official language; the second is a remnant of Dutch influence where Dutch is still official among its polyglot of tongues; and the third is still a dependency—of France. None has a population exceeding 1 million, and all three exhibit social characteristcis and cultural landscapes more representative of the Caribbean than South America. Here British, Dutch, and French colonial powers acquired possessions and established plantations, brought in African and Asian workers, and created economies similar to those that mark a number of the Caribbean islands.

Guyana, with 820,000 people, still has more inhabitants than Suriname and French Guiana combined. When Guyana became independent in 1966, its British rulers left behind an ethnically and culturally divided population in which people of South Asian (Indian) ancestry now make up about 50 percent and those of African heritage (including African-European ancestry) 36 percent. This makes for contentious politics, given the religious mix, which is approximately 58 percent Christian and 37 percent Hindu and Muslim. Guyana remains dominantly rural, and plantation crops continue to figure prominently among exports (gold from the interior is the most valuable single product). Oil may soon become a factor in the economy, though, because a recently discovered reserve that lies offshore from Suriname extends westward beneath Guyana's waters. Still, Guyana is among the realm's poorest and least urbanized countries, and it is strongly affected by the region's narcotics industry. Its thinly populated interior, beyond the reach of antidrug campaigns, has become a staging area for drug distribution to Brazil, North America, and even Europe. A recent official report suggests that drug money amounts to as much as one-fifth of Guyana's total economy.

Venezuela's territorial claim against Guyana may be in abeyance, but the United Nations in 2007 settled a dispute with Suriname over a potentially oil-rich maritime zone in Guyana's favor (Fig. 5B-2). Some geologists believe that this offshore basin may hold more oil than Europe's North Sea, which would—if properly managed—transform Guyana's economy. Exploratory drilling began in 2009.

Suriname actually progressed more rapidly than Guyana did after it was granted independence in 1975, but persistent political instability soon ensued. The Dutch colonists brought South Asians, Indonesians, Africans, and even some Chinese to their colony, making for a fractious nation. More than 100,000 residents—about one-quarter of the entire population—emigrated to the Netherlands, and were it not for support from its former colonial ruler, Suriname would have collapsed. Nonetheless, its rice farms give Suriname self-sufficiency and even allow for some exporting, and plantation crops continue to rank among the exports. Suriname's leading income producer, however, is its bauxite (aluminum ore) mined in a zone across the middle of the country. And the recent oil finds (Fig. 5B-2) could provide important revenues in the years ahead.

Suriname, most of whose population of 480,000 resides along the northern coastal strip, is one of South America's poorest countries, and its prospects remain bleak. Along with Guyana, Suriname is involved in an environmental controversy centering on its luxuriant tropical forests. Timber companies from Asia's Pacific Rim offer lucrative rewards for the right to cut down the magnificent hardwood trees, but conservationists are trying to slow the destruction by buying up concessions before they can be opened to logging.

Suriname's cultural geography is enlivened by its many languages. Dutch may be the official tongue, but other than by officials and in schools it is not heard much. A mixture of Dutch and English, *Sranan Tongo*, serves as a kind of common tongue, but you can also hear Amerindian, Hindi, Chinese, Indonesian, and even some French Creole in the streets. Suriname may have a Dutch past, but today it is a member of CARICOM, the English-speaking, supranational Caribbean economic organization.

French Guiana, the easternmost outpost of the Caribbean North region, is a dependency—mainland South America's only one. This territory is an anomaly in other ways as well. Consider this: it is almost as large as South Korea (population: nearly 50 million) with a population of just over 200,000. Its status is an *Overseas Département* of France, and its official language is French. Nearly half the population resides in the immediate vicinity of the coastal capital, Cayenne.

In 2011, there still was no prospect of independence for this severely underdeveloped relic of the former French Empire. Gold remains the most valuable export, and the small fishing industry sends some exports to France. But what really matters here in French Guiana is the European Space Agency's launch complex at Kourou on the coast, which accounts for more than half of the territory's entire economic activity. From plantation farming to spaceport . . . this must be the ultimate story of globalization.

■■ THE ANDEAN WEST

The second regional grouping of South American states—the Andean West (Fig. 5B-4)—is dominated physiographically by the great Andean mountain chain and historically by Amerindian peoples. This region encompasses Peru, Ecuador, Bolivia, and transitional Paraguay, the last with one foot in the West and the other in the South (Fig. 5B-1). Bolivia and Paraguay also constitute South America's only two landlocked countries. For several decades now, the three main countries in this region, Peru, Ecuador, and Bolivia, have joined in the *Andean Community*, a trading bloc that also includes Colombia.

Spanish conquerors overpowered the Amerindian nations, but they did not reduce them to small minorities as happened to so many indigenous peoples in other parts of the world. Today, roughly 47 percent of the residents of Peru, the region's most populous country, are Amerindian; in Bolivia, Amerindians are in the majority at 57 percent.

About 25 percent of Ecuador's population identifies itself as Amerindian, and in Paraguay the ethnic mix, not regionally clustered as in the other three countries, is overwhelmingly weighted toward Amerindian ancestry.

As the Data Table in Appendix B reports, this is South America's poorest region economically, with lower incomes, higher numbers of subsistence farmers, and fewer opportunities for job-seekers. For a very long time, the urbane lives of the land-owning elite have been worlds apart from the hard-scrabble existence of the landless peonage (the word **peon** is an old Spanish term for an indebted day laborer).

But today, this region, like the realm as a whole, is stirring, and oil and natural gas are part of the story. In Bolivia, the first elected president of Amerindian ancestry is trying to gain control over an energy industry not used to his aggressive tactics. In Ecuador's 2006 elections, a populist gained the presidency on a promise to divert more of the country's oil revenues toward domestic needs. In Peru, where an energy era seems to be opening as new reserves are discovered, the government is under pressure to protect Amazonian peoples and environments, limit foreign involvement, and put domestic needs and rights first.

Peru

Peru straddles the Andean spine for more than 1600 kilometers (1000 mi) and is the largest of the region's four republics in both territory and population (30.4 million). Physiographically and culturally, Peru divides into three components: (1) the desert coast, the European-mestizo subregion; (2) the Andean Highlands or *Sierra*, the Amerindian subregion; and (3) the subregion focused on the eastern slopes, the sparsely populated Amerindian-mestizo interior (Fig. 5B-4).

Three Subregions

Lima (box, p. 248) and its port, Callao, lie at the center of the **desert coast** subregion, and it is symptomatic of the cultural division prevailing in Peru that for nearly 500 years the capital city has been positioned on the periphery, not in a central location in a basin of the Andes. From an economic point of view, however, the Spaniards' choice of a headquarters on the Pacific coast proved to be sound, for the coastal strip has become commercially the most productive part of the country. A thriving fishing industry contributes significantly to the export trade; so do the products of irrigated agriculture in some 40 oases dispersed across the arid coastal plain, which include fruits such as citrus and olives, and vegetables such as asparagus (a big money-maker) and lettuce.

The Andean or *Sierra* subregion occupies just about one-third of the country and is the ancestral home of the largest component in the total population, the speakers of Quechua, the *lingua franca* that emerged during the Inca Empire. Their physical survival during the harsh Spanish colonial regime was made possible by their adaptation to the high-altitude environments they inhabited, but their social fabric was ripped apart by communalization, forced

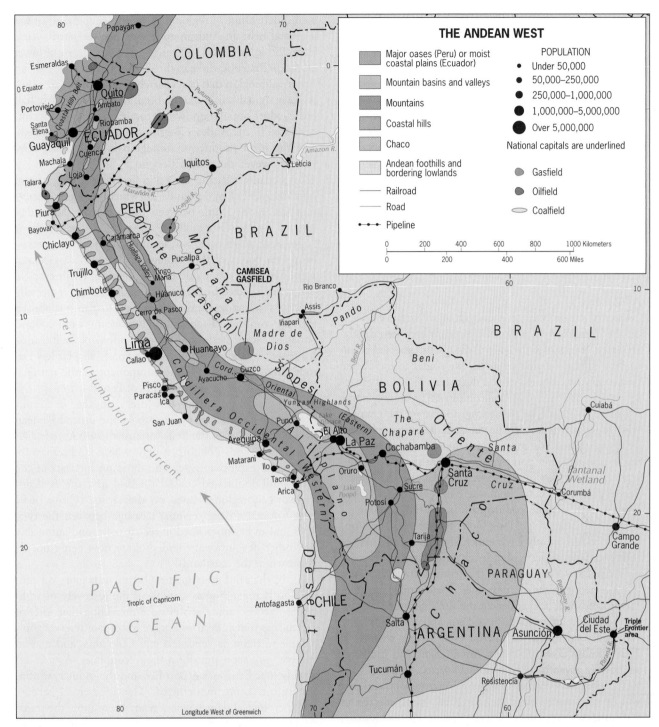

FIGURE 5B-4

© H. J. de Blij, P. O. Muller, and John Wiley & Sons, Inc.

cropping, religious persecution, and outward migration to towns and haciendas where many became serfs.

Although Amerindian people constitute almost half of the population of Peru, their political influence remains slight. Nor is the Andean subregion a major factor in Peru's commercial economy—except, of course, for its enormous mineral storehouse, which yields copper, zinc, and lead from mining centers, the largest of which is Cerro de Pasco. Peru is also the world's fifth-largest producer of

gold, the price of which has been soaring in recent years, contributing to the country's steadily growing economy. The goldmines are located in the Andes north of Lima.

In the high valleys and intermontane basins, the Amerindian population is clustered either in isolated villages, around which people practice a precarious subsistence agriculture, or in the more favorably located and fertile areas where they are tenants, peons on white- or mestizo-owned haciendas. It was a very different story at the time of the

Incas, when the indigenous social structure was still un-harmed and the region provided bountiful harvests. Today, there is a painful contrast between the prosperous coast, booming Lima, and the thriving north on the one hand, and this poverty-plagued southern Andean zone on the other. That division could be a threat to Peru's long-term stability.

Of Peru's three subregions, the East or **Oriente**—the inland slopes of the Andean ranges that lead down to the Amazon-drained, rainforest-covered *montaña*—is the most isolated. The focus of the eastern subregion, in fact, is Iqui-tos, a city that looks east rather than west and can be reached by oceangoing vessels sailing 3700 kilometers (2300 mi) up the Amazon River across northern Brazil. Iquitos grew rap-idly during the Amazon's wild-rubber boom of just over a century ago, and then declined; now it is finally growing again and reflects Peruvian plans to open up the eastern interior.

A New Era

Today, the *Oriente* subregion, and perhaps Peru as a whole, is on the threshold of a new era. Petroleum had been discov-ered west of Iquitos as long ago as the 1970s, when oil began to flow through a pipeline across the Andes to the Pacific port of Bayóvar. But major new discoveries of oil and gas re-serves since the turn of the millennium are promising (some say the verb should be *threatening*) to change the country, with concomitant impacts on the Amerindian peoples still living an isolated existence in this remote, forested East. Al-ready, pipelines transport natural gas from the Camisea re-serve (north of Cuzco) to a conversion plant on the Paracas Peninsula south of Lima, from where an ocean-floor pipeline sends it to an offshore loading platform for tankers taking it to the U.S. market (Fig. 5B-4). Other reserves will come "on-line" later, but already environmentalists and supporters of indigenous rights are raising issues in the interior even as political activists argue against the terms of trade that Peru has accepted with the oil companies. Their complaint is that the proceeds will further benefit the already-favored coastal, northern, and urban residents of Peru, and leave the disad-vantaged residents of the interior even further behind. In 2009, deadly clashes between Amerindians and industry-supporting government forces killed dozens.

Peru has never had an Amerindian president, and that will be the case through at least 2016. Newly-elected in 2011, the country's current leader, populist Ollanta Humala, is, like the others before him, supported by the traditional Hispanic (and mestizo) establishment. Yet Amerindians form a near-majority, they are restive, and they now have models of empowerment not previously seen in this region. The question is what course Peru will take, but with the presently booming economy this may well be the time to bring the country together.

Ecuador

On the map, Ecuador, smallest of the three Andean West republics, appears to be just a northern corner of Peru. But that would be a misrepresentation because Ecuador possesses a complete range of regional variations (Fig. 5B-4). It has a coastal belt; an Andean zone that may be narrow (less than 250 kilometers [150 mi]) but by no means of lower elevation than elsewhere in the region; and an *Oriente*—an eastern subregion that is as sparsely settled and as econom-ically marginalized as that of Peru. As in Peru, nearly half of Ecuador's population (which totals 14.7 million) is con-centrated in the Andean intermontane basins and valleys, and the most productive subregion is the coastal strip. Here, however, the similarities end.

Ecuador's Pacific coastal zone consists of a belt of hills interrupted by lowlands, of which the most impor-tant lies in the south between the hills and the Andes, drained by the Guayas River. Guayaquil—the country's largest city, main port, and leading commercial center—forms the focus of this subregion. Unlike Peru's coastal strip, Ecuador's is not a desert: it consists of fertile tropical plains not afflicted by excessive rainfall. Seafood (especially shrimp) is a leading product, and these lowlands support a thriving commercial agricultural economy built around bananas, cacao, cattle raising, and coffee on the hillsides. Moreover, Ecuador's western subregion is also far less Eu-ropeanized because its white component of the national population is only about half of Peru's.

A greater proportion of whites is engaged in admin-istration and hacienda ownership in the central Andean zone, where most of the Ecuadorians who are Amerindian also reside—and, not surprisingly, where land-tenure re-form is an explosive issue. The differing interests of the Guayaquil-dominated coastal lowland and the Andean-highland subregion focused on the capital (Quito) have long fostered a deep regional cleavage between the two. This schism has intensified in recent years, and autonomy and other devolutionary remedies are now being openly discussed in the coastlands.

In the rainforests of the *Oriente* subregion, oil pro-duction is expanding as a result of the discovery of addi-tional reserves. Some analysts are predicting that interior Ecuador as well as Peru will prove to contain reserves com-parable to those of Venezuela and Colombia, and that an "oil era" will soon transform their economies. As it is, oil already tops Ecuador's export list, and the industry's infra-structure is being modernized. This is essential because much ecological damage has already been done: the trans-Andean pipeline to the port of Esmeraldas, constructed in 1972, was the source of numerous oil spills, and toxic waste was dumped along its route in a series of dreadful environ-mental disasters. A second, more modern pipeline went into operation in 2003, but like Peru, Ecuador faced grow-ing opposition from environmentalists and activists who, in 2005, shut it down for a week. With revenues from oil and gas exports exceeding 50 percent of all exports, Ecua-dor's leaders are hearing an increasingly familiar refrain from its people: demand more from the oil companies and give these funds to those who need it. But it is not as simple as that: oil and gas exploration and exploitation require huge investments, and foreign companies can afford to

From the Field Notes . . .

© H. J. de Blij

"I can't remember being hotter anyplace on Earth, not in Kinshasa, not in Singapore . . . not only are you near the equator here in steamy Guayaquil, but the city lies in a swampy, riverine lowland too far from the Pacific to benefit from any cooling breezes and too far from the Andes foothills to enjoy the benefits of suburban elevation. But Guayaquil is not the disease-ridden backwater it used to be. Its port is modern, its city-center waterfront on the Guayas River has been renovated, its international airport is the hub of a commercial center, and it has grown into a metropolis of 2.8 million. Ecuador's oil revenues have made much of this possible, but from the White Hill (a hill that serves as a cemetery, with the most elaborate vaults near its base and the poorest at the feet of the giant statue of Jesus that tops it) you can see that globalization has not quite arrived here. High-rise development remains limited, international banks, hotels, and other businesses remain comparatively few, and the "middle zone" encircling the city shows little evidence of prosperity (top). Beating the heat and glare is an everyday priority: whole streets have been covered by makeshift tarpaulins and more permanent awnings to protect shoppers (bottom). Talk to the locals, though, and you find that there is another daily concern: the people 'up there in the mountains who rule this country always put us in second place.' Take the 45-minute flight from Guayaquil to cool and comfortable Quito, the capital, and you're in another world, and you quickly forget Guayaquil's problems. That's just what the locals here say the politicians do."

www.conceptcaching.com

Concept Caching

spend what the government's own state company, Petro Ecuador, cannot. That leads to difficult choices because the state needs income to cover its obligations. Clearly, energy riches are a double-edged sword.

Bolivia

Nowhere are the problems so typical of the Andean West more acute than in landlocked, volatile Bolivia. Before reading further, take a careful look at Bolivia's regional geography in both Figures 5B-4 and 5B-5. Bolivia is bounded by remote peripheries of Brazil and Argentina, mountainous Andean highlands and intermontane *altiplanos* [3] (high-elevation basins and valleys) of Peru, and coveted coastal zones of northern Chile. As the maps show, the Andes in this area broaden into a vast mountainous complex some 700 kilometers (450 mi) wide.

On the boundary between Peru and Bolivia, freshwater Lake Titicaca lies at 3700 meters (12,500 ft) above

sea level and helps make the adjacent *Altiplano* (see map, p. 264) livable by ameliorating the coldness in its vicinity, where the snow line lies just above the plateau surface. On the lake's surrounding cultivable land, potatoes and grains have been raised for centuries dating back to pre-Inca times, and the Titicaca Basin still supports a major cluster of Amerindian (Aymara) subsistence farmers. This portion of the *Altiplano* is the heart of modern Bolivia and also contains the capital city, La Paz.

The European/Amerindian Divide

The Bolivian state is the product of the Hispanic impact, and the country's Amerindians (who still make up at least 57 percent of the national population of 10.8 million) no more escaped the loss of their land than did their Peruvian or Ecuadorian counterparts. What made the richest Europeans in Bolivia wealthy, however, was not land but minerals. The town of Potosí in the eastern cordillera became a legend for the immense deposits of silver in its vicinity;

FIGURE 5B-5   © H. J. de Blij, P. O. Muller, and John Wiley & Sons, Inc.

tin, zinc, copper, and several ferroalloys were also discovered there. Amerindian workers were forced to work in the mines under the most appalling conditions.

Today, natural gas and oil, exported to Argentina and Brazil, are leading sources of foreign revenues, and zinc has replaced tin as the leading metal export. But Bolivia's economic prospects will always be impeded by the loss of its outlet to the Pacific Ocean during its war with Chile in the 1880s, despite its transit rights and dedicated port facilities at Antofagasta.

Bolivia's political geography is as divided as its physical geography, the power base of the indigenous population lying in the *altiplanos* and mountains, and that of the mestizo minority centered in the lowlands (see the map above). These photos, taken a day apart in Santa Cruz, where mestizo strength is concentrated, show the crucial difference of opinion: a majority of mestizos want autonomy for their eastern provinces (left); virtually all indigenous people protest the prospect of such autonomy, seen as a prelude to secession and independence, as fatally fracturing their country.

(Left): © AFP/Getty Images, Inc. (Right): © Rodrigo Buendia/AFP/Getty Images, Inc.

More critical than its economic limitations or its landlocked situation is Bolivia's social predicament. The government's history of mistreatment of indigenous people and harsh exploitation of its labor force hangs heavily over a society where about two-thirds of the people, almost all Amerindian, live in dire poverty. In recent years, however, this underrepresented aboriginal majority has been making a growing impact on national affairs. In 2003, violent opposition to a government plan to export natural gas to the United States via a new pipeline to the Chilean coast led to chaos and the government's resignation. In 2005, Bolivian voters elected their first president of Amerindian (Aymara) ancestry, Evo Morales, who proceeded to nationalize the country's natural gas resources.

Departments and Capitals

Landlocked, physiographically bisected, culturally split, and economically divided, Bolivia is a state in trouble. The country's prospects are worsened by its political geography: look again at Figure 5B-5 and you can see that Bolivia's nine provinces are regionally divided between Amerindian-majority *departments* (as subnational units are called here) in the west and those with mestizo majorities in the east. The capital city, La Paz, lies on the Amerindian-majority Andean *Altiplano*, but many mestizo Bolivians do not recognize it as such: historically, the functions of government have been divided between La Paz (the administrative headquarters) and Sucre (which lost most government branches in 1899 during a civil war but retained the Supreme Court, thereafter calling itself the "constitutional capital"). And some mestizo Bolivians even suggest that the eastern city of Santa Cruz should be considered as a candidate for the "compromise capital" of their country.

The Santa Cruz *Department*, like the others in the *Oriente*, stands in sharp contrast to those of the Andes in the west. Here the hacienda system (see Chapter 4A) persists almost unchanged from colonial times, its profitable agriculture supporting a wealthy aristocracy. Nearly 90 percent of Bolivia's agriculturally productive land is still owned by about 50,000 families, much of it taken during the colonization period and the rest expropriated during subsequent military dictatorships. Now this eastern zone of Bolivia proves to contain rich energy resources as well, adding to its economic advantage. So there is talk (and there are frequent public demonstrations) in support of autonomy, even secession here (see photos at left), but neither the haciendas nor the energy industry could function without the Amerindian labor force.

When Evo Morales was elected and began trying to gain control of the energy industry while making moves to alter the division of national wealth, the struggle between east and west took a new turn. When energy prices were high, there was plenty of money to spend on the social programs President Morales had promised to expand. But when, toward the end of the 2000s, the price of natural gas declined, the specter of state failure rose again. The combination of rising expectations among Amerindian citizens and devolutionary hopes among the minority mestizos continues to cast an ominous cloud over Bolivia's future.

Paraguay

Paraguay, Bolivia's landlocked neighbor to the southeast, is one of those transitional countries lying in between regions, exhibiting properties of each (see Fig. 5B-1). Certainly, Paraguay (population: 6.7 million) is not an "Andean" country: it has no highlands of consequence. Its well-watered eastern plains give way to the dry scrub of the *Gran Chaco* in the west. Nor does it have clear, spatially entrenched ethnic divisions between Amerindians, mestizos, and others as do Bolivia and Peru. But indigenous ancestry dominates the ethnic complexion of Paraguay, and Amerindian Guaraní is so widely spoken in the country that this is one of the world's most thoroughly bilingual

These buckets of salt brine sitting in the moonscape of southwestern Bolivia's Uyuni Salt Flats represent one of the country's most promising new development opportunities. Once the liquid evaporates, the residue consists mainly of lithium. This lightest of all metals is used to manufacture the lithium-ion batteries that power the portable electronic devices that dominate our daily lives, such as cellphones, smart phones, and laptops. Larger versions of these batteries are also used to power hybrid motor vehicles, and Bolivia contains as much as half of the world's lithium reserves—enough to support close to 5 billion electric cars. The facility under construction here in this remote part of the *Altiplano* is a pilot lithium extraction plant started operating in 2011. The Bolivian government is the major player in this effort and is determined to keep lithium production a domestic enterprise—despite considerable pressure by foreign companies and investors to become involved. Another fascinating angle is that much of the land of the Salt Flats is the property of Amerindian groups, who are seeking a share in the income to be earned from mineral exploitation. Their claims are consistent with the country's new (2010) constitution, which enables indigenous communities to have control over resources in their territory—and potentially shape the future course of development.

© Martin Bernetti/AFP/GettyImages, Inc.

societies. And continuing protests by landless peasants mirror those taking place elsewhere in the Andean West. Looking south, there is little in Paraguay's economic geography to compare to Argentina, Uruguay, or Chile, as Appendix B confirms. In a sense, Paraguay is a bridge between the Andean West and the Southern Cone, but not yet a heavily traveled one.

And Paraguay is transitional in yet another way: as many as 300,000 Brazilians have crossed the border to settle in eastern Paraguay, where they have created a thriving commercial agricultural economy that produces soybeans, livestock, and other farm products exported to or through Brazil. Brazil, of course, is the giant in the Mercosur/l free-trade zone, but Paraguay is in a geographically difficult position, often complaining that Brazil does not live up to its regional-trade obligations and creates unacceptable difficulties for Paraguayan exporters. Meanwhile, politicians raise fears that increasing Brazilian immigration is creating a foreign enclave within Paraguay, where people speak Portuguese (including in the local schools), the Brazilian rather than the Paraguayan flag flies over public buildings, and a Brazilian cultural landscape is beginning to evolve. Like Bolivia, Paraguay pays heavily for its landlocked weakness.

Paraguay's low GNI resembles that of countries of the Andean West, not the more advantaged Southern Cone region. This is also one of South America's least urbanized states, and poverty dominates the countryside as well as the slums encircling the capital, Asunción, and other towns (Fig. 5B-4). In 2007, nearly two-thirds of the population lived at or below the official poverty level. Even though records are inadequate, research suggests that 1 percent of the population owns about 75 percent of the land. That may be a record for inequality in the South American realm.

In 2008, Paraguay ended its history of ruthless dictatorial rule with the election of a radical, pro-Guaraní priest, Fernando Lugo, who in his days as a missionary supported hacienda invasions by landless peasants and promised land reform and other remedies for the poor. It was a momentous departure from the past. Lugo's election was a "first" in many ways: the first president in half a century who did not come from the establishment Colorado Party; the first who was a cleric (in fact, a Roman Catholic bishop); the first who had been involved in, and inspired by, Liberation Theology (as noted in Chapter 5A, a blend of Christian religion and socialist ideology that interprets the teachings of Christ as a quest to liberate the poor from oppression); and the first who openly declared the relief of poverty of the masses as his overriding objective. No wonder, then, that Lugo's election brought great new hope to much of Paraguay's population.

Money for these remedies will be hard to come by, and it is probably too soon to judge his success. But in 2009 President Lugo scored a major success when he negotiated better terms with Brazil for the sale of Paraguay's share of electricity generated by Itaipu Dam located on the Paraná River between the two countries. When this huge dam's turbines began producing power in 1984, each country (then ruled by dictators) got 50 percent of the electricity, but severely underdeveloped Paraguay needed just a tiny fraction of that. So Paraguay's ruler agreed to sell to Brazil the remainder of his country's share at prices far below market value. Democratically-elected Lugo persuaded Brazil's (also democratic) government to begin paying market rates for the 40 percent of Paraguay's Itaipu-generated electricity it buys each year, thereby adding substantially to the Paraguayan national income. Furthermore, Brazil will assist Paraguay in building a state-of-the-art transmission line from Itaipu to Asunción, to be ready in 2012 and mark the start of a crucial expansion of Paraguay's electrical-power grid.

Another, and quintessentially geographical, problem stemming from Paraguay's long-term weakness and misrule lies in the southeast, where the borders of Brazil, Argentina, and Paraguay converge in a chaotic scene of smuggling, money-laundering, political intrigue, and even terrorist activity, centered on the town of Ciudad del Este. Locals call this the **Triple Frontier [4]**, and warning flags went up when reports of Hizbullah (Iranian-backed terrorist) activity were confirmed by the discovery of maps of the area in a Taliban safe house in Afghanistan. Paraguay's state system decidedly needs strengthening, and not just for domestic reasons.

■■ THE SOUTHERN CONE

This region of South America, consisting of Argentina, Chile, and Uruguay, acquired its name from its tapered, ice-cream-cone shape (Fig. 5B-6). As noted earlier, Paraguay has strong links to this region and in certain ways forms part of it, although social and economic contrasts between Paraguay and those existing in the Southern Cone remain sharp.

Since 1995, the countries of the Southern Cone have been drawing closer together in an economic union named **Mercosur** (*Mercado Común del Sur*), the hemisphere's second-largest trading bloc after NAFTA. Despite setbacks and disputes, Mercosur has expanded and today encompasses Argentina, Uruguay, Paraguay, and Brazil—and (pending ratification) Venezuela; Chile, Bolivia, Peru, Ecuador, and Colombia, participate as associate members. In Portuguese-speaking Brazil, the organization is known as **Mercosul** (*Mercado Comum do Sul*).

Argentina

The largest Southern Cone country by far is Argentina, whose territorial size ranks second only to Brazil in this geographic realm; its population of 41.3 million ranks third after Brazil and Colombia. Argentina exhibits a great deal of physical-environmental variety within its boundaries, and the vast majority of the Argentines are concentrated in the physiographic subregion known as the **Pampa** (a word meaning "plain"). Figure 5A-3 underscores the

FIGURE 5B-6

© H. J. de Blij, P. O. Muller, and John Wiley & Sons, Inc.

degree of clustering of Argentina's inhabitants on the land and in the cities of the Pampa. It also shows the relative emptiness of the other six subregions (which are mapped above): the scrub-forested **Chaco** in the northwest; the mountainous **Andes** in the west, along whose crestline lies the boundary with Chile; the arid plateaus of **Patagonia** south of the Rio Colorado; and the undulating transitional terrain of intermediate **Cuyo, Entre Rios** (also known as "Mesopotamia" because it lies between the Paraná and Uruguay rivers), and the **North**.

The Argentine Pampa is the product of the past 150 years. During the second half of the nineteenth century, when the great grasslands of the world were being opened up (including those of the interior United States, Russia, and Australia), the economy of the long-dormant Pampa began to emerge. The food needs of industrializing Europe grew by leaps and bounds, and the advances of the Industrial Revolution—railroads, more efficient ocean transport, refrigerated ships, and agricultural machinery—helped make large-scale commercial meat and grain production in

From the Field Notes . . .

© H. J. de Blij

"At the heart of Buenos Aires lies the Plaza de San Martín, flanked by impressive buildings but, unusual for such squares in Iberian America, carpeted with extensive lawns shaded by century-old trees. Getting a perspective of the plaza was difficult until I realized that you could get to the top of the 'English Tower' across the avenue you see in the foreground. From there, one can observe the prominent location occupied by the monument to the approximately 700 Argentinian military casualties of the Falklands War of 1982 (center), where an eternal flame behind a brass map of the islands symbolizes Argentina's undiminished determination to wrest these islands from British control."

www.conceptcaching.com

● AMONG THE REALM'S GREAT CITIES . . .

BUENOS AIRES

ITS NAME MEANS "fair winds," which first attracted European mariners to the site of Buenos Aires alongside the broad estuary of the muddy Rio de la Plata. The shipping function has remained paramount, and to this day the city's residents are known throughout Argentina as the *porteños* (the "port dwellers"). Modern Buenos Aires was built on the back of the nearby Pampa's grain and beef industry. It is often likened to Chicago and the Corn Belt in the United States because both cities have thrived as interfaces between their immensely productive agricultural hinterlands and the rest of the world.

Buenos Aires (12.8 million) is yet another classic South American primate metropolis, housing nearly one-third of all Argentines, serving as the capital since 1880, and functioning as the country's economic core. Moreover, Buenos Aires is a cultural center of global standing, a monument-studded city that contains the world's widest boulevard (Avenida 9 de Julio).

During the half-century between 1890 and 1940, the city was known as the "Paris of the South" for its architecture, fashion leadership, book publishing, and performing arts activities (it still has the world's biggest opera house, the Teatro Colón). With the recent restoration of democracy, Buenos Aires is now trying to recapture its golden years. Besides reviving these cultural functions, the city has added a new one: the leading base of the hemisphere's

© H. J. de Blij, P. O. Muller, and John Wiley & Sons, Inc.

motion picture and television industry for Spanish-speaking audiences.

the Pampa not only feasible but also highly profitable. Large haciendas were laid out and farmed by tenant workers; railroads radiated ever farther outward from the booming capital of Buenos Aires and brought the entire Pampa into production.

A Culture Urban and Urbane

Argentina once was one of the richest countries in the world. Its historic affluence is still reflected in its architecturally splendid cities whose plazas and avenues are flanked by ornate public buildings and private mansions. This is true not only of the capital, Buenos Aires, at the head of the Rio de la Plata estuary—it also applies to interior cities such as Mendoza and Córdoba. The cultural imprint is dominantly Spanish, but the cultural landscape was diversified by a massive influx of Italians and smaller but influential numbers of British, French, German, and Lebanese immigrants.

Argentina has long been one of the realm's most urbanized countries: 91 percent of its population is concentrated in cities and towns, a higher percentage even than western Europe or the United States. Almost one-third of all Argentinians live in metropolitan Buenos Aires, by far the leading industrial complex where processing Pampa products dominates. Córdoba has become the second-ranking industrial center and was selected by foreign automobile manufacturers as the car-assembly center for the expanding Mercosur/l market. But what concentrates the urban populations is the processing of products from the vast, sparsely peopled interior: Tucumán (sugar), Mendoza (wines), Santa Fe (forest products), and Salta (livestock). Argentina's product range is enormous. There is even a petroleum reserve near Comodoro Rivadávia on the coast of Patagonia.

Economic Volatility

Despite all these riches, Argentina's economic history is one of boom and bust. With more than 40 million inhabitants, a vast territory with diverse natural resources, adequate infrastructure, and above-average international linkages, Argentina should still be one of the world's wealthiest countries, as it once was. But political infighting and economic mismanagement have combined to ruin a vibrant and varied economy.

Part of the problem, it appears, lies in the country's lopsided political geography. Argentina is a federal state consisting of the Buenos Aires Federal District and 23 provinces (Fig. 5B-7). As would be expected from what we have just learned, the urbanized provinces are populous, while the mainly rural ones have smaller populations—but the gap between Buenos Aires Province (whose capital is La Plata), with 15.7 million, and Tierra del Fuego (capital: Ushuaia), with barely over 100,000, is wide indeed. A dozen other provinces contain less than 750,000 inhabitants, so that the larger ones in addition to dominant Buenos Aires are also disproportionately influential in domestic politics, especially Córdoba and Santa Fe, both with capitals of the same name.

FIGURE 5B-7   © H. J. de Blij, P. O. Muller, and John Wiley & Sons, Inc.

The seemingly never-ending problem for Argentina has been corrupt politics and associated mismanagement, at least since the mid-twentieth century. The darkest episode was the "Dirty War" of 1976–1983, in which a repressive military junta made more than 10,000 (and perhaps as many as 30,000) Argentinians disappear without a trace. Then in 1982, this ruthless military clique launched an invasion of the British-held Falkland Islands (which the Argentines call the Malvinas), resulting in a major defeat for Argentina (Fig. 5B-6). By the time civilian government replaced the discredited junta, inflation and national debt were soaring. Economic revival during the 1990s was followed by another severe down-turn that exposed the flaws in Argentina's fiscal system, including scandalously inefficient tax collection and un-conditional federal handouts to the politically powerful provinces.

The administration in power at the opening of this decade, led by the country's first elected female president, Cristina Fernandez de Kirchner (whose late husband pre-ceded her as president), was primarily focused on return-ing Argentina to fiscal responsibility (controlling inflation) and economic stability. It also had been seeking to heal the festering wounds of the Dirty War through prosecution of the guilty and some form of reconciliation. But the presi-dent's popularity ratings in 2011 remained low, and the majority of Argentines did not show signs of renewed hope. Although Argentina has the highest per capita GNI in the realm, rising inflation and growing labor troubles have cost its leadership much public support. Moreover, foreign investors showed reluctance to assume risks in a country whose government has habitually ignored con-tract terms. And added to that were Argentina's continued uncertainties with regard to its energy supplies.

In 2010, the old Falklands dispute erupted once again when Britain began preparations for oil-drilling op-erations in the potentially rich seabed surrounding these distant South Atlantic islands. Buenos Aires registered strong protests and repeated its claims that the Malvinas should be returned to Argentina.

Chile

For 4000 kilometers (2500 mi) between the crestline of the Andes and the coastline of the Pacific lies the narrow strip of land that is the Republic of Chile (Fig. 5B-8). On aver-age just 150 kilometers (90 mi) wide (and only rarely over 250 kilometers or 150 mi in width), Chile is the world's quintessential example of what **elongation** [5] means to the functioning of a state. Accentuated by its north-south ori-entation, this severe territorial attenuation not only results in Chile extending across numerous environmental zones; it has also contributed to the country's external political, internal administrative, and general economic challenges.

Nevertheless, throughout most of their modern his-tory, the Chileans have made the best of this potentially disruptive centrifugal force: from the beginning, the sea has constituted an avenue of longitudinal communication; the Andes Mountains continue to form a barrier to pre-vent encroachment from the east; and when confronta-tions loomed at the far ends of the country, Chile proved to be quite capable of coping with its northern rivals, Bo-livia and Peru, as well as Argentina in the far south.

Three Subregions

As Figures 5B-6 and 5B-8 indicate, Chile is a three-subregion country. About 90 percent of its 17.4 million people are concentrated in what is called Middle Chile, where Santiago, the capital and largest city, and Valparaíso,

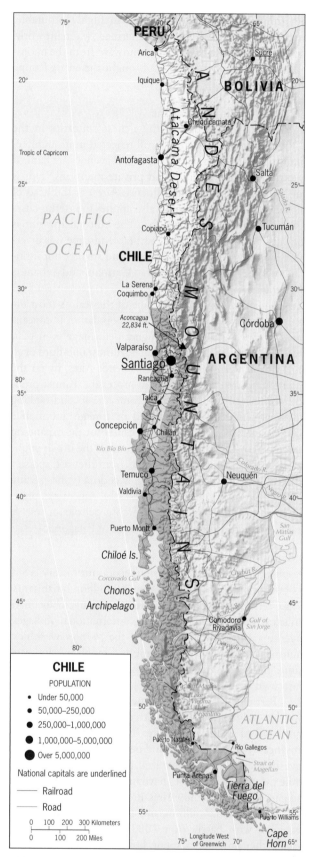

FIGURE 5B-8   © H. J. de Blij, P. O. Muller, and John Wiley & Sons, Inc.

From the Field Notes . . .

"In a small, admission-charge private museum I found a treasure of historic information about this place in Chile—its environmental challenges ranging from droughts to earthquakes, its growth as a center for irrigated agriculture augmented by mining in the hinterland, its religious importance as the site of many churches and convents. But what matters today is tourism and retirement. Not only is La Serena (just inland from Coquimbo) on the international visitor circuit, but it lies at the center of a growing cluster of second-home and retirement complexes, benefiting from its equable climate and attractive scenery, and from the Southern Cone's comparative economic prosperity. So La Serena has become an amusement park, its stores converted to boutique shops, its historic homes on visitor tours, its plaza throbbing with the music of competing boomboxes, its streets clogged by buses. An older resident sitting on a park bench had his doubts. 'All this started when they built this Pan-American Highway right past us,' he said. 'Now you can get here by sea, by road, even by air. Is this your first visit? You came too late.' No doubt about it: what I saw here wasn't what I read in those faded pages in the museum."

© H. J. de Blij

www.conceptcaching.com

the chief port, are located. North of Middle Chile lies the Atacama Desert, wider, drier, and colder than the coastal desert of Peru. To the south of Middle Chile, the coast is punctuated by a plethora of fjords and islands, the topography is mountainous, and the climate—wet and cool near the Pacific—soon turns drier and colder against the Andean interior. South of the latitude of Chiloé Island, there are few permanent overland routes and hardly any settlements.

Not surprisingly, the land in Middle Chile is the most fertile and valuable, with hardly any agriculture in either the North or South (Fig. 5A-5). The three subregions are also apparent on the realm's cultural map, with Europeans dominating the Middle, mestizos prevailing in the North, and Amerindians being the majority in the South (Fig. 5A-4).

Still, prior to the 1990s, the arid Atacama region in the north accounted for more than half of Chile's foreign revenues. It holds the world's largest exploitable deposits of nitrates, which was the country's economic mainstay before the discovery of methods of synthetic nitrate production a century ago. Subsequently, copper became the dominant export (Chile again possesses the world's largest reserves, which in 2010 accounted for more than half of

A view into the gaping maw of northern Chile's Chuquicamata copper mine, the world's largest and second-deepest open-pit mine. For most of the century since mining began here in 1910, 'Chuqui' was the single leading source of copper on Earth. But the increasingly negative environmental impact of its moonscape of slag heaps and toxic materials has now forced the relocation of the town of Chuquicamata and the expected 2013 closure of the surface mine. Nonetheless, huge quantities of copper will continue to be extracted from this still-rich deposit after a major new underground mine is built below this site: it will begin operating in 2018—with plans to keep enlarging it until at least 2060.

© Peter McBride/Aurora Photos, Inc.

all export revenues). It is mined in several places, but the main concentration lies on the eastern margin of the Atacama near Chuquicamata (photo, p. 271), not far from the port of Antofagasta—as the world discovered when 33 trapped miners were miraculously rescued in October 2010. For most of the past decade, commodity prices were high and Chile's export revenues soared; but prices plunged sharply during the recent global recession, underscoring the risks involved—and the continuing need for economic diversification.

Political and Economic Success

Following the withdrawal of its brutal military dictatorship in 1990, Chile embarked on a highly successful program of free-market economic reform that brought stable growth, lowered inflation as well as unemployment, reduced poverty, and attracted massive foreign investment. The last is of particular significance because these new international connections enabled the export-led Chilean economy to diversify and develop in some badly needed new directions. Copper remains the single leading export, but many other mining ventures have been launched. In the agricultural sphere, fruit and vegetable production for export has soared because Chile's harvests coincide with the winter farming lull in the affluent countries of the Northern Hemisphere. Industrial expansion is taking place as well, though at a more leisurely pace, and new manufactures include a modest array of goods that range from basic chemicals to computer software.

Chile's increasingly globalized economy has propelled the country into a prominent role on the international economic scene. The United States, long Chile's leading trade partner, now is in second place compared to the Asian Pacific Rim, where China takes the bulk of Chile's exports followed by Japan and South Korea (Argentina remains Chile's leading source of imports, mostly energy, which is a potential problem given Argentina's own rising needs). Chile's regional commerce also is growing, and Chile has affiliated with Mercosur/l as an associate member.

The 2010 election in Chile saw Sebastián Piñera assume the presidency. Although a member of the Conservative Party whose election marked the end of two decades of center-left government (former President Michelle Bachelet had served the maximum two terms), as his first act he paid tribute to the successful policies that have made Chile South America's richest economy. Even though the global recession of the late 2000s did not leave the Chilean economy unaffected, it looks like Chile's new government will not modify what has been a winning formula for two decades: a pragmatic approach that avoids ideological extremism, prioritizes transparency, and balances economic growth with poverty reduction.

Indeed, the new regime's energies were required to totally focus on the aftermath of another major event of early 2010—one of the most powerful earthquakes in the realm's modern history. This catastrophic temblor and accompanying tsunamis struck the coastal zone of southern Middle Chile on February 27, killing hundreds, causing massive destruction, and disrupting the lives of more than 2 million Chileans. Recovery—tempered by another large quake here at the beginning of 2011—has been steady, but this disaster has clearly slowed progress toward Chile's goal of becoming the first South American country to join the ranks of the world's most developed economies.

Uruguay

Uruguay, in contrast to Argentina or Chile, is compact, small, and rather densely populated. This buffer state [6] of old became a fairly prosperous agricultural country, in effect a smaller-scale Pampa (though possessing less favorable soils and topography). Montevideo, the coastal capital, contains almost 50 percent of the country's population of 3.4 million; from here, railroads and roads radiate outward into the productive agricultural interior (Fig. 5B-6). In the immediate vicinity of Montevideo lies Uruguay's major farming area, producing vegetables and fruits for the metropolis as well as wheat and fodder crops; most of the rest of the country is used for grazing cattle and sheep, with beef products, wool and textile manufactures, and hides dominating the export trade (Fig. 5B-6, inset map). Tourism is another major economic activity as Argentines, Brazilians, and other visitors flock to the Atlantic beaches at Punta del Este and other thriving resort towns.

About the size of Florida but with less than one-fifth the population, Uruguay offers plenty of agricultural potential. However, its government seeks to diversify the economy, and one of its plans—the construction of two large cellulose (paper) factories on the Uruguayan side of the Uruguay River where it forms the border with Argentina—has caused a quarrel between the two Mercosur/l neighbors that has exposed some serious rifts. Montevideo is the administrative headquarters of Mercosur/l, but, like Paraguay, Uruguay has long felt neglected and even obstructed by its much larger neighbors. When Argentinians across the river began demonstrating, barricaded a bridge, and stopped taking vacations on Uruguay's beaches, they severely impacted the Uruguayan economy.

The Argentinians, who in any case dislike major foreign investments (the factories were to be built by Finnish and Spanish companies), argued that the paper mills would accelerate deforestation, cause river pollution, create acid rain, and damage the farming, fishing, and tourist industries. When the presidents of the two countries met to negotiate a solution and Uruguay's leader agreed to halt construction for a "study period," Uruguayan public opinion showed strong opposition to the compromise. This issue reveals just how quickly nationalist feelings can overwhelm the need for international cooperation. It also shows that Mercosur/l partners are far from united on economic matters and that the collaboration implied by Mercosur/l membership is still a distant reality.

BRAZIL: GIANT OF SOUTH AMERICA

The next time you board an airplane, don't be surprised if the aircraft was built in Brazil. When you go to the supermarket to buy provisions, take a look at their sources. Chances are some of them will come from Brazil. When you listen to your car radio, some of the best music you hear may have originated in Brazil. The emergence of Brazil as the regional superpower of South America and an economic superpower in the world at large is going to be one of the main stories of the first quarter of the twenty-first century. Along with Russia, India, and China, Brazil is designated as one of the four biggest emerging markets in the world, the **BRICs [7]** (an acronym based on the first letter of this quartet of countries).

Why is Brazil so upward-bound today? In large measure it is because, after a long period of dictatorial rule by a minority elite that used the military to stay in power, Brazil embraced democratic government in 1989 and has not looked back since. The era of military coups and repeated crushing of civil liberties is over. In some ways, Brazil's political and economic turnaround runs parallel to that of Chile, but because of its sheer size, Brazil matters more to the rest of the realm and, indeed, the world.

The last ten years, in particular, have witnessed spectacular progress in Brazil, and much of this must be attributed to President Luiz Inácio Lula da Silva, popularly known as Lula, who was elected into office in 2002 and completed his second term in 2010. The leader of the Workers' Party, Lula was expected by many to pursue a leftist agenda that would harm business interests (in a show of displeasure at the election result in 2002, the United States did not send a single high-ranking representative to Lula's inauguration—an example of the United States' occasional poor political judgment in this realm). In fact, Lula so skillfully managed Brazil's numerous countervailing forces—thereby unleashing the country's regional and global potential—that many Brazilians dreaded his leaving office in 2010.

By any measure, Brazil is South America's giant. It is so large that it has common boundaries with all the realm's other countries except Ecuador and Chile (Fig. 5B-1). Its tropical and subtropical environments range from the equatorial rainforest of the Amazon Basin to the humid temperate climate of the far south. Territorially as well as in terms of population, Brazil ranks fifth in the world, and on both counts it represents just about half of this entire realm. The Brazilian economy, with its ultramodern industrial base, is now the ninth-largest on Earth—and climbing.

Population and Culture

Brazil's population grew rapidly during the world's twentieth-century population explosion. But over the past three decades, the rate of natural increase has slowed from nearly 3.0 percent to 1.0 percent, and the average number of children born to a Brazilian woman has been more than halved from 4.4 in 1980 to 2.0 in 2010. It is a trend in line with Brazil's overall modernization in recent decades.

The population of just under 200 million is as diverse as that of the United States. In a pattern quite familiar in the Americas, the indigenous inhabitants of the country were decimated following the European invasion (fewer than 200,000 Amerindians now survive deep in the Amazonian interior). Africans came in great numbers, too, and today there are about 13.5 million Afro-Brazilians in Brazil.

Brazil's culture is infused with African themes, a quality that has marked it from the very beginning. Three centuries ago, the Afro-Brazilian sculptor and architect affectionately called Aleijadinho was Brazil's most famous

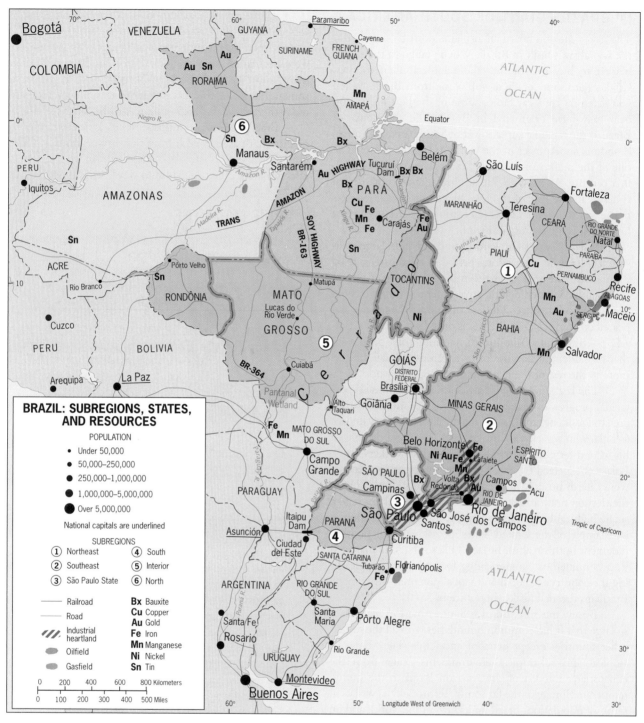

FIGURE 5B-9

© H. J. de Blij, P. O. Muller, and John Wiley & Sons, Inc.

artist. The world-famous composer Heitor Villa-Lobos used numerous Afro-Brazilian folk themes in his music. So many Africans were brought in bondage to the city and hinterland of Salvador in Bahia State (Fig. 5B-9) from what is today Benin (formerly Dahomey) in West Africa that Bahia has become a veritable outpost of African culture.

Significantly, however, there was also much racial mixing, and 87 million Brazilians have combined European,

African, and minor Amerindian ancestries. The remaining 97 million—no longer in the majority at 49 percent—are mainly of European origin, the descendants of immigrants from Portugal, Italy, Germany, and eastern Europe.

Yet another significant, though small, minority began arriving in Brazil in 1908: the Japanese, who today are concentrated in São Paulo State. The more than 1 million Japanese-Brazilians form the largest ethnic Japanese community outside Japan, and in their multicultural environment they have risen to the top ranks of Brazilian society as business leaders, urban professionals, commercial farmers—and even as politicians in the city of São Paulo. Committed to their Brazilian homeland as they are, the Japanese community also retains its contacts with Japan, resulting in many a trade connection.

Brazilian society, to a greater degree than is true elsewhere in the Americas, has made progress in dealing with its racial divisions. To be sure, blacks are still the least advantaged among the country's leading population groups, and community leaders continue to complain about discrimination. But ethnic mixing in Brazil is so pervasive that hardly any group is unaffected, and official census statistics concerning "blacks" and "Europeans" are meaningless.

What the Brazilians do have is a true national culture, expressed in an historic adherence to the Roman Catholic faith (its adherents now constitute about 70 percent of the population, a number that is steadily eroding [from 89 percent in 1980] under Protestant-evangelical and secular pressures); in the universal use of a modified form of Portuguese as the common language ("Brazilian"); and in a set of lifestyles in which soccer, "beach culture," distinctive music and dance, and an intensifying national consciousness and pride are fundamental ingredients.

Inequality and Poverty

For all its accomplishments in multiculturalism, Brazil remains a country of stark, appalling social inequalities

(Fig. 5B-10). Although such inequality is hard to measure precisely, South America is frequently cited as the geographic realm exhibiting the world's sharpest division between affluence and poverty. And within South America, Brazil is reputed to have the widest gap of all.

It is estimated that the richest 10 percent of the population control more than half of Brazil's wealth and that a mere 1 percent of Brazilians own nearly half of all productive land. The poorest 20 percent of the people live in the most squalid conditions prevailing anywhere on the planet, even including the teeming megacities of Africa and Asia. According to UN estimates, in this age of adequate available (but not everywhere affordable) food, about half the population of Brazil suffers some form of malnutrition and, in the northeastern States, even hunger. Packs of young orphans and abandoned children roam the cities, sleeping where they can, stealing or robbing when they must. Some of the world's most magnificent central cities are ringed and sectored by the most wretched *favelas* (shantytowns) where poverty, misery, and crime converge.

But since 2002, government has embarked on a set of policies aimed at bringing relief to the poor while maintaining robust economic growth. It has enacted land reform and increased access to education for Brazil's masses. It also started a subsidy program that had significant results in the poorest States. This *Bolsa Familia* (Family Fund) plan gives families small amounts of cash to keep their children in school and to ensure their vaccinations

FIGURE 5B-10   © H. J. de Blij, P. O. Muller, and John Wiley & Sons, Inc.

against diseases that especially afflict the poor. In just a few years this program has become so successful that it now serves as a model for antipoverty campaigns in many other parts of the world. At the same time, the percentage of people officially living in poverty in Brazil dropped from nearly 30 percent in 2000 to about 15 percent in 2010—with a commensurate decline in the GINI coefficient (see the Introduction chapter) from 0.61 to 0.54.

Development Prospects

Brazil is richly endowed with mineral resources, including enormous iron and aluminum ore reserves, extensive tin and manganese deposits, and highly promising oil- and gasfields (Fig. 5B-9). Other significant energy developments involve massive new hydroelectric facilities and the successful substitution of sugarcane-based alcohol (gasohol) for gasoline, allowing well over half of Brazil's automobiles to use this biofuel instead of more expensive imported petroleum. Besides these natural endowments, Brazil has more arable land than any other country in the world. Brazilian soils sustain a bountiful agricultural output that makes the country the world's biggest exporter of coffee, orange juice, sugar, tobacco, ethanol, beef, and chicken. Its government has set the goal of Brazil becoming the number-one food exporter in the world by 2015, displacing the United States. Commercial agriculture, in fact, is now the fastest growing economic sector, driven by mechanization and the opening of a burgeoning new farming frontier in the fertile grasslands of southwestern Brazil (see photo below).

Large-scale corporate farming of soybeans in Interior Brazil's *cerrado*, one of the world's greatest agricultural frontiers—flat, fertile, well-watered, and ideally suited for mechanized production.

© PauloFridman/SambaPhoto/Getty Images, Inc.

Industrialization has propelled Brazil's rise as a global economic power. Much of the momentum for this continuing development was unleashed in the early 1990s after the government opened the country's long-protected industries to international competition and foreign investment. These new policies proved to be highly effective because productivity has risen by over one-third since 1990 as Brazilian manufacturers attained world-class quality. More than a decade ago, revenues from industrial exports surpassed those from agriculture. Commerce with Argentina made that key member of Mercosur/l one of Brazil's leading trade partners.

On the global stage, Brazil has become a formidable presence in other ways. The country's enormous and easily accessible iron ore deposits, the relatively low wages of its workers, and the mechanized efficiency of its steelmakers enable Brazil to produce that commodity at half the cost of steel made in the United States (which caused American steel producers to demand protectionist measures). The biggest consumer of iron ore and steel, and of many other materials as well, is China. In 2012, the Acu Superport, 400 kilometers (250 mi) northeast of Rio de Janeiro, will open for business as the largest bulk cargo port on Earth, ready to accommodate the biggest ships that sail the oceans—those, too, are Chinese (see photo, p. 251).

A Highly Promising Oil Future

Compared to the other BRICs, Brazil stands out in terms of the enormity and diversity of its natural resources. It has been paying for government programs with revenues mainly derived from commodities—iron ore, soybeans, coffee, orange juice, beef, sugar—that for the past decade have commanded high prices on international markets. Economic geographers note that dependence on a singular commodity is always risky because a country's well-being relies heavily on global market prices for that product (which are, in turn, the combined result of supply and demand across the world). As Brazil diversified its economy after 2000 and increased production in its manufacturing and high-tech sectors, it greatly improved its competitiveness and sustainability. In fact, during the global recession of the late 2000s, Brazil was the last country in and the first one out: by mid-2010, the country was on a roll, way ahead of both the United States and Europe.

On top of this sterling performance, Brazil can now add oil, lots of it, to its portfolio. In the past, a leading concern for

Brazil was energy and its rising costs. Recessions were deepened by high oil and gas prices, and other than some small domestic reserves and those in neighboring Bolivia there were few nearby sources to acquire what was needed. But in 2009 the state-owned (but publicly traded) oil company, Petrobras, confirmed the discovery of an "enormous" oil reserve, quite possibly among the world's largest, capable of yielding an estimated 2 billion barrels. It was proof, said then-President Lula with a broad smile on his face, that "God is Brazilian."

Brazil in the 2010s is already self-sufficient in oil and within the next few years will become a major exporter, thereby adding significantly to its revenues. Moreover, the newfound reserves lie off the coast in a cluster of oil-fields—most fortuitously located near Rio de Janeiro and São Paulo (see Fig. 5B-9)—and are being explored further as you read this. Although these oil deposits are massive and reserve estimates are constantly being revised upward as new discoveries are made, they are buried quite far beneath the ocean floor. Thus exploitation will be costly and require substantial foreign investment as well as state-of-the-art drilling methods, but is likely to take Brazil's technology sector to an even more advanced level.

Brazil's Subregions

Brazil is a federal republic consisting of 26 States and the federal district of the capital, Brasília (Fig. 5B-9). As in the United States, the smallest States lie in the northeast and the larger ones farther west; their populations range from about 475,000 in the northernmost, peripheral Amazon State of Roraima to more than 40 million in burgeoning São Paulo State. Although Brazil is about as large as the 48 contiguous United States, it does not exhibit a clear physiographic regionalism. Even the Amazon Basin, which covers almost 60 percent of the country, is not entirely a plain: between the tributaries of the great river lie low but extensive tablelands. Given this physiographic ambiguity, the six Brazilian subregions discussed next have no absolute or even generally accepted boundaries. In Figure 5B-9, those boundaries have been drawn to coincide with the borders of States, making identifications easier.

The **Northeast** was Brazil's source area, its culture hearth. The plantation economy took root here at an early date, attracting Portuguese planters, who soon imported the country's largest group of African slaves to work in the sugar fields. But the ample rainfall occurring along the coast soon gives way to lower and more variable patterns in the interior, which is home to about half of the region's 50-plus million people. This drier inland backcountry—called the *sertão*—is not only seriously overpopulated but also contains some of the worst poverty to be found anywhere in the Americas. The Northeast produces less than one-sixth of Brazil's gross domestic product, but its inhabitants constitute more than one-fourth of the national population. Given this staggering imbalance, it is not sur-

prising that this subregion contains half of the country's poor, a literacy rate 20 percent below Brazil's mean, and an infant mortality rate twice the national average.

Much of the Northeast's misery is rooted in its unequal system of land tenure. Farms must be at least 100 hectares (250 acres) to be profitable in the hard-scrabble *sertão*, a size that only large landowners can afford. Moreover, the Northeast is tormented by a monumental environmental problem: the recurrence of devastating droughts at least partly attributable to **El Niño [8]** (periodic events of sea-surface warming off the continent's northwestern coast that skew regional weather patterns).

The Northeast today is Brazil's great contradiction. In cities such as Recife and Salvador, hordes of peasants driven from the land constantly arrive to expand the surrounding shantytowns. As yet, few of the generalizations about emerging Brazil apply here, but there are some bright spots. A petrochemical complex has been built near Salvador, creating thousands of jobs and luring foreign investment. Irrigation projects have nurtured a number of productive new commercial agricultural ventures. Tourism is booming along the entire Northeast coast, whose thriving beachside resorts attract tens of thousands of vacationing Europeans. Recife has spawned a budding software industry and a major medical complex. And Fortaleza is the center of new clothing and shoe industries that have already put the city on the global economic map.

The **Southeast** has been modern Brazil's *core area*, with its major cities and leading population clusters. Gold first drew many thousands of settlers, and other mineral finds also contributed to the influx—with Rio de Janeiro itself serving as the terminus of the "Gold Trail" and as the long-time capital of Brazil until 1960. Rio de Janeiro became the cultural capital as well, the country's most international center, *entrepôt*, and tourist hub (see box, p. 246). The third quarter of the twentieth century brought another mineral age to the Southeast, based on the iron ores around Lafaiete carried to the steelmaking complex at Volta Redonda (Fig. 5B-9).

The surrounding State of Minas Gerais (the name means "General Mines") formed the base from which industrial diversification in the Southeast has steadily expanded. The burgeoning metallurgical center of Belo Horizonte paved the way and is now the endpoint of a rapidly developing, ultramodern manufacturing corridor that stretches 500 kilometers (300 mi) southwest to metropolitan São Paulo (Fig. 5B-9, striped zone).

São Paulo State is the leading industrial producer and primary focus of ongoing Brazilian development. This economic-geographic powerhouse accounts for nearly half of the country's GDP, with an economy that today matches Argentina's in overall size. Not surprisingly, this subregion is growing phenomenally (it already contains more than 20 percent of Brazil's population) as a magnet for migrants, especially from the Northeast.

The wealth of São Paulo State was built on its coffee plantations (known as *fazendas*), and Brazil is still the world's leading producer. But coffee today has been eclipsed by other farm commodities. One of them is orange juice concentrate (here, too, Brazil leads the world). São Paulo State now produces more than double the annual output of Florida, thanks to a climate all but devoid of winter freezes, ultramodern processing plants, and a fleet of specially equipped tankers that ship the concentrate to foreign markets. Another leading pursuit

São Paulo may not have an imposing skyline to match New York or a skyscraper to challenge the Sears (now Willis) Tower of Chicago, but what this Brazilian megacity does have is mass. São Paulo is more than just another city—it is a vast conurbation of numerous cities and towns containing more than 26 million inhabitants who constitute the third-largest human agglomeration on Earth. This view shows part of the razor-sharp edge of the CBD, a concrete jungle with little architectural distinction but vibrant urban cultures. In the foreground, juxtaposed against the affluence of downtown, the teeming inner city reflects the opposite end of the social spectrum—a chaotic jumble of *favelas* that mainly house the newest urban migrants, especially from Brazil's hard-pressed Northeast. São Paulo, demographers say, gives us a glimpse of the urban world of the future.

© AP/Wide World Photos.

is soybeans, in which Brazil ranks second among the world's producers.

Matching this agricultural prowess is the State's industrial strength. The revenues derived from the coffee plantations provided the necessary investment capital, ores from Minas Gerais supplied the vital raw materials, São Paulo City's outport of Santos facilitated access to the ocean, and immigration flows from Europe, Japan, and other parts of Brazil contributed the increasingly skilled labor force. As the capacity of the domestic market grew, the advantages of central location and large-scale agglomeration secured São Paulo's primacy. This also resulted in metropolitan São Paulo (profiled to the right) becoming the country's—and South America's—leading industrial complex and megacity (population: 26.2 million).

The **South** consists of three States, whose combined population exceeds 27 million: Paraná, Santa Catarina, and Rio Grande do Sul (Fig. 5B-9). Southernmost Brazil's exceptional agricultural potential has long attracted sizeable numbers of European immigrants. Here the newcomers introduced their advanced farming methods to several areas. Portuguese rice farmers clustered in the valleys of Rio Grande do Sul, where tobacco production has now propelled Brazil to become the world's number-one exporter. The Germans, specialists in raising grain and cattle, occupied the somewhat higher areas to the north and in Santa Catarina. The Italians selected the highest slopes, where they established thriving vineyards. All of these fertile lands proved highly productive, and with growing markets in the mushrooing urban areas to the north, this tri-State subregion became Brazil's most affluent corner.

With the South firmly rooted in the European/commercial agricultural sphere (Figs. 5A-4, 5A-5), European-style standards of living match the diverse Old World heritage that is reflected in the urban centers and countryside (where German and Italian are spoken alongside Portuguese). This has led to hostility against non-European Brazilians, and several communities actively discourage poor job-seeking migrants from the north by offering to pay return bus fares or even blocking their household-goods-laden vehicles. Moreover, extremist groups have proliferated and continue to openly espouse the secession of the South from the rest of Brazil.

Economic development in the South is not limited to the agricultural sector. Coal from Santa Catarina and Rio Grande do Sul is shipped north to the steel plants of Minas Gerais. Local manufacturing is growing as well, especially in Pôrto Alegre and Tubarão. During the 1990s, a major center of the computer software industry was established in Florianópolis, the island-city as well as State capital just off Santa Catarina's coast. Known as *Tecnópolis*, this budding technopole continues to grow by capitalizing on its seaside amenities, skilled labor force, superior air-travel and global communications linkages, and government and private-sector incentives to support new companies.

SÃO PAULO

SÃO PAULO, WHICH lies on a plateau 50 kilometers (30 mi) inland from its Atlantic outport of Santos, possesses no obvious locational advantages. Yet here on this site we find the third-largest metropolis on Earth, whose population has multiplied so uncontrollably that São Paulo has more than doubled in size (from 11 to just over 26 million) during the past three decades.

Founded in 1554 as a Jesuit mission, modern São Paulo was built on the nineteenth-century coffee boom. It has since grown steadily as both an agricultural processing center (soybeans, orange juice concentrate, and sugar besides coffee) and a manufacturing complex (accounting for about half of all of Brazil's industrial jobs). It has also become Brazil's primary focus of commercial and financial activity. Today São Paulo's bustling, high-rise CBD (see photo at left) is the very symbol of urban South America and attracts the continent's largest flow of foreign investment as well as the trade-related activities that befit the city's rise as the business capital of Mercosul (Mercosur).

Nonetheless, even for this biggest industrial metropolis of the Southern Hemisphere, the increasingly global tide of postindustrialism is rolling in and São Paulo is being forced to adapt. To avoid becoming a Detroit-style Rustbelt, the aging automobile-dominated manufacturing zone on the central city's southern fringes is today attracting new industries. Internet companies, in particular, have flocked here in such numbers that they already constitute Brazil's largest cluster of e-commerce activity—and prompt many to now call this area Silicon Village.

Elsewhere in São Paulo's vast urban constellation—whose suburbs now sprawl outward up to 100 kilometers (60 mi) from the CBD—additional opportunities are being exploited. In the outer northeastern sector, new research facilities as well as computer and telecommunications-equipment factories are springing up. And to the west of central São Paulo, lining the ring road that follows the Pinheiros River, is South America's largest suburban office complex replete with a skyline of ultramodern high-rises.

© H. J. de Blij, P. O. Muller, and John Wiley & Sons, Inc.

The huge recent growth of this megacity has also been accompanied by massive problems of overcrowding, pollution, and congestion. Traffic jams here are among the world's worst, with more than twice as many gridlocked motor vehicles on any given day than in Manhattan. Staggering poverty—on a par with Mexico City's—is the most pressing crisis as the ever-expanding belt of shantytowns tightens its grip on much of the metropolis. With its prodigious growth rate expected to persist through the foreseeable future, can anything prevent Greater São Paulo from approaching an unimaginable population of 50 million barely 20 years from today?

The **Interior** subregion—constituted by the States of Goiás, Mato Grosso, and Mato Grosso do Sul—is also known as the *Central-West*. This is the region that Brazil's developers have long sought to make a part of the country's productive heartland, and in 1960 the new capital of Brasília was deliberately situated on its margins (Fig. 5B-9).

By locating the new capital city in the wilderness 650 kilometers (400 mi) inland from its predecessor, Rio de Janeiro, the nation's leaders dramatically signaled the opening of Brazil's development thrust toward the west. Brasília is noteworthy in another regard because it represents what political geographers call a **forward capital [9]**. A state will sometimes relocate its capital to a sensitive area, perhaps near a peripheral zone under dispute with an unfriendly neighbor, in part to confirm its determination to sustain its position in that contested zone. Brasília does not lie near a contested area, but Brazil's interior was an internal frontier to be conquered by a growing nation. Spearheading that drive, the new capital occupied a decidedly forward position.

Despite the subsequent growth of Brasília to 4.1 million inhabitants today (which includes a sizeable ring of peripheral squatter settlements), it was not until the 1990s that the Interior began its economic integration with the rest of Brazil. The catalyst was the exploitation of the vast *cerrado* [10]—the fertile savannas that blanket the Central-West and make it one of the world's most promising agricultural frontiers (at least two-thirds of its arable land still awaits development). As with the U.S. Great Plains, the flat terrain of the *cerrado* is one of its main advantages because it facilitates the large-scale mechanization of farming with a minimal labor force (see photo, p. 276). Another advantage is rainfall, which is more prevalent here than in the Great Plains or Argentine Pampa.

The leading crop is soybeans, whose output per hectare here exceeds even that of the U.S. Corn Belt. Other grains and cotton are also expanding across the farmscape of the *cerrado*, but the current pace of regional development is inhibited by a serious accessibility problem. Thus the Interior's products must travel along poor roads and intermittent railroads to reach the markets and ports of the Atlantic seaboard. Today several projects are finally underway to alleviate these bottlenecks, including the privately financed *Ferronorte* railway that links Santos to the southeastern corner of Mato Grosso State, and the much-improved, so-called ***Soy Highway*** north to the Amazon River port city of Santarém.

The **North** is Brazil's territorially largest and most rapidly developing subregion, which consists of the seven States of the Amazon Basin (Fig. 5B-9). This was the scene of the great rubber boom just over a century ago, when the wild rubber trees in the *selvas* (tropical rainforests) produced huge profits and the central Amazon city of Manaus enjoyed a brief period of wealth and splendor. But the rubber boom ended in 1910, and for most of the seven decades that followed, Amazonia was a stagnant hinterland lying remote from the centers of Brazilian settlement. All that changed quite dramatically during the 1980s as new development began to stir throughout this awakening subregion, which currently is the scene of the world's largest migration into virgin territory. More than 200,000 new settlers arrive each year. The North of Brazil is well known for its high rates of ***deforestation*** (an issue discussed in Chapter 4A). Removing the forest results directly from logging operations, but more of it is a matter of urbanization, land occupation, use by settlers, and the arrival of large-scale agribusiness.

But the worst may be over. From 1995 to 2005, some 20,000 square kilometers (7700 sq mi) were cleared every year. According to the newest data, however, the trend has now shifted downward. In 2008, the Brazilian government pledged to achieve an 80-percent reduction in deforestation by 2020, and to completely end the practice by 2030. At least part of its thinking is that Brazil has huge potential as a "green economy" with its abundance of land,

This is what the Amazon's equatorial rainforest looks like from an orbiting satellite after the human onslaught in preparation for settlement. The colors on this Landsat image emphasize the destruction of the trees, with the dark green of the natural forest contrasted against the pale green and pinks of the leveled forest. The linear branching pattern of deforestation is the preferred approach here in Rondônia State's Highway BR-364 corridor. But farming is not likely to succeed for very long, and much of the cleared land is likely to be abandoned. Then the onslaught will resume to clear additional land.

© NRSC LTD/Photo Researchers, Inc.

water, and sunshine. Today, about 40 percent of the country's energy is already obtained from renewable sources (in contrast to less than 12 percent in the United States), and it becomes easier all the time to envision Brazil as the world's leading environmental power.

Development projects abound in the Amazonian North. One of the most durable is the ***Grande Carajás Project*** in central Pará State, a huge multifaceted scheme centered on one of the world's largest known deposits of iron ore in the hills around Carajás (Fig. 5B-9; photo, p. 244). In addition to a vast mining complex, other new construction here includes the Tucuruí Dam on the nearby Tocantins River and an 850-kilometer (535-mi) railroad to the Atlantic port of São Luis. This ambitious development project further emphasizes the exploitation of additional minerals, cattle raising, crop farming, and forestry. What is taking place here is a manifestation of the growth-pole concept [11]. A growth pole is a location where a set of activities, given a start, will expand and generate widening ripples of development in the surrounding area. In this case, the stimulated hinterland could one day cover one-sixth of all Amazonia.

Understandably, tens of thousands of settlers have descended on this part of the Amazon Basin. Those seeking business opportunities have been in the vanguard, but they have been followed by masses of lower-income laborers and peasant farmers in search of jobs and land ownership. The initial stage of this colossal enterprise has boosted the fortunes of many towns, particularly Manaus northwest of Carajás. Here, a thriving industrial complex (specializing in the production of electronic goods) has emerged in the free-trade zone adjoining the city thanks to the outstanding air-freight facilities and operations at Manaus's ultramodern airport. But many problems have also arisen as the tide of pioneers rolled across central Amazonia. One of the most tragic involved the Yanomami people, whose homeland in Roraima State was overrun by thousands of claim-stakers (in search of newly discovered gold), who triggered violent confrontations that ravaged the fragile aboriginal way of life.

Another leading development scheme, known as the ***Polonoroeste Plan***, is located about 1600 kilometers (1000 mi) to the southwest of Grande Carajás in the 2400-kilometer (1500-mi)-long Highway BR-364 corridor that parallels the Bolivian border and connects the western Brazilian towns of Cuiabá, Pôrto Velho, and Rio Branco (Fig. 5B-9). Even though the government had planned for the penetration of western Amazonia to proceed via the east-west Trans-Amazon Highway, the migrants of the 1980s and 1990s preferred to follow BR-364 and settle within the Basin's southwestern rim zone, mostly in Rondônia State (see photo at left). Agriculture has been the dominant activity here, but in the quest for land, bitter conflicts continue to break out between peasants and landholders as the Brazilian government pursues the volatile issue of land reform.

Brazil is the cornerstone of South America, the dominant economic force in Mercosur/l, the only dimensional counterweight to the United States in the Western Hemisphere, a maturing democracy, and an emerging global giant. Its rapidly intensifying relationship with China, in particular, is a clear sign that South America is no longer in the shadow of the United States. The country's center-left government has successfully steered a course between sustaining growth and reducing poverty. It appears that President Dilma Rousseff, who took office in 2010, will continue on this path. The future of the entire South American realm depends upon Brazil's stability and social as well as economic progress.

POINTS TO PONDER

- Brazil is well on its way to displacing the United States as the world's biggest food exporter; it aims to do so by 2015.
- Much of South America's economic growth in the past ten years is attributed to China's demand for raw materials.
- Watch for Argentina's renewed claims to "their" Malvinas as the British drill for oil near the Falkland Islands.
- If you are of voting age in Brazil, you must vote. It's compulsory. What are the (dis)advantages?

using GEOGRAPHY

1. Most South American countries exhibit similarities in terms of culture (language, religion) and political history (colonialism). Why do you think their economic and political development has diverged so greatly? (Compare, for example, Colombia, Venezuela, and Brazil.) Does geography play a role here, not only within these countries but also at a broader scale, perhaps constraining regional connections and integration?

2. The so-called commodity boom has played a major part in the recent economic rise of South America. But not all countries in South America are equally endowed with natural resources. Taking your cues from trends in market prices in global commodity trading of gold, iron ore, and other minerals, could you predict which countries have shown the most robust growth?

3. Could Brazil be considered the "United States" of South America in terms of its dominance, its history of immigration, its ethnic melting pot, and its colonization of western territory? What are the geographical similarities and differences?

IN THIS CHAPTER

* The cradle of humanity
* Wildlife conservation and sustainable development
* Neocolonial land grabs?
* Africa's varied ethnic mosaic
* The AIDS scourge
* Africa: The latest emerging economy?

CONCEPTS, IDEAS, AND TERMS

FIGURE 6A-1   © H. J. de Blij, P. O. Muller, and John Wiley & Sons, Inc.

The African continent occupies a special place in the physical as well as the human world. This is where **human evolution [1]** began. In Africa we formed our first communities, spoke our first words, and created our first art. From Africa our ancestor hominids spread outward into Eurasia more than 2 million years ago. From Africa our own species emigrated, beginning perhaps 130,000 years ago, northward into present-day Europe and eastward via southern Asia into Australia and, much later, farther afield into the Americas. Disperse our forebears did, but we should remember that, at the source, we are all Africans.

For millions of years, therefore, Africa served as the cradle for humanity's emergence. For tens of thousands of years, Africa was the source of human cultures. For thousands of years, Africa led the world in countless spheres ranging from tool manufacture to plant domestication to forging trade relationships. But in Chapters 6A and 6B we will encounter an Africa that has been struck by a series of disasters ranging from environmental deterioration to human dislocation on a scale unmatched anywhere in the world. When we assess Africa's misfortunes, however, we should remember that they have lasted hundreds, not thousands or millions, of years. Africa's catastrophic interlude will end, and Africa's time and turn will come again. Perhaps, it has already begun.

The focus in these two chapters will be on Africa south of the Sahara, for which the unsatisfactory but convenient name **Subsaharan Africa** has come into use to signify not physically "under" the great desert but directionally south of it. The African continent contains two geographic realms: the African, extending from the southern margins of the Sahara to the Cape of Good Hope, and the western flank of the realm dominated by the Muslim faith and Islamic culture whose heartland lies in the Middle East and the Arabian Peninsula. The great desert forms a formidable barrier between the two, but the powerful influences of Islam crossed it centuries before the first Europeans set foot in West Africa. By that time, the African kingdoms in what is known today as the Sahel had been converted, creating an Islamic foothold all along the northern periphery of the African realm (see Fig. G-3). As we note later, this cultural and ideological penetration had momentous consequences for Subsaharan Africa.

The African continent may be partitioned into two human-geographic realms, but the landmass is indivisible. Before we investigate the human geography of Subsaharan Africa, therefore, we should take note of the entire continent's unique physical geography (Figs. 6A-1 and 6A-2). We have already noted Africa's situation at the center of the planet's land hemisphere; moreover, no other landmass is positioned so squarely astride the equator, reaching almost as far to the south as to the north. This location has much to do with the distribution of Africa's climates, soils, vegetation, agricultural potential, and human population.

AFRICA'S PHYSIOGRAPHY

Africa accounts for about one-fifth of the Earth's entire land surface. The north coast of Tunisia lies 7700 kilometers (4800 mi) from the southernmost coast of South Africa. Coastal Senegal, in West Africa, lies 7200 kilometers (4500 mi) from the tip of the *Horn* in easternmost Somalia.

Village amid farms along the fertile, terraced wall of Kenya's Eastern Rift Valley—a key corridor in the evolution of our human species.

© H. J. de Blij

major geographic qualities of
SUBSAHARAN AFRICA

1. Physiographically, Africa is a plateau continent without a major mountain range, with a set of Great Lakes, several major river basins, variable rainfall, generally low-fertility soils, and mainly savanna and steppe vegetation.

2. Hundreds of distinct ethnic groups make up Subsaharan Africa's culturally rich and varied population. They far outnumber the states in this realm, and their boundaries do not often coincide.

3. Most of Subsaharan Africa's peoples depend on farming for their livelihood.

4. Health and nutritional conditions in Subsaharan Africa need improvement as the incidence of disease remains high and diets are often unbalanced. The AIDS pandemic continues to be a major health crisis in this realm.

5. The realm is rich in raw materials vital to industrialized countries, but many economies continue to rely on primary activities—the extraction of resources—and not the better income-generating activities of assembly and manufacturing.

6. The realm is famous for its wildlife, but many species are threatened. Combining bioconservation and sustainable development is a major challenge.

7. During the Cold War, great-power competition magnified conflicts in several Subsaharan African countries, with results that will be felt for generations.

8. Severe dislocation affects many Subsaharan African countries, from Liberia to Burundi to Zimbabwe. This realm has the largest refugee population in the world today.

9. Government mismanagement and poor leadership afflict the economies of many Subsaharan African countries.

10. In contrast to general patterns in the realm, a small number of countries have witnessed spectacular economic growth in the past decade.

These distances have major environmental implications. Much of Africa is far from maritime sources of moisture. In addition, as Figure G-7 shows, large parts of the landmass lie in latitudes where global atmospheric circulation systems produce arid conditions. The Sahara in the north and the Kalahari in the south form part of this globe-girdling desert zone. Water supply is one of Africa's great problems.

Rifts and Rivers

Africa's topography reveals several properties that are not replicated on other landmasses. Alone among the continents, Africa does not have a mountain backbone; neither the northwestern Atlas nor the far southern Cape Ranges are in the same league as the Andes or Himalayas. Where Africa does have high mountains, as in Ethiopia and South Africa, these are deeply eroded plateaus or, as in East Africa, high snowcapped volcanoes. Furthermore, Africa is one of only two continents containing a cluster of Great Lakes, and the only one whose lakes result from powerful tectonic forces in the Earth's crust. These lakes (with the exception of Lake Victoria) lie in deep trenches called **rift valleys [2]**, which form when huge parallel cracks or faults appear in the Earth's crust and the strips of crust between them sink, or are pushed down, to form great, steep-sided, linear valleys (photo at left). In Figure 6A-2 these rift valleys, which

stretch more than 9600 kilometers (6000 mi) from the Red Sea to Swaziland, are marked by red lines.

Africa's rivers, too, are unusual: their upper courses frequently bear landward, seemingly unrelated to the coast toward which they eventually flow. Several rivers, such as the Nile and the Niger, have inland as well as coastal deltas. Major waterfalls, notably Victoria Falls on the Zambezi, or lengthy systems of cataracts, separate the upper from the lower courses.

Finally, Africa may be described as the plateau continent. Except for some comparatively limited coastal plains, almost the entire continent lies above 300 meters (1000 ft) in elevation, and fully half of it lies over 800 meters (2500 ft) high. As Figure 6A-2 shows, the plateau surface has sagged under the weight of accumulating sediments into a half dozen major basins (three of them in the Sahara). The margins of Africa's plateau are marked by escarpments, often steep and step-like. Most notable among these is the Great Escarpment of southeastern South Africa, marking the eastern edge of the Drakensberg Mountains.

Continental Drift and Plate Tectonics

Africa's remarkable and unusual physiography was a key piece of evidence that geographer Alfred Wegener used to construct his hypothesis of **continental drift [3]**. All of the present-day continents, Wegener reasoned, lay assembled

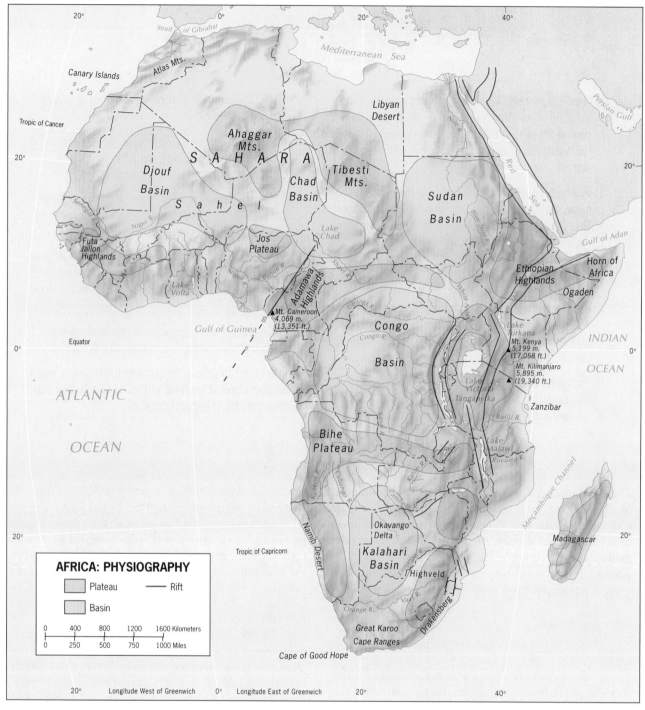

FIGURE 6A-2

© H. J. de Blij, P. O. Muller, and John Wiley & Sons, Inc.

as one giant landmass called **Pangaea** not very long ago in geologic time (220 million years ago). The southern part of this supercontinent was *Gondwana*, of which Africa formed the core (Fig. 6A-3). When, roughly about 200 million years ago, tectonic forces began to split Pangaea apart, Africa (and the other landmasses) acquired their present configurations. That process, now known as **plate tectonics**, continues and is marked by earthquakes and volcanic eruptions. By the time it started, however, Africa's land surface had begun to acquire some of the features that mark it today—and make it unique. The rift valleys, for

example, demarcate the zones where plate movement continues—hence the linear shape of the Red Sea, where the Arabian Plate is separating from the African Plate (Fig. G-4). And yes, the rift valleys of East Africa probably mark the further fragmentation of the African Plate (some geophysicists have already referred to a Somali Plate, which will separate, Madagascar-like, from the rest of Africa).

Africa's ring of escarpments, its rifts, its river systems, its interior basins, and its lack of significant mountains, all relate to the continent's central location in Pangaea, all pieces of the puzzle that led to the plate-tectonics solution.

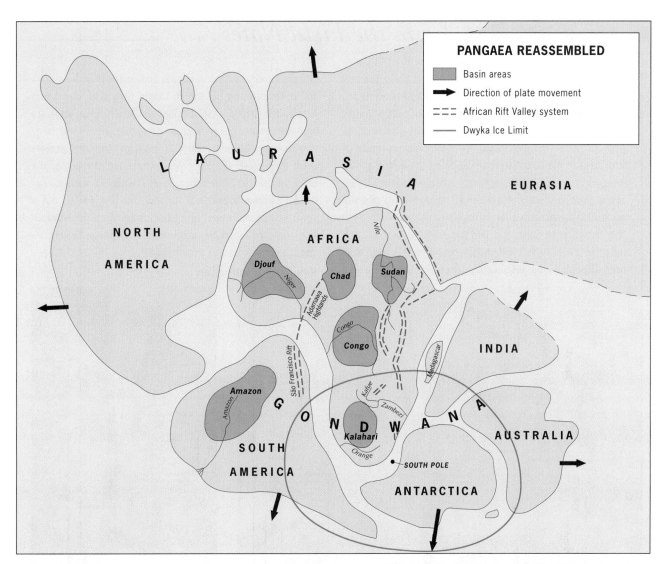

FIGURE 6A-3

© H. J. de Blij, P. O. Muller, and John Wiley & Sons, Inc.

AFRICA'S HISTORICAL GEOGRAPHY

Africa is the cradle of humankind. Archeological research has chronicled 7 million years of transition from Australopithecenes to Hominids to *Homo sapiens*. It is therefore ironic that we know comparatively little about Subsaharan Africa from 5000 to 500 years ago—that is, before the onset of European colonialism. This is partly due to the colonial period itself, during which African history was neglected, numerous African traditions and artifacts were destroyed, and many misconceptions about African cultures and institutions became entrenched. It is also a result of the absence of a written history over most of Africa south of the Sahara until the sixteenth century—and over a large part of it until much later than that.

African Genesis

Africa on the eve of the colonial period was a continent in transition. For several centuries, the habitat in and near one of the continent's most culturally and economically productive areas—West Africa—had been changing. For

2000 years, probably more, Africa had been innovating as well as adopting ideas from outside. In interior West Africa, cities were developing on an impressive scale; in central and Southern Africa, peoples were moving, readjusting, sometimes struggling with each other for territorial supremacy. The Romans had penetrated as far as southern Sudan, North African peoples were trading with West Africans, and Arab *dhows* (wooden boats with triangular sails) were sailing the waters along the eastern coasts, bringing Asian goods in exchange for gold, copper, and a comparatively small number of slaves. These same dhows today are increasingly used in tourism.

It is known that African cultures had been established in all the environmental settings shown in Figure G-7 for thousands of years and thus long before Islamic or European contact. One of these, the Nok culture, endured for over eight centuries on the Benue Plateau (north of the Niger-Benue confluence in present-day Nigeria) from about 500 BC to the third century AD. The Nok people made stone as well as iron tools, and they left behind a treasure of art in the form of clay figurines representing humans and animals.

From the Field Notes . . .

"Got up before dawn this date in the hope of being the first visitor to the Great Zimbabwe ruins, a place I have wanted to see ever since I learned about it in a historical geography class. Climbed the hill and watched the sun rise over the great elliptical 'temple' below, then explored the maze of structures of the so-called fortification on the hilltop above. Next, for an incredible 90 minutes I was alone in the interior of the great oval walls of the temple. What history this site has seen—it was settled for perhaps six centuries before the first stones were hewn and laid here around AD 750, and from the eleventh to the fifteenth centuries this may have been a ceremonial or religious center of a vast empire whose citizens smelted gold and copper and traded them via Sofala and other ports on the Indian Ocean for goods from South Asia and even China. No cement was used: these walls are built with stones cut to fit closely together, and in the valley they rise over 30 feet (nearly 10 m) high. And what secrets Great Zimbabwe still conceals—where are the plans, the tools, the quarries? The remnants of an elaborate water-supply system suggest a practical function for this center, but a decorated conical tower and adjacent platform imply a religious role. Whatever the answers, history hangs heavily, almost tangibly, over this African site."

www.conceptcaching.com

© H. J. de Blij

Early Trade

But we have no evidence that they traded with distant peoples. The opportunities created by environments and technologies still lay ahead.

West Africa, over a north-south span of a few hundred kilometers, displayed an enormous contrast in environments, economic opportunities, modes of life, and products. The peoples of the tropical forest produced and needed goods that were different from the products and requirements of the peoples of the dry, distant north. For example, salt is a prized commodity in the forest, where humidity precludes its formation, but it is plentiful in the desert and semiarid steppe. This enabled the desert peoples to sell salt to the forest peoples in exchange for ivory, spices, and dried foods. Thus there evolved a degree of *regional complementarity* between the peoples of the forest and those of the drylands. And the savanna peoples—those located in between—found themselves in a position to channel and handle the trade (which is always economically profitable).

The markets in which these goods were exchanged prospered and grew, and urban centers arose in the savanna belt of West Africa. One of these old cities, now an epitome of isolation, was once a thriving center of commerce and learning and one of the leading urban places in the world—Timbuktu (now located in Mali). In fact, its university is one of the oldest in the world, with a library that holds valuable documents that are being preserved. Others, predecessors as well as successors of Timbuktu, have declined, some of them into oblivion. Still other savanna cities, such as Kano in the northern part of Nigeria, remain important today.

Early States

Strong and durable states arose inland in West Africa. The oldest state we know anything about is ancient Ghana, located to the northwest of the modern country of Ghana. It covered parts of present-day Mali, Mauritania, and adjacent territory. Ghana lay astride the upper Niger River and included gold-rich streams flowing off the Futa Jallon Highlands, where the Niger River has its origin. Between the ninth and twelfth centuries AD, and perhaps longer, old Ghana managed to weld various groups of people into a stable state. The country had a large capital city complete with marketplaces, suburbs for foreign merchants, religious shrines, and, some distance from the city center, a fortified royal retreat. Taxes were collected from the citizens, and tribute was extracted from subjugated peoples on Ghana's periphery; tolls were levied on goods entering Ghana, and an army maintained control. Muslims from the northern

drylands invaded Ghana in 1067, when it may already have been in decline. Even so, the Ghanaians managed to protect their capital for 14 years. However, the invaders had ruined the farmlands and destroyed the trade links with the north. Ancient Ghana could not survive. It finally broke into smaller units.

Eastward Shift

In the centuries that followed, the focus of politico-territorial organization in this West African *culture hearth* (source area of culture) shifted almost continuously to the east—first to ancient Ghana's successor state of Mali, which was centered on Timbuktu and the middle Niger River Valley, and then to the state of Songhai, whose focus was Gao, a city on the Niger that still exists (Fig. 6A-4). This eastward movement may have been the result of the growing influence and power of Islam. Traditional religions

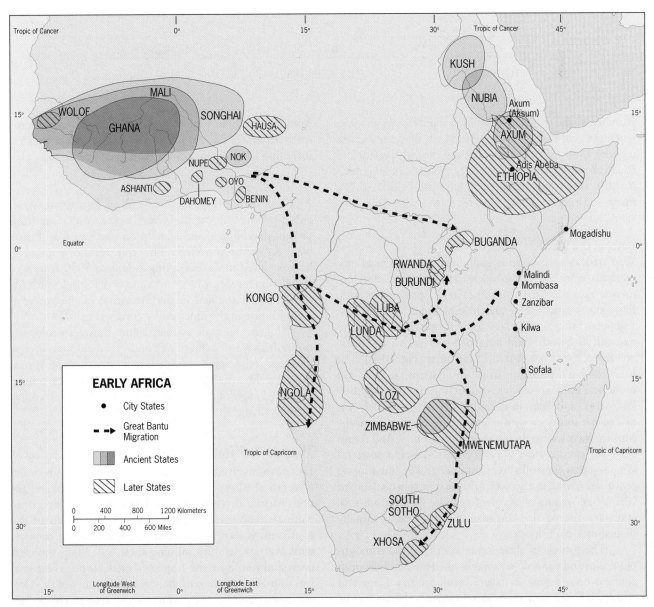

FIGURE 6A-4

© H. J. de Blij, P. O. Muller, and John Wiley & Sons, Inc.

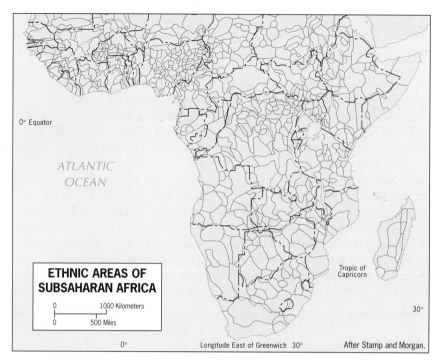

ETHNIC AREAS OF SUBSAHARAN AFRICA

FIGURE 6A-5   © H. J. de Blij, P. O. Muller, and John Wiley & Sons, Inc.

prevailed in ancient Ghana, but Mali and its successor states sent huge, gold-laden pilgrimages to Mecca along the savanna corridor south of the Sahara, passing through present-day Khartoum and Cairo. Of the tens of thousands who participated in these pilgrimages, some remained behind. Today, many Sudanese trace their ancestry to the West African savanna kingdoms.

Beyond the West

West Africa's savanna zone undoubtedly experienced momentous cultural, technological, and economic developments, but other parts of Africa made progress as well. Early states emerged in present-day Sudan, Eritrea, and Ethiopia. Influenced by key innovations from the Egyptian culture hearth, these kingdoms were stable and durable: the oldest, Kush, lasted 23 centuries (Fig. 6A-4). The Kushites built elaborate irrigation systems, forged iron tools, and built impressive structures as the ruins of their long-term capital and industrial center, Meroe, reveal. Nubia, to the southeast of Kush, was Christianized until the Muslim wave overtook it in the eighth century. And Axum was the richest market in northeastern Africa, a powerful kingdom that controlled Red Sea trade and endured for six centuries. Axum, too, was a Christian state that confronted Islam, but Axum's rulers deflected the Muslim advance and gave rise to the Christian dynasty that eventually shaped modern Ethiopia.

The process of **state formation [4]** spread throughout Africa and was still in progress when the first European contacts occurred in the late fifteenth century. Large and effectively organized states developed on the equatorial west coast (notably Kongo) and on the southern plateau from the southern part of the Congo River Basin southeastward to Zimbabwe. East Africa had several city-states, including Mogadishu, Kilwa, Mombasa, and Sofala.

Bantu Migration

A crucial event affected virtually all of Equatorial, West, and Southern Africa: the Great Bantu Migration from present-day Nigeria and Cameroon southward and eastward across the continent. This epic advance appears to have occurred in waves starting as long as 5000 years ago, populating the Great Lakes area and penetrating South Africa, where it resulted in the formation of the powerful Zulu Empire in the nineteenth century (Fig. 6A-4).

All this reminds us that, before European colonization, Africa was a realm of rich and varied cultures, diverse lifestyles, technological progress, and external trade. It was, however, also a highly fragmented realm, its cultural mosaic (Fig. 6A-5) spelling weakness when European intervention came to change the social and political map forever.

The Colonial Transformation

European involvement in Subsaharan Africa began in the fifteenth century. It would interrupt the path of indigenous African development and irreversibly alter the entire cultural, economic, political, and social makeup of the continent. It started quietly in the late fifteenth century, with Portuguese ships groping their way along the west coast and rounding the Cape of Good Hope. Their goal was to find a sea route to the spices and riches of the Orient. Soon other European countries were dispatching their vessels to African waters, and a string of coastal stations and forts sprang up. In West Africa, the nearest part of the continent to European spheres in Middle and South America, the initial impact was strongest. At their coastal control points, the Europeans traded with African intermediaries for the slaves who were destined to work New World plantations, for the gold that had been flowing northward across the desert, and for ivory and spices.

Coastward Reorientation

Suddenly, the centers of activity lay not with the inland cities of the savanna but with the foreign stations on the Atlantic coast. As the interior declined, the coastal peoples thrived. Small forest states gained unprecedented wealth, transferring and selling slaves captured in the interior to the European traders on the coast. Dahomey (now called Benin) and Benin (now part of neighboring Nigeria) were states built on the slave trade. When slavery was eventually abolished in Europe, those who had inherited the power and riches it had brought vigorously opposed abolition on both continents.

Horrors of the Slave Trade

As discussed elsewhere in the book, millions of Africans were forced to migrate from their homelands to the Americas, especially Brazil, the Caribbean Basin, and the United States. The slave trade was one of those African disasters alluded to earlier, and it was facilitated in part by what we may call the peril of proximity. The northeastern tip of Brazil, by far the largest single destination for the millions of Africans forced from their homes in bond-

age, lies about as far from the nearest West African coast as South Carolina lies from Venezuela. This is a short maritime intercontinental journey indeed (it is more than twice as far from West Africa to South Carolina). That proximity facilitated the forced migration of millions of West Africans to Brazil, which in turn contributed to the emergence of an African cultural diaspora in Brazil that is without equal in the New World (see pp. 273–275).

Although slavery was not new to West Africa, the *kind* of slave raiding and trading the Europeans introduced certainly was. In the interior of Africa and within city-states, kings, chiefs, and prominent families traditionally took a few slaves, but the status of those slaves was unlike anything that lay in store for those who were shipped across the Atlantic. In fact, large-scale slave trading had been introduced in East Africa long before the Europeans brought it to West Africa. African intermediaries from the coast raided the interior for able-bodied men and women and marched them in chains to the Arab markets on the coast (the island of Zanzibar was one such slave trading market). There, packed in specially built dhows, they were carried off to Arabia, Persia, and India. When the European slave trade took hold in West Africa, however, its volume was far greater. Europeans, Arabs, and collaborating Africans ravaged the continent, forcing perhaps as many as 30 million persons away from their homelands in captivity (Fig. 6A-6). Families were destroyed, as were whole villages and cultures; those who survived their exile suffered unfathomable misery.

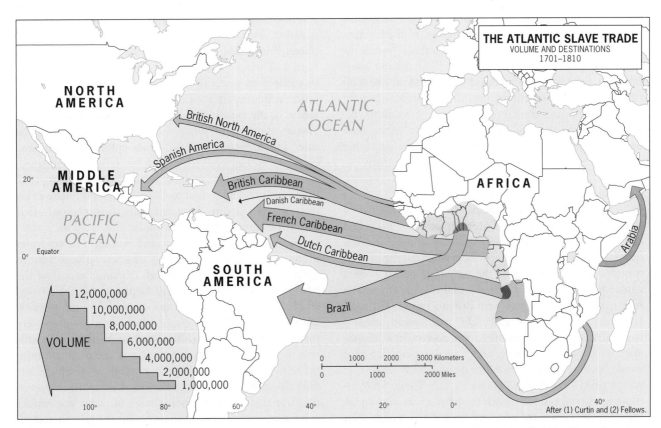

FIGURE 6A-6

© H. J. de Blij, P. O. Muller, and John Wiley & Sons, Inc.

From the Field Notes . . .

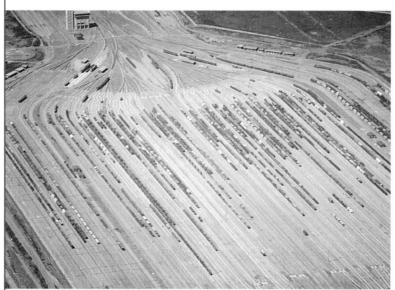

© H. J. de Blij

"Looking down on this enormous railroad complex northeast of Johannesburg, South Africa, we were reminded of the fact that almost an entire continent was turned into a wellspring of raw materials carried from interior to coast and shipped to Europe and other parts of the world. This complex lies near Witbank in the eastern Rand, a huge inventory of freight trains ready to transport ores from the plateau to Durban and Maputo. But at least South Africa acquired a true transport network in the process, ensuring regional interconnections; in most African countries, railroads serve almost entirely to link resources to coastal outlets."

www.conceptcaching.com

Concept Caching

The European presence on the West African coast completely reoriented its trade routes, for it initiated the decline of the interior savanna states and strengthened the coastal forest states. Moreover, the Europeans' insatiable demand for slaves ravaged the population of the interior. But it did not lead to any major European thrust toward the interior or produce colonies overnight. The African intermediaries were well organized and strong, and they held off their European competitors, not just for decades but for centuries. Although the Europeans first appeared in the fifteenth century, they did not carve up West Africa until nearly 400 years later, and they did not conquer many other areas until after the beginning of the twentieth century.

Colonization

In the second half of the nineteenth century, whether or not they had control, the European powers finally laid claim to virtually all of Africa. Colonial competition was intense, and spheres of influence began to overlap. It was time for negotiation among the powerful, and in 1884 a conference was convened in Berlin to divide up Africa. The major colonial contestants were Britain, France, Portugal, Belgium, and Germany itself. On maps spread across a large table, representatives from these powers drew boundaries, exchanged real estate, and forged a new map that would become a liability in Africa decades later (see box on p. 294 titled "The Berlin Conference"). As Figure 6A-7 indicates, when the three-month conference was in progress most of Africa remained under traditional African rule. Not until after 1900 did the colonial powers manage to control all the areas they had marked off on their new maps.

It is important to examine Figure 6A-7 carefully because the colonial powers governed their new dependencies in very different ways, and their contrasting legacies are still in evidence to this day in the countries their colonies spawned. Some colonial powers were democracies at home (Great Britain and France); others were dictatorships (Portugal and Spain). The British established a system of indirect rule over much of their domain, leaving indigenous power structures in place and making local rulers representatives of the British Crown. This was unthinkable in the Portuguese colonies, where harsh, direct control was the rule. The French sought to create culturally assimilated elites that would represent French ideals in the colonies.

Unlike the British, French, or Portuguese, King Leopold II of Belgium sought to claim Congo Free State as his own personal empire. After financing the expeditions that staked Belgium's claim in Berlin, he embarked on a campaign of ruthless exploitation and murder. His enforcers mobilized almost the entire Congolese population to gather rubber (which was in high demand because of the advent of motor vehicles), kill elephants for their ivory, and build public works to improve export routes. For failing to meet production quotas, entire communities were tortured and executed. Killing and maiming became routine in a colony in which horror was the only common denominator (chronicled unforgettably in Joseph Conrad's classic novella, *Heart of Darkness*). After the impact of the slave trade, King Leopold's reign of terror was Africa's most severe demographic disaster. By the time it ended, following growing condemnation from all over the world, as many as 10 million Congolese had been murdered. In

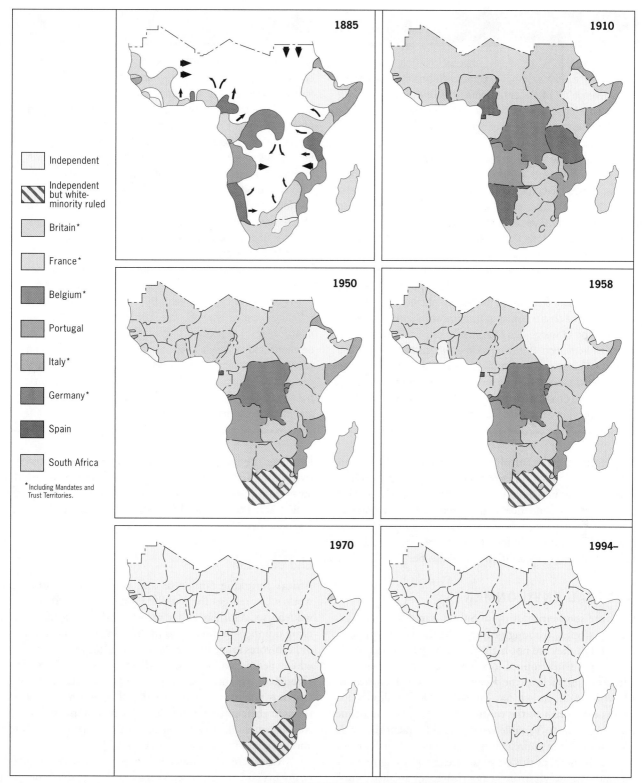

COLONIZATION AND DECOLONIZATION SINCE 1885

FIGURE 6A-7

© H. J. de Blij, P. O. Muller, and John Wiley & Sons, Inc.

1908, the Belgian government took over and began to mirror Belgium's own internal divisions: corporations, government administrators, and the Roman Catholic Church each pursued their sometimes competing interests. But no one thought to change the name of the colonial capital: it

remained Leopoldville until the Belgian Congo achieved independence in 1960.

Colonialism transformed Africa, but in its post-Berlin form it lasted less than a century. In Ghana (then named the Gold Coast), for example, the Ashanti (Asante) Kingdom

The Berlin Conference

IN NOVEMBER 1884, the imperial chancellor and architect of the German Empire, Otto von Bismarck, convened a conference of 14 states (including the United States, which had no African claims) to settle the political partitioning of Africa. Bismarck wanted not only to expand German spheres of influence in Africa but also to play off Germany's colonial rivals against one another to the Germans' advantage. The major colonial contestants in Africa were: (1) the British, who held beachheads along the West, South, and East African coasts; (2) the French, whose main sphere of activity was in the area of the Senegal River as well as north and northwest of the Congo Basin; (3) the Portuguese, who now desired to extend their coastal stations in Angola and Moçambique deep into the interior; (4) King Leopold II of Belgium, who was amassing a personal domain in the Congo Basin; and (5) Germany itself, active in areas where the designs of other colonial powers might be obstructed, as in Togo (between British holdings), Cameroon (a wedge into French spheres), South West Africa (taken from under British noses in a swift strategic move), and East Africa (where German Tanganyika broke the British design for a solid block of territory from the Cape north to Cairo).

When the conference convened in Berlin, more than 80 percent of Africa was still under traditional African rule.

Nonetheless, the colonial powers' representatives drew their boundary lines across the entire map. These lines were drawn through known as well as unknown areas, pieces of territory were haggled over, boundaries were erased and redrawn, and African real estate was exchanged among European governments. In the process, African peoples were divided, unified regions were ripped apart, hostile societies were thrown together, hinterlands were disrupted, and migration routes were closed off (for an idea, see Fig. 6A-5). Not all of this was felt immediately, of course, but these were some of the effects when the colonial powers began to consolidate their holdings and the boundaries on paper became barriers on the African landscape (Fig. 6A-7).

The Berlin Conference was Africa's undoing in more ways than one. The colonial powers superimposed their domains on the African continent. By the time Africa regained its independence after the late 1950s, the realm had acquired a legacy of political fragmentation that could neither be eliminated nor made to operate satisfactorily. The African politico-geographical map is therefore a permanent liability that resulted from three months of ignorant, greedy acquisitiveness during a period when Europe's search for minerals and markets had become insatiable.

still was fighting the British in the early years of the twentieth century; by 1957, Ghana was independent again. These days, much of Subsaharan Africa has already marked half a century of independence, and, in retrospect, the colonial period should be seen as an interlude rather than a paramount chapter in modern African history—but it did leave a deep and lasting imprint on the landscape.

NATURAL ENVIRONMENTS

Only the southernmost tip of Subsaharan Africa lies outside the tropics. Although African elevations are comparatively high, they are not high enough to ward off the heat that comes with tropical location except in especially favored locales such as the Kenya Highlands and parts of Ethiopia. And, as we have noted, Africa's bulky shape means that much of the continent lies far from maritime moisture sources. Variable weather and frequent droughts are among Africa's unrelenting environmental challenges.

It is useful at this point to refer back to Figure G-7. As that map shows, Africa's climatic regions are distributed almost symmetrically about the equator, though more so in the center of the landmass than in the east, where elevation changes the picture. The hot, rainy climate of the Congo Basin merges gradually, both northward and southward, into climates with distinct winter-dry seasons. Winter, however, is marked more by drought than by cold. In parts of the area mapped *Aw* (tropical savanna), the annual seasonal cycle produces two rainy seasons, often referred to

locally as the long rains and the short rains, separated by two winter dry periods. As you go farther north and south away from the moist Congo Basin, the dry season(s) grow longer and the rainfall diminishes and becomes less and less dependable.

End of an Era

Africa's shrinking rainforests and vast savannas form the world's last refuges for wildlife ranging from primates to wildebeests. Gorillas and chimpanzees survive in dwindling numbers in threatened forest habitats, while millions of herbivores range in great herds across the savanna plains where people compete with them for space. European colonizers, who introduced hunting as a "sport" (a practice that was not part of African cultural traditions) and who brought their capacities for mass destruction to animals as well as people in Africa, helped clear vast areas of wildlife and push species to near-extinction.

Later, colonial rulers and subsequent national governments laid out game reserves and other types of conservation areas, but these were not sufficiently large or well enough connected to allow herd animals to follow their seasonal and annual migration routes. The same climatic variability that affects farmers also affects wildlife, and when the rangelands wither, the animals seek better pastures. When the fences of a game reserve wall them off, they cannot survive. When there are no fences, the wildlife invades neighboring farmlands and destroys crops, and the farmers retaliate. After thousands of years of equilibrium, the competition between humans and animals in Africa has taken a new turn.

The statistics are sobering. Overall, the count of "big" animals has gone down by 55 percent over the last four decades, while some species or subspecies are near extinction, such as the northern white rhinoceros. A century ago, the number of rhinos of all types in this realm totaled several hundred thousand; but now only about 2000 remain. As for Africa's lion population, it has dropped from around 400,000 a half-century ago to no more than 20,000 today. It is the end of an era—although it is never too late to save what we can and assume responsible stewardship of nature.

Wildlife Management and Tourism

African governments recognize the importance of nature conservation and benefit from the revenues generated by tens of thousands of tourists who annually come to view Africa's incomparable combination of natural landscapes and varied wildlife. For most countries, the revenues from tourism are (still) very modest, but for some, like Kenya, Tanzania, and especially South Africa, they have become quite significant. The idea here, of course, is that conservation can be economically rewarding and make sense from a monetary point of view.

But managing the sometimes conflicting needs and demands of sedentary local people, migratory animals, and tourist facilities (such as game lodges and artificial water holes to attract animals) is often difficult. The simple notion of fencing animals in and people out obviously does not always work; not only do many wild animals migrate seasonally, but so must the livestock of pastoralists. Maintain a game reserve on ancestral land, and you had better ensure that the local inhabitants have a financial stake in its preservation. In Kenya and Tanzania you can see local herders drive their livestock inside game reserves, across the same pastures that sustain wildlife. And governments have to combat poaching to protect the value of their wildlife heritage.

In some areas, the debates about hunting have become quite contentious. Animal-rights groups and certain conservation agencies strenuously oppose any kind of hunting, yet others argue that *some* hunting would actually make sense. Certainly this is the case with highly abundant species such as oryx or springbok in Namibia (where hunting is now permitted). But it could also apply to elephants because, some say, they are not endangered and they could be "harvested sustainably" (trophy hunting, ivory sales) and become a potential bonanza for local communities. It is a tense debate indeed, loaded with contentious ethical, ecological, and economic issues.

From the Field Notes . . .

"Flying a small propeller plane over East Africa's semiarid grasslands, we had a good view of a Maasai village near the border between Kenya and Tanzania. The Maasai are a pastoral people whose main livelihood is herding cattle, sheep, and goats. I could clearly see the fenced 'kraals' where they keep their animals at night (during the day they are moved around) and the small dwellings that surround them. The more animals a man (!) has, the higher his status. But this way of life is under pressure: many have decided to settle in towns or cities because they offer more and better opportunities. And this area is also the northern extension of Tanzania's famous Serengeti National Park, and here wildlife preservation comes into direct conflict with the Maasai way of life. 'It is really impossible,' said the park guide on the 10-seater plane with us, pointing down at the settlement. 'Of course they want to protect their cattle, but they are killing lions and other animals that should be protected. And we must tell them all the time that they have to stay out of certain areas because the cattle compete for food with the wild animals.' Nodding agreeably, I thought to myself: yes, but isn't this their land? As with so many other places around the world where we want to protect nature, we must involve the local people. That is much easier said than done."

© Jan Nijman

www.conceptcaching.com

It might appear obvious that tourism is a benign industry whose revenues help sustain local ecologies, but there is another side to this. Tourist interest tends to focus on seeing the "big four" animals during short visits, including elephants. But attracting elephant herds through artificial water supply results in habitat change that affects (and can drive away) other wildlife. From ongoing research work by geographers and other scientists, it is clear that ecosystem modification by tree-trampling elephant herds is converting African woodland into shrub or grassland, driving away grazing animals that require particular plant species, and allowing less choosy browsers to take over. Thus the concentration of one species, itself in need of protection and much in tourist demand, negatively impacts wildlife diversity and richness over wide areas. And when ongoing climate change creates severe and prolonged droughts, as parts of Subsaharan Africa have experienced in recent years, the problems of wildlife management become even more difficult. Thus the ability of national governments to protect this natural resource is ever more severely constrained.

People, Farmlands, and Environments

And yet it would seem that there could be space for wildlife as well as humans in Subsaharan Africa. Figure 6A-8 does not present a picture of a densely peopled realm: while there are major clusters of population in West Africa (where Nigeria is the realm's most populous country), East Africa (encircling Lake Victoria), and the Horn (where Ethiopia's highlands sustain a large concentration), most of the rest of the realm seems relatively sparsely populated. Our Data Table in Appendix B bears out this impression: all the countries of Subsaharan Africa *combined* have a population only about two-thirds of China's alone.

But much of this population continues to depend on farming for their livelihood, and we have already noted that African environments are difficult for millions of farmers. Not everywhere, of course: Africa has its areas of good soil, ample water, and high farm productivity. But on the map, these areas are not extensive. The major population concentration in the Great Lakes region reflects the volcanic soils of Mount Kilimanjaro as well as those of the Western Rift Valley zone. The Ethiopian Highlands were

FIGURE 6A-8

© H. J. de Blij, P. O. Muller, and John Wiley & Sons, Inc.

once known as the "breadbasket of northeast Africa" and could be capable of much greater productivity than is presently the case. Higher-latitude, cooler, and moister areas of South Africa are exceptionally productive with a large range of farm produce. And parts of West Africa sustain intensive farming. Taken together, however, these productive areas do not add up to the huge, fertile alluvial basins of China or India, or even the (North African) Nile Valley and Delta. The great majority of African farmers face daunting challenges. These challenges include: (1) climatic variability, (2) the economic policies of national governments, and (3) the difficulties African farmers have in reaching world markets. Look again at Figure G-7, and you can see the relatively short distances involved as the moist tropics of the equatorial Congo Basin give way to the deserts of the north and south. Moreover, as the annual rainfall total declines, its variability increases, so that farmers in the drier zones cannot count on sufficient moisture in any given year.

The economic policies of national governments frequently disadvantage farmers as the prices of their products are kept artificially low to please urban (and politically powerful) consumers. We return to this topic later, but the regional effect on African agriculture has been to inhibit productivity.

And in this era of globalization, African farmers often get a raw deal on world markets. For all the talk of free trade in the wealthy global core, many governments favor their local farmers over those of Africa (and other parts of the global periphery), from rice subsidies in Japan to "market support" in France. Economists estimate that such unfree trade costs African farmers as much as U.S. $200 billion annually.

AFRICANS AND THEIR LAND

Although rainfall is a critical physical criterion for farming, a number of political and economic factors are also influential. Among them are land tenure; commercial (for-profit) vs. subsistence (household-level) agriculture; type of farming system (rotational, shifting cultivation, intercropping, etc.); crop prices; government policies promoting the planting of one crop over another; indigenous agricultural knowledge of particular crops; and the degree of technology adoption and mechanization.

With more than 60 percent of the realm's livelihoods dependent on farming, the issue of land tenure is crucial. Land tenure [5] refers to the way people own, occupy, and use land. African traditions of land tenure are different from those of Europe or the Americas today. In most of Subsaharan Africa, communities, not individuals, customarily hold land. Occupants of the land have temporary, custodial rights to it and cannot sell it. Land may be held by large (extended) families, a village community, or even a traditional chief who holds the land in trust for the people. His subjects may house themselves on it and farm it, but in return they must follow his rules.

Stolen Lands

At the onset of colonialism [6], colonial administrators intended to control the most fertile areas of the occupied colonies through eviction of indigenous peoples. In some cases this was done by physical force (mostly through military conquest), and in other areas by coercion. Prior to colonialism, many traditional African livelihood systems adhered to sustainable land management practices such as leaving the land fallow or rotational grazing. Viewing this land as unused, colonial planners initiated a process of land alienation [7] not unlike what happened in Middle and South America. Many of the most fertile and productive areas were placed under the control of colonial settlers and governments. Over time, these lands were bought and sold and became private legal property. At the end of the colonial era, several newly independent African countries started programs whereby land would revert to traditional forms of ownership and management. However, the legacy of colonialism has been difficult to overcome, and African governments have adopted several neocolonial policies on land management that continue to marginalize farmers.

Rapid population growth, such as Africa experienced early in the postcolonial period, makes access to land even more complicated. Traditional systems of land use, which involve subsistence farming in various forms ranging from shifting cultivation to pastoralism, work best when the population is fairly stable and tenure is communal. Land must be left fallow to recover from cultivation, and pastures must be kept free of livestock so that the grasses can revive. The population explosion that Africa experienced during the mid-twentieth century set into motion a cycle of land overuse. When soils cannot rest and pastures are overgrazed, the land becomes degraded and yields decline.

Persistent Subsistence

Although there is commercial farming in parts of Africa, most African farmers remain subsistence farmers who grow grain crops (maize [corn], millet, sorghum) in drier areas and root crops (yams, manioc [cassava], sweet potatoes) where moisture is ample. Others herd livestock, mostly cattle and goats, as they counter environmental and climatic variability. Farmers and pastoralists alike have been unable to secure access to regular markets and stable prices for their products because of government policies that often promote one particular type of export-oriented crop (for example, peanuts in Gambia, tea and coffee in Kenya, or cacao [cocoa] in Ivory Coast) over other crops that do not fetch high prices on the world market.

Even though African farmers learned to adapt to these problems, farm yields in Africa have remained modest for many years. Governments, in response to World Bank policies, paid greater attention to high-profile industrial projects and as a result have neglected agriculture. The decline of African agriculture has been disproportionally hard on Africa's women who, according to current estimates, produce 75 percent of local food in Subsaharan

Regional ISSUE — The Impact of Colonialism on Subsaharan Africa

COLONIALISM IS THE CULPRIT!

"If you want to see and feel the way colonialism ruined Africa, you don't have to go beyond the borders of my own country, the so-called Democratic Republic of the Congo, for the evidence. I'm a school teacher here in Kikwit, east of Kinshasa, and my whole life I've lived under the dictatorship left us by the Belgians and in the chaos before and after it. All you have to do is look at the map. Who can say that the political map of Africa isn't a terrible burden? Imagine some outside power, say the Chinese, coming into Europe and throwing the Germans and the French together in one country. There would never be peace! Well, that's what they did here. All over Africa they put old enemies together and parceled up our great nations. And now they tell us to get over our animosities and 'tribalism' and live in democratic peace.

"And let me tell you something about tribalism. Before the conquest, we had ethnic groups and clans like all other peoples have, from the Scots to the Siamese. And we certainly fought with each other. But mostly we got along, and there was lots of gray area between and among us. You weren't immediately labeled a Hutu or a Luba or a Bemba. The colonial powers changed all that. Suddenly we were tribalized. Whether you liked it or not, whether you filled the 'profile' or not, you were designated a Tutsi or a Hutu, a Ganda or a Toro. This enabled the European rulers to use the strong to dominate the weak, to use the rich to tax the poor, to police migrants and assemble labor. The tribalization of Africa was a colonial invention. It worsened our divisions and destroyed our common ground. And now we just have to get over it?

"Of course, the colonialists exploited our mineral resources—I tell my pupils how The Congo enriched Europeans and other 'civilized' peoples outside of Africa. But they also grabbed our best land, turning food-producing areas into commercial plantations. Are you surprised that in Zimbabwe Mugabe wanted to chase those white farmers off 'their' land? I'm not, and I was with him all the way. If the colonialists hadn't ruined traditional agriculture, there wouldn't be any food shortages today.

"Look at my country. The war between the Hutu and the Tutsi that started during colonial times has now spread into The Congo. We've got so used to taking sides in 'tribal' conflicts that much of the east is out of control. And who are the losers? The poorest of the poor, who get robbed and killed by the Mai, the Interahamwe, and the Tutsis and their friends. What are the ex-colonial powers doing now to help us with the mess they left behind? Nothing! So don't complain to me about poor African leadership and our failure to carry out economic reforms dictated by some foreign bank. The poisonous legacy of colonialism will hold Africa back for generations to come."

COLONIALISM IS A SCAPEGOAT!

"We Africans have many factors to blame for our troubles, but colonialism isn't one of them. Many African countries have been independent for nearly two generations; most Africans were born long after the colonial era ended. Are we better off today than we were at the end of the colonial period? Are those military dictatorships that ran so many countries into the ground any better than the colonizers who preceded them? Are the ghosts of colonialists past rigging our elections, stifling the media? Is our endemic corruption the fault of lingering colonialism? Did the colonial governors set up Swiss bank accounts to which to divert development funds? Can extinct colonial regimes be blamed for the environmental problems we face? Did the bandits of interior Sierra Leone or eastern areas of The Congo learn their mutilation practices from colonialism? As a nurse in a Uganda hospital for AIDS victims, I say that it's time to stop blaming ancient history for our current failures.

"It's not that we Africans aren't at a disadvantage in this unforgiving world. Our postcolonial population explosion was followed by an HIV-induced implosion. Apartheid South Africa spread its malign influence far beyond its borders. The Cold War pitted regimes against rebels and, to use a colonial term, tribe against tribe. Droughts and other plagues of biblical proportions ravaged our continent when we were more vulnerable than ever. Prices for our raw materials on international markets fell when we most needed a boost. Foreign governments protected their farmers against competition from African producers. Those are legitimate complaints. But colonialism?

"Forget about blaming boundaries, white farmers, tribalism, religion, and other residues of colonialism for our present troubles. Look at what happened here in Uganda. A decade ago we were at the heart of the 'AIDS Belt' in tropical Africa, with one of the worst infection rates on the continent. Then came President Museveni, whose government launched a vigorous self-help campaign, educating people to the risks, urging restraint, distributing condoms, encouraging women to resist unwanted sex. Now Uganda's AIDS incidence is dropping, and we are held up as an example for all of Africa to follow. While South Africa's [former] President Mbeki revealed his confusion about the causes of AIDS, Zimbabwe's Mugabe was too busy chasing white farmers, and Kenya's [ex-president Moi tried] to rig his succession, we in Uganda have leadership in an area where we need it desperately. That certainly beats blaming colonialism for everything.

"We have an African Union. If those colonial boundaries are so bad, let's change them, or do what the European Union is doing and begin to erase them. But first, let's isolate tyrants, nurture open democracy, and combat corruption—and stop blaming colonialism."

Vote your opinion at www.wiley.com/college/deblij

Africa. Development policies often pay too little attention to this situation and its related household dynamics, thereby dooming many well-intended projects.

The **Green Revolution [8]**—the development of more productive, drought-tolerant, pest-resistant, higher-yielding types of grain—has had less impact in Africa than elsewhere (see box titled "A Green Revolution for Africa?"). Where people depend mainly on rice and wheat, the Green Revolution pushed back the prospect of increased hunger. But little research was done on this realm's dominant crops, mostly tubers, so the advances of the Green Revolution were minimal for Africa. And, in any case, the Green Revolution is hardly an unqualified remedy: the poorest farmers, who need help the most, can least afford the more expensive, higher-yielding seeds or the pesticides that are so often required.

Neocolonial Land Grabs?

Low agricultural productivity due to inefficiency and governments' need for revenues have converged in some African countries, and resulted in the selling of huge swaths of public land to the big investors of major agro-industrial concerns. The idea is that if you don't have the means yourself to develop the land, then large-scale commercial farming operations are the best option to increase yields and simultaneously fill government coffers. In 2009 alone, 44 million hectares (110 million acres) were

The cellphone revolution has made a dramatic impact on farmers in many parts of the developing world, especially in Subsaharan Africa where distances can be far, land lines absent, and market information scarce. Now farmers and market women, such as the Kenyan woman pictured here, can be in touch to better gauge when it is best to market crops.

© AP/Wide World Photos

A Green Revolution for Africa?

THE TERM *Green Revolution* refers to the development of higher-yielding types of grains since the 1970s. Worldwide, it has narrowed the gap between population and food production, and it was especially influential in regions where people depend mainly on rice and wheat for their staples. The Green Revolution has had less impact in Subsaharan Africa, however. In part, this relates to the realm's high rate of population growth (which is substantially higher than that of India or China). Other reasons have to do with Africa's staples: rice and wheat support only a small part of the realm's population. Corn (maize) supports many more, along with sorghum, millet, and other grains. In moister areas, root crops, such as the yam and cassava, as well as the plantain (similar to the banana) supply most calories. These crops were not priorities in Green Revolution research.

Lately, there have been a few signs of improvement. Even as terrible famines struck several parts of the continent and people went hungry in places as widely scattered as Moçambique and Mali, scientists worked toward two goals: first, to develop strains of corn and other crops that would be more resistant to Africa's virulent crop diseases, and second, to increase the productivity of those strains. But these efforts faced serious problems. An average African

hectare (2.5 acres) planted with corn, for example, yields only about half a ton of corn, whereas the world average is 1.3 tons. When a virus-resistant variety was developed and distributed throughout Subsaharan Africa in the 1980s, yields rose significantly where farmers could afford to buy the new strain as well as the fertilizers it required. Nigeria saw a near-doubling of its production before it leveled off, but population growth negated the advantage. Hardier types of root crops also raised yields in some areas.

But Africa needs more than this to reverse the trend toward food deficiency. It has been estimated that, for the realm as a whole, food production in recent years has fallen about 1 percent annually even as population grew by more than 2 percent. Lack of capital, inefficient farming methods, inadequate equipment, soil exhaustion, male dominance, apathy, and devastating droughts contributed to this decline—not to mention affordability problems with more expensive higher-yielding seeds and pesticides. And the realm's seemingly endless series of civil conflicts (The Congo, former Sudan, Somalia, Rwanda, Ivory Coast, Chad) have also reduced farm output. The Green Revolution may narrow the gap between production and need, but the battle for food sufficiency and security in Africa is far from won.

involved in such transactions—an area the size of California and West Virginia combined. Most of this land was in former Sudan, Ethiopia, and Moçambique. Some of the investors were domestic, but many others were based abroad (always a contentious issue). When the government of Madagascar in 2009 declared its intention to sell half of all arable public lands to a South Korean corporation, protests ensued and soon escalated to the point that the government was forced to resign.

The impact of these mega land deals is mixed. In some cases, output on the large commercial farms far exceeds previous levels of production. But in others, investors seem to have had mainly speculative intentions and much farmland remains idle. In either situation, many small farmers are uprooted and their villages destroyed. Understandably, there is now considerable mistrust among local populations, with the success of these ventures highly dependent on the way they are handled by the government and the assistance provided to those locals most directly affected. Some of the countries involved in such schemes have a miserable record (former Sudan), but others are showing rapid economic growth (Moçambique).

ENVIRONMENT AND HEALTH

The study of human health in spatial context is the field of medical geography [9], and medical geographers employ modern methods of analysis (including geographic information systems) to track disease outbreaks, identify their sources, detect their carriers, and prevent their recurrence. Alliances between medical personnel, epidemiological researchers, and geographers have already yielded significant results. Doctors know how a disease ravages the body; geographers know how climatic conditions such as wind direction or variations in river flow can affect the distribution and effectiveness of disease carriers. This collaboration helps protect vulnerable populations.

Tropical Africa is the source area of many serious illnesses and has thereby become the focus of much research in medical geography. Investigators look at the carriers (*vectors*) of infectious diseases, the environmental conditions that give rise to them, and also the cultural and social geography of disease dispersion and transmission. Comparing medical, environmental, and social/cultural maps can integrate crucial evidence to help combat the scourge.

In Africa today, hundreds of millions of people carry one or more maladies, often without knowing exactly what ails them. A disease that infects many people (the *hosts*) in a kind of equilibrium, without causing rapid and widespread deaths, is said to be endemic [10] to that population. People affected may not die suddenly or dramatically, but their quality of life and productive capacity are hindered as their overall health is weakened and can deteriorate rapidly when an acute illness strikes. In tropical Africa, hepatitis, venereal diseases, and hookworm are among many public-health threats in this category.

Epidemics and Pandemics

When a disease outbreak has local or regional dimensions, it is called an epidemic [11]. It may claim thousands, even tens of thousands, of lives, but it remains confined to a certain area, perhaps one defined by the range of its vector. In tropical Africa, trypanosomiasis, the disease known as sleeping sickness and vectored by the tsetse fly, has regional dimensions (Figure 6A-9). The extensive herds of savanna wildlife form the *reservoir* of this disease, and the tsetse fly transmits it to livestock and people. It is endemic to wildlife, but it also kills cattle, so Africa's herders try to keep their animals in tsetse-free zones. African sleeping sickness appears to have originated in a West African source area during the fifteenth century, and from there it spread throughout much of tropical Africa. Its epidemic range was limited by that of the tsetse fly: where there are no tsetse flies, there is no sleeping sickness. More than anything else, the tsetse fly has kept substantial parts of Subsaharan Africa's savannas free of livestock and open to wildlife.

When a disease spreads worldwide, it is described as a pandemic [12]. Africa's and the world's most deadly vectored disease is malaria, transmitted by a mosquito and killer of as many as 1 million people each year. Malaria is an ancient affliction. Hippocrates, the Greek physician of the fifth century BC, referred to it in his writings. Apes, monkeys, and several other species also suffer from it. Fever attacks, anemia, and enlargement of the spleen are its key symptoms. Malaria has dispersed around the world and prevails not only in tropical but also in temperate areas.

Eradication campaigns against the mosquito vector have had some success, but always the carrier has come back with renewed vigor. At present, as many as 250 million

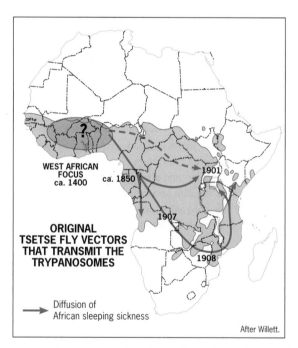

FIGURE 6A-9 © H. J. de Blij, P. O. Muller, and John Wiley & Sons, Inc.

people are affected by malaria globally. Although current efforts at fighting the disease through increased use of mosquito nets appears to be having good results, most of the million annual deaths are African children under the age of five. Indeed, the short life expectancies for the realm's tropical countries listed in Appendix B partly reflect this mortality from malarial infection.

The AIDS Scourge

Malaria remains Africa's most deadly disease to this day, but during the past three decades AIDS has dominated the medical news from this part of the world. AIDS first erupted in Subsaharan Africa and quickly became a global pandemic. No geographic realm has been spared.

AIDS stands for Acquired Immune Deficiency Syndrome, the body's failure to protect itself against a virus. That virus, for want of a better name at the time researchers were trying to identify it, is called the Human Immunodeficiency Virus—HIV. Thus the disease is properly called HIV/AIDS. Since the disease was first recorded in the early 1980s, more than 30 million people worldwide have died of it, between 75 and 80 percent of them Africans.

By the early 1990s, HIV/AIDS had spread most virulently in Equatorial and East Africa, and medical geographers referred to an "AIDS Belt" from The Congo* to Kenya. A decade later, however, the worst-afflicted countries lay in Southern Africa. In 2012, the largest HIV-infected national populations are in South Africa (nearly 6 million) and Nigeria (ca. 2.6 million); ranked next are Moçambique, Tanzania, and Zimbabwe, with each recording about 1.5 million victims. In certain other countries, such as The Congo, we know the outbreak is severe but no statistics are available. When standardized for population size, the dire situations of a number of countries become clearer: around 25 percent of the population aged 15–49 are HIV-positive in Lesotho, Swaziland, and Botswana; 18 percent in South Africa; and approximately 14 percent in Zimbabawe, Zambia, and Namibia.

More than 60 percent of all those infected are women, reflecting cultural and social circumstances. Overall, no part of tropical Africa was spared. Life expectancies plummeted. Children by the millions were orphaned. Companies lost workers and could not replace them. National economies shrank. Associated costs—benefits, treatment, medicines—skyrocketed.

Why is Africa suffering so disproportionately? First, HIV/AIDS originated in tropical African forest margins and quickly spread through all segments of society. Second, the social stigma associated with HIV/AIDS, which is sexually transmitted, makes acknowledging and treating it

especially problematic. Third, life-prolonging medications are expensive and particularly difficult to provide in remote rural areas. Fourth, governmental leadership during the AIDS crisis has varied from highly aggressive and effective (as in Uganda, where political and medical officials cooperated in a massive campaign to distribute free condoms and advocate their use) to catastrophically negligent, as in South Africa where government ministers misled the public and unnecessarily delayed the medical intervention that was critically needed.

The situation remains serious in 2012, but some progress is being made. A new South African government is addressing the AIDS crisis vigorously. Public-health campaigns elsewhere are having a beneficial effect. Lower-cost, generic anti-HIV medicines are becoming more widely available through international help. But the national data in Appendix B remind us that Africa will long suffer from this affliction, only the latest in its long series of death-dealing diseases.

CULTURAL PATTERNS

We tend to think of Africa in terms of its prominent countries and famous cities, its development problems and political dilemmas, but Africans themselves have another perspective. The colonial period created states and capitals, introduced foreign languages to serve as the *linguae francae*, and brought railways and roads. The colonizers stimulated labor movements to the mines they opened, and they disrupted other migrations that had been part of African life for many centuries. But they did not change the ways of life of most of the people. Fully 65 percent of the realm's population still live in, and work near, Africa's hundreds of thousands of villages. They speak one of more than a thousand languages in use in the realm. The villagers' concerns are local; they focus on subsistence, health, and safety. They worry that the conflicts over regional power or political ideology will engulf them, as has happened to millions in Liberia, Sierra Leone, Ethiopia, Rwanda, The Congo, and Angola since the 1970s. Africa's (numerically) largest groups of people are major nations, such as the Yoruba of Nigeria and the Zulu of South Africa. Africa's smallest groups of people number just a few thousand. As a geographic realm, Subsaharan Africa has the most complex cultural mosaic on Earth (see Fig. 6A-5).

African Languages

Africa's linguistic geography is a key component of that cultural intricacy. Most of Subsaharan Africa's more than one thousand languages do not have a written tradition, making classification and mapping difficult. Scholars have attempted to delimit an African language map, and Figure 6A-10 is a composite of their efforts. One feature is common to all language maps of this continent: the geographic realm begins approximately where northern Africa's Afro-Asiatic language

---

*Two countries in Africa have the same short-form name, *Congo*. In this book, we use **The Congo** for the much larger Democratic Republic of the Congo, and **Congo** for the smaller Republic of the Congo just to its northwest.

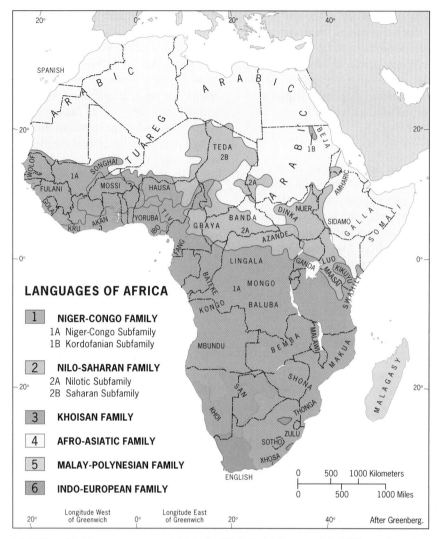

FIGURE 6A-10

© H. J. de Blij, P. O. Muller, and John Wiley & Sons, Inc.

family (mapped above in yellow) gives way, although the pattern is sharper in West Africa than to the east.

In Subsaharan Africa, the dominant language family is the Niger-Congo family, of which the Kordofanian subfamily is a small, historic northeastern outlier (Fig. 6A-10), and the Niger-Congo languages carry the other subfamily's name. This subfamily (mapped in purple) extends across the realm from West to East and Southern Africa. The Bantu language forms the largest branch of this subfamily, but Niger-Congo languages in West Africa, such as Yoruba and Akan, also have millions of speakers. Another important language family is the Nilo-Saharan family (mapped in tan), extending from Maasai in Kenya northwest to Teda in Chad. No other language families are of similar extent or importance: the Khoisan family, of ancient origins, now survives among the dwindling Khoi and San peoples of the Kalahari in the southwest; the small white minority in South Africa speak Indo-European languages; and Malay-Polynesian languages prevail in Madagascar, which was peopled from Southeast Asia before Africans reached it.

The Most Widely Used Languages

About 40 African languages are spoken by 1 million people or more, and a half-dozen by about 10 million or more: Hausa (50 million), Yoruba (23 million), Ibo, Swahili, Lingala, and Zulu. Although English and French have become important *linguae francae* in multilingual countries such as Nigeria and Ivory Coast (where officials even insist on spelling the name of their country *Côte d'Ivoire* in the Francophone way), African languages also serve this purpose. Hausa is a common language across the West African savanna; Swahili is widely used in East Africa. And pidgin languages, mixtures of African and European tongues, are found along West Africa's coast; millions of Pidgin English (called *Wes Kos*) speakers use this medium in Nigeria and Ghana. Not all languages survive. According to UNESCO, Nigeria alone has 15 endangered (near-extinct) languages, and in Africa at large some 46 languages are on the verge of extinction.

Language and Culture

Multilingualism [13] can be a powerful centrifugal force in society, and African governments have tried with varying success to establish national alongside local languages. Nigeria, for instance, made English its official language because none of its 250 languages, not even Hausa, had sufficient internal interregional use. But using a European, colonial language as an official medium invites criticism, and Nigeria remains divided on the issue. On the other hand, making a dominant local language official would invite negative reactions from ethnic minorities. Language remains a potent force in Africa's cultural life.

Religion in Africa

Africans had their own belief systems long before Christians and Muslims arrived to convert them. And for all of Subsaharan Africa's cultural diversity, Africans had a consistent view of their place in nature. Spiritual forces, according to African tradition, are manifest everywhere in the natural environment, not in a supreme deity that exists in some remote place. Thus gods and spirits affect people's daily lives, witnessing every move, rewarding the virtuous and punishing (through injury or crop failure, for example) those who misbehave. Ancestral spirits can inflict misfortune on the living. They are everywhere: in the forest, rivers, and mountains.

As with land tenure, the religious views of Africans clashed fundamentally with those of the colonists. Monotheistic ***Christianity*** first touched Africa in the northeast when Nubia and Axum were converted, and Ethiopia has been a Coptic Christian stronghold since the fourth century AD. But the Christian churches' real invasion did not commence until the onset of colonialism after the turn of the sixteenth century. Christianity's various denominations made inroads in different areas: Roman Catholicism in much of Equatorial Africa, mainly at the behest of the Belgians; the Anglican Church in British colonies; and Presbyterians and others elsewhere. These days, Evangelical churches continue to gain adherents rapidly.

Some of those churches are now far more conservative than their counterparts in Europe and North America, even those appealing to the most conservative congregations in the United States. A split in the U.S. Episcopal (Anglican) Church regarding the treatment of gays and lesbians has some congregations aligning themselves under the Bishop of Uganda. But almost everywhere, Christianity's penetration led to a blending of traditional and Christian beliefs, so that much of Subsaharan Africa is nominally, though not exclusively, Christian (see Fig. G-10). Go to a church in Gabon or Uganda or Zambia, and you may hear drums instead of church bells, sing African music rather than hymns, and see African carvings alongside the usual statuary.

Islam had a different arrival and impact. Today, about 30 percent of the realm's population is Muslim. By the time

The faithful kneel during Friday prayers outside a mosque in Kano, northern Nigeria. The survival of Nigeria as a unified state is an African success story; the Nigerians have overcome strong centrifugal forces in a multiethnic country that is dominantly Muslim in the north, Christian in the south. In the 1990s, some Muslim clerics began calling for an Islamic Republic in Nigeria, and after the death of the dictator Abacha and the election of a non-Muslim president, the Islamic drive intensified. A number of Nigeria's northern States adopted *Sharia* (strict Islamic) law, which led to destructive riots between the majority Muslims and minority Christians who felt threatened by this turn of events (elaborated on pp. 331–332). Can Nigeria avoid the fate of Sudan?

© Pascal Maitre/Cosmos/Aurora Photos, Inc.

of the colonial invasion, Islam had advanced out of Arabia, across the Sahara, and part-way down the coasts of Africa (especially along the Indian Ocean). Muslim clerics converted the rulers of African states and commanded them to convert their subjects. They Islamized the savanna states and penetrated into present-day northern Nigeria, Ghana, and Ivory Coast. They encircled and isolated Ethiopia's Coptic Christians and Islamized the Somali people in Africa's Horn. They established beachheads on the Kenya coast and took over Zanzibar. Arabizing Islam and European Christianity competed for African minds, and Islam proved to be a far more pervasive force. From Senegal to Somalia, the population is overwhelmingly Muslim (see Fig. 6B-9), and Islam's rules dominate everyday life. The Sunni *mullahs* would never allow the kind of marriage between traditional and Christian beliefs seen in much of formerly colonial Africa. This fundamental contradiction between Islamic dogma and Christian accommodation creates a potential for conflict in countries where both religions have adherents.

URBANIZATION AND SOCIAL CHANGE

As can be discerned in Appendix B, Subsaharan Africa remains the least urbanized world realm, but people are now moving to its towns and cities at an increasingly rapid

A typical urban landscape in Subsaharan Africa. This aerial view of the city of Arusha (population: ca. 600,000) in northern Tanzania shows a hallmark feature of the settlement layout of urban Africa: an irregular pattern of low-rise, low-density development. It reflects a general lack of planning, complex land ownership issues, and piecemeal as well as patchwork investment in construction.

© Jan Nijman

time. They live in the most squalid circumstances, in desperate poverty, and governments are unable to assist them. As a result, the squatter-settlement rings around (as well as within) most of Africa's cities are unsafe—miserable, unhealthy slums without adequate shelter, water supply, or basic sanitation. Garbage-strewn (no solid-waste removal here), muddy, and insect-infested during the rainy season, and stifling and smelly during the dry months, they are incubators of disease. Yet few of their residents give up and return to the villages they came from. On the contrary, every new day brings hope.

In our regional discussion in Chapter 6B, we refer to some of Subsaharan Africa's cities, all of which to some degree are stressed by the rate of population influx. Despite the plight of the urban poor and the poverty of Africa's rural areas, some of Africa's capitals remain the strongholds of privileged elites who, dominant in governments, fail to address the needs of the less fortunate and other ethnic groups. Discriminatory policies and artificially low food prices

pace. By the time you read this, the percentage of urban dwellers will have passed 35. This means that more than 300 million people (the size of the U.S. population) now live in towns and cities, of which many were founded and developed by the colonial powers. But the infrastructure of the cities has not kept up with the number of arrivals.

African cities became centers of embryonic national core areas, and of course they served as government headquarters. This **formal sector** [14] of the city used to be the dominant one, with governmental control and regulations affecting civil service, business, industry, and their workers. Today, however, African cities look different. From a distance, the skyline still resembles that of a modern center. But in the streets, on the sidewalks right below the shop windows, there are hawkers, basket weavers, jewelry sellers, garment makers, wood carvers—a second economy, most of it beyond governmental control. This **informal sector** [15] now dominates many African cities. It is peopled by the rural immigrants, who also work as servants, apprentices, construction workers, and in countless other menial jobs.

Millions of urban immigrants, however, cannot find work, at least not for months or even years at a

A recent view over the Jankara Market area on Lagos Island, in the heart of this realm's only megacity, Lagos, Nigeria (11.3 million). In the distance, high-rises stand comparatively isolated (no clustered CBD here); note the minarets of an impressive mosque rising above the cityscape. In the foreground, the street scene is comparatively well-ordered and spacious, in contrast to the city's chaotic ring of shantytowns. Lagos Island, the water-buffered aggregation of office and residential buildings seen here, is relatively modern—but remarkably small for a metropolis with a population as large as that of Greece.

© AFP/Getty Images, Inc.

LAGOS

IN A REALM that is only 35 percent urbanized, Lagos, former capital of federal Nigeria, is the exception: a teeming megacity of 11.3 million, sometimes called the Calcutta of Africa.

Lagos evolved over the past three centuries from a Yoruba fishing village, Portuguese slaving center, and British colonial headquarters into Nigeria's largest city, major port, leading industrial center, and first capital. Situated on the country's southwestern coast, it consists of a group of low-lying islands and sand spits between the swampy shoreline and Lagos Lagoon. The center of the city still lies on Lagos Island, where the high-rises adjoining the Marina overlook Lagos Harbor and, across the water, Apapa Wharf and the Apapa industrial area. The city expanded southeastward onto Ikoyi Island and Victoria Island, but after the 1970s most urban sprawl took place to the north, on the western side of Lagos Lagoon.

Lagos's cityscape is a mixture of modern high-rises, dilapidated residential areas, and squalid slums (see bottom photo at left). From the top of a high-rise one sees a seemingly endless vista of rusting corrugated roofs, the houses built of cement or mud in irregular blocks separated by narrow alleys. On the outskirts lie the shantytowns of the less fortunate, where shelters are made of plywood and cardboard and lack even the most basic facilities.

Traffic in Lagos is as bad as it gets anywhere, with daily commutes of three or more hours rather the rule than the exception. During rush hour (which is almost always) cars move at about 5 kilometers (3 mi) an hour. The only beneficiary of this chaos is the *okada*, the motor-bike taxi that in recent years has taken the city by storm. These motor-bike taxis sometimes carry entire families, weaving swiftly through traffic and violating every rule in the book. The okadas

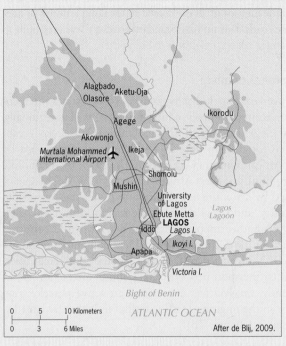

© H. J. de Blij, P. O. Muller, and John Wiley & Sons, Inc.

proliferated so quickly that they themselves became a problem and are now banned from certain roads and locales.

By world standards, Lagos ranks among the most severely polluted, congested, and disorderly cities. Mismanagement and official corruption are endemic. Laws, rules, and regulations, from zoning to traffic, are flouted. The international airport is notorious for its inadequate security and for extortion by immigration and customs officers. In many ways, Lagos is a metropolis out of control.

---

disadvantage farmers and create even greater urban-rural disparities than the colonial period saw. But today the prospect of democracy brings hope that Africa's rural majorities will be heard and heeded in the capitals.

A POSTCOLONIAL REALM

In Subsaharan Africa, more than in any other realm, there is a huge discrepancy between the geography of territorial states (imposed by the former colonial powers) and the geography of national and ethnic identities (look again at Fig. 6A-5). The political map of Subsaharan Africa has 45 states but no nation-states (apart from some microstates and ministates in the islands and in the south). Centrifugal forces are powerful, cultural pluralism prevails, and outside interventions during the Cold War, when communist and anticommunist foreigners took sides in local civil wars, worsened conflict within many African states.

Clearly, the many failings of democracy across the realm cannot be understood without considering this precarious legacy. In 2012, a more or less stable democracy could be found only in South Africa, Namibia, Botswana, Ghana, and the offshore West African states of São Tome-and-Príncipe and Cape Verde. All other countries are ruled by authoritarian regimes or have chronically unstable or flawed political systems. Corruption and incompetence seem rampant, but at a deeper level it is often the ethnic fragmentation and tribal tensions that impede or undermine enlightened governance.

Colonialism's economic legacy was not much better. In Africa, capitals, core areas, port cities, and transport systems were laid out to maximize colonial profit and facilitate the exploitation of minerals and soils; the colonial mosaic inhibited interregional communications except where cooperation enhanced efficiency. Colonial Zambia and Zimbabwe, for instance (then called Northern and Southern Rhodesia, respectively), were landlocked and needed outlets, so rail-roads were built to Portuguese-owned ports. But such routes did little to create intra-African linkages. The modern map reveals the results: in West Africa you can travel from the coast into the interior of all the coastal states along railways or adequate roads. But no high-standard roadway was ever built to link these coastal neighbors to each other.

Supranationalism

To overcome such disadvantages, African states must cooperate internationally, continentwide as well as regionally. The Organization of African Unity (OAU) was established for this purpose in 1963 and was superseded by the African Union (AU) in 2001. In 1975, the Economic Community of West African States (ECOWAS) was established by 15 countries to promote trade, transportation, industry, and social affairs in the region. And in the early 1990s, another important step was taken when 12 countries joined in the Southern African Development Community (SADC), organized to facilitate regional commerce, intercountry transport networks, and political interaction. Most recently, the AU has maintained an active peacekeeping force in the ongoing Darfur conflict in North Sudan (discussed in Chapter 7B).

Beyond Stereotypes: Fast-Growing African Economies

If there is one realm that is subject to stereotypical judgment, it is Subsaharan Africa. All too often, the entire realm is painted as a single, monolithic economic basket case, where dysfunctional government goes hand in hand with economic underperformance, famines, and violence. Such observations are not entirely baseless because, as we have seen, there are abundant problems. But this generalization is off the mark when we acknowledge the remarkable geographic, cultural, and economic diversity exhibited by this realm.

In the following chapter, we highlight the very substantial differences among Subsaharan Africa's regions and countries. For the moment, let us emphasize the enormous economic differences that characterize the realm: according to the International Monetary Fund (IMF), of the top ten countries in the world exhibiting the fastest economic growth between 2001 and 2010, six (!) were located in Subsaharan Africa. Angola ranked first with an average annual GDP growth of 11.1 percent, surpassing even China which came in second. Nigeria, Ethiopia, Chad, Moçambique, and Rwanda ranked, respectively, fourth, fifth, seventh, eighth, and tenth. Moreover, the IMF forecasts that from 2011 to 2015 the number of Subsaharan African countries in the top ten will increase to seven.

Deconstructing the stereotypes even further, consider that Subsaharan Africa as a whole experienced average annual GDP growth rates of 5.7 percent during the 2000s, higher even than those of South America where Brazil leads the way. To be sure, Subsaharan Africa badly needs this kind of development, but in absolute terms the realm's economy has quite a long way to go. Nonetheless, current growth rates are more than double those of the past two decades, while rates of population increase have started to decline. This remarkable economic trend is reason enough for some to speak of Subsaharan Africa as the world's newest emerging market. Perhaps the winds of change are once again gathering speed in this realm, but this time for the better.

POINTS TO PONDER

- Subsaharan Africa remains the most severely AIDS-afflicted realm in the world.

- The realm is home to about four-dozen countries, but the number of languages spoken is estimated to be between 1300 and 2000.

- China's impact on Subsaharan Africa is expanding rapidly through activities ranging from trade and investment to education.

- Of the ten fastest-growing economies in the world during the first decade of this century, six were located in Subsaharan Africa.

REGIONS OF SUBSAHARAN AFRICA

PORTUGAL SPAIN ITALY GREECE TURKEY

15° 15° 45°

0° 30°

Mediterranean Sea

TUNISIA CYPRUS SYRIA

LEBANON IRAQ IR

ISRAEL

MOROCCO JORDAN KUWAIT

30° ALGERIA LIBYA EGYPT SAUDIA BAHRAIN
ARABIA

WESTERN
SAHARA

Red Sea

MAURITANIA NIGER Nile R.

CHAD YEMEN

Niger R. NORTH ERITREA

15° 15° Lake SUDAN DJIBOUTI
SENEGAL MALI Chad
GAMBIA BURKINA
FASO NIGERIA

GUINEA BENIN ETHIOPIA

GUINEA- CENTRAL SOMALIA
BISSAU AFRICAN SOUTH
SIERRA GHANA CAMEROON REPUBLIC SUDAN
LEONE IVORY
LIBERIA COAST TOGO UGANDA

EQUATORIAL KENYA Equator
GUINEA

0° Congo R. 0°
GABON CONGO RWANDA Lake
THE Victoria
CONGO BURUNDI

CABINDA Lake
(Angola) Tanganyika

ATLANTIC TANZANIA INDIA
OCEAN OCEA

ANGOLA MALAWI

15° ZAMBIA 15°
Zambezi R. MOÇAMBIQUE

ZIMBABWE MADAGASCAR

NAMIBIA

BOTSWANA Tropic of Capricorn

SWAZILAND
Vaal R.

30° Orange R. LESOTHO 30°
SOUTH
AFRICA

REGIONS OF
SUBSAHARAN AFRICA

- Southern Africa
- East Africa
- Equatorial Africa
- West Africa
- African Transition Zone
- - - Approximate southern
  boundary of Islam

0 600 1200 Kilometers

0 300 600 Miles

15° 0° 15° 30° Longitude East of Greenwich 45°

▪▪ REGIONS

IN THIS CHAPTER

- Pirates in the Gulf of Adan
- Oil and blood in the Niger Delta
- South Africa after apartheid
- The unforgiving Sahel
- The shadow of Islam
- Madagascar's unique wildlife

CONCEPTS, IDEAS, AND TERMS

FIGURE 6B-1 © H. J. de Blij, P. O. Muller, and John Wiley & Sons, Inc.

© Jan Nijman
Where were these photos taken?
Find out at www.wiley.com/college/deblij

O n the face of it, Africa seems to be so massive, compact, and continuous that any attempt to justify a contemporary regional breakdown is doomed to fail. No deeply penetrating bays or seas create peninsular fragments as they do in Europe. No major islands (other than Madagascar) provide the broad regional contrasts we see in Middle America. Nor does Africa really taper southward to the peninsular proportions of South America. And Africa is not cut by an Andean or a Himalayan mountain barrier. Given Africa's colonial fragmentation and cultural mosaic, is regionalization possible? Indeed it is.

Maps of environmental distributions, ethnic patterns, cultural landscapes, historic culture hearths, and other spatial data yield a four-region structure complicated by a fifth, overlapping zone as shown in Figure 6B-1. Beginning in the south, we identify the following regions:

1. *Southern Africa*, extending from the southern tip of the continent to the northern borders of Angola, Zambia, Malawi, and Moçambique. Ten countries constitute this region, which extends beyond the tropics and whose giant is South Africa.

2. *East Africa*, where equatorial natural environments are moderated by elevation and where plateaus, lakes, and mountains, some carrying permanent snow, define the countryside. Six countries, including the highland part of Ethiopia, comprise this region. The island-state of Madagascar, with Southeast Asian influences, is neither Southern nor East African, but is included here because of its location.

3. *Equatorial Africa*, much of it defined by the basin of the Congo River, where elevations are lower than in East Africa, temperatures are higher and moisture more

© Jan Nijman

City	Population* (in millions)
Lagos, Nigeria	11.3
Kinshasa, The Congo	9.6
Abidjan, Ivory Coast	4.4
Nairobi, Kenya	3.9
Johannesburg, South Africa	3.8
Dar es Salaam, Tanzania	3.8
Cape Town, South Africa	3.5
Adis Abeba, Ethiopia	3.1
Dakar, Senegal	3.0
Ibadan, Nigeria	3.0
Durban, South Africa	2.9
Accra, Ghana	2.5
Harare, Zimbabwe	1.8
Lusaka, Zambia	1.6
Mombasa, Kenya	1.1

*Based on 2012 estimates.

ample, and where most of Africa's surviving rainforests remain. Among the nine countries that make up this region, which includes newly formed South Sudan, The Congo* dominates territorially and demographically.

4. **West Africa**, which includes the countries of the western coast and those on the margins of the Sahara in the interior, a populous region anchored in the southeast by Africa's demographic giant, Nigeria. Fifteen countries form this crucial African region.

5. **The African Transition Zone**, the complicating factor on the regional map of Africa. In Figure 6B-1, note that this striped zone of increasing Islamic influence completely dominates some countries (e.g., Somalia in the east and Senegal in the west) while cutting across others, thereby creating Islamized northern areas and non-Islamic southern zones (e.g., Nigeria, Chad, Ivory Coast) as well as encompassing only a strip of countries such as Ethiopia, Kenya, and Tanzania).

■■ SOUTHERN AFRICA

Southern Africa, as a geographic region, consists of all the countries and territories lying south of Equatorial Africa's The Congo and East Africa's Tanzania (Fig. 6B-2). Thus defined, the region extends from Angola and Moçambique (on the Atlantic and Indian Ocean coasts, respectively) to South Africa and includes a half-dozen landlocked states.

*Two countries in Africa have the same short-form name, Congo. In this book, we use **The Congo** for the much larger Democratic Republic of the Congo, and **Congo** for the smaller Republic of Congo just to its northwest.

Also marking the northern limit of the region are Zambia and Malawi. Zambia is nearly cut in half by a long land extension from The Congo, and Malawi penetrates deeply into Moçambique. The colonial boundary framework, here as elsewhere, produced many liabilities.

Africa's Richest Region

Southern Africa constitutes a geographic region in both physiographic and human terms. Its northern zone marks the southern limit of the Congo Basin in a broad upland that stretches across Angola and into Zambia (the tan corridor extending eastward from the Bihe Plateau in Fig. 6A-2). Lake Malawi is the southernmost of the East African rift-valley lakes; Southern Africa has none of East Africa's volcanic and earthquake activity. Most of the region is plateau country, and the Great Escarpment is much in evidence here. There are two pivotal river systems: the Zambezi (which forms the border between Zambia and Zimbabwe) and the Orange-Vaal (South African rivers that combine to demarcate southern Namibia from South Africa).

Southern Africa is the continent's richest region materially. A great zone of mineral deposits extends through the heart of the region from Zambia's Copperbelt through Zimbabwe's Great Dyke and South Africa's Bushveld Basin and Witwatersrand to the goldfields and diamond mines of Free State and Northern Cape provinces in the heart of South Africa. Ever since these minerals began to be exploited in colonial times, many migrant laborers have come to work in the mines. Southern Africa's agricultural diversity matches its mineral wealth. Vineyards drape the slopes of South Africa's Cape Ranges; tea plantations hug the eastern escarpment slopes of Zimbabwe. Before civil war destroyed its economy, Angola was one of the world's leading coffee producers. South Africa's relatively high latitudes and its range of altitudes create environments for apple orchards, citrus groves, banana plantations, pineapple farms, and many other crops.

Despite this considerable wealth and potential, not all of the ten countries of Southern Africa have prospered. As Figure G-11 shows, four of them remain mired in the low-income category (Malawi, Moçambique, Zambia, and Zimbabwe); even though South Africa, Angola, Lesotho, Swaziland, Namibia, and Botswana fall into the two middle-income ranks, the (desert-dominated) last two rank among the realm's three most sparsely populated countries. A period of rapid population growth, followed by the devastating onslaught of AIDS, civil conflict, political instability, incompetent government, widespread corruption, unfair competition on foreign markets, and environmental problems, have combined to constrain regional development.

Still, with the tragic exception of Zimbabwe (the realm's basket case in 2012), Southern Africa as a region is better off than any other in Subsaharan Africa: as Figure G-11 shows, six countries here have risen above the lowest-income rank among the world's states. Some international

FIGURE 6B-2

© H. J. de Blij, P. O. Muller, and John Wiley & Sons, Inc.

cooperation including a regional association and a customs union are emerging. And South Africa, the realm's most important country by many measures, provides hope for a better future.

South Africa

The Republic of South Africa (RSA) is the giant of Southern Africa, its economy by far the largest of this entire realm. South Africa is an African country that attracts

considerable world attention, a bright ray of hope not only for Africa but for all humankind.

For 40 years, between about 1950 and 1990, multicultural South Africa was in the grip of the world's most notorious racist policy, apartheid [1], the word meaning "apartness" but the practice involving strict racial segregation and severe discrimination. Out of the concept of apartheid grew a notion, promulgated by the white (European) minority then in control of the state, called *separate development*. This would apply apartheid to the entire country, carving it up into racially based entities whose inhabitants would be citizens of those ethnic domains, but not of South Africa as a whole. Predictably, such racist social engineering aroused strong opposition within South Africa and beyond, leading to general condemnation and international sanctions.

South Africa seemed headed for a violent revolution, but disaster was averted by what was, at the time, an almost inconceivable turn of events. A leader of the white-minority government that for decades had ruthlessly pursued its apartheid policies, and a revered leader of the multicultural majority who had for 28 years languished in an island prison not far from Cape Town, struck an accord that, in effect, created a new South Africa virtually overnight. Nelson Mandela walked out of prison a free man in February 1990, and in the RSA's first democratic election of 1994 Mandela became the country's president on a platform representing the African National Congress (ANC), the multiracial anti-apartheid movement that had fought the policy for many years. The ANC's old foes, the architects of apartheid, were seated as part of the loyal opposition in a parliament that was now indeed a rainbow assembly.

South Africa stretches from the warm subtropics in the north to Antarctic-chilled waters in the south. With a land area in excess of 1.2 million square kilometers (470,000 sq mi) and a heterogeneous population of 50.8 million, South Africa contains the bulk of the region's minerals, most of its good farmlands, its largest cities, best ports, most productive factories, and best developed transport networks. Mineral exports from Zambia and Zimbabwe move through South African ports. Workers from as far away as Malawi and as close as Lesotho work in South Africa's mines, factories, and fields.

People and Places in South Africa

South Africa's historical geography differs somewhat from much of the rest of Subsaharan Africa. Its lands were fought over by various African nations before the Europeans arrived and the colonial "scramble for Africa" took place. Peoples migrated southward—first the Khoisan-speakers and then the Bantu peoples—into the South African *cul-de-sac*. The Zulu and Xhosa nations fought over lands at about the time of the first European arrivals. Europeans arrived via the oceans and claimed the southernmost Cape. It is one of the most strategic places on Earth, the gateway from the Atlantic to the Indian Ocean, a source of provisions on the route to Asia's riches. The Dutch East India Company founded Cape Town as early as 1652, and the Hollanders and their descendants, known

South Africa's Cape Town, whose site is dominated by Table Mountain, was founded by the Dutch in 1652 as a waystation for empire building in the East Indies.

© H. J. deBlij

as the **Boers**, have been a part of the South African cultural mosaic ever since. The British took over about 150 years later, and both colonial powers vied for control throughout South Africa's early history as a state. The British came to dominate the Cape, while the Boers trekked into the South African interior and, on the high plateau they called the *highveld*, founded their own republics. By 1910, the Boers and the British had negotiated a power-sharing arrangement, although the Boers eventually achieved hegemony between 1948 and 1994. Having long since shed their European links, they came to call themselves **Afrikaners**, their word for Africans.

The Ethnic Mosaic

In addition to the various native African nations and the Europeans who settled in South Africa are peoples from Asia. The Hollanders brought Southeast Asians to the Cape to serve as domestics and laborers. The British imported laborers from their South Asian colonies to work on sugar-cane plantations, adding cultural diversity to this multiethnic state. Moreover, a substantial population of mixed ancestry, clustered at the Cape, comprises today's so-called *Coloured* sector of the country's citizenry. In the process, South Africa became Africa's most pluralistic and heterogeneous society. Even so, Africans now outnumber non-native Africans by about 4 to 1 (see Table 6B-1).

Although heterogeneity marks the spatial demography of South Africa, regionalism pervades the human mosaic. The Zulu nation still is largely concentrated in today's Kwazulu-Natal Province (Fig. 6B-3). The Xhosa still cluster in the Eastern Cape, from the city of East London to the

FIGURE 6B-3

© H. J. de Blij, P. O. Muller, and John Wiley & Sons, Inc.

Kwazulu-Natal border and below the Great Escarpment. The Tswana still occupy ancestral lands along the Botswana border. Cape Town remains the core area of the Coloured population; Durban still has the strongest South Asian imprint. Travel through South Africa, and you will recognize the diversity of rural cultural landscapes as they change from Swazi to Ndebele to Venda.

This longstanding regionalism was among the factors that led the Afrikaner government to institute its **separate development** scheme, which was linked to **apartheid**, but it could not hold back the tide of urbanization that intermingled the populations. Millions of workers, job-seekers, and illegal migrants converged on the cities, creating vast

shantytowns on their margins. In the legal African townships such as Johannesburg's Soweto (SOuth WEstern TOwnships; right photo, p. 317) and in the squatter settlements the anti-apartheid movement burgeoned, and the strength of the African National Congress (ANC) movement grew, laying the groundwork for its eventual dominance in South Africa's political arena.

Political Change

Largely through the unmatched statesmanship of Nelson Mandela, South Africa did not undergo a massive revolution with great upheaval. Instead it is today governed by a multiracial party, the erstwhile architects of apartheid are

TABLE 6B-1
Demographic Data for South Africa

Population Groups	2012 Estimated Population (in millions)
African nations	**39.9**
Zulu	12.1
Xhosa	8.9
Tswana	4.1
Sotho (N and S)	4.0
Others (6)	10.8
Whites	**4.8**
Afrikaners	3.1
English-speakers	1.6
Others	0.1
Mixed (Coloureds)	**4.7**
African/White	4.3
Malayan	0.4
South Asian	**1.4**
Hindus	1.0
Muslims	0.4
TOTAL	**50.8**

in parliament as a legitimate opposition, and its economy is the largest and healthiest in the realm. Check Appendix B and you will see that, on the African mainland, South Africa's per capita GNI is in a class by itself. By some calculations, South Africa produces 45 percent of all of Subsaharan Africa's GDP.

Together with political change came a change in the administrative structure of the country; in fact, this change occurred before the historic elections of 1994. Before 1994 South Africa was divided into four provinces: the Cape, by far the largest and centered on the legislative capital of Cape Town; Natal, anchored by the port city of Durban; the Transvaal (meaning across the Vaal River), focused on the administrative capital of Pretoria and including the mighty Johannesburg-centered conurbation; and the Orange Free State, the old Boer stronghold with Bloemfontein as its headquarters (see inset map, Fig. 6B-3).

This structure was replaced by nine provinces (Fig. 6B-3), creating a federal arrangement in which each province has its own administration while being represented in the central government. The boundaries of Natal and the Orange Free State remained essentially unchanged, but Natal's name was changed to Kwazulu-Natal, recognizing the Zulu presence there, and the Orange Free State became simply the Free State. But the Cape Province was divided into four new provinces, one of them overlapping into the Transvaal, and the rest of the Transvaal was divided into three, one of which is Gauteng that contains the Johannesburg-Witwatersrand-Pretoria megalopolis.

A New Era Dawns

In 1999, ANC leader Thabo Mbeki, who had served as President Mandela's deputy, became the country's second popularly elected president. In other African states, the succession from heroic founder-of-the-nation to political inheritor-of-the-presidency often has not gone well, but South Africa's new constitution proved its worth. The new president faced several disadvantages: unavoidable comparisons to the incomparable Mandela; the rising tide of AIDS, which Mbeki controversially attributed to causes other than HIV; his awkward stance toward the chaos in Zimbabwe. Nevertheless, Mbeki's economic policies, social programs, and foreign initiatives (The Congo, Rwanda) have served his country well. In 2007, Jacob Zuma succeeded Mbeki as leader of the ANC and, following a brief constitutional crisis, Zuma was elected president in 2009—the RSA's first leader of Zulu ancestry.

How the Economy Evolved: Diamonds and Gold

Ever since diamonds were discovered at Kimberley in the 1860s, South Africa has been synonymous with minerals. The Kimberley finds, made in a remote corner of the Orange Free State, set into motion a new economic geography. Rail lines were laid from the coast to the diamond complex even as fortune seekers, capitalists, and tens of thousands of African workers, many from as far afield as Moçambique, streamed to the site. One of the capitalists was Cecil Rhodes of Rhodes Scholarship fame, who used his fortune to help Britain dominate Southern Africa.

Just 25 years after the diamond finds, prospectors discovered what was long to be the world's greatest goldfield on a ridge called the Witwatersrand (Fig. 6B-3). There, Johannesburg soon became the gold capital of the world, and a new and even larger stream of foreigners arrived, along with a huge influx of African workers. Cheap labor enlarged the profits. Johannesburg grew explosively, satellite towns developed, and black townships mushroomed.

During the twentieth century, South Africa proved to be richer than had been foreseen. Additional goldfields were discovered in the Orange Free State. Coal and iron ore were found in abundance, which gave rise to a major iron and steel industry. Other metallic minerals, including chromium and platinum, yielded large revenues on world markets. Asbestos, manganese, copper, nickel, antimony, and tin were mined and sold; a thriving metallurgical industry developed in South Africa itself. Capital flowed into the country, white immigration expanded, farms and ranches were laid out, and overseas markets multiplied.

Infrastructural Gains During Apartheid

South Africa's cities grew enormously. Johannesburg was no longer just a mining complex: it became an industrial agglomeration and a financial center as well. The old Boer capital, Pretoria, just a short distance north of the Witwatersrand, became the country's administrative center during apartheid. In the Orange Free State, major industrial growth (including oil-from-coal technology) matched

JOHANNESBURG

SUBSAHARAN AFRICA IS urbanizing swiftly, but it still has only one true conurbation, and South Africa's Johannesburg lies at the heart of it. Little more than a century ago, Johannesburg was a small (though rapidly growing) mining town based on the newly discovered gold reserves of the Witwatersrand.

Today Johannesburg forms the focus of a conurbation of more than 7 million, extending from Pretoria in the north to Vereeniging in the south, and from Springs in the east to Krugersdorp in the west. The population of metropolitan Johannesburg itself passed the 3 million mark in 2003 (it now totals 3.8 million), thereby overtaking Cape Town to become South Africa's largest urban agglomeration.

Johannesburg's skyline is the most impressive in all of Africa, a forest of skyscrapers reflecting the wealth generated here over the past hundred years (left photo, p. 317). Look southward from a high vantage point, and you see the huge mounds of yellowish-white slag from the mines of the "Rand," the so-called mine dumps, partly overgrown today, interspersed with suburbs and townships. In a general way, Johannesburg developed as a white city in the north and a black city in the south. Soweto, the black township, lies to the southwest; Houghton and other spacious, upper-class suburbs, formerly exclusively white residential areas, lie to the north.

Johannesburg has neither the scenery of Cape Town nor the climate and beaches of Durban. The city lies at an elevation of 1750 meters above sea level (5750 ft—more than a mile high), and its thin air often is polluted from smog created by motor vehicles, factories, mine-dump dust, and countless cooking fires in the townships and shantytowns that ring the metropolis.

Over the past century, the Johannesburg area produced nearly one-half of all the world's gold by value. But

© H. J. de Blij, P. O. Muller, and John Wiley & Sons, Inc.

today Johannesburg lies at the heart of an industrial, commercial, and financial complex whose name on the new map of South Africa is *Gauteng.*

the expansion of mining. While the core area developed megalopolitan characteristics, most coastal cities also expanded. Durban's port served not only the Witwatersrand but a wider regional hinterland as well. Cape Town was becoming South Africa's second-largest city; its port, industries, and productive agricultural hinterland gave it primacy over a wide area.

The labor force for all this development, from mines to railroads and from farms to highways, came from the African peoples of South Africa (and, indeed, from beyond its borders as well). Even during apartheid, workers from Moçambique, Zimbabwe, Botswana, and the other neighboring countries sought jobs in South Africa while millions of South African villagers also left their homes for the towns and cities. In the process, they built the best infrastructure of any African country. But apartheid ruined South Africa's prospects. The cost of the separate development program was astronomical. Social

unrest during the decade preceding the end of apartheid created a vast educational gap among youngsters. International sanctions against the race-obsessed regime damaged the economy.

The Economy Today
In many respects South Africa is the most important country in Subsaharan Africa, and the entire realm's fortunes are bound up with it. No African country attracts more foreign investment or foreign workers. None has the universities, hospitals, and research facilities. No other has the military forces capable of intervening in African trouble spots. Few have the free press, effective trade unions, independent courts, or financial institutions to match South Africa's. And with a population that has now surpassed 50 million (79 percent black, just under 10 percent white, 9 percent Coloured, 2.8 percent Asian), South Africa has a large, multiracial, and growing middle class.

▲ The leafy suburbs and high-rise CBD of Johannesburg and the shantytown housing of a still poverty-stricken section of crowded Soweto, one of its satellite towns, show the jarring social contrasts in South Africa's ever-more-unequal society. Future political stability will depend on the government's ability to improve living conditions in the poorest areas, rural as well as urban. Although significant advancement has been made in housing, water supply, and electricity hookups, the enormity of the problem, and constraints ranging from political realities to available resources, slow the pace of progress in a country that is in a race against time.

(Left): © Martin Harvey/Gallo Images/Getty Images, Inc. (Right): © Sergio Pitamitz/Robert Harding World Imagery/©Corbis

Despite its successes, South Africa continues to face long-range problems. The economy continues to depend far too heavily on the export of minerals and metals (diamonds, platinum, gold, iron, and steel) at a time when this dependence entails risks at home and abroad. At home, those union-friendly labor laws, including rising wages, are making mining less profitable. Abroad, commodity prices are unreliable. In 1970, South Africa produced nearly 70 percent of the world's gold; today it produces less than 15 percent. And the manufacturing sector of the economy remains too weak.

Even as the black middle class expands, unemployment among blacks may be as high as 50 percent—and the gap between rich and poor is growing. Consider these figures: in 2008, of all people earning under U.S. $7250 per year, 83 percent were black and 7 percent were white; and among those earning more than $108,000 per year, 16 percent were black and 77 percent were white. In addition to this deep and persistent racial inequality, conflicts between locals and immigrants are on the rise. Land reform, an urgent matter in a country where land alienation reached huge proportions, is too slow in the view of many. Add to this the scourge of AIDS and the government's failure to address this crisis effectively, and South Africa's luster is overshadowed by too many serious challenges.

The Middle Tier

Between South Africa's northern border and the region's northern limit lie two groups of states: those with borders adjoining South Africa and those beyond. Five countries form the *Middle Tier*, and all border on South Africa: Zimbabwe, Namibia, Botswana, and the ministates of Lesotho and Swaziland (Fig. 6B-2). As the map shows, four of these five are landlocked. Diamond-exporting (and upper-middle-income) **Botswana** occupies the heart of the Kalahari Desert and surrounding steppe; despite its lucrative diamonds, the majority of its 1.9 million inhabitants are subsistence farmers. In 2011, Botswana remained the most severely AIDS-afflicted country in all of tropical Africa. **Lesotho** and **Swaziland**, both traditional kingdoms, depend very heavily on remittances from their workers in South African mines, fields, and factories.

Southern Africa's youngest independent state, **Namibia** (2.3 million), is a former German colony with a territorial peculiarity: the so-called Caprivi Strip linking it to the right bank of the Zambezi River (Fig. 6B-2), another consequence of colonial partitioning. Administered by South Africa from 1919 to 1990, Namibia is named after one of the world's driest deserts (the Namib, which lines its coast). This state is about as large as Texas and Oklahoma, but only its far north receives enough moisture to facilitate subsistence farming, which is why most of the people live close to the Angolan border. Mining in the Tsumeb area and ranching in the vast steppe country of the south form the leading commercial activities. The capital, Windhoek, is centrally situated opposite Walvis Bay, the main port. German influence still lingers in what used to be called South West Africa, as does an Afrikaner presence from the apartheid period. Although orderly land reform is underway, unresolved issues remain, and much of the population still lives in poverty.

The Tragedy of Zimbabwe

Botswana's other neighbor, Zimbabwe, may be said to lie at the heart of Southern Africa, between the Zambezi River in the north and the Limpopo River in the south, between the escarpment to the east and the desert to the west. Landlocked but endowed with good farmlands, cool uplands, a wide range of mineral resources, and varied natural environments, Zimbabwe (named after historic stone ruins in its interior—see photos, p. 288) had one of Southern Africa's most vibrant economies and seemed to have a bright future. Zimbabwe's core area is defined by the mineral-rich Great Dyke that extends across the heart of the country from the vicinity of its capital, Harare, in the north to the second city, Bulawayo, in the south. Gold, copper, asbestos, chromium, and platinum are among its major exports in normal times. But Zimbabwe is not just an ore-exporting country. Its farms produce tobacco, tea, sugar, cotton, and even cut flowers in addition to staples for the local market.

But these are not normal times for the two nations that constitute most of Zimbabwe's population, the Shona (82 percent) and the Ndebele (14 percent). Following their successful joint campaign to end white-minority rule, the Shona then turned on the Ndebele in a bitter ethnic conflict that established Shona dominance. A leader named Robert Mugabe rose to prominence during this early period, and it seemed that relatively prosperous Zimbabwe would put its conflicts behind it and take its place among Africa's success stories.

During the colonial period, the tiny minority of whites that controlled what was then called Southern Rhodesia (after Cecil Rhodes, the British capitalist of diamond fame and scholarship honors), took the most productive farmlands and organized the agricultural economy. After independence their descendants continued to hold huge estates, only parts of which were being farmed. What was needed was a comprehensive land reform program, but in the 1980s now President Mugabe began to encourage squatters to invade these white farms and a number of owners were killed. The agricultural economy began to collapse as some white farmers held out whereas others abandoned their lands. Corruption in the Mugabe government involved the transfer of land, not to needy squatters but to friends of top officials. Foreign investment in other Zimbabwean enterprises, including the mining industry, started to dry up. By the mid-1990s Zimbabwe was in an economic free fall.

Many Africans within and outside Zimbabwe had little sympathy for the white farmers or corporate powers controlling the means of production, but Mugabe had no compassion for his own people either. He now turned on the informal sector of the economy, which was all most Zimbabweans had left when jobs on farms and in factories and mines disappeared. Shantytown dwellings that were once confined to the outskirts of the cities now appeared in city centers, as did shacks where women repaired garments and men fixed bicycles. Mugabe ordered his henchmen to destroy the dwellings and shacks of some 700,000 of these "informal" urban residents, leaving them in the streets without shelter or livelihood.

As conditions worsened, people by the hundreds of thousands streamed out of Zimbabwe; by 2011, an estimated four-plus million of the country's 12.9 million people were refugees. About 80 percent of the people were jobless. A whole generation of children suffered from malnutrition. But still the Saddam Hussein of Africa, as Mugabe came to be called, would not yield what had become his dictatorial powers. In late 2008 a major outbreak of cholera killed thousands and sent another wave of desperate Zimbabweans across the border, but Mugabe did not express any sympathies; none of the victims was from his elite inner circle. A rapacious ruler had converted an African breadbasket into a basket case.

This terrible tragedy was compounded by the failure of Zimbabwe's key neighbors to intervene. Former president Thabo Mbeki of South Africa proved a weak leader in a time of need; it fell to his predecessor, Nelson Mandela, to express the sentiments of outrage felt by many. Some international food assistance reached desperate Zimbabweans, but the

Although representative government has come to a majority of Subsaharan African states, some of the realm's countries continue to be plagued by dictatorial rule. Zimbabwe is often cited as the worst of the latter: its economy ruined, its people terrorized (perhaps one-quarter of the population lives in forced exile), and its social fabric in tatters. It took great courage to do what these demonstrators did in 2010, calling for free and fair elections and reforms, and supporting a political party known as the Movement for Democratic Change (MDC). This party, headed by Prime Minister Morgan Tsvangirai, opposes the dictatorial ways of President Robert Mugabe. Its members have been subjected to violent repression by the president's clique and no amount of internal negotiation or external mediation has altered Zimbabwe's nightmarish political landscape. All of the Southern Africa region suffers the consequences.

© Philimon/Bulawayo/Reuters/Corbis

international community did not prove a deterrent to Mugabe. In 2008, he was defeated in legislative elections by Morgan Tsvangirai, but through intimidation and violence Mugabe succeeded in preventing a legitimate runoff. Eventually, Tsvangirai agreed to a power-sharing arrangement, but Zimbawe still does not have the representative, competent government it requires. Meanwhile, the people suffer.

The human catastrophe of Zimbabwe stands in sharp contrast to the social achievement of South Africa. The question is whether South Africa's example will ultimately prevail in Zimbabwe, or whether Zimbabwe's affliction will eventually infect its southern neighbor.

The Northern Tier

In the four countries that extend across the *Northern Tier* of Southern Africa—Angola, Zambia, Malawi, and Moçambique—problems abound. Angola (20 million), formerly a Portuguese colony, together with its exclave [2] (outlier) of Cabinda had a thriving economy based on a wide range of mineral and agricultural exports at the time of independence in 1975. But the Portuguese had done a miserable job of preparing the country for self-rule, and soon Angola fell victim to the Cold War, with northern peoples choosing a communist course and southerners falling under the sway of a rebel movement backed by South Africa and the United States. The results included a devastated infrastructure, idle farms, looting of diamonds, hundreds of thousands of casualties, and millions of landmines that continue to kill and maim.

But over the past decade, Angola's economy turned around again, spectacularly so, and it became the fastest growing national economy in the world! Almost all of this growth comes from oil, and the country now vies for first place with Nigeria in Subsaharan Africa in oil production. Yielding about U.S. $5 billion per year, the return of stability is attracting investors to begin rebuilding the ruined country. But the dependence on oil is too great: in 2010, oil accounted for 85 percent of GDP, 95 percent of exports, and 85 percent of government revenues. Yet offshore petroleum exploration is virtually disconnected from the country's other economic sectors, and to make matters worse oil-rich Angola now confronts a secessionist movement in the exclave of Cabinda.

On the opposite coast, the other major former Portuguese dependency, Moçambique (24.5 million), went through a comparable trajectory. Upon independence Moçambique also chose a Marxist course with unfortunate economic and political consequences. Here, too, a rebel movement supported by South Africa caused civil conflict, created famines, and generated a stream of more than a million refugees toward Malawi. Railroad and port facilities lay idle, and Moçambique at one time was ranked by the United Nations as the world's poorest country.

Moçambique has neither oil nor as much agricultural land as Angola. But it does have considerable bauxite (aluminum ore) deposits, and it has a major advantage in terms of relative location. In colonial times its two major ports, Maputo and Beira, handled large volumes of exports and imports for South Africa, Zimbabwe, and Zambia. In recent years, the port traffic has been revived and the country is working with South Africa on a joint Maputo Development Corridor (Fig. 6B-2). In 2011, construction began on a new bridge over the Zambezi River at the interior city of Tete, which will increase road transport from the neighboring landlocked states.

Moçambique's economy has grown quickly over the past decade, but here again we have an African economy that overly depends on a single resource, in this case the bauxite that accounts for one-third of all exports. And, like Angola, the country has done very poorly in distributing the benefits of growth. Late 2010 saw riots on the streets of Maputo to protest against higher food prices. Among Moçambique's masses, poverty still reigns.

Landlocked Zambia (14 million), the product of British colonialism, shares the mineral riches of the Copperbelt with its northern neighbor, The Congo's Katanga Province. Not only have commodity prices on which Zambia depends fluctuated wildly, but Zambia's outlets—Lobito in Angola as well as Beira in Moçambique—and the railways leading there were made inoperative by Cold War conflicts. In recent years, China has taken an interest in Zambia's minerals, and the Chinese are investing in railroad repairs as well as the expansion of mining operations.

Neighboring Malawi (16.3 million) has an almost totally agricultural economic base and is repeatedly challenged by environmental degradation. Malawi's dependence on corn as its food staple, its variable climate, and its severely fragmented land-use pattern combine to create cycles of boom and bust from which the country has yet to escape. But as the 2010s opened, things looked more positive as some bumper harvests were produced and Malawi was able to export its surplus to Zimbabwe.

∷ EAST AFRICA

In only two parts of the world do local people refer to the "Great Lakes" as a geographic landmark: North America and East Africa. A string of elongated lakes, some with colonial-era names, marks the western border of the region we define as East Africa (Fig. 6B-4). Going to school there, you learn these names early: Albert, Edward, Kivu, Tanganyika, Malawi. But in some ways the most important lake of all lies at the heart of this region, not on its border: Lake Victoria. And as the map shows, Lake Victoria is anything but elongated. It is a large body of water lying in a huge basin—but this basin lies high on a plateau.

The equator crosses East Africa and even Lake Victoria itself, but you would hardly think so on the ground here. We tend to associate equatorial environments with rainforests, heat, and moisture, but in East Africa elevation changes everything (as we also note in Chapter 5B for low-latitude, Andean South America). Drought, not incessant rain, is the frequent problem in this region. Nights can be downright cold. Some volcanic peaks even carry snow, most notably Kilimanjaro. Highland East Africa is

FIGURE 6B-4

© H. J. de Blij, P. O. Muller, and John Wiley & Sons, Inc.

as different from lowland Equatorial Africa west of the Great Lakes as Amazonia is from the Andes.

East Africa, as the map shows, extends from the Ethiopian Highlands in the north to Tanzania in the south, and from the Indian Ocean to the eastern margins of the Congo Basin. This is a region of spectacular scenery. Volcanoes, some of them active, rise above the high plains. Canyons carved by rivers rival the one we call Grand; Ethiopia was once called Abyssinia, and not for nothing because steep scarps define *rift valleys* partly filled with water; views from the top are all the more breathtaking when you recall that this is the landscape seen by our earliest ancestors. Herds of wildlife that attract millions of tourists roam vast savannas whose names (Serengeti; Maasai Mara) have become world renowned.

Too many visitors come here to see the wildlife without also learning about East Africa's diversity of cultures. In addition to highland Ethiopia, five other countries form this East African region: Kenya, Tanzania, Uganda, Rwanda, and Burundi. Here the Bantu peoples that make up most of the population met Nilotic peoples from the north, including the Maasai. In the hills of Rwanda and Burundi a stratified society developed in which the minority Tutsi, in their cattle-owning kingdoms, dominated the Hutu peasantry. The Indian Ocean coast was the scene of many historic events: the arrival of Islam from Arabia, the visit of Ming Dynasty Chinese fleets in the 1400s, the quest for power by the Turks, the Arab slave trade, the European colonial competition. Here, too, developed the East African *lingua franca*, Swahili.

Kenya

Kenya (42.2 million) is neither the largest nor the most populous country in East Africa, but over the past half-century it has been the dominant state in the region. Its skyscrapered capital at the heart of its core area, Nairobi, home to 3.9 million, is the region's largest city and hub for many activities; its port, Mombasa, is East Africa's busiest.

After independence, Kenya chose the capitalist path of development, aligning itself with Western interests. Without major known mineral deposits, Kenya depended on exports of coffee and tea as well as other food products, and on a tourist industry based on its magnificent landscapes. Tourism became its largest single earner of foreign exchange, and Kenya prospered, apparently proving the wisdom of its economic planners.

But serious problems soon emerged. Kenya during the 1980s had the highest rate of population increase in the world, and population pressure on farmlands and on the fringes of the wildlife reserves mounted. Poaching became widespread, and tourism declined. During the late 1990s, El Niño-induced rains affected parts of Kenya, causing landslides and washing away large segments of the crucial highway between Nairobi and Mombasa. This disaster was followed by a severe drought lasting several years, bringing famine to the interior. Meanwhile, government corruption siphoned off funds that should have been invested. Democratic principles were violated, and relationships with

Western allies were strained. The AIDS epidemic brought another setback. And then Kenya sustained damage from terrorist attacks in Nairobi and Mombasa, further impacting the tourist industry.

On the surface, Kenya today seems relatively prosperous and exhibits fairly healthy economic statistics. Nairobi's CBD breathes an atmosphere of modernity and worldliness. But the country has failed utterly in building a single nation out of its 41 (!) constituent ethnic groups. Geography, history, and politics have placed the Kikuyu, who account for just over one-sixth of the population, in a position of dominance. But this is contested by several other groups (see Fig. 6B-4). The Luhya, Luo, Kalenjin, and Kamba together constitute just under 50 percent of the population, and on the territorial margins of the country there are peoples such as the Maasai, Turkana, Boran, and Galla.

Kenya is notorious for the corruption of its elected officials. The average Kenyan makes about U.S. $790 per year; a member of the Kenyan parliament earns $145,000. Imagine members of the U.S. Congress getting a salary

Kenya has become one of the world's major exporters of flowers, but not without controversy. Here, workers are picking roses in a huge, 2-hectare (5-acre) greenhouse, part of an industry that currently concentrates around Lake Naivasha in the Eastern Rift Valley. European companies have set up these enterprises, buying large swaths of land and attracting workers far and wide. Locals see wildlife and waters threatened, but the government appreciates the revenues from an industry that, in 2009, accounted for about 20 percent of Kenya's agricultural export income. This is a revealing core-periphery issue: companies growing and exporting the flowers to European markets use pesticides not tolerated in Europe itself, so that the ecological damage prevented by regulations in the global core falls on the periphery, where rules are less stringent and the environment suffers. The burgeoning flower industry around Lake Naivasha also has created social problems as thousands of workers, attracted by rumored job opportunities, have converged on the area and find themselves living in squalid conditions.

© Marta Nascimento/REA/Redux

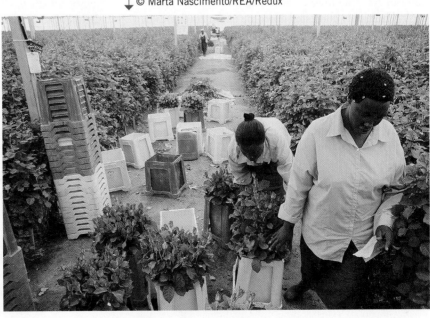

AMONG THE REALM'S GREAT CITIES . . .

NAIROBI

NAIROBI IS THE quintessential colonial legacy: there was no African settlement on this site when, in 1899, the railroad the British were building from the port of Mombasa to the shores of Lake Victoria reached it. However, it had something even more important: water. The fresh stream that crossed the railway line was known to the Maasai cattle herders as Enkare Nairobi (Cold Water). The railroad was extended farther into the interior, but Nairobi grew. Indian traders set up shop. The British established their administrative headquarters here. Not surprisingly, when Kenya became independent in 1963, Nairobi was chosen to be the national capital.

Nairobi owes its primacy to its governmental functions, which ensured its priority through the colonial and independence periods, and to its favorable situation. To the north and northwest lie the Kenya Highlands, the country's leading agricultural area and the historic base of the largest nation in Kenya, the Kikuyu. Beyond the rift valley to the west lie the productive lands of the Luo in the Lake Victoria Basin. To the east, elevations drop rapidly from Nairobi's 1600 meters (5000 ft), so that highland environments make a swift transition to tropical savanna that, in turn, yields to semiarid steppe.

A moderate climate, a modern city center, several major visitor attractions (including Nairobi National Park, on the city's doorstep), and a state-of-the-art airport have

© H. J. de Blij, P. O. Muller, and John Wiley & Sons, Inc.

boosted Nairobi's fortunes as a major tourist destination, though wildlife destruction, security concerns, and political conditions have damaged the industry in recent years.

Nairobi is Kenya's principal commercial, industrial, and educational center. But its growth (to 3.9 million today) has come at a price: its modern central business district stands in stark contrast to the squalor in the shantytowns that house the countless migrants its perceived opportunities attract.

---

184 times that of the average American worker! A recent poll indicated that the typical urban Kenyan pays about 16 bribes per month for anything from accessing running water to avoiding fines.

The highly contested outcome of the 2007 presidential election—the opposition felt it was cheated by allegedly fraudulent results that kept the Kikuyu president in office—uncorked what had been longstanding tensions, and violence erupted that left some 1500 dead. Kenya's continuing challenge is to overcome this still-simmering crisis and sustain a political system that will ensure democracy and represent the interests of Kenya's disparate peoples. New elections are scheduled for 2012.

Tanzania

The name of this country is a hybrid derived from **Tan**ganyika plus **Zan**zibar. This is the biggest and most populous East African state with almost 48 million people; its total area exceeds that of all the region's other countries combined. Tanzania has been described as a country without a core because its clusters of population and zones of productive capacity lie dispersed—mostly on its margins along the east coast (where the capital, Dar es Salaam, is

located), near the shores of Lake Victoria in the northwest, next to Lake Tanganyika in the far west, and facing Lake Malawi in the interior south. This is in sharp contrast to Kenya, which has a well-defined core area in the Kenya Highlands centered on Nairobi in the heart of the country. Moreover, Tanzania consists of many peoples, none numerous enough to dominate the state. About 100 ethnic groups coexist; 30 percent of the population, mainly on the coast, are Muslims.

After receiving its independence Tanzania embarked on a socialist path toward development, including an extensive but poorly planned farm collectivization program. The tourist industry declined sharply, and Tanzania became one of the world's poorest countries. But Tanzania did achieve remarkable political stability and a degree of democracy that none of the other East African states attained. Since 1990, the government has changed course, but the AIDS crisis, problems with the Zanzibar merger, and involvement in the troubles of neighboring countries, including Rwanda, have been costly. Yet today, Tanzania's prospects are improving as the tourist industry has rebounded and political stability continues onward. Tanzania is home to Serengeti National Park, which has now become one of Africa's most popular safari destinations (see photos at the opening of this chapter).

Uganda

When the British colonialists arrived, Uganda (36.1 million) contained this region's most important African state: the Kingdom of Buganda, peopled by the Ganda, located on the northwestern shore of Lake Victoria (Fig. 6B-4; also see Fig. 6A-4). The British established their headquarters near the Ganda capital of Kampala and utilized this ethnic group to control Uganda through indirect rule. Thus the Ganda became the dominant nation in multicultural Uganda, and when the British left, they bequeathed Uganda a complicated federal system designed to perpetuate Ganda supremacy.

That system soon failed, bringing to power one of Africa's most brutal dictators, Idi Amin. Uganda had a strong economy based on coffee, cotton, and other farm exports as well as copper mining; its Asian minority of about 75,000 dominated local commerce. Amin ousted all the Asians, exterminated his opponents, and destroyed the economy. After that, the AIDS epidemic struck Uganda severely. Recovery has been slow since Amin's expulsion, but Uganda has advanced toward more representative government and has made significant gains in the struggle against AIDS. Unfortunately, Ugandan forces played a less constructive role in the conflicts involving its neighbors, Rwanda and The Congo.

As the map shows, Uganda is a **landlocked state [3]** and depends on Kenya for an outlet to the ocean. Its relative location adjacent to such unstable countries as The Congo, Rwanda, and newly independent South Sudan, constitutes a formidable challenge. And in the far north, a rebel group called the Lord's Resistance Army has been terrorizing the local Acholi people for decades.

Rwanda and Burundi

Rwanda and Burundi would seem to occupy Tanzania's northwest corner, and indeed they were part of the German colonial domain forcibly assembled before World War I. But during that war Belgian forces attacked the Germans from their Congo bases and were awarded these territories when the conflict ended in 1918. Local peoples were powerless to prevent this imperial land-grab.

Rwanda (11.0 million) and Burundi (8.9 million), Africa's most densely populated countries, are physiographically part of East Africa, but their cultural geography is linked to the north and west. Here, Tutsi pastoralists from the north subjugated Hutu farmers, who had themselves made serfs of the local Twa (pygmy) population, setting up a conflict that was originally ethnic but became cultural. Certain Hutu were able to advance in the Tutsi-dominated society, becoming to some extent converted to Tutsi ways, leaving subsistence farming behind, and rising within the social hierarchy. These so-called moderate Hutu were—and are—frequently targeted by other Hutus, who

resent their position in society. This longstanding discord, worsened by colonial policies, had repeatedly devastated both countries and in 1994 resulted in the horrific Rwanda genocide. Unfortunately, this conflict continues to simmer and has spilled over into The Congo. As many as 4 million people have perished as Hutu, Tutsi, Ugandan, and Congolese rebel forces have fought for control over the eastern areas of The Congo. Only massive international intervention could stabilize the situation, but the world has turned a blind eye to this region's woes—again.

Ethiopia

The highland zone of Ethopia also forms part of East Africa. Adis Abeba, the historic capital, was the headquarters of a Coptic Christian, Amharic empire that held its own against the colonial intrusion except for a brief period from 1935 to 1941, when the Italians defeated it. Indeed, the Ethiopians, based in their mountain fortress (Adis Abeba lies more than 2400 meters/8000 ft above sea level), became colonizers themselves, taking control of much of the Islamic part of the African Horn to the east.

This country's natural outlets are toward the Gulf of Adan and the Red Sea, but its government was forced to yield independence to Eritrea (part of the African Transition Zone) and Ethiopia is now effectively landlocked. A bitter border war between the two countries, starting in 1998, cost some 100,000 lives. Physiographically and culturally, however, Ethiopia is part of East Africa, and the Amhara and Oromo peoples are neither Arabized nor Muslim: they are Africans. However, Figure 6B-4 shows that functional linkages between Ethiopia and East Africa remain weak—but this is likely to change in the future.

In the past decade, Ethiopia has made considerable economic progress and ranked as the fifth-fastest-growing economy in the world (but starting from a low base). Much of this growth has been attributed to new governmental initiatives, such as selling large tracts of publicly owned arable land to foreign agribusinesses and courting Chinese foreign investment. By 2011, Chinese investment in Ethiopian infrastructure projects (roads, bridges, hydroelectric facilities) had risen to U.S. $2.5 billion.

Ethiopia's diverse environments, spectacular scenery, and archeological fame sustain an economy that depends primarily on farm products (coffee, tea, and spices) and tourism. But the overwhelming majority of Ethiopia's 90 million citizens are subsistence farmers, and the country still requires food imports from relief agencies.

Not all of Ethiopia is properly considered part of East Africa: its northern parts lie in the African Transition Zone and as such face challenges of a completely different regional order. We will return to these issues later in the chapter.

Madagascar

Only 400 kilometers (250 mi) off Africa's central east coast lies the world's fourth-largest island, Madagascar (Fig. 6B-5).

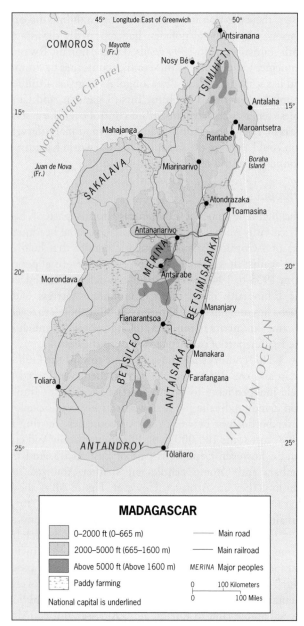

FIGURE 6B-5　　© H. J. de Blij, P. O. Muller, and John Wiley & Sons, Inc.

organized and resisted colonial conquest. Eventually, Madagascar became part of France's empire, and French became the *lingua franca* of the educated elite.

Because of its Southeast Asian imprint, Madagascar's staple food is rice, not corn. It has some minerals, including chromite, iron ore, and bauxite, but the economy is weak, damaged by long-term political turmoil and burdened by rapid population growth. The infrastructure has crumbled; the "main road" from the capital to the nearest port (Fig. 6B-5) is now a potholed 250 kilometers (150 mi) that takes 10 hours for a truck to navigate.

Meanwhile, Madagascar's unique flora and fauna retreated before the human onslaught. Madagascar's long-term isolation kept evolution here so distinct that the island is a discrete zoogeographic realm. Primates living on the island are found nowhere else; 33 varieties of lemurs are unique to Madagascar. Many species of birds, amphibians, and reptiles are also exclusive to this island. Their home, the rainforest, covered 168,000 square kilometers (65,000 sq mi) in 1950, but today less than one-third of it is left. Logging, introduced by the colonists, damaged it; slash-and-burn agriculture continues to destroy it; and severe droughts since 1980 have intensified the impact. Obviously, Madagascar should be a global conservation priority, but funds are limited and the needs are enormous. Malnutrition and poverty are powerful forces when survival is at stake for villages and families.

Madagascar's cultural landscape retains its Southeast Asian imprints, in the towns as well as the paddies. The capital, Antananarivo, is the country's primate city, its architecture and atmosphere combining traces of Asia and Africa. Poverty dominates the townscape here, too, and there is little to attract in-migrants (Madagascar is only 31 percent urbanized). But perhaps the most ominous statistic is the high natural increase rate (2.7 percent) of Madagascar's population.

■■ EQUATORIAL AFRICA

The term *equatorial* is not just locational but also environmental. The equator bisects Africa, but only the western two-thirds of central Africa features the conditions associated with the low-elevation tropics: intense heat, high rainfall and extreme humidity, little seasonal variation, rainforest and monsoon-forest vegetation, and enormous biodiversity. To the east, beyond the Western Rift Valley, elevations rise, and cooler, more seasonal climatic regimes prevail. As a result, we recognize two regions in these lowest latitudes: Equatorial Africa to the west and East Africa to the east.

Equatorial Africa is physiographically dominated by the bowl-shaped Congo Basin. The Adamawa Highlands separate this region from West Africa; rising elevations and climatic change mark its southern limits (see the *Cwa* boundary in Fig. G-7). Its political geography consists of nine states, of which The Congo (formerly Zaïre) is by far the largest in both territory and population (Fig. 6B-6).

But Madagascar is not part of either East Africa or Southern Africa. About 2000 years ago, the first human settlers arrived here—not from Africa but from distant Southeast Asia. A powerful Malay kingdom of the Merina came to flourish in the highlands, and its language, Malagasy, became the indigenous tongue of the entire island (see Fig. 6A-10).

The Malay immigrants later brought Africans to their island, but today the Merina (5.5 million) and the Betsimisaraka (3.2 million) remain the largest of about 20 ethnic groups in the population of 21.2 million. Like mainland Africa, Madagascar experienced colonial invasion and competition. Portuguese, British, and French colonists appeared after 1500, but the Merina were well

From the Field Notes . . .

"Taking an early morning stroll along the beach near the town of Maroantsetra on Madagascar's northeastern coast,

© Jan Nijman

I stopped to observe a group of fishermen at work. It was a very basic way of fishing, with minimal equipment. Using a couple of small rowboats, they dropped their net about 20 meters (65 ft) from shore and then pulled it in toward the beach. It was a meager haul. Most of this catch, they said, was intended for consumption by the men's families and the rest would be sold informally. More than 70 percent of the people in this country live below the national poverty level, and in rural and coastal areas the percentage is even higher. These economic problems stand in sharp contrast to the country's natural beauty—or what remains of it after years of severe environmental degradation."

www.conceptcaching.com

Five of the other eight states—Gabon, Cameroon, Congo, Equatorial Guinea, and São Tomé and Príncipe— all have coastlines on the Atlantic Ocean. The Central African Republic, the southern part of Chad that belongs to this region, and newly-formed South Sudan are landlocked. In many ways, vast and complex Equatorial Africa is the most troubled region in the entire realm.

The Congo

As the map shows, The Congo (known officially as the Democratic Republic of the Congo) has only a tiny window (37 kilometers/23 mi) on the Atlantic Ocean, barely enough to accommodate the wide mouth of the Congo River. Oceangoing ships can reach the port of Matadi, inland from which falls and rapids make it necessary to move goods by road or rail to the capital, Kinshasa. Lack of navigability also necessitates transshipment between Kisangani and Ubundu, and at Kindu. Follow the railroad south from Kindu, and you reach another narrow corridor of territory at the city of Lubumbashi in the southeastern corner of the country. That vital part of Katanga Province contains most of The Congo's major mineral resources, including copper and cobalt.

With a territory not much smaller than the United States east of the Mississippi, a population of just under 72 million, a rich and varied mineral base, and much serviceable agricultural land, The Congo would seem to have all the ingredients needed to lead this region and, indeed, all of Subsaharan Africa. But powerful centrifugal forces, arising from its physiography and cultural geography, pull The Congo apart. The immense forested heart of this basin-shaped country creates communication and transportation barriers between east and west, north and south. Many of The Congo's productive

areas lie along its periphery, separated by considerable distances. These areas tend to look across the border, to one or more of The Congo's nine neighbors, for outlets, markets, and often ethnic kinship as well.

Crisis in the Interior

The Congo's civil wars of the 1990s started in one such neighbor, Rwanda, and spilled over into what was then still known as Zaïre. Rwanda has for centuries been the scene of conflict between sedentary Hutu farmers and invading Tutsi pastoralists. Colonial borders and practices exacerbated the situation, and after independence in the early 1960s a series of terrible crises followed. In the mid-1990s, the latest of these crises generated one of the largest refugee streams ever seen in the world, and that conflict engulfed eastern (and later northern and western) Congo. The death toll will never be known, but estimates range from 5 to 6 million, a calamity that was still not enough to propel the international community into concerted peacemaking action. By 2004 a combination of power transfer in the capital, Kinshasa, negotiation among various rebel groups and the African states involved in the conflict, UN assistance, and general exhaustion had produced a semblance of stability in all but certain eastern areas of The Congo (Fig. 6B-6). But in 2007 civil unrest erupted again and continues intermittently, adding to the ever-rising human toll (see photo, p. 327).

Across the River

To the west and north of the Congo and Ubangi rivers lie seven of Equatorial Africa's other eight countries (Fig. 6B-6). Two of them are landlocked. Chad, straddling the African Transition Zone as well as the regional boundary

FIGURE 6B-6

© H. J. de Blij, P. O. Muller, and John Wiley & Sons, Inc.

with West Africa, is one of Africa's most remote countries, although recent oil discoveries in the south and assistance from China in exploiting those reserves are today changing its status. The **Central African Republic**, chronically unstable and deeply mired in poverty, never was able to convert its agricultural potential and mineral resources (diamonds, uranium) into real progress. And one country consists of two small, densely forested volcanic islands: **São Tomé and Príncipe**, a ministate with a

population of only 220,000 whose economy is being transformed by recent oil discoveries.

The four coastal states present a different picture. All possess oil reserves and share the Congo Basin's equatorial forests; thus petroleum and timber figure prominently among their exports. In **Gabon**, this combination has produced an upper-middle-income economy (see Fig. G-11). Of the four, coastal Gabon also has the largest proven mineral resources, including manganese, uranium, and iron ore.

For decades the easternmost areas of The Congo, especially the provinces of North Kivu and South Kivu (see Fig. 6B-6), have endured recurrent violence as internal conflicts are worsened by civil strife in neighboring countries spilling across the borders. Government forces, rebel groups, militiamen and their followers, warlords, and thieves fight each other, often at the behest of outsiders, driving millions from their homes and sowing death and destruction across the countryside. This photo, taken late in 2008, shows thousands of refugees walking along a road north of the town of Goma, hoping to return to their looted homes near Kibumba, scene of a battle among militias seeking control but driven out by Congolese government forces.

© AP/Wide World Photos

Its capital, Libreville (the only coastal capital in the region), reflects all this in its high-rise downtown, bustling port, and mushrooming squatter settlements.

Cameroon, not as well endowed with oil or other raw materials, has the region's strongest agricultural sector by virtue of its higher-latitude location and higher-relief topography. Western Cameroon is one of the more developed parts of Equatorial Africa and includes the capital, Yaoundé, and the port of Douala.

With five neighbors, Congo (the "other" Congo) could be a major transit hub for this region, especially for The Congo (DRC) if it recovers from civil war. Its capital, Brazzaville, lies across the Congo River from Kinshasa and is linked to the port of Pointe Noire by road and railway. But devastating power struggles have negated Congo's geographic advantages.

As Figure 6B-6 shows, Equatorial Guinea consists of a rectangle of mainland territory and the island of Bioko, where the capital of Malabo is located. A former Spanish colony that remained one of Africa's least-developed territories, Equatorial Guinea has now been so dramatically affected by increased oil production and exporting that it has become Africa's only high-income economy (Fig. G-11). However, as in so many other oil-rich countries, this bounty has not notably increased the incomes of most of the people.

One other coastal territory would also seem to be a part of Equatorial Africa: *Cabinda*, wedged between the two Congos just to the north of the Congo River's mouth. But Cabinda is one of those colonial legacies on the African map—it belonged to the Portuguese and was administered as part of Angola. Today it is an exclave of independent Angola, and a valuable one: it contains major oil reserves. But, as noted above, problems of secession loom here.

South Sudan

The ninth state of Equatorial Africa lies in the region's northeast corner: South Sudan. This newest country to appear on the world political map officially became independent in mid-2011. Its birth came in the aftermath of six disastrous decades of postcolonial strife within the former state of Sudan, a brutal conflict magnified by the bisection of that country by the religious divide between Islam and Christianity-animism that marks the African Transition Zone. The Zone will be discussed as the final region of Subsaharan Africa in this chapter; the remainder of former Sudan—now renamed North Sudan—is discussed in Chapter 7B.

The British effectively ruled Sudan from the 1890s until independence in 1956. One of the roots of the country's long-running internal conflict, which peaked during the past three decades, lies in the decision of the British colonial administration to combine northern Sudan, which was Arabized and Islamized, with a large area to the south that was African and where many villagers had been Christianized. As soon as the British departed, the Khartoum-based regime in the north sought to impose its Islamic rule on southern Sudan, and a bitter civil war immediately broke out. After dragging on for about 15 years, things quieted down between the early1970s and the mid-1980s. But by 1985 the war resumed with a ferocity that lasted until a peace agreement was finally brokered in 2005. About 1.5 million people are believed to have died in the protracted violence, and another half million succumbed to famine and disease triggered by the hostilities.

The drawn out civil war impoverished the country, and per capita income in Sudan ranked among the world's lowest. But that changed in the 1990s when oil was discovered in Kordofan Province and elsewhere, including major deposits in the embattled south (Fig. 6B-6). Leaders in the south saw oil as a ticket to self-sufficiency, and the north felt that it too would walk away with control of several oilfields.

The 2005 peace agreement stipulated a division of the oil revenues between north and south, and also laid the groundwork for a referendum in the south on secession and its reconstitution as an independent country. That referendum was held in the southern provinces in January 2011, with 99 percent of its population voting for independence. This was followed by the official declaration of independence of the new Republic of South Sudan on July 9, 2011.

South Sudan has oil and minerals as well as ample arable land, and the upper White Nile brings plenty of water, much of which evaporates as the flow is dispersed across the flatlands of the Sudd (Arabic for "barrier"), the gigantic swamp filled with floating vegetation atop the inland delta where the ancient river drained into the shallow Sudan Basin (Figs. 6B-6, 6A-2). But the new country offers a landscape of grinding poverty. Pre-2011 governance there had been so deficient that the size of the overall population is not even known—it can only be estimated at somewhere between 8 and 14 million people. Most of the people are cattle herders, stuck in a traditional livelihood in which individual and family status depends on the size of one's herd. Most people are illiterate, and child mortality rates are among the highest in Africa. Less than one-third of the population has access to any form of health care, and clean drinking water is hard to find. There are only about 100 kilometers (60 mi) of paved road in the entire country; nearly half of that is in the capital, Juba, and most of the remainder services the Chinese-run oilfields.

And these are just the economic challenges, with life along the African Transition Zone tending to make things even more complicated. In the run-up to independence amid pressures from the north, one of the provinces astride the new border between North and South Sudan, Abyei, did not participate in the referendum—much against the wishes of the south. Abyei includes a major oilfield, and its population is split between Christian-animist Africans and Muslim Arabs. Violence erupted here swiftly following the 2011 referendum: hundreds died, tens of thousands fled the province, and some feared the 2005 peace treaty would come apart. As we shall see later in the chapter, the African Transition Zone contains a number of very sharp fault lines, and Abyei sits right on top of one of them. South Sudan urgently needs good governance, but most of all it needs a lasting peace with North Sudan.

■■ WEST AFRICA

West Africa occupies most of Africa's Bulge, extending south from the margins of the Sahara to the Gulf of Guinea coast and from Lake Chad west to Senegal (Fig. 6B-7). Politically, the broadest definition of this region includes all those states that lie to the south of Western Sahara, Algeria, and Libya and to the west of Chad (itself sometimes included) and Cameroon. Within West Africa, a rough division is sometimes made between the large, mostly steppe and desert states that extend across the southern Sahara (Chad could also be included here) and the smaller, better-watered coastal states.

Apart from once-Portuguese Guinea-Bissau and long-independent Liberia, West Africa encompasses four former

VOICE FROM THE

Region

YOUNG PEOPLE AND POLITICS IN NIGERIA

© Victoria Okoye
Victoria Okoye, West Africa

"My family is from Nigeria, and although I've lived in Abuja and Lagos, I was born in the U.S. I'm an urban planner working in Accra, Ghana, and I plan to work in Nigeria in a few years. My connection to Nigeria has given me a deep interest in the nation's political development. Democracy in Nigeria is slowly improving, but many young people still don't see the political system as accountable or as a tool to improve conditions in the country. Instead, politics remains an avenue for personal gain and those in government seem disconnected from the everyday struggles of average citizens. Problems are often attributed to Nigeria's ethnic diversity, but this diversity is part of the richness of Nigeria's culture. To me, this is one of the most prominent examples of the failure of politics and government, which leads people to fall back on their social, cultural, familial, and religious ties. We young people are eager for change, but we find it hard not to become cynical. Many Nigerians, including some of my friends and relatives, have become disillusioned. They were not interested in voting in the 2011 elections because of their lack of faith in the system. In some parts of the country, there was voter intimidation, violence, and fraud at polling stations. But there are numerous grassroots initiatives and projects aimed at improving the political system, and engaging youth is part of this. Through on-the-ground community mobilizing as well as the Internet, Facebook, and mobile phones, young people are playing a key role in creating and working with organizations aimed at real political change to make the politicians and government more responsive to citizens' needs. There is a big generational shift happening in that sense. Lasting change will take time, but I'm very hopeful."

British and nine former French dependencies. The British-influenced countries (Nigeria, Ghana, Sierra Leone, and Gambia) lie separated from one another, whereas Francophone West Africa is contiguous. As Figure 6B-7 shows, political boundaries extend perpendicularly from the coast into the interior, so that from Mauritania to Nigeria, the West African habitat is parceled out among parallel, coast-oriented states. Across these boundaries, especially across those between former British and former French territories, there is only limited interaction. For example, in terms of value, Nigeria's trade with Britain is about 100 times as great

FIGURE 6B-7

© H. J. de Blij, P. O. Muller, and John Wiley & Sons, Inc.

as its trade with nearby Ghana. The countries of West Africa are not interdependent economically, and their incomes are largely derived from the sale of their products on the non-African international market.

In light of these cross-currents of subdivision within West Africa, why are we justified in speaking of a single West African region? First, this part of the realm has remarkable cultural and historical momentum. The colonial interlude failed to extinguish West African vitality, expressed not only by the old states and empires of the savanna and the cities of the forest, but also by the vigor and entrepreneurship, the achievements in sculpture, music, and dance, of peoples from Senegal to Nigeria's southeastern Iboland. Second, West Africa contains a set of parallel east-west environmental belts, very clearly reflected in Figure G-7, whose role in the development of the region is pervasive. As the transport-route pattern on the map of West Africa indicates, overland connections within each of these belts, from country to country, are poor; no coastal

or interior railroad ever connected this tier of countries. Yet spatial interaction is stronger across these belts, and some north-south economic exchange does take place, notably in the coastal consumption of meat from cattle raised in the northern savannas. And third, West Africa received an early and crucial imprint from European colonialism, which—with its maritime commerce and slave trade—transformed the region from one end to the other. This impact reached into the heart of the Sahara and set the stage for the reorientation of the entire area, from which emerged the present patchwork of states.

West Africa today is Subsaharan Africa's most populous region (Fig. 6A-4). In these terms, Nigeria (whose census results are in doubt, but with an estimated population of 166 million) is Africa's largest state; Ghana (25.1 million) and Ivory Coast (23.0 million) rank prominently as well. The southern half of the region, understandably, is home to most of the people. Mauritania, Mali, and Niger include too much of the unproductive belt of steppe

From the Field Notes . . .

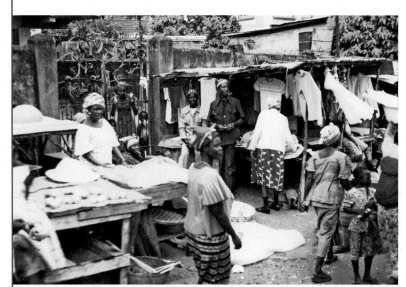

© H. J. de Blij

"Wherever you go in West Africa, you are struck by the bright colors, especially in the markets. And Freetown, Sierra Leone on this Friday seems to be one huge market, the streets lined with stalls, young and old crowding intersections in huge numbers. Most of the goods were for local trade; compared to what you see in East Africa, where tourism plays a bigger role, these markets still have a strong domestic atmosphere. Music plays everywhere, in shops, out in the open; and the whole city seems suffused with the aromas of spicy foods."

www.conceptcaching.com

environment known as the Sahel as well as the arid Sahara to its north to sustain populations comparable to those of Nigeria, Ghana, or Ivory Coast.

The peoples along the coast reflect the modern era that the colonial powers introduced: they prospered in their newfound roles as intermediaries in the coastward trade. Later, they experienced the changes of the colonial period; in education, religion, urbanization, agriculture, politics, health, and many other endeavors, they adopted new ways. In contrast, the peoples of the interior retained their ties with a different era in African history. Distant and aloof from the main theater of European colonial activity and often drawn into the Islamic orbit, they experienced a significantly different kind of change. But the map reminds us that Africa's boundaries were not drawn to accommodate such contrasts. Both Nigeria and Ghana possess population clusters representing the interior as well as the coastal peoples, and in both countries the wide cultural gap between north and south has produced political problems.

Nigeria

Nigeria, the region's cornerstone, is home to just over 166 million people, by far the largest population of any African country. When Nigeria achieved full independence from Britain in 1960, its new government was faced with the daunting task of administering a European political creation containing three major nations and nearly 250 other peoples ranging from several million to a few thousand in number.

For reasons obvious from the map (Fig. 6B-8), Britain's colonial imprint always was stronger in the two southern subregions than in the north. Christianity became

the dominant faith in the south, and southerners, especially the Yoruba, took a lead role in the transition from colony to independent state. The choice of Lagos, the port of the Yoruba-dominated southwest, as the capital of a federal Nigeria (and not one of the cities in the more populous north) reflected British desires for the country's future. A three-region federation, two of which lay in the south, would ensure the primacy of the non-Islamic part of the state. But this framework did not last long. In 1967, the Ibo-dominated Eastern Region declared its independence as the Republic of Biafra, leading to a three-year civil war at a cost of about 1 million lives. Since then, Nigeria's federal system has been modified repeatedly; today there are 36 States, and the capital has been moved from Lagos (now a megacity, profiled on page 305) to more centrally located Abuja (Fig. 6B-8).

Fateful Oil

Large oilfields were discovered beneath the Niger Delta during the 1950s, when Nigeria's agricultural sector produced most of its exports (peanuts, palm oil, cocoa, cotton) and farming still had priority in national and State development plans. Soon, revenues from oil production dwarfed all other sources, bringing the country a brief period of prosperity and promise.

But before long, Nigeria's petroleum riches brought more bust than boom. Misguided development plans now focused on grand, ill-founded industrial schemes and expensive luxuries such as a national airline; the continuing mainstay of the vast majority of Nigerians, agriculture, fell into neglect. Worse, poor management, corruption, outright theft of oil revenues during military misrule, and

FIGURE 6B-8

© H. J. de Blij, P. O. Muller, and John Wiley & Sons, Inc.

excessive borrowing against future oil income led to economic disaster. The country's infrastructure collapsed. In the cities, basic services broke down. In the rural areas, clinics, schools, water supplies, and roads to markets crumbled. Across the country, 80 percent of the people still earn less than U.S. $2 a day.

The Niger Delta itself has become conflict-ridden, and with every spasm of violence, oil production is interrupted and so are revenues. Local people, beneath whose land the oil was being exploited, demanded a share of the revenues and reparations for ecological damage; the military regime under General Abacha responded by arresting and executing nine of their leaders. In 2010, thousands of insurgents agreed to lay down their arms in return for a promise of amnesty and a stipend. But peace remains

tenuous, and more than 50 years of oil production has decimated many of the Delta's rural occupations in agriculture and fisheries.

Nigeria's GDP growth over the past decade is among the highest on the African continent, But on global indices of national well-being, Nigeria has sunk to the lowest rungs even as its output has ranked it as high as the world's tenth-largest oil producer, with the United States its chief customer.

Islam Ascendant

In 1999, Nigeria's hopes were raised when, for the first time since 1983, a democratically elected president was sworn into office. But Nigeria's problems (which now also included a deepening AIDS crisis) worsened when northern States, beginning with Zamfara, proclaimed *Sharia* (strict

To the people living in Nigeria's "middle belt," the Islamic Front is much more than a geographic concept: it is like living in an earthquake zone where things can fall apart in a moment. In Figure 6B-8, you can see Plateau State bordering Islamic-law Kaduna and Bauchi States, with a northward-pointing land extension separating these two. In that protrusion is where Jos, the State capital, is situated, and here Muslim and Christian tensions are never far from the surface. In late November 2008 local elections led to rumors that the mainly Christian Peoples Democratic Party had won over the mainly Muslim All Nigerian Peoples Party. Riots broke out, and in a few days hundreds died and much property destruction took place. Mosques and churches, as well as more than 3000 shops in one market alone, were destroyed before the government could restore order. This photo, taken in early December 2008, shows the Nigerian military separating an advancing group of protesters from their targets by barricading a road in central Jos while local elders try to calm the people. It took the national army and State law enforcement several more days to control the violence, but it will not be the last time such strife will erupt along Nigeria's Islamic Front (see Fig. 6B-9).

© AFP/Getty Images, Inc.

region of West Africa. Four of these states, comprising a huge territory on the Sahara's margins but containing small populations, are landlocked: Mali, Burkina Faso, Niger, and Chad. Figure G-7 shows clearly how steppe and desert conditions dominate the natural environments of these four interior states. Figure 6A-8 reveals the concentration of population within the steppe zone and along the ribbon of water provided by the Niger River. Scattered oases form the remaining settlements and anchor regional trade as well. But even the coastal states do not escape the dominance of the desert over West Africa. Mauritania's environment is almost entirely desert. Senegal, as Figure 6B-7 shows, is mostly a semiarid Sahel country; and not only northern Nigeria but also northern Benin, Togo, and Ghana have interior steppe zones. The loss of pastures to *desertification* (human-induced desert expansion) is a constant worry for the livestock herders there.

West Africa's countries all share the effects of the environmental zonation depicted in Figure G-7, but they also exhibit distinct regional geographies. **Benin**, Nigeria's neighbor with a population of 10.4 million, has a growing cultural and economic link with the Brazilian State of Bahía, where many of its people were taken in bondage and where elements of West African culture have survived. Benin's geography is comparable to that of neighboring **Togo**: both are markedly elongated narrow political units with savannas in the north and humid tropical lowlands near the coast, but without any significant rivers.

Burkina Faso is a poor landlocked state with undeveloped reserves of gold; its economy relies on exports of cotton. With a population of 17.3 million that is almost 60 percent Muslim, this country lies fully within the Islamic orbit.

Ghana, once known as the Gold Coast, was the first West African state to achieve independence (1957), with a sound economy based on cocoa exports. Two grandiose post-independence schemes can be seen on its map: the port of Tema, which was to serve a vast West African hinterland, and Lake Volta, which resulted from the region's largest dam project. When neither fulfilled expectations, Ghana's economy collapsed. But during the 1990s, Ghana's military regime was replaced by a democratic and stable government in the capital, Accra, and by the 2000s Ghana had become a model for Africa. Even ethnic strife in the northern region in 2002 did not disrupt the country's progress, and its agriculture-dominated export economy, based on cocoa, received an unexpected boost when chaos descended on its chief West African competitor,

Islamic) **law** [4]. When Kaduna State followed suit, riots between Christians and Muslims devastated the venerable capital city of Kaduna. There, and in 11 other northern States (Fig. 6B-8), the imposition of *Sharia* law led to the departure of thousands of Christians, intensifying the cultural fault line that threatens the cohesion of the country. Although Kaduna State temporarily repealed its decision, the Islamic revivalism now exhibited in the north raises the prospect that Nigeria, West Africa's cornerstone and one of this realm's most important states, may succumb to devolutionary forces arising from its location in the African Transition Zone. For Africa, that would be a calamity.

Coast and Interior

Nigeria is one of 17 countries (counting Chad and offshore Cape Verde, not shown in Fig. 6B-7) that constitute the neighboring Ivory Coast. In 2006, Ghana authorized the world's largest mining conglomerate to begin exploiting gold reserves northwest of Kumasi, the historic Ashanti capital, under strict compensatory and environmental conditions that set an example for the rest of the realm and, indeed, the world.

In 2007, Ghana celebrated 50 years of independence, its democracy maturing, its economy robust, corruption persistent but declining, and its international stature rising. The same year also brought news of a major discovery of oil reserves off Ghana's coast and caused a surge of optimism. This came only a few months after (former) President Kufuor had signed a nearly U.S. $550 million aid package from the Millennium Challenge Corporation in the United States to expand commercial agriculture, further improve infrastructure, and combat poverty, a grant made in recognition of Ghana's achievements that, in late 2008, included a democratic, closely-contested, and peaceful presidential election.

Ivory Coast (which, as we have stated, officially still goes by its French name, *Côte d'Ivoire*) translated three decades of autocratic but stable rule into economic progress that gave it lower-middle-income status based mainly on cocoa and coffee sales. Continued French involvement in the country's affairs after independence in 1960 contributed to this prosperity; the capital, Abidjan, reflected its comparative well-being. But then Ivory Coast's president-for-life engineered the transfer of the capital to his home village, Yamoussoukro, spending tens of millions of dollars to build a Roman Catholic basilica there to rival that of St. Peter's in Rome. It was dedicated just as the country's economy was slowing and its social conditions worsened.

By the late 1990s, Ivory Coast's economy had reverted to the low-income category (Fig. G-11). Worse, regional strife intensified during the run-up to the presidential election of 2000, when southern politicians objected to the ethnic origins and alleged Muslim sympathies of a northern candidate. In 2002, a series of coordinated attacks by northern rebels failed to take the largest city and former capital, Abidjan, but succeeded in capturing major northern centers. The attacks cost hundreds of lives and confirmed the worst fears of citizens in this long-stable and comparatively prosperous country, where immigrants had contributed importantly for decades to the economy and society.

At the turn of this century, the population included about 2 million Burkinabes (immigrants from Burkina Faso to the north), as many as 2 million Nigerians, about one million Malians, half a million Senegalese, and smaller groups of Ghanaians, Guineans, and Liberians. This cultural mix had been kept stable by a strong central government and a growing economy, but when both of these mainstays failed, disorder followed. In 2004, French troops, sent to Ivory Coast to stabilize the situation and to facilitate negotiations as well as protect the French expatriate community, suffered a deadly attack by the Ivoirian Air Force. This resulted in retaliatory action and led to assaults on French citizens in Abidjan and elsewhere. Most of the expatriate community departed, leaving Ivory Coast, its cocoa-based economy devastated, a West African success story that took a tragic turn.

In late 2010, then-President Gbagbo lost the election to his rival Alassane Ouattara, but Gbagbo refused to concede and allow a transfer of power. The voting patterns reflected the geographic divisions: the northern (mainly Islamic) half of the country voted for Ouattara and most (but not all) of the south cast its vote for Gbagbo. After four months of ferocious hostilities, Gbagbo was finally ousted in April 2011—but the country remains deeply troubled as it now tries to rebuild its shattered society and economy.

All of West Africa has been affected by what has happened in Liberia and Sierra Leone. **Liberia**, a country founded in 1822 by freed U.S. slaves who returned to Africa with the help of American colonization societies, was ruled by their descendants for more than six generations. Rubber plantations and iron mines made life comfortable for the "Americo-Liberians," but simmering resentment among the local peoples finally exploded in 1980. A military coup was later followed in 1989 by full-scale civil war that pitted ethnic groups against each other and drove hundreds of thousands of refugees into neighboring countries, including Ivory Coast, Guinea, and Sierra Leone. Monrovia, the capital (named after U.S. president James Monroe), was devastated; an estimated 230,000 people, almost 10 percent of the population, perished. In 1997, one of the rebel leaders, Charles Taylor, became president, but he was indicted by a UN war-crimes tribunal in 2003 just as he was negotiating an end to the civil war with his opponents. Nigerian forces and UN peacekeeping troops finally entered Monrovia in 2005 and a semblance of stability returned to the country. Later that year, Liberia became the first African state to elect a female president when Ellen Johnson Sirleaf won the first post–civil war contest for that office.

Among the countries in the West African region most severely affected by Liberia's turmoil was **Sierra Leone**, its coastal neighbor to the northwest. Like Liberia, Sierra Leone also was founded as a haven for freed slaves, in this case by the British in 1787. Independent since 1961, Sierra Leone went the all-too-familiar route from self-governing British Commonwealth member to republic to one-party state to military dictatorship. But in the 1990s a civil war brought untold horror to this small country even as refugees from Liberia's war were arriving. Rebels enabled by "blood" diamond sales fought supporters of the legitimate government in a struggle that devastated towns and countryside alike and killed and mutilated tens of thousands. Eventually a combination of West African, United Nations, and British forces intervened in the conflict and resurrected a semblance of representative government, but not before Sierra Leone had sunk to dead last on the world's list of national well-being.

As Figure 6B-7 shows, **Guinea** borders all three of the troubled countries just discussed, and it has involved itself in the affairs of all of them—receiving in return a stream of refugees in its border areas. Guinea, with a population of 11.5 million, has been under dictatorial rule ever since its founding president, Sekou Touré, turned from father of the nation to usurper of the people. As Figure G-11 shows, Guinea is one of Africa's lowest-income states, but representative government and less corrupt administration could have made it a far better place. Guinea has significant gold deposits, may possess as much as one-third of the world's bauxite (aluminum ore) reserves, and can produce

far more coffee and cotton than it does; offshore lie productive fishing grounds. As the map shows, this is no ministate like neighboring **Guinea-Bissau**, the former Portuguese colony. Guinea's capital, Conakry, has a substantial hinterland reaching well into West Africa's interior, giving the country a wide range of environments.

Figure 6B-7 shows that **Senegal** has a wedge-shaped territory situated between forested Guinea to the south and desert Mauritania to the north. Senegal's environmental problems notwithstanding, this country is far ahead of its neighbors in almost every respect. Its great advantage, a legacy from colonial times, is the city of Dakar, headquarters for France's West African empire and now the oversized capital of a state with 13.2 million people, more than half of them farmers. Dakar anchors an unusually well-defined core area, but its surface connections with a much larger West African hinterland are not as favorable as its large commercial and fishing port would suggest. Senegal's economy remains largely agricultural, but the export sector now depends mostly on fish, fertilizers, converted petroleum products, and some iron ore. Peanuts, once a big revenue earner, now figure less prominently.

In many ways Senegal's most valuable asset is its tradition of representative government. In 2010, therefore, this country could not only celebrate 50 years of independence (from France) but also more than four decades of democracy. Although Senegal is 96 percent Muslim, its cultural mosaic is varied: the Wolof, concentrated in and around the capital, constitute nearly half the population; the Fulani farmers of the interior about one-quarter; and the Serer about one-seventh. In general, Senegal's leaders have maintained close relationships with France, which remains the country's chief financial supporter in times of need. Without significant oil reserves, diamonds, or other riches, and with a mainly subsistence-farming population,

Even in the poorer countries of the world, you see something that has become a phenomenon of globalization: gated communities. Widening wealth differences, security concerns, and real estate markets in societies formerly characterized by traditional forms of land ownership combined to produce this new element in the cultural landscape. This is the entrance to Golden Gate, the first private gated-community development in Accra, Ghana, started in 1993 as a joint venture between a Texas-based construction company and a Ghanaian industrial partner.

© Richard Grant

Senegal is still a low-income economy but with GNI levels that exceed the region's average (see the Data Table in Appendix B). Here is evidence that reasonably representative government and stability are greater assets than gold, liquid or otherwise.

A closer look at Figure 6B-7 reveals that Senegal's territory surrounds that of another state, the English-speaking enclave [5] of **Gambia**. Efforts to achieve unification as "Senegambia" have failed, and Senegal has even faced a secession movement in the Casamance District lying south of Gambia. Yet none of this has threatened the Senegalese state, whose strong cultural traditions and steady democracy are a model for French-speaking West Africa.

Saharan Shadows

The huge western bulge of the African landmass is environmentally dominated by the Sahara, the great desert whose dust is widely spread by *Harmattan* "winter" winds across the entire region and far out to sea. Under nature's regime, the Sahara expands and contracts over time, its semiarid southern margin or **Sahel** (an Arabic word meaning "border") shifting north and south. Figure 6B-7 shows the Sahel in an average position, but people and their animals are increasingly affecting its configuration through the process of **desertification**. Herds of livestock quickly occupy the desert-adjoining steppes when they turn temporarily green, but their grazing in numbers far beyond the carrying capacity of this land removes vast areas of this grass as they yank it out by the roots and trample it into the fragile soil layer, making the now-dusty surface highly vulnerable to wind erosion. Inevitably, drought soon returns, the great desert advances, and the people again retreat into the moister savannas to the south, where they invade farmlands and confront sedentary inhabitants. It is an ancient cycle intensified by unrelenting population growth and political partitioning, and it challenges virtually every state in this key zone of Africa.

Much of West Africa's countryside retains its traditional African forms of farming and herding, which, as noted earlier, reflect skillful adaptations to environmental limitations and social constraints. Visit villages south of the Sahel, and you may find that there is not enough commerce to justify daily markets everywhere. This has led to an economic tradition called the **periodic market [6]**, involving a circulation system whereby particular village markets are not open every day, but only every third or fourth day or in some other rotation adjusted to factors of product supply, distance, and population distribution. This system ensures that all villages of a certain size participate in the exchange network. Cultural traditions of this kind endure throughout West Africa, even while the towns and cities beckon the farmers and burst at the seams. Look to the north of the Sahel, and even the vast Sahara is compartmentalized into states, from Mauritania in the west to at least as far as Chad in the east. As we will see in Chapters 7A and 7B, Islam dominates in these sparsely

peopled desert countries, but here too population pressures and environmental issues have intensified. Within the West African region and beyond, the overriding challenges are economic survival and nation-building, complicated by a boundary framework as burdensome as any on Earth.

■■ THE AFRICAN TRANSITION ZONE

From Mauritania and Senegal in the far west to the Horn of Africa in the east, a semiarid belt of land, about 500 kilometers (300 mi) wide, lies between the bone-dry Sahara to the north and the more humid savanna zone to the south. Here in the Sahel, steppe (**BS**) climates separate desert (**BW**) from tropical (**Aw**) savanna climates (see Fig. G-7). Life in this stressed belt of steppelands is unpredictable. During the nine-month dry season, from October to June, there is virtually no rain and temperatures rise to 50°C (122°F). When the rains finally come, they are erratic and some areas get a little while others are flooded. Some years the harvest is good, sometimes it is almost nonexistent. According to a recent UNICEF report, malnutrition kills 225,000 children a year in the five western Sahelian countries (those excluding former Sudan and the African Horn). But the problems of the Sahel go far beyond climate.

As Figures 6B-9 and 6B-1 show, the African Transition Zone is unlike the other four regions of Subsaharan Africa. This is where the realm's cultures intersect with those of the Muslim world. This Islamic gateway into Subsaharan Africa encompasses some countries in their entirety that are also part of the formal regions on the map (such as Senegal and Burkina Faso); it divides others into Muslim and non-Muslim sectors (such as Nigeria, Ivory Coast, and Chad); and in Africa's Horn, it contains countries that do not form parts of other regions (Eritrea, Djibouti, Somalia). As elsewhere in the world where certain geographic realms meet and overlap, complications mark the African Transition Zone. In some areas, the transition from Muslim to non-Muslim society is gradual, as it remains in the northern parts of Ivory Coast and Nigeria despite the recent cultural problems that have erupted in these important states. Elsewhere the break is considerably sharper, as in eastern Ethiopia and newly-divided Sudan where a traditional border between Christian/animist African cultures and Islamic African communities is decidedly abrupt. The southern border of the African Transition Zone, the religious frontier occasionally referred to as Africa's **Islamic Front [7]**, is therefore neither static nor everywhere the same.

Conflict marks much of the African Islamic Front (mapped in Fig. 6B-9): long term in Sudan, where a decades-long war for independence by non-Islamic Africans in the south against the Arabized, Muslim north was successful but cost millions of lives; intermittent in Nigeria, where Islamic revivalism is currently clouding that country's prospects; and recent in Ivory Coast, where political rivalries have spurred

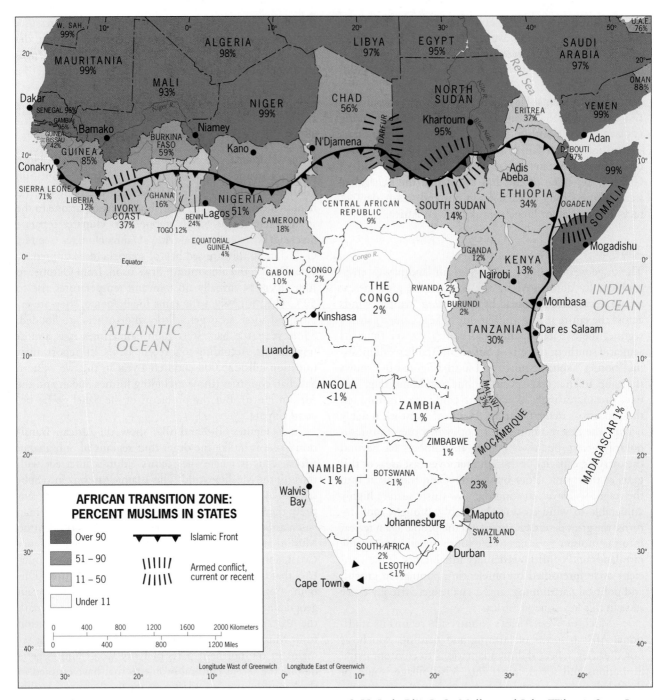

FIGURE 6B-9

© H. J. de Blij, P. O. Muller, and John Wiley & Sons, Inc.

religious strife in a once stable country. Nonetheless, the most conflict-prone part of the African Transition Zone lies in the east, involving not only now-bifurcated Sudan but also the historic Christian state of Ethiopia, nearly encircled by dominantly Muslim societies.

The Horn of Africa

Africa's Horn (Fig. 6B-10) is an especially volatile subregion of the African Transition Zone, which contains Ethiopia, Eritrea, Djibouti, and Somalia. We will concentrate our attention here on the two largest of these countries. But note

that the ministate of **Djibouti**, fully 97 percent Muslim, has taken on added importance since it lies directly across from Yemen and overlooks the narrow entry to the Red Sea, the Bab el Mandeb Strait, a **choke point [8]** in international commerce. **Eritrea** came into existence in 1993 when it split off from Ethiopia. Ongoing boundary conflicts with Ethiopia have damaged the economies of both countries.

Ethiopia

Earlier we noted that Highland Ethiopia where the country's core area, capital (Adis Abeba), and Christian heartland lie, is virtually encircled by dominantly Muslim societies. One poor-quality road, along which bandits rob bus passengers and law enforcement is virtually nonexistent, forms the only direct surface link between Adis Abeba and Nairobi, the capital of Kenya far to the south (Fig. 6B-10). But, as Figure 6B-9 reminds us, Ethiopia's eastern Ogaden is traditional Somali country and almost entirely Islamic, a vestige of a time when the rulers in Adis Abeba extended their power from their highland fortress over the plains below. Today, 34 percent of Ethiopia's population of

89.7 million is Muslim, and the Islamic Front is quite sharply defined (Fig. 6B-9).

With the realm's second-largest population, Ethiopia is an important African cornerstone state, but its economy remains weak, its politics unrepresentative, and its relations with neighbors adversarial. Landlocked and fragmented by the African Transition Zone, Ethiopia forfeited international support when its disputed parliamentary election in 2005 was followed by the imprisonment of opposition leaders and

FIGURE 6B-10

© H. J. de Blij, P. O. Muller, J. Nijman, and John Wiley & Sons, Inc.

journalists. The World Bank, for example, withheld budgetary subsidies, while human rights organizations criticized Ethiopian actions in its southern Oromo and eastern Somali areas. In 2007, an attack by the Ogaden National Liberation Front (a terrorist organization) on an oil rig in eastern Ethiopia killed 68 Ethiopians and 9 Chinese. Ethiopian forces, with U.S. approval, tried to intervene in the chaotic situation prevailing in southern Somalia, but the campaign proved to be beyond their capabilities.

The secession of Eritrea in 1993 had the effect of landlocking Ethiopia, which had made Assab its key Red Sea port while it controlled Eritrea. As the Muslim sector of Eritrea's population of more than 5 million grew faster than the Christian sector, relations between Ethiopia and Eritrea worsened, and in 1998 the two countries launched a seemingly senseless war over their common border. This war flared up repeatedly, cost billions of dollars, and required foreign mediation to resolve, but a negotiated agreement reached in 2000 was not enough to halt it altogether. By 2009, more than 100,000 soldiers on both sides had died in old-style combat over a strip of rocky land and a few abandoned villages. But by that time Eritrea had become a magnet for Islamic militants eager to join a campaign against Christian Ethiopia. As millions went hungry and the once-promising Eritrean economy collapsed, it was a reminder that the Islamic Front is far more than a line on a map.

Somalia

The key component in the eastern sector of the African Transition Zone is Somalia, where 10 million people, virtually all Muslim, live at the mercy of a desert-dominated climate that requires cross-border migration into Ethiopia's Ogaden area in pursuit of seasonal pastures. As many as 3 to 5 million Somalis live permanently on the Ethiopian side of the border, but this is not the only division the Somali "nation" faces. In reality, the Somali people embody an assemblage of 5 major ethnic groups fragmented into hundreds of clans engaged in an endless struggle for power as well as survival.

Early in this new century's first decade, Somalia's condition as a **failed state** [9] led to the country's fragmentation into three subdivisions (Fig. 6B-10). In the north, the sector called *Somaliland*, which had proclaimed its independence in the 1990s and which remains by far the most stable of the three, functioned essentially as an African state, although the "international community" will not recognize it as such. In the east, a conclave of local chiefs declared their territory to be separate

from the rest of Somalia, named it *Puntland*, and asserted an unspecified degree of autonomy. In the south, where the official capital, Mogadishu, is located on the Indian Ocean coast, local secular warlords and Islamic militias continued their grim contest for supremacy, the warlords supported by U.S. funding. In 2006, the militias stormed into the capital and took control, ousting the warlords and proclaiming their determination to create an Islamic state.

As noted above, Ethiopian forces intervened with the unheralded support of the United States. They were joined by an African Union peacekeeping force, but before long both came under continuous and increasing attack from Somali militias. Hundreds of thousands of refugees streamed out of Mogadishu, where the fighting was worst, and many others left their villages in the embattled zone. Recent UN estimates show that more than one million Somalis were stranded between the Kenya border (where many were turned back) and the sea, and that starvation was widespread.

Meanwhile, Indian Ocean waters off Somalia became an arena of action for a growing number of pirates based along the country's lengthy coastline and with increasing capabilities. Piracy had long been occurring here, but most of the incidents involved smaller boats and did not make the news. But then Somali pirates began to target larger ships; they captured a Ukrainian freighter loaded with tanks and

This boat photographed near Hobyo, Somalia is one of about 100 speedboats operated by the so-called Central Regional Coast Guard, a pirate group formed in 2005 with about 400 men in its ranks today. As recently as 2008, this organization attacked 29 ships in the waters off eastern Africa, eventually extorting a total of U.S. $10 million from their owners. In this decade, stronger protection of commercial shipping in the area has reduced the pirates' success rate, and some pirates have been killed by defensive actions. But the threat continues.

© Veronique de Viguerie/Getty Images, Inc.

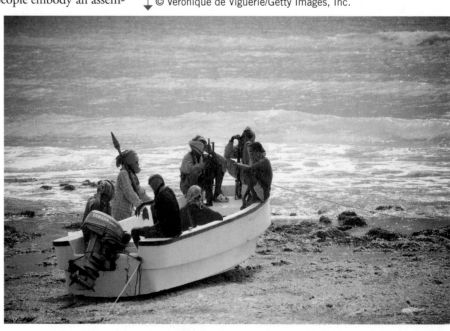

other weapons, a 300,000-ton oil tanker, and even tried to board cruise ships. Their sphere of activity expanded from the Gulf of Adan to waters south of the latitude of Kenya's port of Mombasa, and their impact was global. Warships from Russia, India, and the United States were among vessels trying to secure the vast maritime region off Somalia, but the "industry" could not easily be closed down. The combination of a failed state in Africa's Horn, Islamic fundamentalism driven by local militants, al-Qaeda in the mix, and pirates operating ever more boldly off the coast continues to make this segment of the African Transition Zone a focus of heightened international concern.

As this chapter has underscored, Africa is a continent of infinite social diversity, and Subsaharan Africa is a realm of matchless cultural history. Relatively recent environmental change has widened the Sahara, which separates north from south. The push of Islam turned the north to Mecca and left the south open to Christian proselytism. In colonial times, millions of Africans were carried away in bondage; the cruelties of imperialism subjugated those left to the interests of Europe. Africa's equatorial heart is an incubator for debilitating diseases second to none in the world, but Africa's health problems have, until recently, not been a top priority in the medical world. When the European powers finally abandoned their African territories, African peoples found themselves with state borders that were incompatible with ethnic identities, and with economies that were unable to compete on world markets. We said at the beginning of Chapter 6A that Africa's time and turn will come again, and that perhaps it has already turned the page—but it will take far more than what is being done today to rectify half a millennium's malefactions.

POINTS TO PONDER

- The newly-independent state of South Sudan is by most measures the poorest country in Subsaharan Africa.
- Even though South Africa was freed from apartheid nearly two decades ago, current data report that it is still the most unequal society on Earth.
- In 2010, Angola was China's top oil supplier, having surpassed both Saudi Arabia and Iran.
- Some African countries are among the poorest in the world, but their GDP growth rates are quite high—what should their governments do with the increased revenues?

using GEOGRAPHY

- Imagine you were on a well-equipped expedition using 4-wheel drive vehicles, driving from Lagos, Nigeria due north to southern Algeria. Your task was to keep a diary about the changing landscape in terms of vegetation, major landforms, human habitats, weather, soils, economic activity, cultural identities, population density, and so on. What would some of the main observations in your diary be, from beginning to end?

- The International Monetary Fund expects that from 2011 to 2015, Ethiopia, Moçambique, and Tanzania will rank among the five fastest-growing economies in the world (along with China and India). On the basis of a geographic analysis (resources, infrastructure, etc.), which of these three Subsaharan countries do you think will have the healthiest, most diversified, and most sustainable growth?

Elevation (m)

3000
1500
600
300
150
0
below sea level
0
-150
-1500
-3000

Roads
Railroads

NORWAY   SWEDEN
Bergen   Oslo   Västerås   Uppsala
Stavanger   Göteborg   Stockholm

Shetland Islands (U.K.)   North Sea   Skagerrak   Aalborg   Gotland (SWEDEN)   Baltic Sea
Orkney Islands   DENMARK   Gdansk
Glasgow   Edinburgh   Newcastle   COPENHAGEN   Malmö   Odense   Kaliningrad
IRELAND   Belfast   UNITED   Leeds   HAMBURG   Bremen   Szczecin   Bydgoszcz
Dublin   Liverpool   KINGDOM   Amsterdam   Hannover   BERLIN   Poznan   POLAND
Cork   Birmingham   Manchester   NETH.   Essen   Leipzig   Dresden   Wroclaw
LONDON   The Hague   Rotterdam   Cologne   PRAGUE   CZECH REP.   Krak
Plymouth   Bristol   Antwerp   BRUSSELS   BELG.   Frankfurt   Brno
Channel Islands (U.K.)   English Channel   LUX.   Luxembourg   VIENNA   SLOVAK
Celtic   Le   Brest   Rouen   GERMANY   Stuttgart   MUNICH   LIECH.   AUS.   Bratislava   BUDA
Sea   Havre   PARIS   Strasbourg   Zürich   Salzburg   Graz   HUNGARY
Nantes   Tours   Geneva   SWITZ.   Innsbruck   LJUBLJANA   Zagreb
FRANCE   Lyon   MASSIF   Turin   MILAN   SLOV.   BOSNIA   SERB
Bordeaux   CENTRAL   Genoa   Venice   Rijeka   CROATIA   Sarajevo
Pico de Aneto   Toulouse   Nice   MONACO   Florence   SAN   MONTENEGRO
(3,404 m,   PYRENEES   Marseille   Corsica   MARINO   Podgorica   KOS
11,168 ft.)   ANDORRA   (FRANCE)   Bologna   ROME   MAC
La Coruña   Gijón   Bilbao   Elba   Adriatic   Tirane   ALBANIA
CANTABRIAN MOUNTAINS   Zaragoza   Balearic   Naples   Bari   Sea
Vigo   Porto   MADRID   Sea   Palma de   Sardinia   ITALY
IBERIAN   Barcelona   Majorca   (ITALY)   Tyrrhenian   Ionian
PORTUGAL   SPAIN   Valencia   Cagliari   Palermo   Sea   Sea
Lisbon   PENINSULA   Murcia   Mediter   Sicily
Seville   ALGIERS   Valletta
Strait of Gibraltar   Gibraltar (U.K.)   Oran   MALTA
Tangier   Alboran Sea   Mostaganem   Setif   Annaba   Tunis
RABAT   Oujda   Tlemcen   Batna   Constantine   Sousse
Madeira Is.   CASABLANCA   Fez   Taza   Djelfa   Biskra   Tebessa   Qairouan   TUNISIA
(PORTUGAL)   Funchal   Safi   Khouribga   Meknes   Gafsa   Sfax
MOROCCO   Beni Mellal   Laghouat   Gabes
Salvage Is.   Marrakech   Jebel Toubkal   Ouargla   Zuwarah
(PORTUGAL)   (4,165 m, 13,665 ft.)   Bechar   Ghardaia   TRIPOLI   Derna
La Palma   Canary Islands   Agadir   Benghazi   CYR
Santa Cruz de Tenerife   (SPAIN)   Ghadamis   Gulf of Sidra   Ajdabi
Hierro   Tenerife   Lanzarote   GREAT ERG OCCIDENTAL   Sidra   TRIPOLITANIA
Gran   Las   Fuerteventura   TADEMAIT   GREATER ERG ORIENTAL
Canaria   Palmas   Tan Tan   PLATEAU   HAMADA DE TINRHERT
El Aaiún   Adrar   LIBYA
Tropic of Cancer   Western Sahara   ERG IGUIDI   ALGERIA   I-n-Salah   Sabha
Smara   ERG CHECH   TASSILI-N-AJJER   FEZZAN
Ad Dakhla   TANEZROUFT   Al Birkah   MARZUQ   Al Birkah
Nouadhibou   Fderik   DESERT
Cape Blanc   ERG   AHAGGAR MTS.   Mt. Greboun   TIBESTI
Atar   DJOUF   Tamanrasset   (1,944 m, 6,378 ft.)   MTS.
MAURITANIA   Akjoujt   Mt. Koussi
Nouakchott   S   A   H   A   R   A   (3,415 m, 11,204 ft.)
Tidjikja   AIR   BORKOU
Cape Verde Is.   Arlit   Bilma   Ouniang
Santo   Mindelo   Saint Louis   Rosso   Nema   Tomboutou   Faya-Largeau
Antão   São   Aleg   Kiffa   (Timbuktu)   Agadez
Nicolau   CAPE VERDE   Kayes   Nioro   MALI   NIGER   CHAD   ENI
Fogo   Boa Vista   DAKAR   Thies   Diourbel   Gao   Tahoua   Mao   Biltine
Praia   Cape Verde   SENEGAL   Kiffa   Mopti   Niamey   Zinder   Lake   Abéché
Santiago   Kaolack   Nioro   Segou   Dori   Dosso   Sokoto   Maradi   Chad   N'Djamena
GAMBIA   Diourbel   Ouahigouya   Katsina   Diffa   Kousséri
Banjul   Ziguinchor   Bissau   BAMAKO   Niger   BURKINA FASO   KANO   MAIDUGURI
GUINEA-   Bafata   Koutiala   Koudougou   Ouagadougou   ZARIA   Mokolo   Maroua
BISSAU   Boke   Siguiri   Sikasso   Bobo   KADUNA   Bauchi   Jos
GUINEA   Labe   Kankan   Dioulasso   Korhogo   Tamale   Minna   Abuja   Garoua
Kindia   FUTA JALLON   Gambaga   Oshogbo   Ilorin   Makurdi   Kélo
CONAKRY   HIGHLANDS   IVORY   Sokode   Benue   Moundou   Doba
SIERRA   Nzerekore   COAST   Bouake   BENIN   Oshogbo   ADAMAWA   CHAÎNE
Freetown   LEONE   Bo   Man   Lake   GHANA   Abeokuta   NIGERIA   HIGHLANDS   DES   BON
Kenema   Kossou   KUMASI   Ibadan   Enugu   Umuahia   Aba   CENTRAL
LIBERIA   Yamoussoukro   Koforidua   BENIN CITY   Ngaoundéré   AFRI
Monrovia   Divo   ACCRA   Lomé   Porto   Bamenda   REPUBLI
Buchanan   Harper   ABIDJAN   Cotonou   Novo   LAGOS   Bafoussam   Bossangoa
River Cess   Cape Palmas   Sekondi   Cape Coast   Bight of Benin   CAMEROON   Bouar   Bertoua
Equator   San Pedro   Takoradi   Cape   PORT   DOUALA   Bangui
Three Points   HARCOURT   Bight   Bambari
ATLANTIC   OCEAN   Gulf of   Malabo

IN THIS CHAPTER

- The cradle of civilization
- Terror in the name of Islam
- Oil: A mixed blessing
- The revolutions of 2011
- The demise of Usama bin Laden

CONCEPTS, IDEAS, AND TERMS

Cultural geography 1

Culture hearth 2

Cultural diffusion 3

Cultural landscape 4

Hydraulic civilization 5

Climate change 6

Spatial diffusion 7

Expansion diffusion 8

Relocation diffusion 9

Contagious diffusion 10

Hierarchical diffusion 11

Islamization 12

Stateless nation 13

Choke point 14

Fragmented modernization 15

League of Nations 16

Religious revivalism 17

Jihad 18

Wahhabism 19

Domino effect 20

FIGURE 7A-1 © H. J. de Blij, P. O. Muller, and John Wiley & Sons, Inc.

From Morocco on the shores of the Atlantic to the mountains of Afghanistan, and from the Horn of Africa to the steppes of inner Asia, lies a vast geographic realm of enormous cultural complexity. It stands at the crossroads where Europe, Asia, and Africa meet, and it is part of all three (Fig. 7A-1). Throughout history, its influences have radiated across Eurasia and Africa and reached practically every other part of the world as well. This is one of humankind's primary source areas. On the Mesopotamian Plain between the Tigris and Euphrates rivers (in modern-day Iraq) and on the banks of the Egyptian Nile arose several of the world's earliest civilizations. In its soils, plants were domesticated that are now grown from the Americas to Australia. Along its paths walked prophets whose religious teachings are still followed by hundreds of millions. And early in the second decade of the twenty-first century, the heart of this realm is beset by some of the most bitter and dangerous conflicts in the world.

major geographic qualities of
NORTH AFRICA/SOUTHWEST ASIA

1. North Africa and Southwest Asia were the scene of several of the world's great ancient civilizations, based in its river valleys and basins.

2. From this realm's culture hearths diffused ideas, innovations, and technologies that changed the world.

3. The North Africa/Southwest Asia realm is the source of three world religions: Judaism, Christianity, and Islam.

4. Islam, the most recent of the major religions to arise in this realm, transformed, unified, and energized a vast domain extending from Europe to Southeast Asia and from Russia to East Africa.

5. Drought and unreliable precipitation dominate natural environments in this realm. Population clusters exist where water supply is adequate to marginal.

6. Certain countries of this realm have enormous reserves of oil and natural gas, creating great wealth for some but doing little to raise the living standards of the majority.

7. The boundaries of the North Africa/Southwest Asia realm consist of volatile transition zones in several places in Africa and Asia.

8. Conflict over water sources and supplies is a constant threat in this realm, where population growth rates are high by world standards.

9. The Middle East, as a region, lies at the heart of this realm; and Israel lies at the center of the Middle East conflict.

10. The broad commonalities and interrelationships across much of the realm were demonstrated in the recent diffusion of political uprisings against autocratic regimes.

NAMING THIS PIVOTAL REALM

As you may conclude from its long and somewhat cumbersome name—North Africa/Southwest Asia—this is a sprawling, geographically complex realm. In our era of high-speed communication you will sometimes see it referred to as *NASWA*, the first letters of its regional deployment. And it is tempting to refer to it in other kinds of geographic shorthand, based on some of its dominant features. But such generalizations can be misleading, and some examples follow.

A "Dry World"?

This realm is, for instance, sometimes called the Dry World, containing as it does the vast Sahara as well as the Arabian Desert. But most of the realm's people live where there is water—along the Nile River, along the hilly Mediterranean coastal strip or *tell* (meaning mound in Arabic) of northwesternmost Africa, along the Asian eastern and northeastern shores of the Mediterranean Sea, in most of the Tigris-Euphrates Basin, in far-flung desert oases, and along the lower mountain slopes of Iran south of the Caspian Sea as well as Turkestan to the northeast. Figure 7A-2 reflects this covarying relationship between population distribution and available water. Quite a few of this realm's prominent geographic features—river valleys, basins, and deltas; moist coastlines; adequately-watered mountain basins—are virtually defined by the clusters you see on this map.

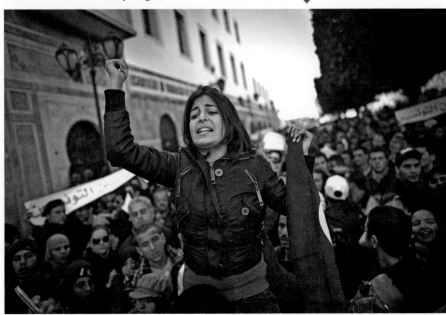

Tunis, January 22, 2011: Holding the Tunisian flag, a woman raises her fist in a demonstration to demand democratic reforms. Women played a key role in efforts to force regime change in Tunisia in early 2011—which swiftly triggered a wave of major political upheavals that swept across the realm.

© Martin Bureau/AFP/GettyImages, Inc.

FIGURE 7A-2

© H. J. de Blij, P. O. Muller, and John Wiley & Sons, Inc.

So we know this world realm as a place where water is almost always at a premium, where consumers (such as Israelis and Palestinians) compete for limited supplies, where peasants often struggle to make soil and moisture yield a small harvest, where nomadic peoples and their livestock still migrate across dust-blown flatlands, where oases are green islands in the desert that sustain local farming, supply weary travelers, and support trade across vast expanses of aridity.

But this is also the land of the Nile, the lifeline of Egypt, the crop-covered *tell* of coastal Algeria, the verdant shores of western Turkey, and the meltwater-fed valleys of Central Asia. Compare Figure 7A-1 to Figure G-7, and the dominance of *B* climates becomes evident, underscoring just how water-dependent this realm's clustered population is.

Is *This* the "Middle East"?

This realm is commonly referred to as the Middle East. That must sound odd to someone in, say, India, who might think of a Middle West rather than a Middle East! The name, of course, reflects the biases of its source: the Western world, which saw a Near East in Turkey; a Middle East in Egypt, Arabia, and Iraq; and a Far East in China and Japan. Still, the term has taken hold, and it can be seen and heard in everyday usage by scholars, journalists, and members of the United Nations. Even so, we feel that it should only be applied to one of the regions of this sprawling realm, not to the realm as a whole.

An "Arab World"?

Another shorthand label often used for North Africa/Southwest Asia is the Arab World. This term implies a uniformity

that does not actually exist. First, the name *Arab* is applied loosely to the peoples of this area who speak Arabic and related languages, but ethnologists normally restrict it to certain occupants of the Arabian Peninsula—the Arab "source." In any case, Turks are not Arabs, and neither are most Iranians or Israelis. Moreover, although the Arabic language prevails from Mauritania in the west across all of North Africa to the Arabian Peninsula, Syria, and Iraq in the east, it is not spoken in other parts of this realm. In Turkey, for example, Turkish is the major language, and it has Ural-Altaic rather than Arabic's Semitic or Hamitic roots; the Iranian language is a member of the Indo-European linguistic family (see Fig. G-9). Other Arab World languages that have separate ethnological identities are spoken by the Jews of Israel, the Tuareg people of the Sahara, the Berbers of northwestern Africa, and the peoples of the wide transition zone between North Africa and Subsaharan Africa to the south.

An "Islamic World"?

Finally, and perhaps most significantly, this realm is routinely referred to as the Islamic World. Indeed, the prophet Muhammad was born on the Arabian Peninsula in AD 571, and Islam, the religion to which he gave rise, spread across this realm and beyond after his death in 632. This great saga of Arab conquest saw Islamic armies, caravans, and fleets carry Muhammad's teachings across deserts, mountains, and oceans, penetrating Europe, converting the kings of West African states, threatening the Christian strongholds of Ethiopia, invading Central Asia, conquering most of what is now India, and even reaching the peninsulas and islands of Southeast Asia. Today, the world's largest

Muslim state is Indonesia, and the Islamic faith extends far outside the realm under discussion (Fig. G-10). In this context, the term Islamic World is also misleading in that it suggests that there is no Islam beyond NASWA's borders.

In any case, this World of Islam is not entirely Muslim either. Christian minorities continue to survive in all the regions and many of the countries of the North Africa/Southwest Asia realm. Judaism has its base in the Middle East region; and smaller religious communities, such as Lebanon's Druze and Iran's Bahais, many of them under pressure from Islamic regimes, continue to diversify the religious mosaic.

Nonetheless, Islam's impact on this realm's cultural geography is pervasive. It ranges from dominant in such countries as Iran, Saudi Arabia, and North Sudan to laminate (that is, more of a veneer) in several of the countries of Central Asia's Turkestan, where the former Soviet rulers discouraged the faith and where its comeback is still in progress—one of the criteria that allow us to identify Turkestan as a discrete region within this realm. The rise, impact, and continuing evolution of Islam lie at the core of the story of this chapter.

States and Nations

Although Islam and its cultural expressions dominate this geographic realm, its political and social geographies are divided and often fractious. We will encounter states with strong internal divisions (Lebanon, Iraq), nations as yet without their own states (Palestinians, Kurds, Berbers), and territories still in the process of integration (Western Sahara, Palestine). No single state dominates this realm as Mexico eclipses the rest of Middle America or Brazil looms over South America. Of its nearly 30 states and territories, the three largest and in many ways the most important are Egypt in North Africa, Turkey on the threshold of Europe, and Iran at the margins of Turkestan. All three have populations between 75 and 85 million, modest by world standards. But at the other end of this continuum there are ministates such as Bahrain, Qatar, and Djibouti, each with fewer than two million inhabitants. It is quite important to study Figure 7A-1 attentively, because the boundary framework—much of it inherited from the colonial era—lies at the root of some of the realm's current problems.

It is also useful to take a careful look at Figure 7A-2, because it clearly shows how differently populations are distributed within the countries we will examine. Compare the three largest countries just mentioned: Egypt's population forms a river-valley ribbon and delta cluster; Turkey's, between the Black Sea and the Mediterranean Sea, is far more evenly dispersed, although its west is more populous than the east; and in Iran you can draw a line down its center from the Caspian Sea to the Persian Gulf that separates the comparatively crowded west, where the capital is located, from the drier and far more sparsely peopled east. In North Africa, almost everyone seems to live along or near the Mediterranean coast, but no such coastal concentrations mark the Arabian Peninsula except in its far south, where Yemen's inhabitants cluster near the Red Sea. Certainly the historic role of water can be seen in the population patterns displayed in Figure 7A-2.

HEARTHS OF CULTURES

This geographic realm occupies a pivotal part of the world: here Africa, the source of humankind, meets Eurasia, crucible of human cultures. Two million years ago, the ancestors of our species walked from East Africa into North Africa and Arabia and spread all across Asia. Less than one hundred thousand years ago, *Homo sapiens* crossed these lands on their way to Europe, Australia, and, eventually, the Americas. Ten thousand years ago, human communities in what we now call the Middle East began to domesticate plants and animals, learned to irrigate their fields, enlarged their settlements into towns, and formed the earliest states. The world's dominant monotheistic religions originated here: first Judaism, then Christianity, and most recently the heart of the realm was stirred and mobilized by the teachings of Muhammad and the Quran (Koran), and Islam began. Today this realm is a cauldron of religious and political activity and turmoil: complicated by foreign interventions; weakened by conflict; empowered and enriched by oil in some places; plagued by poverty in many others.

Dimensions of Culture

In the introductory chapter, we discussed the concept of culture and its regional expression in the cultural landscape. Cultural geography [1], we noted, is a wide-ranging and comprehensive field that studies spatial aspects of human cultures, focusing not only on cultural landscapes but also on culture hearths [2]—the crucibles of civilization, the sources of ideas, innovations, and ideologies that changed regions and realms. Those ideas and innovations spread far and wide through a set of processes that we study under the rubric of cultural diffusion [3]. Because we understand these processes better today, we can reconstruct many of the ancient routes by which the knowledge and achievements of culture hearths spread—that is, *diffused*—to other areas.

Another aspect of cultural geography, particularly noteworthy in the context of the North Africa/Southwest Asia realm, is the study of cultural landscapes [4] that a dominant culture creates. Human cultures exist in long-term accommodation with (and adaptation to) their natural environments, exploiting opportunities that these environments present and coping with the extremes they can impose. In the process, they fuse their physical landscape and cultural landscape into an interacting unity.

Rivers and Communities

In the basins of the major rivers of this realm (the Tigris-Euphrates system of modern-day Turkey, Syria, and Iraq as well as the Nile of Egypt) lay two of the world's earliest culture hearths (Fig. 7A-3). Mesopotamia (land amidst the

FIGURE 7A-3

© H. J. de Blij, P. O. Muller, and John Wiley & Sons, Inc.

rivers) had fertile alluvial soils, abundant sunshine, ample water, and animals and plants that could be domesticated. Here, in the Tigris-Euphrates lowland between the Persian Gulf and the uplands of present-day Turkey, arose one of humanity's first culture hearths, a cluster of communities that grew into larger societies and, eventually, into the world's first states. (Early state development probably occurred simultaneously in East Asia's river basins.)

Mesopotamia

Mesopotamians were innovative farmers who domesticated crops such as wheat and knew when to sow and harvest various crops, water their fields, and store their surpluses. Their knowledge spread to villages near and far, and a *Fertile Crescent*, a region of significant agricultural productivity, evolved, extending from Mesopotamia across southeastern Turkey into Syria and the eastern Mediterranean coast beyond (Fig. 7A-3).

Irrigation was the key to prosperity and power in Mesopotamia, and urbanization was its major reward. Among many settlements in the Fertile Crescent, some thrived, grew, enlarged their hinterlands, and diversified socially and occupationally; others failed. What determined success? One theory, the hydraulic civilization theory [5], holds that cities that could control irrigated farming over large hinterlands held power over others, used food as a weapon, and thrived. One such city, Babylon on the Euphrates River, endured for nearly 4000 years (from 4100 BC). A busy port, its walled

and fortified center endowed with temples, towers, and palaces, Babylon for a time was the world's largest city.

Egypt and the Nile

Egypt's cultural evolution may have started even earlier than Mesopotamia's, and its focus lay upstream from (south of) the Nile Delta and downstream from (north of) the first of the Nile's series of rapids, or cataracts. This part of the Nile Valley is surrounded by inhospitable desert, and unlike Mesopotamia (which lay open to all comers), the Nile provided a natural fortress here. The ancient Egyptians converted their security into progress. The Nile was their highway of trade and interaction; it also supported agriculture through irrigation. The Nile's cyclical ebb and flow was much more predictable than that of the Tigris-Euphrates river system. By the time Egypt fell victim to outside invaders (about 1700 BC), a full-scale urban civilization had emerged. Ancient Egypt's artist-engineers left a magnificent legacy of massive stone monuments, some of them containing treasure-filled crypts of god-kings called Pharaohs. These tombs have enabled archeologists to reconstruct the ancient history of this culture hearth.

Today, the world continues to benefit from the accomplishments of the ancient Mesopotamians and Egyptians. They domesticated cereals (wheat, rye, barley), vegetables (peas, beans), fruits (grapes, apples, peaches), and many animals (horses, pigs, sheep). They also advanced the study of the calendar, mathematics, astronomy, government,

From the Field Notes . . .

"Flying north over central Egypt from Luxor's ruins toward Cairo along the Nile's ribbon of green, I pondered the famous remark by the ancient Greek historian Herodotus that

"Egypt is the gift of the River Nile." The Nile brings water from the interior of Africa that crosses the vast northern deserts and is crucially important to a country with as little rainfall as Egypt. About 95 percent of the Egyptian population lives within 20 kilometers (12 mi) of the river's banks. Almost all of Egypt's agriculture has always depended on irrigation, which is how this narrow strip of desert turned fertile. Notice the razor-sharp boundaries between irrigated and nonirrigated land. Without a doubt, the Nile was the lifeline that supported one of the ancient world's major civilizations."

www.conceptcaching.com

© Jan Nijman

engineering, metallurgy, and a host of other skills and technologies. In time, many of their innovations were adopted and then modified by other cultures in the Old World and eventually in the New World as well. Europe was the greatest beneficiary of these legacies of Mesopotamia and ancient Egypt, whose achievements constituted the foundations of Western civilization.

Decline and Decay

As Figure G-7 reminds us, many of the early cities of this culture hearth lay in what is today desert territory. Assuming that there were no good reasons to build large settlements in the middle of deserts, we may hypothesize that **climate change [6]** sweeping over this region, not a monopoly over irrigation techniques, gave certain cities in the ancient Fertile Crescent an advantage over others. Climate change, closely associated with shifting environmental zones after the final Pleistocene glacial retreat, may have destroyed the last of the old civilizations. Perhaps overpopulation and human destruction of the natural vegetation contributed to the process. Indeed, some cultural geographers suggest that the momentous innovations in agricultural planning and irrigation technology were not "taught" by the seasonal flooding of the rivers but were forced on the inhabitants as they tried to survive changing environmental conditions.

The scenario is not difficult to imagine. As outlying areas began to fall dry and farmlands were destroyed, people congregated in the already crowded river valleys—and made every effort to increase the productivity of the land that could still be watered. Eventually overpopulation, destruction of the watershed, and perhaps reduced rainfall in the rivers' headwater areas dealt the final blow. Towns were abandoned to the encroaching desert; irrigation channels filled with drifting sand; croplands dried up. Those who could migrated to areas that were still reputed to be productive. Others stayed, their numbers dwindling, increasingly reduced to subsistence.

As old societies disintegrated, power emerged elsewhere. First the Persians, then the Greeks, and later the Romans imposed their imperial designs on the tenuous lands and disconnected peoples of North Africa/Southwest Asia. Roman technicians converted North Africa's farmlands into irrigated plantations whose products went by the boatload to Roman Mediterranean shores. Thousands of people were carried off as slaves to the cities of the new conquerors. Egypt was quickly colonized, as was the area we now call the Middle East. One region that lay distant, and therefore remote from these invasions, was the Arabian Peninsula, where no major culture hearth or large cities had emerged and where the turmoil had not affected Arab settlements and nomadic routes.

STAGE FOR ISLAM

In a remote place on the Arabian Peninsula, where the foreign invasions of the Middle East had had little effect on the Arab communities, an event occurred early in the seventh century that was to change history and affect the destinies of people in many parts of the world. In a town called Mecca (Makkah), about 70 kilometers (45 mi) from the Red Sea coast in the Jabal Mountains, a man named Muhammad in the year AD 611 began to receive revelations from Allah (God). Muhammad (571–632) was then in his early forties. Convinced after some initial self-doubt that he was indeed chosen to be a prophet, Muhammad committed his life to fulfilling the divine commands he believed he had received. Arab society was in social and cultural disarray, but Muhammad forcefully taught Allah's lessons and began to transform his culture. His personal power soon attracted enemies, and in 622 he fled from Mecca to the safer haven of Medina (Al Madinah), where he continued his work until his death. This moment, the *hejira* (meaning migration), marks the starting date of the Muslim era, Year 1 on Islam's calendar. Mecca, of course, later became Islam's holiest place.

The Faith

The precepts of Islam in many ways constituted a revision and embellishment of Judaic and Christian beliefs and traditions. All of these faiths have but one god, who occasionally communicates with humankind through prophets; Islam acknowledges that Moses and Jesus were such prophets but considers Muhammad to be the final and greatest prophet. What is earthly and worldly is profane; only Allah is pure. Allah's will is absolute; Allah is omnipotent and omniscient. All humans live in a world that Allah created for their use but only to await a final judgment day.

Islam brought to the Arab World not only the unifying religious faith it had lacked but also a new set of values, a new way of life, a new individual and collective dignity. Islam dictated observance of the Five Pillars: (1) repeated expressions of the basic creed, (2) daily prayer, (3) a month each year of daytime fasting (Ramadan), (4) the giving of alms, and (5) at least one pilgrimage to Mecca in each Muslim's lifetime (the *hajj*). Islam prescribed and proscribed in other spheres of life as well. It forbade alcohol, smoking, and gambling. It tolerated polygamy, although it acknowledged the virtues of monogamy. Mosques appeared

FIGURE 7A-4

© H. J. de Blij, P. O. Muller, and John Wiley & Sons, Inc.

in Arab settlements, not only for the (Friday) sabbath prayer, but also as social gathering places to knit communities closer together. Mecca became the spiritual center for a divided, widely dispersed people for whom a collective focus was something new.

The Arab-Islamic Empire

Muhammad provided such a powerful stimulus that Arab society was mobilized almost overnight. The prophet died in 632, but his faith and fame spread like wildfire. Arab armies carrying the banner of Islam formed, invaded, conquered, and converted wherever they went. As Figure 7A-4 shows, by AD 700 Islam had reached far into North Africa, into Transcaucasia, and into most of Southwest Asia. In the centuries that followed, it penetrated southern and eastern Europe, Central Asia's Turkestan, West Africa, East Africa, and South and Southeast Asia, even reaching China by AD 1000.

Routes of Diffusion

The spread of Islam provides a good illustration of a series of processes known as spatial diffusion [7] that focus on the way ideas, inventions, and cultural practices propagate through a population in space and time. In 1952, the Swedish geographer named Torsten Hägerstrand published his fundamental study of spatial diffusion entitled *The Propagation of Innovation Waves.* He reported that diffusion takes place in two forms: expansion diffusion [8], when propagation waves originate in a strong and durable source area and spread outward, affecting an ever larger region and population; and relocation diffusion [9], in which migrants carry an innovation, idea, or (for example) a virus from the source to distant locations and it diffuses from there. Islam spread mostly through expansion diffusion (the worldwide spread of AIDS is a case of relocation diffusion).

Islam on the March

Both expansion and relocation diffusion include several types of processes. The dissemination of Islam, as Figure 7A-4 shows, initially proceeded via a form of expansion diffusion called contagious diffusion [10] as the faith moved from village to village across the Arabian Peninsula, North Africa, and Southwest Asia. But Islam got powerful boosts when kings, chiefs, and other high officials were converted, who in turn propagated the faith downward through their bureaucracies to far-flung subjects. This form of expansion diffusion, called hierarchical diffusion [11], served Islam well.

But the map leaves no doubt that Islam later spread by relocation diffusion as well, notably to the Ganges Delta in South Asia, to present-day Indonesia, and to East Africa. That process continues to this day, but the heart of Islam remains in Southwest Asia. There, Islam became the cornerstone of an Arab Empire with Medina as its first capital. As the empire grew by expansion diffusion, its headquarters were relocated from Medina to Damascus (in what is now Syria) and later to Baghdad on the Tigris River (in modern Iraq). And it prospered. In architecture, mathematics, and science, the Arabs overshadowed their European contemporaries. The Arabs established institutions of higher learning in many cities including Cairo, Baghdad, and Toledo (Spain), and their distinctive cultural landscapes united their far-flung domain (see box titled "The Flowering of Islamic Culture"). Non-Arab societies in the path of the Muslim drive were not only Islamized, but also Arabized, adopting other Arab traditions as well. Islam had spawned a culture; it still lies at the heart of that culture today.

As we have noted, Islam's expansion eventually was checked in Europe, Russia, and elsewhere. But a map showing the total area under Muslim sway in Eurasia and Africa reveals the enormous dimensions of the domain affected by Islamization [12] at one time or another (Fig. 7A-5). Islam continues to expand, now mainly by relocation diffusion. There are Islamic communities in cities as widely scattered as Vienna, Singapore, and Cape Town, South Africa; Islam is also growing rapidly in the United States. With about 1.7 billion adherents today (24 percent of humankind), Islam is a vigorous, burgeoning, and still-growing cultural force around the world.

Islam and Other Religions

Two additional major faiths also had their sources in the **Levant** (the area extending from Greece eastward along the Mediterranean coast to northern Egypt), Judaism and Christianity, and both are older than Islam. Islam's rise submerged many smaller Jewish communities, but the Christians, not the Jews, waged centuries of holy war against the Muslims, seeking, through the Crusades, not only to drive Islam back but to reestablish Christian communities where they had dominated before the Islamic expansion. The aftermath of that campaign still marks the realm's cultural landscape today. A substantial Christian minority (about 40 percent of the population) remains in Lebanon, and Christian minorities also survive in Israel, Egypt, Syria, and Jordan. Strained relations between the long-dominant Christian minority in Lebanon and the Muslim majority (itself divided into five sects) contributed to the disastrous armed struggle that engulfed the country during the 1970s and 1980s. The conflict reemerged in 2006, and Lebanon has once again fallen into turmoil.

By far the most intense conflict in modern times has pitted the Jewish state, Israel, against its Islamic neighbors near and far. Israel's United Nations-sponsored creation in 1948 has precipitated more than six decades of intermittent strife and attempts at mediation; it also caused friction among Islamic states in the region. Jerusalem—holy city for Judaism, Christianity, and Islam—lies in the crucible of this confrontation.

From the Field Notes . . .

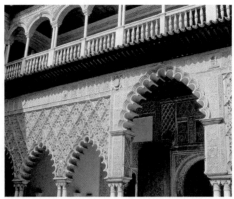

"Seville was a Muslim capital of Iberia, and walking the streets of the old city center is an adventure in cultural and historical geography. Seville fell to the Muslim invaders in 711, and the Muslims built a large mosque in the heart of town. Later the Christians, who ousted the Muslims in 1248, destroyed most of the mosque and built a huge cathedral on the site, but they saved the ornate minaret, made it the bell tower, and called it the Giralda (left). This minaret had been built by the Muslims to match the 67-meter (220-ft) minaret of the Koutoubia Mosque in Marrakech (in present-day Morocco). But the finest Islamic legacy in Seville surely is the Alcazar Palace (above), begun in the late twelfth century and embellished and finished by the Christians after their victory. Its arched façades and intricately carved walls remain a monument to the Muslim architects and artists who designed and created them."

www.conceptcaching.com

© H. J. de Blij

The Flowering of Islamic Culture

THE CONVERSION OF a vast realm to Islam was not only a religious conquest: it was also accompanied by a glorious explosion of Arab culture. In science, the arts, architecture, and other fields, Arab society far outshone European society. While the western European remnants of the Roman Empire languished, Arab energies soared.

When the wave of Islamic diffusion reached the Maghreb (western North Africa), the Arabs saw, on the other side of the narrow Strait of Gibraltar, an Iberia ripe for conquest and ready for renewal. An Arab-Berber alliance, called the Moors, invaded Spain in 711 and controlled all but northern Castile and Catalonia before the end of the eighth century.

It took seven centuries for Catholic armies to recapture all of the Arabs' Iberian holdings, but by then the Muslims had made an indelible imprint on the Spanish and Portuguese cultural landscape. The Arabs brought unity and imposed the rule of Baghdad, and their works soon overshadowed what the Romans had wrought.

Al-Andalus, as this westernmost outpost of Islam was called, was endowed with thousands of magnificent castles, mosques, schools, gardens, and public buildings. The ultimately victorious Christians destroyed most of the less durable art (pottery, textiles, furniture, sculpture) and burned the contents of Islamic libraries, but the great Islamic structures survived, including the Alhambra in Granada, the Giralda in Seville, and the Great Mosque of Córdoba, three of the world's greatest architectural achievements. While Spanish and Portuguese culture became Hispanic-Islamic culture, the Muslims were transforming their cities from Turkestan to the Maghreb in the image of Baghdad. The lost greatness of a past era still graces those townscapes today.

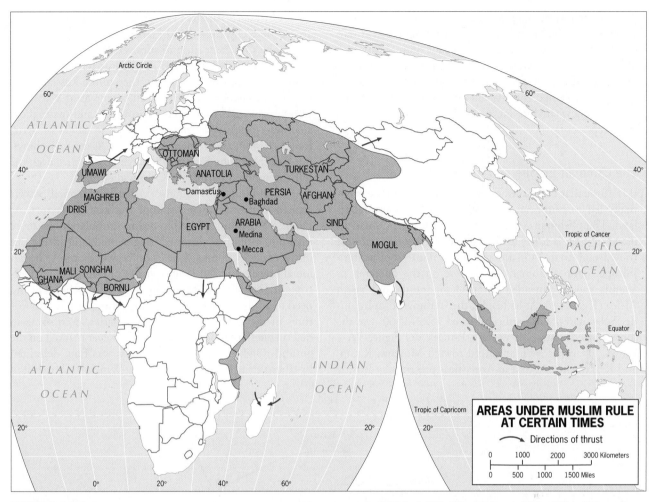

FIGURE 7A-5

© H. J. de Blij, P. O. Muller, and John Wiley & Sons, Inc.

ISLAM DIVIDED

For all its vigor and success, Islam is still fragmented into sects. The earliest and most consequential division arose after Muhammad's death. Who should be his legitimate successor? Some believed that only a blood relative should follow the prophet as leader of Islam. Others, a majority, felt that any devout follower of Muhammad was qualified to take over. The first chosen successor was the father of Muhammad's wife (and thus not a blood relative). But this did not satisfy those who wanted to see a man named Ali, a cousin of Muhammad, made *caliph* (successor). When Ali's turn came, his followers, the **Shi'ites**, proclaimed that Muhammad finally had a legitimate successor. This offended the **Sunnis**, those who did not see a blood relationship as necessary for the succession. From the beginning of this disagreement, the numbers of Muslims who took the Sunni side far exceeded those who regarded themselves as Shia (followers) of Ali. The great expansion of Islam was largely propelled by the Sunnis; the Shi'ites survived as minorities scattered throughout the realm. Today, about 85 percent of all Muslims are Sunnis, but the Shia–Sunni schism is at the root of various conflicts in this realm.

The Strength of Shi'ism

The minority Shi'ites have always vigorously promoted their version of the faith. In the early sixteenth century their work paid off: the royal house of Persia (present-day Iran) made Shi'ism the only legal religion throughout its vast empire. That domain extended from Persia into lower Mesopotamia (modern Iraq), into Azerbaijan, and into western Afghanistan and Pakistan. As the world map of religions (Fig. G-10) shows, this created for Shi'ism a large culture region and gave the faith unprecedented strength. Iran remains the bastion of Shi'ism in the realm today, and the appeal of Shi'ism continues to radiate into neighboring countries and even farther afield.

During the late twentieth century, Shi'ism gained unprecedented influence in the realm. In its heartland, Iran, a *shah* (king) tried to secularize the country and to limit the power of the *imams* (mosque officials); he provoked a revolution that cost him the throne and made Iran a Shi'ite Islamic republic. Before long, Iran was at war with neighboring Sunni-ruled but Shi'ite-majority Iraq, and Shi'ite parties and communities elsewhere were invigorated by the newfound power of Shi'ism. From Arabia to

Africa's northwestern corner, Sunni-ruled countries warily watched their Shi'ite minorities, newly imbued with religious fervor. Mecca, the sacred place for both Sunnis and Shi'ites, became a battleground during the week of the annual pilgrimage, and for a time the (Sunni) Saudi Arabian government denied entry to Shi'ite pilgrims. That rift has healed somewhat, but intra-Islamic sectarian differences run deep.

The Ottoman Empire and Its Aftermath

It is ironic that Islam's last great advance into Europe eventually led to European occupation of Islam's very heartland. The Ottomans (named after their leader, Osman I), based in what is today Turkey, conquered Constantinople (now Istanbul) in 1453 and then pushed into southeastern Europe. Soon Ottoman forces were on the doorstep of Vienna; they also invaded Persia, Mesopotamia, and North Africa (Fig. 7A-6). At its height, the Ottoman Empire under Suleyman the Magnificent (ruled 1522–1560) was the most powerful state in western Eurasia. As Figure 7A-7 shows, its armies even advanced toward Moscow, Kazan, and Krakow from a base at Azov (near present-day Rostov

in southwestern Russia). The Turks also launched marine attacks on Sicily, Spain, and France.

The Ottoman Empire survived for more than four centuries (it ended in 1923), but it lost territory as time went on, first to the Hungarians, then to the Russians, and later to the Greeks and Serbs until, after World War I, the European powers took over its provinces and made them colonies—colonies we now know by the names of Syria, Iraq, Lebanon, and Yemen (Fig. 7A-7). As the map shows, the French and the British took large possessions; even the Italians annexed part of the Ottoman domain.

The boundary framework that the colonial powers created to delimit their holdings was not satisfactory. As Figure 7A-2 illustrates, this realm's population of more than 600 million is clustered, fragmented, and strung out in river valleys, coastal zones, and crowded oases. The colonial powers laid out long stretches of boundary as ruler-straight lines across uninhabited territory; they saw no need to adjust these boundaries to cultural or physical features in the landscape. Other boundaries, even some in desert zones, were poorly defined and never marked on the ground. One product of this process was Iraq, delimited by (then) British Colonial Secretary Winston Churchill in

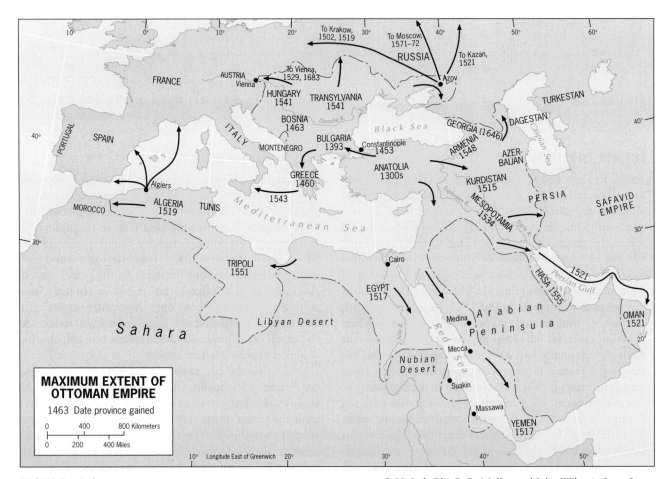

FIGURE 7A-6

© H. J. de Blij, P. O. Muller, and John Wiley & Sons, Inc.

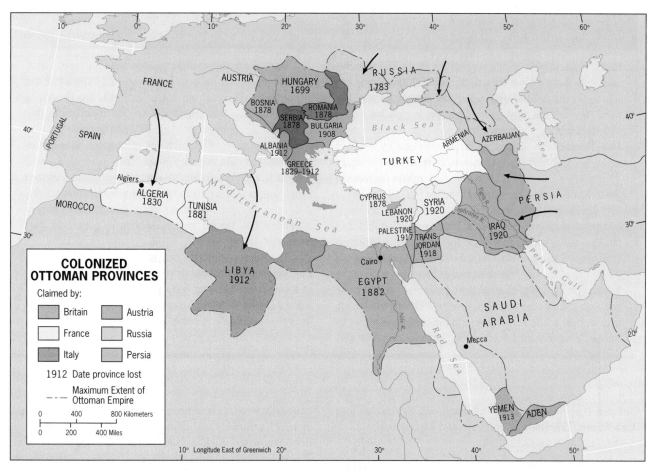

FIGURE 7A-7

© H. J. de Blij, P. O. Muller, and John Wiley & Sons, Inc.

1921 as a kingdom with Baghdad as its capital; another was Jordan, centered on Amman. Later, when the colonies had become independent states, such boundaries led to quarrels, even armed conflicts, among neighboring Muslim states. Moreover, certain national populations were entirely excluded from the new politico-geographical matrix, and nearly a century later they still aspire to statehood (see box titled "A Future Kurdistan?").

THE POWER AND PERIL OF OIL

About 30 of the world's countries have significant oil reserves. But five of these countries, all located in the North Africa/Southwest Asia realm, in combination possess larger reserves (roughly 57 percent of the world total) than all of the rest combined. The estimated reserves of this Big Five in 2011, averaged from several sources including the governments of the countries themselves and reported in billions of barrels, were as follows: (1) Saudi Arabia, 264; (2) Iran, 150; (3) Iraq, 143; (4) Kuwait, 102; and (5) the United Arab Emirates, 98. The next-ranking country, Venezuela, had known reserves just above 96 billion barrels; Russia about 79; Libya, Nigeria, Kazakhstan, and Qatar between 25 and 50; and the United States just above 21.

No other country's known reserves surpassed 16 billion barrels as of mid-2011. It should also be noted that Canada may possess the largest oil reserves of all—but most of this supply is contained within Alberta's *tar sands* (as noted in Chapters 3A and 3B), which requires complex and expensive recovery and refining processes. Yet with oil prices at unprecedented high levels, Canada's production is likely to grow. And finally, as we reported in Chapter 5B, spectacular new discoveries are being made in Brazil's offshore oilfields; if as-yet-unproven estimates of these massive reserves are confirmed, that country is poised to surpass the United States and become a major oil exporter before the end of this decade.

In general terms, oil (and associated natural gas) exists in this realm in three discontinuous zones (Fig. 7A-8). The most productive of these zones extends from the southern and southeastern part of the Arabian Peninsula northwestward around the rim of the Persian Gulf, reaching into Iran and continuing northward into Iraq, Syria, and southeastern Turkey, where it peters out. The second zone lies across North Africa and extends from north-central Algeria eastward across northern Libya to Egypt's Sinai Peninsula, where it ends. The third zone begins on the margins of the realm in eastern Azerbaijan, continues eastward under the Caspian

A Future Kurdistan?

POLITICAL MAPS OF the heart of the North Africa/Southwest Asia realm do not show it, but where Turkey, Iraq, and Iran meet, the cultural landscape is not Turkish, Iraqi, or Iranian. Here live the Kurds, a fractious and fragmented nation of approximately 35 million (their numbers are uncertain). More Kurds live in Turkey than in any other country (perhaps as many as 15 million); possibly as many as 6.5 million in Iraq; another 5 or so million in Iran; and smaller numbers in Syria, Armenia, and even Azerbaijan (see Fig. 7B-4).

The Kurds have occupied this isolated, mountainous, frontier zone for over 3000 years. They are a nation, but they have no state; nor do they enjoy the international attention that peoples of other **stateless nations [13]** (such as the Palestinians) receive. Turkish as well as Iraqi repression of the Kurds, and Iranian betrayal of their aspirations, briefly make the news but are soon forgotten. Relative location has much to do with this: spatial remoteness and the obstacles created by the ruling regimes inhibit access to their landlocked domain.

Many Kurds dream of a day when their fractured homeland will become a nation-state. Most would agree that the city of Diyarbakir, now in southeastern Turkey, would become the capital. It is the closest any Kurdish town comes to a primate city, although the largest urban concentration of Kurds today is in the shantytowns of Istanbul, where more than 3 million have migrated. In their heartland, meanwhile, the Kurds, their shared goals notwithstanding, are a divided people whose intense disunity has thwarted their objectives; and, as in the case of Europe's Basques, a small minority of extremists has tended to use violence in pursuit of their aims.

The Kurds—without a seacoast, without a powerful patron, without a global public relations machine—are victims of a conspiracy between history and geography. One of the world's largest stateless nations keeps hoping for a deliverance that is unlikely to come.

Sea into Turkmenistan and Kazakhstan, and also reaches into Uzbekistan, Tajikistan, Kyrgyzstan, and Afghanistan.

Producers and Consumers

Saudi Arabia is the world's largest oil exporter, but in recent years Russia has risen to second place. As we noted in Chapters 2A and 2B, oil and natural gas are by far Russia's most valuable commodities, and the Russian state direly needs the foreign revenues they produce. Thus Russia, with modest (though expanding) known reserves, is vigorously exporting to international markets and is now a leading force in the global energy picture. In combination, however, production by the countries of the North Africa/Southwest Asia realm far exceeds that from all other sources. The United States, not surprisingly, is the world's leading importer even while it consumes almost all of its own production.

As Figure G-11 indicates, the production and export of oil and gas has elevated several of this realm's countries into the higher-income categories. But petroleum wealth also has enmeshed these Islamic societies and their governments in global strategic affairs. When regional conflicts create instability in producing countries that have the potential to disrupt supply lines, powerful consumers are tempted to intervene—and have done so.

Colonial Legacy

When the colonial powers laid down the boundaries that partitioned this realm among themselves, no one knew about the riches that lay beneath the ground. A few wells had been drilled, and production in Iran had begun as early as 1908 and in Egypt's Sinai Peninsula in 1913. But the

major discoveries came later, in some cases after the colonial powers had already withdrawn. Some of the newly independent countries, such as Libya, Iraq, and Kuwait, found themselves with wealth undreamed of when the Turkish Ottoman Empire collapsed. As Figure 7A-8 shows, however, others were less fortunate. A few countries had (and still have) potential. The smaller, weaker emirates and sheikdoms on the Arabian Peninsula always feared that powerful neighbors would try to annex them (Kuwait faced this prospect in 1990 when Iraq invaded it). The unevenly distributed oil wealth, therefore, created another source of division and distrust among Islamic neighbors.

A Foreign Invasion

The oil-rich countries of the realm found themselves with a coveted energy source but lacking in the skills, capital, or equipment to exploit it. These had to come from the Western world and entailed what many tradition-bound Muslims feared most: a strong foreign presence on Islamic soil, foreign intervention in political as well as economic affairs, and penetration of Islamic societies by the vulgarities of Western ways. The transport of oil from these countries to destinations around the world also raised the importance of strategic arteries connecting this realm to the rest of the world (see box titled "Choke Points: Danger on the Sea Lanes").

Many Muslims' worst fears came true in Iran during the 1950s, when its government tried to control British oil exploitation centered on Abadan and the Persian Gulf. The imperious ways of the British, the luxurious life of the European

Choke Points: Danger on the Sea Lanes

The Hormuz Strait at the outlet of the Persian Gulf between the Arabian Peninsula and Iran (see Fig. 7A-1) is only one of dozens of maritime choke points that have taken on added significance during this time of terrorist challenges. A **choke point [14]** is defined as the narrowing of an international waterway causing marine traffic congestion, requiring reduced speeds and/or sharp turns, and increasing the risk of collision as well as vulnerability to attack.

For many years, Southeast Asia's Strait of Malacca between the Malay Peninsula and the Indonesian island of Sumatera (Sumatra) has been a notorious choke point because reduced speeds in the Strait gave pirates the opportunity to board vessels and plunder them or worse, kill their crews, and take them over. Since the 1980s, hundreds of acts of piracy have made this one of the world's least safe waterways, especially for smaller ships.

One of the world's busiest choke points is the Strait of Gibraltar between Spain and Morocco, which was in the news a few years ago when an al-Qaeda document was found that referred to plans for an attack on Western ships slowing down through this 60-kilometer (35-mi)-long funnel that narrows to 13 kilometers (8 mi) in width and is subject to high winds and fast-flowing currents.

Another particularly risky choke point these days is the Gulf of Adan that leads to the narrow Bab el Mandeb Strait between Africa and the Arabian Peninsula, or, more specifically,

between Djibouti and Yemen, at the southern entrance to the Red Sea. Barim (Perim) Island partially blocks this 30-kilometer (18-mi)-wide strait. Hundreds of acts of piracy committed by Somali bandits, based on the beaches of the nearby Horn of Africa, have imperiled shipping in these sea lanes (see photo, page 338). In 2008, among other vessels they seized a freighter loaded with heavy weapons and a fully-loaded 300,000-ton oil tanker; but their attempt to hijack the *Maersk Alabama* and its American crew in 2009 triggered a successful counterattack by the U.S. Navy and strengthened international resolve to end this scourge.

The English Channel between Britain and France, Turkey's narrow Bosphorus and Dardanelles between the Black and Mediterranean seas, Indonesia's Sunda Strait between the Java Sea and the Indian Ocean, the Strait of Magellan for ships rounding the southern end of South America, and the Hainan Strait between southernmost China's Leizhou Peninsula and Hainan Island are just some of the many other choke points on the world map.

And some choke points are artificial. The Suez Canal between the Red Sea and the Mediterranean Sea as well as the Panama Canal between Atlantic (Caribbean) and Pacific waters are the busiest and most prominent, but many other route-shortening canals are vulnerable choke points at a time when the risk of hostile action looms larger than the risk of collision.

expatriates compared to the abject poverty of Iranian workers, and the unfair terms of the concession under which the oil was exported all contributed to the rise of nationalist sentiment in Iran, whose leaders appealed to the United States for help in negotiating a better deal with the British. American President Harry Truman had some sympathy for the Iranians, but his successor Dwight Eisenhower worried about Iran turning toward communism. Capitalizing on political disputes within Iran, CIA operatives engineered the overthrow of elected Premier Mohammad Mosaddeq, restoring to power the young shah and setting into motion a sequence of events that would eventually lead to the downfall of Iran's monarchy and the proclamation of an Islamic Republic in 1979.

Even when foreign intervention was more subtle, it intensified clashes between the traditional and modern. Oil revenues created cultural landscapes in which gleaming skyscrapers towered over ornate mosques and historic neighborhoods. The social chasms between the rich, well-connected elites and less fortunate citizens bred resentment. Involvement of Western armies in regional conflicts produced, in the eyes of some, the unthinkable: foreign armies on Arabia's hallowed soil. To some Muslims, all this violated the most basic tenets of their faith, and some

retaliated with violence of their own. That violence, in the form we define as terrorism, has become one of the major challenges of our time.

The Geography of Oil's Impact

As is true of every natural resource found across the planet, oil has completely transformed cultural landscapes in certain parts of the North Africa/Southwest Asia realm and left others virtually unchanged. For hundreds of millions of this realm's inhabitants, a patch of tillable soil and a source of water still mean more to daily life than all the oil in OPEC.* Countrysides in Arab as well as non-Arab regions in this part of the world continue to carry the imprints of centuries of cultural tradition, not decades of oil-driven modernization. Nonetheless, oil and natural gas—

---

*OPEC, the Organization of Petroleum Exporting Countries, is the international oil cartel (syndicate) formed by 12 producing countries to promote their common economic interests through the setting of joint pricing policies and the limitation of market options for consumers. Its 8 NASWA members are Algeria, Iran, Iraq, Kuwait, Libya, Qatar, Saudi Arabia, and the United Arab Emirates.

FIGURE 7A-8

© H. J. de Blij, P. O. Muller, and John Wiley & Sons, Inc.

their location, production, transportation, and sale—have produced massive changes that include the following.

1. *Urban Transformation.* Undoubtedly the most visible manifestation of oil wealth is the modernization of cities. By far the tallest building in the world rises above Dubai in a state called the United Arab Emirates, but it is only one in a forest of glass-encased skyscrapers, many of which test the limits of design and engineer-

ing. Besides the capitals (such as Saudi Arabia's Riyadh) a number of other cities on the Arabian Peninsula also reflect the riches oil has brought—yes, even Mecca. And, as we note in Chapter 7B, entirely new cities are springing up in the deserts and along the coasts.

2. *Variable Incomes.* Oil and natural gas prices fluctuate on world markets (in 2008 oil sold for more than [U.S.] $140 per barrel before plunging to less than

Map of the Middle East / Central Asia showing oil fields, pipelines, and national boundaries. Countries labeled include RUSSIA, KAZAKHSTAN, UZBEKISTAN, TURKMENISTAN, AZERBAIJAN, GEORGIA, TURKEY, SYRIA, LEBANON, IRAQ, IRAN, KUWAIT, SAUDI ARABIA, QATAR, BAHRAIN, UNITED ARAB EMIRATES, OMAN, YEMEN, TAJIKISTAN, KYRGYZSTAN, AFGHANISTAN, ERITREA, ETHIOPIA, JORDAN. Water bodies: Black Sea, Caspian Sea, Aral Sea, Red Sea, Persian Gulf, Gulf of Oman, Arabian Sea, Strait of Hormuz. Cities: Novorossiysk, Ceyhan, Baku, Basra, Abadan, Kuwait, Khark I., Ad Dammam, Dhahran, Doha, Mutrah, Yanbu, Al Hudayadah.

Bar chart: WORLD NATURAL GAS PRODUCTION — RUSSIA, UNITED STATES, CANADA, IRAN, ALG., NORWAY, NETH., SAUDI AR., INDIA, CHINA, TURK., UZBEK., MALAYSIA, QATAR, INDON., U.A.E., MEXICO, OTHER N.A./S.W. ASIA, OTHER.

Bar chart: WORLD NATURAL GAS RESERVES — RUSSIA, IRAN, QATAR, SAUDI AR., U.S., U.A.E., NIG., VENEZ., ALG., IRAQ, INDON., OTHER N.A./S.W. ASIA, OTHER.

$40, but by mid-2011 the price of a barrel was once again near $100). When energy prices are high, several states in this realm rank among the highest-income societies in the world. Even when they decline, many petroleum-exporting countries manage to remain at least in the upper-middle-income category (Fig. G-11).

3. **Infrastructure.** Massive spending on airports, seaports, bridges, tunnels, four-lane highways, public buildings, shopping malls, recreational facilities, and other components of national infrastructures creates an image

of comfort and affluence quite unlike that prevailing in countries without major oil or gas revenues. Saudi Arabia is now engaged in a vast modernization project extending from coast to coast.

4. **Industrialization.** A number of far-sighted governments among those with oil wealth, realizing that reserves will not last forever, are investing some of their income in industries that will outlast the era of massive oil exporting. Petrochemical manufacturing using domestic supplies, aluminum, steel, and fertilizers are

among those industries, although others potentially more promising, for example those in high-technology fields, are not yet a significant part of this important initiative.

5. *Regional Disparities.* Oil wealth, like other high-value resources, tends to create strong regional contrasts both within and among countries. Saudi Arabia's ultramodern east coast is a world apart from most of its interior, which remains a land of barren deserts, widely scattered oases, vast distances, isolated settlements, and slow change. This is not unique to countries in this realm: sharp regional disparities mark oil-rich countries from Algeria to Azerbaijan.

6. *Foreign Investment.* Governments and private entrepreneurs have invested massive quantities of oil-generated wealth in foreign countries, buying financial securities as well as acquiring prestigious hotels, famous stores, and other high-cachet properties. These investments have created a network of international linkages that not only connects NASWA states and individuals to the national economies of other realms but also links them to the growing Islamic communities in those countries.

7. *Foreign Involvement.* To many inhabitants of this realm, especially those with strong Islamic-revivalist (fundamentalist) convictions, the inevitable presence of foreigners (including businesspeople, politicians, architects, engineers, and even armed forces) on Islamic soil is an unwelcome byproduct of the energy era. In fact, public opinion in Saudi Arabia forced its ruling regime to negotiate the 2003 departure of American troops from the country's territory.

8. *Intra-Realm Migration.* Oil wealth allows governments, industrialists, and private individuals to hire workers from less-favored parts of the realm to labor in the oilfields, ports, and numerous menial service occupations. This has brought many Shi'ites to the countries of the eastern zone of the Arabian Peninsula, where they now form significant sectors of national populations; hundreds of thousands of Palestinian Arabs also sought temporary work here in local industries. In 2010 it was estimated that Saudi Arabia, with a population of just over 30 million, hosted more than 5 million foreign workers from elsewhere in the realm.

9. *Migration from Other Realms.* The labor market in such places as Dubai and Abu Dhabi in the United Arab Emirates has also attracted workers from beyond the realm's borders during recent periods of rapid growth. Wages in such countries as Pakistan, India, Sri Lanka, and Bangladesh are even lower than those paid by the building industry or private employers in oil-boom-driven NASWA states. These workers serve mainly as domestics, gardeners, trash collectors, and the like. In 2010, about 85 percent of the total population of the United Arab Emirates consisted of foreign employees! Over the past few years, substandard working conditions and wage issues have also led to protests that exposed the harsh circumstances faced by many guest workers throughout this realm.

10. *Diffusion of Revivalism.* Oil and gas revenues are utilized by Islamist regimes to support Muslim communities as well as their mosques and cultural centers throughout the world. No NASWA country has spent more money on such causes than Saudi Arabia, and thousands of mosques from England to Indonesia prosper as a result. This example of relocation diffusion creates nodes of recruitment for the faith—and ensures the dissemination of revivalist principles.

FRAGMENTED MODERNIZATION

More than any other realm, North Africa/Southwest Asia is marked by a human spatial mosaic of **fragmented modernization [15]**: in some respects it is highly modern and prosperous but in more ways it remains traditional, stagnant, and quite poor. These disparities that characterize so many of the realm's societies are best explained by the uneven impact of oil, the dominance of conservative autocratic regimes, and the tenuous role of religion. It explains why much of the realm often seems on edge, and it provides a vital backdrop to the recent revolutionary turmoil that surged from Tunisia to Yemen to Syria and engulfed several countries in between.

The Uneven Impact of Oil

Oil brought this realm into contact with the outside world in ways unforeseen just a century ago. Oil has strengthened and empowered some of its peoples, but has dislocated and imperiled others. It has truly been a double-edged sword.

Before we turn to the regions of this realm in Chapter 7B, we should remind ourselves that the great majority of NASWA's inhabitants are not, in their daily lives, directly affected by the changes the energy era has brought. Most Moroccans, Tunisians, Egyptians, Jordanians, Yemenis, Lebanese, and many millions of others—Kurds, Palestinians, Berbers, Tuaregs—make ends meet by trading or farming or working at jobs their parents and grandparents performed. Take the case of Iran, which as a country earns about two-thirds of its income from oil and natural gas. Only one-half of one percent of that country's workforce (just over 100,000 out of more than 20 million workers) earn a salary from energy or energy-related work. By far the largest number in any single occupation are the farmers (5 million). And for all their oil, Iranians in 2010 earned a mere (U.S.) $10,840 per person—less than in Turkey and only 23 percent of the average in Singapore. As we shall discover, cultural-geographic rather than economic-geographic forces mainly shape the regionalization of this realm.

The nearly 30 states of the North African/Southwest Asia geographic realm display an often stunning degree of variety and diversity. Oil and natural gas have generated wealth, but also severe inequality and disparity—not only *within* the countries with large reserves, but also *between* the haves and the have-nots (see photo below). Ultramodern city skylines on the Arabian Peninsula stand in sharp contrast to Nile Valley villages virtually unchanged for millennia. Super-highways cross deserts still traversed by nomadic traders riding camels. Social geographies vary just as greatly: for instance, the female literacy rate in Lebanon is more than 80 percent, but only 25 percent in Yemen. And the demographic range is also enormous: the annual growth of oil-poor Tunisia's population equals the world average (which is far below the realm's average), while in oil-rich Saudi Arabia it is more than twice the global rate.

Autocratic Regimes

Democracy has been a rarity across this realm; although specific reasons vary, the majority of countries share a history of autocratic, conservative regimes, usually backed strongly by their military forces. Most countries were part of the Ottoman Empire before World War I (see Fig. 7A-7), and only became independent after a period of (attempted) European rule that was, in some cases, sanctioned by the **League of Nations [16]** (the forerunner of the United Nations). Accordingly, Egypt, Iraq, Palestine, Transjordan, and parts of Yemen were administered by Britain; Lebanon, Syria, Tunisia, Morocco, and Algeria were ruled by France; and Libya was controlled by Italy.

In most cases, the Europeans faced major and tenacious opposition from local populations but were reluctant to let go. Independence was generally preceded by intense conflict or even war, and none of these countries can be said to have been prepared to function as West European-style democracies. Indeed, virtually everywhere the newly independent countries fell into the hands of the military (e.g., Egypt) or restored monarchs with strong military support (e.g., Yemen).

After World War II, these autocratic regimes were often cemented into place through foreign involvement. Mostly, this was about oil: foreign powers needed oil, showed little interest in domestic politics, and provided local governments with abundant revenues to build up their military power, secure their political position, and maintain order. Saudi Arabia, the world's richest petrostate, is the quintessential example. In other cases, it was Cold War politics that effectively supported the realm's nondemocratic governments. For instance, the United States and the Soviet Union competed to incorporate Egypt into their spheres of influence. Egypt was first championed by the Soviet Union from the mid-1950s through the early 1970s, and then by the United States since 1973 (Egypt has for many years been the second-biggest recipient of U.S. foreign aid after Israel).

The governments in this realm vary: some are republics and others are monarchies; most are secular (nonreligious), but a few others function as Islamic states. For example, Egypt, Tunisia, and Syria are secular republics; Iran and Afghanistan are Islamic republics; Saudi Arabia and Qatar (to some extent) are Islamic monarchies; and Jordan and Yemen are secular monarchies. Yet all these countries share traditions of autocratic government: even though they have embraced some notions of Western modernization, democratic freedoms have been scarce. Frequently, government leaders have been in power for many years, and their main concern has been to preserve their control and the status quo. Their reigns have been generally conservative, top-down with little concern for the "people on the street," and sometimes marked by repression and violence. Thus, across much of this realm, the common people have been deprived of economic advancement as well as political and religious freedoms. For how long, you might wonder, could the masses put up with that kind of situation?

Religious Revivalism

Another schism in this realm relates to the resurgence of religious fundamentalism, or as Muslims refer to it, **religious revivalism [17]**. During the 1970s, the imams in Shi'ite Iran sought to reverse the shah's moves toward liberalization and secularization: they wanted to (and did) recast society in a traditional, revivalist Islamic mold. An *ayatollah* (leader under Allah) replaced the shah in 1979; Islamic rules and punishments were instituted. Urban women, many of whom had been considerably liberated and educated during the shah's regime, were forced to return to more traditional roles. Vestiges of Westernization, encouraged by the shah, disappeared. Under these new conditions, even the war against neighboring Iraq (1980–1990), which began as

The geography of inequality is literally visible anywhere in this realm—even here just blocks away from some of the world's highest real-estate values along Dubai's main axis, Sheikh Zayed Road. In astonishing contrast to the city's glittering, ultramodern skyline, these ramshackle structures sitting atop the desert surface house poverty-stricken migrants whose meager shacks are all the protection they have against searing summer heat and nocturnal winter cold in this virtually rain-free climate.

© Picture Contact BV/Alamy Limited

a conflict over territory, became a holy crusade that cost more than a million lives.

Islamic revivalist fundamentalism did not rise in Iran alone, nor was it only confined to Shi'ite communities. Many Muslims—Sunnis as well as Shi'ites—in all parts of the realm disapproved of the erosion of traditional Islamic values, the corruption of society by European colonialists and later by Western modernizers (that brought economic benefits only to the elites), and the declining power of the faith within the secular state. As noted above, most governments in this realm are both secular *and* repressive, and they do not provide the economic growth their citizens yearn for. In that sense, it is not so hard to see why many people would seek solace in religion or even turn fundamentalist. Think of religion as a way to regain hope and dignity.

But some have taken their return to the faith quite far and imbued it with a fanaticism that is alien to most Muslims—and this has set Muslim against Muslim in many parts of the realm. Revivalists fired the faith with a new militancy, challenging the status quo from Afghanistan to Algeria. The militants forced their governments to ban "blasphemous" books, to re-segregate the schools according to gender, to enforce traditional dress codes, to legitimize religious-political parties, and to heed the wishes of the *mullahs* (teachers of Islamic ways). Militant Muslims proclaimed that Western ideas about modernization, inherited from colonialists and adopted by Arab nationalists, were incompatible with the dictates of the Quran.

Terror in the Name of Islam

The great majority of Muslims are not fundamentalists; not all fundamentalists are militants; and not all militants are terrorists. Most religious scholars point out that Islam is no more predisposed toward militancy than other religions. We should also note that terrorism is a tool of war that has been used around the world, and indeed throughout modern history, by people of different beliefs and ethnic backgrounds.

But nowadays the religion of Islam, in particular, has proven to be a significant vehicle for political mobilization and a powerful source of inspiration for militants with terrorist agendas. To be sure, their extreme interpretation of the Quran is contested by many; but at the same time, there can be little doubt that their actions are at least partly inspired by their particular fundamentalist beliefs. Theirs is a "holy war"—a *jihad* [18]—and as such it is a deeply reactionary movement that looks to the past rather than the future.

In Afghanistan, the so-called **Taliban** ("seekers of religion"), a kind of Islamist militia, control large parts of the country. They adhere to **Wahhabism [19]**, an orthodox form of Sunni Islam similar to what is practiced in Saudi Arabia. The Taliban impose a very rigid interpretation of Islamic law and have decreed a reversion to what in many ways resembles a premodern society. For example, criminals are executed in public, men are jailed for not growing beards, and education for women is outlawed. The Taliban also exert a strong influence across the border in parts of northern Pakistan and are well known for harboring **al-Qaeda** terrorists.

If the Taliban are mainly concerned with regional territorial control (they emerged from rebel groups that opposed the Soviet occupation of Afghanistan during the 1980s), al-Qaeda has a much broader global agenda. *Al-Qaeda* means "base" or "foundation," and it is better understood as a multinational organization or network. It possesses a tightly knit core but is difficult to define along its edges, with participants and sympathizers in several countries. Al-Qaeda's stated aim is to establish Islamic rule across the realm and to banish all foreign influence. To them, all means are justified. Their actions have killed thousands of people, mostly innocent bystanders, inside and outside the realm, with the most infamous being the 9/11 attacks of 2001 in the United States. The organization was co-founded in Afghanistan in the early 1990s by Usama bin Laden, the wealthy Saudi who especially resented the presence of U.S. military personnel in his home country. Bin Laden was a primary target of American retaliation for well over a decade before he was finally found and killed by U.S. Special Forces in Abbottabad, Pakistan on May 2, 2011. The impact of his demise on al-Qaeda, and more generally on militant Islam, remained a matter of debate at press time. Both the Taliban and al-Qaeda are discussed further in the section on Afghanistan in Chapter 7B.

THE POPULAR UPRISINGS OF 2011: AN ARAB SPRING?

On December 17, 2010, 26-year-old Tunis fruit vendor Mohammed Bouazizi became so distraught when police in the Tunisian capital, after months of routine harassment, confiscated his unlicensed produce stand, that he set himself on fire. His act of desperation resonated with countless Tunisians whose frustration with their government had reached a boiling point. Bouazizi died about two weeks later, but his self-immolation quickly came to symbolize Tunisia's "Jasmine Revolution." By mid-January of 2011, popular opposition to the government of Zine El Abidine Ben Ali had grown so fierce that the president was forced to step down and flee to Saudi Arabia after 23 years in power.

Ben Ali's departure was the climax to an unprecedented sequence of events in this realm, where autocratic regimes routinely cling to power for decades with the silent acquiescence of the masses. In the Arab World, conventional wisdom suggested, democratic government or even the *desire* for democracy was nonexistent. But that latter myth appeared to explode as Tunisians of all backgrounds, including large numbers of women and youths (see photo, p. 343), demanded an end to cronyism, corruption, repression, and economic mismanagement. The Tunisian people now insisted on being heard—and their voices would soon echo throughout the realm.

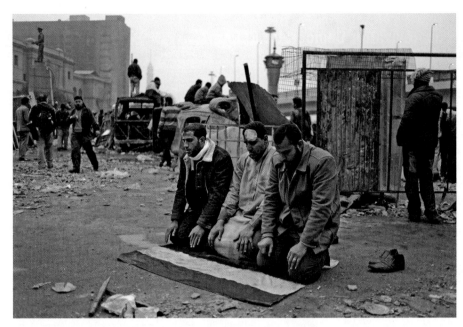

Tahrir Square, Cairo, February 5, 2011: Demonstrators praying over an Egyptian flag at the height of the anti-Mubarak uprising. The protesters demanded the ouster of Mubarak's regime as well as democratic reforms while simultaneously expressing their love of country and their Islamic beliefs. The upheaval in Egypt and elsewhere in the realm clearly reflected the deep popular resentment of long-established regimes and, as such, constituted a massive democratic breakthrough. But in mid-2011 it remained unclear as to what the political outcome would be.

© Samuel Aranda/© Corbis

Revolutionary Dominoes

What began in Tunisia, as 2010 turned into 2011, rapidly diffused across North Africa and into adjacent parts of Southwest Asia. Within a few months, unrest had spread to Egypt, Libya, Syria, Yemen, and Bahrain. The conditions in all these countries were similar: they were ruled by long established autocratic regimes that failed to bring economic progress, repressed their own people, and had lost touch (especially with the younger generation). Across the realm, people followed the events in Tunisia via television and the Internet, and revolutionary fervor spread like wildfire. It may well have been the most spectacular **domino effect [20]** in recent history.

Undoubtedly, the most important of these "falling dominoes" was Egypt. The most heavily populated, historically prominent, and centrally located country in the Arab World, Egypt had for decades suffered from stunted economic growth and an out-of-touch, repressive government. In Cairo and elsewhere, the masses took to the streets and demanded the ouster of President Mubarak, even though protesters were repeatedly confronted by harsh and bloody encounters with the police and military forces (see the photo above). In Egypt alone, an estimated 850 people died in these clashes and thousands were wounded; nonetheless, after 18 days of demonstrations Mubarak, too, was compelled to resign.

The events in Egypt generated shockwaves throughout the realm. Whereas Tunisia is a small and comparatively peripheral state and Ben Ali was not that well known outside of his country, Egypt lies at the epicenter of the Arab World and Mubarak was widely regarded as unassailable as a pharaoh. If the tide could be turned here, it could happen anywhere. At least that seemed to be the conviction of fearless protesters in surrounding countries.

However, the more despotic regimes in Libya and Syria responded with stubborn resistance and iron fists. Libya quickly degenerated into civil war, where Moammar Gadhafi's unleashing of military forces against his unarmed people was so brutal that a makeshift alliance of France, the United Kingdom, and the United States (later subsumed by NATO) intervened with air attacks to protect the Libyans from annihilation by their own government. As the spring of 2011 turned to summer (and press time), the standoff continued, with the protesters maintaining control of northeastern Libya. The rebellion in Syria began in earnest in early April and after two months of particularly ruthless repression by government forces, the death toll approached 1000. Nevertheless, the struggle in Syria continued with demonstrations in numerous cities maintaining pressure on the regime.

A New Generation

In most countries of the North Africa/Southwest Asia realm, more than half of the population is under 25 years of age. This enormous demographic cohort has grown up in a world in which information flows almost uncontrollably, so these young people know their home countries deny them the opportunities that exist elsewhere. Moreover, most have known no other national leader than the autocrat already in power at the time of their birth.

The youthfulness of these populations stands in stark contrast to the antiquated nature of their governments. Consider this list of despotic longevity at the outset of 2011: Tunisia's Ben Ali, 75 years old, 24 years in office; Egypt's Mubarak, 82/30; Yemen's Abdullah Saleh, 69/33; Libya's Moammar Gadhafi, 69/42; and Syria's Al-Assad father-and-son rulers, in office for 41 years.

The 2011 uprisings were predominantly (although not exclusively) led by the younger generation, from teenagers to thirty-somethings. In Tunisia, Egypt, and Syria, students took the lead, using the Internet (Facebook, Twitter, blogging) to organize demonstrations. This emphasized one crucial aspect of the "Arab Spring": the disconnect between the realm's aging autocrats and their young, globally aware subjects.

Regional ISSUE Religious Revival or Democratic Reform?

ISLAMIC REVIVAL IS THE ONLY WAY

"We Muslims must return to, and protect, the fundamental tenets of our faith. Too many of us have strayed, lured and seduced by the vulgarities of Western and other foreign ways. Consider our history. The Prophet set us on course, the Quran was and is the source of all truth, and the message took root from Mecca to Morocco to Malaya. Don't forget: Allah's revelations came many centuries after other religions—Hinduism, Buddhism, Christianity—had captured the minds of millions. But many saw that Islam is the only way, and today practicing Muslims (which means every believer) outnumber Christians, millions of whom never see the inside of a church and do not live pious lives.

"As a devout Muslim living in Egypt and having suffered the consequences of my devotion, let me say this: we Muslims have tried to live harmoniously with other faiths. We not only brought science and enlightenment to the world we converted, but we lived in peace with non-Arabs. In Islamic Andalusia, Jewish communities were safe and secure. It was the Christian Catholics who invented the horrors of the Inquisition. From West Africa to Central Asia, we proved our willingness to live in peace with infidels. So what did we get in return? The 'three c's'—the crusades, colonialism, and commercial exploitation! From my American friends I understand that there's an expression, 'three strikes and you're out.' Well, you non-Muslims are out. And so are those Muslims who still cooperate with foreigners, including former president Mubarak of my own country (his reward for being an infidel!) and the entire Saudi royal family. I agree with Sheikh Hamoud: whoever supports the infidel against Muslims is himself an infidel. When you sell your soul for oil dollars or foreign aid, you're subject to a fatwa.

"Yes, I'm a revivalist, or what you people call a fundamentalist. Our salvation lies in a return to the strictest rules of Islam. Every Muslim must adhere to every tenet of the faith. Youngsters should learn the Quran by heart: it is the source of all truth beyond which nothing else matters. Women should know and keep their place as it is accorded to them by the Quran. Sharia law should be imposed by all Islamic countries, so that the punishment of transgressors will serve as an example to others. We should dress conservatively, resist the degenerate temptations of other cultures, and follow the teachings of our religious leaders. Let me ask you this: are we better off today than when Sheikh al-Wahhab and Ibn Saud were forging the Islamic state of Arabia? I would say not. We have to go back to our roots to undo the damage that has been done to us.

"Ayatollah Khomeini's Islamic revolution failed in Iran, and our Algerian brothers were prevented from making Algeria an Islamic state, but these were just harbingers of things to come. Our goal is Saudi Arabia, where it all started fourteen centuries ago—and where we will make the new start that will transform the world."

ISLAMIC COUNTRIES NEED DEMOCRATIC REFORM

"How is it that we Muslims, with our rich cultural and scientific heritage, wealth of oil, history of interaction with other societies, and central location in the world have been held back so long by autocratic, corrupt, and incompetent governments, interfering and medieval clerics, stultifying traditions, and internal discord? As a practicing Tunisian Muslim who took part in the Jasmine Revolution that ousted our corrupt former regime, I can tell you that we need democratic reforms, not the religious revivalism about which I keep hearing.

"Let's be objective about this. Look at the world's social and economic statistics, and you'll see that countries where Islam is the dominant religion tend to rank near the bottom—not always, but far too often for this to be a coincidence. The globe is in the grip of a revolution in information and communication technology, but not here. The planet has a mosaic of religions, each professing to be the real path to salvation, and each bedeviled by a fundamentalist fringe, but not here. Here the fundamentalists are a constituency, not a fringe. And they are using tactics ranging from intimidation to terrorism to maintain and increase their clout.

"The problem is we are caught between despotic and self-serving governments on one side and extremist revivalists on the other side. And they put us in a downward spiral. Repressive governments that persecute revivalists only make the problem worse and Islamic extremism only invites the government to become even more repressive. This has to end.

"It all boils down to this: freedom. We need the freedom to debate the issues that confront us without fear of fatwas or imprisonment. So we need reform on both sides, political and religious. Here in Tunisia we have rid ourselves of a government that delivered nothing to the people while it persecuted revivalists. We need a government that allows people to practice the religion they want and practice it the way they want. And, most importantly, we need a government that is democratically elected and that works for the people, that cares about our youth and helps them get a proper education and jobs.

"At the same time, people must understand that religious revivalism is not a solution but is like stepping back in time. It will only bring more darkness. If anything, Islam needs a reformation, a reinterpretation of its inspired writings in light of modern times. And the mullahs need to get out of the business of controlling government, which is their constant aim. The separation of mosque and state is indispensable to our progress.

"And our reformation will have to address the plight of women. As a woman, it infuriates me that half of all women in the Arab World can neither read nor write, that they have unequal citizenship, and that they play virtually no role in government. Nowhere in the world is the domination of men more absolute, and I refuse to believe that the Prophet would approve of this. We need a democratic revolution, not a religious revival."

Vote your opinion at www.wiley.com/college/deblij

Another dimension of the upheaval, however, has a darker side. Earlier revolutionary movements, including the "Prague Spring" that led to reforms in communist former Czechoslovakia in the late 1960s, the overthrow of dictatorships in Portugal (1970s) and South Korea (1980s), and the nonviolent democratic transformation of South Africa (1990s), occurred in countries and regions where certain social and political conditions already prevailed and could be used and modified to channel the process. Not one of the Arab Spring states produced an opposition leader who could articulate the demands of the reformers and negotiate a transition—i.e., no Arab Nelson Mandela emerged. Not a single country had genuinely free media that could carry the messages of the oppressed. No country had even a flawed judiciary that could be converted into a dependable and impartial one. No country had police who impartially enforced the law. The odds against the Arab Spring were daunting. In July 2011, there still was no clear indication that it would succeed in any of the affected countries. And the most consequential domino of them all—Saudi Arabia—remained essentially unaffected.

The Arab Spring also should be seen in the context of the Sunni-Shi'ite split discussed earlier. Superimposed on the conflicts in several Arab states is the role of Shi'ite minorities in propelling the uprisings—or of crushing them. We deal with this issue in our regional discussion in Chapter 7B, but let us note here that Shi'ites played major roles in the uprisings in Bahrain (where they are in the majority, ruled by a Sunni monarchy), in Yemen (where their rebellion started even before Tunisia fell apart), and, paradoxically, in Syria, where a Shi'ite sect, the Alawites, form the ruling minority whose murderous response to the people's demands drew global condemnation.

Religion and Revolution

The Arab Spring arose from a combination of populist grievances ranging from economic issues to religious re-pression. Obviously, religious revivalists saw their opportunity as part of the wider revolt that was aimed primarily at the removal of repressive regimes: those aging autocrats had relentlessly persecuted fundamentalists of all stripes. In several countries, including Egypt and Syria, Muslim Brotherhoods had tenuous, conflict-ridden relationships with the ruling regimes. But at least in the early stages of the Arab Spring, religious revivalists did not play leading roles in the popular uprisings. In fact, the young rebels and their notions of democracy ran counter to the fundamentalist views of reactionary Islamists, a divide that adds still another layer to the complex social strata marking this fractious part of the world.

Whatever the outcome of the events roiling the Arab regions of this unruly realm, they will mark a social transformation the outlines of which were still hazy almost a year after they began. The question is whether the reformers can build civil societies even as they compete for the power vacuums they have created.

POINTS TO PONDER

- The expansion of Islam in Africa and Central Asia is changing the cultural geography of the margins of this realm.
- Oil-rich countries tend to have nondiversified economies and maintain a wide gap between rich and poor.
- In Afghanistan, the Taliban outlawed education for females; in Tunisia, women played a leading role in the "Jasmine Revolution."
- If you were a student in Egypt or Syria or Libya, would you have participated in the demonstrations for democratic reform?

POLITICAL UNITS AND GEOGRAPHIC REGIONS OF NORTH AFRICA/SOUTHWEST ASIA

Egypt and the Lower Nile Basin
The Maghreb and Its Neighbors
Middle East
Arabian Peninsula
The Empire States
Turkestan
African Transition Zone

POPULATION
- Under 50,000
- 50,000–250,000
- 250,000–1,000,000
- 1,000,000–5,000,000
- Over 5,000,000

National capitals are underlined
Railroad
Road
Canal

0   400   800   1200   1600 Kilometers
0   200   400   600   800   1000 Miles

ATLANTIC OCEAN

NORWAY   Oslo   FINLAND
SWEDEN   Helsinki   St. Petersburg
Stockholm   ESTONIA   Volga R.
DENMARK   LATVIA   Moscow
UNITED KINGDOM   LITHUANIA   RUSSIA
London   NETH.   Berlin   BELARUS
BELG.   GERMANY   Warsaw   Mensk   Kiev
LUX.   POLAND   UKRAINE
Paris   CZECH REPUBLIC   Dnieper R.   Don R.
FRANCE   SWITZ.   Vienna   SLOVAKIA
AUSTRIA   HUNGARY   Budapest   ROMANIA   MOLDOVA
Milan   SLO.   CRO.   Rostо
North Sea   BOSNIA-HERZEGOVINA   Belgrade   Bucharest
SPAIN   ITALY   MONTENEGRO   SERBIA   BULGARIA   Black Sea
Barcelona   Rome   KOS.   GEORG.
Naples   ALBANIA   MAC.   Istanbul   ARME
GREECE   Ankara   EMPI
MALTA   Izmir   Bursa   TURKEY   Gaziantep
Gibraltar (U.K.)   Algiers   Annaba   Athens   Adana   Aleppo
Tangier   Oran   Tunis   CYPRUS   SYRIA   MIDDLE EAST
Rabat   Constantine   Nicosia   LEBANON   Damascus
Casablanca   TUNISIA   Sfax   Beirut
Marrakech   MAGHREB   Tripoli   Mediterranean Sea   ISRAEL   Amman
MOROCCO   Bechar   Benghazi   Tel Aviv-Jaffa   Jerusalem
Canary Is. (Sp.)   Alexandria   Port Said   JORDAN
El Aaiún   ALGERIA   Cairo
WESTERN SAHARA   LIBYA   Libyan Desert   EGYPT   ARA
Tropic of Cancer   Sahara   Thebes   Nile R.
Nouakchott   Aswan   Lake Nasser
MAURITANIA   EGYPT AND THE LOWER NILE BASIN   Jiddah
MALI   Red Se
Dakar   NIGER   CHAD   NORTH SUDAN ZONE   Port Sudan
GAMBIA   SENEGAL   Khartoum   ERITREA
GUINEA-BISSAU   Ségou   Mopti   AFRICAN   TRANSITION   Wad Madani   As
GUINEA   Niamey   Kano   N'Djamena
Conakry   Bamako   BURKINA FASO
SIERRA LEONE   Kaduna   Adis Abeba
Freetown   BENIN   NIGERIA   SUDD   ETHI
IVORY COAST   TOGO   Abuja
LIBERIA   GHANA   Ibadan   Benue R.   SOUTH SUDAN
Monrovia   Yamoussoukro   Lagos   CENTRAL AFRICAN REPUBLIC   Juba
Abidjan   Accra   Yaoundé   Bangui   UGANDA   KENYA
Equator   EQ. GUINEA   CAMEROON   Kampala
ATLANTIC OCEAN   SÃO TOMÉ & PRÍNCIPE   GABON   CONGO   THE CONGO   Lake Victoria   Nairo
Congo R.   TANZANIA
Brazzaville   Kinshasa

Longitude West of Greenwich   0°

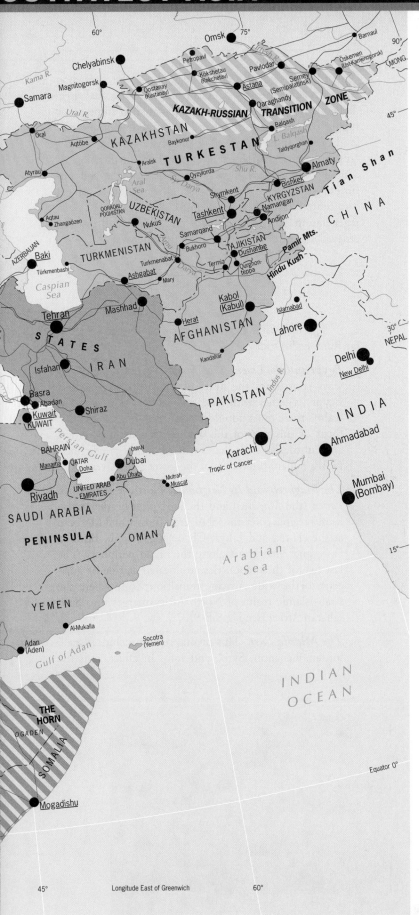

REGIONS

Egypt and the Lower Nile Basin
The Maghreb and Its Neighbors
The Middle East
The Arabian Peninsula
The Empire States
Turkestan

IN THIS CHAPTER

- Tunisia's Jasmine Revolution
- The ouster of Egyptian President Mubarak
- Civil war in Libya
- The rise and fall of Usama bin Laden
- Israel and the Palestinians . . . again
- Dubai's fantasyland and Qatar's World Cup

CONCEPTS, IDEAS, AND TERMS

Cultural revival	1
Turkish Model	2
Rain shadow effect	3
Police state	4
Stateless nation	5
Choke point	6
Qanat	7
Forward capital	8
Buffer state	9
Taliban	10
al-Qaeda	11

FIGURE 7B-1 © H. J. de Blij, P. O. Muller, and John Wiley & Sons, Inc.

© Jan Nijman

Where were these photos taken?
Find out at www.wiley.com/college/deblij

dentifying and delimiting regions in the sprawling North Africa/Southwest Asia realm is quite a challenge. Population clusters are widely scattered in some areas, highly concentrated in others. Cultural transitions and landscapes—internal as well as peripheral—make it difficult to discern a regional framework. This, furthermore, is a highly changeable realm and always has been. Several centuries ago it extended into eastern Europe; now it reaches into central Asia, where an Islamic **cultural revival [1]** (the regeneration of a long-dormant culture through internal renewal and external infusion) is well under way.

The following are the regional components of this far-flung realm today (Fig. 7B-1):

1. **Egypt and the Lower Nile Basin.** This region in many ways constitutes the heart of the realm as a whole. Egypt (together with Iran and Turkey) is one of the realm's three most populous countries. It is the historic focus of this part of the world and a major political and cultural force. It shares with its southern neighbor, North Sudan, the waters of the lower Nile River.

2. **The Maghreb and Its Neighbors.** Northwestern North Africa (the *Maghreb*) and the areas that border it also form a region, consisting of Algeria, Tunisia, and Morocco at the center, and Libya, Chad, Niger, Mali, Burkina Faso, and Mauritania along the broad periphery. The last five of these countries are located astride or adjacent to the wide African Transition Zone where the Arab-Islamic realm of North Africa merges into Subsaharan Africa (see Fig. 6B-9).

3. **The Middle East.** This pivotal region includes Israel, Jordan, Lebanon, Syria, and Iraq. In effect, it is the

© Jan Nijman

crescent-shaped zone of countries that extends from the eastern Mediterranean coast to the head of the Persian Gulf.

4. ***The Arabian Peninsula.*** Dominated by the enormous territory of Saudi Arabia, the Arabian Peninsula also includes the United Arab Emirates (UAE), Kuwait, Bahrain, Qatar, Oman, and Yemen. Here lies the source and focus of Islam, the holy city of Mecca; here, too, lie many of the world's greatest oil deposits.

5. ***The Empire States.*** We refer to this region as the Empire States because two of the realm's giants, both the centers of major historic empires, dominate its geography. Turkey, the heart of the Ottoman Empire, now is the realm's most secular state and (still) aspires to joining the European Union. Shi'ite Iran, the core of the erstwhile Persian Empire, has become an Islamic republic. To the north, Azerbaijan, once a part of the Persian Empire, lies in the turbulent, Muslim-infused Transcaucasian Transition Zone. To the south, the island of Cyprus is divided between Turkish and Greek spheres, the Turkish sector being another remnant of imperial times.

6. ***Turkestan.*** Turkish influence ranged far and wide in southwestern Asia, and following the collapse of the Soviet Union in 1991 that influence turned out to be durable and strong. In the five former Soviet Central Asian republics the strength and potency of Islam vary, and their post-Soviet governments deal warily (sometimes forcefully) with Islamic revivalists. Russian influence continues, democratic institutions remain weak, and Western (chiefly American) involvement has increased since the campaign against terrorism began and

Afghanistan—which also forms part of this region—became a major target of that campaign.

▪▪ EGYPT AND THE LOWER NILE BASIN

Egypt occupies a pivotal location in the heart of a realm that extends over 9600 kilometers (6000 mi) longitudinally and some 6400 kilometers (4000 mi) latitudinally. At the northern end of the Nile and of the Red Sea, at the eastern end of the Mediterranean Sea, in the northeastern corner of Africa across from Turkey to the north and Saudi Arabia to the east, adjacent to Israel, to North Sudan, and to Libya, Egypt lies in the crucible of this realm. Because it owns the Sinai Peninsula, Egypt, alone among states on the African continent, has a foothold in Asia. This foothold gives it a coast overlooking the strategic Gulf of Aqaba (the northeasternmost arm of the Red Sea). Egypt also controls the Suez Canal, the vital link between the Indian and Atlantic oceans and lifeline of Europe. The capital, Cairo (Al Qahira), is the realm's largest city and a leading center of Islamic civilization. We hardly need to further justify Egypt's designation (together with North Sudan) as a discrete region.

Egypt has six subregions, mapped in Figure 7B-2. Most Egyptians live and work in Lower (i.e., northern) and Middle Egypt, subregions ① and ②, the country's core area anchored by Cairo and flanked by the leading port and second-largest manufacturing center, Alexandria. The other four regions are remote and sparsely populated, especially Upper Egypt ③, the Western Desert ④, and the Sinai ⑥. The northern stretches of the Eastern Desert ⑤ are now in the path of Cairo's encroaching urban sprawl. Oil discoveries in Sinai and in the Western Desert have benefited the economy, so that Egypt now is self-sufficient and even exports some petroleum products.

Gift of the Nile

Egypt's Nile is the aggregate of two great branches upstream: the White Nile, which originates in the streams that feed Lake Victoria in East Africa, and the Blue Nile, whose source lies in Lake Tana in Ethiopia's highlands. The two Niles converge at Khartoum, capital of present-day North Sudan. About 95 percent of Egypt's 84 million people live within 20 kilometers (12 mi) of the great river's banks or in its delta (Figs. 7B-2, 7A-2).

Before dams were constructed on the Nile, the ancient Egyptians used *basin irrigation*, building fields with earthen ridges and trapping the annual floodwaters with their fertile silt, to grow their crops. That practice continued for thousands of years until, during the nineteenth century, the construction of permanent dams made it possible to irrigate Egypt's farmlands year round. These dams, with locks for navigation, controlled the Nile's annual flood, expanded the country's cultivable area, and allowed farmers to harvest more than one crop per year on the same field. In a single century, all of Egypt's farmland was brought under this *perennial irrigation*.

● MAJOR CITIES OF THE REALM

City	Population* (in millions)
Cairo, Egypt	11.3
Istanbul, Turkey	10.8
Tehran, Iran	7.4
Baghdad, Iraq	6.2
Khartoum, North Sudan	5.5
Riyadh, Saudi Arabia	5.1
Kabol (Kabul), Afghanistan	4.1
Tel Aviv-Jaffa, Israel	3.4
Casablanca, Morocco	3.4
Algiers, Algeria	2.9
Damascus, Syria	2.8
Tashkent, Uzbekistan	2.3
Beirut, Lebanon	2.0
Almaty, Kazakhstan	1.4
Tripoli, Libya	1.2
Jerusalem, Israel	0.8

*Based on 2012 estimates.

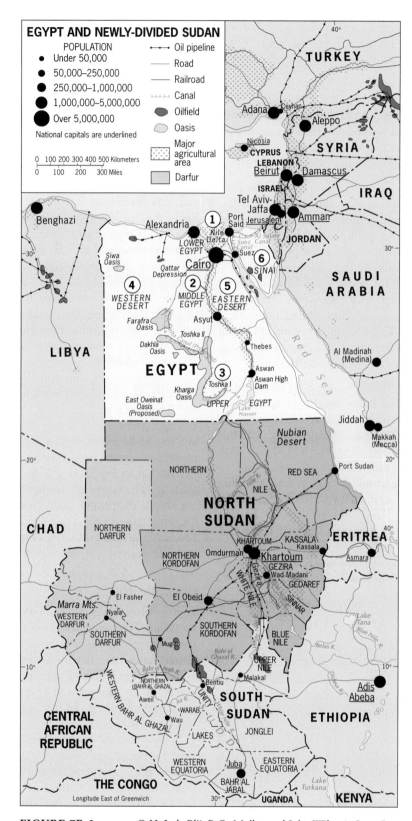

FIGURE 7B-2 © H. J. de Blij, P. O. Muller, and John Wiley & Sons, Inc.

From the Field Notes . . .

"Taking off from Cairo International Airport on the northeast edge of this sprawling megacity, we flew over one of Cairo's newest residential developments at the edge of the northern reaches of the Eastern Desert. Look at this density—in a desert! With a birthrate twice that of the United States, Egypt's critical problem is the provision of a sufficient water supply—which right now is steadily decreasing. Earlier on this day in November 2009, we took a field trip through this area led by a local geographer. These are middle- and upper-middle-class residential complexes. All these new apartment towers will need water, and so will anything green planted here. Where will these people work and how will they get there? Is this sustainable urban development?

© Jan Nijman

www.conceptcaching.com

The largest of these dams, the Aswan High Dam, began operating in 1968, controlling water and sediments as well as bringing reliability to what had been an uncertain flood regime. It created Lake Nasser, which extends southward 150 kilometers (100 mi) into North Sudan, where 50,000 people had to be resettled to make way for it. The Aswan High Dam increased Egypt's irrigable land by nearly 50 percent and today provides the country with more than 10 percent of its electricity.

Farther upstream, in the western highlands of Ethiopia where the Blue Nile originates, the Ethiopian government intends to build the so-called Grand Millennium Dam. This project would quintuple Ethiopia's electrical power production—and would provide significant leverage over downstream North Sudan and Egypt. Ethiopia's shortage of investment capital may derail the project, although there is talk of Chinese funding to carry it through.

Valley and Delta

Egypt's elongated oasis along the Nile, just 5 to 25 kilometers (3 to 15 mi) wide (see photo, p. 347), broadens north of Cairo across a delta anchored in the west by the historic city of Alexandria and in the east by Port Said, gateway to the Suez Canal. The delta contains extensive farmlands, but it is a troubled area today. The ever more intensive use of the Nile's water and silt upstream is depriving the delta of much needed replenishment. And the low-lying delta is also at risk due to geological subsidence and sea-level rise, increasing the danger of salt-water intrusion from the Mediterranean that can severely damage soils.

Egypt's millions of subsistence farmers, the *fellaheen*, struggle daily to make their living off the land, as did the peasants of the Egypt of five millennia ago. Some rural landscapes seem barely to have changed; ancient farming methods are still used, and dwellings remain rudimentary. Poverty, disease, high infant mortality rates, and low incomes prevail among Egypt's masses. The government has announced grandiose plans to expand irrigated acreage, but almost nothing has taken place so far.

Economic and Political Discontent

Egypt has experienced steady economic growth over the past decade at a yearly rate of about 5 percent, and the economy has further diversified. At the same time, its annual rate of population growth has declined, while literacy and per capita consumption have increased markedly. Egypt now exports fruits, vegetables, rice, and textiles in addition to its longtime staple, cotton; it also has a sizeable tourist industry (which is frequently damaged by the actions of Islamist extremists and lately by major social unrest). Since it has begun to exploit its domestic resources, the country's expenditures on oil imports have substantially declined.

Even though the Egyptian economy has been improving in recent years, the primary beneficiaries are the wealthy and the upper-middle class. The poor are as deprived and entrenched in poverty as ever. As the population continues to grow, the gap between food supply and demand widens, and Egypt must now import about (U.S.) $6 billion worth of food every year. More than half the population is under 25 years of age, and the young, most

⊙ AMONG THE REALM'S GREAT CITIES . . .

CAIRO

© H. J. de Blij, P. O. Muller, and John Wiley & Sons, Inc.

STAND ON THE roof of one of the high-rise hotels in the heart of Cairo, and in the distance you can see the great pyramids, monumental proof of the longevity of human settlement in this area. But the present city was not founded until Muslim Arabs chose the site as the center of their new empire in AD 969. Cairo became and remains Egypt's primate city, situated where the Nile River opens onto its delta, home today to nearly one-seventh of the entire country's population.

Cairo at 11.3 million ranks among the world's 20 largest urban agglomerations, and it shares with other cities of the less advantaged realms the staggering problems of crowding, inadequate sanitation, crumbling infrastructure, and substandard housing. But even among such cities, Cairo is noteworthy for its stunning social contrasts. Along the Nile waterfront, elegant skyscrapers rise above carefully manicured, Parisian-looking surroundings. But look toward the east and the metropolitan landscape extends gray, dusty, almost featureless as far as the eye can see. Not far away, more than a million squatters live in the sprawling cemetery known as the City of the Dead (see photo below). On the urban outskirts, millions more survive in overcrowded shantytowns of mud huts and hovels, while exclusive new residential developments for the middle and upper classes sprawl well into the desert beyond (see From the Field Notes box, p. 369).

Nonetheless, Cairo is not only the dominant city of Egypt but also serves as the cultural capital of the entire Arab World, with centers of higher learning, splendid museums, world-class theater and music, as well as magnificent mosques and Islamic learning centers. Although Cairo has always primarily been a focus of government, administration, and religion, it also is a river port and an industrial complex, a commercial center, and, as it sometimes seems, one giant bazaar. Cairo is truly the heart and soul of the Arab World, a creation of its geography and a repository of its history.

From the Field Notes . . .

© H. J. de Blij

"On the eastern edge of central Cairo we saw what looked like a combination of miniature mosques and elaborate memorials. Here lie buried the rich and the prominent of times past in what locals call the 'City of the Dead.' But we found it to be anything but a dead part of the city. Many of the tombs here are so large and spacious that squatters have occupied them. Thus the City of the Dead is now an inhabited graveyard, home to at least one million people. The exact numbers are impossible to determine; indeed, whereas metropolitan Cairo in 2012 has an officially estimated population of 11.3 million, many knowledgeable observers believe that 17 million is closer to the mark."

www.conceptcaching.com

of whom lack economic opportunities, are restless. After the Mubarak regime came into office following the assassination of President Anwar Sadat in 1981, it failed to connect with Egypt's new generation, and in the end was unable to justify its three-decade lock on power. All along, this government repressed Islamic groups such as the Muslim Brotherhood, but discontent had also been widening across broader swaths of the population, especially the secular middle classes and the young.

Political Upheaval

As we noted in Chapter 7A, when Tunisia's longtime president was forced to resign in January 2011 after massive, relentless demonstrations in Tunis, it was the spark that ignited an "Arab Spring" of political upheaval all across the realm, most notably in Egypt. By February 1, hundreds of thousands of people gathered at Cairo's Tahrir Square were demanding the removal of President Mubarak. Following initial efforts by the authorities to crush the protests, and after more than a week of violent confrontations in which an estimated 850 people perished, the regime caved in and Mubarak resigned on February 11. This outcome would have been unthinkable just a few weeks earlier, and many Egyptians—even those risking their lives at Tahrir Square—could hardly believe they had been victorious.

When Mubarak stepped down, military leaders took control and quickly organized a referendum for constitutional change on March 19, responding to some of the key demands of the protesters. Constitutional amendments (approved overwhelmingly) included term limits for the president and judicial oversight of regular free elections. Parliamentary elections were planned for the summer of 2011. The Muslim Brotherhood, a longtime major organization that most would describe as moderate, expressed a preference for future governance according to the **Turkish Model [2]**—a multi-party democracy that has a place for, but is not dominated by, Islamic parties.

When U.S. President Obama hailed Mubarak's decision to resign, he said that Egypt would "never be the same." To be sure, the country's popular uprising and revolution were unprecedented. However, as so often occurs with revolutions, the removal of the old regime is only the beginning and does not clearly indicate what will replace it. Egypt in mid-2011 was at a crossroads in more ways than one. The country's new political direction will recast the role of Islamist forces, but it will also involve choices as to how to deal with the disadvantaged masses. Democratic breakthroughs must, before too long, be followed by economic policies that recognize the needs of the poor and the young—and simultaneously foster a vibrant growing economy. These are colossal challenges for any government in this part of the world.

North Sudan

As Figure 7B-1 shows, Egypt is flanked by two countries that have posed challenges to its leadership: North Sudan to the south and Libya to the west. North Sudan, almost

twice the size of Egypt but with less than 40 percent of the population, lies centered on the confluence of the White Nile (from Uganda) and the Blue Nile (from Ethiopia). Here the twin capital, Khartoum–Omdurman, anchors a sizeable agricultural area where cotton was planted during colonial times.

Almost all of North Sudan is desert, with irrigated agriculture along the banks of the While Nile and the Blue Nile. The country also has a 500-kilometer (300-mi) coastline on the Red Sea, where Port Sudan lies almost directly across from Jiddah and Makkah (Mecca) in Saudi Arabia. For many years after independence, the economy was typical of the energy-poor global periphery, exchanging sheep, cotton, and sugar for oil, with Saudi Arabia the main trading partner. Moreover, decades of civil war following the end of colonial rule had impoverished the Islamic regime in Khartoum, and per capita income in then-Sudan was one of the lowest in the world. All that, however, changed in the 1990s.

Oil in the Transition Zone

During the 1980s it became clear that significant oil reserves lie in the country, and during the 1990s discoveries were made in Kordofan Province and elsewhere in central Sudan, including major deposits in the embattled south. Sudan quickly became an exporter instead of an importer of oil. Foreign companies, particularly from China, which was only too willing to overlook human rights violations by the Khartoum regime, soon arrived to explore and exploit. The government then saw its income rise, allowing it to purchase more weapons. Local peoples living on top of or near the newly found deposits were forcibly relocated. Ultramodern buildings in Khartoum reflected the newfound wealth for the few. But the discovery of oil in the south also fueled the drive for independence in that part of former Sudan and culminated in the breakup of 2011.

As a single political entity, Sudan had been a fixture on the map of northeastern Africa for over a century; but on July 9, 2011, the country (independent since 1956) split in two. North Sudan is the much bigger, Islamic, northern part, and South Sudan is a typical Equatorial African country dominated by Christianity and African religions (see pp. 329–330). The breakup followed six decades of bitter and bloody conflict with an estimated 1.5 million deaths. North Sudan is now a more homogeneously Islamic state, with a Muslim population of more than 90 percent (compared to the 68 percent of pre-breakup Sudan). The new boundary between North and South puts the majority of the former Sudan's oil reserves in the South, although the pipelines for export lie in the North. Ominously, a smoldering, unresolved conflict continues over Abyei Province on the border between North and South Sudan. Abyei, which lies directly on the Islamic Front in the heart of the African Transition Zone (see Fig. 6B-9), has a mixed population and also sits atop an important oilfield.

Tragedy in Darfur

Even as the issue of Southern independence heated up, a terrible crisis engulfed Sudan's western provinces of Darfur (Fig. 7B-2), where the local people, the Fur, have for centuries lived as Arabized pastoralists in the north and as settled farmers in the south. It is not clear how the conflict began, but a number of officials of the Khartoum regime suspected the southern Fur of sympathizing with rebels in the far south who were opposed to Islamic rule. In 2003 northern pastoralist militias, encouraged by the regime, rode into the villages and fields of the southern Fur, burning

The "international community" has been slow to respond to the crisis in Darfur, arguing over the question of whether the assault by Arabized Muslim militias on sedentary African farmers in these western provinces of Sudan technically constitutes genocide, and relying on an undermanned, poorly equipped, and inadequately mandated African Union force to halt the violence. The result can be seen in this photograph of the Northern Darfur village of Bandago, where virtually every dwelling has been burned by *janjaweed* militias empowered by Khartoum's Islamic regime. By the middle of 2011, about 3 million people had been driven from their homes, leaving behind charred remains like this; as many as 400,000 had died; and this part of the African Transition Zone faced famine and dislocation for many years to come, whatever the result of the power struggles afflicting them.

© AP/Wide World Photos

homes, destroying crops, and killing thousands. Perhaps as many as 3 million people were driven from their homes, and a combination of violence and resultant disease took a toll that by now may exceed 400,000.

North and South Sudan are locked in an uneasy embrace dictated by the location of their shared energy resources, but marred by the memories of decades of deathly conflict. The final chapter in this geopolitical saga has yet to be written.

■■ THE MAGHREB AND ITS NEIGHBORS

The countries of northwestern Africa are collectively called the *Maghreb*, but the Arab name for them is more elaborate than that: *Jezira-al-Maghreb*, or "Isle of the West," in recognition of the great Atlas Mountain range rising like a huge island from the Mediterranean Sea to the north and the sandy flatlands of the immense Sahara to the south.

The countries of the Maghreb (sometimes spelled *Maghrib*) include Morocco, last of the North African kingdoms; Algeria, a secular republic beset by the religious-political problems we noted earlier; and Tunisia, smallest and most Westernized of the three (Fig. 7B-3). Neighboring Libya, facing the Mediterranean between the Maghreb and

FIGURE 7B-3

© H. J. de Blij, P. O. Muller, and John Wiley & Sons, Inc.

Egypt, is unlike any other North African country: an oil-rich desert state, embroiled in a violent civil war in 2011, whose population is almost entirely clustered in settlements along the coast.

Atlas Mountains

Whereas Egypt is the gift of the Nile, the Atlas Mountains facilitate the settled Maghreb. These high ranges wrest from the rising air enough orographic rainfall to sustain life in the intervening valleys, where good soils support productive farming. From the vicinity of Algiers eastward along the coast into Tunisia, annual rainfall averages more than 75 centimeters (30 in), a total more than three times as high as that recorded for Alexandria in Egypt's delta. Even 240 kilometers (150 mi) inland, the slopes of the Atlas still receive over 25 centimeters (10 in) of rainfall. The effect of the topography can even be read on the world climate map (Fig. G-7): where the highlands of the Atlas terminate, aridity immediately takes over.

The Atlas Mountains are aligned southwest-northeast and begin within Morocco as the High Atlas, with elevations close to 4000 meters (13,000 ft). Eastward, two major ranges dominate the landscapes of Algeria proper: the Tell Atlas to the north, facing the Mediterranean, and the Saharan Atlas to the south, overlooking the great desert. Between these two mountain chains, with each consisting of several parallel ranges and foothills, lies a series of intermontane basins markedly drier than the northward-facing slopes of the Tell Atlas. Within these valleys, the **rain shadow effect [3]** of the Tell Atlas is reflected not only in the steppe-like natural vegetation but also in land-use patterns: pastoralism replaces cultivation, and stands of short grass and shrubs blanket the countryside.

Colonial Impact

During the colonial era, which began in Algeria in 1830 and lasted until the early 1960s, well over a million Europeans came to settle in North Africa—most of them French, and a substantial majority bound for Algeria—and these immigrants soon dominated commercial life. They stimulated the renewed development of the region's towns; Casablanca, Algiers, Oran, and Tunis became the urban foci of the colonized territories. Although the Europeans dominated trade and commerce and integrated the North African countries with France and the European Mediterranean world, they did not confine themselves to the cities and towns. They recognized the agricultural possibilities of the favored parts of the *tell* (the lower Tell Atlas slopes and narrow coastal plains that face the Mediterranean) and established prosperous farms. Not surprisingly, agriculture here is Mediterranean. Algeria soon became known for its vineyards and wines, citrus groves, and dates; Tunisia has long been one of the world's leading exporters of olive oil; and Moroccan oranges went to many European markets.

From the Field Notes . . .

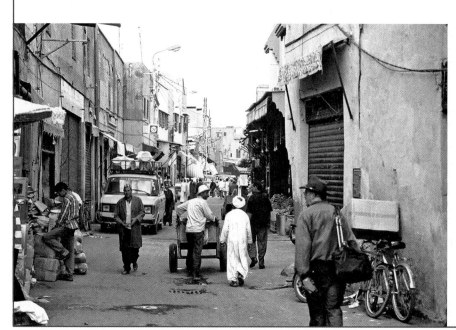

"The medina (old city) in Marrakech, Morocco is lively, and the central square is crowded with locals and visitors alike. But the neighborhoods and streets and alleys beyond the tourist destinations reveal a different aspect of this sprawling city where modernization and tradition coexist. You'll look in vain here for signs in English—or, for that matter, French . . ."

www.conceptcaching.com

© H. J. de Blij

The Maghreb Countries

Between desert and sea, the Maghreb states display considerable geographic diversity (Fig. 7B-3). **Morocco**, a conservative kingdom in a revolution-marked region, is tradition-bound and weak economically. Its core area lies in the north, anchored by four major cities; but the Moroccans' political attention is focused on the south, where the government is seeking to absorb its neighbor, ***Western Sahara*** (a former Spanish dependency with almost 500,000 inhabitants, many of them immigrants from Morocco). Even if this campaign is successful, it will do little to improve the lives of most of Morocco's 33 million people. Hundreds of thousands have migrated to Europe, many of them via Spain's two tiny exclaves on the Moroccan coast, Ceuta and Melilla (see box, p. 85), or via rickety boats to the Canary Islands in the nearby Atlantic.

Algeria, France's former colony whose agricultural potential drew more than a million European settlers, now has an economy based primarily on its substantial oil and gas reserves. Its bitter war of liberation (1954–1962) was followed by recurrent conflict between Islamists and military-backed secular authorities, costing over 100,000 lives and resulting in a long-delayed compromise still punctuated by occasional skirmishes. The country's primate city, Algiers, is centrally situated on the Mediterranean coast and contains nearly 10 percent of Algeria's 37 million people. But even more Algerians—officially estimated between 1.5 and 2 million—have emigrated to France, where they form one of Europe's largest Muslim population sectors.

The smallest of the Maghreb states, **Tunisia**, lies at the eastern end of the region. As Appendix B shows, Tunisia in many ways outranks surrounding countries: it has a higher urbanization level, higher social indicators, and a much lower growth rate (among its population of almost 11 million) than elsewhere in the Maghreb. Most of the country's productive capacity lies in the north, in the hinterland of its historic capital, Tunis.

Nonetheless, in January 2011 Tunisia took center stage as the launching pad of a wave of revolt that swept through the Arab World and beyond. Despite its rather modern appearance, Tunisia's repressive and corrupt government had become intolerable to the people. Their response was the *Jasmine Revolution*, a sudden and explosive outburst of street protests that forced out the regime of President Ben Ali, who fled the country to Saudi Arabia. It was an amazing feat—the dictator of a **police state [4]** entrenched for a generation ousted in a matter of weeks by a spontaneous uprising.

All across the Arab World, the events in Tunisia were monitored closely by excited populations and nervous governments. But in Tunisia, as in Egypt shortly thereafter, the ouster of the regime gave way to a power vacuum and the government system was rendered dysfunctional. Free elections were scheduled for the summer of 2011, raising hope but also concern that democratization could prove to be a "Trojan horse" for radical Islam (polls suggested that the religious *Ennahda* Party [the word means Renaissance] would have more support than any other). Still, Tunisia's Jasmine Revolution was incredibly swift and relatively bloodless; but when the spirit of the revolution soon spilled into neighboring Libya, things turned out differently and far more violently.

Libya

Almost rectangular in shape, Libya (with a population of 6.7 million), lies on the Mediterranean Sea between the Maghreb states and Egypt (Fig. 7B-3). What limited agricultural possibilities exist lie in the district known as Tripolitania in the northwest, centered on the capital, Tripoli, and in the northeast in Cyrenaica, where Benghazi is the urban focus. But it is oil, not farming, that drives the economy of this highly urbanized country. The oilfields are located well inland from the central Gulf of Sidra, linked by pipelines to coastal terminals. Libya's two interior corners, the desert Fezzan district in the southwest and the Kufra oasis in the southeast, are connected to the coast by two-lane roads subject to frequent sandstorms.

Colonel Moammar el-Qadhafi had been one of the Arab World's most violent despots, clinging to power for more than four decades. His iron-fisted response to anti-government demonstrations in several Libyan cities that began in late January 2011 did not come as a surprise. This rebellion was different in some ways because of Libya's large number of tribes (140) and resulting tribal divisions as well as the regime's systematic repression of several tribes that had withheld support for Gadhafi's reign. The latter were among the most tenacious protesters and played a leading role in the ensuing civil war.

By early summer of 2011, rebel forces held on to a number of cities including their main base of Benghazi (photo above) as well as Tobruk in the northeast; Misrata, about 200 kilometers (125 mi) east of Tripoli; and a few smaller towns in the far northwest near the Tunisian border. They could not have done so without the intervention of air power supplied by a coalition of NATO countries (including the United States), which enforced a no-fly zone for Qadhafi's air force throughout northern Libya and carried out a series of bombardments on key military targets and government ground troops bent on laying siege to rebel-held cities. If Tunisia's and Egypt's challenge was to overcome a power vacuum, Libya's predicament was a vicious dictator who unhesitatingly attacked his own people and seemed incapable of even considering resignation.

Adjoining Saharan Africa

Between the coastal states of the Maghreb/Libya and the southern margin of the Sahara lies a tier of five states, all but one of them landlocked, dominated by the desert, sustained by ribbons of water, and under the sway of Islam. Figure 7B-1 best displays this chain of countries whose populations are not insignificant by African standards, totaling some 66 million.

Rebel volunteers in the city of Benghazi in northeastern Libya receive military training as the country's civil war escalates in March 2011. The second-largest city in the country after Tripoli, Benghazi was a crucial stronghold for the anti-Gadhafi forces. An estimated 2000 people, mostly civilians, were killed during the first four months of fighting, and the death toll would have been much higher without NATO's air support for the rebels.
© al-mak/Demotix/Demotix/Corbis

Only coastal **Mauritania** has a minuscule population even in this company: less than 4 million in a territory the size of Texas and New Mexico combined, half of it concentrated in and around the coastal capital, Nouakchott, base of a small fishing fleet. Overwhelmingly Muslim and experiencing its first democratic presidential election as recently as 2007, Mauritania has been infamous for tolerating slavery despite its official ban in 1981; following the 2007 election, its parliament voted to impose prison terms on slave-owners continuing the practice.

By contrast, neighboring **Mali** (16.2 million), is a multi-party, democratic state with a multi-cultural population that includes a sizeable minority adhering to traditional beliefs as well as a small Christian minority. Mali depends on the waters of the upper Niger (pronounced NYE-jer), the river on which its capital and irrigated farming zone lie, concentrating the population in the southwest—within the African Transition Zone. The fabled city of Timbuktu, today a mere shadow of its former greatness, lies downstream from the capital, Bamako. Mali shares with several neighbors an increasingly assertive minority of Tuaregs, a nomadic people of the interior Sahara (see Fig. 7B-3) who are demanding a share of local mineral revenues, including uranium.

Uranium is now the leading export of **Niger**, which is pronounced nee-ZHAIR. As in Mali, the Niger River is the lifeline of this country of 17 million, as reflected by Figure 7A-2, and Niamey, the capital, lies on its banks in the far southwest. We should remember that these land-locked and low-income countries emerged more than half a century ago as independent states from French imperial rule, and when their boundaries were laid out nobody ever imagined that these colonial dependencies would become sovereign states. This explains why there is a country called **Burkina Faso**—lying south of Mali and southwest of Niger—where there is no Niger River, no coastline, and little economic opportunity (this is one of the world's poorest countries, with nearly half the population of 17.3 million subsisting on one U.S. dollar a day or less). Environmental variability here is marked by destructive floods alternating with severe droughts, the latter driving people and their livestock southward into Ivory Coast, resulting in deadly conflicts. Despite all this, a representative government in the capital of Ouagadougou manages to play a regional and international role in this part of the continent.

The fifth member of this tier of desert-margin states, **Chad**, lies east of Niger and, as Figure 7B-1 shows, is also bisected by the wide African Transition Zone. Here, divisions between Chad's Islamized north and Christian/animist south are strong, and the map shows its capital, N'Djamena, lies directly on the Islamic Front. Since 2000, this country of 12 million has been in turmoil, the crisis in Sudan's Darfur provinces spilling over into its eastern territorial margin, the south experiencing an oil boom (see Fig. 6B-6), and the capital the focus of a struggle among rebel factions trying to overthrow the government.

Islamic militancy, commodities of growing value, and restive minorities are today drawing the international attention of global powers to a broad, remote periphery of North Africa long in the shadow of more prominent coastal neighbors to the north and south.

▪▪ THE MIDDLE EAST: CRUCIBLE OF CONFLICT

The regional term **Middle East**, we noted earlier, is not satisfactory, but it is so common and generally used that avoiding it creates more problems than it solves. It originated when Europe was the world's dominant realm and when places were *near*, *middle*, and *far* from Europe: hence a Near East (Turkey), a Far East (China, Japan, and other countries of East Asia), and a Middle East (Egypt, Arabia, Iraq). If you check definitions used in the past, you will see that these terms were applied inconsistently: Syria, Lebanon, Palestine, and even Jordan sometimes were included in the Near East, and Persia and Afghanistan in the Middle East.

Today, the geographic designation *Middle East* has a more specific meaning. And at least half of it has merit: this region, more than any other, lies at the middle of the far-flung Islamic realm (Fig. 7B-1). To the north and east of it, respectively, lie Turkey and Iran, with Muslim Turkestan beyond the latter. To the south lies the Arabian Peninsula.

And to the west lie the Mediterranean Sea and Egypt, and the rest of North Africa. This, then, is the key region of the realm, its very heart.

Five countries comprise the Middle East (Fig. 7B-4): Iraq, largest in population and territorial size, facing the Persian Gulf; Syria, next in both categories and fronting the Mediterranean; Jordan, linked by the narrow Gulf of Aqaba to the Red Sea; Lebanon, whose survival as a unified state has come into question; and Israel, Jewish nation in the crucible of the Muslim world. Because of the extraordinary importance of this region in world affairs, we focus in some detail on issues of cultural, economic, and political geography.

Iraq's Enduring Importance

Figure 7B-4 explains, even at a glance, why Iraq is pivotal among the states of the Middle East. With about 60 percent of the region's total area, more than 40 percent of its predominantly Arab population, and most of its valuable energy and agricultural resources, Iraq is key to the Middle East's fortunes. Iraq also is heir to the early Mesopotamian states and empires that emerged in the basin of its twin rivers, the Tigris-Euphrates, and the country is studded with matchless archeological sites and museum collections.

Much of this heritage was disastrously damaged in 2003 when the United States led a military invasion of Iraq in the aftermath of the 9/11 terrorist attacks in America. The map indicates why this attack had (and continues to have) critical consequences for the region and beyond: Iraq has six neighbors, all of them affected in serious ways. To the west lie Syria, Iraq's staunch ally during dictator Saddam Hussein's despotic rule, and Jordan, whose leaders chose Iraq's side during an early 1990s conflict with southern neighbor Kuwait. To the south, Iraq possesses a lengthy border with Saudi Arabia. And to the north, Iraq has borders with Iran, against which it fought a bitter war during the 1980s, and with Turkey, source of its crucial river lifelines.

But perhaps the most obvious feature in Figure 7B-4 is the striped zone that covers northeastern Iraq, large parts of Turkey and Iran, and smaller parts of other countries. This is the area where the majority of residents are not Arabs, but Kurds, a people who have been discriminated against by all the countries in which they live. More than 6 million Kurds inhabit the hilly north of Iraq; adjacent southeastern Turkey is the traditional home of 15 million more, and another 5 million live in western Iran. Together, they form one of the world's largest **stateless nations [5]**, divided and often exploited by those who rule over them (see box, p. 354). The American invasion of Iraq gave the Kurds something they had not experienced anywhere else in their domain: a chance at regional self-government and a large measure of cultural freedom.

With so many land neighbors, it is no surprise that Iraq is nearly landlocked. Figure 7B-4 shows how short and congested its single outlet to the Persian Gulf is; this

FIGURE 7B-4

© H. J. de Blij, P. O. Muller, and John Wiley & Sons, Inc.

was the main reason that Saddam tried to conquer and annex neighboring Kuwait in 1990 (Kuwait has significant oil reserves as well). As a result, a network of pipelines across Iraq's neighbors must carry much of Iraq's oil export volume to coastal terminals in other countries—another reason these countries have an interest in what happens to Iraq in the years ahead.

Discordant Cultural and Political Geographies

Let us now turn to the larger-scale map of Iraq (Fig. 7B-5), which shows how Iraq is divided into three cultural domains in which religion, tradition, and custom form the basis of its political geography. The largest population sector (within the national total of 33.2 million) is that of the 20 million Shi'ites in the southeast, whose religious affinities are with neighboring Iran; the next largest is the

Sunni minority of about 10 million in the north and west, which ruled the country before the 2003 invasion; and the smallest of Iraq's major components is that of the Kurds (most of whom are Sunnis), roughly 6.5 million in their northeastern stronghold. Even at this scale, however, the map is a generalization of a considerably more intricate mosaic. The borders between the major cultural domains in reality are transition zones, most of them without sharp definition (see the Kurd-Sunni frontier). Also not shown at this scale are smaller minorities of Turkmen, Assyrians, and others clustering in the north. And most important of all, the population of the capital, Baghdad (now 6.2 million), has grown larger than either the Sunni or the Kurdish domains (inset map), a microcosm of the divided state.

After the U.S. invasion in 2003, the multipurpose war in Iraq unleashed terrible sectarian violence and enabled

IRAQ

POPULATION
- Under 50,000
- 50,000–250,000
- 250,000–1,000,000
- 1,000,000–5,000,000
- Over 5,000,000

Oilfield
Road
Railroad
Oil pipeline
Province boundary
T Turkmen Minority
Su Sunni Minority

- - - Limit of strong Kurdish
regional influence

National capitals are underlined

Domains depict prevailing majorities; minorities (Turkmen, Assyrians)
not shown. Borderlands between domains have mixed populations.

0   40   80   120   160   200 Kilometers
0   20   40   60   80   100 Miles

CARTOGRAM OF POPULATION GROUPS IN IRAQ

TAMIM
NINEVEH
SALAHUDDIN
BAGHDAD
BASRA

Shia   Kur
Sunni   Mix

TURKEY

Naxcivan
Lake Van
Tatvan   Van
Salmas
Orumiyeh
Hakkari   Lake
Orumiyeh

Gaziantep   Sanliurfa
Kilis
Zakhu
Qamishli
Aleppo
Lake
Asad

Mosul   **T**
Irbil   **Kurdish
Domain**

S u n n i   Kirkuk (Tamim) **T**   Sulaymaniyah   Sanandaj

Dayr az Zawr   **D o m a i n**   **T**   Hamadan

S Y R I A   Euphrates River   Kermanshah   Arak   Borujerd

Qadisiya
Lake   Tikrit   **Z**   Khorramabad
Tharthar   Samarra   **a**
Lake   Baqubah   Mandali   **g**
Falluja   Baghdad   **r**   **I R A N**
Ramadi   **o**
Rutbah   Karbala   **s**
Razzaza   Hillah   Kut   Dezful
L.   **M**
J O R D A N   **D**   **I   R   A   Q**   Najaf   Diwaniyah   **Shia**   'Amarah   **t**   Ahvaz
e   Nukhayb   **s**   **Domain**   **KHUZESTAN
PROVINCE**
Nasiriyah
Badanah   **e**   **Su**
Basra
r   **Su**   Abadan
Su   Rumaylah Oilfield
Rafha   **Persian**
Kuwait   **Gulf**
KUWAIT
Hafar al Batin   Subayhiyah

S A U D I   A R A B I A

PROVINCES OF IRAQ

0   100   200 Kilometers
0   50   100 Miles

TURKEY
Dahuk
SYRIA   Arbil
Ninawa
(Nineveh)   Tamim
(Kirkuk)   Sulaymaniyah
Salah ad Din
(Saladin)   IRAN
Diyala
Baghdad
Anbar
Babil   Wasit
Karbala
Qadisyah   Maysan
JORDAN   Najaf   Dhi Qar
Muthanna   Basra
SAUDI
ARABIA   KUWAIT

Longitude East of Greenwich

39°   42°   45°   48°

al-Qaeda terrorists, who had not previously been active here, to worsen the security situation. Even so, Iraq's people repeatedly voted in elections that advanced the country toward representative government, and some middle-class Iraqis (among the hundreds of thousands who fled the country in the mid-2000s) began to return to their homeland. The allied presence also helped to create a much more open economy.

In 2010, the official combat role of U.S. troops in Iraq was declared to be at an end, following a protracted war that claimed an estimated 150,000 Iraqi lives and caused the death of some 5000 allied soldiers. By the end of 2011, even as suicide bombers continued to regularly attack targets from religious pilgrims to government facilities, American troops were scheduled to withdraw from Iraqi soil. It remained uncertain whether Iraq would be able to achieve anything resembling structural stability and political democracy.

Syria

Like Iraq before 2003, Syria is ruled by a minority. Although Syria's population of 23.6 million is about 75 percent Sunni Muslim, the ruling elite comes from a smaller Shi'ite Islamic sect based there, the Alawites who account for about 10 percent of the population. Leaders of this powerful minority have retained control over the country for decades, at times by ruthless suppression of dissent. In 2000, president-for-life Hafez al-Assad died and was succeeded by his son, Bashar, signaling a continuation of the political status quo. For 25 years, part of this status quo was the occupation and control of neighboring Lebanon, but in 2005 the Syrians finally withdrew.

Like Lebanon and Israel, Syria has a Mediterranean coastline where crops can be raised without irrigation. Behind this densely populated coastal zone, Syria has a much larger interior than its neighbors, but its areas of productive capacity are widely dispersed. Damascus, in the southwest corner of the country, was built on an oasis and is considered to be one of the world's oldest continuously inhabited cities. It is now the capital of Syria, with a population of close to 3 million.

The far northwest is anchored by Aleppo, the focus of cotton- and wheat-growing areas in the shadow of the Turkish border. Here the Orontes River is the chief source of irrigation water, but in the eastern part of the country the Euphrates Valley is the crucial lifeline. It is in Syria's interest to develop its northeastern provinces, now enhanced by recent discoveries of oil there, but the war in neighboring Iraq has triggered problems along Syria's eastern border that will delay development plans.

Syria's ruthless dictatorship at first seemed immune to the wave of unrest that cascaded across much of the Arab World in early 2011. But soon demonstrations broke out in a number of cities, and the protest movement in-tensified rapidly in response to the regime's especially harsh crackdown, with many of the brutalities focused on the western parts of the country, particularly the cities of Aleppo, Homs, Latakia, and Dara'a. By mid-2011, more than 1000 people were estimated to have been killed in the violence while several thousand were reported to be in custody or missing.

Popular resentment in Syria, as elsewhere, reflected the workings of government that robbed its population of freedom while failing to deliver economic progress. The demonstrations here, too, expressed a long-simmering discontent, most notably among the young. The resolution of the conflict in Syria is critically important to the future direction of the Middle East and the realm at large (more important, arguably, than what happens in Tunisia, Libya, or Yemen). Syria is a hub of influence in the region, particularly regarding its immediate neighbors Israel, Lebanon, Jordan, Iraq, and Turkey. A peaceful democratic turn in Syria could be immensely beneficial for all—but it seemed that Syria's "Spring" had a very long way to go before "Summer" would arrive.

Jordan

Syria's southern neighbor, Jordan is a classic case (and victim) of changing relative location. This desert kingdom was a product of the Ottoman collapse after World War I, but when Israel was created in 1948 it lost its window on the Mediterranean Sea as well as the (now Israeli) port of Haifa, previously in the British-administered Mandate of Palestine. Following Jordan's independence in 1946 with about 400,000 inhabitants, newly-created Israel soon bequeathed the country some 500,000 West Bank Palestinians and, later, a huge inflow of refugees. Today Palestinians outnumber original residents by more than two to one in the population of almost 7 million. It may be said that the 47-year rule by King Hussein, ending in 1999, was the key centripetal force that held the country together.

With a poverty-stricken capital (Amman), lacking in oil reserves, and possessing only a small and remote outlet to the Gulf of Aqaba, Jordan has survived with U.S., British, and other aid. It lost its West Bank territory in the 1967 war with Israel, including its sector of Jerusalem (then the kingdom's second-largest city). No third country has a greater stake in a settlement between Israel and the Palestinians than does Jordan.

Jordan did not experience the kind of upheaval that shook many other Arab countries in 2011 because the monarchy has enjoyed widespread support and government policies lacked the level of repression found in Syria or Tunisia. Yet there were some demonstrations, and King Abdullah II had to dismiss his cabinet in February to quell the protests. Jordan's long-term challenges are economic, although continued deprivation can readily turn into political discontent.

FIGURE 7B-5 © H. J. de Blij, P. O. Muller, and John Wiley & Sons, Inc.

Lebanon

The map suggests that Lebanon has significant geographic advantages in this region: a lengthy coastline on the Mediterranean Sea; a well-situated capital, Beirut, on its shoreline; oil terminals along its coast; and a major capital (Syria's Damascus) in its hinterland. The map at the scale of Figure 7B-4 cannot reveal yet another asset: the fertile, agriculturally productive Bekaa Valley in the eastern interior.

French colonialism following the Ottoman period created a territory, which in 1930 was about equally Muslim and Christian. Beirut soon became known as the Paris of the Middle East. After independence in 1946, Lebanon functioned as a democracy and did well economically as the region's leading banking and commercial center. But the Muslim sector of the population grew much faster than the Christian one, and in the late 1950s the Arabs launched their first rebellion against the established order. Following the first of several waves of influx by Palestinian refugees, Lebanon fell apart in 1975 in a civil war that wrecked Beirut, devastated the economy, and left the country at the mercy of Syrian forces, which had entered to take control.

For so small a country (4.4 million, including more than 400,000 naturalized and noncitizen Palestinians), Lebanon is permeated with religious and ethnic factions and is prone to sectarian breakdown. The Lebanese were unhappy with what they felt to be a Syrian occupation, yet that situation persisted for several decades. In the meantime, an Iran-sponsored terrorist movement, Hizbollah, came into being and even became a political force. The Syrians were ousted in 2005 under United Nations auspices, and a stable new Lebanese government coalesced around a power-sharing agreement. However, the peace lasted only a few years. In 2006, a Hizbollah kidnapping of Israeli soldiers provoked an Israeli attack, shattering the reconstructed physical and political infrastructure. Lebanon, sometimes aptly described as a "garden without a fence" because of its vulnerability to outside interference, remains in precarious condition.

Israel and the Palestinian Territories

Israel, the Jewish state, lies in the center of the Arab World (Fig. 7B-1). Since 1948, when Israel was created as a homeland for the Jewish people on the recommendation of a United Nations commission, the Arab-Israeli conflict has affected all else in the region.

Figure 7B-6 helps us understand the complex issues involved here. In 1946, the British, who had administered this area in post-Ottoman times, granted independence to what was then called Transjordan, the kingdom east of the Jordan River. In 1948, the orange-colored area became the UN-sponsored state of Israel—including, of course, land that had long belonged to Arabs in this territory called Palestine.

As soon as Israel proclaimed its independence, neighboring Arab states attacked it. Israel, however, not only held its own but pushed the Arab forces back beyond its borders, gaining the green areas in 1949 (Fig. 7B-6). Meanwhile, Transjordanian armies crossed the Jordan River and annexed the yellow-colored area named the West Bank (of the Jordan River), including part of the city of Jerusalem. The king called his newly enlarged country Jordan.

More conflict soon followed. In 1967, a six-day-long war produced a major Israeli victory: Israel took the Golan Heights from Syria, the West Bank from Jordan, and the Gaza Strip from Egypt, and conquered the entire Sinai Peninsula all the way to Egypt's Suez Canal. In later peace agreements, Israel returned the Sinai but not the Gaza Strip to Egypt.

All this strife produced a huge outflow of Palestinian Arab refugees and displaced persons. The Palestinian Arabs constitute another of this realm's *stateless nations*; about 1.6 million continue to live as Israeli citizens within the borders of Israel, but more than 2.3 million are in the West Bank and another 1.5 million in the Gaza Strip (the

Israel's Security Barrier takes several forms—a concrete wall, an iron fence—and its effect also has multiple dimensions. This section of the barrier separates the village of Abu Dis on the outskirts of Jerusalem (left) from the West Bank (right). While Israel defines this demarcation as a security issue, the Arabs, citing its location inside Palestinian territory, argue that it is motivated primarily by politics.

© AP/Wide World Photos

FIGURE 7B-6

© H. J. de Blij, P. O. Muller, and John Wiley & Sons, Inc.

Golan Heights population is comparatively insignificant). Many Palestinians live in neighboring and nearby countries, including Jordan (2.8 million), Syria (630,000), Lebanon (405,000), and Saudi Arabia (280,000); another 400,000 or so reside in Iraq, Egypt, and Kuwait; and some live in other countries around the world, including the United States (ca. 250,000). Many in this Palestinian Diaspora have been assimilated into their host societies, but tens of thousands continue to live in refugee camps. The estimated size of the total, stateless Palestinian population in 2012 approached 12 million.

Israel is approximately the size of Massachusetts and has a population of just under 8 million (including its 1.6 million Arab citizens), but because of its location amid adversaries and its strong international links, these data do not reflect Israel's importance. Israel has built a powerful military even as its regional Arab neighbors have grown stronger, and its policies arouse Arab and Muslim passions. As a democracy with strong Western ties, Israel has been a recipient of massive U.S. financial aid, and American foreign policy has been to seek an accommodation between Jews and Palestinians, as well as between Israel and its Arab neighbors. Washington is tenaciously pushing toward a two-state solution that would create a Palestinian state alongside Israel, but as always there are thorny issues to be dealt with and a need for political will and accommodation on both sides that does not appear to be forthcoming.

FIGURE 7B-7

© H. J. de Blij, P. O. Muller, and John Wiley & Sons, Inc.

The geographic obstacles to this kind of an accommodation include the following:

1. *The West Bank.* Even after its capture by Israel in 1967, the West Bank might have become a Palestinian homeland (and possibly a state), but Jewish immigration to the area made such a future problematic. In 1977, only 5000 Jews lived in the West Bank, including East Jerusalem. By 2012 there were about 350,000, making up more than 12 percent of the population and creating a seemingly inextricable jigsaw of Jewish and Arab settlements (Fig. 7B-7), making the two-state solution difficult.

2. *The Golan Heights.* The right inset map in Figure 7B-6 suggests how difficult the Golan Heights issue is. The Heights overlook a large area of northern Israel, and they flank the Jordan River and crucial Lake Kinneret (the Sea of Galilee), the main water reservoir for Israel. Relations with Syria are not likely to become normalized until the Golan Heights are returned, but in democratic Israel the political climate may make the ceding of this territory impossible.

3. *Jerusalem.* The United Nations intended Tel Aviv to be Israel's capital, and Jerusalem an international city. But the Arab attack and the 1948–1949 war allowed Israel to drive toward Jerusalem (see Fig. 7B-6, the green wedge into the West Bank). By the time a cease-fire was arranged, Israel held the western part of the city, and Arab forces the eastern sector. But in this eastern sector lay major Jewish historic sites, including the Western Wall. Still, in 1950 Israel declared the western sector of Jerusalem its capital, making this, in effect, a *forward capital.* Figure 7B-8 shows the position of the (black) armistice line, leaving most of the Old City in Jordanian hands. But then, in the 1967 war, Israel conquered all of the West Bank, including East Jerusalem; in 1980, the Jewish state reaffirmed Jerusalem's status as capital, calling on all nations to move their embassies from Tel Aviv. Meanwhile, the government redrew the map of the ancient city, building Jewish settlements in a ring around East Jerusalem that would terminate the old distinction between Jewish west and Arab east. This enraged Palestinian leaders, who still view Jerusalem as the eventual headquarters of a hoped-for Palestinian state. Renewed U.S. efforts in 2011 to persuade Israel to halt the construction of new settlements in East Jerusalem were in vain, despite promises of increased American aid to Israel.

4. *The Security Fence.* In response to the infiltration of suicide terrorists, Israel has now walled off most of the West Bank along the Security Barrier border shown in Figure 7B-7 and the photo on page 380. The new wall does not conform to the de facto post-1967 boundary of the West Bank, cutting off an additional 10 percent of its territory and in effect annexing it (and its inhabitants) to Israel. This reinforcement of the West Bank border imposes a hardship on Palestinians, and for some it runs between their homes and their farm fields. Palestinians demand that the wall be taken down; Israelis reply that the Palestinian Authority must control the terrorists.

5. *The Gaza Strip.* Israel in 2005 decided to withdraw from the Gaza Strip, removing all Jewish settlements and yielding control of the area to the (then) Palestinian Authority (Fig. 7B-6, left inset map). A power struggle ensued between the two main Palestinian political factions, Fatah and the more militant Hamas, and Hamas won a disputed 2006 election. As rockets smuggled in from Egypt through tunnels dug beneath the Gaza-Egyptian border fell on Israel causing some deaths, injuries, and damage, Israel warned the Hamas regime of retaliation. Hamas authorities either would not or could not stop the rockets fired from Gaza, and in 2009 Israel launched a massive attack, killing about 1300 Gaza residents including many innocent civilians and inflicting heavy infrastructure damage, including a UN facility. Condemnation of Israel's assault as disproportionate was worldwide, but a majority of Israeli citizens approved of it.

Israel lies in the path of a fast-moving geopolitical storm. The issues raised above are only part of the overall problem: others involve water rights, compensation for expropriated land, Arabs' "right of return" to pre-refugee abodes, and the exact form a Palestinian state would take. Meanwhile, an Iranian president continues his aggressive declaratory stand toward Israel while his country appears

Jerusalem's urban landscape is rife with cultural contrasts—and is also home to the holy ground that means so much to Jews, Christians, and Muslims.

© H. J. de Blij

FIGURE 7B-8

© H. J. de Blij, P. O. Muller, and John Wiley & Sons, Inc.

on its way to nuclear weapons capability and already has the means to deliver them. In this age of nuclear, chemical, and biological weapons, Israel's search for an accommodation with its Arab neighbors is a race against time.

Israel's recent resolve may partly be due to new discoveries of major offshore gasfields in the Mediterranean (see Fig. 7B-6). Exploration has already begun in some fields, while test drilling proceeds in others. These gas reserves presage a major boost to Israel's income, but they also heighten political tensions. The largest field is also the furthest north and has generated a dispute with Lebanon as to its exact location relative to maritime boundaries. And the southernmost field lies off the shore of the Gaza Strip. Although the Palestinian authorities have been eager to commence exploitation that could provide a badly needed lift to their impoverished economy, Israel has vetoed the project on security grounds.

▰▰ THE ARABIAN PENINSULA

The regional identity of the Arabian Peninsula is clear: south of Jordan and Iraq, the entire peninsula is encircled by water. This is a region of old-style sheikdoms made

fabulously wealthy by oil, modern-looking emirates, and the site where Islam originated.

Saudi Arabia

Saudi Arabia itself has only 30.7 million residents (plus about 8 million expatriate workers) in its vast territory, but we can see the kingdom's importance in Figure 7A-8: the Arabian Peninsula contains the world's largest concentration of known petroleum reserves. Saudi Arabia occupies most of this area and may possess more than one-fifth of the world's liquid oil deposits. As Figure 7B-9 shows, these reserves lie in the eastern part of the country, particularly along the Persian Gulf coast, and extend southward into the Rub al Khali (Empty Quarter).

As a region, the Arabian Peninsula is environmentally dominated by a desert habitat and politically dominated by the Kingdom of Saudi Arabia (Fig. 7B-9). Containing 2,150,000 square kilometers (830,000 sq mi), Saudi Arabia is the realm's third-biggest state; only Kazakhstan and Algeria are larger. On the peninsula, Saudi Arabia's neighbors (proceeding clockwise from the head of the Persian Gulf) are Kuwait, Bahrain, Qatar, the United Arab

ARABIAN PENINSULA

POPULATION

- Under 50,000
- 50,000–250,000
- 250,000–1,000,000
- 1,000,000–5,000,000
- Over 5,000,000

(E) Cities targeted for expansion

National capitals are underlined

Oilfield
Gasfield
4-lane road
Major road
Road
Oil pipeline
Gas pipeline

LAND USE

Dry farming
Desert
Permanent crops and arable lands

0 100 200 300 400 500 Kilometers
0   100   200   300 Miles

FIGURE 7B-9

© H. J. de Blij, P. O. Muller, and John Wiley & Sons, Inc.

Emirates (UAE), the Sultanate of Oman, and the Republic of Yemen. Together, these countries on the eastern and southern fringes of the peninsula contain 40 million inhabitants; the largest by far is Yemen, with 25 million. Figure 7B-9 reveals that most of the economic activities in Saudi Arabia are concentrated in a wide belt across the "waist" of the peninsula, from the oil boomtown of Dhahran on the Persian Gulf through the national capital of Riyadh in the interior to the Mecca–Medina area near the Red Sea. A modern transportation and communications network has recently been completed, but in the more remote zones of the interior Bedouin nomads still ply their ancient caravan routes across the vast deserts. For decades, Saudi Arabia's royal families were virtually the sole beneficiaries of their country's wealth, and there was hardly any impact on the lives of villagers and nomads. When the oil boom arrived in the 1950s, foreign laborers were brought in (today there are roughly 8 million, many of them Shi'ite) to work in the oilfields, ports, factories, and as domestics. The east boomed, but the rest of the country has lagged behind.

Disparities and Uncertainties

Efforts to reduce Saudi Arabia's regional economic disparities have been impeded by the enormous cost of bringing water from deep-seated sources to the desert surface to stimulate agriculture, by the sheer size of the country, and by the high rate of population growth (2.6 percent). Still, housing, health care, and education have seen major improvement, and industrialization also has been stimulated in such places as Jubail on the Persian Gulf and Yanbu, near Medina, on the Red Sea.

Saudi Arabia's conservative monarchic rule, official friendship with the West, and huge social contrasts resulting from its economic growth have raised political opposition, for which no adequate channels exist. Relations with the United States were affected by the role of Saudi suicide terrorists in 9/11, Saudi financial support for Wahhabist extremists, and additional issues arising in the wake of the events of 2001.

The Saudi regime watched nervously as the uprisings diffused across the realm in early 2011, but things remained relatively quiet here. It was quick to announce a new jobs program involving billions of government dollars while beefing up police presence on the streets. With the disruption of Libya's oil production as its rebellion-triggered civil war raged, the Saudis increased their own output to help stabilize global market prices (which, for a variety of reasons, were already fluctuating around record-high levels).

On the Peninsular Periphery

Five of Saudi Arabia's six neighbors on the Arabian Peninsula face the Persian and Oman gulfs (Fig. 7B-9) and are monarchies in the Islamic tradition. All five also derive substantial revenues from oil. Their populations range from 1.3 to 5.5 million in addition to hundreds of thousands of foreign workers; these are not strong or influential states, but they do display considerable geographic diversity.

Kuwait, at the head of the Persian Gulf, almost cuts Iraq off from the open sea, an issue the 1990–1991 Gulf War did not settle. Ministate Kuwait's oil reserves are among the world's five largest, but even Kuwait feels the need to diversify its economy. A major oil refinery, the largest of its kind in the realm, is being built near the capital. Like other wealthy countries in the GCC (Gulf Cooperation Council), Kuwait has attracted a large foreign workforce; in this mostly Sunni emirate, where 80 percent of the workers in private industry are foreigners, this has produced a Shia minority of greater than 25 percent. As in other GCC states, the working conditions for foreigners in this Arab country are often dismal. On the other hand, women were allowed to register and vote for the first time in the 2007 parliamentary elections.

Bahrain, an island-state ruled by a royal family since the eighteenth century, is a tiny territory with dwindling oil reserves. It has become an important banking center for the region and has some of the more progressive social policies in this part of the world. Its 1.3 million people are split between Sunnis (30 percent) and Shi'ites (70 percent); nearly two-thirds of its labor force is foreign. When faced with popular unrest and demonstrations in early 2011 (mainly by the marginalized Shi'ite majority), the government assumed an increasingly tough stance and invited in extra troops from Saudi Arabia to suppress demonstrators and maintain order. Over the years, Bahrain has maintained such friendly relations with the United States that the U.S. Navy's Fifth Fleet is headquartered there. Thus the country's police-state actions during the 2011 uprising posed a special problem for the Americans, whose criticism had to be tempered by the need to maintain long-term military cooperation.

Neighboring **Qatar** consists of a peninsula jutting out into the Persian Gulf that has capitalized on its oil reserves and its increasing natural gas reserves. With a modest population of 1.7 million but a strong Sunni majority, this emirate is trying to establish itself as not only an Arab World focus for international economic, social, and political reform, but as a global center for financial deliberation. Its capital, Doha, is the site of annual "Doha Round" meetings of the world's leading economic powers. The political unrest that coursed through much of the realm in 2011 bypassed Qatar, which was the first Arab country to grant full recognition to Libya's rebels. Qatar will be hosting soccer's World Cup tournament in 2022, its winning bid a surprise to most observers. This small nation has no soccer tradition to speak of, and its summer temperatures can reach 50°C (122°F). The local organizers, hardly lacking monetary resources, have proposed the building of several indoor, fully air-conditioned stadiums.

The **United Arab Emirates (UAE)**, a federation of seven emirates, faces the Persian Gulf between Qatar and Oman. A reigning sheik is the absolute monarch in each of the emirates, and the seven sheiks together form the Supreme Council of Rulers. In terms of oil revenues, however, there is no equality: two emirates—Abu Dhabi and Dubai—have most of the reserves. As the photograph

The Burj Khalifa, centerpiece of the ultramodern city of Dubai in the United Arab Emirates. Completed in 2010, the 160-story Burj is the world's tallest building, topping out at 828 meters (2716 ft)—just over a *half mile* high, almost twice the height of New York's Empire State Building.

© Jose Fuste Raga/Corbis

architecture, state-of-the-art engineering feats, and world-class sports, leisure, and entertainment complexes.

The eastern corner of the Arabian Peninsula is occupied by the Sultanate of **Oman**, another absolute monarchy, centered on the capital, Muscat. Figure 7B-9 shows that Oman is fragmented into two parts: the large eastern corner of the peninsula, and a small but highly critical cape to the north, the Musandam Peninsula, that protrudes into the Persian Gulf to form a narrow channel or **choke point [6]**—the Hormuz Strait (Iran lies on the opposite shore). Tankers leaving the other Gulf states must negotiate this slender, twisting channel at slow speed, and during politically tense times warships have needed to protect them. Iran's claim to several small islands near the Strait that are owned by the UAE remains a potential source of dispute.

This brings us to what is, in so many ways, Saudi Arabia's most substantial peninsular neighbor: **Yemen**. The boundary between Saudi Arabia and Yemen has only recently been satisfactorily delimited in an area where there may be substantial oil reserves. Another boundary, between former North Yemen and South Yemen, was erased in 1990, when the two countries joined to form the present state. San'a, formerly the capital of North Yemen, retained that status; Adan (Aden), the only major

above reveals, Dubai (Dubayy) has converted its favorable geographic location together with its oil revenues into a booming economy (although staggered by the recent global recession) in which tourism, international transit functions, banking, and trade are symbolized by cutting-edge

From the Field Notes . . .

© H. J. de Blij

"The port of Mutrah, Oman, like the capital of Muscat nearby, lies wedged between water and rock, the former encroaching by erosion, the latter crumbling as a result of tectonic plate movement. From across the bay one can see how limited Mutrah's living space is, and one of the dangers here is the frequent falling and downhill sliding of large pieces of rock. It took about five hours to walk from Mutrah to Muscat; it was extremely hot under the desert sun but the cultural landscape was fascinating. Oil also drives Oman's economy, but here you do not find the total transformation seen in Kuwait or Dubai. Townscapes (as in Mutrah) retain their Arab-Islamic qualities; modern highways, hotels, and residential areas have been built, but not at the cost of the older and the traditional. Oman's authoritarian government is slowly opening the country to the outside world after long-term isolation."

www.conceptcaching.com

port along a lengthy stretch of the peninsula, anchors the south.

Yemen, with a population of 25 million—roughly 53 percent Sunni and 45 percent Shi'ite—initially made significant progress toward representative government and political stability, but centrifugal forces and terrorist activity caused later setbacks. A Shi'ite rebellion against the central government has destabilized the north even as a secessionist movement flares in the south, where oilfields encourage such aspirations. In 2000, a terrorist attack on a U.S. warship docked in Adan killed 17 sailors, nearly sank the vessel, and ushered in a series of subsequent attacks on foreigners, government facilities, and other targets for which an organization called *Al-Qaeda of the Arabian Peninsula* claimed responsibility.

Figure 7B-9 underscores the geographic attractions Yemen has for al-Qaeda and other Islamic terrorist groups.

A sea of Yemeni protesters on May 13, 2011 carry an enormous flag during a demonstration in San'a to demand the ouster of President Saleh. Here and in most other countries touched by the Arab Spring, demonstrators often defiantly displayed their national flag to confirm their nationalistic feelings—but also as a gesture of scorn to remind their government how badly it had failed the country and its people.

© YAHYA ARHAB/epa/Corbis

Not only does Yemen lie at the back door of Saudi Arabia, its southern tip overlooks another of the world's busiest choke points at the entrance to the Red Sea (the Bab el Mandeb ["Gate of Grief"] Strait), where ships converge and risk capture by pirates. Across the Gulf of Adan lies another terrorist haven (and leading pirate base), Somalia, and across the Red Sea lies Eritrea, a likely target for extremists of the al-Qaeda stripe. From Yemen, too, these militants can stoke strife all along Africa's Islamic Front (see Fig. 6B-9).

Yemen's precarious position became even more complicated when the country was inundated by the passions unleashed during the Arab Spring of 2011. The closely followed events in Tunisia and Egypt quickly sparked mass demonstrations against the 33-year-old regime of President Ali Abdullah Saleh. The government responded with a mix of conciliatory measures and violent confrontation, but the courageous protesters, many students among them, defied the brutalities inflicted by police and returned to Yemen's streets again and again. The result was utter chaos. At several points, an agreement appeared to have been reached between the government and the protesters—only to be revoked by one side or the other at the last moment. But pressure on the regime steadily mounted, especially when certain government officials resigned and a number of high-ranking military leaders expressed support for the demonstrators. By early summer, Saleh, seriously injured in an attack and recuperating in Saudi Arabia, had still not resigned, and Yemen was sliding toward anarchy.

■■ THE EMPIRE STATES

Two major states, both with imperial histories, dominate the region that lies immediately to the north of the Middle East and Persian Gulf (Fig. 7B-1), where Arab ethnicity gives way but Islamic culture endures—Turkey and Iran. Although they share a short border and are both Muslim countries, they display significant differences as well. Even their versions of Islam are different: Turkey is an officially secular but dominantly Sunni state, whereas Iran is the heartland of Shi'ism. Turkey's leaders have worked to establish satisfactory relations with Israel; Iran's president promised to "wipe Israel off the map." In recent years, Turkey has been trying to negotiate entry into the European Union. Iran's goals have been quite different as that country has defied international efforts to constrain its nuclear ambitions. Two smaller countries are inextricably bound up with these two regional powers: island *Cyprus* to the southwest, still divided today between Turks and Greeks in a way that threatens Turkey's European ambitions (discussed in Chapter 1B); and oil-rich, Caspian Sea-bordering *Azerbaijan* to the northwest, with ethnic and religious ties to Iran but economic links elsewhere (covered in Chapter 2B).

Turkey

As Figure 7B-10 below indicates, Turkey is a mountainous country of generally moderate relief; it also tends to exhibit considerable environmental diversity, ranging from steppe to highland (see Fig. G-7). On central Turkey's dry Anatolian Plateau, villages tend to be small and subsistence farmers grow cereals and raise livestock. Coastal plains are not large, but they are productive as well as densely populated. Textiles (from home-grown cotton) and farm products dominate the export economy, but Turkey also has substantial mineral reserves, some oil in the southeast as well as massive dam-building projects on the Tigris and Euphrates rivers, and a small steel industry based on domestic raw materials.

In Chapter 7A we chronicled the historical geography of the Ottoman Empire, its expansion, cultural domination, and collapse. By the beginning of the twentieth century, the country we now know as Turkey lay at the center of this disintegrating state, ripe for revolution and renewal. This occurred in the 1920s and thrust into prominence a leader who became known as the father of modern Turkey: Mustafa Kemal, known after 1933 as Atatürk, meaning Father of the Turks.

Capitals New and Old

The ancient capital of Turkey was Constantinople (now Istanbul), located on the Bosphorus, part of the strategic straits connecting the Black and Mediterranean seas. But the struggle for Turkey's survival had been waged from the heart of the country, the Anatolian Plateau, and it was here that Atatürk decided to place his seat of government (Fig. 7B-10). Ankara, the new capital, possessed certain advantages: it would remind the Turks that they were (as Atatürk always said) Anatolians; it lay nearer the center of the country than Istanbul; and it could therefore act as a stronger unifier. Istanbul lies on the threshold of Europe, with the minarets and mosques of this largest and most varied Turkish city rising above a townscape that resembles those in eastern Europe (see photo next page).

Even though Atatürk moved the capital eastward and inward, his orientation was westward and outward. To implement his plans for Turkey's modernization, he initiated reforms in almost every sphere of life within the country. Islam, formerly the state religion, lost its official status, and Turkey became a secular state whose army ensured that the Islamists would not take over again. The state took over most of the religious schools that had controlled education. The

FIGURE 7B-10

© H. J. de Blij, P. O. Muller, and John Wiley & Sons, Inc.

AMONG THE REALM'S GREAT CITIES . . .

ISTANBUL

© H. J. de Blij, P. O. Muller, and John Wiley & Sons, Inc.

FROM BOTH SHORES of the Sea of Marmara, northward along the narrow Bosphorus toward the Black Sea, sprawls the fabulous city of Istanbul, known for centuries as Constantinople, headquarters of the Byzantine Empire, capital of the Ottoman Empire, and, until Atatürk moved the seat of government to Ankara in 1923, capital of the modern Republic of Turkey as well.

Istanbul's site and situation are incomparable. Situated where Europe meets Asia and where the Black Sea joins the Mediterranean, the city was built on the requisite seven hills (as Rome's successor), rising over a deep waterway that enters the Bosphorus from the west, the famous Golden Horn. A sequence of empires and religions endowed the city with a host of architectural marvels (see photo below) that give it, when approached from the water, an almost surreal appearance—and what is seen today is a mere remnant of history's accretion, survivals of earthquakes, fires, and war.

Turkey's political capital may have moved to Ankara, but Istanbul remains its cultural and commercial headquarters. It also is the country's leading urban magnet, luring millions from the poverty-stricken countryside. In the heart of the city known as Stamboul and in the "foreign" area north of the Golden Horn called Beyoğlu, modern buildings vie for space and harbor views (see photo, p. 35), blocking the vistas that long made Istanbul's cultural landscape unique. On the outskirts, shantytowns emerge so rapidly that Istanbul's population is doubling every 15 years at current rates, having reached 10.8 million in 2012. Istanbul's infrastructure is crumbling under this unprecedented influx, its growth is virtually without planning controls, and its legacy of two millennia of cultural landscapes is threatened.

From the Field Notes . . .

© H. J. de Blij

"Having spent the night in Istanbul's Beyoğlu District on the European side, I walked across the Atatürk (Unkapani) Bridge to the Eminönü District on the south side of the Golden Horn, intending to make it to the upscale Fatih District later. But getting a clear view of the Hagia Sophia 'Museum' was not easy from this angle, amid the congested buildup in this hilly area. In a line of taxis at a taxi stand one had a sign saying 'Speak English' so I asked the driver about getting to a good vantage point, and he showed me on my map how to find a lookout point on the terrace of a local restaurant. From there, I could see the bridge I had crossed to my left, the Hagia Sophia Church with its later minarets in the distance, the Sultan Haseki Hürrem Baths to the right, and in the foreground a simple, local mosque. 'You should know the rules about mosques,' he said. 'The number of minarets reveals the importance of any mosque. One minaret, and it's local, neighborhood. Two, and it's likely to be more prosperous, perhaps the work of a very successful citizen or community. Any more than two, and the builders have to have not only religious sanction but also state permission to build. And if they get that, they go for four.' Thousands of mosques, many of them architectural treasures, grace the urban landscape of Istanbul, whose population today, according to local geographers, is really 21 million."

www.conceptcaching.com

Roman alphabet replaced the Arabic. A modified Western code supplemented Islamic law. Symbols of old—growing beards, wearing the fez—were prohibited. Monogamy was made law, and the emancipation of women was begun. The new government emphasized Turkey's separateness from the Arab World, and in many ways it has remained aloof from the affairs that engage other Islamic states.

Turkey and Its Neighbors

From before Atatürk's time, Turkey has had a history of intolerance of minorities. Soon after the outbreak of the First World War, the (pre-Atatürk) regime decided to expel all the Armenians, concentrated in the country's northeast. Nearly 2 million Turkish Armenians were uprooted and brutally forced out; an estimated 600,000 died in a campaign that still arouses anti-Turkish emotions among Armenians today.

In modern times, Turkey also has been criticized for its harsh treatment of its large and regionally concentrated Kurdish population. About one-fifth of Turkey's population of 75.4 million is Kurdish, and successive Turkish governments have mishandled relationships with this minority nation, even prohibiting the use of Kurdish speech and music in public places during one especially repressive period. The historic Kurdish homeland lies in the southeast of Turkey, centered on Diyarbakir (Fig. 7B-10), but millions of Kurds have moved to the shantytowns around Istanbul—and to jobs in the countries of the European Union. With Kurdish nationalism rising across the border in embattled Iraq, Turkey has responded in two principal ways: by suppressing Kurdish insurgent movements both in Turkey itself and across the border in Iraq, and by awarding more rights and freedoms to those Kurds not involved in armed resistance. Thus Turkey improved its human rights record, a key stumbling block in its aspirations to join the European Union. Simultaneously, Ankara made tentative but significant moves to rehabilitate its relations with Armenia, addressing an erosive, longstanding problem.

Turkey and the EU

With its diversified and growing economy, secular government, and independent posture in international affairs, Turkey is the key, predominantly Islamic state potentially eligible for admission into the European Union (Albania and Kosovo may precede it). Turkey's soccer team already plays in the European championships, and for a long time Turkey's westward orientation, membership in NATO, its many migrants, and its solid economic ties to that realm made this a distinct possibility. Already in a customs union with the EU since 1995, Turkey in 1999 was officially declared an accession candidate. Formal negotiations toward admission began in 2005.

However, major obstacles lie in the way of this becoming a reality. The Cyprus and Kurdish issues have both negatively affected Turkey's prospects for joining the EU. Admitting Turkey would put Europe's boundary on the doorstep of a dangerous neighborhood, which is a concern to some, though a positive aspect for others. Another consideration

is Turkey's human rights record, which remains inadequate. In particular, Turkey's numerous "honor killings," usually of girls and women by men, arouse European anger—not only because they take place (heinous murders occur in non-Islamic societies as well) but because they constitute fully half of all murders committed in the country—and because they continue to be tolerated through ineffective gender-biased law enforcement and insufficient public condemnation.

Nonetheless, Turkey has made considerable progress during the past decade. The state is now a functioning democracy, and its secular values are constitutionally protected so that even an Islamic party majority in government does not portend radicalization even as it may encourage symbols of piety. For instance, after the largest Islamic party won a majority in Turkey's 2007 elections, good relations with Israel persisted even as the new government lifted the long-existing ban on women wearing head scarves. As noted above, the new government also continued to seek accommodation with its large Kurdish minority as well as with neighboring (Christian) Armenia.

Even though Turkey might be closing in on European Union targets and requirements, enthusiasm among Turks for joining the EU is diminishing. The economy has been doing so well in recent years that the need for EU membership is not as alluring as it once looked—especially as some established EU members (Greece among them) effectively went bankrupt in the early 2010s. In addition, many Turks take offense at the unrestrained public debate concerning their candidacy going on in Europe, often including criticisms of Turkish customs and traditions not only in Turkey itself but also in Turkish communities already established in Europe. So we have now reached a point that if the EU eventually does extend an invitation to Turkey, it is by no means certain that a future Turkish government would accept it. Polls indicate that popular support for EU membership dropped from 75 percent in 2000 to a mere 25 percent in 2010.

Turkey is a crucial country in a unique geographical position that is trying to steer an independent and singular course in one volatile realm while accommodating the demands of another. Its importance was confirmed in dramatic fashion when the Turkish Model (see p. 371) was repeatedly held up by protesters across the Arab World in 2011 as the ideal future for their countries: a mildly Islamist government that does not alienate the religiously devout, and prioritizes democratic reform while pursuing vigorous economic growth. For the EU to pass on Turkey's admission, many observers say, looks increasingly imprudent.

Iran

Iran has its own history of conquest and empire, collapse and revival. Oil-rich and vulnerable, Iran already was just a remnant of a once-vast regional empire that waxed and waned, from the Danube to the Indus, over more than two millennia. Long known to the outside world as *Persia* (although the people had called themselves Iranians for centuries), the state was officially renamed *Iran* by the reigning shah (king) in 1935. His intention was to stress his country's Indo-European heritage as opposed to its Arab and Mongol infusions.

As we noted earlier, democracy was stirring in oil-rich Iran during the 1950s when the United States undermined a popular prime minister and helped reinstate the degenerating monarchy. In 1971, the then-ruling shah and his family celebrated the 2500th anniversary of Persia's royal rule with unmatched splendor. By 1979, however, revolution had engulfed Iran, and Shi'ite fundamentalists, with widespread public support, drove the shah from power. The monarchy was replaced by an Islamic republic ruled by an ayatollah (leader under Allah), and a frightful wave of retribution followed. Today, despite a veneer of democracy, Islamists continue to tightly control the state. The disputed 2009 presidential election—that led to a wave of street protests, violence, reprisals, and murder—dramatically exposed these divisions.

A Crucial Location and Dangerous Terrain

As Figure 7B-1 shows, Iran occupies a critical position in this turbulent realm. It controls the entire corridor between the Caspian Sea and the Persian Gulf. To the west it adjoins Turkey and Iraq, both historic enemies. To the north (west of the Caspian Sea) Iran borders both Azerbaijan and Armenia, where once again Islam confronts Christianity. To the east Iran meets Afghanistan and Pakistan, and to the northeast lies similarly volatile Turkmenistan.

Iran, as Figure 7B-11 demonstrates, is a country of mountains and deserts. The heart of it is a sizeable upland, the Iranian Plateau, that lies surrounded by even higher mountains, including the Zagros in the west, the Elburz in the north along the Caspian Sea coast, and the mountains of Khurasan to the northeast. This mountainous topography signals geologic danger: here the Eurasian and Arabian tectonic plates converge, causing major and often devastating earthquakes. The Iranian Plateau therefore is actually a huge highland basin marked by salt flats and wide expanses of sand and rock. The highlands wrest some moisture from the air, but elsewhere only oases interrupt the arid monotony—oases that for countless centuries have been stops on the area's caravan routes.

City and Countryside

In ancient times, Persepolis in southern Iran (located near the modern city of Shiraz) was the focus of a powerful Persian kingdom, a city dependent on **qanats** [7], underground tunnels carrying water from moist mountain slopes to dry flatlands some distance away. Today, Iran's population of 77.1 million is 69 percent urban, and the capital, Tehran, lies far to the north, on the southern, interior slopes of the Elburz Mountains. This mushrooming metropolis of seven and a half million, lying at the heart of modern Iran's core area, still depends in part on the same kinds of qanats that sustained Persepolis more than 2000 years ago. As such, Tehran symbolizes the internal contradictions of Iran: a country in which modernization

has taken hold in the cities, but little has changed in the vast countryside, where a generation ago the mullahs led their peasant followers in a revolution that overthrew a monarchy and installed a theocracy.

The Ethnic-Cultural Map

Iranians dominate Iran's affairs, but the country is no more an ethnically unified state than Turkey is. Figure 7B-11 shows that three corners of Iran are inhabited by non-Iranians, two in the west and one in the southeast. Iran's northwestern corner is part of greater Kurdistan, the multistate region in which Kurds are in the majority and of which southeastern Turkey also is a part (Fig. 7B-4). Between the Kurdish area and the Caspian Sea the majority population is Azeri, who number more than 18 million and also form the majority in the neighboring state of Azerbaijan to the north.

In the southwest, the minority is Arab and is concentrated in Khuzestan Province. During the Iran-Iraq War of the 1980s, Saddam Hussein had his eye on this part of Iran, whose Arab minority often complained of mistreatment by the Iranian regime, and where a large part of Iran's huge oil reserve is located (note the situation of the export terminal of Abadan in Fig. 7B-11). The Arab minority here is among Iran's poorest and most restive, in a province whose stability is key to the country's economy.

The third non-Iranian corner of the country lies in the southeast, where a coastal Arab minority focused on the key port of Bandar-e-Abbas on the Strait of Hormuz gives way eastward to a scattered and still-nomadic population of Baluchis who have ties to fellow ethnics in neighboring Pakistan (see Fig. 7B-13). On the Iranian side of the border, this is a remote and currently inconsequential area, but in Pakistani Baluchistan the government faces a persistent rebellion by tribes who feel slighted by official neglect. Iran's ethnic-cultural map, therefore, is anything but simple.

Energy and Conflict

As Figure 7A-8 shows, Iran possesses the realm's second-largest concentration of oil reserves, which amount to just over half of those in Saudi Arabia. Petroleum and natural gas production provide about 90 percent of the country's income. The reserves lie in a zone along the southwestern periphery of Iran's territory, and Abadan became its petroleum capital near the head of the Persian Gulf. But Iran is a large and populous country, and the wealth oil generated could not transform it in the way the last shah intended, a transformation that had it occurred might have staved off the revolution. Modernization remained but a veneer: in the villages distant from Tehran's polluted air, the holy men continued to dominate the lives of ordinary Iranians. As elsewhere in the Islamic world, urbanites, villagers, and nomads remained enmeshed in a web of production and profiteering, serfdom and indebtedness that has always characterized traditional society here.

The Islamic revolution swept this system away, but it did not improve the lot of Iran's millions: the economic failure of the shah was followed by the economic failure of the

FIGURE 7B-11

© H. J. de Blij, P. O. Muller, and John Wiley & Sons, Inc.

ayatollahs. A devastating war with Iraq (1980–1990), into which Iran ruthlessly poured hundreds of thousands of its young men, sapped both the coffers and energies of the state. When it was over, Iran was left poorer, weaker, and aimless, its revolution spent on unproductive pursuits. In the early years of the twenty-first century, evidence abounded that the people of Iran remained divided between conservatives determined to protect the power of the mullahs and reformers seeking to modernize and liberalize Iranian society.

In Iran's 2005 election, many reformist candidates were disqualified by religious conservatives who hold ultimate authority, and a president who had proved himself to be mildly reformist was succeeded by the far more extreme Mahmoud Ahmadinejad, who, shortly after taking office, proclaimed that the state of Israel should be "wiped off the map." In 2009, Ahmadinejad's reelection elicited charges of fraud and produced widespread and deadly riots.

In the meantime, Iran was pursuing a course toward nuclear-power-generating capability with implications that trouble many countries. In addition, the Iranians want to construct a natural gas pipeline across Pakistan to India, making India highly dependent on the Iranian supply. The deal is currently stalled, with Iran's price too high for India and the United States trying to terminate the plan.

Iran's rise as a nuclear power and its regional ambitions should not surprise anyone. Although revolutionary Iran disavowed the imperial designs of its Persian predecessors, this does not mean that its national interests now end at its borders. Tehran has a major stake in developments in Iraq and Afghanistan, both neighbors; it has an already-nuclear and unstable Pakistan on its eastern border; it has a strong interest in the fate of Shi'ite majorities and minorities elsewhere in the realm; it has long been an avowed enemy of Israel and a strong supporter of Palestinian causes; and it has funded organizations labeled terrorist whose actions have even reached across the Atlantic. Iran's revolution ousted the shah and terminated a monarchy, but it did not extinguish all ties to an imperial past. Iran's economy is now in dreadful shape, but the Arab Spring may have presented Iran with a renewed geostrategic advantage: the wave of political turmoil in 2011 weakened Egypt, Syria, and Jordan, and at the very least gave pause to Saudi Arabia. Iran might well use this opportunity to increase its traditional support of Shi'ite factions in those countries.

▪▪ TURKESTAN: THE SIX STATES OF CENTRAL ASIA

For many centuries, Turkish (Turkic) peoples spread outward across an enormous Central Asian domain that extended from Mongolia and Siberia to the Black Sea. Propelled by population growth and energized by Islam, they penetrated Iran, defeated the Byzantine Empire in present-day Turkey, and colonized much of eastern Europe. Eventually, their power declined as Mongols, Chinese, and then Russians invaded their strongholds. However, these conquerors could not expunge them, as the names on the modern map attest (Fig. 7B-12). The latest conquerors, the Chinese and particularly the Russian czars and their communist successors, created five Soviet Socialist Republics named after the majority peoples within their borders. Thus the Kazakhs, Turkmen, Kyrgyz, and other Turkic peoples retained some geographic identity in what was then Soviet Central Asia.

Central Asia—*Turkestan*—is a still-evolving region. In certain sectors, the cultural landscapes of neighboring realms extend into it, for example, in northern (Russian) Kazakhstan. In other areas, Turkestan extends into adjoining realms, as in the Xinjiang Uyghur Autonomous Region of westernmost China. Some areas once penetrated by Turkic peoples are no longer dominated by them—for instance, Afghanistan. And certain peoples now living in Turkestan are not of Turkic ancestry, notably the Tajiks. This is a fractious region in sometimes turbulent transition.

The underlying cultural-geographic reason for this contentiousness—not only in Turkestan but in neighboring regions in Southwest and South Asia as well—is illustrated in Figure 7B-13. This detailed map of ethnolinguistic groups actually is a generalization of an even more complex mosaic of peoples and cultures. Every cultural domain on the map also includes minorities that cannot

be mapped at this scale, so that people of different faiths, languages, and ways of life rub shoulders everywhere. Often this results in friction, and at times such friction escalates into ethnic conflict.

As we define it here, Turkestan includes six states: (1) *Kazakhstan*, territorially larger than the other five combined but situated astride a major ethnic transition zone; (2) *Turkmenistan*, relatively closed and dictatorial, with important frontage on the Caspian Sea and bordering both Iran and Afghanistan; (3) *Uzbekistan*, the most populous state and situated at the heart of the region; (4) *Kyrgyzstan*, wedged between powerful neighbors and chronically unstable; (5) *Tajikistan*, regionally and culturally divided as well as strife-torn; and (6) *Afghanistan*, engulfed in almost continuous warfare since it was invaded by Soviet forces in 1979.

The States of Former Soviet Central Asia

During their hegemony over Central Asia, the Soviets tried to suppress Islam and install secular regimes (this was their objective when they invaded Afghanistan as well), but today Islam's revival is one of the defining qualities of this region. From Almaty to Samarqand, mosques are being repaired and revived, and Islamic dress again is part of the cultural landscape. National leaders make high-profile visits to Mecca; most are sworn into office on the Quran. All of Central Asia's countries now observe Islamic holidays. In other ways, too, Turkestan reflects the norms of the NASWA realm: in its dry-world environments and the clustering of its population; its mountain-fed streams irrigating farms and fields; its sectarian conflicts; and its oil-based economies. It also is a region where democratic government remains an elusive goal.

The two influences of resurgent Islam and residual Russian both contribute significantly to the shaping of post-Soviet Turkestan. In recent years, Russia's leaders have worked to revive Moscow's links with this part of the Near Abroad and to exert Russian influence here. For instance, the government of Kyrgyzstan, which had granted the United States permission to lease an air base in the country for the military operations in Afghanistan, came under Russian pressure in 2009 to end that agreement and to oust the Americans, which it ultimately resisted. But with more than 250,000 Kyrgyz citizens working in Russia and more than one-third of the impoverished country's GDP consisting of remittances sent back by workers based outside their homeland, Moscow held a strong hand. The leaders of Turkestan's states tend to have Russian or Russified names, not Turkic ones (Kazakhstan's Nazarbayev, Uzbekistan's Karimov, Turkmenistan's Berdimuhamedov, Kyrgyzstan's Otunbayeva, and Tajikistan's Rakhmon), and are inheritors of the power vested in pro-Russian elites during Soviet rule in the region. More importantly, the concentration of authority prevailing in communist times continues today: representative government has but a weak foothold in Turkestan. The personality cult and absolute power that

FIGURE 7B-12

© H. J. de Blij, P. O. Muller, and John Wiley & Sons, Inc.

marked Saparmurat Niyazov's protracted rule of Turkmenistan (1990–2006), for example, was on a par with North Korea's Kim Il Sung and Kim Jong Il.

Turkestan's intricate ethnolinguistic mosaic is vividly represented in Figure 7B-13. Although states in this region tend to be named after their largest ethnic population sectors, political boundaries do not match the distribution of cultures. Perhaps the most complex cultural fabric is that of Afghanistan, populated by Pushtuns (occasionally spelled Pashtuns), Tajiks, Hazaras, Uzbeks, Turkmen, Baluchis, and several lesser groups. Add to this the region's rugged relief, vast deserts, and limited surface communications, and you can see why national integration is a particularly difficult challenge here.

SOUTH-CENTRAL EURASIA

After National Geographic.

FIGURE 7B-13

© H. J. de Blij, P. O. Muller, and John Wiley & Sons, Inc.

REGIONAL CONFLICTS

A Abkhazia – Georgia
B South Ossetia – Georgia
C North Ossetia-Alaniya – Ingushetiya
D Chechnya – Russia
E Turkey – Kurdish separatists
F Nagorno-Karabakh – Azerbaijan
G Afghan war
H Iraq war

MAJOR ETHNOLINGUISTIC GROUPS

ALTAIC
1 Azerbaijani
2 Kalmyk
3 Kazakh
4 Kyrgyz
5 Turk
6 Turkmen
7 Uzbek

CAUCASIAN
8 Chechen–Ingush
9 Dagestani
10 Georgian

AFRO-ASIATIC
11 Arabic

INDO-EUROPEAN
12 Armenian
13 Bakhtiari
14 Baluchi
15 Chahar Aimak
16 Hazara
17 Kurdish
18 Lur
19 Persian
20 Punjabi
21 Pushtun
22 Russian
23 Sindhi
24 Tajik

25 Other
26 Uninhabited or sparsely populated

0   200 km
0   200 mi

Kazakhstan is the region's giant and borders two even greater territorial titans: Russia and China (Fig. 7B-12). During the Soviet period, northernmost Kazakhstan was heavily Russified and become, in effect, part of Russia's Southeastern Frontier (see Fig. 2B-1). Rail and road links crossed the area mapped as the Kazakh–Russian Transition Zone, connecting the north to Russia. The Soviets made Almaty, in the heart of the Kazakh domain, the territory's capital. Today the Kazakhs are in control, and they in turn have moved the capital to Astana, right in the heart of the Transition Zone, where over 4 million Russians (who constitute about 25 percent of the total national population of 17 million) still live. Clearly, Astana is another **forward capital [8]**.

Figure 7B-12 further reveals Kazakhstan's situation as a corridor between the huge oil reserves of the Caspian Basin and China. Oil and gas pipelines nearing completion across Kazakhstan will eliminate, or greatly reduce, China's dependence on oil carried by tankers along distant international sea lanes.

Uzbekistan occupies the heart of Turkestan and borders every other state in it. Uzbeks not only make up 80 percent of the population of 29 million, but also form substantial minorities in several neighboring states. The capital, Tashkent, lies in the eastern core area of the country, where most of the people live in towns and farm villages, and the crowded Fergana Valley is the focus. In the far west lies the shrunken Aral Sea, whose feeder streams were diverted into cotton fields and croplands during the Soviet occupation; heavy use of pesticides contaminated the groundwater, and countless thousands of local people continue to suffer horrific medical problems as a result.

In the post-Soviet period, Islamic (Sunni) revivalism has become a problem for Uzbekistan. Wahhabism, the particularly virulent form of Sunni fundamentalism, took root in eastern Uzbekistan; government efforts to suppress it have only increased its strength in the countryside. In turn, the country's regime has taken on increasingly dictatorial habits, and public demonstrations have been put down with many casualties.

Turkmenistan, the autocratic desert republic that extends all the way from the Caspian Sea to the borders of Afghanistan, has a population of little more than 5 million, of which more than 80 percent are ethnic Turkmen. During the Soviet era, the communist planners began work on a massive project: the Garagum (Kara Kum) Canal designed to bring water from Turkestan's eastern mountains into the heart of the desert. Today the canal is 1100 kilometers (700 mi) long, and it has enabled the cultivation of some 1.2 million hectares (3 million acres) of cotton, vegetables, and fruits. The plan is to extend the canal all the way to the Caspian Sea, but meanwhile Turkmenistan has hopes of greatly expanding its oil and gas output from Caspian coastal and offshore reserves. But take another look at Figure 7B-12: this country's relative location is clearly not advantageous for export routes.

Kyrgyzstan's topography and political geography are reminiscent of the Caucasus. Mapped in yellow in Figure 7B-12, Kyrgyzstan lies intertwined with Uzbekistan and Tajikistan to the point of having exclaves and enclaves along their common borders. The Kyrgyz, for whom the Soviets established this "republic," constitute about 70 percent of the population of 5.5 million. Uzbeks and other minorities form a complex cultural geography; mountains and valleys isolate communities and complicate nation-building. The agricultural economy is generally weak, consisting of pastoralism in the mountains and farming in the valleys.

About seven-eighths of the people profess allegiance to Islam, and Wahhabism has gained a strong foothold here as well. The town of Osh is often referred

If you were asked where this photo might have been taken, Russia would seem to be a good guess, or perhaps Belarus (although the English word "bar" might not be expected there). Certainly the bleak apartment building in the background seems to be of Soviet-socialist vintage. But this is the entrance to a casino on the main shopping street of Almaty, the largest city and former capital of Kazakhstan. The Soviets and their Kazakh allies administered Kazakhstan from Almaty (then known as Alma-Ata) for some 70 years, and the Russian imprint remains strong in this city. But, as Figure 7B-12 shows, the Russian imprint is even stronger in the country's north, and in 1998 the president and his party, over the objections of the mainly Kazakh opposition, moved the national capital from scenic and warm Almaty to drab and cold Astana. Yet Almaty retains its attractions: more than 40 casinos as well as countless bars and clubs coexist with Sunni Islam, Kazakhstan's leading religion.

© Abbie Trayler-Smith/Panos Pictures

AFGHANISTAN

Highest relief
Moderate relief
Low relief
Main opium-poppy growing areas
Forest
Steppe, grassland

Road
The Ring Road
Railroad
International boundary
Provincial boundary
TAJIK/T Ethnic group

National capitals are underlined

0   50   100   150   200   250 Kilometers
0   30   60   90   120   150 Miles

TURKMENISTAN

UZBEKISTAN

TAJIKISTAN

CHI

Samarqand

Dushanbe

Amu Darya

JOWZJAN

Termiz

Mazar-e Sharif

Kholm

Sheberghan

BALKH

Kondoz

KUNDUZ

Talogan

TAKHAR

Khanabad

BADAKHSHAN

Feyzabad

Wakhjir Pass

Baroghil Pass

T A J I K S

Vakhan Corridor

Meymaneh

UZBEKS

Sar-e Pol

Aybak

SAMANGAN

T

FARYAB

Pol-e Khomri

BADGHIS

Qal'eh-ye Now

SAR-E POL

BAGHLAN

H i n d u K u s h

KASHMIR

Sabzak Pass

T

Paropamisus Range

Hari Rud River

Herat

T

HAZARAS

Nil Pass

Bamian

BAMIAN

Shebar Pass

PANJSHIR

Panjshir Valley

NURESTAN

P

Mingora

HERAT

T

Shotor Khun Pass

Chaghcharan

Ghow Gardan Pass

PARVAN

Charikar

KAPISA

Mahmud-e Raqi

Kabul

(Kabul)

LAGHMAN

Mehtar Lam

KONAR

Asadabad

Farah River

GHOWR

DAYKUNDI

VARDAK

Meydan Shahr

KABUL

SarowbL

Jalalabad

NANGARHAR

Tora Bora

Khyber Pass

Mardan

Abbottabad

Islamabad

Srir

Pol-e Alam

LOWGAR

Peywar Kandaw Pass

Meydan Khvolah

Peshawar

Farah

FARAH

P

GHAZNI

Ghazni

PAKTIA

Gardiz

KHOWST

Khost

TRIBAL AREAS

Bannu

Rawalpindi

ORUZGAN

Tarin Kowt

Sharan

Gujran

P

ZABOL

PAKTIKA

Lahore

Lashkar Gah

Kandahar

Qalat

Zhob

PAKISTAN

Faisalabad

P

HELMAND

KANDAHAR

Khojak Pass

Zaranj

NIMRUZ

Rigestan Desert

Helmand River

Quetta

Multan

IRAN

Zahedan

IND

AFGHANISTAN
PROVINCES AND ETHNIC AREAS
(including adjoining Pakistan)

TURKMENISTAN

UZBEK.

TAJIKISTAN

JOWZJAN

BALKH

KUNDUZ

TAKHAR

BADAKHSHAN

FARYAB

SAR-E POL

SAMANGAN

BAGHLAN

PANJSHIR

NURESTAN

BADGHIS

BAMIAN

PARVAN

KAPISA

LAGHMAN

KONAR

HERAT

GHOWR

VARDAK

KABUL

NANGARHAR

LOWGAR

DAYKUNDI

GHAZNI

PAKTIA

KHOWST

FARAH

ORUZGAN

ZABOL

PAKTIKA

INDIA

IND

HELMAND

KANDAHAR

PAKISTAN

IRAN

NIMRUZ

Pushtun   Tajik
Uzbek   Hazara
Turkmen   Chahar
Baluchi   Mixed

0   100   200 Kilometers
0   50   100 Miles

to as the headquarters of the movement in Turkestan. In 2010, ethnic rioting erupted in Osh between the Kyrgyz majority and the Uzbek minority, killing at least 400 people and putting the country in jeopardy of breaking apart. Tens of thousands of Uzbeks, who make up only 14 percent of the total population but form a prosperous merchant class, fled their homes. About 2000 houses were destroyed by arson.

Tajikistan's mountainous scenery is even more spectacular than Kyrgyzstan's, and here as well topography is a barrier to the integration of a multicultural society. The Tajiks, who constitute more than three-quarters of the population of 8 million, are ethnically Persian (Iranian), not Turkic, and speak a language related to Persian (an Indo-European tongue). Most Tajiks, despite their Persian affinities, are Sunni Muslims, not Shi'ites. Small though Tajikistan is, regionalism plagues the state: the government in Dushanbe is often at odds with the barely connected northern area (Fig. 7B-12), a hotbed not only of Islamic revivalism but also of anti-Tajik, Uzbek activism.

Taliban fighters in an "undisclosed location" inside Afghanistan brandish their weapons and practice their threatening glares. The Taliban still control a significant part of northeast Afghanistan and exert influence in northern Pakistan.
Reuters/© Corbis

Fractious Afghanistan

Afghanistan, the southernmost country in this region, exists because the British and Russians, competing for hegemony in this area during the nineteenth century, agreed to tolerate it as a cushion, or **buffer state [9]**, between them. This is how Afghanistan acquired the narrow extension leading from the main territory eastward to the Chinese border—the Vakhan Corridor (Fig. 7B-14). As the colonialists delimited it, Afghanistan adjoined the domains of the Turkmen, Uzbeks, and Tajiks to the north, Persia (now Iran) to the west, and the western flank of British India (now Pakistan) to the east.

Landlocked and Fractured

Geography and history seem to have conspired to divide Afghanistan. As Figure 7B-14 shows, the towering Hindu Kush range dominates the center of the country, creating three broad environmental zones: the relatively well-watered and fertile northern plains and basins; the rugged, earthquake-prone central highlands; and the desert-dominated southern plateaus. Kabol (Kabul), the capital, lies on the southeastern slopes of the Hindu Kush, linked by narrow passes to the northern plains and by the famous Khyber Pass to Pakistan.

Across this rugged variegated landscape moved countless peoples: Greeks, Turks, Arabs, Mongols, and others. Some settled here, their descendants today speaking Persian, Turkic, and other languages. Others left archeological remains or no trace at all. The present population of Afghanistan (just over 30 million) has no ethnic majority. This is a country of minorities in which the Pushtuns of the east are the most numerous but make up only 42 percent of the total. The second-largest minority are the Tajiks (27 percent), a world away across the Hindu Kush, concentrated in the

zone near Afghanistan's border with Tajikistan. The Hazaras (9 percent) of the central highlands and the south, the Uzbeks (9 percent) and Turkmen (3 percent) in the northern border areas, the Baluchis (1.8 percent) of the southern deserts, and a handful of smaller groups scattered across this country create one of the world's most complicated cultural mosaics (Figs. 7B-14 inset, 7B-13). Two major languages, Pushtun and Dari (the local variant of Persian), plus several others create a veritable Tower of Babel here.

The Cold War and the Emergence of the Taliban and al-Qaeda

Episodes of conflict have marked the history of Afghanistan, but none was as costly as its involvement in the Cold War (1945–1990). Following the Soviet intervention of 1979, the United States supported the Muslim opposition, the *mujahideen* (strugglers), with up-to-date weapons and money, and a decade later the Soviets were forced to withdraw. Soon the factions that had been united during the anti-Soviet campaign were in conflict, delaying the return of some 4 million refugees who had fled to Pakistan and Iran. The situation resembled the pre-Soviet past: a feudal country with a weak and ineffectual government in Kabol.

In 1994, what at first seemed to be just another warring faction appeared on the scene: the **Taliban [10]** (seekers of religion), from religious schools (*madrassas*) in Pakistan. Their avowed aim was to end Afghanistan's chronic factionalism and endemic corruption by instituting strict Islamic law. Popular support in the war-weary country, especially among the Pushtuns, led to a series of successes, and by 1996 the Taliban had taken Kabol.

The Taliban's imposition of Islamic (*Sharia*) law was so uncompromising and severe that Islamic as well as

FIGURE 7B-14   © H. J. de Blij, P. O. Muller, and John Wiley & Sons, Inc.

non-Islamic countries objected. Restrictions on the activities of women ended their professional education, employment, and freedom of movement, and had a devastating impact on children as well. Public amputations and stonings enforced the Taliban's code. In the process, Afghanistan became a haven for groups of revolutionaries whose goals went far beyond those of the Taliban: they plotted attacks on Western interests throughout the realm and threatened Arab regimes they deemed compliant with Western priorities. Taliban-ruled, cave-riddled, remote and isolated Afghanistan was an ideal locale for these outlaws. Already in possession of arms and ammunition (Soviet as well as American) left over from the Cold War, they also benefited from Afghanistan's huge, thriving, illicit opium trade. It is estimated that, in a typical year since 2000, Afghanistan produces over 90 percent of the world's opium. Much of the revenue found its way into the coffers of the conspirators.

With such resources, the militants were able to organize and launch attacks against several Western targets in the realm and elsewhere, but events took a fateful turn in 1996. A terrorist organization named al-Qaeda [11] took root in the country, an expanding global network that would further the aims of the revolutionaries once loosely allied. Afghanistan now became al-Qaeda's headquarters, and Usama bin Laden, a notorious Saudi renegade, its director. Al-Qaeda's exploits were funded by numerous Islamic sources and ranged from terrorist training in local camps to lethal attacks on American targets, including a warship in Yemen's port of Adan and two U.S. embassies in East Africa.

Al-Qaeda's most fateful operation came with the suicide attacks of September 11, 2001 that destroyed New York's Twin Towers, killed thousands, caused billions of dollars in physical destruction, and inflicted an untold amount of psychological damage. Several weeks later, United States and British forces, with the acquiescence of Pakistan, attacked both the Taliban regime and the al-Qaeda infrastructure in Afghanistan.

The Taliban and al-Qaeda leaderships may not have counted on Pakistan's compliance with Western demands, but they surely found out what the consequences of 9/11 were for Afghanistan and its people. Once again a foreign power (the United States and its allies) invaded the country and set its political course, the associated upheaval enabling rapacious warlords to exploit the peoples of border provinces far from the capital.

The Rise and Fall of Usama bin Laden

The child of a Saudi billionaire, Usama bin Laden graduated from a university in Jiddah in 1979 and headed for Afghanistan with an inherited fortune estimated at U.S. $300 million to help the anti-Soviet campaign. Well connected in Saudi Arabia and now a pivotal figure in Afghanistan, bin Laden saw his fame as well as his war chest grow as the Soviet intervention collapsed at the end of the 1980s. Following the withdrawal of the communist forces, he returned to Saudi Arabia, where he denounced his government for allowing U.S. troops on Saudi soil during the 1990–1991 Gulf War. The Saudi regime responded

by stripping him of his citizenship and expelling him, and bin Laden fled to Sudan. There he set up several legitimate businesses to facilitate his now-global financial transactions, but he also established terrorist training camps. Under heavy international pressure, the Khartoum regime ousted him in 1996, and bin Laden returned to a country he knew would welcome him again—Afghanistan.

It was from his base in Afghanistan that bin Laden directed the 9/11 attacks as well as other acts of terrorism, while overseeing a steady recruitment of al-Qaeda warriors and trainees from madrassas in poverty-stricken areas in Afghanistan, Pakistan, and elsewhere. Proof of bin Laden's and al-Qaeda's complicity in the September 11 assault was discovered in videotape and documentary form after the American-led invasion of Afghanistan, and successive U.S. governments declared it a priority to bring him to justice. For nearly a decade, he was thought to be hiding in the rugged, cave-riddled, Taliban-controlled highlands in the remote Afghanistan-Pakistan border zone.

He was finally found in the city of Abbottabad in the central flatlands of northern Pakistan. Just after midnight local time on Monday, May 2, 2011, a helicopter-supported unit of U.S. Navy SEALs stormed a house inside a fortified compound and shot bin Laden. They took his body and, after retaining DNA samples, swiftly buried him at sea in the nearby Indian Ocean. The role of Pakistan's military and intelligence services in the hiding of bin Laden remained unclear, but people in the United States overwhelmingly received the news of his death as justified retribution. However, the effect of his demise on al-Qaeda's future is still a matter of debate.

Clouded Future

In the immediate aftermath of al-Qaeda's 9/11 attacks, the American response provided reasons for optimism: the Taliban were quickly defeated, a more representative government was forming, warlords were co-opted or sidelined, Pushtun and other refugees were streaming back into Afghanistan, and life in Kabol and other urban centers returned to some semblance of normal. In 2004, Afghanistan held elections that produced a representative government headed by President Hamid Karzai.

But already the seeds of disaster had been sown because the U.S. invasion of Iraq in 2003 had begun to divert attention and resources from Afghanistan. In a pattern recurring throughout Afghanistan's history, opponents of modernization in any form resumed their attacks on foreign forces and local facilities (such as reopened girls' schools). From their mountain hideaways in the Afghanistan-Pakistan border zone that had long protected Usama bin Laden from capture, Taliban warriors not only challenged Western ground forces in Afghanistan but also staged a major attack in northern Pakistan, where they briefly held a large area in which they carried out their hallmark public executions, destroyed schools, kidnapped prominent citizens, and terrorized shopkeepers before Pakistani forces beat them back.

By 2010, Afghanistan seemed to have regained its prominence among American priorities: even as the war

As U.S. armed forces withdraw from Iraq, an even greater challenge looms: the pacification and stabilization of Afghanistan. This challenge is most urgent (and difficult) in Afghanistan's eastern and southern provinces, where high relief, distance, and poor surface communications favor the Taliban. This image shows U.S. soldiers on patrol near a bridge at Nishagam in Konar province (see Fig. 7B-14), not far from the border with the Federally Administered Tribal Areas of Pakistan, a known Taliban stronghold.

© Liu Jin/AFP/Getty Images, Inc.

North Africa/Southwest Asia realm. The most important commonality among the realm's states is the predominance of Islam and its varying tendencies toward politicization and militancy. The Arab Spring of 2011 exposed yet another dimension: widespread simmering discontent with repressive regimes that failed to bring economic progress, a grievance expressed in powerful demands for democratic reform. The political upheaval affected many countries in the realm, although not in the same manner. It also caused the United States to rethink its policies toward a number of countries in the realm: it could no longer engage in geopolitical treaties with countries like Yemen, Bahrain, or Egypt while closing its eyes to the undemocratic imperfections of those regimes—not when pro-democracy protesters were being mowed down on the streets. In mid-2011, the long-term implications of the Arab Spring were impossible to predict even though events in several countries appeared to have reached a point of no return. As never before, this realm is now at a historic crossroads.

in Iraq was winding down, the U.S. government approved a significant increase in combat troops and support personnel to augment its inadequate forces. But the moment of opportunity may have come and gone. Afghanistan's physical geography, social complexity, experience with foreigners and resultant warrior culture, Islamic intensity and traditional male dominance, endemic corruption, and highly limited subsistence opportunities combine to create an arena in which no military power has ever prevailed over the long term. In 2011, the United States and the Afghan regime reached agreement that by 2014 the entire country would come under the control of government forces, implying that U.S. troops would be withdrawn at that time.

This chapter opened by emphasizing the diversity and variety among the cultures, countries, and regions of the

POINTS TO PONDER

- The popular uprisings across the Arab World in 2011 involved disproportionate numbers of young people and were in no small part facilitated through the use of the Internet and cell-phone communication.

- The Arab Spring has elevated the stature of Turkey as the country showing the way to the future for the Islamic world.

- Women played a major role in the revolution in Tunisia but far less so in Libya or Yemen.

- Should the Palestinian Authority have the right to exploit the natural gasfield off the shore of the Gaza Strip?

using
GEOGRAPHY

1. Which of the realm's countries that witnessed democratic uprisings in 2011 have shown the most progress to date, and which have shown the least? Comparing these countries, what do you think are the main reasons for the different outcomes?

2. Find out where the United States presently maintains military bases in this realm. Rate those countries in terms of their democratic functioning. Should this be a factor in locating U.S. military bases?

IN THIS CHAPTER

- South Asia as a birthplace of religions
- Backward agriculture and cutting-edge IT
- Two nuclear powers quarrel over Kashmir
- The Indian Ocean basin as a new geopolitical arena
- South Asia's missing girls

CONCEPTS, IDEAS, AND TERMS

FIGURE 8A-1 © H. J. de Blij, P. O. Muller, and John Wiley & Sons, Inc.

South Asia is a realm of almost magical geographic names: Mount Everest, Kashmir, the Khyber Pass, the Ganges River. There was a time when this realm was legendary and prized. Remember that it was "India" and its fabled wealth that the European explorers were after, from Vasco da Gama to Columbus to Magellan. Before them, the fourteenth-century North African geographer Ibn Battuta had traveled overland to South Asia, and his writings about its riches were met with astonishment and even disbelief. From the sixteenth century onward, European trading companies derived enormous profits from commerce in this realm.

However, by the late nineteenth century South Asia seemed to have become remote from the affairs of the world—hungry, weak, exploited, the epitome of the global periphery. Even after independence in 1947, India as well as the other countries of the realm long remained among the world's poorest. For decades, population growth outstripped economic expansion.

Today, for a number of reasons, South Asia commands the world's attention once again. It became the most populous geographic realm on Earth in 2011 (see the Data Table in Appendix B). Two of its states, India and Pakistan, often find themselves in conflict and both are nuclear powers. In the remote mountain hideaways of Pakistan, a terrorist organization's leaders planned attacks that changed the skyline of New York and altered the battleground of Iraq. In the ports of India, a growing navy reflects the emergence of the Indian Ocean Basin as a new global geopolitical arena in which China, too, is asserting itself. Meanwhile, outsourcing by U.S. companies to India has become a hot topic, and India's spectacular rise in information technology has changed that industry. Our daily lives will increasingly be affected by what happens in this crowded and restive part of the world.

THE GEOGRAPHIC PANORAMA

The Eurasian landmass incorporates all or part of six of the world's geographic realms, and of these six none is more clearly defined by nature than the one we call South Asia. Figure 8A-1 shows us why: the huge triangular subcontinent that divides the northern Indian Ocean between the Arabian Sea and the Bay of Bengal is so sharply demarcated by mountain walls and desert wastes that you could take a pen and mark its boundary, from the Naga Hills in the east through the Great Himalaya and Karakoram in the north to the Hindu Kush and the Iran-bordering wastelands of Baluchistan in the west. Note how short the distances are over which the green of habitable lowlands turns to the dark brown of massive, snowcapped mountain ranges.

South Asia's kaleidoscope of cultures may be the most diverse in the world, proving that neither formidable mountains nor forbidding deserts could prevent foreign influences from further diversifying an already variegated realm. We will encounter many of these influences in this chapter and the next, but South Asia also possessed one unifying force of sorts: the British Empire, which in its late-nineteenth-century heyday came to hold sway over all of it in a **raj** (period of rule) that endured through the mid-twentieth. When, in the immediate aftermath of World War II, the British wanted to transfer their authority to a single regional government, local objections swiftly nullified this notion. That regional government would have been Hindu-dominated, but Muslims concentrated in the realm's eastern and western flanks refused, as did a pair of small kingdoms in the mountainous north as well as the Buddhist-dominated southern island then called Ceylon (now Sri Lanka). Negotiations and compromises produced partition and the political boundaries seen in Figure 8A-1. As a result, India, the

One of India's many famous Hindu temples: 15th-century Chamunda Devi Mandir, in the desert city of Jodhpur.

© Jan Nijman

major geographic qualities
SOUTH ASIA

1. South Asia is clearly defined physiographically, and much of the realm's boundary is marked by mountains, deserts, and the Indian Ocean.

2. South Asia's great rivers, especially the Ganges, have for tens of thousands of years supported huge population clusters.

3. South Asia, and especially northern India, was the birthplace of major religions that include Hinduism and Buddhism.

4. Due to the realm's natural boundaries, foreign influences in premodern South Asia came mainly via a narrow passage in the northwest (Khyber Pass).

5. South Asia covers just over 3 percent of the Earth's land area but contains nearly 26 percent of the world's human population.

6. South Asia's annual monsoon continues to dominate life for hundreds of millions of subsistence and commercial farmers. Failure of the monsoon cycle spells economic crisis.

7. Certain remote areas in the realm's northern mountain perimeter are a dangerous source of friction between India and both Pakistan and China.

8. South Asia is still predominantly rural with hundreds of thousands of small villages; but it is also home to some of the biggest cities in the world.

---

realm's giant, is flanked by six countries (in clockwise order, Pakistan, Nepal, Bhutan, Bangladesh, Sri Lanka, and the Maldives) as well as a remaining disputed territory, Kashmir.

Since Islam is Pakistan's official religion (India has none) and that faith is a key criterion in defining the realm we designated as North Africa/Southwest Asia, should Pakistan be included in the latter? The answer lies in several aspects of Pakistan's historical geography. One criterion is ethnic continuity, which links Pakistan to India rather than to Afghanistan or Iran. Another factor involves language: although Urdu is Pakistan's official language, English is the *lingua franca*, as it is in India. Still another factor, of course, is Pakistan's evolution as part of the British South Asian Empire. Furthermore, the boundary between Pakistan and India does not signify the eastern frontier of Islam in Asia. As we shall see, more than 180 million of India's 1.227 billion citizens are Muslims (which is just about as many as there are in all of Pakistan), and millions live very close to the Indian side of the border whose creation cost so many lives in 1947. And not only are Pakistan and India linked in the cultural-historical arena: they are locked in a deadly and dangerous embrace in embattled Kashmir.

The close integration of Pakistan with South Asia will not surprise you after you study the realm's physiography in Figure 8A-1: the natural boundary in this part of the realm lies west of the Indus River, not east of it. Pakistan today remains part of a realm that changes not in the Punjab, but at the Khyber Pass, the highland gateway to Afghanistan.

SOUTH ASIA'S PHYSIOGRAPHY

From snowcapped peaks to tropical forests and from bone-dry deserts to lush farmlands, this part of the world presents a virtually endless array of environments and ecologies, a diversity that is matched by its cultural mosaic. The broad outlines of this realm's physiography are best understood against the backdrop of its fascinating geologic past.

A Tectonic Encounter

As Figure 8A-2 shows, the spectacular relief in the north of this realm resulted initially from the collision of two of the Earth's great tectonic plates (see Fig. G-4). About 10 million years ago, after a long geologic journey following the breakup of the supercontinent Pangaea (Fig. 6A-3), the Indian Plate encountered Eurasia. In this slow-motion, accordion-like collision, parts of the crust were pushed upward, thereby creating the mighty Himalaya mountain range. This process is still going on—at the rate of 5 millimeters (0.2 in) per year. One major outcome was that the northern margins of the South Asian realm were thrust upward to elevations where permanent snow and ice make the landscape look polar. The march of the seasons melts enough of this snow in spring and summer to sustain the great rivers below, providing water for farmlands that support hundreds of millions of people. The Ganges, Indus, and Brahmaputra all have their origins in the Himalaya. Only south of the Ganges Basin does the massive plateau begin that marks the much older geologic core of the Indian Plate as it drifted northeast toward Eurasia.

The Monsoon

Physical geography, therefore, is crucial here in South Asia—but not just on and below the ground. What happens in the atmosphere is critical as well. The name "South Asia" is almost synonymous with the term **monsoon [1]** because the annual rains that come with its onset, usually in June,

TECTONIC EVOLUTION OF SOUTH ASIA

EURASIAN PLATE

Himalayas

INDIA Today

10 million years ago

SRI LANKA

38 million years ago

Equator

55 million years ago

INDIAN OCEAN

71 million years ago

"INDIA" landmass

```
0    400    800 Kilometers
0   200   400 Miles
```

SRI LANKA

Modified after USGS drawing.

FIGURE 8A-2   © H. J. de Blij, P. O. Muller, J. Nijman, and John Wiley & Sons, Inc.

rises and condensing large amounts of rainfall. The other branch of the wet monsoon originates in the Bay of Bengal and gets caught up in the convection (rising hot air) over northeastern India and Bangladesh ②. Seemingly endless rain now inundates a much larger area, including the entire North Indian Plain. The Himalaya mountain wall blocks the onshore airstream from spreading into the Asian interior and the rain from dissipating ③. Thus the moist airflow is steered westward, drying out as it advances toward Pakistan ④. After persisting for weeks, this pattern finally breaks down and the wet monsoon gives way to periodic rains and, eventually, another dry season. Then the anxious wait begins for the next year's summer monsoon, for without it India would face disaster. In much of rural India, life can hang by a meteorological thread.

Physiographic Regions

Figure 8A-3 underscores South Asia's overall division into three physiographic zones: northern mountains, southern plateaus, and, in between, a wide crescent of river lowlands.

The ***northern mountains*** extend from the Hindu Kush and Karakoram ranges in the northwest through the Himalaya in the center (Everest, the world's tallest peak, lies on the crestline that forms the Nepal-China border) to the ranges of Bhutan and the Indian State of Arunachal Pradesh in the east. Dry and barren in the west on the Afghanistan border, the ranges become green and tree-studded in Kashmir, forested in the lower-lying sections of Nepal, and even more densely vegetated in Arunachal

The arrival of the annual rains of the wet monsoon transforms the Indian countryside. By the end of May, the paddies lie parched and brown, dust chokes the air, and it seems that nothing will revive the land. Then the rains begin, and blankets of dust turn into layers of mud. Soon the first patches of green appear on the soil, and by the time the monsoon ends all is green. The photograph on the left, taken just before the onset of the wet monsoon in the State of Goa, shows the paddies before the rains begin; three months later this same countryside looks as on the right.

© Steve McCurry/Magnum Photos, Inc.

are indispensable to subsistence as well as commercial agriculture in the realm's key country, India.

Figure 8A-3 shows how the monsoon works. As the South Asian landmass heats up during the spring, a huge low-pressure system forms above it. This low-pressure system begins to draw in vast volumes of air from over the ocean onto the subcontinent. When the inflow of moist oceanic air reaches critical mass in early June, the ***wet monsoon*** has arrived. It may rain for 60 days or more. The countryside turns green, the paddies fill, and another dry season's dust and dirt are washed away. The region is reborn (see photo pair at right). The moisture-laden air flowing onshore from the Arabian Sea is forced upward against the Western Ghats ①, cooling as it

FIGURE 8A-3

© H. J. de Blij, P. O. Muller, and John Wiley & Sons, Inc.

Pradesh. Transitional foothills, with many deeply eroded valleys cut by rushing meltwater, lead to the river basins below.

The belt of *river lowlands* extends eastward from Pakistan's lower Indus Valley (the area known as Sindh) through India's wide Gangetic Plain and then on across the great double delta of the Ganges and Brahmaputra in Bangladesh (Fig. 8A-3). In the east, this physiographic region is often called the North Indian Plain. To the west lies the lowland of the Indus River, which rises in Tibet, crosses

Kashmir, and then bends southward to receive its major tributaries from the Punjab ("Land of Five Rivers").

With so much of the realm, and so many of its people, depending on the water carried down from the High Himalaya by these great rivers, the melting of glaciers due to global warming is a serious concern. The Brahmaputra, for example, relies on meltwater for 21 percent of its volume, the remainder coming from precipitation. The impact of global warming is actually hard to determine even if we know the ice is melting at an increasingly rapid rate. In the short term, increased melting will actually add to the river's volume, but as the glaciers shrink there will be a tipping point at which meltwater will suddenly and rapidly decrease. Some models indicate this is likely to happen in about 40 years, with potentially devastating consequences for populations at lower elevations that have depended on this water for millennia.

Peninsular India is mainly plateau country, dominated by the massive *Deccan*, a tableland built of basalt that poured out when India separated from Africa during the breakup of Pangaea. The Deccan (meaning "South") tilts to the east, so that its highest areas are in the west and most of the rivers flow into the Bay of Bengal. North of the Deccan lie two other plateaus, the Central Indian Plateau to the west and the Chota Nagpur Plateau to the east (Fig. 8A-3). On the map, also note the Eastern and Western Ghats: "ghat" means step, and it connotes the descent from Deccan-Plateau elevations to the narrow coastal plains below. The onshore winds of the annual wet monsoon bring ample precipitation to the Western Ghats; as a result, here lies one of India's most productive farming areas and one of southern India's largest population concentrations.

BIRTHPLACE OF CIVILIZATIONS

South Asia has a special place in the historical geography of humankind. First, recent research indicates that the Ganges Basin was a crucial stop on the great out-of-Africa migration that carried our species from the shores of the Red Sea to the coast of Australia. Here, perhaps as far back as 70,000 years ago, environmental conditions enabled groups of hunter-gatherers to stabilize and expand. From the Indian subcontinent, they went on to Southeast Asia and Australia, and, scholars conclude, emigrated westward to the Middle East and Europe as well. The Ganges Basin may have been a crucible of humanity, but that was only the beginning. Subsequently, the great river basins of South Asia gave rise to two separate world-class culture hearths (see Fig. 7A-3), and continue to nurture what has now become the largest population concentration on Earth (Fig. G-8).

Indus Valley Civilization

Whatever the distant past, we know that a complex and technologically advanced civilization had emerged in the Indus Valley by about 2500 BC, simultaneous with other Bronze Age "urban revolutions" in Egypt and Mesopotamia. The Indus Valley civilization was centered on two major cities, Harappa and Mohenjo-Daro, which may have been capitals during different periods of its history (Fig. 8A-4); in addition, there were more than 100 smaller urban settlements. The locals apparently called their state *Sindhu*, and both *Indus* (for the river) and *India* (for the later state) may derive from this name. Although the influence of this civilization extended as far east as present-day Delhi, it did not last because of environmental change

and, perhaps, because the political center of gravity shifted southeastward into the Ganges Basin.

Aryans and the Origins of Hinduism

Around 1500 BC, northern India was invaded by the *Aryans* (peoples speaking Indo-European languages based in what is today Iran). As the Iron Age dawned in India, acculturated Aryans began the process of welding the Ganges Basin's isolated tribes and villages into a new organized system, and urbanization made a comeback. The Aryans brought their language (Sanskrit, related to Old Persian) and a new social order to the vast riverine flatlands of northern India. Their settlement here was also accompanied by the emergence of a religious belief system, *Vedism*. Out of the texts of Vedism and local creeds there arose a new religion—*Hinduism*—and with it a new way of life.

It is thought that the arrival and accommodation of Aryans in this new society forged a system of social stratification [2] that would solidify the powerful position of the Aryans and be legitimized through religion. Starting about 3500 years ago, a combination of regional integration, the organization of villages into controlled networks, and the emergence of numerous small city-states produced a hierarchy of power among the people, a ranking from the very powerful (Brahmins—highest-order priests) to the weakest. Hinduism's class-based *caste system* is highly controversial in the West (and among groups of Indians too) because of its rigidity and the ways in which it justifies structural inequality. Those in the lowest castes, deemed to be there because they deserved it given their past lives, are worst off, without hope of advancement and at the mercy of those higher up on the social ladder. In recent years, the caste system appears to be eroding due to the combined effects of globalization, economic growth, and urbanization; this is true for India's bigger cities, but much less so for the majority of people in the country who still reside in rural areas. (The caste system is examined in more detail in Chapter 8B.)

What is really most striking about Hindu civilization is that, after more than three millennia, it is still around and functioning with more or less the same beliefs, the same pantheon of gods, the same caste system, the same temples, and the same religious practices. You can visit Varanasi or any other holy place on the Ganges River today and witness *poojas* (prayers, offerings) and rituals that have been performed in similar fashion for some 3000 years. Nowhere in the world has a major religious culture been sustained for so long. It helps explain why the South Asian realm has such a powerful spiritual appeal to many Westerners, even if they are dismissive of the caste system.

Although Hinduism spread across South Asia and even reached the Southeast Asian realm (especially Cambodia and Indonesia), Indo-European languages never took hold in the southern part of the subcontinent. As

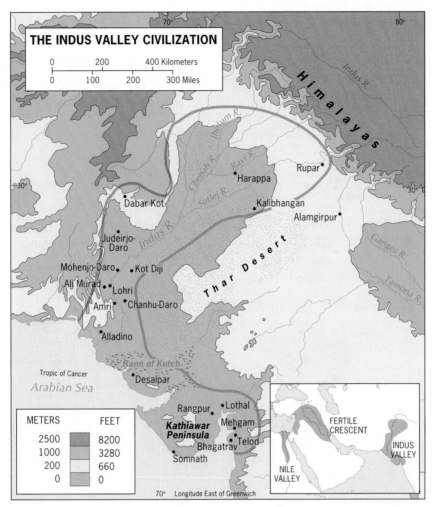

FIGURE 8A-4   © H. J. de Blij, P. O. Muller, and John Wiley & Sons, Inc.

Figure 8A-5 shows, **Indo-European languages [3]** (several of which are rooted in Sanskrit) dominate the western and northern parts of the realm, whereas the southern languages belong to the **Dravidian [4]** family—languages that were indigenous to the realm even before the arrival of the Aryans. But these are not fossil languages: they remain vibrant today and have long literary histories. Telugu, Tamil, Kanarese (Kannada), and Malayalam are spoken by some 265 million people. In India's northern and northeastern fringes, Sino-Tibetan languages predominate, and smaller pockets of Austro-Asiatic speakers can be found in eastern India and neighboring Bangladesh.

Buddhism and Other Indigenous Religions

Hinduism is not the only religion that emerged in this realm. Around 500 BC, **Buddhism** arose in the eastern Ganges Basin in what is today the Indian State of Bihar. The famous story of the "enlightenment" of the Prince Siddhartha (the Buddha) took place in the town of Bodh Gaya, and his following soon expanded in all directions. The appeal of Buddhism was (and is) especially strong among lower-caste Hindus, and substantial numbers have

converted through the ages. Even some prominent ancient Hindu kings were known to have turned to Buddhism, prompting their subjects to follow suit. Interestingly, Buddhism emerged inside India, but its ultimate influence was felt beyond the realm in Southeast and East Asia. Today, less than 1 percent of the population of India adheres to Buddhism (81 percent are Hindu), but it is the state religion in Bhutan and a large majority (more than 70 percent) of Sri Lankans are Buddhist as well.

Another (much smaller) indigenous religion that has evolved alongside Hinduism since ancient times is **Jainism**, which is often described as a more purist, principled, and deeply spiritual form of Hinduism. It is especially well known for its uncompromising stand on nonviolence and vegetarianism. Jains today comprise under 1 percent of the population in India. Finally, we should take note of **Sikhism** as another of the realm's indigenous religions, a blend of sorts of Islamic and Hindu beliefs. This religion, practiced by about 2 percent of the population, is of course much younger; it emerged around AD 1500, a few centuries after Islam became a dominating force in much of the South Asian realm.

FIGURE 8A-5

© H. J. de Blij, P. O. Muller, J. Nijman, and John Wiley & Sons, Inc.

FOREIGN INVADERS

The Reach of Islam

In the late tenth century, Islam came rolling like a giant tide across South Asia, spreading across Persia and Afghanistan, through the high mountain passes, into the Indus Valley, across the Punjab, and into the Ganges Basin, converting virtually everybody in the Indus Valley and foreshadowing the emergence, many centuries later, of the Islamic Republic of Pakistan. By the early thirteenth century, the Muslims had established the long-surviving and powerful *Delhi Sultanate*, which expanded across much of the northern tier of the peninsula. The Muslims also came by sea, arriving at the Ganges-Brahmaputra Delta and spreading their faith from the east as well as the west, in the process laying the foundation of today's predominantly Islamic state of Bangladesh.

Why did millions of Hindus convert to Islam? It was a combination of compulsion and attraction. The princes of India's local states faced the choice of cooperation or annihilation, and thus the elites of the native states

found it prudent to convert. And, as was the case with Buddhism, Islam was a welcome alternative for Hindus of lower caste. Thus Islam was the faith of the rulers and of the disadvantaged, a powerful force in the heartland of Hinduism.

The Muslims, not unlike other conquerors, also fought among themselves. In the early 1500s, a descendant of Genghis Khan named Babur placed his forces in control of Kabol (Kabul) in Afghanistan, and from that base he penetrated the Punjab and challenged the Delhi Sultanate. In the 1520s, his Islamicized Mongol armies ousted the Delhi rulers and established the *Mughal (Mogul) Empire*.

By most accounts, Mughal rule was at times remarkably enlightened, especially under the leadership of Babur's grandson Akbar, who expanded the empire by force but adopted tolerant policies toward Hindus under his sway; Akbar's grandson, Shah Jahan, made his mark on India's cultural landscape through such magnificent architectural creations as the Taj Mahal in the city of Agra (bottom photo, p. 428).

Nonetheless, by the early eighteenth century the Mughal Empire was in decline. Maratha, a Hindu state in the west, expanded not only into the peninsular south but also northward toward Delhi, capturing the allegiance of local rulers and weakening Islam's hold. Fractured India now lay open to still another foreign intrusion, this time from Europe.

Reflecting on more than seven centuries of Islamic rule in South Asia, it is remarkable that Islam never achieved proportional dominance over the realm as a whole. Whereas Pakistan is more than 96 percent Muslim and Bangladesh 90 percent, India—where the Delhi Sultanate and the Mughal Empire were centered—remains less than 15 percent Muslim today. Islam may have arrived like a giant tide, but Hinduism stayed afloat and outlasted the invasion. The accompanying caste system has many critics and understandably so, but it is hard to deny the ingenuity of the Hindu religion as it withstood first the Islamic incursions and then the European onslaught that culminated in the incorporation of the entire realm into the British Empire.

The European Intrusion

By the middle of the eighteenth century, the British had taken over much of the trade in South Asia. Their power was imposed through the East India Company (EIC), which represented the empire but whose main purpose was economic control. The British took advantage of the weakened and fragmented power of the Mughals and followed a strategy commonly known as "indirect rule." They left local rulers in place as long as they extracted the desired trading arrangements. In fact, thanks to arrangements with the British, many local maharajas became wealthier than ever before: from northern to southern India, you can find beautiful palaces (now often converted to either museums or hotels) that were built by these rulers, often as recently as the late nineteenth or even early twentieth century.

The EIC not only controlled trade with Europe in spices, cotton, and silk goods, but also India's longstanding commerce with Southeast Asia, which until then was in the hands of Indian, Arab, and Chinese merchants. This system worked well (for the British and their Indian trade partners) for almost a century, but by then political developments and heightened tensions were making it inevitable that the British government itself would take over from the EIC and assume responsibility. Thus "East India" became part of the British colonial empire in 1857—a *raj* that would endure for the next 90 years—and Queen Victoria officially became its empress 20 years later.

From the Field Notes . . .

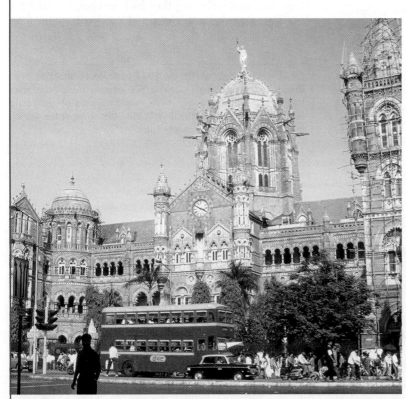

© H. J. de Blij

"More than a half-century after the end of British rule, the centers of India's great cities continue to be dominated by the Victorian-Gothic buildings the colonizers constructed here. Here is evidence of a previous era of globalization, when European imprints transformed urban landscapes. Walking the streets of some parts of Mumbai (the British called it Bombay) you can turn a corner and be forgiven for mistaking the scene for London, double-decker buses and all. One of the British planners' major achievements was the construction of a nationwide railroad system, and railway stations were given great prominence in the urban architecture. I had walked up Naoroji Road, having learned to dodge the wild traffic around the circles in the Fort area, and watched the throngs passing through Victoria (now Chhatrapati Shivaji) Station. Inside, the facility is badly worn, but the trains continue to run, bulging with passengers hanging out of doors and windows."

www.conceptcaching.com

Concept Caching

Colonial Transformation

British colonialism in South Asia coincided with the Industrial Revolution in Europe, and the impact of Britain on the realm must be understood in that context. South Asia became, in large part, a supplier of raw materials needed to keep the factories going in Manchester, Birmingham, and other industrial centers in Britain. For instance, when the supply of cotton from the American South came to a halt during the U.S. Civil War in the early 1860s, the British quickly encouraged (and enforced) cotton production in what is today western India.

When the British took power in South Asia, this was a realm with already considerable industrial development (notably in metal goods and textiles) and an active trade with both Southeast and Southwest Asia. The colonialists saw this as competition, and soon India was exporting raw materials and importing manufactured products—from Europe, of course. Local industries declined, and Indian merchants lost their markets.

Colonialism did produce assets for India. The country was bequeathed one of the most extensive transport networks of the colonial era, particularly the railroad system—even though the network focused on interior-to-seaport linkages rather than fully interconnecting the various parts of the country. British engineers laid out irrigation canals through which millions of hectares of land were brought into cultivation. Coastal settlements that had been founded by Britain developed into major cities and bustling ports, led by Bombay (now renamed Mumbai), Calcutta (now Kolkata), and Madras (now Chennai). These three cities still rank among India's largest urban centers, and their cityscapes bear the unmistakable imprint of colonialism (photo, previous page).

British rule also produced a new elite among the South Asian natives. They had access to education and schools that combined English and Indian traditions, and their Westernization was reinforced through university education in Britain. This elite drew from Hindu and Muslim communities, and it was to play a major part in the rising demands for self-rule and independence. These demands started to gather momentum in the early twentieth century and could no longer be denied when World War II came to an end in the mid-1940s.

THE GEOPOLITICS OF MODERN SOUTH ASIA

Partition and Independence

Even before the British government decided to yield to demands for independence, it was clear that British India would not survive the coming of self-rule as a single political entity. As early as the 1930s, Muslim activists were promoting the idea of a separate state. As the colony moved toward independence, a major political crisis developed that eventually resulted in the separation of India and Pakistan. But **partition [5]** was no simple matter. True, Muslims were in the majority in the western and eastern sectors of British India, but smaller Islamic clusters were scattered throughout the realm. Furthermore, the new boundaries between Hindu and Muslim communities had to be drawn right through areas where both sides coexisted—thereby displacing millions.

The consequences of this migration for the social geography of India were especially far reaching. Comparing the country's 1931 and 1951 distributions of Muslims in Figure 8A-6, you can see the impact on the Indian Punjab and in what is today the State of Rajasthan. (Since Kashmir was mapped as three entities before the partition and as one afterward, the change there represents an administrative, not a major numerical, alteration.) Even in the east a Muslim exodus occurred, as reflected on the map by the State of West Bengal, adjacent to Bangladesh, where the Islamic component in the population declined substantially.

The world has seen many **refugee [6]** migrations but none involving so many people in so short a time as the one resulting from British India's partition (which occurred on Independence Day, August 15, 1947). Scholars who study the refugee phenomenon differentiate between "forced" and "voluntary" migrations, but as this case underscores, it is not always possible to separate the two. Many Muslims, believing they had no choice, feared for their future in the new India and joined the stampede.

Flight was one response to the 1947 partition of what had been British India, resulting in one of the greatest mass population transfers in human history. Here, two trainloads of eastbound Hindu refugees fleeing (then) West Pakistan arrive at the station in Amritsar, the first city inside India.
© Corbis-Bettmann

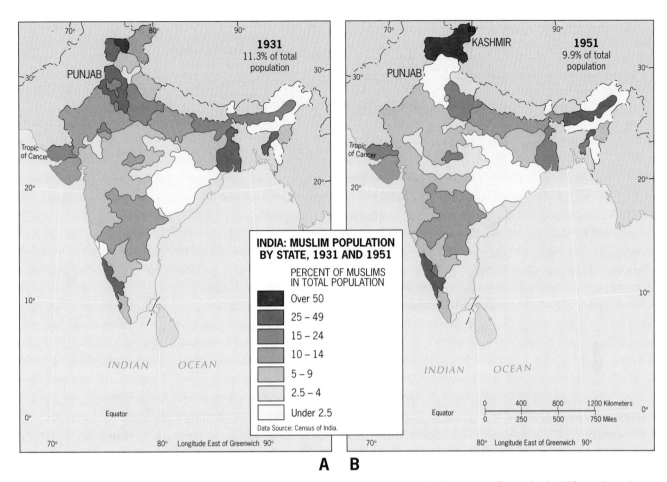

FIGURE 8A-6

© H. J. de Blij, P. O. Muller, and John Wiley & Sons, Inc.

Others had the means and the ability to make a decision to stay or leave, but even these better-off migrants undoubtedly sensed a threat.

The great majority of Hindus who lived on the "wrong" side of the border moved as well. The Hindu component of present-day Pakistan may have neared 16 percent in 1947 but is only about 2 percent today; in Bangladesh, which was named East Pakistan at the time of partition, it declined from 30 percent to around 10 percent today. Partition therefore created a new cultural and geopolitical landscape in South Asia.

India–Pakistan

From the moment of their separate creation, India and Pakistan have had a tenuous relationship. Upon independence, present-day Pakistan was united with present-day Bangladesh, and the two countries were respectively called West Pakistan and East Pakistan. As we have noted, the basis for this scheme was Islam: in both Pakistan and Bangladesh, Islam is the state religion. Between the two Islamic wings of Pakistan lay Hindu India. But there was little else to unify the Muslim easterners and westerners, and their union lasted less than 25 years. In 1971, a costly war of secession, in which India supported East Pakistan, led to the collapse of this unusual arrangement. East

Pakistan, upon its "second independence" in 1971, took the name Bangladesh; and since there was no longer any need for a "West" Pakistan, that qualifier was dropped and the name Pakistan remained on the map.

India's encouragement of independence for Bangladesh emphasized the continuing tension between Pakistan and India, which had already led to war in 1965, to further conflict during the 1970s over Jammu and Kashmir, and to periodic flare-ups over other issues. During the Cold War, India tilted toward Moscow, while Pakistan found favor in Washington because of its strategic location adjacent to Afghanistan. Armed conflict between the two South Asian countries seemed to be a regional matter—until the early 1990s, when their arms race took on ominous nuclear proportions. Since then, the specter of nuclear war has hung over the conflicts that continue to embroil Pakistan and India, a concern not just for the South Asian realm but for the world as a whole. No longer merely a decolonized, divided, and disadvantaged country trying to survive, Pakistan has taken a crucial place in the political geography of a geographic realm in turbulent transition.

The relationship between India and Pakistan is especially sensitive because there are still so many Muslims in India. Massive as the 1947 refugee movement was, it left far more Muslims in India than those who had departed.

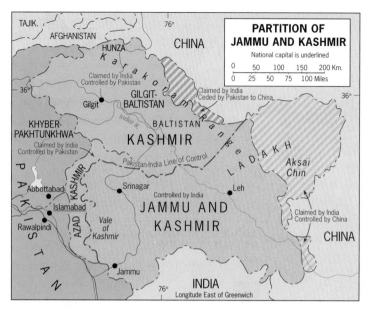

FIGURE 8A-7   © H. J. de Blij, P. O. Muller, and John Wiley
& Sons, Inc.

The number of Muslims in India declined sharply, but it remained a huge minority, one that was growing rapidly to boot. By 2012, it surpassed 180 million, more than 14 percent of the total population—the largest cultural minority in the world and almost as large as Pakistan's entire population (193 million).

What this means is that a sizeable portion of India's population tends to have "natural" sympathies vis-à-vis Pakistan. Their presence works at times as a brake on hawkish Indian policies toward Islamabad. On the other hand, conflict with Pakistan can have detrimental effects on Hindu-Muslim relations inside India, and over the years this has led to communal violence and deadly clashes. This issue is further complicated today by the alleged role of Indian Muslims in terrorist activities in India, orchestrated from Pakistan.

Contested Kashmir

When Pakistan became an independent state following the partition of British India in 1947, its capital was Karachi on the south coast, near the western end of the Indus Delta. As the map shows, however, the present capital is Islamabad. By moving the capital from the "safe" coast to the embattled interior, and by placing it on the doorstep of the contested territory of Kashmir, Pakistan announced its intent to stake a claim to its northern frontiers. And by naming the city Islamabad, Pakistan proclaimed its Muslim foundation here in the face of the Hindu challenge. This politico-geographical usage of a national capital can be assertive, and as such Islamabad exemplifies the principle of the **forward capital** [7].

Kashmir is a territory of high mountains surrounded by Pakistan, India, China, and, along more than 50 kilometers (30 mi) in the far north, Afghanistan (Fig. 8A-7). Although known simply as Kashmir, the area actually consists of several political divisions, including the Indian State properly referred to as Jammu and Kashmir, a major bone of contention between India and Pakistan.

When partition took place in 1947, the existing States of British India were asked to decide whether they wanted to be incorporated into India or Pakistan. In most of the States, the local ruler made this decision, but Kashmir was an unusual case. It had about 5 million inhabitants at the time, nearly three-quarters of them Muslims, but the maharajah of Kashmir himself was a Hindu. When he decided not to join Pakistan and instead aimed to retain autonomous status, this was answered with a Muslim uprising supported by Pakistan. The maharajah, in turn, called for help from India. After more than a year's fighting and through the intervention of the United Nations, a cease-fire line left most of Jammu and Kashmir (including nearly four-fifths of the territory's population) in Indian hands. Eventually, this line—now known as the Line of Control—began to appear on maps as the final boundary settlement, and Indian governments have proposed that it be so recognized.

With about two-thirds of the inhabitants still Muslim, Pakistan has long demanded a referendum in Jammu and Kashmir in which the people can decide for themselves to remain with India or become part of Pakistan. India has refused, arguing that there is a place for Muslims in secular India but not for Hindus in the Islamic Republic of Pakistan. This stalemate has continued for over 60 years

Regional ISSUE   Who Should Govern Kashmir?

KASHMIR SHOULD BE PART OF PAKISTAN!

"I don't know why we're even debating this. Kashmir should and would have been made part of Pakistan in 1947 if that colonial commission hadn't stopped mapping the Pakistan–India boundary before they got to the Chinese border. And the reason they stopped was clear to everybody then and there: instead of carrying on according to their own rules, separating Muslims from Hindus, they reverted to that old colonial habit of recognizing "traditional" States. And what was more traditional than some Hindu potentate and his minority clique ruling over a powerless majority of Muslims? It happened all over India, and when they saw it here in the mountains they couldn't bring themselves to do the right thing. So India gets Jammu and Kashmir and its several million Muslims, and Pakistan loses again. The whole boundary scheme was rigged in favor of the Hindus anyway, so what do you expect?

"Here's the key question the Indians won't answer. Why not have a referendum to test the will of all the people in Kashmir? India claims to be such a democratic example to the world. Doesn't that mean that the will of the majority prevails? But India has never allowed the will of the majority even to be expressed in Kashmir. We all know why. About two-thirds of the voters would favor union with Pakistan. Muslims want to live under an Islamic government. So people like me, a Muslim carpenter here in Srinagar, can vote for a Muslim collaborator in the Kashmir government, but we can't vote against the whole idea of Indian occupation.

"Life isn't easy here in Srinagar. It used to be a peaceful place with boats full of tourists floating on beautiful lakes. But now it's a violent place with shootings and bombings. Of course we Muslims get the blame, but what do you expect when the wishes of a religious majority are ignored? So don't be surprised at the support our cause gets from Pakistan across the border. The Indians call them terrorists and they accuse them of causing the 60,000 deaths this dispute has already cost, but here's a question: why does it take an Indian army of 600,000 to keep control of a territory in which they claim the people prefer Indian rule?

"Now this so-called War on Terror has made things even worse for us. Pakistan has been forced into the American camp, and of course you can't be against 'terrorism' in Afghanistan while supporting it in Kashmir. So our compatriots on the Pakistani side of the Line of Control have to stay quiet and bide their time. But don't underestimate the power of Islam. The people of Pakistan will free themselves of collaborators and infidels, and then they will be back to defend our cause in Kashmir."

KASHMIR BELONGS TO INDIA!

"Let's get something straight. This stuff about that British boundary commission giving up and yielding to a maharajah is nonsense. Kashmir (all of it, the Pakistani as well as the Indian side) had been governed by a maharajah for a century prior to partition. What the maharajah in 1947 wanted was to be ruled by neither India nor Pakistan. He wanted independence, and he might have gotten it if Pakistanis hadn't invaded and forced him to join India in return for military help. As a matter of fact, our Prime Minister Nehru prevailed on the United Nations to call on Pakistan to withdraw its forces, which of course it never did. As to a referendum, let me remind you that a Kashmir-wide referendum was (and still is) contingent on Pakistan's withdrawal from the area of Kashmir it grabbed. And as for Muslim 'collaborators', in the 1950s the preeminent leader on the Indian side of Kashmir was Sheikh Muhammad Abdullah (get it?), a Muslim who disliked Pakistan's Muslim extremism even more than he disliked the maharajah's rule. What he wanted, and many on the Indian side still do, is autonomy for Kashmir, not incorporation.

"In any case, Muslim states do not do well by their minorities, and we in India generally do. As far as I am concerned, Pakistan is disqualified from ruling Kashmir by the failure of its democracy and the extremism of its Islamic ideology. Let me remind you that Indian Kashmir is not just a population of Hindus and Muslims. There are other minorities—for example, the Ladakh Buddhists—who are very satisfied with India's administration but who are terrified at the prospect of incorporation into Islamic Pakistan. You already know what Sunnis do to Shi'ites in Pakistan. You are aware of what happened to ancient Buddhist monuments in Taliban Afghanistan (and let's not forget where the Taliban came from). Can you imagine the takeover of multicultural Kashmir by Islamabad?

"To the Muslim citizens of Indian Kashmir, I, as a civil servant in the Srinagar government, say this: look around you, look at the country of which you are a subject. Muslims in India are more free, have more opportunities to participate in all spheres of life, are better educated, have more political power and influence than Muslims do in Islamic states. Traditional law in India accommodates Muslim needs. Women in Muslim-Indian society are far better off than they are in many Islamic states. Is it worth three wars, 60,000 lives, and a possible nuclear conflict to reject participation in one of the world's greatest democratic experiments?

"Kashmir belongs to India. All inhabitants of Kashmir benefit from Indian governance. What is good for all of India is good for Kashmir."

Vote your opinion at www.wiley.com/college/deblij

and, if anything, has intensified during the past few years. In 2010, Kashmir witnessed the third consecutive "summer of rage" in which the Muslim population rebelled against the Indian "occupation"—in that year alone it cost 50 Muslim lives at the hands of Indian troops. The sad truth seems to be that decades of efforts by Indian authorities to win the hearts and minds of the people in this territory have failed. In the words of a local Muslim cleric in mid-2010, the people felt "angry, humiliated, and willing to face death." Given the specter of terrorism in India and the dangerous precedent of a concession given India's enormous ethnic and regional diversity, the Kashmir conflict is left to smolder and is unlikely to be resolved for years to come.

The Specter of Terrorism

Comparatively successful as the integration of India's Muslim communities into the fabric of the Indian state has been, the risk of Islamic violence, directed against Indian society in general, is also rising. A brief review of some of the major terrorist attacks in just the past decade underscores the severity of the situation. In 2001, Islamic terrorists with roots in Pakistan assaulted the Indian Parliament in New Delhi and brought the two countries to the brink of war. In 2003, a series of bombs set off in different parts of Mumbai killed more than 100 people and injured many more. In 2006, a series of simultaneous, coordinated attacks struck several trains along a commuter line in Mumbai, killing about 200. The most daring attack to date followed in 2008, when terrorists targeted Mumbai's two most upscale, Westernized hotels. Nearly 200 people died and hundreds were wounded. Live pictures of smoke billowing from the famous Taj Mahal Hotel in southern Mumbai were seen around the world (photo at right). And in 2010, a bombing in Pune, another major city not far from Mumbai, claimed 11 more lives.

The group responsible for the two most recent attacks was *Lashkar-e-Taiba* (the Party of the Righteous), a Pakistan-based organization that among other things aims to return Kashmir to Islamic rule. These events have significant implications for India, where the overwhelming majority of Muslim citizens have remained uninvolved in extremist causes. It seems that a small number of Indian Muslims are joining local terrorist cells with links to Pakistan and perhaps other Islamic countries. Their terrorist acts lead to investigations and a reactionary climate that offend ordinary and peaceful Indian Muslims, radicalizing a number of them and expanding the market for Islamic militancy. It is still too early to gauge the potential impact of this development on a country that has long and justly prided itself on its multicultural democracy, but the portents for India's political, social, and economic geography are obviously serious.

In the meantime, Pakistan's northwestern frontier is effectively managed by Taliban forces that provide shelter and training grounds to the al-Qaeda terrorist network. This border zone with Afghanistan is well beyond the control of the Pakistani government. U.S. efforts to defeat the Taliban in Afghanistan continue, but are thwarted by the Taliban's ability to move back and forth across this border. Thus the United States is increasing pressure on Islamabad to confront the Taliban on the Pakistani side of the border in this remote mountain refuge.

It is a delicate geopolitical chess game. Pakistan is careful not to alienate its Islamic base even if it despises the northern extremists, and it is fearful of an economically stronger India and its growing intimacy with the United States. India is deeply concerned about Pakistan's role in terrorism on Indian soil and, even worse, the possibility of fundamentalists taking control of Pakistan's nuclear capabilities; at the same time, it must also guard against increased

Smoke billows from the landmark Taj Mahal Hotel in Mumbai on November 29th, 2008. The Taj was one of several sites in southern Mumbai that were simultaneously targeted for attack by Islamic militants. The siege of the Taj lasted four days and almost 200 people died in the violence. Indian commandos killed all of the terrorists but one, who was captured. It was not the first time that terrorism of this magnitude had struck Mumbai—unfortunately, as the attacks of July 13, 2011 showed, it was not the last.

© AP/Wide World Photos

tensions between Hindus and Muslims inside India. India is also impatient with American reluctance to choose its side in the Kashmir conflict. The United States, in turn, is sympathetic to the world's largest democracy but needs Pakistan to be an ally in the global counterterrorism campaign. Each of these parties is walking a tightrope, where the slightest mistake could have deadly consequences.

Chinese Border Claims

An overview of this realm's geopolitical landscape would not be complete without noting the powerful presence of China, South Asia's northeastern neighbor, and that country's territorial claims. As Figure 8A-7 shows, the northeastern extension of Jammu and Kashmir State is claimed by China. This conflict has been fairly quiet in recent years, but officially neither China nor India shows the slightest sign of conceding. China is also part of the equation because of its strategic support for Pakistan, expressed for example in the building of nuclear power plants. Not surprisingly, this draws heavy criticism from India.

Another, historically more volatile dispute lies in India's far northeast, where China claims the bulk of the territory of the poetically named Indian State of Arunachal Pradesh (land of the dawn-lit mountains). Here India borders Tibet, and back in 1914, long before the Chinese took control, independent Tibet agreed to this boundary. Tibet (now called Xizang by the Chinese) figures prominently in this conflict, especially because the exiled Dalai Lama has called India his second home since the 1960s. A brief war between China and India in 1962 was fought over this border, but the matter of Tibet was never far removed. Today, this issue remains unresolved and is compounded by Chinese plans to construct dams in the Tibetan tributaries of the upper Brahmaputra River, potentially jeopardizing the supply of water to northeastern India (as well as Bangladesh).

Indian Ocean Geopolitics

It has been labeled the "New Great Game of Asia," the emergence of a highly strategic competition for control of the Indian Ocean Basin that involves India, China, and the United States as the leading players. The U.S. role here today is comparable to that of the British during the nineteenth century in Central Asia (Turkestan), the theater of the original Great Game that featured Russia as the other main protagonist.

China needs access to markets for its products and supplies of raw materials to sustain its rapidly growing industrial production, and a good part of this access runs through the Indian Ocean. Along the way, it is extending Chinese political and military power through the expansion of its navy in the Indian Ocean and building bases in Pakistan, Myanmar, and Bangladesh. India, increasingly concerned about Chinese intentions, has responded by building its own new alliances with such Southeast Asian states as Indonesia and Vietnam.

From a broader pan-Asian perspective (South and East), geopolitical developments are increasingly a matter of U.S.-China-India relations in which China is asserting itself in the Indian Ocean Basin as well as along the northern Indian border; in which India's economic rise has given it a new assertiveness in the political arena; and in which the United States is ideologically inclined to sympathize with India while the imperatives of political reality motivate it to steer toward maintaining the traditional balance of power politics between the two Asian giants. The future could be a U.S.-Chinese-Indian condominium or, more likely, a continued informal alliance of the United States and India seeking to put the brakes on China's assertiveness. U.S. support for Indian membership in the United Nations Security Council is but one manifestation of this evolving U.S.-India relationship.

Growing economic interdependence will help to keep things in check. China wants to penetrate India's vast and growing consumer markets. China-India trade is growing rapidly, reaching U.S. $60 billion in 2010 ($40 billion from China to India and half of that in the other direction). China runs a significant surplus, and this reflects, up to now, the fact that the **terms of trade [8]** are to China's advantage: India supplies mainly raw materials while China sells it finished goods with a higher added value. In December 2010, the two pledged to increase their trade to $100 billion by 2015, with India hoping to reduce its trade deficit.

EMERGING MARKETS AND FRAGMENTED MODERNIZATION

In recent years, optimistic television news reports and articles in the popular press have been proclaiming a new era for South Asia, marked by rising growth rates for the realm's national economies, rewards from globalization and modernization, and increasing integration into the global economy. India, obviously the key to the realm, has even been described as "India Shining" during this wave of enthusiasm.

And indeed, a combination of circumstances, ranging from America's involvement with Pakistan in the campaign against terrorism to the real estate and stock market booms in India, suggest that a new era has arrived. But consider this: well over half of India's 1.227 billion people continue to live in poverty-stricken rural areas, their villages and lives virtually untouched by what is happening in the cities (where, by the way, tens of millions of urban dwellers inhabit some of the world's poorest slums). Fully one-third of Pakistan's population lives in abject poverty; female literacy is below 30 percent. Half of the people of Bangladesh, and nearly half of those in the realm as a whole, live on the equivalent of one U.S. dollar per day or less. It is estimated that half the children in South Asia are malnourished and underweight, a majority of them girls—this at a time when the world is able to provide adequate calories for all its inhabitants, if not sufficiently balanced daily meals. It still remains to be seen if the benefits of newfound economic growth can be spread around widely enough for the good of South Asia's masses.

AMONG THE REALM'S GREAT CITIES . . .

DELHI NEW AND OLD

© H. J. de Blij, P. O. Muller, and John Wiley & Sons, Inc.

FLY DIRECTLY OVER the Delhi–New Delhi conurbation into its new international airport (opened in 2010), and you may not see the place at all. A combination of smog and dust creates an atmospheric soup that can limit visibility to a few hundred meters for weeks on end. Relief comes when the rains arrive, but Delhi's climate is mostly dry. The tail-end of the wet monsoon reaches here during late June or July, but altogether the city only gets about 60 centimeters (25 in) of rain a year. When the British colonial government decided to leave Calcutta and build a new capital city adjacent to Delhi, conditions were different. South of the old city lay a hill about 15 meters (50 ft) above the surrounding countryside, on the right bank of the southward-flowing Yamuna River, a tributary of the Ganges. Compared to Calcutta's hot, swampy environment, Delhi's was agreeable. In 1912 it was not yet a megacity. Skies were mostly clear. Raisina Hill became the site of a New Delhi.

This was not the first time rulers chose Delhi as the seat of empire. Ruins of numerous palaces mark the passing of powerful kingdoms. But none brought to the Delhi area the transformation the British did. In 1947, the Indian government decided to keep its headquarters here. In 1970, the metropolitan-area population exceeded 4 million. By 2012, it was a staggering 22.9 million, and India's largest urban region.

Delhi is popular as a seat of government for the same reason as its ongoing expansion: the city has a fortuitous relative location. The regional topography creates a narrow corridor through which all land routes from northwestern India to the North Indian Plain must pass, and Delhi lies in this gateway. Thus the twin cities not only contain the government functions; they also anchor the core area of this populous country.

Old Delhi was once a small, traditional, homogeneous town. Today Old and New Delhi form a multicultural, multifunctional urban giant. From above, the Delhi conurbation looks like an inkblot, a near concentric region that has steadily expanded in all directions. The fastest growth has been to the south, where formerly separate towns and satellite cities such as Faridabad are now part of this sprawling urban region. Gurgaon is especially well-known, an activity hub south of the airport that has witnessed explosive growth over the past 15 years or so, an agglomeration of calling centers, IT companies, and new middle-class residential developments. Another sign of Delhi's modernization is the construction of a new, heavy-rail transit system, aimed at providing some relief to the metropolitan area's massive congestion and pollution problems.

Economic Liberalization

Most countries in this realm have liberalized their economies since the late 1980s as part of a worldwide turn toward **neoliberalism** [9]. This involves privatization of state-run companies, lowering of international trade tariffs, reduction of government subsidies, cutting of corporate taxes, and overall deregulation to spur business activity. It was an important change from previous times in which markets were tightly controlled by large central governments that since independence had espoused ideologies opposed to unrestrained capitalism. This change of direction was unavoidable. The ineffective policies of the past, continuing grinding poverty, and near fiscal bankruptcy demanded support from the International Monetary Fund; the IMF, in turn, insisted on structural reforms.

The results of these reforms have been especially noticeable in India, Bangladesh, and Pakistan, where economic growth rates climbed to levels never seen before (Fig. 8A-8). Most of this growth is in manufacturing, services, finance, and, in India, information technology (IT). A more open economy has attracted increased foreign investment, and during the past two decades a new (urban) middle class has emerged. This new class may account for only 25 percent or so of the population, but in South Asia that translates into fully 300 million people—a huge new consumer market for an array of products ranging from cars to smart phones. Nevertheless, that still leaves a staggering 1 billion South Asians who have *not* attained middle-class status, for whom little has changed, who remain overwhelmingly dependent on agriculture, and who are not likely to log on to the new information economy anytime soon.

One striking feature of the South Asian realm, then, is that even if the majority of the people live traditional lives in rural villages, there are also a fair number of

SOUTH ASIA: ECONOMIC GROWTH PER CAPITA, 1984–86 AND 2004–06

Percent growth, 1984–86
5%
10%
Percent growth, 2004–06

PAKISTAN
NEPAL
BHUTAN
INDIA
BANG.
SRI LANKA

0 250 500 Kilometers
0 125 250 Miles

Data Source: World Resources Institute

FIGURE 8A-8   © H. J. de Blij, P. O. Muller, J. Nijman, and John Wiley & Sons, Inc.

megacities in which social and economic change is the order of the day. The urban regions of Mumbai, Delhi–New Delhi, Kolkata, Dhaka, and Karachi all contain populations larger than 13 million that have grown rapidly over the past two decades. Their density is often overwhelming, environmental conditions are poor, and the contrast between rich and poor can be painfully sharp. But almost always it is better to be poor in the city than in the countryside, simply because cities offer opportunities that do not exist in the villages—which explains the steady streams of rural-to-urban migration that have helped to drive the growth of towns and cities all across South Asia.

The Significance of Agriculture

Across South Asia, more than half of the entire workforce is employed in agriculture, ranging from about 40 percent in Pakistan to about 80 percent in Nepal. But overall productivity is low, and the contribution from agriculture to the overall economy is only around 20 percent. Incomes in rural areas are much lower than in major cities, and the same is true for the standard of living. About 70 percent of South Asia's population is rural, and even those who do not work in agriculture tend to rely on it indirectly.

Millions of lives every year depend on a good harvest. As Figure 8A-3 shows, the wet monsoon brings life-giving rains to the southwestern (Malabar) coast, and a second branch from the Bay of Bengal sweeps across north-central India toward Pakistan, losing strength (and moisture) as it proceeds. This means that amply watered eastern India and Bangladesh as well as the southwestern coastal strip grow rice as their staple crop; but drier northwestern India and Pakistan raise wheat.

Farmers' fortunes tend to vary with geography, and this can be illustrated in the situation on either side of India's Western Ghats upland. The monsoon rains are generally plentiful along the coast-paralleling, western slopes of this linear highland, from the southern tip of the subcontinent as far north as the vicinity of Mumbai. Here you can see the hillside vegetation assume all shades of fresh

From the Field Notes . . .

"During a period of field research in burgeoning Mumbai a few years ago, I took a road trip of about 300 kilometers (200 mi) into the interior of Maharashtra State for a change of scenery. These remote rural areas present a different world, where life has not changed much for centuries. On this drier interior side of the Western Ghats, the wet monsoon is less predictable and so is the harvest. Farms are small and worked by extended families, mostly without machinery. Millet is the main crop in these parts. Here a grandfather pumps irrigation water from a well with the use of a bullock while the rest of the family does the more exacting work in the surrounding fields. That year the rains had been abundant and the water level in the well was still high. Just as the bullock goes around and around, day after day, life tends to repeat itself in a cyclical pattern for generation after generation in India's myriad small villages."

© Jan Nijman

www.conceptcaching.com

● AMONG THE REALM'S GREAT CITIES . . .

KARACHI

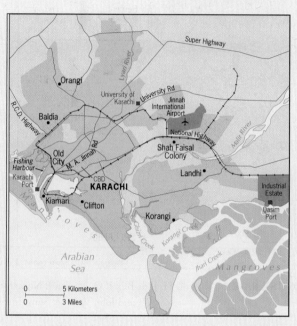

© H. J. de Blij, P. O. Muller, J. Nijman, and John Wiley & Sons, Inc.

LOCATED IN SOUTHERNMOST Pakistan next to the Indus Delta and on the Arabian Sea, Karachi is the country's biggest city as well as its economic and financial center. In former days, it was the capital too, but that function was shifted to Islamabad in the 1960s (see p. 414). Until independence in 1947, Karachi was a typical colonial port city, designed to serve British interests. This function can still be seen in the urban landscape: the city was organized around the port, and this is where you can still find the main railroad terminals. Raw materials and agricultural goods were brought in from the hinterland to be loaded onto ships headed for England's industrial complexes. The old city center lies just north of the port, and today's main business district is located just to the east of it.

Karachi has grown enormously over the years, from around 100,000 at the beginning of the twentieth century to about half a million in 1950 to 13.7 million today. This undoubtedly is Pakistan's "world city" because it dominates the country's linkages to the global economy (even more so than Mumbai does for India). Karachi is home to most of the foreign multinational firms, accounts for about three-quarters of international trade, and produces about 20 percent of Pakistan's GDP. It has the busiest airport and two seaports that together handle more than 90 percent of all transoceanic trade (the newer port of Qasim was built east of the metropolis in the 1970s). The city's economic importance is manifest in a major central business district (Saddar) with an imposing skyline.

But this is a city besieged with problems as well. Rampant and unplanned growth has resulted in congestion and pollution. Many people live in abject poverty, and the contrast with the small but very wealthy urban elite is stark. Compared to most other cities in this realm, street crime is considerable. Karachi sits near the bottom in rankings of the world's "most liveable" cities. The great majority of its residents are first- or second-generation immigrants from a wide variety of ethnic backgrounds (a large number of them originally hailing from northern India prior to the 1947 partition), and recent years have witnessed growing communal tension and violent conflict (press reports indicate that more than 1000 people were killed in such violence during 2010 alone). In addition, the city is widely described as a nurturing ground for Islamic fundamentalists and an ideal haven for Taliban leaders and followers. Clearly, Karachi's image as Pakistan's most cosmopolitan center is being seriously threatened.

green come the month of June, a sure sign that the harvest will be bountiful. But on the eastern (rain shadow) side and deeper into the interior of the Deccan Plateau, it is a different story. The rains don't come as often and don't last as long. Farming becomes a gamble with nature, and life becomes precarious. Many of the farmers here are members of lower castes, landless and indebted, and have the hardest time making ends meet. Almost every year, Maharashtra's inland districts report several thousand (!) farmer suicides as desperately poor peasants end their lives because they can no longer provide for their families.

It is clear that the majority of people in this realm depend on agriculture and that governments must aim their economic policies at improving agricultural productivity to raise the standard of living in rural areas. But they have a long way to go. The demands on governments are many, and they often seem distracted by economic sectors that can make a faster and greater contribution to the tax base, such as manufacturing, financial services, and IT—economic activities that mostly take place in the big cities far from the impoverished countryside.

SOUTH ASIA'S POPULATION GEOGRAPHY

Given its enormous human content, the South Asian realm's areal size is relatively quite small. It totals less than two-fifths the size of equally populous East Asia. Comparing the world's two giants shows that China's area is almost three times as large as India's. The total population of Subsaharan Africa is less than half of South Asia's, in a territory almost five times as large. Adjectives such as "teeming," "crowded," and "jammed" are often used to describe the realm's habitable living space, and with good reason. South Asia's intricate cultural mosaic is tightly packed, with only the deserts in the west and the mountain fringe in the north displaying large empty spaces (Fig. 8A-9); in fact, you can

FIGURE 8A-9

© H. J. de Blij, P. O. Muller, and John Wiley & Sons, Inc.

even see the outlines of the densely populated river basins in the dot patterns of this population distribution map.

But what is the significance of the realm's enormous population size and density, and how are South Asia's population dynamics related to issues of development? The field of **population geography [10]** focuses on the characteristics, distribution, growth, and other aspects of spatial demography in a country, region, or realm as this relates to soils, climates, land ownership, social conditions, economic development, and other factors. In the South Asian context, it is useful to concentrate on four demographic

dimensions: the role of density; the demographic transition; age distributions and economics; and the gender bias in birth rates. As we shall see, population issues are often more complicated than they first appear.

Population Density and the Question of Overpopulation

Population density [11] measures the number of people per unit area (such as a square kilometer or square mile) in a country, province, or, as is the case in Figure 8A-9, an entire realm. We distinguish between two types of measures. *Arithmetic density* is simply the number of people per area, usually a country. **Physiologic density [12]** is a more meaningful measure because it takes into account only land that is arable, that can be used for food production. Please take a careful look at the data displayed for South Asia in Appendix B, and you will see that, for example in Pakistan, the two measures are quite different due to that country's large deserts and inhospitable mountain ranges.

Until recently, South Asia's persistent poverty was often related to its enormous and rapidly growing population and its high population densities. The idea was that there were simply "too many mouths to feed": the realm was "overpopulated." The notion of *overpopulation* can be compelling and seems to make sense at an intuitive level because every country or region can be thought of having a limited "carrying capacity."

But things are more complex than that. If you look again at Appendix B, you will find that some countries with high densities, such as the Netherlands or Japan, are doing very well, and it is not necessarily because they have such impressive natural resources (they do not). The point is that high density in itself is not always a problem and that, in certain circumstances, population can be considered as a *human resource*. If productivity is high, there does not appear to be a problem, but if productivity is low, then large populations can be a drain on the economy. Countries with high education levels, institutional efficiency, and technological know-how are able to use their natural resources more efficiently.

Thus in South Asia, with large numbers of people still illiterate and undereducated, population tends to function as a *burden* rather than a resource. The problem is not so much that there are too many, but that too many are not sufficiently productive. The good news is that, as a result of higher economic growth rates stemming from economic reforms, there is more money to invest in education. The bad news so far is that not enough of the money is actually spent that way.

The Demographic Transition

The relevance of population issues to development go far beyond density, which is really just a snapshot of the population pattern at a moment in time. It gets more interesting—and more complicated—when we relate population change to economic trends. For example, for a considerable time South Asia's population grew faster than the realm's economy. Clearly that was a problem because more and more people had to survive on less. Today, fortunately, it is the other way around: in most of the realm, the economy is growing faster than the population.

The term **demographic transition [13]** refers to a structural change in birth and death rates resulting, first, in rapid population increase and, subsequently, in declining growth rates and a stable population (Fig. 8A-10). The United States and other highly developed countries had already passed through this transition by the mid-twentieth century, and most countries in the South Asian realm are in the third stage today. Note that Stage 2 and part of Stage 3, with high birth rates and low death rates (due to medical advances), entail a population expansion. In South Asia, this expansion occurred from the 1950s through the 1970s. The key issue, of course, is for the birth rates to come down so that overall growth rates will drop and the population will stabilize. This is happening today, but the process is not yet complete.

Figure 8A-11 shows how **fertility rates [14]** (the number of births per woman) have dropped across the realm over the past quarter-century. Only Sri Lanka seems to have

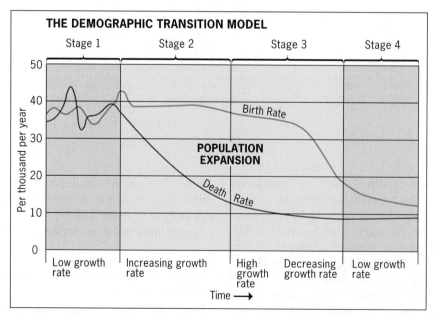

FIGURE 8A-10 © H. J. de Blij, P. O. Muller, and John Wiley & Sons, Inc.

FIGURE 8A-11

© H. J. de Blij, P. O. Muller, J. Nijman, and John Wiley & Sons, Inc.

completed the transition, although Bhutan is now very close. Elsewhere, fertility rates are still too high (India by itself has been adding about 15 million people *per year* during the past decade), but at least they are moving in the right direction.

Demographic Burdens

The immediate significance of demography to economics lies in what we call the **demographic burden [15]**. This term refers to the proportion of the population that is either too old or too young to be productive and that must be cared for by the productive population. Typically, the most productive population in developing countries is represented in the age cohorts between 20 to 50 years. A country with low death rates and high birth rates will have a relatively large share of old and young people, and thus a large demographic burden. Obviously, the way to reduce this burden is to lower birth rates.

Let us now turn to examine Figure 8A-12 and compare today's **population pyramids [16]** (diagrams showing the age–sex structure) for India and China. The latter has been more successful in curtailing births since 1980, so China now faces a smaller demographic burden than India. But, interestingly, what is advantageous today can become

POPULATION PYRAMIDS: INDIA AND CHINA, 2012–2037

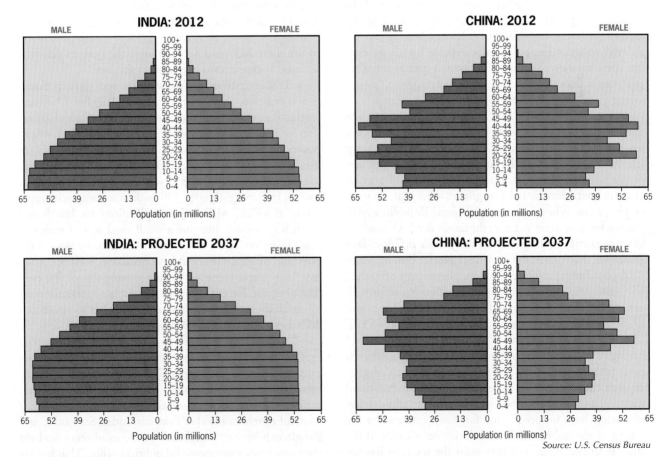

Source: U.S. Census Bureau

FIGURE 8A-12

© H. J. de Blij, P. O. Muller, J. Nijman, and John Wiley & Sons, Inc.

a disadvantage tomorrow. Look at Figure 8A-12 again and see what the population profiles are projected to be 25 years from now. Assuming that India will be able to further reduce its birth rate in the coming years, its demographic burden will be less than China's one generation from now. In China, today's productive cohort will have moved on to old age, thereby adding to its national demographic burden. It is another reason for India boosters to be optimistic about the future. Yet some other major hurdles remain, and certain forms of birth control are as morally reprehensible as they are economically counterproductive.

The Missing Girls

Issues of family planning and birth control shed some interesting light on the "fragmented modernization" of South Asia. As we saw, the realm finds itself in an advanced stage of the demographic transition wherein birth rates have begun to decrease. But take a close look at India's population pyramid for 2012 and note that among young children boys far outnumber girls. In fact, males outnumber females well into middle age.

Traditionally, boys are valued more than girls because they are thought to be more productive income-earners, because they are entitled to land and inheritance, and because they do not require a dowry at the time of marriage. When a couple gets married (often arranged, and at a young age), the bride comes into the care of the groom's family, where she also contributes her work in and around the house. For this, the bride's family must provide a dowry that can impose a major expense on her parents. For these reasons, the birth of a boy is a greater cause for celebration than that of a girl. "Raising a daughter," as one saying goes, "is like watering your neighbor's garden."

One reason for the high fertility rates in the past was that families would continue to have children until there were enough sons to take care of the parents in their old age (the girls, after all, would be taking care of their future husband's parents). Hence, this gender bias is in itself a major factor in South Asia's population growth—but that is not the only problem. When a poor couple repeatedly produces girls and not boys, in some instances the family decides to end the life of the newborn daughter. It is this *female infanticide* or *gendercide* that causes the unnatural gender bias in South Asia's population profiles (the same applies to parts of China).

But why—as the economic situation has improved, as birth rates have receded, and as modernization has begun to set in—do we still observe this skewed **sex ratio [17]**? The answer is that with fewer children, the importance of having at least one boy has for many families become even more pressing. And here is where "modernization" throws another curve: newly available technologies of ultrasound scanning and rising incomes (i.e., the growing affordability of a scan) have induced many families to determine the gender of the unborn child and decide on abortion if the child is female. Thus in recent years the sex ratio has become more, not less, skewed, and the most extreme ratios

A group of girls in a park in Delhi showing off their henna tattoos. "Mehndi" is a centuries-old festive tradition involving decorative painting on the hands and wrists. The designs usually wear off in a few weeks' time.

© Jan Nijman

are now found in some of the most developed parts of the realm, such as the Indian States of Punjab and Haryana.

In the long run, however, this leads to a shortage of females, which becomes particularly apparent at marriage age. In some areas, families now face a problem in finding brides for their sons, and this "bachelor angst," in turn, is leading to a change in attitudes. Look at India's population pyramid for 2037 and note that over the next 25 years the sex ratio is expected to become less skewed.

Discrimination against women and girls in South Asia is expressed in a variety of ways in daily life. One matter that is now receiving more attention is the lack of sanitation for females. In many workplaces, schools, slum neighborhoods, and public spaces, women still have no (or insufficient) access to toilets. For safety and personal reasons women need more privacy, but often that is not to be had. In India, an estimated 330 million females lack proper access to toilets, with major implications for health and social functioning. Imagine a small rural school with one toilet (if indeed there is one): it will be used by boys at the exclusion of girls. Girls will have no access at all the entire day. This simple fact leads to girls missing school and even dropping out. Recent efforts by development organizations to install toilets for girls have so far only led to significant decreases in female-student dropout rates.

Still, it can be difficult to make generalizations about gender relations across this populous realm, in part because of religious and regional diversity as well as rural-urban differences. To be sure, these are in many respects male-dominated societies, especially at a young age. But it is useful to remember that Pakistan, India, Sri Lanka, and Bangladesh have all had female prime ministers who held their countries' most powerful political office. That has yet to happen in the United States.

From the Field Notes . . .

"Strolling the grounds of one of India's leading high-tech companies, Wipro, in the southern city of Bengaluru (formerly Bangalore), I was struck by the thought that their use of the word 'campus' was right on target: the great majority of the 33,000 employees on this site (120,000 in total) are in their mid-20s and almost all have a college degree; a large

© Jan Nijman

number of them are involved in research and product development; and the layout of the premises could easily be mistaken for a (well-funded) university. This Silicon Valley-like campus is the heart of the company's booming outsourcing and consulting business. The company is active in almost all major Indian cities and has a sizeable global presence. While some skilled workers still emigrate to Europe, Australia, and the United States, where their skills earn them more than at home, many now choose to stay here and benefit from the higher local income the IT manufacturers pay. I was shown around by Rohit, a 26-year-old with a bachelor's degree in engineering from an Indian college and an MBA from Singapore National University. He reveled in the opportunity: 'I joined the company just four months ago in their marketing division and feel extremely fortunate. Everybody wants this job! It's a great professional opportunity and the salary is very good. In addition, Wipro is an IT leader in the world and we are proud of it.' And Bengaluru is a pleasant place to live: at 900 meters (3000 ft) above sea level, even the summer weather is tolerable. Bengaluru, however, is growing so rapidly that its infrastructure cannot keep up, so that congestion, traffic jams, and commuting times are all growing as well."

www.conceptcaching.com

FUTURE PROSPECTS

South Asia is a realm in transition—politically, economically, and demographically. It is a realm that seems clearly bounded by nature, yet it is vitally linked to Southwest Asia and, increasingly, the entire world. It is also a realm that at times is difficult to read. India–Pakistan tensions continue to be a cause for concern, and the specter of terrorism haunts those who wish only to live in peace. This is not just in the hands of the governments of the two biggest nations in the realm. Religious movements (Muslim and Hindu) and the way they engage politics are crucially important, and the United States and China have major roles to play as well.

Economically, there is no question that India's rise will increasingly demand the world's attention. Indian transnational corporations will continue to penetrate the global economy, and the growing Indian middle class with its appetite for consumption will increasingly draw interest from producers around the world. That English is the subcontinent's *lingua franca* and that IT is a leading economic sector give it an enormous advantage into the future. And the fact that India, the realm's giant, can claim to be the world's biggest democracy gives it tremendous credibility.

When, during the next several decades, South Asia passes through the demographic transition; when it keeps the peace; when it continues its leading role in the global IT sector; when its economic growth is used to educate and empower the masses; and when the reorganization of agriculture allows more productive and prosperous lives— and these are all real possibilities—then this populous and wondrous South Asian realm may yet turn out to be the biggest story of the twenty-first century.

► POINTS TO PONDER

- The South Asian realm contains three of the world's mightiest rivers and the world's largest human concentration.

- After more than 60 years, the conflict over Kashmir is still unresolved.

- The most skewed sex ratios in the population occur in the most prosperous parts of the realm.

- The Indian Ocean Basin may be the world's most crucial geopolitical arena of the twenty-first century.

- Hundreds of millions of females in South Asia suffer from a lack of access to basic human needs, including sanitary facilities.

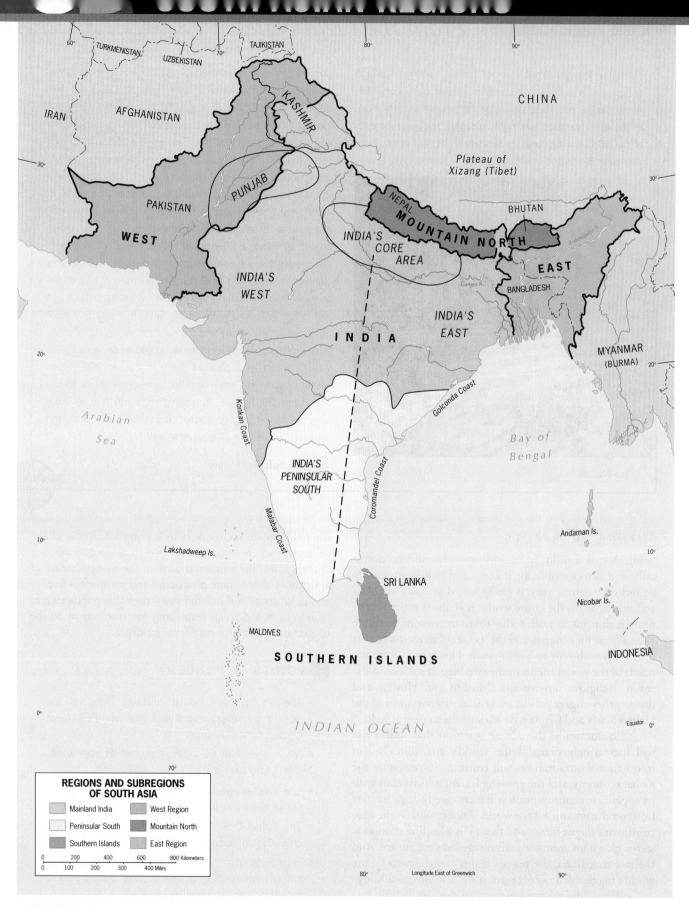

REGIONS AND SUBREGIONS OF SOUTH ASIA

- Mainland India
- Peninsular South
- Southern Islands
- West Region
- Mountain North
- East Region

0   200   400   600   800 Kilometers
0   100   200   300   400 Miles

FIGURE 8B-1

© H. J. de Blij, P. O. Muller, and John Wiley & Sons, Inc.

REGIONS

Pakistan

India

Bangladesh

Mountainous North

Southern Islands

IN THIS CHAPTER

- Pakistan: A hard place to govern
- Caste in modern India
- India's new consumers
- Bangladesh: Survival in the double delta
- Sri Lanka: Stability at last?

CONCEPTS, IDEAS, AND TERMS

Developments in the South Asian realm, as discussed in Chapter 8A, demand analysis at multiple scales. Realmwide as well as international geopolitical issues, from Kashmir to the Indian Ocean, are closely tied to various regional and national trends such as Pakistan's dealings with the Taliban, India's economic rise, or the role of the Communist Party in Nepal. This chapter provides a better understanding of these regional and national developments, all in geographic context.

India, the cornerstone of South Asia, has a core area centered on the broad basin of the Ganges River, the historic heart of this realm (Fig. 8B-1). India is a region as well as a state, with key subregions based on cultural and other criteria to be discussed later in this chapter. To the west lies Islamic Pakistan, whose lifeline, the Indus River and its tributaries, creates a core area in the Punjab. But note that the Punjab subregion extends across the border with India. What you see here is the aftermath of a human drama. When independence approached, Muslim leaders in the west demanded the creation of an Islamic state. A boundary drawn hastily in 1947 between a future Muslim West Pakistan and a dominantly Hindu India precipitated the sudden migration of millions of people in the border zone—Muslims westward and Hindus eastward—with countless casualties and innumerable personal tragedies.

In South Asia's east, the British had earlier drawn a border between (Hindu) Bengal and (dominantly Muslim) East Bengal, and that border became the boundary of what became known as East Pakistan, the political connection based on shared Islamic values. In 1971, however, East Pakistan severed its link with West Pakistan (which thereupon renamed itself Pakistan) and became independent Bangladesh.

© Jan Nijman
Where were these photos taken?
Find out at www.wiley.com/college/deblij

© Jan Nijman

● MAJOR CITIES OF THE REALM

City	Population* (in millions)
Delhi–New Delhi, India	22.9
Mumbai (Bombay), India	20.7
Kolkata (Calcutta), India	16.1
Dhaka, Bangladesh	15.4
Karachi, Pakistan	13.7
Lahore, Pakistan	7.5
Chennai (Madras), India	7.2
Hyderabad, India	6.9
Bengaluru (Bangalore), India	6.8
Ahmadabad, India	5.9
Varanasi, India	1.5
Kathmandu, Nepal	1.1
Colombo, Sri Lanka	0.7

*Based on 2012 estimates.

As Figure 8B-1 shows, three separated entities make up the region defined as the Mountainous North. From east to west, these are the traditional kingdom of Bhutan, as isolated a country as the world has today; the troubled state of Nepal, where royal rule failed but the struggle to establish representative government continues; and, at the western end, the disputed territory of Kashmir, where the British boundary-making effort failed to resolve a complicated cultural and political situation, and where India and Pakistan have repeatedly clashed in armed conflict.

Finally, the region mapped as the Southern Islands consists of two very different countries: dominantly Buddhist Sri Lanka, where a quarter-century-long civil war ended in 2009, and the ministate Republic of the Maldives, whose official religion is Islam.

Physiographically, India mainly consists of plateau-and-basin country, but its neighbors to the north, Nepal and Bhutan, straddle the southern slopes of the mighty Himalaya range and its foothills. The waters of the Indian Ocean separate India from neighbors Sri Lanka and the Maldives in the south. Thus there are good geographic reasons for using the political framework to designate the regions of this realm. Culture and nature constitute formidable regionalizing factors here, even within India itself. The cultural forces that broke British India apart at the time of independence continue to exert enormous power in still-evolving South Asia.

██ PAKISTAN: ON SOUTH ASIA'S WESTERN FLANK

Gift of the Indus

If, as is so often said, Egypt is the gift of the Nile, then Pakistan is the gift of the Indus. The Indus River and its principal tributary, the Sutlej, nourish the ribbons of life that form the heart of this populous country (Fig. 8B-2).

Territorially, Pakistan is not large by Asian standards; its area is about the same as that of Texas plus Louisiana. But Pakistan's population of 193 million makes it one of the world's ten most populous states. Among Muslim countries (its official name is the Islamic Republic of Pakistan), only Southeast Asia's Indonesia is larger.

Pakistan lies like a gigantic wedge between Iran and Afghanistan to the west and India to the east. Here in Pakistan lay South Asia's earliest urban civilizations, whose innovations radiated southeastward into the triangular peninsula. Here lies South Asia's Muslim frontier, contiguous to the great Islamic realm to the west and irrevocably linked to the enormous Muslim minority to its east.

Pakistan's cultural landscapes bear witness to its transitional location. Teeming, disorderly Karachi is the typical South Asian city; as in India, the largest urban center lies on the coast. Historic, architecturally Islamic Lahore is reminiscent of the scholarly centers of Muslim Southwest Asia. In Pakistan's east, the 1947 partition boundary divides a Punjab subregion that is otherwise continuous—a land of villages, wheatfields, and irrigation ditches. In the northwest, Pakistan resembles Afghanistan in its huge migrant populations and its mountainous frontier. And in the far north, Pakistan and India are locked in a deadly and intractable conflict over Kashmir (discussed in Chapter 8A). A legacy of the times of partition, this territory is claimed by Pakistan because the majority of the inhabitants are Muslim, while India refuses to give up control because it claims that the Hindu minority would have no future inside Pakistan. This issue has plagued relations between the two countries for decades and is not likely to be resolved in the foreseeable future.

A Hard Place to Govern

At independence (West) Pakistan had a bounded national territory, a capital, a cultural core, and a population—but few centripetal forces to bind state and nation. The disparate subregions of Pakistan shared the Islamic faith and an aversion for Hindu India, but little else. Karachi and the coastal south, the desert of Baluchistan, the city of Lahore and the Punjab, the rugged northwest along Afghanistan's border, and the mountainous far north remain worlds apart. Urdu is the official language, and English is still the *lingua franca* of the elite. Yet several other major languages prevail in different areas (see Fig. 8A-5), and ways of life vary enormously.

Successive Pakistani governments, civilian as well as military, turned to Islam to provide the common bond that history and geography had denied this nation. In the process, Pakistan became one of the world's most theocratic states. But even Islam itself is not unified in restive Pakistan. About 77 percent of the people are Sunni Muslims, and the Shia minority numbers approximately 20 percent. Sunni fanatics intermittently attack Shi'ites, leading to retaliation and establishing grounds for subsequent revenge.

To govern so diverse and fractious a country would challenge any system, and so far Pakistan has failed the test. Democratically elected governments have repeatedly

FIGURE 8B-2

© H. J. de Blij, P. O. Muller, and John Wiley & Sons, Inc.

squandered their opportunities, only to be overthrown by military coups. Pakistan's recent economic boom has not filtered down to the poor; literacy rates are not rising; health conditions are not improving significantly; national institutions are weak (for instance, there are only about 2 million registered taxpayers in a country of nearly 200 million); and one consequence of the global antiterrorism campaign is that Pakistanis, who used to be overwhelmingly secular in their political choices, are now increasingly joining Islamic parties.

Meanwhile, too little is being done to confront a growing water-supply crisis, an insurgency festers in Baluchistan, the army is incapable of establishing control over mountainous Waziristan (where al-Qaeda and Taliban groups maintain hideouts), the issue of Kashmir costs Pakistan dearly, and relations with neighboring India (which if satisfactory would bring enormous benefits) remain conflicted.

Adding to Pakistan's woes, the late summer of 2010 witnessed the worst floods in the country's history, a result of the wettest monsoon in decades. The Indus River, the life-

line that traverses the entire country from north to south, swelled beyond capacity, backed up many of its tributaries, and flooded fully one-quarter of the country. The worst-hit areas were in the northwest, especially in Khyber Pakhtunkhwa (formerly North West Frontier) Province, but as the river surged southward other parts of the country met with disaster as well. In all, an estimated 1800 people were killed, about 700,000 homes were destroyed and many more damaged, and several million people were directly affected. And not only was the 2010 harvest decimated: because of damage to water-control structures and irrigation equipment, 2011 crop yields were expected to be sharply curtailed as well. This disaster seemed to push Pakistan to the brink as the government, already facing so many major problems, was not in a position to provide adequate relief. Asked if the crisis would threaten the current government, President Zardari commented that "I don't think anybody in their right mind would want to take over Pakistan right now." A hard country to govern, indeed.

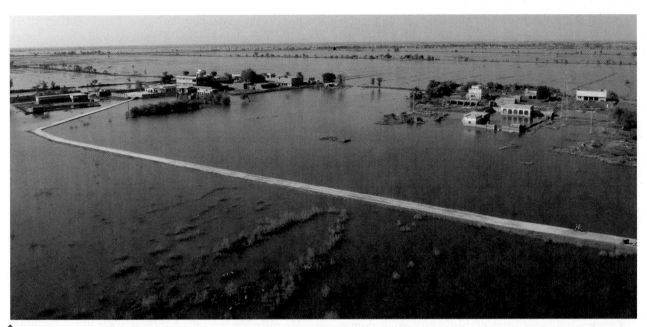

The floods that plagued Pakistan in the late summer and fall of 2010 occurred almost everywhere along the mighty Indus River, from Khyber-Pakhtunkhwa Province in the north to Sindh in the south. This photo, taken on October 30th, shows the extent of the flooded area around Jacobabad (located about 50 kilometers [30 miles] from the river's main channel on the wide Indus floodplain), near the border between Sindh and Baluchistan just north of Shikarpur (see Fig. 8B-2). From this medium-sized city alone, some 20,000 people were forced to flee for their lives. This massive flood also destroyed much of the year's harvest, further adding to Pakistan's mounting problems.

© ASIF HASSAN/AFP/Getty Images, Inc.

Subregions of Pakistan

Punjab

Pakistan's core area is the Punjab (Fig. 8B-2), the Muslim heartland across which the post-independence boundary between Pakistan and India was superimposed. (As a result, India also has a State named Punjab, sometimes spelled Panjab there.) Pakistan's Punjab is home to about 55 percent of the country's population; in the triangle formed by the Indus River and its tributary, the Sutlej, live more than 100 million people. Punjabi is the language here, and wheat farming is the mainstay.

Three cities anchor this core area: Lahore, the outstanding center of Islamic culture in the realm; Faisalabad; and Multan. Lahore, now home to 7.5 million people, lies close to the India–Pakistan border. Founded around 2000 years ago, Lahore was situated favorably to become a great Muslim center during the Mughal period, when the Punjab was the main corridor into India. After partition in 1947, the city received hundreds of thousands of refugees and grew rapidly. Lahore did lose its eastern hinterland, but its new role in independent Pakistan sustained its growth. Punjab's relationship with Pakistan's other three provinces is one of the country's weak points. Both the governments and residents of those provinces feel uneasy about the dominance of Punjab, the populous, powerful core of the country from which most of the army is drawn.

Sindh

The lower Indus River is the key to life in Sindh (Fig. 8B-2), but the Punjab controls the waters upstream, which is one of the issues dividing Pakistan. When the Punjab-dominated regime proposes to build dams across the Indus and its major tributaries, Sindhis (who make up almost one-fourth of the national population) are reminded of their underrepresentation in government and talk of greater autonomy. Nationalist, anti-Pakistan rage swept Sindh following the 2007 assassination of Sindh's presidential candidate, Benazir Bhutto.

The ribbon of fertile, irrigated, alluvial land along the lower Indus, where the British laid out irrigation systems, makes Sindh a Pakistani breadbasket for wheat and rice. Commercially, cotton is king here, supplying textile factories in the cities and towns (textiles account for more than half of Pakistan's exports by value).

But the dominant presence in southern Sindh is the chaotic, crime-ridden megacity of Karachi (profiled in Chapter 8A), with its stock market, dangerous streets, crowded beaches, and poverty-stricken shantytowns, a place of searing contrasts under a broiling sun. Karachi grew explosively during and after partition in 1947, when refugee Muhajirs from across the new border with India streamed into this urban area, setting off riots and gang battles whose aftermath still simmers. With little effective law enforcement, Karachi has today become a hotbed of terrorist activity, but somehow the city still functions as Pakistan's (and Afghanistan's) major maritime outlet and the seat of Sindh's provincial government.

Khyber Pakhtunkhwa

As Figure 8B-2 shows, this province lies wedged between the powerful Punjab to the east and troubled Afghanistan to the west, with the territory long known as the Tribal

Areas intervening in the south. The name "Pakhtunkhwa" connotes "belonging to the Pushtuns," the Afghan-associated tribes that inhabit this region.

Why is there a zone designated "Tribal Areas" in a country otherwise subdivided into provinces? The designation began during colonial times when the British assigned a special status to the obstinate Pushtun villages and their chiefs living in the remote hills and mountains on the Afghan border. The "Tribal Agencies" were the responsibility of an official "agent" who exercised control over the local chiefs through ample reward and harsh punishment—and little or no accountability. There were seven of them, some very small, several large enough to appear on a relatively small-scale map such as Figure 8B-2. North and South Waziristan were among the largest.

The Tribal Areas have had a certain degree of autonomy ever since British times, and the Pakistani government's reach into these parts is quite limited to this day. The province's mountainous physical geography reflects its remoteness and the isolation of many of its people. Mountain passes lead to Afghanistan; the Khyber Pass, already noted as the historic route of invaders, is legendary (see photo this page). Coming from Afghanistan, the Khyber's road leads directly to the provincial capital, Peshawar, which lies in a broad, alluvium-filled, fertile valley where wheat and corn drape the countryside.

During the Soviet occupation of Afghanistan in the 1980s and later during the Taliban regime, several million Pushtun refugees streamed through the Khyber and other passes into refugee camps in this area. Following the ouster of the Taliban from Afghanistan in late 2001, the great majority of Pushtuns returned home. Khyber Pakhtunkhwa remains a conservative, deeply religious, militant province, where Islamic political parties and movements are proportionately stronger than in any other part of the country and where the obstruction of national policies (including antiterrorist operations) is a common goal.

Pakistan's northwest frontier is critical to the U.S. war in Afghanistan as well as the wider campaign against Islamic terrorism, because it is extremely difficult to monitor the cross-border movements of Taliban and al-Qaeda forces. The area is a hotbed of terrorist activity and training camps. The United States constantly prods Pakistan to try to secure the province militarily, but it is not clear to what extent Pakistan is willing—or able—to comply. This does not stop America from flexing its own muscles. In April 2010, bombings by CIA-operated drones (small unmanned planes)

killed dozens of militants in North Waziristan (Fig. 8B-2), but the Taliban immediately retaliated with a brazen attack on the U.S. consulate in Peshawar. The most spectacular American intervention on Pakistani soil, of course, occurred one year later on the night of May 1-2, 2011, when U.S. special forces attacked Usama bin Laden's hiding place in Abbottabad and killed the al-Qaeda mastermind. The extent of participation by Pakistani authorities is likely to remain unclear.

Baluchistan

As Figure 8B-2 indicates, Baluchistan is by far the largest of Pakistan's four provinces, accounting for (not including Kashmir) nearly half of the national territory but inhabited only by an estimated 8.8 million people or barely 5 percent of the country's population. For a sense of its terrain, take a look at Figure 8A-1; much of this vast territory is desert, with mostly barren mountains that wrest some moisture from the air only in the northeast. Sheep raising is the leading livelihood here, and wool the primary export. In the far north, Baluchistan abuts the Tribal Areas and Afghanistan. The provincial capital, Quetta, lies in this zone. This province could easily be called the "South West Frontier Province."

However, Baluchistan is not unimportant. Beneath its parched surface lie possibly substantial reserves of oil as well as coal, and already the province produces most of Pakistan's natural gas. The construction of a major port and energy terminal is under way at Gwadar on the southwestern coast. The port will serve Baluchistan but also become a

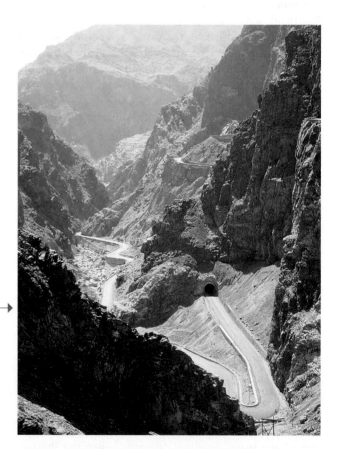

The Khyber Pass that links Afghanistan and Pakistan across the Hindu Kush. One of the most strategic passes in the world, it was no easy passage for invading armies in times past. Nowadays, the roads and tunnels facilitate the movement of refugees, drugs, and arms, and they are used by militant separatist forces from both sides of the border. Pakistan's North West Frontier Province [*now renamed Khyber Pakhtunkhwa*], and especially the city of Peshawar, are hotbeds of these activities."

© Barbara A. Weightman

transshipment port for oil and gas from Iran and the Caspian Basin, destined for markets in East Asia. China is a major investor in this project and has dispatched engineers to plan and build the port with a loan to Pakistan. But Baluchis complain that they were left out of the process. Note that 90 percent of local residents have no energy supply themselves. And there are other persistent problems. Even though there is groundwater in several localities, eight out of ten Baluchis do not have access to clean water. For decades, short-lived local rebellions have signaled dissatisfaction with the government, but more recently the Baluchistan Liberation Army (BLA) has instigated a more serious and more durable insurgency. Indeed, three Chinese engineers were killed and over a dozen wounded in a terrorist attack at Gwadar.

Pakistan's Prospects

Pakistan is a country of enormous cultural contrasts, where the modern and medieval exist side by side, where you hear residents of one province call those of another (but not themselves) "Pakistanis," and where a sense of nationhood is still elusive among many people whose loyalties to their family, clan, and village are stronger.

Pakistan has experienced so many cycles of progress and failure that confidence is in short supply. And yet, against all odds, there are areas of progress: a combination of expanded irrigation and Green Revolution farming techniques has allowed Pakistan during the past decade to export some rice (although wheat imports continue), and the country's manufacturing and services sectors have shown substantial growth. Exports include not only cotton-based textiles but also carpets, tapestries, and leather goods. Domestic manufacturing remains limited, but Pakistan has built its own steel mill near Karachi. Meanwhile, the authorities struggle to control Pakistan's growing production and trade in opium and hashish. Neighboring Afghanistan remains the world's dominant source of this illegal commerce, and the spillover effect continues; but Pakistan has many remote corners where poppy fields yield high returns and trade routes are well established. In this as in so many other spheres, Pakistan displays the contradictory symptoms of a state in transition.

This western flank of South Asia is the realm's most critical region, today more so than ever before. Islamic Pakistan's coherence and stability have become crucial at a time when the global struggle against terrorism is entangling its leaders with Western power and priorities. The role of Pakistan's government in this struggle is disputed and resented by some of its own people, who express their distaste by voting for militant Islamic parties, voicing support for the resurgent Taliban movement, and giving refuge to fugitive al-Qaeda operatives in their homes. Militancy and instability are no longer confined to the northwest or the far north: the cities of Lahore and Karachi were targeted multiple times in 2010 and 2011 by terrorists, with the number of casualties ranging from a handful to nearly 100. The government blames al-Qaeda and the Taliban.

Even if Islamic militants are not a majority, their recourse to violence and their politics of intimidation are casting a shadow over Pakistan. In early 2011, the former governor of Punjab, Salman Taseer, who had returned to his legal profession, was shot dead on a street in Islamabad—because he had spoken out against the country's pernicious blasphemy laws. He had done so in connection with the case against his client, a poor Christian woman whom a court had condemned to death for blasphemy (swearing). The killer, one of Mr. Taseer's own security guards, was later showered with rose petals by groups of lawyers (!) outside the courthouse in Islamabad. Liberal critics in Pakistan were quick to voice their fears of a wholesale revival of religious intolerance across the country.

Pakistan faces many difficult challenges, and it is easy to think of a number of ways in which the country could descend into disastrous turmoil. Pakistan's future hangs in the balance and with it the stability of South (as well as Southwest) Asia.

INDIA: GIANT OF THE REALM

If you have been reading the press and watching television over the past few years, you have seen the increasing attention being paid to India—not just in North America but around the world. Examples of news coverage include the purported "strategic partnership" between the United States and India, the outsourcing of American jobs to India, the rapid rise of a new Indian middle class, and the emergence of large Indian companies that are making their presence felt around the world. Undoubtedly, India is on the move—but will it really become the next economic superpower?

Certainly India has the dimensions to make the world take notice. Not only does it occupy three-quarters of the great land triangle of South Asia: India also is poised to overtake China to become the most populous country on Earth before the middle of this century. According to some estimates, India will have the third biggest economy in the world by 2020 (after the United States and China). Already, India is the world's largest democracy, a federation of 28 States and several additional Territories with a population exceeding 1.2 billion.

That India has endured as a unified state is a politico-geographical miracle. The country contains a cultural mosaic of immense ethnic, religious, linguistic, and economic diversity and contrast; it is truly a state of many nations. Upon independence in 1947, India adopted a democratic, secular, federal system of government, giving regions and peoples some autonomy and identity, and allowing others to aspire to such status.

India's pluralist and democratic complexion cannot be fully understood without considering the huge influence of Mahatma Gandhi, the great spiritual and political leader who played a central role in the achievement of independence from the British. More than anyone else, he symbolized tolerance, reason, nonviolence, and perseverance—the basic principles on which the Indian freedom struggle and the new state were founded. Jawaharlal Nehru, India's first prime minister, also exerted enormous influence through his strong, unshakable beliefs in democracy and secular government. Interestingly, those beliefs and principles can be said to be firmly embedded in Hindu culture. After all, Hinduism is in some ways an extremely open, diverse, and introspective religion and way of life. Indeed, in his book *The Argumentative Indian*, Harvard's Nobel laureate economist Amartya Sen emphasizes this very idea—that Indian civilization, despite its spiritual tendencies, has an inherent penchant for democracy.

Political Geography

A Federation of States and Peoples

The map of India's political geography shows a federation of 28 States, 6 Union Territories (UTs), and 1 National Capital Territory (NCT) (Fig. 8B-3). The federal government retains direct authority over the UTs, all of which are

FIGURE 8B-3

© H. J. de Blij, P. O. Muller, and John Wiley & Sons, Inc.

small in both territory and population. The NCT, however, includes most of the Delhi/New Delhi urban region, and contains more than 17 million inhabitants.

This form of political spatial organization is mainly the product of India's restructuring following independence from Britain. Its State boundaries reflect the broad outlines of the country's cultural mosaic: insofar as possible, the system recognizes languages, religions, and cultural traditions. Indians speak 14 major and numerous minor languages, and while Hindi is the official language, it is by no means universal (see Fig. 8A-5). At the time of independence, Hindi was the most widely spoken native language, but it was the mother tongue for only about one-third of the population. It is telling that the famous Midnight Speech of Prime Minister Nehru at the moment of independence in 1947 was given in English—Hindi or any other language would have been divisive and politically impossible. English soon became the country's *lingua franca*.

As Figure 8B-3 shows, the territorially largest States lie in the heart of the country and on the great southward-pointing peninsula. Uttar Pradesh (197 million, according to the 2011 population estimate) and Bihar (96 million) constitute much of the Ganges Basin and also form the core area of modern India. Maharashtra (111 million), anchored by the great coastal city of Mumbai (called Bombay before 1996), also has a population larger than that of most countries. West Bengal, the State that adjoins Bangladesh, contains 89 million residents, about 16 million of whom live in its urban focus, Kolkata (known as Calcutta before 2000).

These are staggering numbers, and they do not decline much toward the south. Southern India consists of four States linked by a discrete history and by their distinct Dravidian languages. Facing the Bay of Bengal are Andhra Pradesh (84 million) and Tamil Nadu (67 million), both part of the hinterland of the city of Chennai (formerly Madras) and located on the coast near their joint border. Facing the Arabian Sea are Karnataka (almost 59 million) and Kerala (34 million).

As Figure 8B-3 indicates, India's smaller States lie mainly in the northeast, on the far side of Bangladesh, and in the northwest, toward Jammu and Kashmir. North of Delhi, physical and cultural landscapes change from the flatlands of the Ganges to the hills and mountains of spurs of the Himalaya. In the State of Himachal Pradesh, forests cover the hillslopes and high relief reduces living space; less than 7 million people live here, many in small, comparatively isolated clusters.

The map becomes even more complicated in the distant northeast, beyond the narrow corridor between Bhutan and Bangladesh. The dominant State here is Asom (Assam), home to just over 30 million, famed for its tea plantations and important because its petroleum and gas production amounts to more than 40 percent of India's total. In the Brahmaputra Valley, Asom resembles the India of the Ganges. But in almost all directions from Asom, things change.

North of Asom, in sparsely populated Arunachal Pradesh (1.2 million), we are in the Himalayan offshoots once again. To the east, in Nagaland (2.2 million), Manipur (2.4 million), and Mizoram (1.0 million), lie the forested and terraced hills that separate India from Myanmar (Burma). This is an area of numerous ethnic groups (more than a dozen in Nagaland alone) and of frequent rebellion against Delhi's government. And to the south, the States of Meghalaya (2.6 million) and Tripura (3.5 million), hilly and still wooded, border the teeming floodplains of Bangladesh. Here in the country's northeast, where peoples are restive and where population growth is still soaring, India confronts one of its strongest regional challenges.

India: Population by State

State	2001 Census	Estimated 2011
Andhra Pradesh	75,728,000	84,075,000
Arunachal Pradesh	1,091,000	1,244,000
Asom (Assam)	26,638,000	30,187,000
Bihar	82,879,000	96,422,000
Chhattisgarh*	20,796,000	23,886,000
Goa	1,344,000	1,704,000
Gujarat	50,597,000	58,229,000
Haryana	21,083,000	24,963,000
Himachal Pradesh	6,077,000	6,692,000
Jammu and Kashmir	10,070,000	11,633,000
Jharkhand*	26,909,000	30,977,000
Karnataka	52,734,000	58,815,000
Kerala	31,839,000	34,256,000
Madhya Pradesh	60,385,000	71,141,000
Maharashtra	96,752,000	111,138,000
Manipur	2,167,000	2,447,000
Meghalaya	2,306,000	2,639,000
Mizoram	891,000	994,000
Nagaland	1,989,000	2,241,000
Orissa	36,707,000	40,433,000
Punjab	24,289,000	27,388,000
Rajasthan	56,473,000	66,795,000
Sikkim	541,000	604,000
Tamil Nadu	62,111,000	67,039,000
Tripura	3,191,000	3,536,000
Uttar Pradesh	166,053,000	197,319,000
Uttarakhand*	8,480,000	9,767,000
West Bengal	80,221,000	88,724,000
National Capital Territory**	13,783,000	17,116,000
Union Territories	2,670,000	3,512,000

* Established 2000

** Established 1993

India's Ever-Changing Map

In view of the country's enormous diversity and its federal democratic structure, it should not surprise us that India's political map is the product of endless compromise. The newly devised framework in 1947, based on the major regional languages, proved to be unsatisfactory to many communities in India. In the first place, many more languages are in use than the 14 that had been officially recognized. Demands for additional States soon arose. As early as 1960, the State of Bombay was divided into two language-based States, Gujarat and Maharashtra. In 1966, Hindu-dominated Haryana was carved out of Punjab to give Sikhs (and Punjabi speakers) more self-control.

As noted above, India's northeast harbors numerous ethnic groups in a varied, forest-clad topography. The Naga, a cluster of peoples whose domain had been incorporated into Asom State, rebelled soon after India's independence. A protracted war brought federal troops into the area; after a truce and lengthy negotiations, Nagaland was proclaimed a State in 1961. That led the way for other politico-geographical changes in India's problematic northeastern wing. In the State of Manipur, separatist groups have challenged Indian authority for more than three decades (few tourists ever come here, more than 1600 kilometers [1000 mi] from New Delhi and on the doorstep of Myanmar). In 2009 alone, this conflict cost more than 400 lives.

Devolutionary pressures have continued throughout India's existence as an independent country. In certain cases, the federal government and the military come down hard on the insurgents, but in other cases the government is more inclined to negotiate. In 2000, for instance, three new States were recognized: Jharkhand, carved out of southern Bihar State on behalf of 18 poverty-stricken districts there; Chhattisgarh, where tribal peoples had been agitating since the 1930s for separation from the State of Madhya Pradesh; and Uttarakhand (originally named Uttaranchal), which split from India's most populous, Ganges Basin, core-area State of Uttar Pradesh on the basis of its highland character and lifeways (Fig. 8B-3).

As recently as the end of 2009, India's central government appointed a committee to consider the creation of a new State of Telangana out of the northwestern sector of Andhra Pradesh. Telangana is an inland area that sees itself as marginalized by coastal elites. The proposed State counts

FIGURE 8B-4   © H. J. de Blij, P. O. Muller, J. Nijman, and John Wiley & Sons, Inc.

about 31 million people and covers approximately one-third of Andhra Pradesh. The issue is fiercely contested, and remained unresolved at press time. Some other demands for independent statehood have gone unrewarded for many decades and flare up occasionally, such as the calls for secession of "Gorkhaland" from the State of West Bengal.

Over the past several years, there have been troubling signs of still another challenge to India's federal government: communist- (avowedly Maoist-) inspired rebellions that seem to be transforming into a coordinated revolutionary campaign. It is known in India as the **Naxalite** movement, named after a town in the State of West Bengal where it was founded in the 1960s. Mainly active in India's poorest and most disaffected States—such as Bihar, Jharkhand, and Andhra Pradesh—the Naxalites appeal to the poor and other minorities (especially tribal people), whose plight is blamed on India's elites and neoliberal economic policies. According to Indian observers, they remain active in one-third of India's more than 600 districts (the administrative unit below the State) and maintain a violent presence in about 90 districts (Fig. 8B-4). They blow up railroad tracks, attack police stations, and intimidate villagers. In early 2010 alone, they killed 24 police officers in West Bengal, murdered 12 villagers in Bihar, and ambushed and killed 76 Indian soldiers in the eastern State of Chhattisgarh. The death toll for all of 2010 reached nearly 1000. The Indian government has declared the Naxalites to be the country's greatest internal security threat. Stability is key to India's economic prospects, and this is what the Naxalites try to undermine.

Although the Maoist Naxalite rebels denied involvement, Indian authorities blamed their leadership for planning and executing the carefully timed destruction of two trains in May 2010, their most audacious terrorist act up to that time. In this dramatic photo, the cars of the rust-colored train, struck by an explosion on the track about 150 kilometers (90 mi) west of Kolkata on the way to Mumbai, lie crushed against those of the oncoming blue train, which could not stop to avert even greater carnage. More than 150 passengers were killed and many more injured. The State of West Bengal, where this tragedy happened, is a hotbed of Naxalite antigovernment activity.

© AP/Wide World Photos

Communal Tensions

The term **communal tension** [1] (communal violence) refers to the several different kinds of conflicts that recur among India's highly diverse sociocultural groups. Most commonly, these conflicts have a base in (politicized) religion, but they can also be caste-based.

The Sikhs

The Sikhs (the word means "disciples") adhere to a religion that was created about five centuries ago to unite warring Hindus and Muslims into a single faith. They rejected what they considered to be the negative aspects of Hinduism and Islam, and Sikhism gained millions of followers in the Punjab and adjacent areas. During the colonial period, many Sikhs supported the British administration of India, and by doing so they won the respect and trust of the colonialists, who employed tens of thousands

of Sikhs as soldiers and policemen. By 1947, there was a large Sikh middle class in the Punjab. Today they still exert a strong influence over Indian affairs, far in excess of the less than 2 percent of the population (about 21 million) they constitute.

For a time after India's independence, the Sikhs created India's most serious separatist problem, demanding the formation of an autonomous State they wanted to call *Khalistan*. The government sought to defuse the situation in 1966 by dividing the original State of Punjab into a Sikh-dominated northwest—which retained the name Punjab—and a Hindu-dominated southeast (Haryana; see Fig. 8B-3). But the conflict intensified again in the 1970s when Prime Minister Indira Gandhi declared a "national emergency" and imposed strong centralized rule across the federation. Tensions culminated in an attack by federal military forces on Sikh rebels holed up in the Golden Temple in the city of Amritsar, Sikhdom's holiest site. In the wake of this attack, Mrs. Gandhi was assassinated by her own Sikh bodyguards in 1984 and more violence followed. Since the 1980s, both sides have sought to gain better understanding.

The Muslims

When India became a sovereign state and the great population shifts across its borders had ended, the country was left with a Muslim minority of about 35 million widely dispersed throughout the country. By 1991, that minority had grown to nearly 100 million, representing 11.7 percent of the total population. The current Muslim population is estimated to be 180 million, about 14.7 percent, and clearly it continues to grow faster than the Indian population in general. Muslims are an actual majority in Jammu and Kashmir, Asom, and West Bengal, while their largest absolute numbers are in the bigger states of Uttar Pradesh, Bihar, and Maharashtra. To India's great geographic advantage, the Muslim minority is not regionally concentrated, avoiding what otherwise could lead to a dangerous secession movement.

The position of Muslims in Indian society is not easy to separate from India's relations with Pakistan, from the issue of Kashmir, and from acts of Islamic terrorism that have haunted India in recent years (especially the attacks on Parliament in 2001 and Mumbai's upscale hotels in 2008). At present, India holds the unenviable distinction of being the most frequently targeted country in the world for terrorist attacks. As a result, relations between Muslims and Hindus are worsening. Concern over linkages between foreign (read Pakistani) terrorists and local Muslim subversives is changing Hindu attitudes and tactics, a topic elaborated in the next section.

The most serious threat to Muslim integration into Indian society arises from their comparatively low level of education and poor economic standing. Official statistics show that less than 4 percent of Muslims nationwide have completed secondary school. And according to one recent government report, Muslims have now become as poor as the members of India's lowest-ranked Hindu castes. Even as small cadres of Muslim university graduates have become successful professionals and have increased their representation among high-profile politicians, film stars, and sports icons, the vast majority appear to be stuck near the bottom of the social ladder.

Hindutva

Hindu extremism or fundamentalism may seem like a contradiction in terms, but a movement has come to the fore in recent decades that seeks to remake India as a society in which Hindu principles prevail. *Hindutva* [2], or Hinduness, is variously expressed as Hindu nationalism, Hindu heritage, or Hindu patriotism. It has been the guiding agenda for the Bharatiya Janata Party (BJP), a powerful force in national politics and in big States like Maharashtra, Gujarat, and Madhya Pradesh. *Hindutva* fanatics want to impose a Hindu curriculum on schools, change the flexible family law in ways that would make it unacceptable to Muslims, inhibit the activities of non-Hindu religious proselytizers, and forge an India in which non-Hindus are essentially outsiders.

This naturally worries Muslims as well as other minorities, but it also concerns those who understand that India's secularism—its separation of religion and state—is indispensable to the survival of democracy. Moderate Hindus and non-Hindus in India oppose such notions, which are as divisive as any India has faced. It is easy to see how Hindu hardliners and Islamic fundamentalists can fuel each other's agendas and maintain a vicious cycle of conflict.

From the Field Notes . . .

© Jan Nijman, Photo by Zach Woodward

"It is a Friday afternoon as we make our way through a main thoroughfare in one of Mumbai's biggest slums. Dharavi 'houses' approximately half a million people on less than 2 square kilometers! About 20 percent are Muslims and the overwhelming majority of all others are dalits. Increasingly, the Muslims live spatially segregated in their own tightly knit neighborhoods. There are several mosques but there is little space inside them. So the men, taking a break from work nearby, place their mats alongside the road and prepare for prayer outdoors. Having a faith is important to people living and working in Dharavi's filthy and impoverished environs, especially for the Muslim minority."

www.conceptcaching.com

Recent State and national elections have seen a decline in the fortunes of the BJP and, more importantly, an internal party struggle between moderates and *hindutva* hardliners in which the moderates have seemed to prevail. Overall, Indian voters have not rushed to embrace this radicalization of Hinduism, a confirmation of the continuing robustness and vitality of India's democracy.

The Persistence of Caste

Hinduism's benign and admirable properties combine with a system of social stratification that is generally derided in the West (and by a large number of Indians as well). Under Hindu dogma, castes are fixed layers in society whose ranks are based on ancestries, family ties, and occupations. The **caste system** [3] has its origins in the early social divisions into priests and warriors, merchants and farmers, and it is also thought to have had a racial basis (the Sanskrit term for caste, *varna*, means color). More specifically, caste became associated with specific professions and, over the centuries, its complexity grew until India had thousands of castes, some of them with a few hundred members and others containing millions.

Hindus believe in reincarnation, and a person is thought to be born into a particular caste based on his or her actions in a previous existence. Hence it would not be appropriate to counter such ordained caste assignments by permitting movement (or even contact) from a lower caste to a higher one. Persons of a particular caste could perform only certain jobs, wear only certain clothes, and worship only in prescribed ways at particular places. They or their children could not eat, play, or even walk with people of a higher social status.

The *untouchables* occupying the lowest tier of all were the most debased, wretched members of this rigidly structured social system. Indeed, the term "untouchable" acquired such negative connotations that it was replaced several times. Mahatma Gandhi, a powerful critic of the system, introduced the term *harijans*, meaning children of God. But more recently this label was thought to have a condescending connotation, and it gave way to the term *dalits* (the oppressed), indicating a greater sense of awareness and assertiveness among them.

In the isolated villages deep in the countryside, dalits still suffer from severe discrimination and harsh treatment by higher castes. Children often are made to sit on the floor of their classroom (if they go to school at all); dalits are not allowed to draw water from the village well because they might pollute it; they must take off

The Holy Ganges

STAND ON THE banks of the Ganges River in Varanasi, Hinduism's holiest city, and you will see people bathing in holy water, drinking it, and praying as they stand in it. The Ganges (*Ganga*, as the Indians call it) is Hinduism's most sacred river. At Varanasi, Prayagraj (Allahabad), and other cities and towns along the Ganges, the riverbanks are lined with Hindu temples, decaying ornate palaces, and dozens of wide stone staircases called *ghats*. These stepped platforms lead down to the water, enabling thousands of bathers to enter the river. They come from the city and from afar, many of them pilgrims in need of the healing and spiritual powers of the water. It is estimated that more than a million people enter the river somewhere along its 2600-kilometer (1600-mi) course every day. During religious festivals, the number may be ten times as large.

The river's ceaseless flow and spiritual healing power are revered by Hindus, its waters believed to purify the soul. Therefore, tradition has it, the river's water is immaculate, and no amount of human (or other) waste can pollute its sacred quality. But at a nonspiritual level, this is one of the most toxic of the world's major rivers. By the time the fresh, clear glacial meltwaters from the Himalaya reach Varanasi, they have been used up almost entirely in upstream industrial cities like Delhi and Kanpur, and replaced with effluents. As far back as 1986, then-Prime Minister Rajiv Gandhi launched a major scheme to reduce the level of pollution in the river, a decade-long construction program

The stone steps leading into the Ganges' waters in Varanasi. This is India's holiest city, and millions descend these ghats every year.

© Jan Nijman

of sewage treatment plants and associated facilities. At the same time, however, industrial waste and sewage have steadily increased as well.

To Hindu believers, this is all immaterial; to them it is all about the divinity of the *Ganga*. Their congregation on the ghats, every single morning of the year, is an occasion for joy and a spectacle to watch.

their shoes, if they own any, when they pass higher-caste houses; and they cannot take jobs in professions other than the one they were born into (sweeping, cleaning, and the like).

Today, dalits are estimated to constitute more than 15 percent of all Hindus while Brahmins, the highest caste, account for a comparable share. The rest of the population occupies a wide range of in-between castes. The Indian government officially abolished castes at independence, but the system has proven hard to dismantle. Many dalits have chosen to convert to other religions, most notably Buddhism. It is estimated that nearly 90 percent of Buddhists in India were originally dalits, but it appears that even after converting they do not lose this stigma.

Successive Indian governments have now introduced an elaborate system of affirmative action on behalf of *Scheduled Castes* (the official government label for dalits). This effort has had more effect in urban than in rural areas of India. In any case, dalits now have reserved for them places in the schools, a fixed percentage of State and federal government jobs, and a quota of seats in national and State legislatures. These jobs are often highly desirable because they tend to be white-collar and much better paid than those generally available to dalits.

Because of the great number of minorities in India, affirmative action schemes have become increasingly complex over time and are at the center of seemingly never-ending debates. In addition to the existing quota of 15 percent of federal jobs for dalits, the central government is now considering whether it should reserve public-sector jobs for Muslims and other religious minorities such as Sikhs. In another proposal that is presently being debated, one-third of all Parliament seats would go to women. This was protested by some Muslim organizations that feared there would not be enough eligible Muslim women, and they subsequently demanded a subquota for them. In the world's largest and most diverse democracy, politics can be a highly complicated matter!

Just how far India's political pendulum can swing was shown in provincial legislative elections in Uttar Pradesh State in 2007, where a dalit party won an absolute majority and where a woman named Mayawati Kumari was the first dalit to become a chief minister of one of India's major States. It was a stunning victory that made headlines throughout the country and shook up the political establishment, revealing the growing power of the lowest castes and marking a turning point for India's representative government. But most lower-caste Indians are still faced with very limited opportunities, widespread discrimination, and abject poverty. In traditional India, caste provided stability and continuity; in contemporary India, it constitutes an often painful and difficult legacy.

Economic Geography

East and West

The most commonly cited, and most clearly evident, regional division of India is between north and south. The north is India's heartland, the south its Dravidian appendage; the north speaks Hindi as its *lingua franca*, the south prefers English over Hindi; the north is bustling and testy, the south seems slower and less agitated.

But there is another, as yet less obvious, but potentially more significant divide across India. In Figure 8B-5 draw a line from Lucknow, on the Ganges River, south to Chennai, near the northern tip of Tamil Nadu State (cf. Fig. 8B-1). To the west of this line, India is showing signs of economic progress, the kind of productive activity that has brought Pacific Rim countries such as Thailand and Indonesia a new life. To the east, India has more in common with the less-promising countries that also face the Bay of Bengal: Bangladesh and Myanmar (Burma). The manufacturing map may seem to suggest that much of India's industrial strength lies near Kolkata, but the heavy industries built here by the state in the 1950s are now outdated, uncompetitive, and in steady decline. The State of Bihar represents the stagnation that afflicts much of India east of our line: by several measures it ranks among the poorest of the 28 States.

Compare this to western India. The State of Maharashtra, the hinterland of Mumbai, leads India in many categories, and Mumbai leads Maharashtra. Many smaller, private industries have emerged here, manufacturing goods ranging from umbrellas to satellite dishes and from toys to textiles. Across the Arabian Sea lie the oil-rich economies of the Persian Gulf. Hundreds of thousands of workers from western India have found jobs there, sending remittances back to families from Punjab to Kerala. More importantly, many have used their foreign incomes to establish service industries back home. Outward-looking western India, in sharp contrast to the inward-looking east, has begun to establish additional ties to the outside world. Satellite links have propelled Bengaluru (formerly Bangalore) to become the center of a growing software-producing complex that reaches world markets. Maharashtra's economic success also has spilled over into Gujarat, its northern neighbor, and even landlocked Rajasthan (the next State to the north beyond Gujarat) is experiencing the beginnings of what, by Indian standards, is a boom.

Life Is in the Harvest

Agriculture provides more jobs in India than any other economic sector, and India's fortunes (and misfortunes) remain strongly tied to farming. More than two-thirds of the population still lives on (and from) the land, spread in and around the country's 600,000 villages. There, traditional farming methods persist, yields per hectare remain among the world's lowest, and hunger and malnutrition still afflict millions even as grain surpluses accrue. The relatively few areas of modernization, as in the wheatlands of the Punjab, are islands in a sea of stagnation. Land reform has essentially failed; roughly one-quarter of India's entire cultivated area, including much of the best land, is still owned by less than 5 percent of the country's landholders, who have much political influence and obstruct redistribution.

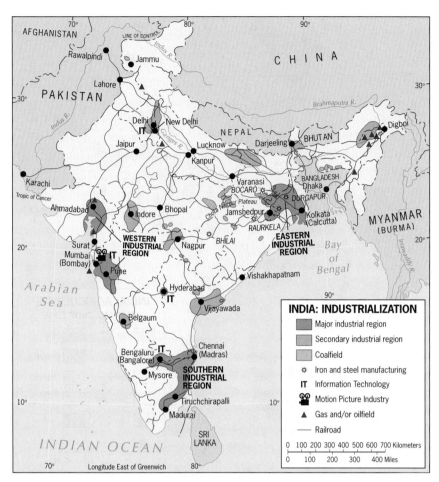

FIGURE 8B-5   © H. J. de Blij, P. O. Muller, and John Wiley & Sons, Inc.

Perhaps half of all rural families own less than 1 hectare (2.5 acres) or no land at all; and an estimated 175 million live and work as tenants, always uncertain of their fate.

As everywhere, agriculture depends strongly on physiography and climate, and agricultural fortunes vary significantly across the country. In India, the monsoon is absolutely critical, its abundance and timing a reliable predictor of the harvest. Figure 8B-6 maps India's agricultural geography and reflects the rainfall patterns and monsoonal cycle depicted in Figure 8A-3. Rice dominates along the Arabian Sea-facing southwestern coast (on the rainy side of the Western Ghats) and in the monsoon-drenched peninsular northeast; where drier conditions develop, wheat and other grains prevail.

Even though the country has a good variety of crops, yields are often too low as a result of inefficient land ownership and small farmers' lack of access to inputs such as fertilizer, irrigation equipment, and machinery. If there is any to sell, getting produce to market is yet another struggle for millions of farmers. In 2010, almost half of India's 600,000 villages could not be accessed by truck let alone automobile, and in this era of modern transportation animal-drawn carts still far outnumber motor vehicles nationwide.

Manufacturing and Information Technology

India's industrial sectors have significantly improved over the past two decades, but the pace of change is still too slow to meet the country's needs. The map of manufacturing geography (Fig. 8B-5) is a legacy of colonial times, with coastal Mumbai, Kolkata, and Chennai anchoring major industrial zones, and textiles—the entry-level industry of disadvantaged countries—dominating the manufacturing scene. Other industries, such as steel, machinery, and building materials, have become much stronger, and some of the companies involved have even become major global players. For example, Tata Steel and Mittal Steel (now officially called ArcelorMittal), rank among the half-dozen biggest steel producers in the world.

These changes were made possible through major policy changes in the early 1990s, when India liberalized its economy following four decades of stifling overregulation and inefficiency. Along with many other developing countries, India embraced neoliberalism: trade barriers came down, state companies were privatized, and foreign investment was encouraged. Although it is not yet clear what the impact will be on India's massive working class, there is no doubt that many companies became far more competitive and profitable. This, in turn, increased the government's tax revenues, allowing it to invest in, among other things, badly needed infrastructure projects. Since 2000, India's economy has grown faster than ever before. Much of this growth is related to the information technology sector.

FIGURE 8B-6   © H. J. de Blij, P. O. Muller, and John Wiley & Sons, Inc.

From the Field Notes . . .

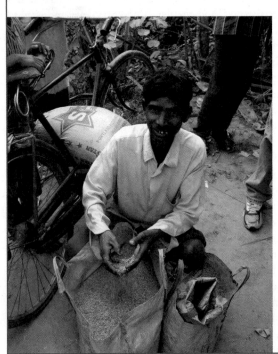

© Jan Nijman

"During my travels in northeastern India in early 2011, I visited some agricultural areas in the Brahmaputra Valley that benefit from plentiful irrigation. Tea and rice are two major crops. Because big companies run the large tea plantations, many of the rice farmers have to make do with small plots of land. Here a farmer in Asom State shows off part of his rice harvest. I encountered him and several others at a small machine-operated mill shop. Note that the rice is brown—milling it removes the bran layer and makes the rice white. 'We have the best rice in all of India,' he proudly said, as he slowly poured it from his hand back into the bag. Having enjoyed some excellent meals in the area, I could not disagree."

www.conceptcaching.com

From the Field Notes . . .

"With the help of a real estate agent, I toured some of the new residential developments in India's biggest city, Mumbai. The new urban middle class wants better and more spacious housing, but given the lack of space in the city it has to be a high-rise far away from the center. From a new construction site I looked out onto the northeastern edge of the city, marked by an already completed condominium complex. These homes are of a much better quality than the average dwelling in Mumbai, containing most of the luxuries we take for granted in the United States. Two-bedroom apartments sell for about U.S.$75,000 and up—a huge sum of money even for India's upwardly mobile classes and totally out of reach for the city's masses. Half of Mumbai's 20-plus million people live in substandard housing, and many in the kind of slums you see in the foreground, nestled up against the coveted middle-class residences. As the new middle class takes off, the contrasts are getting sharper."

www.conceptcaching.com

© Jan Nijman

India's information-technology (IT) industries, clustered in and around Bengaluru (Bangalore), Hyderabad, Mumbai, and New Delhi, draw much international attention, in part because of the large-scale **outsourcing [4]** of U.S. jobs to India. Leading Indian software companies such as Infosys, Wipro, and Satyam are now household names all across urban India. The growth of software and IT services has been spectacular and now accounts for about 8 percent of GDP and no less than one-quarter of merchandise exports. But this sector employs only around 2 percent of the workforce, with about ten times as many people holding government jobs!

What India needs far more, in terms of employment, are manufacturing industries competitively selling goods on world markets, putting tens of millions to work, and transforming the economy. Clearly, India is very different from China. The two may have comparably-sized populations, but the Chinese economy employs 12 times as many workers in manufacturing and it has a much larger urban middle class. India, on the other hand, has shown spectacular growth; but too much of it is confined to IT, and too much benefits only a small, highly educated urban elite.

Urbanization

India is famous for its great and teeming cities, but the country is not yet an urbanized society. Only 29 percent of the population currently lives in cities and towns—compared to an average of about 80 percent across the developed world.

But whereas rural-urban migration has long stabilized at a low level in the West, the rate of urbanization in India is much higher. People by the hundreds of thousands are arriving in the already overcrowded cities, swelling urban India by about 3 percent annually—twice as fast as the country's overall population growth. Not only do the cities attract as they do everywhere, but many villagers are also driven off the land by desperate conditions in the countryside. As these migrants manage to establish themselves in Mumbai or Kolkata or Chennai, they help their relatives and friends to join them in squatter settlements that often are populated by newcomers from the same locality, bringing their language and customs with them and cushioning the stress of the move.

As a result, India's cities display staggering social contrasts. Squatter shacks without any amenities crowd cheek-by-jowl against the walls of modern high-rise apartment buildings and condominiums (see photo above). Hundreds of thousands of homeless people roam the streets and sleep in parks, under bridges, and even on sidewalks. As crowding intensifies, social stresses multiply.

Figure 8B-3 displays the distribution of major urban centers in India. Except for Delhi–New Delhi, the largest cities have coastal locations: Kolkata dominates the east, Mumbai the west, and Chennai the south. The overriding influence of these coastal cities is a colonial legacy. But urbanization also has expanded in the interior, notably in the core area. In 2010, even though less than 30 percent of

AMONG THE REALM'S GREAT CITIES . . .

MUMBAI (BOMBAY)

IN SOME WAYS, Mumbai is a microcosm of India, a burgeoning, crowded, chaotic, fast-moving agglomeration of humanity. Shrines, mosques, temples, and churches evince the pervasive power of religion in this multicultural society. Street signs come in a bewildering variety of scripts and alphabets. The Victorian-Gothic architecture of the city center is a legacy of the British colonial period (see photo, p. 411). Creaking double-decker buses compete with ox-wagons and handcarts on the congested roadways. The throngs on the sidewalks—businesspeople, holy men, sari-clad women, beggars, clerks, homeless wanderers—spill over into the streets.

In precolonial times, fishing folk living on the seven small islands at the entrance to this harbor named the place after their local Hindu goddess, Mumbai Devi. The Portuguese, first to colonize it, called it Bom Bahia, "Beautiful Bay." The British, who came next, corrupted both names to form *Bombay*, and so it remained for more than three centuries. In 1996, the government of the State of Maharashtra, with support from the federal authorities, changed the city's name back to Mumbai.

With nearly 21 million people spread across the metropolis, this is India's second-largest urban region. Think of Mumbai as a combination of New York and Los Angeles—India-style. This is the country's commercial and financial center with most of the banks, the leading stock exchange, and most foreign multinationals. And it is also India's most glamorous city, owing in no small part to the presence of Bollywood, center of the Hindi film industry that produces more movies (for more viewers) than anywhere else on Earth.

The 2009 movie *Slumdog Millionaire*, a box office hit around the world, focused on the darker side of this megacity, the approximately 8 million slum dwellers who try to make a living alongside the movie stars and the burgeoning middle classes. In part because its peninsular geography imposes strict limits, Mumbai is an extremely crowded city with about seven times the density of major cities in the United States—and contains slums marked by much higher densities than that. Anyone visiting the city is bound to be overwhelmed by the sheer scale of the poverty, which stands in sharp contrast to the lavish, hugely expensive homes of the rich.

Mumbai is very much a product of colonial times. During the Industrial Revolution, England needed raw materials at low prices and India was a main supplier, especially of

© H. J. de Blij, P. O. Muller, and John Wiley & Sons, Inc.

cotton. Much of the raw cotton was processed in the textile mills of Bombay and then shipped to London, Manchester, or Liverpool to be converted into finished products. From the 1860s on, Bombay became one of the most important colonial cities and a vital node in the British Empire. The imprint of that era is still visible in the architecture of the Fort district in the south, the port, the railways, and the textile mills lining the rail corridors farther north.

Today not much is left of the textile industry in the urban economy. Instead, Mumbai has become the country's most important world city with its growing sectors of finance and producer services that include accounting, advertising, and consulting.

its citizens were urbanites, India had 46 metropolitan areas containing populations in excess of 1 million.

When you arrive in any Indian city, you are struck by the number of small shops everywhere—tiny businesses wedged into every available space in virtually every non-public building along every street. Even the upper, walk-up floors are occupied by shops, their advertising signs suspended from windows and balconies. According to a study by the Indian government's Department of Consumer Affairs, India has the highest density of retail outlets of any country in the world: a total of 15 million shops (compared to well below 1 million in the United States, where the marketplace is 13 times richer). After farming, the retail sector is India's largest provider of jobs.

What keeps all these small stores in business? Most of them earn very little and can afford to stay open only because they are part of what economic geographers call the **informal sector [5]**—they are essentially unregistered, pay no rent and probably no taxes, use family labor, have been handed down through generations, and keep going because India's economic geography, bound by longstanding protective government regulation, has been slow to change. When you are in India you may wonder how so many shops can survive, but the answer is in the throngs of people on the sidewalks (spilling over into the clogged traffic in the streets). Not many of these people are wealthy, but all of them need basic goods and some can afford small luxuries, and so the shops tend to be busy all day.

This bustling retail scene notwithstanding, the newest statistics confirm that change is under way, especially in the cities. India's middle class, now estimated to include some 300 million people (equal to the entire U.S. population), is expanding rapidly, and something quite new is beginning to appear in the urban landscape: shopping malls (see photo above). As recently as 2000, this vast country containing more than one-sixth of humankind did not have a single shopping center! But by 2005 the 100th had opened its doors, and by 2010 more than 500 were in operation. You will see American fast-food restaurants among the establishments in these malls as well as the brand names of numerous other foreign companies, proving that globalization has already breached India's walls. Nonetheless, if 300 million middle-class Indians can afford to visit a mall, more than 700 million Indians cannot, and the gap is widening by the day. Can India's economic transformation be accomplished without severe social dislocation?

Infrastructural Challenges

As the second decade of the twenty-first century opened, India was in the midst of constructing a nationwide four-lane superhighway—the *Golden Quadrilateral*—that is going to link the four anchors of the urban system (Delhi, Mumbai,

Sandesh Kadur–wildlife photographer and filmmaker, Bengaluru, India Photo Courtesy Sandesh Kadur/www.felis.in

"I was born and raised in Bangalore [Bengaluru], India. My passion in life is natural history. Most people in the world associate wildlife with Africa and when you mention India they think of a huge human population, or of poverty, or the IT businesses. But India is also home to the largest population of Asian lions, elephants, tigers and countless other species, many of which are only found here. Indians traditionally revere nature; maybe that is why we still have so much wilderness and wildlife despite there being over a billion people. Mahatma Gandhi once said, "Nature can provide for every man's need, but not for every man's greed." And this is what I believe. We must be careful to protect nature from our fast growing cities, expanding road systems, and pollution. There are many beautiful nature reserves in India but my favorite place is

Greater One-Horned Rhino with Egret, Kaziranga National Park, India Photo Courtesy Sandesh Kadur/www.felis.in

Kaziranga in Assam [Asom]. It has some of the highest wildlife densities in the world. For the past 10 years, I have been a professional wildlife photographer and filmmaker, having worked with the BBC, National Geographic, and other organizations. My goal is to further nature education through powerful photography and film documentaries."

From the Field Notes . . .

© Jan Nijman

"As I set foot into this spacious (and partially air-conditioned) mall from the steamy and crowded Andheri Link Road in Mumbai, I thought of the implications—is this the future in a land of small family-owned shops where 'retail' has a different and historic meaning? Infinity Mall, an upscale shopping center opened in 2005 in suburban Mumbai, caters to the megacity's new middle class. My Indian friend Pankaj thought it was all for the better. 'On the weekends my wife and I often come here. There is a nice food court and the shopping is great. These are the things you Americans want and these are the things we Indians want.' Malls are now springing up across India's major cities, transforming urban landscapes as well as consumer behavior."

www.conceptcaching.com

KOLKATA (CALCUTTA)

THE NAME CALCUTTA seems to invoke all that can go wrong in large cities: poverty, disease, pollution, congestion, and lack of planning. The city was renamed Kolkata in 2001, one in a series of cities in India whose names were "decolo-nized" and returned to vernacular pronunciation. But that may have done little to change its reputation. It wasn't always so.

Calcutta's heyday began in 1772 when the British made it their colonial capital. The British chose the site, 130 kilometers (80 mi) up the Hooghly River from the Bay of Bengal and less than 9 meters (30 ft) above sea level, not far from some unhealthy marshes but well placed for com-merce and defense as well as accessibility to the famous tea plantations in the northeast. When the British East India Company was granted freedom of trade in this populous hin-terland, the British sector of the city was drained and raised, and so much wealth accumulated here that Calcutta became known as the "city of palaces." Outside the British town, rich Indian merchants built magnificent mansions.

But even in those times the city knew incredible pov-erty and deprivation. Beyond the "palaces" lay neighborhoods that were often based on occupational caste, whose names are still on the map today (such as Kumartuli, the potters' district). Almost everywhere, on both banks of the Hooghly, lay the huts and hovels of the poorest of the poor. Searing social contrasts characterized Calcutta and the city was no-torious for its lack of hygiene and frequency of epidemics.

The twentieth century was not kind to Calcutta. In 1912, the British moved their colonial capital to New Delhi. The 1947 partition that created Pakistan also created then-East Pakistan (now Bangladesh), cutting off a large part of Calcutta's hinterland and burdening the city with a flood of refugees. The Indian part of the city had arisen virtually without any urban planning, and the influx created almost unimaginable conditions.

Kolkata today counts 16.1 million residents, and the city is in some ways left behind in India's modernization. It was surpassed by Bombay (now Mumbai) in the industrial

© H. J. de Blij, P. O. Muller, and John Wiley & Sons, Inc.

era, and the current information age belongs to Bengaluru (formerly Bangalore) and Hyderabad. But despite its poverty and urban problems, Kolkata still has a reputation as India's true cultural capital, home of the poet Rabindranath Tagore (who in 1913 became the first non-European Nobel laure-ate), the well-known 1950s art-film maker Satyajit Ray, and the great musician Ravi Shankar (legendary maestro of the stringed instrument known as the sitar).

Chennai, and Kolkata) and in the process connect 15 other major cities along this all-important route (Fig. 8B-3). The impacts of this project are multiple: it is expanding urban hin-terlands; commuters are using it to travel farther to work than ever before; many once-remote rural areas now have a link to markets; and it is accelerating the rural-to-urban migration flow that will transform India in the decades to come.

Improving India's infrastructure, however, is not enough to overcome all of the serious impediments to the country's economic advancement. Take another look at Fig-ure 8B-3 and note that the Golden Quadrilateral crosses numerous State boundaries. In the United States, we are used to seeing thousands of trucks on the interstate highways,

crossing from one State into another without slowing down; there are truck-weighing stations along these expressways, but in the newest ones all the truck has to do is slow down enough for electronic surveillance to record its passing.

In India, truck drivers face a very different expe-rience. It can take as long as nine days, including more than 30 hours waiting at State-border checkpoints and tollbooths, for a loaded truck to travel from Kolkata to Mumbai via Chennai, or less than the distance from Los Angeles to New Orleans on Interstate-10. Drivers are sub-ject to daunting piles of paperwork and repeated demands for bribes, and when mechanical problems occur it may take days to get a truck back on the road. It will take more than expanding India's highway network to improve the circulation the country so desperately needs.

The Energy Problem

If you have a friend in (or from) India, you know someone who is familiar with power outages. They are a way of life in India, where electricity demand routinely exceeds the available supply, where governments cannot bring themselves to require customers to pay for the actual cost of the power they consume, and where power grids, generating equipment, and other associated infrastructure are in bad shape. On top of all this, keep in mind that hundreds of millions of villagers still have no electrical supply at all.

And yet electrical power is key to India's modernization. Already, some foreign companies doing business in India are importing their own generators, and others are discouraged from investing in factories and other facilities because the power supply is so unreliable. Most of India's electricity is generated in thermal plants burning coal, oil, or natural gas; about 25 percent comes from hydroelectric sources, and about 3 percent from nuclear plants. The problems are many: India has substantial coal deposits, but the railroads cannot handle the transport to power plants. So India must import coal, but the ports do not have the required capacity. And India possesses only limited oil and natural gas reserves. Add to this an increasingly inadequate national power-supply grid in a country with a still-exploding population and even faster-growing demand, and you have trouble.

A major remedy, of course, lies in increased oil and gas imports, but here India runs into geopolitical problems. Although the U.S. government was eager to give India leeway in the nuclear arena, the United States made it clear that it does not like India's plan to buy Iranian natural gas via a pipeline across Pakistan. India's other options lie in interior Asia, but those sources are more distant and pipeline construction would involve further diplomatic complications. Once again, the other alternative—importing oil and gas via tankers and ports—is constrained by inadequate infrastructure. Energy will remain a problem for some time to come.

India's Prospects

So what lies in store for India? Some economic geographers suggest that India might leapfrog China and move quickly from an "underdeveloped" to a "postindustrial" information-based economy. Certainly India has the requisite intellectual clout. But to reach India's hundreds of millions of potential wage-earners, India also needs a vigorous expansion of its secondary industries, those that make goods (other than textiles) and sell them at home and abroad. Here's how it may happen: when an economy churns along the way China's has over the past three decades, labor and production costs tend to rise. That causes manufacturers to look for places where cheaper labor will reduce such costs. India, with its long history of local manufacturing, huge domestic market, and vast reservoir of capable labor, would then take its turn on the world stage.

But there are plenty of obstacles. In 2008, a division of one of India's largest companies, Tata Motors, introduced a tiny automobile (the Nano) that would sell for little more than U.S. $2000. Experts internationally praised this vehicle as a marvel of Indian ingenuity and engineering, and when Tata proposed to build the factory to produce it in one of India's poorest States, West Bengal, it planned to boost the job market where this seemed to be most needed. But Tata's planners failed to reckon with the farmers in West Bengal and their political party. First, urged on by political leaders, these farmers refused to negotiate over the price of their land. Next, the company faced rallies and blockades as it tried to persuade locals that the thousands of prospective jobs would more than compensate for the approximately 400 hectares (1000 acres) of land it sought. Then, suddenly, the company withdrew its plans and began considering offers from other States. But the implications were clear: if the Tata Group, an Indian corporation, could not build a factory in an Indian State, would foreign investors looking for cheap labor want to risk similar problems?

Nonetheless, India's prospects are brightening, if not yet in the dramatic terms used in popular-media hype. A middle class of more than 300 million demands goods that range from cell phones (over the past few years, 2.5 million new subscribers were signing up every month) to motor bikes (10,000 per day were being sold in 2009). Even if Tata's mini-car was not going to be built in West Bengal State, Chennai was already becoming an Indian automobile center: Korean and German cars are coming off its assembly lines not only to be marketed in India but also to be shipped to foreign consumers.

India's economy today ranks as the world's sixth-largest, and many economists suggest that, whatever happens, India by 2020 may be in third place. Over the past several years, the economy has grown by an average of 7 percent—less than China's, but far ahead of growth rates in most other parts of the world. As we will see in Chapters 9A and 9B, China's dramatic growth resulted from decisions at the top, a transformation planned and implemented in controlled detail. In India the economy is growing from the bottom up, with all the traditional chaos that makes it a country like no other. As with China's provinces, some of India's States will advance ahead of others, and the country's already incredible socioeconomic contrasts will intensify. But over time, India could achieve what China hitherto has not: an economic and cultural geography of consensus.

■■ BANGLADESH: CHALLENGES OLD AND NEW

On the map of South Asia, Bangladesh looks like another State of India: the country occupies the area of the **double delta [6]** of India's great Ganges and Brahmaputra rivers, and India almost completely surrounds it on its landward side (Fig. 8B-7). But Bangladesh is an independent country, born in 1971 after its brief war for independence against Pakistan, with a territory about the size of Wisconsin. Today

FIGURE 8B-7   © H. J. de Blij, P. O. Muller, and John Wiley & Sons, Inc.

it remains one of the poorest and least developed countries on Earth, with a population of 170 million that is growing at an annual rate of 1.5 percent.

A Vulnerable Territory

Not only is Bangladesh a poor country; it also is highly susceptible to damage from natural hazards. During the twentieth century, eight of the ten deadliest natural disasters in the entire world struck this single country. The most recent of these occurred in 1991 when a cyclone (as hurricanes are called in this part of the world) killed more than 150,000 people. Smaller cyclones strike several times every year and almost routinely kill anywhere between dozens and thousands of people.

The reasons for Bangladesh's vulnerability can be deduced from Figures 8B-7 and 8A-1. Southern Bangladesh consists of the deltaic plain of the combined Ganges–Brahmaputra river system, integrating fertile alluvial (river-deposited silt) soils that attract farmers with the low elevations that endanger them when water rises. The shape of the Bay of Bengal forms a funnel that steers cyclones and their storm surges of wind-whipped water barreling into the delta coast. Without money to build seawalls, floodgates, elevated shelters in sufficient numbers, or adequate escape routes, hundreds of thousands of people are at continuous risk, with deadly consequences.

Limits to Opportunity

Bangladesh remains a nation of subsistence farmers. Only about one-fourth of the population lives in urban settlements, and more than half of the workforce is engaged in agriculture. Dhaka, the megacity capital, and the cities of Chittagong, Rangpur, Khulna, and Rajshahi are the only urban centers of consequence. Moreover, Bangladesh has one of the highest **physiologic densities [7]** in the world (1867 people per square kilometer/4835 per square mile), and only

● AMONG THE REALM'S GREAT CITIES . . .

DHAKA

DHAKA (formerly spelled Dacca) is situated in the heart of Bangladesh in the northern sector of the double delta formed by the Ganges and Brahmaputra. It is the economic, political, and cultural focus of the country, and with a population of some 15.4 million it is by the far the biggest city.

Known as the "city of mosques and muslin" (a type of cotton), the modern history of Dhaka began in the early seventeenth century when the area came under the rule of the Islamic Mughals, along with what is today northern India. Dhaka occupied an advantageous position on the main waterways, and to this day the local Buriganga River is at the center of the city's activities. The oldest section of Dhaka lines the north bank of the waterfront. The British turned the place into a provincial capital of sorts, and their legacy can still be seen in some of the colonial architecture near the old center alongside the remaining Mughal structures. At independence in 1947 the city became the administrative capital of East Pakistan, and in 1971 it became the seat of government for the newly formed state of Bangladesh.

As part of the great double delta, Dhaka lies close to sea level and is prone to flooding during the summer monsoon. Every so often, major cyclones barrel in from the Bay of Bengal and cause massive human and material devastation. The greatest disaster in modern times was triggered by the Bhola cyclone of 1970 that killed as many as 500,000 of the country's citizens, destroyed hundreds of thousands of homes, and flooded most of the city.

Dhaka has grown rapidly in recent decades as it attracted large numbers of migrants from rural areas, and the city gradually expanded northward. At the time of partition, many Hindus left and Muslims arrived, and now more than 90 percent of the population adheres to Islam. Religious

© H. J. de Blij, P. O. Muller, J. Nijman, and John Wiley & Sons, Inc.

fanaticism is not as widespread as it is in Pakistan. The main schism here is between the tiny rich minority and the huge poverty-stricken majority. Dhaka is also known as the bicycle-riksha capital of the world: some 400,000 colorfully painted rikshas criss-cross the thoroughfares of this city on a daily basis—surely the cheapest mode of urban transportation on the planet.

higher-yielding varieties of rice and the introduction of wheat in the crop rotation (where climate permits) have improved diets and food security. But diets remain poorly balanced and, overall, the country's nutrition level is unsatisfactory.

Bangladesh's birth rate has come down substantially in the past few decades but is still rather high. No less than 70 percent of the population is under 35 years of age, which translates into a significant demographic burden. At the same time, per capita incomes have risen markedly, yet remain well below those in India or Pakistan. The troubled textile industry provides most of the country's foreign revenues, along with remittances from Bangladeshis working abroad, but the once-thriving jute industry continues its decline. The discovery of a natural gas reserve is now the subject of a national debate: home consumption or money-making export?

Bangladesh is an overwhelmingly Muslim society, though not one dominated by revivalists (for example, 30 seats in the national legislature are reserved for women).

However, Bangladesh's politics are chaotic, patronage-driven, corrupt, and seemingly beyond redemption. Foreign aid, essential to the country's survival, pours in, but the country remains mired among the world's lowest-income societies. Its relations with neighboring India have at times been strained over water resources (India controls the Ganges, which is Bangladesh's lifeline), cross-border migration (one-tenth of the population is Hindu), and transit between parts of India across Bangladesh's north (refer to Fig. 8B-5 to see the reason). Bangladesh faces an uphill battle to improve people's livelihoods. For most Bangladeshis, survival is the leading industry and all else is luxury.

▪▪ THE MOUNTAINOUS NORTH

As Figure 8B-1 shows, a tier of landlocked countries and territories lies across the mountainous zone that walls India off from China. One of them, Kashmir, is in a state of near-war. Another, Sikkim, was absorbed by India in 1975 and made into one of its federal States. But Nepal and Bhutan retain their independence.

Nepal

A **buffer state** [8] between India and China, Nepal exhibits an internal politics that is increasingly a reflection of the competitive influences of these two giant neighbors. Culturally and, until recently, economically, Nepal lies within India's sphere of influence, but China is increasingly making its presence felt. Maoist groups are supported by China but loathed by India, resulting in fractious and contentious governance. China, as it does just about everywhere in the world nowadays, is using its economic clout as leverage for growing political influence.

Nepal, located just northeast of India's Hindu core, has a population of 29 million and is the size of Illinois. It has three geographic zones (Fig. 8B-8): a southern, subtropical, fertile lowland called the Terai; a central belt of Himalayan foothills with swiftly flowing streams and deep valleys; and the spectacular high Himalaya itself (topped by Mount Everest) in the north. The capital, Kathmandu, lies in the east-central part of the country in an open valley of the central hill zone.

Nepal, birthplace of Buddha, is materially poor but culturally rich. The Nepalese are a people of many origins, including India, Tibet, and interior Asia. About 80 percent are Hindu, and Hinduism is the country's official religion; but Nepal's Hinduism is a unique blend of Hindu and Buddhist ideals. Thousands of temples and pagodas ranging from the simple to the ornate grace the cultural landscape, especially in the valley of Kathmandu, the country's core area. Although well over a dozen languages are spoken, 90 percent of the people also speak Nepali, a language related to Indian Hindi.

This is a troubled country suffering from severe underdevelopment, with the lowest GNI in the realm. It is the only country in the realm with a declining income per capita over the past two decades. It also faces strong centrifugal social and political forces. Environmental degradation, crowded farmlands, and soil erosion, as well as deforestation, scar the countryside. The Himalayan peaks form a world-renowned tourist attraction, but tourist spending in Nepal, always relatively modest, has been cut back because of the recurrent disorder associated with a Maoist-communist insurrection and the collapse of the country's monarchy.

Nepal's cultural and political geographies have long been in turmoil. Strong regionalism divides the country both north-south and east-west. The southern Terai with its tropical lowlands is a world apart from the foothills of the Himalaya in the central interior, and the peoples of the west have origins and traditions different from those in the east. An absolute (Hindu) monarchy held Nepal together until 1990, when antigovernment demonstrations and resulting casualties persuaded the king to lift the ban on political parties. But chaos soon resumed, the king was assassinated, and by 2002 Maoist rebels controlled

FIGURE 8B-8

© H. J. de Blij, P. O. Muller, and John Wiley & Sons, Inc.

nearly half of the country, turning Nepal into a *failed state*. The ruling king (the brother of the former king) was effectively forced from office in 2006, the insurgents were brought into the political system, and in the following year a new constitution formalized the abolition of the monarchy and the reinvention of Nepal as a secular federal republic. Peaceful elections in 2008 confirmed the capacity of Nepal's new system to accommodate the still-powerful centrifugal forces in the country, but it remains to be seen if the new government can deliver on the demands of the country's highly diverse communities and constituencies.

Bhutan

Mountainous Bhutan, with a population of 740,000, lies wedged between India and China's Tibet (Fig. 8B-8), the only other buffer between Asia's giants. In landlocked, fortress-like Bhutan, time seems to have stood still. Bhutan long was a constitutional monarchy ruled by a king whose absolute power was unquestioned by his subjects. But in 2007 the reigning king, who had recently succeeded his father, and perhaps with an eye on what had happened to the monarchy in nearby Nepal, decided to order his subjects to vote for a political party in a newly created democracy. Thus in 2008 Bhutan went from an absolute monarchy to a multiparty democracy on the orders of its monarch, and Thimphu, the capital, became the seat of a new National Assembly.

In the mountainous countryside, the symbols of Buddhism, the state religion, dominate the cultural landscape, and the government, in its development policies, stresses the importance of spiritual fulfillment alongside the satisfaction of material needs. Still, there is social tension driven by the presence of a Nepalese (Hindu) minority, and a Bhutanese refugee population remains housed in camps in eastern Nepal. Add to this a still-unresolved boundary issue with China and the possibility of rising material expectations, and newly democratic Bhutan faces some important challenges.

Regarding economic geography, forestry, hydroelectric power, and tourism all have much potential here, and Bhutan possesses considerable mineral resources. But isolation and inaccessibility preserve traditional ways of life in this mountainous nation.

██ THE SOUTHERN ISLANDS

As Figure 8B-1 shows, South Asia's subcontinental landmass is flanked by several sets of islands: Sri Lanka just off the southern tip of India, the Maldives in the Indian Ocean to the southwest, and the Andaman Islands (which belong to India) marking the eastern edge of the Bay of Bengal.

The Maldives

The Maldives consists of more than 1000 tiny islands whose combined area is just 300 square kilometers (115 sq mi) and

whose highest elevation is less than 2 meters (6 ft) above sea level. Its population of 330,000 from Dravidian and Sri Lankan sources is now virtually 100 percent Muslim, with one-quarter of it concentrated on the capital island named Maale. The Maldives might be unremarkable, except that, as the table in Appendix B shows, this country has the realm's highest GNI per capita. The locals have translated their palm-studded, beach-fringed islands into a tourist mecca that attracts tens of thousands of mainly European visitors annually. One key question is how well the Maldives' economy can withstand the effects of the global downturn that began in 2008, because tourism, accounting for nearly 30 percent of its income, is a vulnerable industry.

The Maldives' low elevation (80 percent of the country lies less than one meter [3 ft] above sea level) has been singled out repeatedly in assessments of the future impact of global warming, which would produce rising sea levels. Those risks became sudden reality on December 26, 2004 when a devastating Indian Ocean tsunami generated near Indonesia swept over the islands, killing more than 100 residents as well as tourists and destroying resort facilities along the shore as well as inland.

Sri Lanka: Paradise Lost and Regained?

Sri Lanka (known as Ceylon before 1972), the compact, pear-shaped island located just 35 kilometers (22 mi) across the Palk Strait from southernmost India, became independent from Britain in 1948 (Fig. 8B-9). There were good reasons to create a separate sovereignty for Sri Lanka. This is neither a Hindu nor a Muslim country: the majority of its population of 21.2 million people—some 70 percent—are Buddhists. Furthermore, unlike Pakistan or India, plantations dominate Sri Lanka, and commercial farming still is the mainstay of the agricultural economy.

The great majority of Sri Lanka's population is descended from migrants who came to this island from northwestern India beginning about 2500 years ago. Those migrants introduced the advanced culture of their source area, building towns and irrigation systems and bringing Buddhism. Today, their descendants, known as the Sinhalese, speak a language (Sinhala) that belongs to the Indo-European language family of northern India.

The Dravidians who lived on the mainland, just across the Palk Strait, came much later. During the nineteenth century the British brought hundreds of thousands of Tamils to labor on their tea plantations, and soon a small minority became a substantial component of Ceylonese society. The Tamils brought their Dravidian tongue to the island and also introduced their Hindu faith. At the time of independence, they constituted more than 15 percent of the population; today they total about 18 percent.

When Ceylon became independent, it was one of the great hopes of the postcolonial world. The country had a sound economy plus a democratic government, and it was renowned for its tropical beauty. Its reputation soared

FIGURE 8B-9

© H. J. de Blij, P. O. Muller, and John Wiley & Sons, Inc.

From the Field Notes . . .

© H. J. de Blij

"Sri Lanka is a dominantly Buddhist country, the religion of the majority Sinhalese. Most areas of the capital, Colombo, have numerous reminders of this in the form of architecture and statuary: shrines to the Buddha, large and small, abound. But walk into the Tamil parts of town, and the cultural landscape changes drastically. This might as well be a street in Chennai or Madurai: elaborate Hindu shrines vie for space with storefronts, and Buddhist symbols are absent. The people here seemed to be less than enthused about the Tamil Tigers' campaign for an independent state. 'This would never be a part of it anyway,' said the fellow walking toward me as I took this photograph. 'We're here for better or worse, and for us the situation up north makes it worse.' But, he added, Sri Lankan governments of the past had helped create the situation by discriminating against Tamils."

www.conceptcaching.com

Concept Caching

when a massive effort succeeded in eradicating malaria and when family-planning campaigns reduced population growth while the rest of the realm was experiencing a population explosion. Rivers from the cooler, forested interior highlands fed the paddies that provided ample rice; crops from the moist southwest paid the bills, and the capital, Colombo, grew to reflect the optimism that prevailed.

In the midst of this glowing scenario, the seeds of disaster were already being sown. Sri Lanka's Tamil minority soon began proclaiming its sense of exclusion, demanding better treatment from the Sinhalese majority. Although the government recognized Tamil as a "national language" in 1978, sporadic violence marked the Tamil campaign, and in 1983 full-scale civil war broke out. Now many in

This is the tragic end of an insurgent state overpowered by a national counteroffensive. Here in northeastern Sri Lanka, the Tamil Tigers made their last stand before their defeat by the Sri Lankan military in May 2009 (this photo was taken exactly one year later). Thousands of civilians, used as human shields in the final stage of the battle, died or were wounded; as many as 200,000 were displaced by the fighting. In the aftermath of this civil war, the Sinhalese-dominated government must find ways to reintegrate the Tamils and others swept up in the insurgents' campaign, in order to avoid a return to the social conditions that gave rise to the insurgency in the first place.

© STR/Reuters/© Corbis

the Tamil community demanded a separate Tamil state to encompass the north and east of the country (see Fig. 8B-9 inset map), and a rebel army calling itself the Liberation Tigers of Tamil Eelam (LTTE) confronted Sri Lanka's national forces.

The Tamil Tigers first took the Jaffna Peninsula in the far north (Fig. 8B-9), next they extended their control over the Vanni Region just to its south, and soon they started publishing maps that revealed their ultimate demand: the entire northern and eastern periphery of Sri Lanka would become the secessionist state of *Eelam* (shown by the striped zone in the inset map). Not only had an ***insurgent state*** (a concept discussed in Chapter 5B) formed in northeastern Sri Lanka: the "Tigers" backed up these demands through terrorist attacks in the capital, Colombo, and unleashed combat in many parts of the state they were seeking to establish.

Eventually, the town of Kilinochchi near the mainland entry to the Jaffna Peninsula became the LTTE "capital," with all the hallmarks of an insurgent state. The Tigers collected taxes, administered banks and issued currency, organized police and court systems, established schools, and even ran hotels and guest houses for visitors. By 2002, the government of Sri Lanka had agreed to a cease-fire, and negotiations leading to the country's partitioning—at least from the LTTE perspective—began.

But in 2007 the Colombo government decided to embark on the counteroffensive that marks the third and decisive stage in the success or failure of an insurgent state. After five years of intermittent fighting, cease-fires, and fruitless negotiations, it concluded that the only way to settle the conflict was through an all-out military campaign. On January 2, 2009, President Mahinda Rajapaksa announced the capture of Kilinochchi and proclaimed that the threat of partition in Sri Lanka had ended.

In the spring of 2009, the final battle of the counteroffensive took place in the Vanni Region amid terrible losses among civilians, who were used as human shields by the doomed LTTE even as government forces impeded their escape. The leader of the insurgency was killed, and the Tamil Tigers' insurgent state disappeared from the map. An international outcry over the tactics of the Sri Lankan military was followed by expressions of concern over the future of Sinhalese-Tamil relations in this still-fractured country.

Reconciliation is urgently needed, but even though several million Tamils had never wanted anything to do with an armed insurrection, the Colombo government seemed to focus more on retribution than pacification. More than three decades ago the LTTE insurgency arose because of Hindu-Tamil anger over the government's Buddhist-Sinhalese cultural dominance and social policies. Therefore, only more even-handed management of national affairs can stave off the resumption of the cycle that transformed South Asia's great success story into a dreadful tragedy whose after-effects will be felt for decades. Divided Sri Lanka still awaits a turn toward peaceful cooperation.

POINTS TO PONDER

- Pakistan is crucially important to the stability of both Southwest and South Asia, yet it is becoming increasingly difficult to govern and faces growing religious fanaticism and political turmoil.

- India's IT sector accounts for one-quarter of all exports, but employs only 2 percent of the national workforce.

- If India goes ahead with plans to reserve 33 percent of its Parliament seats for women, it will have almost twice as much female representation as the U.S. Congress.

- Nepal, struggling to emerge from failed-state status, is the only country in South Asia in which per capita incomes have declined in recent years.

using GEOGRAPHY

1. One of the current issues in the United States involves the so-called outsourcing of jobs: the transfer of salaried employment from American to foreign locations. Many economists say that this is a logical outgrowth of globalization, that it involves comparatively few jobs, and that it helps U.S. companies survive competition to which they might otherwise succumb—costing more jobs than the outsourcing did in the first place. India has become a major player in the outsourcing of certain work. What are these jobs? Where in India do they cluster and why? Do language and/or time-zone location have anything to do with India's success, and if so, how?

2. Six and a half decades after India and Pakistan ceased to be part of the British colonial empire, the contrasts between India and Pakistan—demographic, economic, social, political—could hardly be greater despite the fact that they are neighbors. What, in your view, are the key differences between the two South Asian states in these contexts? Can you identify crucial geographic factors that would account for such enormous dissimilarities? Given their shared histories, do you see any common ground between Pakistan and India today that might form a basis for their future convergence and cooperation?

FIGURE 9A-1   © H. J. de Blij, P. O. Muller, and John Wiley & Sons, Inc.

East Asia is a geographic realm like no other. At its heart lies the world's most populous country, the product of what may be the world's oldest continuous civilization. On its Pacific mainland shores an economic transformation has taken shape with no parallel in world history. Its offshore islands witnessed the first use of atomic weapons on civilian populations and the postwar emergence of one of the world's most powerful economies. Few lives in this world were left unaffected, directly or indirectly, by the momentous events that occurred in East Asia over the past two generations. Just look around you. Japanese-designed automobiles, Chinese-made televisions (and so much more), South Korean motorbikes, Taiwanese computers—from toys to textiles and from hardware to software, East Asian products fill streets and stores, homes and hotels.

China is the world's biggest exporter and home to three of the five busiest container ports in the world: Shanghai, Shenzhen, and Hong Kong (photo). The other two are in Asia as well: Singapore and Busan (South Korea)

© Jan Nijman

It has all happened with astonishing speed. Some of us can remember the time when you were no more likely to find anything useful made in China than you were to buy anything from Russia. But after the 1960s ended, things changed rapidly. Japan led the way, turning its World War II defeat into postwar economic triumph. By the mid-1970s, Japan's economic growth—compared to China's seemingly total stagnation—seemed to justify Japan's recognition as a discrete geographic realm, an economic engine for the world and a strong competitor for

major geographic qualities of
EAST ASIA

1. East Asia is encircled by snowcapped mountains, vast deserts, cold climates, and Pacific waters.

2. East Asia was one of the world's earliest culture hearths, and China is one of the world's oldest continuous civilizations.

3. East Asia is the second most populous geographic realm, but its population remains strongly concentrated in its eastern regions.

4. China, the world's largest state demographically, is the current rendition of an empire that has expanded and contracted, fragmented, and unified many times during its long existence.

5. China's sparsely peopled western regions are strategically important to the state, but they lie exposed to minority pressures and Islamic influences.

6. Along China's east coast, an economic transformation launched three decades ago is now rapidly expanding westward.

7. Increasing regional disparities and fast-changing cultural landscapes are straining East Asian societies.

8. Japan, one of the economic giants of the East Asian realm, has a history of colonial expansion and wartime conduct that still affects international relations here.

9. East Asia is home to the world's newest superpower as China's economic and political influence is increasingly projected around the globe.

10. The political geography of East Asia contains a number of flashpoints that can generate conflict, including Taiwan, North Korea, and several island groups in the realm's seas.

the top-ranked U.S. economy. But then Hong Kong, South Korea, and autonomous Taiwan began to show the world what other peoples in East Asia were capable of. When the ruler of communist China, the realm's giant, opened the door to an American president in 1972, a pivotal moment in global history (and geography) had arrived. Before the end of the century, China had taken its place as modern East Asia's core. That event in 1972 signified a new direction in Chinese political thinking, and it would facilitate major revisions in economic policy. Today, East Asia is the most dynamic realm in the global economy: led by China, it is changing the world, and in the process is itself undergoing a profound transformation.

THE GEOGRAPHIC PANORAMA

As Figure 9A-1 shows, the East Asian geographic realm forms a roughly triangular wedge between the vast expanses of Russia to the north and the populous countries of South and Southeast Asia to the south, its edges often marked by high mountain ranges or remote deserts. The darker brown on the map designates the highest mountains and plateaus, which create a vast arc north of the Himalayas before bending southward and becoming lower (tan shading) toward Myanmar, Laos, and Vietnam in Southeast Asia. Here in the southwest, where Tibet is located, mountains and plateaus alike are covered by permanent ice and snow, the soaring ranges crumpled up like the folds of an accordion. Three major rivers, their valleys parallel for hundreds of kilometers, disclose the orientation of this high-relief topography. Northward, note how rapidly the mountains give way to broad, flat deserts whose names appear prominently: the Takla Makan in the far west, the Gobi where China meets Mongolia, the Ordos in the embrace of what looks like a huge meander of a river we will learn more about later, the Huang He (Yellow River).

One basin, lodged between high mountains to its west and lower ones elsewhere, and of special interest, is the Sichuan (Red) Basin. It looks rather small on the map, but is home to more than 100 million people and, we will discover, is important for other reasons as well. In this country of mountains and deserts, living space is at a premium. And speaking of living space, the green areas on the map, which have the lowest relief and (often) the most fertile soils, are home to the vast majority of this realm's population. Here the great rivers that come from the melting ice and snow in the interior highlands have been depositing their sediment for eons, and when humans domesticated plants and started to grow crops, this was the place to be. That was thousands of years ago—perhaps as long as 10,000 years—and ever since, this has been the largest cluster of humanity on Earth.

But the East Asian realm is not confined to the mainland of mountains and river basins. You can imagine how its offshore islands were populated: the Korean Peninsula seems to form part of a bridge pointing toward the southernmost island of what is today Japan, and from there it seems likely that the early migrants moved farther north until they reached Hokkaido. In warmer times, they may even have ventured beyond, onto the Kurile Islands. And in

Places and Names

AS IN OTHER parts of the world, European colonists in China wrote down the names of places and people as they heard them—and often got them wrong. The Wade-Giles system put such place names as Peking, Canton, and Tientsin on the map, but that's not how the Chinese knew them. In 1958, the communist regime replaced the foreign version with the *pinyin* system, based on the pronunciation of Chinese characters in Northern Mandarin, which was to become the standard form of the Chinese language throughout China. The world now became familiar with these same three cities written as Beijing, Guangzhou, and Tianjin. Among prominent Chinese geographic names, only one has not gained universal recognition: Tibet, called Xizang by China but referred to by its old name on many maps for reasons that will become obvious later in this chapter.

Pinyin usage may be standard in China today, but China remains a country of many languages and dialects. Not only do minorities speak many different languages, but Mandarin-speakers use numerous dialects—so that when you sit down with some Chinese friends who have just met, it may take just one sentence for one to say to the other "ah, you're from Shanghai!" Migration is changing this, of course. China's population is on the move, and a growing number of residents of Shanghai today were not born in that city. But China's language map is anything but simple. And not just place names, but personal names, too, were revised under pinyin rules. The communist leader who used to be called Mao Tse-tung in the Western media is now Mao Zedong. Also remember that the Chinese write their last name first: President Hu Jintao is Jintao to friends and Mr. Hu to others. To the Chinese, it is Obama Barack, not the other way around.

the south, Taiwan lies even closer to mainland China than Japan does to Korea, whereas tropical Hainan Island, the realm's southernmost extremity, is almost—but not quite—connected to the small peninsula that reaches toward it.

In terms of total land area, though, East Asia is mostly mainland—but the islands and their peoples have played leading roles in forging this realm's regional geography. The waters between mainland and islands (the Taiwan Strait, the South China Sea, the East China Sea, the Yellow Sea, the Korea Strait, and others) also figure prominently in the geographic saga of this realm. Today, the Japanese and the Chinese are arguing over the ownership of small islands in these waters, specks of land with large oil reserves and fishing grounds claimed by both sides. So the realm's map is considerably more complicated than that wedge-shaped triangle in Figure 9A-1 initially suggests.

A GEOPOLITICAL REALM

It is all too easy to refer to China when you mean East Asia, because China is the latter's dominant country, contains more than 85 percent of the realm's population, and has taken an increasingly prominent role on the world stage. But there are five other political entities on East Asia's map: Japan, South Korea, North Korea, Mongolia, and Taiwan. Note that we refer here to *political entities* rather than *states*. In this realm, the distinction is important. Taiwan refers to itself as the Republic of China (ROC), but it is not recognized as a sovereign state by most members of the international community. The communist administration in Beijing, capital of the People's Republic of China (PRC), regards Taiwan as part of China and as a temporarily wayward province. And North Korea is widely viewed as a *failed state*, whose (communist)

regime has one of the world's most dreadful human rights records; North Korea is not even a fully functional member of the United Nations.

Nevertheless, China is the realm's dominant entity: demographically, economically, and politically. It is important to keep in mind, as you read Chapters 9A and 9B, that portions of what we map today as regional components of China were not part of the country in the past, and that other areas now lying outside China are regarded by many Chinese as Chinese property (for example, a large sector of the Indian State of Arunachal Pradesh, and sections of the Russian Far East). As we will see, China's imperial past saw the state expand and contract and expand again, leaving unfinished business on land as well as at sea. On such issues, Chinese emotions can run quite deeply.

Six hundred years ago China already was the largest nation on the planet, its history spanning thousands of years, its unmatched fleets exploring lands and peoples in South Asia and East Africa, its technologies unequaled. But even the Chinese could not curb the growing presence of Europe's colonial powers, and eventually the British, French, Russians, and Germans accomplished the unthinkable, taking control over most of the Chinese state. When the Japanese emulated the Europeans and conquered their own colonial empire, much of it at China's expense, humiliation was complete—and never forgotten. To this day, the sense that China's borders (including those with Russia and India) were imposed by outsiders is never far from the surface.

ENVIRONMENT AND POPULATION

To understand the complex physical geography of East Asia shown in Figure 9A-1, it is useful to refer back to Figures G-4 and G-5 in the introductory chapter. The

1

111

1

high, snowcapped mountain ranges of the realm's south-western interior result from the gigantic collision of the Indian and Eurasian tectonic plates (Fig. 8A-2), pushing the Earth's crust upward and creating not only the mountain ranges of which the Himalaya is the most famous, but also the dome-like Qinghai-Xizang (Tibetan) Plateau. In Figure G-5, note the high incidence of earthquakes associated with this collision, converging on a narrow zone of instability that crosses southwestern China and leads into Southeast Asia. The devastating May 12, 2008 earthquake (magnitude 7.9) in Sichuan Province that killed more than 70,000 originated in this danger zone and is only one in an endless series that will continue to take its toll.

From Figure G-5 it is also obvious that the Pacific Ring of Fire, with its lethal combination of volcanism and earthquakes, endangers Japan far more than it does China. This threat became horrific reality on March 11, 2011 when a monstrous 9.0-magnitude earthquake struck off Japan's northeastern coast near the city of Sendai. It was the most powerful quake ever recorded in that country, and one of the five biggest in the world since 1900. Even worse, the massive temblor and its torrent of violent aftershocks triggered a series of devastating **tsunamis [1]** (seismic sea waves) that swept across the narrow, densely populated coastal plains that hug the shoreline of Honshu (Japan's largest island) north of the Tokyo area. And as if the destruction and death left behind was not enough, the leakage of radioactivity from a heavily damaged nuclear-power complex near Fukushima south of Sendai (Fig. 9A-2) may have caused serious health problems in the local population. Now officially known as the Tohoku earthquake, its death toll three months later was finalized at about 23,000. The vulnerability of Japan to such disasters results from a dangerous combination of circumstances: it is located in a particularly active tectonic-plate collision zone, and many of the country's habitable (and therefore most densely populated) areas are confined, low-lying plains on the islands' east coast open to flooding by Pacific tsunamis (see box titled "Gambling with Nature"). Not surprisingly, the location of most Japanese nuclear power plants along these susceptible shorelines has now become a hotly debated issue.

Looking at the map of East Asian climates (Fig. 9A-3, left map), we should not be surprised that the western and northern sectors of this realm are dominated by conditions that do not favor substantial population clusters. Permanent snow and ice cover much of the area mapped as *H* (highland climes) including Tibet (Xizang) and Qinghai. Northward, the *B* climates (desert and steppe) prevail because this vast expanse lies about as far from maritime influences (and air masses) as you can get in Asia. Mongolia is one of the driest countries in the world, but even here—and even in frosty Tibet—there are places where people manage to eke out a living. But the map leaves no

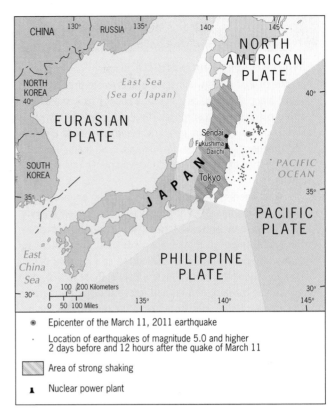

FIGURE 9A-2   © H. J. de Blij, P. O. Muller, J. Nijman, and John Wiley & Sons, Inc.

doubt as to why the great majority of East Asians live in the eastern part of this realm.

When we compare the climates prevailing in the East Asian realm to those familiar to us in North America

Houses and other buildings being washed away by massive tsunami waves minutes after the gigantic earthquake of March 11, 2011. Virtually the entire area around the city of Sendai was destroyed, including this satellite town of Natori. It was a truly apocalyptic event: first the magnitude-9.0 quake, then the tsunamis, and finally the radioactive pollution emitted by the damaged reactors of the Fukushima Daiichi nuclear complex 95 kilometers (60 mi) south of Sendai (see map above).

© Kyodo/XinHua/Xinhua Press/©Corbis

Gambling With Nature

THE CATASTROPHIC NATURAL disaster that hit Japan in March 2011 overwhelmed the northeastern coast of its main island (Honshu) as well as the country as a whole—but it did not come as a surprise. The Earth's crustal dynamics in this part of the realm guarantees earthquakes at a relentless frequency. As Figure 9A-2 shows, it all stems from the configuration of the underlying tectonic plates that converge on Japan.

The Pacific Plate moves westward at an average of 7–10 centimeters (3–4 in) per year, and the only way it can do so is by pressing forward (or subducting) *beneath* the North American Plate (Japan sits atop the western tip of the latter). The earthquake occurred at a moment of violent and sudden movement as the Pacific Plate was blocked from going forward, and instead forced the North American Plate upward. With the epicenter of the quake 110 kilometers (70 mi) offshore, this vertical thrust released enormous amounts of energy and lifted a vast dome of seawater whose gravity-induced dispersal triggered the tsunamis. In fact, so much energy was released that tsunami waves reached speeds of 800 kph (500 mph). Thus within 30 minutes, waves over 8 meters (27 ft) high repeatedly crashed onto Honshu's low-lying northeast coastlands, causing unimaginable devastation (photo p. 461).

Take another look at Figure 9A-2 and consider the distribution of all those smaller quakes before and after the big one. Any one of them could have been the epicenter, and it is obvious that the entire east coast of Japan lies in the danger zone, including all of the gigantic conurbation centered on Tokyo. So even the misfortunes that befell Japan on March 11, 2011 could have been far worse.

All Japanese, and especially the nearly 30 million living in and around Tokyo, know about the "70-year" rule: over the past three and a half centuries, the Tokyo area has been struck by major earthquakes roughly every 70 years—in 1633, 1703, 1782, 1853, and 1923. The Great Kanto Earthquake of 1923 set off a firestorm that swept over the city and took the lives of an estimated 143,000 people. Tokyo Bay virtually emptied of water; then a killer tsunami roared back in, sweeping everything before it.

Today, metropolitan Tokyo is decidedly more vulnerable than it was ninety years ago. True, building regulations are stricter and civilian preparedness is better. But entire expanses of industries have been built on landfill that will liquefy; the city is honeycombed by underground gas lines that will rupture and stoke countless fires; congestion in the area's maze of narrow streets will hamper rescue operations; and many older high-rise buildings do not have the structural integrity that has lately emboldened builders to erect skyscrapers in excess of 50 stories. Add to this the burgeoning population of the greater Kanto Plain—which surpasses 40 million on what may well be the most dangerous 4 percent of Japan's territory—and we realize that the next major earthquake along northern Japan's Pacific coast could produce an ever bigger disaster than the historic one of 2011.

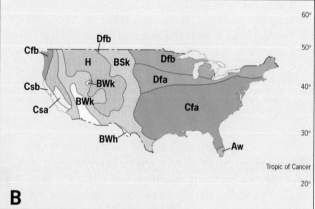

CLIMATES OF EAST ASIA AND THE CONTERMINOUS UNITED STATES
After Köppen-Geiger

A HUMID EQUATORIAL CLIMATE	B DRY CLIMATE	C HUMID TEMPERATE CLIMATE	D HUMID COLD CLIMATE	H HIGHLAND CLIMATE
Am Short dry season	**BS** Semiarid	**Cf** No dry season	**Df** No dry season	**H** Unclassified highlands
Aw Dry winter	**BW** Arid	**Cw** Dry winter	**Dw** Dry winter	
	h=hot **k**=cold	**Cs** Dry summer **a**=hot summer **b**=cool summer **c**=short, cool summer		

FIGURE 9A-3

© H. J. de Blij, P. O. Muller, and John Wiley & Sons, Inc.

From the Field Notes . . .

"From the train, traveling from Beijing to Xian, we had a memorable view of the Loess Plateau. Loess is a fine-grained dust formed from rocks pulverized by glacial action, blown away by persistent winds, and deposited in sometimes well-defined locales. Loess covers much of the North China Plain, which is what makes it so fertile, but there it is not as thick as in the Loess Plateau in the middle basin of the Huang He (Yellow River). Here the loess averages 75 meters (250 ft) in thickness and in places reaches as much as 180 meters (600 ft). Because loess has some very distinctive physical properties, it tends to create unusual landscapes. Through a complicated physical process following deposition, loess develops the capacity to stand upright in walls and columns, and resists collapse when it is excavated. As a result the landscape looks terraced: streams cut deep and steep-sided valleys. The Loess Plateau is a physiographic region, but it is a cultural region too. Hundreds of thousands of people have literally dug their homes into vertical faces of the kind shown here, creating cave-like, multi-room dwellings with wooden exterior doors."

www.conceptcaching.com

© H. J. de Blij

(Fig. 9A-3, right map), it is immediately obvious that the *C* or humid-temperate climates are more extensive in the United States than in East Asia. Note especially the comparative location of the milder *Cfa* climate, which in the United States extends beyond 40 degrees North latitude up to New England, but which in China yields to colder *D* climates at a latitude equivalent to Virginia's. Thus the capital, Beijing, has the warm summers but long, bitterly cold winters characteristic of *D* climates. Take a closer look at Figure 9A-3 and it is clear that both the Korean Peninsula and the Japanese islands lie astride this transition from *C* to *D* climates. South Korea is significantly milder and moister than North Korea, and Japan's largest island (Honshu) is temperate in the south but cold in the north. Not surprisingly, the least densely populated island of Japan is northernmost Hokkaido, where the climate is rather like that of northern Wisconsin.

Comparing just the United States and China, note that whereas *C* or *D* climates prevail over more than half of the United States, these climates predominate over less than one-third of China—even though the United States has only 313 million people compared to China's 1.36 billion. That is what makes the population distribution map (Fig. 9A-4) so noteworthy: the overwhelming majority of East Asia's people reside in the easternmost one-third of the realm's territory, creating the biggest, most densely settled population cluster on Earth, which mainly depends on the limited green-colored area in Figure 9A-1.

Several times in earlier chapters of this book we have noted the capacity of humans to live under virtually all environmental circumstances; technological developments enable the survival of year-round communities in Antarctica, on oil platforms at sea, in the driest of deserts, and on the highest of plateaus. But the world population distribution map (Fig. G-8) still reminds us how we got our modern start—through crop-raising and herding. The fertile river basins and coastal plains of East Asia supported ever-larger farming populations, whose descendants still live on that same land: for all its current industrialization and urbanization, more than half the population of China remains rural to this day. Environment, in the form of elevation, relief, water supply, soil fertility, and climate played the crucial role in the evolution of the population distribution displayed in Figure 9A-4—and will continue to do so for centuries to come.

The Great Rivers

China is in some respects the product of four great river systems and their basins, valleys, deltas, and estuaries. These rivers and their tributaries are visible, but do not stand out as clearly in Figure 9A-1 as they do on the physiographic

FIGURE 9A-4

© H. J. de Blij, P. O. Muller, and John Wiley & Sons, Inc.

map (Fig. 9A-5). So important are they as designators of the realm's regional geography that we should acquaint ourselves with them here. Of the four, the two in the middle are in many ways the most important: the ***Huang He*** (Yellow River) that makes a huge loop around the Ordos Desert and then flows into the Bohai Gulf, and the ***Yangzi*** River, probably the most famous river in China's historical geography, called the ***Chang Jiang*** (Long River) upstream.

As the map shows, the Yellow River and its tributaries form the sources of water vital to the historic core area of China, the North China Plain, where the capital, Beijing, is located. The Yangzi River is the major artery of the Lower Chang Basin; at its mouth lies China's largest city, Shanghai, and in its middle course the water flow is controlled by the world's biggest dam (see photo at right). Both the Huang and the Yangzi rivers originate in the snowy mountains of the Qinghai-Xizang Plateau, a reminder that these remote environments are critical to hundreds of millions of people thousands of kilometers away.

The other two rivers have much shorter courses, but the ***Pearl*** River outlet of the one in the south, the ***Xi*** River, forms an estuary that has become China's (and East Asia's) greatest hub of globalization—an ongoing saga discussed

in Chapter 9B. This is where you find Hong Kong and, right next to it, the fastest-growing major city in the history of humankind, Shenzhen.

Finally, the northernmost of China's four major rivers, the ***Liao*** River, originates near the margin of the Gobi Desert and then forms an elbow as it crosses the Northeast China Plain to reach the Bohai Gulf flowing southward. Here the climate is colder, low relief scarcer, and population smaller than in the more southerly river basins, but mining and industry create opportunities lacking for agriculture. As is obvious, each of East Asia's river-based clusters has its own combination of problems and potentials.

Along the Coast

East Asia's Pacific margin is a jumble of peninsulas and islands. The Korean Peninsula looks like a near-bridge from Asia to Japan, and indeed it has served as such in the past. The Liaodong and Shandong peninsulas protrude into the Yellow Sea, which continues to silt up from the sediments of the Huang and Liao rivers. Off the mainland lie the islands that have played such a crucial role in the modern human geography of Asia and, indeed, the world: Japan,

FIGURE 9A-5

© H. J. de Blij, P. O. Muller, and John Wiley & Sons, Inc.

Three Gorges Dam is emblematic of China's modern "era of the mega-project." The dam wall rises 180 meters (600 ft) above the now-inundated valley floor of the Chang/Yangzi River. It is 2 kilometers (1.2 mi) wide and creates a reservoir—largest of its kind in the world—that extends more than 600 kilometers (400 mi) upstream.

© Wen Zhenxiao/XinHua/XinhuaPress/©Corbis

Taiwan, and Hainan. Japan's environmental range is expressed by cold northern Hokkaido and warm southern Kyushu, but Japan's core area lies on its main island, Honshu. As Figure 9A-1 shows, myriad smaller islands flank the mainland and dot the East and South China seas. As we will discover, some of these smaller islands significantly affect the human geography of this realm.

NATURAL RESOURCES

Given that the East Asian realm contains nearly one-fourth of the world's population, it is not difficult to imagine the magnitude of the demand for natural resources here. The world first received notice of this about a hundred years ago when Japan's imperial expansion was in part driven by the need for raw materials to feed its rapidly expanding industrialization. Today, East Asia's demand for natural resources is expressed not in imperialism but in the global marketplace: it is driving a commodity boom all over the world, from Russia to Australia to Brazil.

While China was moribund under its early communist administration, and before post–World War II Japan embarked on its headlong rush to world economic power, East Asia's requirements remained modest by global standards. But then Japan's economic success, followed by China's swift adoption of market economics, created unimagined and unprecedented needs.

Japan, about the size of Montana but with a population of more than 100 million, showed what lay ahead. With few domestic resources to support large-scale manufacturing, the Japanese set up global networks through which flowed commodities ranging from oil and natural gas to iron ore and chemicals. Urbanizing and modernizing populations demand ever more consumer goods, and Japanese products poured onto domestic as well as foreign markets. Japanese-owned fleets of freighters plied the oceans, and for a while Japan even became remote Australia's top-ranked customer for commodities.

When China took off in the 1980s, economic geographers cast a wary eye on the geologic map. Until then, China's biggest resource had been its fertile, river-deposited alluvium (silt): despite the communist regime's best efforts, state-run industries planned to satisfy domestic needs were no match for globalizing Japan. Most Chinese were farmers. Staving off famines was a never-ending preoccupation. But when China opened its doors to the world, and its cities burgeoned even as its industries proliferated, its needs—for oil, gas, metals, food, electricity, water—multiplied. Before long, China had replaced Japan as Australia's number-one customer. And Chinese manufacturers and suppliers searched for commodities from Indonesia to Iraq and from Tanzania to Brazil. If China holds an advantage, it is in the so-called *rare earth* elements not commonly known, such as thulium (used in lasers), praseodymium (aircraft parts), lanthanum (electric automobiles), and promethium (X-ray equipment). By some measures, East Asia contains almost 95 percent of the known deposits of these minerals, which are increasingly utilized in missile technology and "green" energy applications.

But China needs the world because East Asia's storehouse of other known resources is not encouraging (Fig. 9A-6). Coal reserves are widespread and can satisfy the

FIGURE 9A-6

© H. J. de Blij, P. O. Muller, and John Wiley & Sons, Inc.

In China's headlong rush to become a global economic power, environmental protection has always taken a back seat. This scene of industrial production in Chongqing, the Sichuan Basin's largest city, shows the consequences.

© Yann Layma/The Image Bank/Getty Images, Inc.

expanding coal-fired energy system (at the cost of thousands of miners' lives every year). But oil reserves, while also widespread, tend to be rather modest and diminishing. Deposits in northeastern and far-western China are the largest; moreover, exploration is proceeding offshore. Nothing in China compares with the massive iron ore deposits and alloy-metal sources available to Russia when it industrialized to meet the Nazi challenge; and nothing in China can match the oil and gas reserves of Southwest Asia's Persian Gulf region. Hence China competes with Japan for energy pipelines from Russia and with other sources for its industrial raw materials. So in just a few short years, China has become the world's biggest customer as well as a massive exporter.

East Asia's ascent to the globalizing world's principal stage has put it in the forefront of regional development, but at a high environmental cost in terms of atmospheric pollution and water contamination (see photo above). The East Asian geographic realm is in the midst of an industrial, social, and political revolution the outcome of which is far from clear.

UNFOLDING THE CULTURAL MAP

We in the Western world often take it for granted that the pivotal events that led to the domestication of animals and the selective farming of plants, the herding of livestock and the harvesting of crops, the storage of produce and the growth of villages into towns, all began in what we now call the Middle East. The story of rivers ebbing and flowing, the "lessons" of natural irrigation, the planned harvesting of grains, the rise of cities, and the organization of the first states is the story of "Western" civilization, the saga of Mesopotamia and ancient Egypt, the Tigris and Euphrates,

and the Nile. From the Fertile Crescent, we were taught, these revolutionary changes diffused to other parts of Eurasia and then to the rest of the world.

The Chinese have long taken a different view, and we now know from extensive archeological research that East Asia was one of the few places in the world where the process of state formation [2] occurred independently thousands of years ago, and that the modern Chinese state could trace its roots to that ancient time, long before there was a Greece or a Rome.

Ancient China

For a long time, it was assumed that the Chinese state first arose on the North China Plain, in the lower basin of the Yellow River. All early states (wherever they arose) were ruled by elites, small groups that took power in growing towns, but in East Asia their political history was chronicled in dynasties [3] because a succession of rulers came from the same line of male descent, sometimes enduring for centuries. And the earliest dynasty about which much was known with certainty was the Shang Dynasty, which lasted from around 1766 BC until about 1080 BC.

But then new discoveries began to overturn established knowledge. First, evidence for an even older dynasty, the Xia Dynasty, was found in the area where the Yellow River is joined by its Wei tributary, and the remains of its capital gave it the name *Erlitou* culture, dated from ca. 2200 BC to 1770 BC. And then the whole notion that the cradle of Chinese civilization lay in this middle Yellow River Basin was overturned by the discovery of even more ancient and complex cultures in other areas, hundreds of kilometers away. Now we know about such cultures as the Yangshao (6000 BC), Hongshan (4500 BC), and Liangzhu (3500 BC), and others that long predated even the Xia city builders. And it is clear that such crops as millet were domesticated in China nearly 10,000 years ago, rice about 9000 years ago, and wheat some 5000 years ago. The early Chinese did not need Middle Eastern guidance when it came to farming, urbanization, or state formation.

Geographically, the most consequential result of these discoveries is the knowledge that East Asia, as a realm, was not the product of one dominant culture that emerged on the North China Plain and spread outward, but was forged from numerous cultures that flourished in several areas including the Lower Chang Basin to the south and even beyond. During this Neolithic [4] ("New Stone Age") period, these regional cultures specialized in skills and arts that diffused far afield. Now we know that certain manufactured products important in East Asia to this day are of early Neolithic vintage: two 8000-year-old pots in the shape of a silkworm cocoon were found in Hebei Province near Beijing. Don't be surprised when, visiting China, you hear students suggest that maybe the inventions we

Westerners think diffused from west to east instead spread from east to west!

Beyond the River Basins

Eventually, the most powerful dynasties did establish themselves on the North China Plain, but their influence was not confined to present-day China. Even as East Asia's cultures matured and dynasties endured for centuries, their domains expanded and their innovations reached ever-more-distant communities. The Korean Peninsula, where tradition has it that the first king ascended the throne more than 4000 years ago and where the modern period is said to date from 57 BC, witnessed growing Chinese influence; in the late seventh century AD, the Chinese even helped one Korean kingdom defeat two competitors. Later, the Chinese helped the Koreans repel Japanese invaders, but it was not until China's final dynasty, the Qing (Manchu), that the Chinese forced the Koreans to acknowledge China's regional supremacy. That authority, as it turned out, would not be permanent: during much of the nineteenth century the Korean Peninsula was in turmoil as European, Chinese, and Japanese interests competed for influence, a contest that ended in 1895 when the Japanese ousted the Chinese and made Korea their colony.

In Japan, where local tradition places the country's beginnings in 660 BC, the ruling elites did not confront a Chinese presence but borrowed heavily from Chinese culture, including town and city plans, building styles, legal models, and even writing systems. Thus the cultural geography of the East Asian realm acquired a common base that, despite long-term Korean independence as well as Japanese modifications and adaptations to the Chinese norms they had borrowed, acquired an overall Chinese denomination. Buddhism matured in China and diffused to Japan; **Confucianism [5]** (see box titled "Confucius") infused Korean kingdoms. The realm was being defined in numerous ways.

Peoples of the East Asian Realm

If we were studying the East Asian realm six centuries ago, during China's Ming Dynasty, we would have had no doubt as to the location of core and periphery. China was the core. Encircling the Chinese state in the periphery were not only the Koreans and the Japanese but also the

---

Confucius

CONFUCIUS (*Kongfuzi* in pinyin) was China's most influential philosopher and teacher. His ideas dominated Chinese life and thought for over 20 centuries. Confucius was born in 551 BC and died in 479 BC. Appalled at the suffering of ordinary people during the Zhou Dynasty, he urged the poor to assert themselves and demand explanations for their harsh treatment by the feudal lords. He tutored the indigent as well as the privileged, giving the poor an education that had hitherto been denied them and ending the aristocracy's exclusive access to the knowledge that constituted power.

Confucius's revolutionary ideas extended to the rulers as well as the ruled. He abhorred supernatural mysticism and cast doubt on the divine ancestries of China's aristocratic rulers. Human virtues, not godly connections, should determine a person's place in society, he taught. Accordingly, he proposed that the dynastic rulers turn over the reins of state to ministers chosen for their competence and merit.

His earthly philosophies notwithstanding, Confucius took on the mantle of a spiritual leader after his death. His thoughts, distilled from the mass of philosophical writing (including Daoism) that poured forth during his lifetime, became the guiding principles of the formative Han Dynasty. The state, he said, should not exist just for the power and pleasure of the elite; it should be a cooperative system for the well-being and happiness of the people.

With time, a mass of writings evolved, much of which Confucius never wrote. At the heart of this body of literature lay the *Confucian Classics*, 13 texts that became the basis for education in China for 2000 years. From government to morality and from law to religion, the *Classics* were Chinese civilization's guide. The entire national system of education (including the state examinations through which everyone, poor or privileged, could enter the civil service and achieve political power) was based on the *Classics*. Confucius championed the family as the foundation of Chinese culture, and the *Classics* prescribe a respect for the aged that was a hallmark of Chinese society.

But his philosophies also were conservative and rigid, and when the colonial powers penetrated China, his ideas came face to face with practical Western education. For the first time, some Chinese leaders began to call for reform and modernization, especially of teaching. Confucian principles, they said, could guide an isolated China, but not China in the new age of competition.

The communists who took power in 1949 attacked Confucian thought on all fronts. The *Classics* were abandoned, indoctrination pervaded education, and, for a time, even the family was viewed as an institution of the past. Here the communists miscalculated. It proved impossible to eradicate two millennia of cultural conditioning in a few decades. As soon as China entered its post-Mao period in 1976, public interest in Confucius surged, and the shelves of bookstores again sagged under the weight of his writings. The spirit of Confucius will pervade physical and mental landscapes in China for generations to come.

FIGURE 9A-7

© H. J. de Blij, P. O. Muller, and John Wiley & Sons, Inc.

Mongols and Tatars to the north and northwest; the Kazakhs, Kyrgyz, Tajiks, and Uyghurs to the west; the Tibetans, Nepalese, and others to the southwest; and peoples too numerous to identify individually to the south, including majorities (Burmans, Thais, Vietnamese) and minorities in states that were still forming in South and Southeast Asia. Like all empires, the China of the Ming emperors expanded and contracted over time, but during the next (and final) dynasty, that of the Qing, many of the states and peoples just mentioned fell under Chinese rule. But it was the emperors' last hurrah: European and Japanese imperialists challenged Qing rule, took control over most of the state's core area, ousted the Chinese from much of the periphery, and left the country in chaos, bringing about the end of five millennia of dynastic rule in 1911.

But by that time, the ethnic geography of East Asia had already become very complex. Even after losing control

over peoples from Korea to Vietnam and from Mongolia to Burma (now called Myanmar), the Chinese state of the twentieth century still governed numerous minorities. East Asia today, therefore, remains a mosaic of ethnicities and languages.

Figure 9A-7 highlights the cultural diversity of a realm in which China dominates numerically, but where infusions from elsewhere are evident. We already know of the Mongols, the Muslim Uyghurs, and the Buddhist Tibetans; but the most varied and most numerous minority groups inhabit the southeastern corner of this realm, from the island of Hainan to the mouth of the Yangzi River. For example, the Yue language, middle green on the map (it used to be called Cantonese), is the common language in the pivotal Pearl River Estuary. Many of the other minority tongues shown here have links to Southeast Asian languages.

And do not be misled by what the map seems to show about Han Chinese (or "Mandarin"). True, the elites and the highly educated speak standard Mandarin, with those in the south somewhat less true to the Beijing-area version than those in the north. But ordinary people in villages a few kilometers apart may not be able to understand each other at all. What they are able to do is read the characters in which standard Chinese (locally called *Putonghua*) is written. And so, wherever you are in China, when you watch an anchorperson read the news on television, you will see a ticker scroll below showing in characters what is being said on-screen. Viewers in Sichuan may not understand the newsreader in Beijing, but still will be able to get the news by reading the crawl.

The ethnolinguistic mosaic of the East Asian realm is paralleled by regional variations in belief systems that are discussed in Chapter 9B, because beliefs—their origins, diffusion, and current regional expression—all reflect the diverse ways that cultural and political geographies are interconnected in this populous realm.

CHINA'S HISTORICAL ROLE IN EAST ASIA

Chinese Empires and Dynasties

China's great antiquity, as we noted earlier, is chronicled in dynasties, the earliest of which are still shrouded in mystery. But, as with any long-existing state, some of China's dynastic rulers proved more productive than others. Some

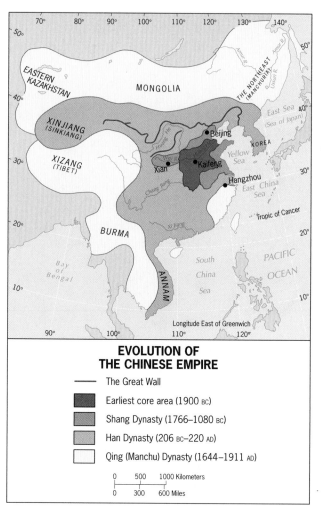

EVOLUTION OF THE CHINESE EMPIRE

—— The Great Wall

■ Earliest core area (1900 BC)

▨ Shang Dynasty (1766–1080 BC)

▨ Han Dynasty (206 BC–220 AD)

□ Qing (Manchu) Dynasty (1644–1911 AD)

```
0        500      1000 Kilometers
0    300      600 Miles
```

FIGURE 9A-8   © H. J. de Blij, P. O. Muller, and John Wiley & Sons, Inc.

The Great Wall of China, built and rebuilt from the fifth to the sixteenth century AD in the effort to fend off Mongol invaders, extends for more than 6000 kilometers (3700 mi) across northern China (see Fig. 9A-8). Some parts are completely gone but other sections have been renovated in recent times, such as this one just north of Beijing.

© Jan Nijman

dynasties bequeathed long-term geographic legacies to the state. Others molded the Chinese nation. The Han Dynasty did both, which is why we speak to this day of the Chinese as the ***People of Han***.

But the Zhou Dynasty, centuries before the Han, witnessed the arrival of Buddhism in China, saw Confucius walk the pathways of the north, started the building of the Great Wall (photo at left), and, in another context, spread the use of chopsticks for eating. Confucianism was to become China's guiding philosophy for two millennia.

More than a thousand years after the Han, something happened that the Russians would understand. The Mongols drove the local rulers from power and took over the state. But instead of imprinting Mongol rules on their Chinese subjects, the Mongols underwent **Sinicization [6]** or, as it is sometimes said, **Hanification [7]**—that is, they adopted many of the ways of the People of Han. Hostility between the Mongols and the Chinese

XIAN

THE CITY KNOWN today as Xian in Shaanxi Province is the site of one of the world's oldest urban centers. It may have been a settlement during the Shang-Yin Dynasty more than 3000 years ago; it was a town during the Zhou Dynasty, and the Qin emperor was buried here along with 6000 life-sized terracotta soldiers and horses, reflecting the city's importance. During the Han Dynasty (around 200 BC to AD 200) the city, then called Chang'an, was one of the greatest centers of the ancient world, the Rome of ancient China. Chang'an formed the eastern terminus of the Silk Route, a storehouse of enormous wealth. Its architecture was unrivaled, from its ornamental defensive wall with elaborately sculpted gates to the magnificent public buildings and gardens at its center.

Situated on the fertile loess plain of the upper Wei River, Chang'an was the focus of ancient China during crucial formative periods. After two centuries of Han rule, political strife led to a period of decline, but the Sui emperors rebuilt and expanded Chang'an when they made it their capital. During the Tang Dynasty, Chang'an again became a magnificent city with three districts: the ornate Palace City; the impressive Imperial City, which housed the national administration; and the busy Outer City containing the homes and markets of artisans and merchants.

After its Tang heyday the city again declined, although it remained a bustling trade center. During the Ming Dynasty it was endowed with some of its architectural landmarks, including the Great Mosque marking the arrival of Islam; the older Big Wild Goose Pagoda dates from the influx of Buddhism. After the Ming period, Chang'an's name was changed to Xian (meaning "Western Peace"), then to Siking, and in 1943 back to Xian again.

Having been a gateway for Buddhism and Islam, Xian in the 1920s became a center of Soviet communist

© H. J. de Blij, P. O. Muller, and John Wiley & Sons, Inc.

ideology. The Nationalists, during the struggle against the Japanese, moved industries from the vulnerable east coast to Xian, and when the communists took power they enlarged Xian's industrial base still further. The present city (population: 4.9 million) lies southwest of the famed tombs, its cultural landscape now dominated by a large industrial complex that includes a steel mill, textile factories, chemical plants, and machine-making facilities. Little remains (other than some prominent historic landmarks) of the splendor of times past, but Xian's location on the main railroad line to the vast western frontier of China sustains its long-term role as one of the country's key gateways.

never diminished, however. Marco Polo saw it first-hand during his early medieval travels to "Cathay" and reported on it when he returned to Europe.

Mongol (Yuan Dynasty) rule ended in 1368, and it is amazing that, in the five and a half centuries dynastic China was subsequently to endure, only two dynasties were to rule: the Ming (1368 to 1644) and the Qing (1644 until the final collapse in 1911). The Ming, ruling their empire from the Forbidden City in Beijing (where ordinary mortals were forbidden entry), started out triumphantly: they annexed northern Korea, Mongolia, even Myanmar (Burma). They dispatched huge fleets into Pacific and Indian Ocean waters to explore the wider world. But it all came crashing down under conditions we talk a great deal about today: climate change. A major environmental shift called the *Little Ice Age* that had struck Europe in the previous century now afflicted Ming China. Wheat-

fields lay barren. With 100 million mouths to feed, the emperors worried about revolution. They burned the fleets, built barges, and extended the Grand Canal from the Lower Chang Basin to transport rice to the hungry north. The Ming Dynasty never recovered.

By the mid-1600s, the Manchus, a northern people with historic Mongol-Tatar links, saw their opportunity and took control in Beijing. The People of Han are still shaking their heads about it—a people numbering about 1 million managed to grab the reins of power over a nation of several hundred million. The invaders adopted Chinese ways, kept the Ming systems of administration and education, called themselves Qing, and set about expanding the empire. It was, as we noted earlier, the emperors' last hurrah, although a long drawn out one. The Qing Empire was the largest Chinese empire ever (Fig. 9A-9), but it was also weak. Our map shows the maximum dimensions the Qing

FIGURE 9A-9

© H. J. de Blij, P. O. Muller, and John Wiley & Sons, Inc.

Empire acquired as well as the colonial domains the Europeans, Russians, and Japanese appropriated. Five (or more) millennia of dynastic rule over China ended in war, revolution, and final collapse in 1911, just over a century ago.

China in Disarray

Figure 9A-9 provides only a mere hint of China's disarray. Chinese emperors (and their subjects) had come to regard the state as invincible, but the colonial powers demonstrated otherwise. Economically, the colonialists imported cheap manufactures into China, whose handicraft industries succumbed in the face of unbeatable competition. In addition, British merchants imported large quantities of opium from British India into China, and soon this addictive intoxicant was destroying the very fabric of Chinese cultural life. When the Qing government tried to resist this influx, the First Opium War (1839–1842) proved disastrous. The victorious British forced China's rulers to acquiesce to opium importation; the breakdown

Extraterritoriality

DURING THE NINETEENTH century, as China weakened and European colonial invaders entered China's coastal cities and sailed up its rivers, the Europeans forced China to accept a European doctrine of international law—*extraterritoriality*. Under this doctrine, foreign states and their representatives are immune from the jurisdiction of the country in which they are based. Today this applies to embassies and diplomatic personnel. But in Qing (Manchu) China, it went far beyond that.

The European, Russian, and Japanese invaders established as many as 90 treaty ports—extraterritorial enclaves in China's cities under unequal treaties enforced by gunboat diplomacy. In their "concessions," diplomats and traders were exempt from Chinese law. Not only port areas but also the best residential suburbs of large cities were declared to be "extraterritorial" and made inaccessible to Chinese citizens. In the city of Guangzhou (Canton of colonial times), Sha Mian Island in the Pearl River was a favorite extraterritorial enclave. A sign at the only bridge to the island stated, in English and Cantonese, "No Dogs or Chinese."

of Chinese sovereignty was under way. When the Chinese tried that approach again, the Second Opium War 15 years later led to the Chinese being forced to allow the cultivation of the opium poppy in China itself. Chinese society was disintegrating beneath a narcotic tide that proved uncontrollable.

Meanwhile, the ethnic complex of Figure 9A-7 was in the making. The Beijing court was forced to grant concessions and leases to foreign merchants; China ceded Hong Kong to the British even as British ships sailed up the Yangzi to consolidate a huge sphere of influence in China's midsection. Portugal took Macau, Germany established itself on the Shandong Peninsula, the French encroached on China from their Asian colonies farther south, and the Russians entered the Northeast, then still called *Manchuria*. The Japanese subsequently invaded Korea, annexed the Ryukyu Islands (they still possess them, including Okinawa, today), and colonized Formosa (now called Taiwan) in 1895.

All over China, under the doctrine of **extraterritoriality [8]**, the British and other colonial powers carved out ***concessions*** in which Europeans, as well as Russians and Japanese, were immune from local Chinese law (see box titled "Extraterritoriality"). Some of these urban enclaves, business as well as residential, were even off-limits to Chinese citizens. In many other buildings, parks, and additional facilities, Chinese found themselves unable to enter without permission from foreigners. This contributed to the loss of face that helped provoke the so-called Boxer Rebellion of 1900, when bands of revolutionaries roamed cities and countryside, attacking and killing not only foreigners but also their Chinese collaborators. It was a portentous uprising, put down with great loss of life by a multinational force consisting of British, Russian, French, Italian, German, Japanese, and American soldiers. China had no friends to call upon.

Revolutionary China

Even as China's public order continued to disintegrate, a better-organized revolutionary movement arose: the so-called Nationalist movement under a prominent leader named Sun Yat-sen. Blaming the sclerotic dynastic regime for China's troubles, these Nationalists in 1911 mounted a fierce attack on the emperor's garrisons all over China. Within a few months the 267-year-old Qing Dynasty was overthrown, and with it the system that had survived for thousands of years at China's helm.

The Nationalists, however, faced insurmountable problems in their own efforts to impose a new order on chaotic China. They did negotiate an end to the extraterritorial treaties, and, because of their well-orchestrated military campaign, they did acquire greater legitimacy than the previous regime had when confronting colonial interests. But the Nationalist government set up its base in the southern city of Canton (now Guangzhou), leaving another would-be government to attempt to rule from Peking (Beijing), the old imperial headquarters. Meanwhile, a group of intellectuals in Shanghai founded the Chinese Communist Party; one of its prominent members was a young man named Mao Zedong.

During the chaotic 1920s, the Nationalists and the Communist Party at first cooperated, with the remaining foreign presence their joint target. After Sun Yat-sen's death in 1925, Chiang Kai-shek became the Nationalists' leader, and by 1927 the foreigners were on the run, escaping by boat and train or falling victim to rampaging Nationalist forces. But soon the Nationalists began purging communists even as they pursued foreigners, and in 1928, when Chiang established his Nationalist capital in the city of Nanjing on the banks of the Yangzi, it appeared that the Nationalists would emerge victorious from their campaigns. They had driven the communists ever deeper into the interior, and by 1933 the Nationalist armies were on the verge of encircling the last communist stronghold in the area around Ruijin in Jiangxi Province. This led to a momentous event in Chinese history: the ***Long March***. Nearly 100,000 people—soldiers, peasants, leaders—marched westward from Ruijin in 1934, a communist column that included Mao Zedong and Zhou Enlai. Nationalist forces rained down attack after attack on the marchers, and of the original 100,000, about three-quarters were eliminated. But new sympathizers joined along the way (see the route marked in Fig. 9A-9), and the 20,000 survivors found a refuge in the remote mountainous interior of Shaanxi Province, 3200 kilometers (2000 miles) away. There, they prepared for a renewed campaign that would eventually bring them to power.

JAPAN'S HISTORICAL ROLE IN EAST ASIA

From Isolationism to Imperialism

Since the seventeenth century, Japan had cultivated and enforced a strict policy of isolationism [9]. Foreign influences were shunned, the Japanese people were not allowed to travel outside of Japan, and foreigners were not tolerated on Japanese soil. Contact with the outside world was extremely limited, therefore, and even trade was heavily restricted. For some time during the seventeenth and eighteenth centuries, the Dutch East India Company obtained the exclusive right to trade with Japan, but Company personnel were prohibited from setting foot on Japanese soil. So the Japanese built a small artificial island in Nagasaki Bay named Deshima where Dutch ships could dock and sailors stretch their legs. Japanese workers would cross the bridge to the island to haul the cargo ashore, but the foreigners were strictly confined to Deshima.

But as Western imperial powers (especially the British and the French) imposed their designs on Asia, isolationism became an increasingly unrealistic option for Japan because it did not want to end up itself one day as a colonial prize of the Europeans. The Industrial Revolution had yet to reach Japan, and its ability to wage war was premodern compared to the more advanced military technology of the Western powers (e.g., steamships). Prodded by a local show of force by the U.S. navy in the 1850s, the so-called *Meiji Restoration* (the return of "enlightened rule" focused on the Emperor Meiji) in 1868 introduced a wholesale change of Japanese foreign policy: away from isolationism and aimed at rapid modernization. Japan had decided to emulate the West.

After these modernizers took control of Japan, they turned to Britain for guidance in reforming their nation and its economy. During the decades that followed, the British advised the Japanese on the layout of cities and the construction of a railroad network, the location of industrial plants, and the organization of education. The British influence still is visible in the Japanese cultural landscape: the Japanese, like the British, drive on the left side of the road. (Consider how this affects the effort to open the Japanese market to U.S.-made automobiles!)

The Japanese reformers of the late nineteenth century undoubtedly saw many geographic similarities between Britain and Japan. At that time, most of what mattered in Japan was concentrated on the country's largest island, Honshu (literally, *mainland*). The ancient capital, Kyoto, lay in the interior, but the modernizers wanted a coastal, outward-looking headquarters. So they chose the town of Edo, situated on a large bay where Honshu's eastern coastline bends sharply (Fig. 9A-1). They renamed the place *Tokyo* (meaning eastern capital), and little more than a century later it was the largest urban agglomeration on Earth. Honshu's coasts were also close to mainland Asia, where raw materials and potential markets for Japanese products could be found. Importantly, the notion of a greater Japanese empire, too, was inspired in part by British example.

But the Japanese reformers who oversaw this process of modernization [10] managed to build on, not replace, Japanese cultural traditions. We in the Western world tend to equate modernization with Westernization [11]: urbanization, the spread of transport and communications facilities, the establishment of a market (money) economy, the breakdown of local traditional communities, the proliferation of formal schooling, and the acceptance and adoption of foreign innovations. In Japan, and later in other East and Southeast Asian countries, modernization was viewed as progress mainly confined to the introduction of new technologies and new ways of production. Modernization served to increase efficiency, from producing goods to running railroads to fighting war. But modernization need not affect culture and cultural norms and values. In their view, a society can be modernized without being Westernized.

In this context, Japan's modernization was a novelty. Having long resisted foreign intrusion, the Japanese did not achieve the transformation of their society by importing a Trojan horse; it was guided by Japanese planners, building on the existing Japanese infrastructure, to fulfill Japanese objectives. Certainly, the Japanese imported foreign technologies and adopted innovations from the British and others, but the Japan that was built, a unique combination of modern and traditional elements, was basically an indigenous achievement.

Japan in China

By the end of the 1880s, Japan had already built up the strongest military force in East Asia and was prepared to use it. The so-called First Sino-Japanese War of 1894–1895 resulted in a Japanese victory: a long-term Japanese presence on the mainland in the Northeast, Japanese control of Korea, and the effective Japanese annexation of Taiwan (renamed Formosa). If it was not yet clear that power in East Asia had shifted to Japan, this would become obvious when the Japanese defeated Russia in their war of 1904–1905, both powers having had imperialist designs on northeastern China (the Russians were especially interested in Port Arthur, the Yellow Sea port now known as Dalian). Along the way, Japan became ever more aggressive in its foreign policy: in 1910, it annexed Korea; in 1931, it took firm control of Manchuria (Northeast China); in 1940, it invaded French Indochina, the U.S.-controlled Philippines, the Dutch East Indies (now

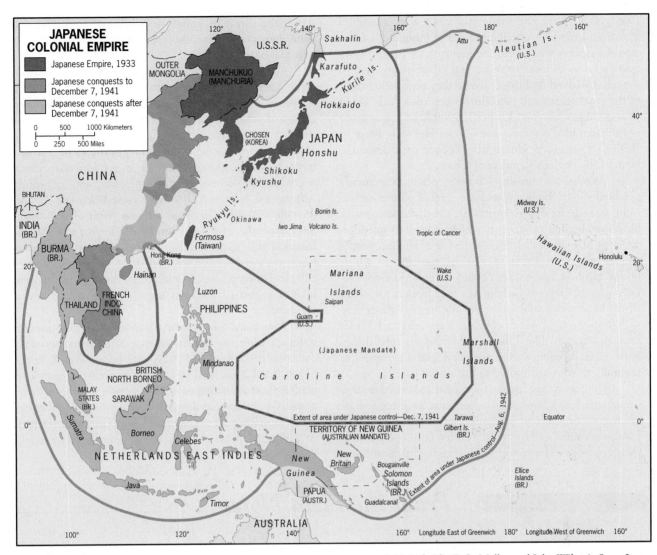

FIGURE 9A-10

© H. J. de Blij, P. O. Muller, and John Wiley & Sons, Inc.

Indonesia), and the British colonies of Burma and modern-day Malaysia; and finally, on December 7, 1941, it ignited World War II in the Pacific by attacking the United States at Pearl Harbor in Hawai'i (Fig. 9A-10).

The Japanese-Chinese conflict during this era had exposed the weakness of China, and that in turn fueled a drive for change (and rehabilitation) among the Chinese. Although many foreigners fled China during the 1920s and 1930s, others took advantage of the opportunities presented by the contest between the Nationalists and communists. The Japanese took control over the Northeast, and when the Nationalists proved unable to dislodge them, they established a puppet state there, appointed a Manchu ruler to represent them, and named their possession *Manchukuo*.

The inevitable full-scale war between the Chinese and the Japanese broke out in 1937, with the Nationalists bearing the brunt of it (providing the communists with even more time and room to regroup). Figures 9A-9 and 9A-10 show how much of China the Japanese con-

quered. The Nationalists moved their capital inland to Chongqing, and the communists controlled the territory centered on Yanan to the north. China had been broken into three pieces.

POSTWAR EAST ASIA

Communist China

After the U.S.-led Western allies defeated Japan in 1945, the civil war in China quickly resumed. The United States, hoping for a stable and friendly government in China, attempted to mediate the conflict but at the same time recognized the Nationalists as the legitimate government. The United States also aided the Nationalists militarily, destroying any chance of genuine and impartial mediation. By 1948, it was clear that Mao Zedong's well-organized militias would defeat Chiang Kai-shek's forces. Chiang kept moving his capital—back to Guangzhou, seat of Sun Yat-sen's first Nationalist government, then again to

Chongqing. In the late summer of 1949, after a series of disastrous defeats in which hundreds of thousands of Nationalist troops were killed, the remnants of Chiang's faction gathered Chinese treasures and valuables and fled to the nearby island of Taiwan. There they acquired control of the government and proclaimed their own Republic of China. Meanwhile, on October 1, 1949, standing in front of the assembled masses at the Gate of Heavenly Peace in Beijing's Tiananmen Square, Mao Zedong proclaimed the birth of the People's Republic of China.

Under communism, Chinese society was totally overhauled and the dynasties soon seemed part of an ancient past. Benevolent or otherwise, the dynastic rulers of old China headed a country in which—for all its splendor, strength, and cultural richness—the fate of landless people and of serfs often was indescribably miserable; in which floods, famines, and diseases could decimate the populations of entire regions without any help from the state; in which local lords could (and often did) repress the people with impunity; in which children were sold and brides were purchased. The European intrusion made things even worse, bringing slums, starvation, and deprivation to the millions who had moved to the cities.

The communist regime, dictatorial and brutal though it was, attacked China's weaknesses on several fronts, mobilizing virtually every able-bodied citizen in the process. Land was confiscated from the wealthy; farms were collectivized; dams and levees were built with the hands of thousands; the threat of hunger for millions receded; health conditions improved; child labor was reduced; literacy was encouraged.

At the same time, the regime committed colossal errors and engaged in systematic repression of its own people at an unheard of scale. The so-called **Great Leap Forward** (the propaganda term of what this was supposed to be) became what was perhaps the worst human-engineered catastrophe in the history of the world. A combination of delusional attempts at revolutionary glory, overwhelming economic incompetence, and unlimited ruthlessness is estimated to have caused between 30 and 45 million deaths—most as a direct result of starvation, but many others falling victim to murder, torture, and exhaustion due to forced labor. The idea was to enforce labor-intensive industrialization through the compulsory enlisting of large rural populations. But the results were inferior industrial products and a horrific disruption of agriculture.

Mao ruled China from 1949 to 1976, long enough to leave lasting imprints on the state. Another aspect of his communist ideology had to do with population. Like the Soviets (and influenced by a horde of Soviet advisors and planners), Mao refused to impose or even recommend any population policy, arguing that such a policy would represent a capitalist plot to constrain China's human resources. As a result, China's population mushroomed explosively during his rule.

From the Field Notes . . .

"Visiting the Peace Memorial Park in Hiroshima is a difficult experience. Over this site on August 6, 1945 began the era of nuclear weapons use, and the horror arising from that moment in history, displayed searingly in the museum, is an object lesson in this time of nuclear proliferation and enhanced risk. In the museum is a model of the city immediately after the explosion (the red ball marks where the detonation occurred), showing the total annihilation of the entire area at an immediate loss of more than 80,000 people and the death from radiation of many more subsequently. In the park outside, the Atomic Bomb Memorial Dome, the only building to partially survive the blast, has become the symbol of Hiroshima's devastation and of the dread of nuclear war."

www.conceptcaching.com

© H. J. de Blij

The Wages of War

JAPAN AND THE former Soviet Union never signed a peace treaty to end their World War II conflict. Why? There are four reasons, and all of them are on the map just to the northeast of Japan's northernmost large island, Hokkaido (see Fig. 9A-1). Their names are Habomai, Shikotan, Kunashiri, and Etorofu. The Japanese call these rocky specks at the southern end of the Kurile Island chain their "Northern Territories." The Soviets occupied them late in the war and never gave them back to Japan. Now they are part of Russia, and the Russians have not given them back either.

The islands themselves are no great prize. During World War II, the Japanese brought 40,000 forced laborers, most of them Koreans, to mine the minerals there. When the Red Army overran them in 1945, the Japanese were ordered out, and most of the Koreans fled. Today the population of about 50,000 is mostly Russian, many of them members of the military based on the islands and their families. At their closest point the islands are only 5 kilometers (3 mi) from Japanese soil, a constant and visible reminder of Japan's defeat and loss of land. Moreover, territorial waters bring Russia even closer, so that the islands' geostrategic importance far exceeds their economic potential.

All attempts to resolve this issue have failed. In 1956, Moscow offered to return the two smallest, Shikotan and Habomai, but the Japanese declined, demanding all four islands back. In 1989, then-Soviet President Mikhail Gorbachev visited Tokyo in the hope of securing an agreement. The Japanese, it was widely reported, offered an aid-and-development package worth U.S. $26 billion to develop Russia's eastern zone—its Pacific Rim and the vast resources of the eastern Siberian interior. This would have begun the transformation of Russia's Far East, stimulated the ports of Nakhodka and Vladivostok, and made Russia a participant in the spectacular growth boom of the western Pacific Rim.

But it was not to be. Subsequently, Russian presidents Yeltsin, Putin, and Medvedev also were unable to come to terms with Japan on this issue, facing opposition from the islands' inhabitants and from their own governments in Moscow. And so the Second World War, nearly 70 years after its conclusion, continues to cast a shadow over this northernmost segment of the Asian Pacific Rim.

Yet another calamitous episode of Mao's rule was the so-called *Great Proletarian Cultural Revolution*, launched by Mao Zedong during his final decade in power (1966–1976). Fearful that Maoist communism was becoming contaminated by Soviet "deviationism" and concerned about his own stature as its revolutionary architect, Mao unleashed a vicious campaign against what he viewed as emerging elitism in society. He mobilized young people living in cities and towns into cadres known as Red Guards and ordered them to attack "bourgeois" elements throughout China, criticize Communist Party officials, and root out "opponents" of the system. He shut down all of China's schools, persecuted untrustworthy intellectuals, and encouraged the Red Guards to engage in what he called a renewed revolutionary experience. The results were disastrous: Red Guard factions took to fighting among themselves, and anarchy, terror, and economic paralysis followed. Thousands of China's leading intellectuals died, moderate leaders were purged, and teachers, elderly citizens, and older revolutionaries were tortured to make them confess to crimes they did not commit. As the economy suffered, food and industrial production diminished. Violence and famine killed as many as 30 million people as the Cultural Revolution spun out of control. One of those who survived was a Communist Party leader who had himself been purged and thereafter been reinstated—Deng Xiaoping. Deng was destined to lead the country in the post-Mao period of economic transformation.

Japan's Defeat and Recovery

During World War II, Japan had expanded its domain farther than the architects of the 1868 modernization could ever have imagined (Fig. 9A-10). But by 1945, when American nuclear bombs devastated two Japanese cities and produced surrender shortly thereafter, the expansionist era was over and the country lay in ruins. But once again, aided this time by an enlightened U.S. postwar administration, Japan surmounted disaster. Japan was forced to accept a new constitution, and territorial adjustments were imposed (see box entitled "The Wages of War"). The new constitution stipulated that the country could not spend more than 1 percent of its GDP on the military and it had to accept the permanent stationing of U.S. troops on its soil. This was meant to constrain Japan in terms of any possible future expansionist urges—and it probably did. It also induced Japan to shift its focus from military might to economic prowess. And once again everything turned out to be extraordinarily successful.

EAST ASIA'S ECONOMIC TRANSFORMATION

Japan's Economic Miracle

Japan's accelerated economic recovery and rise to the status of world economic superpower was one of the greatest success stories of the second half of the twentieth century. Japan had lost a war and an empire, but now was scoring many economic victories in a new global arena. Japan became an industrial giant, a technological pacesetter, a fully urbanized society, a political power, and one of the most affluent nations on Earth.

In the postwar world, Japan once again proved adept at emulating the West: by 1980 the U.S. automobile industry had effectively been competed out of the market by the likes of Toyota, Nissan, and Honda. The same thing happened in the domain of consumer electronics and other high-technology products. Japan did it better and cheaper than anyone else. Cities everywhere have reliable Japanese cars on their streets; tourists across the globe shoot pictures with Japanese cameras; laboratories the world over use Japanese optical equipment. From microwave ovens to the latest smartphones, from oceangoing ships to state-of-the-art plasma TVs, Japanese-designed products flood global markets.

Nonetheless, for the past 20 years Japan's economy has been in a stall. It is still the third-biggest economy in the world, but it has lost much of its dynamism and momentum. We elaborate on Japan's current challenges in Chapter 9B.

The Asian Tigers

The Japanese economic miracle was replicated, in turn, by the four **Asian Tigers [12]**: Hong Kong, South Korea, Taiwan, and Singapore. In the 1960s and 1970s, they embarked on similar strategies that resulted in rapid industrialization propelled by the attraction of foreign investment and the creation of export processing zones for the manufacturing of **high value-added goods [13]**, including computers, mobile phones, kitchen appliances,

and a plethora of electronic devices. The Tigers quickly became trading nations, and they particularly oriented themselves to the rich Western markets. Today, it is no coincidence that most of the largest ports in the world are located in East Asia (see photo at the beginning of this chapter), with their exports shipped mainly to North America and Europe.

Note that the Asian Tigers (as well as Japan) share yet another characteristic: while they all implemented significant market liberalization measures (e.g., reducing tariffs or limits on foreign investment), they also maintained strong central governments that exerted a major, sometimes authoritarian, influence on their economies. It might even be argued that this is a reflection of "Asian values" reminiscent of Confucius—especially loyalty to the family, the corporation, and the nation together with the suppression of individual desires.

The Chinese Economic Miracle

After Mao Zedong's death in 1976, a political struggle among China's leaders pitted the so-called purists (communist bosses and intellectuals who wanted to adhere strictly to Marxist principles) against those who called themselves pragmatists, who envisaged a China that would continue to be politically and administratively communist but would open itself economically to free-market forces. Deng Xiaoping and his pragmatist allies soon prevailed, and in 1979 China began its historic transformation that was to have a global impact.

The essence of China's transformation was comparable to what the Asian Tigers had achieved earlier: the creation of a favorable environment for foreign investment to support the growth of a manufacturing sector mostly geared toward exports. Chinese wages, at least initially, were kept low, and training programs were aimed at constantly upgrading the skills of the local workforce. At the same time, political conditions remained stable because in China, more than anywhere else, the government maintains tight controls. After all, this was (and is) a communist state—but one that proved to be extremely adept at understanding how global capitalism works and how to put it to use. Pragmatism, it seems, has become the hallmark of Chinese policies.

It was as if Japan and its Tiger emulators had merely been warm-ups for East Asia's main economic act: the rise of colossal China. In less than a quarter of a century, China has emerged as the most dynamic and fastest-growing component of the world economy. To be sure, Japan and the Tigers had experienced double-digit growth rates as well, but it is a different story altogether when the leading character is a country of more than a billion people. In 2010, China surpassed

Hong Kong Island (left photo), long the focus of a British Crown Colony (until 1997) but always closely linked to adjacent mainland China, is one of the most densely built-up areas in the world, with its high-rise towers crammed together cheek by jowl. This former city-state has a major, globally-oriented central business district and offers a high-end residential lifestyle for those who can afford it. English is still the chief language and many transnational companies are located here. Just across the water to the north lies Hong Kong's port and Kowloon (right photo), a decidedly more "Chinese" area, vibrant but with a lower cost of living and catering to ethnic Chinese rather than foreign residents.

© Jan Nijman

From the Field Notes . . .

"I could not quite believe my eyes when I came upon this Starbucks in the middle of Beijing's Forbidden City: this icon of American consumerism, wrapped in ancient Chinese architecture, in the heart of this revered complex, in a country that calls itself communist? It was back in 2004 that the Chinese authorities had agreed to allow the Starbucks café on the premises in return for a hefty financial contribution to ongoing restoration efforts. But not everybody agreed this was a good idea. Several Chinese groups protested against this "'erosion" of Chinese culture, and by 2007 the government felt compelled to reverse its decision. The deal was rescinded and Starbucks was evicted from the Forbidden City. It was a splendid illustration of the pragmatism that marks the culture of the Chinese and permeates their economic policies. Call it capitalism or authoritarianism, but the Chinese prefer to do whatever works given the circumstances, and they have little time for ideological principles."

© Jan Nijman

www.conceptcaching.com

Japan to become the second-largest economy in the world, and its rise steadily continues in this decade.

GEOPOLITICS IN EAST ASIA

Sino-Japanese Relations

It should not come as a surprise that Sino-Japanese relations are problematic. A hundred years ago, China was in disarray and Japan was taking East Asia by storm—and by force. The contrast was stark: one was a wounded civilization, the other was beating the Western powers at their own game (at least for a time). Japan proudly proclaimed to stand for pan-Asian ideals, and that is how it legitimized the occupation of China. But the Japanese committed unspeakable atrocities during their campaign in China. Millions of Chinese citizens were shot, burned, drowned, subjected to gruesome chemical and biological experiments, and otherwise wantonly victimized. When you ask the Japanese, they say World War II began in 1941 when the United States declared war right after Pearl Harbor. When you ask the Chinese, they say World War II began in 1931 when the Japanese invaded Manchuria.

Years later, when China's economic reforms of the 1980s and 1990s led to a renewed Japanese presence in China, the Chinese public and its leaders called for Japan to acknowledge and apologize for these wartime abuses. The unqualified apology the Chinese desire, in word and deed, has not been forthcoming. Some Japanese history textbooks still do not acknowledge what happened to the satisfaction of the Chinese, and surveys indicate strong public sentiment on this still-sensitive issue.

But now that China has surpassed Japan in terms of economic size; now that China is the biggest exporter in the world; and now that Japan's economy has been stagnating for over two decades—this time it is China that is full of confidence and on the front foot. Although the two countries have close economic ties, their diplomatic relations are strained by clashing interests, historical memory, and cultural friction.

One of the most recent flashpoints to arise involves the Sino-Japanese-Taiwanese dispute over the Senkaku Islands in the East China Sea (Fig. 9A-11). These tiny uninhabited islands were seized by Japan in 1895 but are claimed by the Chinese, who call them the Diaoyu Islands. And they are also claimed by Taiwan, which refers to them as the Diaoyutai Islands. It is not so much the intrinsic value of the islands that is at stake here: it is far more a matter of national pride and entitlement. Moreover, some oil and gas deposits have recently been discovered in this area, and ownership of the islands grants rights to their territorial waters and what lies beneath them. It may well be that China has the most compelling claims to these islands, because eighteenth- and nineteenth-century maps (from China and Japan) seem to corroborate that they were in Chinese hands before the Japanese takeover.

EAST ASIA: GEOPOLITICAL FLASHPOINTS

FIGURE 9A-11   © H. J. de Blij, P. O. Muller, J. Nijman, and John Wiley & Sons, Inc.

The Korea Factor

Throughout history, the Koreans have been divided, partitioned, colonized, and occupied many times. Even when outsiders were not involved, their indigenous kingdoms struggled for supremacy. During the Qing Dynasty, Chinese emperors intervened at will. As China fell apart, the Japanese conquered the Korean Peninsula and annexed it as their colony in 1910.

When Japan was defeated in 1945, the Allied powers divided Korea for "administrative" purposes. The territory north of the 38th parallel was placed under the control of Soviet forces; south of this latitude, the United States was in control. In 1950, communist forces from North Korea invaded the South in a forced-unification drive, unleashing a devastating conflict that first swept southward across the peninsula, then back northward across the 38th parallel, and finally drew in Chinese armed forces who pushed the front southward again. By 1953, a military stalemate halted the hostilities, and the Korean War ended at the Cease-Fire Line not far from where the 38th parallel had marked the original 1945 boundary. Ever since, a heavily armed demilitarized zone (DMZ) has more or less hermetically sealed North from South—with the two Koreas, having grown apart, still in danger of renewed conflict.

But this is not just a bilateral issue. The Korean conflict has long had realmwide and even global dimensions. One reason is that North Korea's nuclear capability is in the hands of perhaps the most repressive and archaic regime on

Earth—unreliable, unpredictable, and without scruples. North Korea's nuclear missiles may be intended primarily for South Korea, but (theoretically at least) can also reach China, Russia, Japan—and Alaska. In late 2009, the North Koreans announced they had carried out their second successful nuclear weapons test, thereby putting East Asia on the verge of a nerve-wracking arms race.

The North Korea issue also deepens divisiveness in the realm. South Korea and Japan are diametrically opposed to the regime based in Pyongyang, whereas China takes a more neutral position, at times even seeming to lend the North Korean regime support. China appears to use North Korea in its dealings with Japan and the United States, since it is widely thought that China is crucial to containing the North Korean threat.

The North Koreans, in turn, are aware of latent Chinese support, and this seems to encourage them to test the waters. In March 2010, a North Korean submarine torpedoed a South Korean naval vessel patrolling off the west coast of South Korean near the boundary with the North, killing 46 (see both Fig. 9A-11 and the photo to the right). South Korea, the United States, Japan, and much of the world strongly condemned this act of violence, but China pointedly did not. In July 2010, South Korea and the United States engaged in a joint naval exercise in the Yellow Sea so as to remind North Korea of the wider dangerous implications of their behavior. In November 2010, North Korea shelled a small island on the South Korean side of their marine boundary, Yeonpyeong, killing two military personnel (Fig. 9A-11). This again earned them strong condemnation from the across the realm and from the United States, but once again China remained silent (adding tension to U.S.-China relations).

Taiwan: The Other China

Mention the island of Taiwan in China, and you are likely to be greeted with a frown and a headshake. Taiwan, your host may tell you, is a problem foreigners do not understand. Virtually all of the 23 million people of Taiwan are Chinese. Taiwan was part of China during the Qing Dynasty. Taiwan was stolen from China by Japanese imperialists in 1895, when it was known as Formosa. Then, when communists and "nationalists" were fighting each other for control of mainland China right after World War II, and the communists were winning, the Nationalists in 1949 fled by plane and boat to Taiwan, where they overpowered the locals. Even as Mao Zedong was proclaiming the People's Republic of China (PRC) in Beijing, the loser, Chiang Kai-shek, named his regime in Taiwan the Republic of China (ROC)—and told the world that the ROC was China's "legitimate" government.

Of course, the PRC never agreed to this, but the ROC had powerful friends, including America. Chiang Kai-shek's regime was installed in the United Nations in China's seat. Washington sent massive aid to support the island's economic recovery and weapons to ensure its security. While

As Figure 9A-11 shows, the Cease-Fire Line marking the DMZ that forms the land boundary on the Korean Peninsula extends westward into the Yellow Sea, making Paengnyong Island part of South Korea. Offshore waters have long been a source of dispute between the two Koreas, with a dangerous escalation in March 2010, when the South Korean warship named *Cheonan* suffered an explosion, broke in half, and sank with a loss of 46 sailors. North Korea denied all involvement, but recovery of the two major parts of the ship (the stern is raised in this photo) enabled experts to reconstruct what happened. A submarine-launched torpedo, not an on-board accident or a wayward mine, destroyed the vessel in what many South Koreans saw as an act of war. Once again, state terrorism brought the Korean Peninsula perilously close to open conflict.

© AFP/Getty Images, Inc.

the PRC languished under communist rule, Taiwan (the name commonly used for the ROC) grew economically. And over time, Taiwan's political system matured into a functioning (if turbulent) democracy. Chiang's followers were even voted out of office, but not by a party that wanted to reunite Taiwan with the PRC. What the Taiwanese wanted was economic prosperity, and they got it.

But the ROC could not match its economic prowess with political progress. The PRC was adamant: things might be going quite well in Taiwan, but Taiwan was still a "wayward province" that must be reunited with the motherland. And when U.S. President Nixon arrived in Beijing in 1972 for a visit that was to change the world, Taiwan was a bargaining chip. Soon, the ROC's United Nations delegation was dismissed and representatives of the Beijing government were seated in its place. Many countries around the world that had recognized Taiwan as the legitimate heir to China's leadership suddenly changed their tune. Beijing's leaders set about trying to isolate the ROC, and to a large extent they succeeded. When Deng Xiaoping and his pragmatists set up China's coastal development zones, some of them right across from Taiwan, it looked like Taiwan's days as a political entity and as an economic power were numbered.

Geography, however, was to intervene. With billions of dollars in reserves and with good connections to Overseas

Chinese in Southeast Asia, Taiwan had some strong cards to play—and the Beijing regime could not afford to deny Taiwanese companies permission to exploit opportunities in the development zones. And so, via the "back door" of Hong Kong, Taiwanese entrepreneurs built thousands of factories in mainland China, many of them located directly across the Taiwan Strait. Taiwanese businesspeople now funneled hundreds of millions of dollars into China's development boom. To be sure, the economies of Taiwan and the PRC are tightly integrated. In 2011, an estimated 40 percent of Taiwan's exports went to China along with two-thirds of all Taiwanese "foreign" investment; and at least 800,000 Taiwanese now live and work (temporarily) on the mainland.

Today, even in difficult economic times, per capita annual income in Taiwan exceeds U.S. $18,500—quadruple that of China and on a par with such countries as South Korea and Saudi Arabia. Even though a large majority of Taiwanese oppose reunification, few would want to see their economic well-being imperiled by political adventures.

China Goes Global

Napoleon once famously described China as a sleeping giant that would shake the world when it finally awoke. China today is indeed awake and changing the world; but it is changing itself and the East Asian realm even more. A closer look at these changes within the framework of East Asian's regions is provided in Chapter 9B.

> ### ▶ POINTS TO PONDER
>
> - The fortunes of China and Japan have always seemed to be inversely related: when one was rising, the other was in decline.
> - East Asia now constitutes the single biggest production complex in the world, but the realm's natural resources are limited.
> - North Korea is said to have received technological aid from Pakistan in acquiring nuclear capabilities; Pakistan is said to have received nuclear technological support from China.
> - China in 2010 objected vigorously to the sale of military aircraft by the United States to Taiwan because China views Taiwan as a temporarily wayward province to be returned to Chinese control.

REGIONS OF EAST ASIA

POPULATION

- • 50,000–250,000
- ● 250,000–1,000,000
- ⬤ 1,000,000–5,000,000
- ⬤ Over 5,000,000

National capitals are underlined

China's Coastal Core
China's Interior

—— Road
—— Railroad

| 0 | 300 | 600 | 900 | Kilometers |
| 0 | 100 | 200 | 300 | 400 | 500 | Miles |

FIGURE 9B-1

© H. J. de Blij, P. O. Muller, J. Nijman, and John Wiley & Sons, Inc.

REGIONS

China's Coastal Core

China's Interior

China's Western Periphery

Mongolia

The Korean Peninsula

Japan

Taiwan

IN THIS CHAPTER

- One-Child Policy under fire
- The miracle of Shenzhen
- China's growing regional disparities
- The demise of Tibetan culture
- Taiwan's future
- Japan's series of misfortunes

CONCEPTS, IDEAS, AND TERMS

In Chapter 9A we defined the boundaries of the East Asian geographic realm, using its politico-geographical layout as a frame of reference. But when it comes to the regions that constitute this physiographically and culturally diverse realm, things get rather more complicated. One problem is the speed with which East Asia is changing. In the introductory chapter we emphasized that regions are always subject to change, but East Asia over the past half-century has been a special case. The spectacular takeoff and subsequent faltering of Japan; the meteoric rise of eastern China; the transformation of South Korea from moribund dictatorship to burgeoning (and high-income) democracy—these are just some of the changes affecting East Asia. So what follows reflects the situation at the beginning of our century's second decade, and we can only imagine what the regional geography of this realm will look like at the opening of the third.

© Jan Nijman
Where were these photos taken?
Find out at www.wiley.com/college/deblij

© Jan Nijman

Half a century ago, some geographers described East Asia as a realm with an "empty heart"—not because China, its once and future core area, was depopulated, but because China lay isolated, with very limited contact with the outside world and less influence over its Eurasian neighbors. Imagine this: after the Chinese communists quarreled with their Soviet advisors in the 1950s and threw them out, just a few *dozen* foreigners were left in all of China. What little contact there was with the world beyond its borders went via Hong Kong. China's cities were huge but showed little evidence of modernization. The old, colonial-era bank and other corporate buildings were still the newest structures (other than the gray, faceless tenements the Soviets introduced) in their CBDs. Today, hundreds of thousands of foreigners do business, teach, study, and live long-term in Chinese cities that now have skylines to match those of any city in the world. And China is again—as it

● MAJOR CITIES OF THE REALM

City	Population* (in millions)
Tokyo, Japan	26.8
Shanghai (Shi), China	17.1
Beijing (Shi), China	12.8
Osaka, Japan	11.3
Seoul, South Korea	9.8
Chongqing (Shi), China	9.6
Guangzhou, China	9.3
Shenzhen, China	9.3
Tianjin (Shi), China	8.2
Wuhan, China	8.0
Hong Kong (Xianggang) SAR, China	7.3
Shenyang, China	5.5
Chengdu, China	5.2
Xian, China	4.9
Nanjing, China	4.7
Harbin, China	4.4
Busan, South Korea	3.4
Taipei, Taiwan	2.7
Macau SAR, China	0.6

*Based on 2012 estimates.

was during its dynastic period—the core of the East Asian realm. Indeed, this time around China is asserting itself as a full-fledged global power.

The colonial powers employed the name China Proper to denote the "real" China, the China of those who call themselves **the people of Han**. They do so because the Han Dynasty (206 BC–AD 220) is seen as the time when China became the real China, when the state adopted Confucian principles, culture flourished as never before, cities burgeoned (present-day Xian was the "Rome of China"), armies conquered distant lands, trading goods moved across the heart of Eurasia via the Silk Road to and from the Roman Empire, and population grew steadily. Ever since, the ethnic Chinese majority has referred to itself and the country as *Han China*. In the eyes of foreigners, Han China was China Proper.

Han China corresponds approximately to the extensive ethnolinguistic area mapped as Han Chinese in Figure 9A-7 and, as Figure 9B-1 shows, it consists of two major regions. The core of Han China, and of China overall, lies in the eastern coastal provinces where the great cities of Beijing, Shanghai, Hong Kong, and Guangzhou are located. Today, this is not only East Asia's most influential region but also the most dynamic spatial component of the global economic system. The economic success story that began only a little more than 30 years ago in a few designated port cities on the coast has steadily expanded to encompass a zone of development that now reaches as far west as Chengdu in Sichuan Province. But farther west still, in that vast territory

centered on Lanzhou, is quite another China. This region we call the Interior. Here the Han are rural, poor, and far removed—geographically as well as functionally—from the booming Pacific seaboard. The contrasts between the increasingly affluent coastal core and the much poorer and still-rural remainder of the country are a matter of growing concern to the Chinese themselves. Will the core fracture? Can China bridge the gap between coast and interior?

Beyond the Interior, as Figure 9B-1 shows, lies the Western Periphery that includes blue-colored Tibet (which the Chinese call Xizang) and Xinjiang (colored brown) under Chinese control, and independent Mongolia (green) outside of it. And in the east, seaward of the mainland shores, is a part of the realm that could not be more different from the landlocked Western Periphery: here lie North Korea, where China has significant influence; South Korea, now a strong trading partner of China; Japan, a longtime rival of China; and Taiwan, which is claimed by mainland China as a wayward province.

To summarize, we will discuss East Asia's regions in the following sequence:

1. China's Coastal Core

2. China's Interior

3. China's Western Periphery

4. Mongolia

5. The Korean Peninsula

6. Japan

7. Taiwan

But in order to set the stage, we need to begin with a general overview of China as well as a profile of its administrative structure, because that is critical to the way in which this huge country is governed.

■■ THE PEOPLE'S REPUBLIC OF CHINA

Citizens of the People's Republic of China (PRC) have no doubt: not only does their country still have the largest population of any state in the world ("still" because India is closing in), but theirs is the oldest continuous civilization on the planet. Yet China's territory does not match its demography. For all its 1.36 billion people, China is just a bit larger than the conterminous United States (Fig. 9B-2). Distances in China are comparable to those in the Lower-48. It is about the same distance from Beijing to Shanghai as it is from New York to Chicago. And as the map shows, east-west distances also are similar. Latitudinally, however, China extends farther north as well as south of the contiguous United States, creating a greater range of natural environments. In the far northeast, China comes close to Siberian cold. The south exhibits Caribbean warmth—with typhoons (as hurricanes are called in Pacific Asia) to match.

CHINA AND THE CONTERMINOUS UNITED STATES

Political and Administrative Divisions

China and the United States may be approximately the same size, but they have very different government structures: the United States is a federation whereas China is a highly centralized unitary state. For administrative purposes, China is divided into the following units (Fig. 9B-3):

4 Central-government-controlled Municipalities (each is known as a *Shi*)

22 Provinces

2 Special Administrative Regions (SARs)

5 Autonomous Regions (ARs)

The four central-government-controlled ***Municipalities*** (or ***Shi's***) are the capital, Beijing; the nearby port city of Tianjin; China's largest metropolis, Shanghai; and the Chang River port of Chongqing in the Sichuan Basin. These *Shi's* form the cores of China's most populous and important

FIGURE 9B-2   © H. J. de Blij, P. O. Muller, and John Wiley & Sons, Inc.

POLITICAL DIVISIONS OF CHINA

- ─·─·─   International boundary
- ─ ─ ─   Province boundary

National capital is underlined

FIGURE 9B-3   © H. J. de Blij, P. O. Muller, and John Wiley & Sons, Inc.

BEIJING

BEIJING (population 12.8 million), capital of China, lies at the northern apex of the roughly triangular North China Plain, just over 160 kilometers (100 mi) from its port, Tianjin, on the Bohai Gulf. Urban sprawl has now reached the hills and mountains that bound the Plain on the north, a defendable natural barrier fortified by the builders of the Great Wall. Today you can drive to the Great Wall from central Beijing in about an hour.

Although settlement on the site of Beijing began thousands of years ago, the city's rise to national prominence began during the Mongol (Yuan) Dynasty, more than seven centuries ago. The Mongols, preferring a capital close to their historic heartland, endowed the city with walls and palaces. Following the Mongols, later rulers at times moved their capital southward, but the government always returned to Beijing. From the third Ming emperor onward, Beijing was China's focus; it was ideally situated for the Manchus when they took over in 1644. During the twentieth century, China's Nationalists again chose a southern capital, but when the communists prevailed in 1949 they immediately reestablished Beijing as their (and China's) headquarters.

Ruthless destruction of historic monuments carried out from the time of the Mongols to the communists (and now during China's modernization) has diminished, but not totally destroyed, Beijing's heritage. With fifteenth-century monuments such as the Forbidden City of the Manchu emperors (left photo below) and the Temple of Heaven, Beijing

© H. J. de Blij, P. O. Muller, and John Wiley & Sons, Inc.

remains an outdoor museum and the cultural focus of China. Successive emperors and aristocrats, moreover, bequeathed the city numerous parks and additional recreational spaces that other cities lack. Most recently, Beijing has been transformed by China's new economic policies. A forest of high-rises now towers above the retreating traditional cityscape, avenues have been widened, and expressways have been built. A new era has arrived in this historic capital (right photo below).

Beijing, old and new: the Forbidden City and Chang'an Avenue.
© Jan Nijman

subregions, and direct control over them from the capital entrenches the central government's power.

We should keep in mind that the administrative map of China continues to change—and to pose problems for geographers. The city of Chongqing was made a *shi* in 1996, and its municipal area was enlarged to incorporate not only the central urban area but an enormous hinterland covering all of eastern Sichuan Province. As a result, the urban population of Chongqing is now officially 30 million, making this the world's "largest metropolis"—but in truth, the central urban area contains

less than 10 million inhabitants. And because Chongqing's population is officially *not* part of the province that borders it to the west (Fig. 9B-3), the official population of Sichuan dropped by 30 million when the *Chongqing Shi* was created.

China's 22 **Provinces**, like U.S. States, tend to be smallest in the east and largest toward the west. The territorially smallest are the three easternmost provinces on China's coastal bulge: Zhejiang, Jiangsu, and Fujian. The two largest are Qinghai, flanked by Xizang (Tibet), and Sichuan, China's Midwest.

As with all large countries, some provinces are more important than others. Hebei Province nearly surrounds Beijing. Shaanxi Province is centered on the great ancient city of Xian. In the southeast, momentous economic developments are occurring in Guangdong Province, whose urban focus is Guangzhou. When, in the pages that follow, we refer to a particular province or other administrative unit, Figure 9B-3 is a useful locational guide.

In 1997, the British dependency of Hong Kong was taken over by China and became the country's inaugural *Special Administrative Region (SAR)*. In 1999, Portugal similarly transferred Macau to Chinese control, and this former colony, situated opposite Hong Kong on the Pearl River Estuary, became the second SAR under Beijing's administration.

The five *Autonomous Regions (ARs)* were created in order to recognize the non-Han ethnic minorities living there. Some laws that apply to the Han Chinese do not apply to certain minorities. As we saw in the case of the former Soviet Union, however, demographic changes and population movements affect such regions, and the policies of the 1940s may not work in the twenty-first century. Han Chinese immigrants now outnumber several minorities in their own ARs. The label "autonomous" should not be taken literally here, and more than anything else it is a political gesture; Beijing maintains firm control in every corner of China. The five Autonomous Regions are: (1) Nei Mongol AR (Inner Mongolia); (2) Ningxia Hui AR (adjacent to Inner Mongolia); (3) Xinjiang Uyghur AR (China's northwest corner); (4) Guangxi Zhuang AR (far south, bordering Vietnam); and (5) Xizang AR (Tibet, anchoring the southwest).

China's "Capitalist" Turn

It was not hard to see why China needed to change course in the late 1970s: agriculture had picked up from the disastrous effects of the "Great Leap Forward" two decades earlier, but meaningful industrial progress continued to elude Beijing's economic planners. China was now falling further behind in terms of technology, a liability made abundantly clear in any comparison to the miraculous ascent of Japan. And even that "wayward province" of Taiwan was outperforming the mainland—and at a Tigerish pace to boot. Something had to change.

Taking their cues from the accelerated development of Japan and the four trailing Asian Tigers, the new leadership saw opportunities that could not even be mentioned during Mao's reign (1949–1976). For example, the Lower Chang Basin—the subregion anchored by Shanghai and networked by the Yangzi and its tributaries—had retained some of its historic identity and energy. Drab, teeming, and decrepit though it had become, Shanghai had not completely lost its intellectual vigor, artistic individuality, risk-taking entrepreneurs, or even its opponents of communist dogma. And Shanghai's geography still offered immense opportunities. From day one, Deng Xiaoping

and his pragmatists in Beijing saw in Shanghai what they could not yet foresee in their own backyard. Here was a vibrant hub on the Pacific coast right at the mouth of the realm's greatest river, loaded with talent and accustomed to taking chances. And in its vast hinterland lay much more than a farmscape of flood-prone wheatfields: the Sichuan Basin alone had a population (at the time) of more than 100 million, growing everything from rice to tea and from fruits to spices.

If China's planners needed encouragement, all they had to do was look to the southernmost subregion of eastern China: the Pearl River Estuary, outlet of the Xi River. There was Hong Kong, about as vivid a contrast between success and failure as China could present. When Deng took charge, Hong Kong was a thriving port city importing raw materials by the shipload and disgorging manufactured products that sold on markets all around the world. Hong Kong may have been a British crown colony, but it was Chinese managers and Chinese workers who propelled its economy. If you were a visitor to Hong Kong in the 1970s, they took you to a place on a nearby hillside from where you could peer across the fortified border into "Red China"—and observe some villages with duck ponds and paddies and wooden fishing boats. The contrast was crystal clear.

China's newly formulated policies of the late 1970s did not entail an official departure from communism. This is still the People's Republic of China, the Communist Party is still in control, the Politburo still calls the shots, Mao is still officially revered like a deity, and freedom of speech remains anything but. Yet the Chinese leadership fundamentally reorganized the way the economy operates in China—at least in certain vital parts of it. China created the "perfect" conditions for foreign investors from around the globe with a docile, skilled, hardworking labor force; efficient facilities; a good infrastructure; fiscal advantages; and, not least, superb accessibility vis-à-vis the rest of East Asia—already the fastest-growing regional component in the world economy.

Population Issues

China has been the world's most populous state for many centuries, and during the population explosion of the twentieth century it was also one of the fastest-growing. During Mao's rule, when China still was a dominantly agricultural society, families were encouraged to produce numerous children, and China grew at a rate approaching 3 percent per year. That was the policy inherited by Deng and his reformers, who immediately grasped that slowing down China's rate of growth was critical to its economic future. Accordingly, the new regime embarked on a vigorous population-control program that imposed on Han Chinese families (but not minorities) a one-child limit. Enforced by sometimes draconian means, this policy had the desired effect: by the mid-1980s China's annual population growth was down to 1.2 percent, and by 2010 the rate had dipped to 0.5 percent, less than half the average for the world as a whole (1.2 percent).

The one-child policy had its desired economic impact, but its other results were less salutary. China's is a dominantly patriarchal society, where male children are much preferred; rates of female fetus abortion, infanticide, and abandonment skyrocketed, resulting in a gender imbalance that raises concern for the future. China's administration estimates that, if the policy is not modified, the society will be short some 30 million brides (the reported gender imbalance in births for 2010 was 120 boys for every 100 girls). This is already increasing the trafficking of women both within China and from neighboring countries, where public resentment of Chinese males pursuing (and sometimes abducting) local females is growing.

The one-child policy had other outcomes as well. An important one was that China's became an aging population as its proportion of youngsters shrank while the older age cohorts mushroomed. That raised the specter of a population implosion (discussed in the chapters on Europe and Russia): would there be a sufficient number of workers in the younger age groups to support the ever-expanding older population? For China, this concern was magnified by the prospect that, unlike in European countries and Japan, China might grow old before it grew rich.

In 2009, a rare event occurred in the PRC: official toleration of public debate regarding Politburo policy. In the English-language newspaper *China Daily* and other media, brief articles mentioned that a national debate over the one-child policy was "tolerable," signaling open discussion of it at high levels in the Communist Party. Should China abandon the policy, demographers will need to revise their projected world population totals for the twenty-first century.

China's Urban Transformation

As recently as 1980, four out of ten Chinese people were farmers; today, almost exactly half live in towns and cities. China's urban transformation is without precedent and on a scale the world has never seen before. The urban population increased from 18 percent in 1978 to 26 percent in 1990 to 48 percent in 2011. Much of this growth has been planned and controlled by the government, and that by itself is an incredible feat. But it has brought deep changes to Chinese society, uprooting millions, infusing awareness even in remote rural areas, creating inequality within cities and between city and countryside, and creating a huge **floating population [1]** that consists of temporary urban dwellers with restricted residency rights.

The so-called *hukou* system **[2]** is based on residency permits that indicate where individuals are from and where they may exercise particular rights such as education, health care, and housing. *Hukou* dates back to ancient China and is not uncommon in other parts of Asia including Japan and Vietnam. In Maoist times, this residency-permit system became far more rigid and was used as a tool of central

VOICE FROM THE
Region

© Zheng Yue
Zheng Yue, Beijing

HUKOU AND LIFE IN CHINA

"*Hukou* is the term we use in China for a person's residential status: it tells you where you have residency rights such as health care or housing or education and how many social benefits you can enjoy. In the past, during the time of the Planned Economy, *hukou* was very important because it basically determined your entire course of life, even your access to food. When I was a little kid, my family's *hukou* entitled us to 250 grams of pork per month for all three of us— my parents usually let me eat most of it. It was the same with sugar, rice, and other things. Lots of people always want to move to the cities because everything is better there, but it was impossible for most people to get their *hukou* changed. Today, *hukou* no longer determines access to food but it still decides where you can go to school, if you are allowed to buy a car, a house, and other important things. It is still difficult to change your *hukou*. The government has made moving to selected cities easier because more workers are needed there. They will adjust your *hukou* so that you can get a house. Sometimes the government gives certain big state-owned companies the right to provide a local-city *hukou* to attract new talent from other parts of the country. I was born in Xian, and that's where I had my first *hukou*. When I was 9 years old, my parents moved to Shenzhen to work, and so I had my second *hukou* from there. When I was 27, I settled in Beijing after getting my Masters degree in the USA. Beijing provides a *hukou* for those with degrees from good overseas universities, so I got my third *hukou* from there; otherwise, it would be very difficult to obtain a Beijing *hukou*. Now, because of my status, my son has a Beijing *hukou* as well. I won't say *hukou* is a headache for everyone, but it still affects the lives of a lot of people—you'd better plan ahead if you can! And *hukou* changes from time to time. When my baby son grows up, I wonder how it will affect him and his generation."

planning to restrict the movement of people. More recently, *hukou* has been employed to manage the needed migration to China's cities—enough migrants to meet the demand for factory workers (but not too many), and always with the option to enforce a return to the originating rural area if the need for urban workers should decline.

⬤ AMONG THE REALM'S GREAT CITIES . . .

SHANGHAI

SAIL INTO THE mouth of the great Yangzi River, and you see little to prepare you for your encounter with China's largest city (population: 17.1 million). For that, you turn left into the narrow Huangpu River, and for the next several hours you will be spellbound. To starboard lies a fleet of Chinese warships. On the port side you pass oil refineries, factories, and crowded neighborhoods. Next you see an ultramodern container facility, white buildings, and rust-free cranes, built by Singapore. Soon rusty tankers and freighters line both sides of the stream. High-rise tenements tower behind the cluttered, chaotic waterfront where cranes, sheds, boatyards, piles of rusting scrap iron, mounds of coal, and stacks of cargo vie for space. In the river, your boat competes with ferries, barge trains, and cargo ships. Large vessels, anchored in midstream, are being offloaded by dozens of lighters tied up to them in rows. The air is acrid with pollution. The noise—bells, horns, whistles—is deafening.

What strikes you is the vastness and sameness of Shanghai's cityscape, until you pass beneath the first of two gigantic suspension bridges. Suddenly, everything changes. To the left, or east, lies **Pudong**, an ultramodern district with the space-age Oriental Pearl Television Tower rising like a rocket on its launchpad (see photo at right) above a forest of modern, glass-and-chrome skyscrapers that make the Huangpu look like Hong Kong's Victoria Harbor. To the right, along the waterfront Bund, stand the remnants of Victorian buildings, monuments to the British colonialists who made Shanghai a treaty port and started the city on its way to greatness. Everywhere, construction cranes rise above the cityscape. Shanghai looks like one vast construction zone, complete with a 460-meter (1510-ft) office tower—named the World Financial Center—that is one of the world's tallest buildings.

The Chinese spent heavily to improve infrastructure in the Shanghai area, including brand-new Pudong International Airport, connected to the city by the world's first maglev (magnetic levitation) train—the fastest anywhere at 430 kph (267 mph). Pudong itself has become a magnet

© H. J. de Blij, P. O. Muller, and John Wiley & Sons, Inc.

for foreign investment, attracting as much as 10 percent of the country's annual total. Shanghai's income is rising much faster than China's, and the Yangzi River Delta is becoming a counterweight to the massive Pearl River Estuary development in China's South. Among other things, Shanghai is becoming China's "motor city," complete with a Formula One racetrack seating 200,000 at the center of an automobile complex where all levels of the industry, from manufacturing to sales, are being concentrated.

In 2010, Shanghai hosted World Expo on the banks of the Huangpu River (the maglev route was extended to the site), and the city's planners, who have already transformed the place into a quintessential symbol of the New China, enjoyed the opportunity to show the world what has been accomplished here in just a single generation.

China now counts nearly 100 cities containing at least one million people, twice as many as the United States. By 2025, there could be 200. But the future is uncertain. Will there continue to be enough jobs for all these rural-urban migrants? Can cities that grow at such an accelerated pace be managed in an environmentally sustainable way? Will the Communist Party be able to maintain control over the new masses of urbanites with their changing lifestyles, identities, and demands? For

how long can the government justify the ever-widening income inequalities within the major cities (this is, after all, *communist* China), and can they meet the steadily increasing demands of industrial workers? Could these enormous economic and social changes at some point force a political transformation—and would that be for the better? Clearly, there is no blueprint for China as it tries to chart its course into the next stage of its remarkable odyssey.

From the Field Notes . . .

"Shanghai in April 2009 showed no sign of any economic slowdown and gave much evidence of the environmental consequences. From my vantage point on the top floor of the Jin Mao Tower in Pudong District, buildings beyond the Huangpu River disappeared in a dense smog on a hot spring day that locals described to me as 'pretty clear.' China's prodigious economic growth has made the People's Republic the world's largest polluter of the atmosphere, surpassing the United States. Some projections suggest that, by 2020, China may be emitting twice as much pollution as America, but since China has more than four times the population, the U.S. would still lead in terms of pollutants per capita. That's just one reason the United States shouldn't invoke the 'China excuse' to weaken its own efforts to reduce pollution from motor vehicles, factories, and other sources. 'Enjoy this view of the Oriental Pearl Television Tower,' Li Sheng told me. 'We're building the Shanghai Tower right down there where you see the construction site, and it'll be in the way. And it'll be half as tall again as your Empire State Building.' Well, let's hope the view from the top will be clearer."

www.conceptcaching.com

© H. J. de Blij

■■ CHINA'S COASTAL CORE

Many observers in the late 1970s thought that the new Chinese leadership would enact change cautiously and slowly, so nobody really foresaw the massive shift ahead. China's economic transformation was first introduced in a few carefully selected places. By setting new economic rules that would initially apply only to selected urban regions on the Pacific coast, most of the rest of the country and the bulk of the population would remain comparatively unaffected, at first. Figure 9B-4 shows how it was planned to work.

Special Economic Zones

The government introduced a three-tier system of **Special Economic Zones (SEZs) [3]**, so-called ***Open Cities***, and ***Open Coastal Areas***, which would attract technologies and investments from abroad and reshape the economic geography of eastern China. In these economic zones, investors were offered numerous incentives that included low taxes, eased import-export regulations, and simplified land leases. The hiring of labor under contract was permitted. Products made in the economic zones could be sold on foreign markets and, under some restrictions, in China as well. Even Taiwanese enterprises could operate here. And profits earned were allowed to be sent back to the investors' home countries.

When Deng's government made the decisions that would reorient China's economic geography, location was a prime consideration. Beijing wanted China to participate in the global market economy, but it also wanted to produce as little impact on interior China as possible—at least in the initial stages. The obvious answer was to position the Special Economic Zones along the coast. Initially, four SEZs were established in 1980, all with particular geographical properties (Fig. 9B-4):

1. ***Shenzhen***, adjacent to then-booming British Hong Kong on the Pearl River Estuary in Guangdong Province

2. ***Zhuhai***, across from (then still Portuguese) Macau, also on Guangdong's Pearl Estuary

FIGURE 9B-4   © H. J. de Blij, P. O. Muller, and John Wiley & Sons, Inc.

3. *Shantou*, opposite southern Taiwan, a colonial treaty port, also in Guangdong Province, source of many expatriate Chinese now residing in Thailand

4. *Xiamen*, on the Taiwan Strait, another colonial treaty port, in Fujian Province, source of many Chinese now based in Singapore, Indonesia, and Malaysia

In 1988 and 1990, respectively, two additional SEZs were proclaimed:

5. *Hainan Island*, declared an SEZ in its entirety, its potential success linked to its forward location vis-à-vis Southeast Asia

6. *Pudong*, across the Huangpu River from Shanghai, China's largest city, different from other SEZs because it was a giant state-financed project designed to attract large multinational corporations

And the process continues, with the newest SEZ authorized in 2006:

7. *Binhai New Area*, the coastal zone of the northern port city of Tianjin, a long-established Open City with

considerable foreign investment, now elevated to SEZ status, projected to outperform even Shanghai–Pudong and Shenzhen itself

The grand design of China's economic planners, therefore, was to stimulate economic growth in the coastal provinces and to capitalize on the exchange opportunities created by location; the availability of funding; the proximity of foreign investors in Southeast Asia, Taiwan, Japan, and, importantly, still-British Hong Kong; the presence of cheap labor; and the promise of world markets eager for low-cost Chinese products.

Remember that hillside view from Hong Kong in the 1970s? In those days the name *Shenzhen* was hardly known. But during the following decade, those rice paddies and duck ponds gave way to the fastest-growing city in the history of humanity—from 80,000 (mainly fishermen and farmers) in the late 1970s to over 9 million today (photo next page). Virtually everyone who lives in Shenzhen comes from somewhere else, and because all are really outsiders they speak to each other in Mandarin Chinese rather than the Cantonese that prevails all around them.

All of us who witnessed the incredible conversion of a fishing village with duck ponds and paddies into a metropolis of millions in a few years of frenzied economic and demographic growth wondered, inevitably, what kind of city Shenzhen would be. After all, shouldn't cities grow slowly, have histories etched in their architecture, develop traditional neighborhoods, and negotiate by consensus the pitfalls of planning and zoning? But look at Shenzhen only 30 years later, and it seems far more orderly, established, and accommodating than could have been imagined. Certainly Shenzhen has its richer and poorer neighborhoods—not everyone can afford to live in the CBD shown here. And when China was buffeted by the global economic downturn, Shenzhen was among SEZs where factories closed and workers had to leave. But this near-megacity is here to stay, an icon of the new China.

© Photolibrary

Meanwhile, the entire Pearl River Estuary was rapidly evolving into a world-class industrial powerhouse (Fig. 9B-5). What made it happen? When China opened the Shenzhen SEZ, hundreds of companies moved their factories from neighboring Hong Kong to Shenzhen, capitalizing on lower wages and fewer taxes, less environmental regulation, and weaker official oversight (Mao—let alone Marx—would surely have spun in his coffin). At the same time, the Chinese government also built business-friendly infrastructure; expanded seaport facilities, roads, railroads, and airports; and constructed countless apartment buildings to accommodate hundreds of thousands of workers. Soon

FIGURE 9B-5　　　　© H. J. de Blij, P. O. Muller, and John Wiley & Sons, Inc.

Shenzhen was the shining example of Deng's Open China policy, to be emulated (inasmuch as possible) by the other SEZs from Zhuhai (next to Macau) to Pudong (Shanghai) to Binhai (Tianjin).

Think about it: Shenzhen, a city barely 30 years old, now has more inhabitants than New York City, and Guangdong Province has an industrial workforce larger than that of the entire United States. The Pearl River Estuary is the most productive economy of its size in the world, accounting for just about half of all of China's exports. If you have a computer, an iPod, a mobile phone, or any similar device, it is certain to contain parts that came via the Pearl Estuary's massive manufacturing complex.

Regional Engines of Growth

None of this would hurt Hong Kong's economy. In fact, it ushered in a new and even more prosperous era for the then-still crown colony. Hong Kong now engaged less in manufacturing but far more in banking, finance, fiscal management, and business services, taking advantage of the greater skills already known to exist among its workforce during earlier economic transitions. Such transitions are the concern of the field of **economic geography [4]**, which focuses on raw-material distributions, environmental issues, historical-cultural factors (such as the lingering effects of colonialism), environmental issues, and particularly the role of spatial economic networks in regional development.

Chinese strategies were based on the understanding that rapid economic growth can be achieved and managed if it is spatially concentrated, because local conditions can be manipulated and controlled. Once growth takes off, it can have a stimulating effect on surrounding places, and the benefits can be spread around by a strong central government. China devised a truly remarkable strategy combining pure capitalist principles with powerful government involvement.

Deng and his pragmatists answered the question "How can we best utilize our resources in our effort to speed economic development?" by deliberately favoring the most accessible (i.e., coastal) provinces and cities; by opening its entire Pacific Rim to free-market forces; by

CHINA: EXPORTS BY PROVINCE, 2009, IN U.S. $BILLION

- More than 200
- 50–199
- 10–49
- 5–9
- Less than 5

FIGURE 9B-6

© H. J. de Blij, P. O. Muller, J. Nijman, and John Wiley & Sons, Inc.

exploiting its huge working class; by encouraging millions of **Overseas Chinese [5]**—most of whom had left China from those very coastal provinces to settle (and often thrive) in other countries—to invest their money in their ancestral homeland; and by attracting ever more substantial **foreign direct investment [6]** from Japan, the United States, and other key countries. Figure 9B-6 shows in which provinces China's export production is concentrated; note especially those containing the Special Economic Zones.

The Expanding Core

China's economic miracle began with the Special Economic Zones, and the government continues to use this strategy to launch new regional engines of growth through massive but concentrated investment plans. But economic development has also spilled over into adjacent areas and spread to other provinces, so an increasing number and variety of places now carry a "special" designation of one sort or another.

We should not get too fixated on the SEZs: some have received much less attention and investment than others. Hainan, for instance, has only experienced modest increases in manufacturing, which mainly involves the processing of agricultural products. Furthermore, it does not necessarily require the official status of SEZ to obtain investment (particularly domestic investment) and to achieve economic growth. Thus the entire corridor of seaboard provinces from Guangdong Province in the south to Liaoning in the north now forms the centerpiece of a highly productive regional economy, and the Core region itself has expanded over the past few years as far west as Chengdu in the Sichuan Basin (Fig. 9B-1). Export industries continue to be clustered along the Pacific Rim, but the newest economic development has decidedly been channeled toward the interior (Fig. 9B-7) to form growing manufacturing centers around such cities as Zhengzhou, Xian, Shenyang, Wuhan, Chongqing, and Chengdu (Fig. 9B-8). Therefore, today's main contrasts, geographically speaking, are between the cities and their surrounding rural areas at the metropolitan scale, and between the Coastal Core and the Interior at the regional level.

FIGURE 9B-7

© H. J. de Blij, P. O. Muller, J. Nijman, and John Wiley & Sons, Inc.

FIGURE 9B-8

© H. J. de Blij, P. O. Muller, and John Wiley & Sons, Inc.

The city of Chengdu, capital of Sichuan Province, has grown and modernized at a dizzying pace since 2000. Not long ago, this was part of the Interior; now the city has been drawn into China's Core region (Fig. 9B-1). Traditionally known as one of the country's most laid-back and livable cities, it has quickly acquired a new reputation as a leading economic center specializing in information technology, finance, and biopharmaceutical industries.

© Jan Nijman

▪▪ CHINA'S INTERIOR

China's Interior region offers a stark contrast to the burgeoning environs of Shenzhen, Shanghai, or Beijing. Situated between the Coastal Core and Western Periphery, it stretches in a vast arc from the Russian border in the far northeast to the city of Yumen (gateway to China's far west) in the northwest to the Vietnamese border in the farthest reaches of the south (Fig. 9B-1). The Interior can be divided into three subregions: the Northeast, Central China, and the South.

The Northeast

Most of this subregion, formerly known as Manchuria, is covered by Heilongjiang Province and the northeastern part of the Inner Mongol Autonomous Region (Figure 9B-3). It is relatively well endowed with coal and metallic mineral reserves (Fig. 9B-9), and

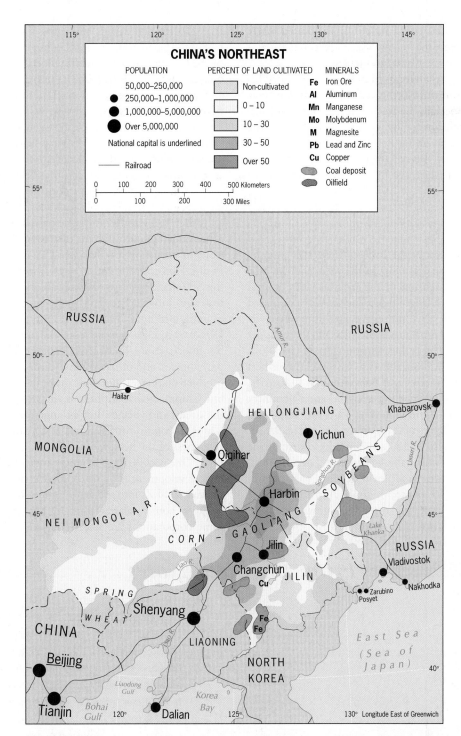

FIGURE 9B-9 © H. J. de Blij, P. O. Muller, and John Wiley & Sons, Inc.

this is where the Japanese built substantial industrial infrastructure during their colonial occupation. In 1949, the Northeast contained half of China's railroad mileage.

The communist regime, too, made the industrial redevelopment of the Northeast a priority, and cities and towns grew exponentially. But even though this subregion in the late 1950s produced more than one-quarter of China's industrial output, its state-run factories were inefficient, workers' perks were unaffordable in the long run, and the cost of its products was, to put it mildly, uncompetitive.

There was no way the Northeast would be an asset in an economically open China.

Nonetheless, this part of China is steadily changing as well. Liaoning Province, once the focus of the old regional economy, has undergone rapid transformation and is now fully integrated into the Coastal Core region. Undoubtedly, proximity to Beijing along with superior coastal access were key factors in this transformation. Its leading city, Shenyang (population: 5.5 million), today has a diversified economy that, besides heavy industry, includes aerospace, electronics, and banking. Its primary port city of Dalian, located where the Bohai Gulf meets the Yellow Sea, has become a major export zone vis-à-vis the Koreas and Japan.

Farther north, change has been slower to occur. This more remote area of the Northeast derives its significance from oilfields (especially between Qiqihar and Harbin), agriculture (though limited by a colder climate), and the potential of the zone lining its long border with Russia (Heilongjiang means "Black Dragon River," the Chinese name for the Amur River that marks much of the boundary) (Fig. 9B-9).

Central China

The Central subregion of the Interior borders Mongolia in the north, Xinjiang and Tibet in the west, and the protruding westward extension of the Coastal Core around Chengdu in the south (Fig. 9B-1). The sparsely populated western margins of this subregion are dominated by Tibetan pastoralists. Lanzhou, centrally located in the upper basin of the Huang He, is the subregion's main city. With a population of 2.4 million, this key crossroads city is best known for its oil refineries and petrochemical industries. Much of the central Interior is desert-and-steppe country that includes parts of the Gobi and Ordos deserts and, farther west, the salt lakes of the Qaidam Basin (Fig. 9A-5). But this subregion also includes the middle course of the Huang with its agriculturally productive Loess Plateau (*loess* consists of fertile, windblown deposits of rock pulverized by glaciers [see photo, p. 463]). Add the water of the middle Huang Basin and an adequate growing season, and a sizeable population has materialized.

To the south, this subregion includes the northern half of the Sichuan Basin, another highly fertile area, crossed by the Chang Jiang. Despite its vulnerability to major earthquakes (most recently in 2008), this huge natural amphitheater has supported human communities for a very long time, and you can readily discern its population cluster in Figure 9A-4. The Sichuan Basin, encircled as it is by mountains, is also one of the world's most sharply defined physiographic regions, and the concentration of its approximately 125 million inhabitants reflects that delineation.

Although there are minorities in the Interior, the great majority of the people are Han Chinese. Understandably, they feel entitled to participate in China's incredible achievements, but hundreds of millions of them still continue to subsist on less than five U.S. dollars a day. China's Pacific Rim transformation drew many millions more to the SEZ factories as well as urban and regional construction sites in what came to be the largest internal national migration in human history. Yet those who stayed behind found themselves remote from the opportunities brought by globalization—and often exploited by local mayors, party bosses, and others when it came to land, property, or produce markets. It does not make the international news, but China's inner periphery is rife with public protests in the streets and villages. It is a dangerous situation, and the Chinese communist regime is showing an increasing awareness of the need for mitigation.

One way Beijing's planners are responding to this uneven development is by spreading the privileges enjoyed by the coastal SEZs into the urban centers of the Interior. With workers' hourly salaries rising in the seaboard provinces, some corporations are finding it profitable to move their

From the Field Notes . . .

"It is about 8 in the morning when, after a long hike through the cloud forest, we reach the first major monastery on Mount Emei in Sichuan Province. The Shaolin temple appears magically from the mist as we get closer. A UNESCO World Heritage Site, Mount Emei is the highest (elevation 3099 meters/ 10,168 ft) of China's Four Sacred Buddhist Mountains. This is where the first Buddhist temple was built in China during the first century AD and where Shaolin martial arts originated in the sixteenth century. Most of China's famous old temples are located well away from the coast and major cities, undisturbed by the atheist ideologies of the communist state. Mount Emei is a major destination for Buddhist pilgrims from all over China. It is hard to imagine a more serene venue, a more perfect place for spiritual dedication."

© Jan Nijman

www.conceptcaching.com

A symbolic event took place in China during July 2006: the inaugural journey of a train linking China's capital, Beijing, with Lhasa, the capital of Xizang (Tibet). Laid by 100,000 construction workers across icy mountains, the final 1150 kilometers (700 mi) of the track constitutes an engineering marvel, climbing as high as 5072 meters (16,640 ft) across permafrost-ridden terrain. No other railway in the world reaches such altitudes; the carriages of the "Sky Train," with a total capacity of 900 passengers, are sealed and pressurized like the fuselage of an airplane. But the train has its detractors. This $4.2 billion project, they say, will facilitate Chinese penetration of Xizang and enhance Beijing's economic and military control over the remote territory while Tibet's infrastructure needs, such as road building and electrification, are neglected. It is evidence that China will spend whatever it takes to bring in larger numbers of Han Chinese to overwhelm the local Buddhist population, suppress Tibetan separatism, and promote "modernization." As such, the train to Xizang symbolizes the future as seen in Beijing, not Lhasa.

© AP/World Wide Photo

minorities here as well, especially near the borders with Myanmar, Laos, and Cambodia (see Fig. 9A-7). Indeed, bordering Vietnam in the southeast we find the Guangxi Zhuang Autonomous Region, which was specially created with these minorities in mind (Fig. 9B-3).

The Yunnan Plateau is the source of the streams that feed the Xi River and its Pearl Estuary outlet. Much of southeastern China has comparatively high relief; it is hilly and in certain places even mountainous. Among other things, this challenging terrain has inhibited overland contact between China and Southeast Asia. Yunnan Province itself, in addition to its obvious geopolitical significance, has important mineral deposits. This is also an area of great natural beauty and biodiversity that contains more than half of China's plant species.

■■ CHINA'S WESTERN PERIPHERY

Historically, empires have expanded and contracted, acquiring and losing territory as well as subjects, and leaving their imprint where they once were. Russians in Central Asia, British in East Africa, French in Indochina, Japanese in Taiwan all came and went, leaving behind languages, religions, infrastructures, and traditions.

And so it was with China. Figure 9A-9 reminds us just how vast China's Qing Dynasty empire was and how much it lost—but Figure 9B-1 reveals that China still is an empire today. Its beyond-the-Han domain includes a pair

Tibetan Buddhist pilgrims at Jokhang Temple, Lhasa.
© Jan Nijman

production facilities into the periphery. To encourage them to do so, new economic zones have been established in such deep-interior cities as Kunming and Lanzhou. Another incentive is the construction of massive public works in the central Interior, such as the diversion of part of the upper course of the Chang across the north to supply more water to drier areas. Moreover, to improve linkages between Coastal Core and Interior, China is constructing tens of thousands of kilometers of four-lane highways; a far-flung, high-speed passenger rail network (Fig. 9B-8, inset map); and hundreds of new or expanded airports.

The South

The southern subregion of the Interior, like its two counterparts, is dominated by Han Chinese, but there are substantial clusters of of large Autonomous Regions and a third one rapidly being integrated into Han China itself. Two other, smaller Autonomous Regions in the inner periphery are similarly misnamed (see Fig. 9B-3). There is nothing "autonomous" about China's minority entities: just ask the Tibetans of Xizang and the Uyghurs of Xinjiang.

Xizang (Tibet)

Tibet (called *Xizang* by the Han) is the icebound heartland of Tibetan-Buddhist culture that extends northeastward into Qinghai Province and, importantly, into a corner of far northeastern India in the Indian State of Arunachal Pradesh. Tibet shook off Chinese domination at the end of the nineteenth century, but in 1950, almost immediately after the communist regime took control in Beijing, it ordered the Red Army into Xizang to recapture the territory. Tibetan society had been organized around the fortress-like monasteries of Buddhist monks who paid allegiance to their supreme leader, the Dalai Lama.

The Chinese wanted to modernize this feudal system, but the Tibetans clung to their traditions, and in 1959 the army crushed an uprising. With the Dalai Lama now exiled and the people powerless, the Chinese destroyed much of Tibet's cultural heritage, looting its religious treasures and works of art. Under Deng's administration, part of this heritage was returned to Xizang, but Beijing also encouraged Han Chinese to move to Tibet and built "the world's highest railroad" (photo, p. 499) across the Tibetan Plateau to the capital of Lhasa. While the Dalai Lama travels the world making

the 3 million Tibetans' case and asking not for independence but for genuine autonomy, Hanification continues and resulted in another uprising in 2008 when the world was focused on China in the run-up to the Beijing Olympics. Most Tibetans alive today were born after the Chinese took control; they have gone through Chinese-mandated schools and, for the young especially, the Chinese presence has become a fact of life. A leading casualty is Tibetan culture, which is steadily being extinguished.

Tibet is now firmly controlled by the Chinese, not only for geopolitical reasons (e.g., the border with India) but also because many critical parts of China depend on the water provided by the great rivers that rise in this soaring highland. In fact, the Chinese government refers to the Tibetan Plateau as "China's water tower" and has launched several canal projects to divert some of Tibet's surplus river waters to the populated areas of the Western Periphery's arid northern interior.

Xinjiang

Xinjiang constitutes the westernmost margin of China's modern empire and is even larger than Xizang. With 22 million people, about half of them Han Chinese, the Xinjiang-Uyghur AR is even more important than its Buddhist, Tibetan neighbor. Figure 9B-10 shows why: here China meets the peoples of Turkestan and the faith of Islam; here China has significant energy reserves; and here, in the remote, clear-sky, desert-dominated far west, China built its original space program. Moreover, only a single country lies between this inner Asian outpost and the much larger

From the Field Notes . . .

© Jan Nijman

"Tibet is one of the most colorful and spiritual places on Earth, high in the Himalaya with its steely blue skies and dry, crisp air. Since 1950, Tibet has been controlled by the People's Republic of China. The Chinese military seem omnipresent, and it is hard to escape a sense of forceful occupation, especially in the capital city, Lhasa. In late 2009, I explored the busy streets around the famous Jokhang Temple, a major pilgrimage destination for Tibetan Buddhists. There, too, the Chinese were conspicuously present, with observation posts perched atop buildings as if to remind ordinary Tibetans of the political order of the day."

www.conceptcaching.com

FIGURE 9B-10

© H. J. de Blij, P. O. Muller, and John Wiley & Sons, Inc.

oil and gas reserves of the Caspian Sea Basin: Kazakhstan. As noted in Chapter 7B, connecting pipelines are already in place and more are under construction.

But Xinjiang's Autonomous Region is no more autonomous than Tibet. During the Qing Dynasty, the Muslim peoples here—including Uyghurs, Kazakhs, Tajiks, Kyrgyz, and others—fell under Chinese control; but even when the communist regime came to power in Beijing, fewer than one in 20 inhabitants of Xinjiang was Han Chinese, and the area was named the "East Turkestan Republic." Today, barely six decades later, nearly 50 percent are Han, and Xinjiang has become a two-tiered society.

The modern, regimented, and urbanized component of Xinjiang is focused on the revitalized capital of Ürümqi and the nearby model city of Shihezi, located in the northern, Junggar Basin; the traditional, religious, and rural component is anchored by the historic town of Kashgar (Kashi on Chinese maps) in the extreme western corner of the southern, Tarim Basin (Fig. 9B-10).

Where the Uyghurs, who form the majority among the local peoples, meet the Han, as occurs in the factories of Ürümqi and the streets of Kashgar, the results are sometimes tragic. In 2009, a skirmish between Uyghur workers and Han managers at a plant on the outskirts of Ürümqi spun out of control and led to deadly rioting in which some 200 Han Chinese citizens were killed and some 1,700 were wounded. For years, Uyghur militants have pursued an intermittent and fruitless campaign of violence against Han Chinese domination.

In the meantime, Beijing is tightening its grip on the Western Periphery for economic as well as geopolitical reasons. In 2010, the city of Kasghar was designated an "economic development zone," the first in this subregion. At the same time, the government declared Kashgar and Shenzhen to be "sister-cities," which supposedly means that Guangdong Province is going to "pay" about half of all the initial development costs in Kashgar and that the latter will be modeled after Shenzhen. It is a promise of accelerated progress, held up constantly to the local population.

Consider the geographic significance of Kashgar: it has a long history as an important waystation on the ancient Silk Road; it is close to the borders with Pakistan, Afghanistan, Kyrgyzstan, Uzbekistan, and Tajikistan; and it is already the home of Central Asia's largest marketplace. The city is by far the most effective link between China and this sector of Central Asia, and is to be swiftly developed into a major trading and logistical bridge. Clearly, China today is also looking closely at its far west as it continues to concentrate on its westward-expanding Core region.

Both Xizang and Xinjiang exhibit the properties of peripheries: local cultures and traditional economies are overpowered by national interests and global systems, widening disparities not only between core and periphery but also between and among societies within the periphery itself. When, as in China's case, the political system affords inadequate opportunity for the expression of grievances and the representation of local interests, the symptoms of marginalization become entrenched.

We now turn to the East Asia lying beyond China's borders. Some of what we encounter will be reminiscent of China's Interior, but elsewhere the mainland and island entities of this realm display all the properties of core areas—not just the East Asian core, but the global core. Here, in fact, was where the East Asian transformation began that in short order reached the PRC and changed the world.

▪▪ MONGOLIA

That East Asian transformation, however, has not yet reached Mongolia. This immense, landlocked, isolated country wedged between China and Russia, with an area larger than Alaska but a sparse population of less than 3 million, suggests a steppe-and-desert-dominated vacuum between two of the world's most powerful countries (Fig. 9B-1).

This used to be the domain of a powerful people who swept westward to challenge the Russians and southeastward to rule China, but today it is a weak and vulnerable country whose 800,000 herders and their millions of sheep follow nomadic tracks along the fenceless fringes of the immense Gobi Desert, where Siberian cold periodically causes severe human and livestock losses. During the Soviet era, the location of the capital of Ulaanbaatar symbolized Mongolia's security against Chinese encroachment; today, however, Chinese investment and involvement are growing. As a historic **buffer state [7]** within interior Asia, Mongolia has few options.

But change may be on the horizon now that large deposits of gold, copper, and coal have been found and exploration by large foreign companies has begun. The International Monetary Fund predicts double-digit annual growth for years to come. The main buyer of these commodities, to no great surprise, will be China, and Chinese investments have been pouring in, mainly for infrastructural projects. It is not clear whether ordinary Mongolians, among the poorest people in Asia, will get to benefit from this bonanza any time soon.

▪▪ THE KOREAN PENINSULA

If you were able to cross the tightly sealed border between China and North Korea, it would seem (as in the case of Mongolia) that you have once again gone from core to periphery. No evidence here of the economic success story we tend to associate with the geographic term "Korea." But if you could travel freely to North Korea's other border with South Korea and cross it, you would have no doubt you were still in the realm's core. That half of the Korean Peninsula was and continues to be the scene of what economic geographers refer to as an economic miracle. Once a sclerotic dictatorship with a moribund economy, South Korea now has a dynamic economy typical of one of the most prominent burgeoning *economic tigers* (modernizing, industrialized, fast-developing beehive countries) in East Asia.

The Korean Peninsula is not that large: its total territory is about the size of Idaho, and much of it (especially the north) is mountainous and rugged. Nonetheless, its population stands at 72.3 million (Idaho's is only 1.6 million). The map (Fig. 9B-11) suggests that the Korean Peninsula is very nearly a bridge connecting East Asia's mainland to the islands of Japan, and in ancient times when sea levels were lower that link was even closer. As a result, Korea has long been a stage for the contest between Chinese and Japanese cultural influences. More recently it became the scene of devastating great-power conflict between communist and anticommunist armies. The ceasefire line you see on the map marks the armistice that ended the Korean War of 1950–1953. But unlike the long-gone Berlin Wall and Iron Curtain, it is still there. Thus North and South Korea still await reunification.

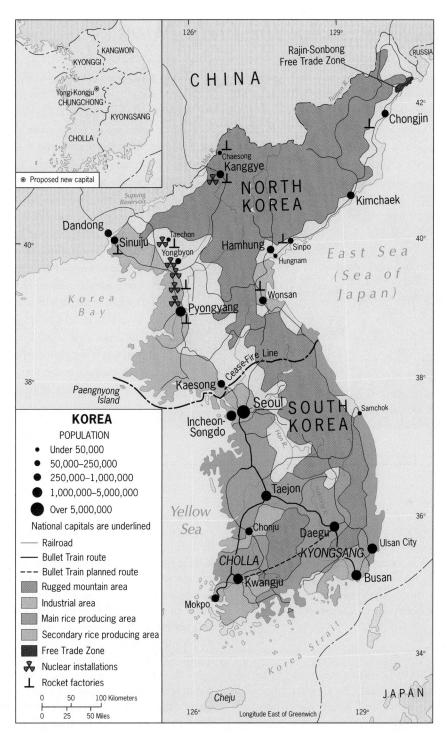

FIGURE 9B-11   © H. J. de Blij, P. O. Muller, and John Wiley & Sons, Inc.

North Korea

Almost seven decades of the harshest kind of communist rule have turned North Korea (23 million) into one of the poorest, hungriest, and most regimented countries on Earth. A regime that imprisons its citizens for the slightest infractions, operates a gulag of prison camps reputedly worse than that of the Soviets, starves its people as punishment, and isolates its subjects from the rest of the world nevertheless finds sufficient resources to achieve missile technology, nuclear capability, as well as associated weaponry with which to blackmail its neighbors and infect the world. Refugees' and escapees' stories tell of extreme poverty and misery, but China, North Korea's ideological ally, has sealed its lengthy border in the interest of "stability" rather than pressuring the

From the Field Notes . . .

"From the observation platform atop the Seoul Tower one would be able to see into North Korea except for the range of hills in the background: the capital lies in the shadow of the DMZ (demilitarized zone), relic of one of the hot conflicts of the Cold War. The vulnerable Seoul-Incheon metropolitan area ranks among the realm's largest, its population approaching 10 million. I asked my colleague, a professor of geography at Seoul National University, why the central business district, the cluster of buildings in the center of this photo, does not seem to reflect the economic power of this city (note the sector of traditional buildings in the foreground, some with blue roofs, the culture's favorite color). 'Turn around and look across the [Han] River,' he said. 'That's the new Seoul, and there the skyline matches that of Singapore, Beijing, or Tokyo.' Height restrictions, disputes over land ownership, and congestion are among the factors that slowed growth in this part of Seoul and caused many companies to build in the new Samsung area. Looking in that direction, you cannot miss the large U.S. military facility, right in the heart of the urban area, on some of the most valuable real estate and right next to an upscale shopping district. In the local press, the debate over the presence of American forces dominated the letters page day after day."

© H. J. de Blij

www.conceptcaching.com

rulers in the capital of Pyongyang to reconsider their policies. Recent tentative moves toward change—a few cross-border visits by South Koreans to families in the North, some tourists allowed into the North, several (failed) industrial cooperatives, an American symphony orchestra visiting the capital—have essentially gone nowhere.

South Korea

The tragedy of the Korean Peninsula is especially painful when put in geographic context. As all environmental maps indicate, North Korea and South Korea actually need each other: the North has raw materials required by industries in the South, and South Korea produces food needed in the North. This is a condition known as **regional complementarity [8]**, but in Korea the political situation prevents it from becoming operational to mutual advantage.

Therefore, the narrow demilitarized zone (DMZ) that divides the Koreas marks the ultimate core-periphery border at this peninsular scale. South Korea (49.3 million) emerged from the ravages of the Korean War as an unstable, dictatorial, politically and economically corrupt nation, its initial **state capitalism [9]** propelled by powerful industrial conglomerates in cahoots with the politicians.

But then democracy took hold, corruption was confronted (if not altogether tamed), and the economy took off. South Korea, ever more connected to the world, became an early economic tiger on the Pacific Rim, the world's largest shipbuilding nation, a major automobile manufacturer, a producer of goods ranging from iron and steel to electronics and chemicals. Seoul, the capital, lies at the center of an urban-industrial complex with 9.8 million inhabitants facing the Yellow Sea across from China.

But there is much more to South Korea than its capital-dominated core area. The country's historic, traditional subregions still figure prominently in its contemporary economic geography (Fig. 9B-11). In the southeast, *Kyongsang*, located on the Korea Strait opposite Japan's main island and centered on the city of Busan, has become South Korea's second-ranking manufacturing agglomeration (the first Hyundai cars were made at nearby Ulsan). And in the far southwest lies *Cholla*, anchored by the city of Kwangju, an area historically marginalized by the powers in Seoul but now finally coming into its own.

The contrasts between North and South on the Korean Peninsula are startling—and dangerous. Take a look at the comparative statistics for North and South in Appendix B and you will see why North Koreans flee to the South whenever an opportunity arises.

● AMONG THE REALM'S GREAT CITIES . . .

SEOUL

SEOUL (population: 9.8 million), located on the Han River, is ideally situated to be the capital of all of Korea, North and South (its name means capital in the Korean language). Indeed, it served as such from the late fourteenth century until the early twentieth, but events in that century changed its role. Today the city lies in the northwest corner of South Korea, for which it serves as capital; not far to the north lies the tense DMZ that contains the cease-fire line with North Korea. That line cuts across the mouth of the Han River, thereby depriving Seoul of its inland waterborne traffic. Its bustling ocean port, Incheon, has emerged as a result (Fig. 9B-11).

Seoul's undisciplined growth, attended by a series of major accidents including the failure of a key bridge over the Han and the collapse of a six-story department store, reflects the unbridled expansion of the South Korean economy as a whole, as well as the political struggles that carried the country from autocracy to democracy. Central Seoul lies in a basin surrounded by hills to an elevation of about 300 meters (1000 ft), and this megacity has sprawled outward in all directions, even toward the DMZ (see photo at left). An urban plan designed in the 1960s was overwhelmed by the immigrant inflow.

During the period of Japanese colonial control, Seoul's surface links to other parts of the Korean Peninsula were improved, and this infrastructure played a role in the city's later

© H. J. de Blij, P. O. Muller, and John Wiley & Sons, Inc.

success. Seoul is not only the capital but also the leading industrial center of South Korea, exporting sizeable quantities of textiles, clothing, footwear, and (increasingly) electronic goods. South Korea continues to thrive as an Asian-Pacific economic tiger, and Seoul is its heart.

▪▪ JAPAN

Japan consists of four main islands—Honshu, the largest; Hokkaido to the north; and Shikoku and Kyushu to the south—as well as numerous small islands and islets (Fig. 9B-12). Most of the country is mountainous and steep-sloped, geologically young, earthquake-prone, and studded with volcanoes. A mere 18 percent of the country is considered habitable. Japan's high-relief topography has constrained its economic development. Except for the ancient capital of Kyoto, all of Japan's major cities are perched along the coast, and virtually all lie partly on artificial land claimed from the sea. Sailing into Kobe harbor near Osaka (the country's second-largest city), one passes artificial islands designed for high-volume shipping and connected to the mainland by automatic, space-age trains. Enter Tokyo Bay, and the refineries and factories to the east and west stand on huge expanses of landfill that have pushed the bay's shoreline outward.

With just over 127 million people, four-fifths of whom reside in towns and cities, Japan uses its habitable living space very intensively—and expands it wherever possible. As Figure 9B-12 shows, farmland in Japan is both limited and regionally fragmented. Urban sprawl has invaded much of the cultivable land. In the hinterland

of Tokyo, around Osaka, and surrounding Nagoya, major farming zones are under relentless urban pressure. All three of these plains lie within Japan's fragmented but well-defined core area (delimited by the red line on the map), the heart of Japan's prodigious manufacturing complex.

Coastal Development

Figure 9B-12 also tells us a great deal about the nature of Japan's external orientation and its dependence on foreign trade. All of the country's primary and secondary regions lie on the coast. Dominant among these regions is the ***Kanto Plain***, the heart of Japan's core area, which is focused on metropolitan Tokyo and contains about one-third of the country's population. Among its advantages are an unusually extensive area of low relief, a fine natural harbor at Yokohama, a relatively mild and moist climate, and a central location with respect to the country as a whole. (Its principal disadvantage lies in its vulnerability to major earthquakes.)

The second-ranking economic region within Japan is named the ***Kansai District*** (Fig. 9B-12); it contains the Osaka-Kobe-Kyoto triangle and is located at the east end

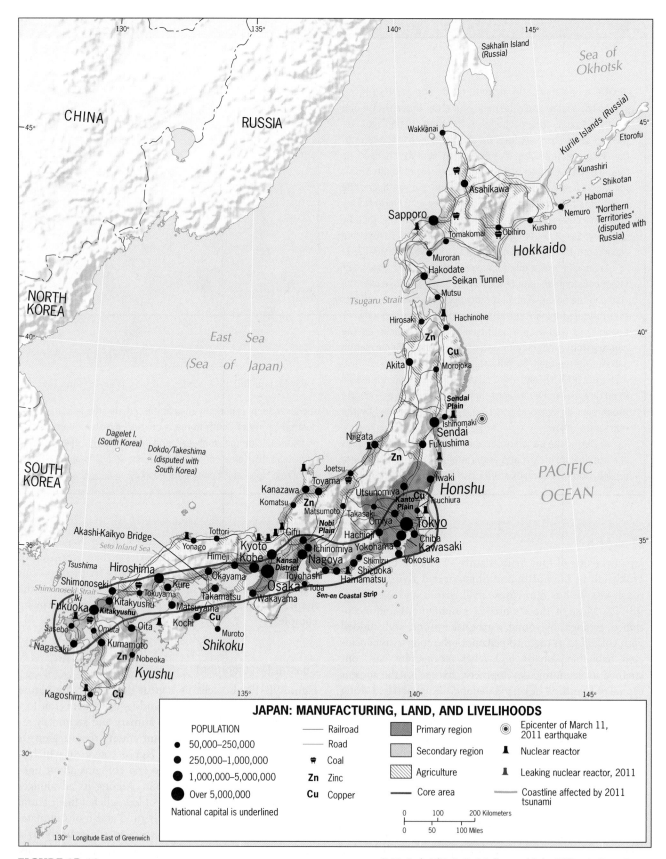

JAPAN: MANUFACTURING, LAND, AND LIVELIHOODS

POPULATION

- ● 50,000–250,000
- ● 250,000–1,000,000
- ● 1,000,000–5,000,000
- ● Over 5,000,000

National capital is underlined

— Railroad
— Road
⛏ Coal
Zn Zinc
Cu Copper

▨ Primary region
▨ Secondary region
▨ Agriculture
— Core area

◎ Epicenter of March 11, 2011 earthquake
⚒ Nuclear reactor
⚒ Leaking nuclear reactor, 2011
— Coastline affected by 2011 tsunami

0   100   200 Kilometers
0   50   100 Miles

FIGURE 9B-12

© H. J. de Blij, P. O. Muller, and John Wiley & Sons, Inc.

TOKYO

MANY URBAN AGGLOMERATIONS are named after the city that lies at their heart, and so it is with the second largest of all: Tokyo (26.8 million). Even its longer name—Tokyo–Yokohama–Kawasaki—does not begin to describe the congregation of cities and towns that form this massive, crowded metropolitan area that encircles the head of Tokyo Bay and continues to grow, outward and upward.

Near the waterfront, some of Tokyo's neighborhoods are laid out in a grid pattern. But the urban area has sprawled over hills and valleys, and much of it is a maze of narrow winding streets and alleys. Circulation is slow, and traffic jams are legendary. The train and subway systems, however, are models of efficiency—although during rush hours you must get used to the *shirioshi* pushing you into the cars to get the doors closed.

At the heart of Tokyo lies the Imperial Palace with its moats and private parks. Across the street, buildings retain a respectful low profile, but farther away Tokyo's skyscrapers seem to ignore the peril of earthquakes. Nearby lies one of the world's most famous avenues, the Ginza, lined by department stores and luxury shops. In the distance you can see an edifice that looks like the Eiffel Tower, only taller: this is the Tokyo Tower (see photo, p. 508), a multipurpose structure designed to test lighter Japanese steel, transmit television and cellphone signals, detect Earth tremors, monitor air pollution, and attract tourists.

Tokyo is the epitome of modernization, but Buddhist temples, Shinto shrines, historic bridges, and serene gardens

© H. J. de Blij, P. O. Muller, and John Wiley & Sons, Inc.

still grace this burgeoning urban complex, a cultural landscape that reflects Japan's successful marriage of the modern and the traditional.

of the Seto Inland Sea between Honshu and Shikoku. Osaka and Kobe are major industrial centers and busy ports, but the Kansai District also yields large harvests of rice, Japan's staple food. Between the Kanto Plain and the Kansai District lies the ***Nobi Plain*** (Fig. 9B-12), where Nagoya is the key city. And, as the map shows, the Japanese core area is anchored in the west by the **conurbation [10]** centered on ***Kitakyushu***, situated not on Honshu Island but in the northwestern corner of Kyushu, Japan's southernmost major island. This still-expanding edge of the core area is particularly favored by its location relative to South Korea and China.

A Trading Nation

Japan is a remarkable country that despite its poor natural endowments (except for its advantageous location) has done extremely well. As noted in Chapter 9A, it became the dominant power of East and Southeast Asia through military force prior to World War II, and its postwar economic miracle astounded the world. In the 1970s and 1980s, people referred to this corner of the world as the "Yen Bloc," one that was dominated by the Japanese economy.

The country defeated the odds by calling on old Japanese virtues: organizational efficacy, massive productivity, dedication to quality, and adherence to common goals. Note that this has always been an exceptionally homogeneous, collectively oriented, and consensus-minded nation with a strong work ethic and a deep-seated conviction that common interests override those of the individual. It is hard to overestimate this cultural context in Japan's enormous achievements—even if it can turn into a liability, as when collective rigidity impedes adaptation to new circumstances.

More concretely, Japan dealt with its own limited geographic opportunities by engaging the alternatives. Look at Figures 9A-6 as well as 9B-12 and note Japan's heavy reliance on nuclear energy—a strategy not without major risks in this earthquake-prone region, as became so painfully clear in the aftermath of the massive 2011 quake/tsunami. The importance of land reclamation in and around the country's major cities has already been underscored. But most important of all, Japan had no choice but to engage other countries, near and far, in order to compensate for what it does not have itself. Thus Japan became wealthy through trade, imported raw materials,

Tokyo, at the center of one of the largest metropolises in the world, continues to change. Land-filling and bridge-building in the bay continue; skyscrapers sprout amid low-rise neighborhoods in this earthquake-prone area; traffic congestion worsens. The red-painted Tokyo Tower, a beacon in this part of the city, was modeled after the Eiffel Tower in Paris but, as a billboard at its base announces, is an improvement over the original: lighter steel, greater strength, and less weight.

© Yann Arthus-Bertrand/Photo Researchers, Inc.

the efficient production and processing of high-value-added goods, and the ability to turn the last into lucrative exports.

Japan's Two Lost Decades

At the end of the 1980s, it seemed that Japan's meteoric rise had been so enormous that it would only be a matter of time before it would surpass even the United States' economy—but it all came to a rather sudden halt in the early 1990s. And, unexpectedly, Japan has been stuck in a rut ever since. This now more than 20-year-long economic slowdown had many causes, ranging from government mismanagement and inefficiency to growing international competition (such as from South Korean cars and Taiwanese high-technology products). Fully globalized Japan, linked to so many other economies, suffered unduly when cyclical downturns affected its trading partners. Japanese corporations and workers, long accustomed to relationships that guaranteed lifelong employment and comfortable retirement, turned stagnant. And during the lengthy pre-1990 economic boom, Japan's busy investors had bought up all kinds of properties, from landmark buildings in

From the Field Notes . . .

"The city of Kyoto, chronologically Japan's second capital (after Nara; before Tokyo), is the country's principal center

© H. J. de Blij.

of culture and religion, education and the arts. Tree-lined streets lead past hundreds of Buddhist temples; tranquil gardens provide solace from the bustle of the city. I rode the bullet train from Tokyo and spent my first day following a walking route recommended by a colleague, but got only part of the way because I felt compelled to enter so many of the temple grounds and gardens. And not only Buddhism, but also Shinto makes its mark on the cultural landscape. I passed under a *torii*—a gateway usually formed by two wooden posts topped by two horizontal beams turned up at their ends—which signals that you have left the secular and entered the sacred, and found this beautiful Shinto shrine with its orange trim and olive-green glazed tiles."

www.conceptcaching.com

New York and movie studios in Hollywood to department stores in Europe and resorts in Hawai'i. When the values of those assets plummeted, significant economic damage resulted.

One of the main problems was that Japan turned out to lack adaptability, so its culturally embedded economic system could not cope with rapidly changing circumstances. When the economy turned sour, there was no vigorous response as one might expect in a true free market. With bankruptcy a near-taboo, too many weak firms were kept alive, and this kind of economic sclerosis greatly inhibits the creation of new companies (the United States in any given year has three times the number of start-ups as Japan and twice the number of bankruptcies).

Japan's conservative culture became a liability. Whereas women constitute an ever greater share of Japan's highly educated workforce, only 8 percent of all managers are female (compared to 40 percent in the United States and 20 percent in China). For the first time in decades, Japan suffered a crisis of confidence and, more tangibly, something unfamiliar to the burgeoning Japan of the late twentieth century—rising unemployment. Faced with much leaner economic times, men especially seemed to seek refuge in the past as growing numbers preferred to stay with the same lifeless big companies instead of starting their own businesses. Even in the realm of romance, Japan seems to be in the doldrums: out of step with women's emancipation and evincing poor economic prospects, Japanese young men are sometimes referred to as "herbivores" (not the kind of vigorous, ambitious, eligible bachelors that most women would be looking for). Accordingly, the number of marriages is down and female marriage age is up—far from an ideal situation in a country hoping for a much needed increase in its birth rate.

Japan is a rapidly aging society, projected to decline from 127.4 million today to barely 95 million in 2050 and only about 65 million by the end of this century. It faces an enormous and increasing **demographic burden [11]**: by 2025, it is projected there will be only two workers for every retiree! Ethnically homogeneous Japan has historically resisted immigration, so when the government tried to recruit ethnic Japanese living in Brazil (and elsewhere) to return home, the experiment was not very successful. Thus a shrinking base of qualified workers threatens the government's capacity to sustain social programs for all.

Still another set of problems has to do with Japan's international relations. Japan never signed a peace treaty with the (then) Soviet Union after World War II because the Russians had occupied and refused to return four small island groups in the Kurile chain northeast of Hokkaido (Fig. 9B-12). Failed negotiations for the return of these "Northern Territories," as the Japanese call them, have cost Japan the opportunity to play a key role in the economic development of the Russian Far East, where crucial energy as well as mineral resources abound.

Furthermore, relations with both South and North Korea are troubled—with the South over ownership of a small island group in the East Sea (Sea of Japan) and over memories of Japanese misconduct during World War II, and with the North over the kidnapping of Japanese citizens by North Korean agents. Moreover, Japan has obvious reasons to feel particularly threatened by North Korea's nuclear weapons development. And add to all this a number of lingering issues with China over Japanese actions during the colonial and wartime periods, and it is clear that Japan has to contend with discomforts other than just economic ones.

Ever since the end of the Second World War in 1945, Japan has adhered to a constitution that essentially forbids its rearmament and to a relationship with the United States that has involved the stationing of tens of thousands of American armed forces on Japanese soil. But Japan's 2009 election campaign brought to the fore an unusually forceful reappraisal of these issues, and public opinion has been shifting toward a stronger military posture and the ouster of U.S. troops. One result of the international community's failure to constrain North Korea's nuclear aspirations may be Japan's military revival. Another consequence may be a diminution of America's capacity to maintain the stability of a potentially volatile region.

Japan's Challenges

The catastrophic earthquake of March 11, 2011 shook Japan, literally and figuratively: at least 23,000 people died, entire communities in Honshu's northeastern prefectures were destroyed, many thousands lost their loved ones, and economic damages were initially estimated at U.S. $300 billion. Besides the immediate human suffering, the disaster added a huge burden to a country already straining to return its economy to a healthier state. Before the quake, Japan had already amassed a debt exceeding twice its GDP, bigger than that of any other rich country (the United States, in comparison, has a national debt that about equals its GDP and many consider that hugely problematic). The staggering costs of recovery will unavoidably push that deficit even deeper into financial distress.

Most experts would say that Japan's earthquake preparedness has always been much better than any other nation's. And the Japanese have had much experience with lesser quakes to upgrade their plans for the big one. Perhaps so, but the prudence of building nuclear reactors in some of the most dangerous areas—let alone the country's massive reliance on nuclear power in the first place—is now being seriously questioned.

To be sure, Japan has fared poorly since 1990 in adjusting to new economic realities, and this problem has now become even more acute. Quite apart from the earthquake disaster of 2011, to get its house back in order Japan must also deal with: (1) its demographic burden, by increasing both births and foreign immigration; (2) its moribund economy, by allowing inadequately performing companies to fail and letting the creative destruction of capitalism do its work; (3) the issues of encouraging new upstart businesses by rewarding creativity and risk; (4) abandoning its conservative tendencies with respect to gender; and (5) the settlement of several geopolitical disputes that impede potentially rewarding new economic relationships, especially with Russia and China.

■■ TAIWAN

When we see where Taiwan is located, it would seem to be inextricably bound up with China because it lies a mere 200 kilometers (125 mi) from the mainland. Not surprisingly, Taiwan has a history of immigration from coastal Chinese provinces. As discussed in Chapter 9A, China's rulers view Taiwan as a "wayward province" that must one day return to the People's Republic. Taiwan lost its seat in the United Nations in 1971 when it was replaced by the PRC. The United States at the time also rescinded full-fledged recognition of Taiwan in favor of good relations with mainland China.

Taiwan's Island Geography

Taiwan, as Figure 9B-13 shows, is not a large island. It is smaller than Switzerland but has a population much larger

(23.3 million), most of it concentrated in a crescent lining the western and northern coasts. The Chungyang Mountains, an area of high elevations (some over 3000 meters [10,000 ft]), steep slopes, and dense forests dominate the eastern half of the island. Westward, these mountains yield to a zone of hilly topography and, facing the Taiwan Strait, a substantial coastal plain. Streams from the mountains irrigate the paddyfields, and farm production has more than doubled since 1950 even though hundreds of thousands of farmers left the land for employment in Taiwan's expanding industries.

Today, the lowland urban-industrial corridor of western Taiwan is anchored by the capital, Taipei, at the island's northern end and rapidly growing Kaohsiung in the far south. The Japanese developed Chilung, Taipei's port, to export nearby coal, but now the raw materials flow the other way. Taiwan imports raw cotton for its textile industry, bauxite (for aluminum) from Indonesia, oil from Brunei, and iron ore from Africa. Taiwan has a developing iron and steel industry, nuclear power plants, shipyards, a substantial chemical industry, and modern transport networks. Increasingly, however, it has been exporting high-tech products: personal computers, telecommunications equipment, and precision electronic instruments. Taiwan offers enormous brainpower resources, and many foreign

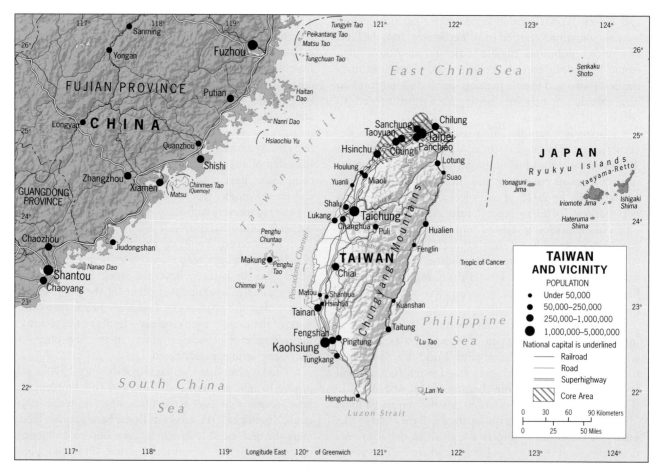

FIGURE 9B-13

© H. J. de Blij, P. O. Muller, and John Wiley & Sons, Inc.

Regional ISSUE   Chinese Rule or Independence for Taiwan?

SHOULD CHINA RULE TAIWAN?

"I am a graduate student in political geography here at East China Normal University, and I've been listening to your lecture about Chinese-American relations and the disagreement we have over Taiwan. You talk about this as though you have a right to stand between two parts of China, and with the power you have right now, I guess you can. You said that you hoped that mainland Chinese would become more 'reasonable' in their attitude toward the island province, but let me (as my professor would say) erect a hypothetical.

"Imagine that America wakes up one morning and discovers that a band of thugs have flown several 747s, loaded with weapons of all kinds, including those of mass destruction ready for use, into the San Juan, Puerto Rico airport. After taking the airport amid many casualties, they move into the city and enter administration buildings, threatening mass annihilation unless the local government resigns. By evening, Puerto Rico is under their control, American and Puerto Rican flags have been lowered and replaced by the banner of an 'independent' Puerto Rico.

"In Washington, your Congress meets and, realizing that the entire population of the island is at risk, makes the only move it can under the circumstances: it orders your armed forces to assemble a fleet to head for Puerto Rico, to blockade the island so as to create a basis for negotiation and a platform for possible military action. The fleet assembles near Miami and heads eastward through the channel between the Bahamas and Cuba.

"A day before reaching Puerto Rican waters, the path of the hastily assembled American fleet is blocked by a larger fleet from China, which has sailed from its base at Colón near the northern end of the Panama Canal. The commander of the American fleet communicates with his Chinese counterpart. 'We cannot let you pass,' states the Chinese admiral. 'We have not yet determined whether the invaders of Puerto Rico were invited by the people there to liberate them. We may have common ground with them. Please avoid an unnecessary conflict by going back to your bases.'

"How, sir, do you suppose the American people would feel toward China if a version of this scenario took place? They would be outraged at China's interference in a domestic matter. They would order your navy to attack. Well, we in China have had something like this happen with Taiwan. Thugs took it over. American aircraft carriers appeared in the Taiwan Strait. We've 'avoided conflict' for many decades, but that won't last forever. One way or another, Taiwan will submit to Beijing's rule. The Chinese people demand it."

TAIWAN SHOULD BE INDEPENDENT

"Instead of harping constantly on the 'reunification' issue, why don't Beijing's rulers get their own house in order first? Their political system is straight out of the discarded past, they're apparently allergic to democracy, and they can't seem to accept or accommodate any disobedience or dissent. I represent a Taiwanese high-tech corporation, and I fly frequently between Taipei and Hong Kong and visit the so-called People's Republic via that route—and I can tell you, the Chinese could learn a lot about democracy, openness, freedom, transparency, and honesty from the Taiwanese. Politically, that 'One Country, Two Systems' promise in Hong Kong is eroding as we speak. Economically, the Chinese ought to acknowledge that a lot of the 'Chinese miracle' on the Pacific Rim is actually a 'Taiwanese miracle.' Without our money, skills, discipline, and rigor there wouldn't be much of a miracle, especially if you also subtract what Southeast Asia's Chinese have contributed.

"Of course, Taiwan is a great rallying point for Beijing's autocrats. Gives them a chance to rail against the Western world, against America, against social excesses (we even have freedom of religion in Taiwan, imagine that!), against international conspiracies. But the truth is that the voters of Taiwan long ago ousted the 'thugs' who took us over, forged a thriving economy and a vibrant democracy, and made this island an example of what might have been across the Taiwan Strait.

"Could I say a few words about the 'historic' justification for China's claim to Taiwan? First of all, this island was peopled by Southeast Asian groups long before any Chinese arrived here, so if it's a matter of who got here first, we're last on the list. Beijing didn't integrate Taiwan politically until the 1680s, after groups of hungry citizens started to cross the Strait. That's pretty late in China's vaunted 5000-year history. Taiwan didn't become a province until 1886, which lasted exactly nine years. From 1895 to 1945 Taiwan belonged to Japan as part of the war settlement, and then Beijing took it back only to lose it again four years later to the 'thugs' under Chiang Kai-shek. That makes 13 years as part of the mainland system and 103 years of something else.

"For a country that has expanded and contracted for millennia, has gained and lost territory from Tibet to Taiwan, you would think that China could live with an ethnically compatible, economically successful, politically stable, democratic and independent neighbor. Those stiff bureaucrats in Beijing should have better things to do than threaten force to 'reincorporate' their exemplary neighbor."

Vote your opinion at www.wiley.com/college/deblij

firms join in the research and development carried on in such places as Hsinchu in the north, where the government helped to establish a **technopole [12]** centered on the microelectronics and personal computer industries. In the south, the "science city" of Tainan specializes in microsystems and information technology.

The establishment of China's Special Economic Zones went a long way toward mitigating longstanding problems between Beijing and Taipei. China needed investment capital and Taiwan had it, and in the almost-anything-goes environment of the SEZs, Taiwanese entrepreneurs could buy or build factories just as "real" foreigners were able to do. Importantly, as we noted earlier, the Xiamen SEZ was laid out directly across from Taiwan for this purpose (people there will tell you that 80 percent of locals have relatives across the Taiwan Strait). As a result, thousands of Taiwanese-owned factories began operating in China, their owners and investors having a strong interest in avoiding violent conflict over political issues.

Taiwan's Future

Taiwan's unresolved status entails risks, but there are signs that all parties want to find a solution. Given that the PRC will never allow Taiwan to attain independence, and the growing prospect that the United States, Taiwan's chief guarantor, could not secure Taiwan against Chinese armed intervention, many Taiwanese as well as their allies and adversaries are seeking a long-range, negotiated solution. Among the options is one that appears to be working in Hong Kong, where the principle called *One Nation, Two Systems* has functioned better than many observers antici-

pated. As a Special Administrative Region (SAR), Hong Kong's economy has thrived, and freedoms not available in the Shi's, Provinces, or Autonomous Regions continue to prevail. In time, the Taiwanese may negotiate a comparable status, internationally as well as domestically, that would benefit all concerned.

POINTS TO PONDER

- In the two decades between 1990 and 2010, China's economy vaulted from tenth place in the world to second.

- The area surrounding China's Pearl River Estuary has a manufacturing workforce larger than that of the entire United States.

- The increasing social inequalities resulting from China's rapid economic growth are becoming ever more difficult to justify by China's self-proclaimed communist government.

- The massive earthquake and resulting tsunami of March 11, 2011 that struck northeastern Japan's Honshu Island took a huge human toll and continues to have social, economic, and political consequences reaching far beyond Japan itself.

- North Korea remains the greatest single threat to geopolitical stability in East Asia. In May 2011, China hailed visiting North Korean President Kim Jong-Il for giving "top priority to improving people's lives."

using GEOGRAPHY

1. Four of the five busiest (container) ports in the world are located in this realm, a reflection of the economic powerhouse that is East Asia. Where do all the exports go? What are the destination ports of most ships leaving from Shenzhen, Shanghai, Hong Kong, and Busan? What about incoming cargo, where does it originate? Use export and import data and/or data provided by these four ports to trace the most significant trade routes and trade

partners. What will these findings teach us about the geography of today's global economy?

2. Consider Japan's geography and the distribution of its many nuclear energy plants. What has been the rationale for locating reactors at these sites? In view of the earthquake/tsunami disaster of 2011, should the government reconsider some of these locations?

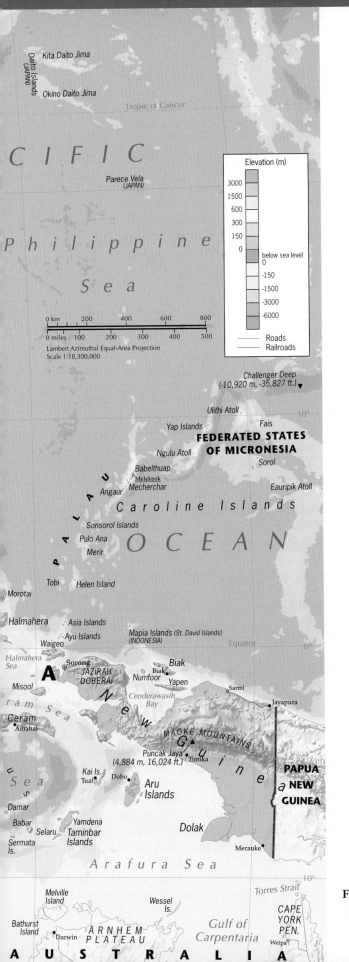

IN THIS CHAPTER

- ◆ Earthquakes, tsunamis, and volcanoes
- ◆ A religious crossroads
- ◆ The Overseas Chinese and the connection to East Asia
- ◆ Singapore's centrality
- ◆ ASEAN to counter Chinese dominance
- ◆ State territorial morphology: Why the shape of a country matters

CONCEPTS, IDEAS, AND TERMS

FIGURE 10A-1 © H. J. de Blij, P. O. Muller, and John Wiley & Sons, Inc.

Southeast Asia is a realm of peninsulas and islands, a corner of Asia bounded by India on the northwest and China on the northeast (Fig. 10A-1). Its western coasts are washed by the Indian Ocean, and to the east stretches the vast Pacific. From all these directions, Southeast Asia has been penetrated by outside forces. From India came traders; from China, settlers; from across the Indian Ocean, Arabs to engage in commerce and Europeans to build empires; and from across the Pacific, Americans. Southeast Asia has been the scene of countless contests for power and primacy—the competitors have come from near and far.

Southeast Asia's geography in some ways resembles that of eastern Europe. It is a mosaic of smaller countries on the periphery of two of the world's largest states. It has been a **buffer zone [1]** between powerful adversaries. It is a **shatter belt [2]** in which stresses and pressures from without and within have fractured the political geography. Like eastern Europe, Southeast Asia exhibits great cultural diversity. This is a realm of hundreds of cultures and ethnicities, numerous languages and dialects, global as well as local religions, and diverse national economies ranging from high- to low-income.

A GEOGRAPHIC OVERVIEW

Figure 10A-1 displays the relative location and dimensions of the Southeast Asian geographic realm, an assemblage of 11 countries situated on the Asian mainland and on thousands of islands, large and small, extending from Sumatera* in the west to New Guinea in the east, and from Luzon in the north to Timor in the south. We will become familiar with only the largest and most populous of these islands, but sailing the seas of this part of the world you would pass dozens of smaller ones every day, and if you could stop you would find even the tiniest populated islands to have their own character resulting from the cultural sources of their residents, the environmental challenges they face and the opportunities they exploit, their modes of dress, and the structure of their dwellings, even the vivid colors with which they often decorate their boats.

The giant of this realm, in terms of both territory and population, is the far-flung archipelago of Indonesia, labeled appropriately on the map by the largest letters. In the east, the state of Indonesia extends beyond the Southeast Asian realm into the Pacific Realm because it controls the western half of an island—New Guinea—whose indigenous peoples are not Southeast Asian. We focus on this unusual situation later (and on New Guinea as a whole in Chapter 12), but note it here because unusual borders and divided islands are a hallmark of this realm.

In mainland Asia, as noted above, Southeast Asia is bordered by India and China, both sources of immigrants, cultural infusions, economic initiatives, and other relationships evident in the realm's cultural landscapes. Also originating there are rivers that play a central role in the lives of many millions of people south of the realm border. As we will find, the core areas of several of Southeast Asia's most populous countries are located in the basins of major rivers whose sources lie inside China.

Before we get started, it is helpful to get acquainted with the key states of this realm, some of which will already be familiar. North of Indonesia lies the Philippines, well known to Americans because this was once an American colony and is still a source of many immigrants to the United States. Also north of Indonesia is Malaysia, easy to find on the map because its core area lies on the long peninsula that almost connects mainland Asia with Indonesia. At the tip of that peninsula lies another famous geographic locality: Singapore, the spectacular economic success story of Southeast Asia and a "global city" by any measure.

On the Asian mainland, the name Vietnam still resonates in America, which fought a bitter and costly war there in the 1960s and 1970s. As the map shows, Vietnam looks like a sliver of land extending from its border with China to the delta of the greatest of all Southeast Asian rivers, the Mekong. Looking westward, neighboring Laos and Cambodia may not be so familiar, but next comes centrally positioned Thailand, the dream (and reality) of millions of tourists and one of the world's most interesting countries. And then, on Southeast Asia's western margin, we come to Myanmar, one of the planet's most tragic places; here human potential and talent as well as natural endowment and opportunity were ruined and wasted by

---

*As in Africa and South Asia, names and spellings have changed with independence. In this chapter we will use contemporary spellings, except when we refer to the colonial era. Thus Indonesia's four major islands are Jawa, Sumatera, Kalimantan (the Indonesian part of Borneo), and Sulawesi. Prior to independence they were respectively known as Java, Sumatra, Dutch Borneo, and Celebes.

major geographic qualities of
SOUTHEAST ASIA

1. Southeast Asia extends from the peninsular mainland to the archipelagos offshore. Because Indonesia controls part of New Guinea, its functional region reaches into the neighboring Pacific geographic realm.

2. Southeast Asia, like eastern Europe, has been a shatter belt between powerful adversaries and has a fractured cultural and political geography shaped by foreign intervention.

3. Southeast Asia's physiography is dominated by high relief, crustal instability marked by volcanic activity and earthquakes, and tropical climates.

4. A majority of Southeast Asia's more than 600 million people live on the islands of just two countries: Indonesia, with the world's fourth-largest population, and the Philippines. The rate of population increase in the Insular region of Southeast Asia exceeds that of the Mainland region.

5. Although the great majority of Southeast Asians have the same ancestry, cultural divisions and local traditions abound, which the realm's divisive physiography sustains.

6. The legacies of powerful foreign influences, Asian as well as non-Asian, continue to affect the cultural landscapes of Southeast Asia.

7. Southeast Asia's political geography exhibits a variety of boundary types and several categories of state territorial morphology.

8. The Mekong River, Southeast Asia's Danube, has its source in China and borders or crosses five Southeast Asian countries, sustaining tens of millions of farmers, fishing people, and boat owners.

9. Singapore is the leading world city in Southeast Asia and lies at the realm's center of trade and business relations.

10. The realm contains a number of rapidly emerging markets and fast-growing economies, resulting in widening income disparities and uneven development.

11. Economic relations within the realm have intensified in recent years, while China's influence has been growing steadily.

The stupas at Jawa's Borobudur, one of the world's greatest Buddhist monuments.

© H. J. de Blij

power struggles and the force of arms, resulting in a state so devastated by poverty that time seems to have stood still.

SOUTHEAST ASIA'S PHYSICAL GEOGRAPHY

In certain ways, Southeast Asia is reminiscent of Middle America, a fractured realm of islands and peninsulas flanking a populous mainland studded with soaring mountains and deep valleys. It is useful to look back at Figures G-4 and G-5 to see why: both Southeast Asia and Middle America are dangerous places, where the Earth's crust is unstable as tectonic plates are in collision, earthquakes are a never-ending threat, volcanic eruptions take their toll, tropical cyclones lash sea and land, and floods, landslides, and other natural hazards make life riskier than in most other parts of the world.

In physiographic terms, Southeast Asia forms part of the Pacific Ring of Fire and all the hazards this designation brings with it. In 2004, an undersea earthquake (not unlike the one that struck near the coast of Japan in 2011) off westernmost Indonesia triggered a **tsunami [3]** in the Indian Ocean that killed more than 300,000 people along coastlines from Sumatera to Somalia. It was one in an endless succession of natural disasters in this realm. In 1883, the Krakatau volcano between Sumatera and Jawa exploded with a death toll estimated at 36,000. In 1815, the Tambora volcano in the chain of islands east of Jawa known as the Lesser Sunda Islands blew up, darkening skies throughout the world and affecting climates around the planet (the year that followed is still known as the "year without a summer" when crops failed, economies faltered, and millions went hungry as far away as Egypt, France, and New England).

Research on what may have been the most calamitous of all such eruptions suggests that, about 73,000 years ago, the Toba volcano on Sumatera exploded with such force that its ash and soot not only darkened skies and affected weather, but changed global climate for perhaps as long as 20 years and threatened the very survival of the human population, then still small in number and widely dispersed. Some scientists postulate that this eruption caused such widespread casualties that human genetic diversity was significantly diminished.

The most recent hazardous event occurred in late 2010, when Mount Merapi, one of Jawa's most active volcanoes (near the city of Yogyakarta), erupted intermittently for nearly two weeks. The Indonesian Geological Agency estimates that during this time period Merapi disgorged some 50 million cubic meters of volcanic material, accompanied by explosions that released huge gas clouds and ash plumes

November 3, 2010: After more than 10 days, the eruption of Merapi volcano in central Jawa had left its immediate surroundings razed by fire and buried in ash. In areas of abundant rainfall, surfaces inundated by new lava can recover rapidly, with new vegetation beginning to appear less than a year after an eruption. In the longer run, measured in decades and centuries, the erosion of such volcanic materials can produce highly fertile soils. This is what draws large numbers of people to inhabit volcanic landscapes—but hardly without risk.

© HADI SUSANTO/epa/Corbis Images

that forced the closure of Jawa's biggest airport at Jakarta, more than 400 kilometers (250 mi) away. By the end of this episode, nearly 200 people were dead and hundreds of thousands homeless. It was Merapi's worst eruption since 1930.

It therefore goes without saying that high relief dominates Southeast Asia, from the Arakan Mountains in western Myanmar to the glaciers (yes, glaciers!) of New Guinea. Take a close look at Figure 10A-1 and you can see how many elevations approach or exceed 3000 meters (10,000 ft), and, using the elevation guide in the upper right-hand corner, how mountainous and hilly much of the realm's topography is. Lengthy ranges form the backbone not only of islands such as Sulawesi and Sumatera, but also of the Malay Peninsula, most of Vietnam, and of the border zone between Thailand and Myanmar. And in Figure G-5, it is even possible to trace these volcano-studded mountain ranges by their earthquake epicenters and volcanic eruptions.

Exceptional Borneo

But among the islands there is one significant exception. The bulky island named Borneo in Figure 10A-1 has high elevations (Mount Kinabalu in the north reaches 4101 meters [13,455 ft]), yet has no volcanoes and negligible Earth tremors. This island has been called a stable "mini-continent" amid a mass of volcanic activity, a slab of ancient crust that long ago was pushed high above sea level by tectonic forces and was subsequently eroded into its present landscapes. Borneo's soils are not nearly as fertile as those of the volcanic islands, so that an equatorial rainforest developed here that long survived the human population explosion, giving sanctuary to countless plant and animal species including Southeast Asia's great ape, the orangutan. That era is now ending as human encroachment on Borneo's tropical habitat is accelerating, logging is destroying the remaining forest, and roads and farms are penetrating its interior. Borneo, along with other parts of Indonesia and Malaysia, has also experienced the rapid expansion of palm-oil plantations at the expense of tropical woodlands (see box titled "Palm-Oil Plantations and Deforestation").

Borneo's tropical forests are a remnant of a much larger stand of equatorial rainforest that once covered most of this realm. Besides Borneo, eastern Sumatera still contains limited expanses of rainforest, including a few orangutan sanctuaries. So does less populous and more remote New Guinea, which was never reached by these great apes. As we noted in the introductory chapter, equatorial rainforests can still be found today in three low-latitude areas of the world: Equatorial Africa, the Amazon Basin of South America, and here in Southeast Asia. The combination of climatic conditions that sustains these forests—consistently warm temperatures and year-round rainfall—produces the biologically richest and ecologically most complex vegetation regions on Earth. Huge numbers and varieties of trees and other plants grow in very close proximity, vying for space and sunlight both horizontally and vertically. Yet despite all this luxuriant growth, the soils beneath rainforests are nutrient poor. Most of the surface nutrients are derived

Palm-Oil Plantations and Deforestation

IN THE MID-NINETEENTH century, it was discovered that the reddish fruit of a species of palm tree in West Africa contained rich oils that could be used for everything from making soap to lubricating steam engines. The European colonial powers, who liked to experiment with different crops throughout their empires to obtain the highest and most profitable yields, eventually learned that these palm trees grew especially well in Southeast Asia. Beginning in the 1930s, oil-palm plantations were established across this realm, particularly in what is now Malaysia and Indonesia. Today, palm oil is used in an enormous range of products; in fact, it is so widely utilized in processed foods that up to half of all products sold in supermarkets contain it.

This form of agriculture is also quite lucrative, because the yields of oil palms per hectare (2.5 acres) are much higher than those of such competing oil crops as soybeans and sunflowers. Moreover, the price of this commodity on global markets has risen steadily over the past few years, triggering successive rounds of production increases. Malaysia and Indonesia have benefited the most, and by 2010 together accounted for fully 90 percent of all the world's palm oil.

The achievement of this success, however, has come at a high price. According to UN agencies, the expansion of palm-oil plantations now constitutes the single biggest threat to tropical forest preservation in the two countries. Most at risk for survival are such key wildlife species as orangutans, pygmy elephants, Sumatran rhinos, and tigers. In 2010, the swiftly expanding closed-canopy palm oil plantations were conservatively estimated to already cover an area the size of Austria, with about half located in Sumatera and the rest divided about evenly between Borneo and peninsular Malaysia (Fig. 10A-2 below).

The palm-oil industry is now being pressured by various groups to adhere to more sustainable practices, mainly by avoiding start-ups of new plantations in areas of high conservation value, and by introducing a "certified" label for oil produced in a sustainable manner. But progress on this front is slow, and as long as global demand remains high, the temptation to expand production in suitable areas will be hard to resist.

A section of a vast palm-oil plantation in Sarawak State in the southern part of Malaysian Borneo.

© G. Bowater/Corbis Images

FIGURE 10A-2

© H. J. de Blij, P. O. Muller, J. Nijman, and John Wiley & Sons, Inc.

---

from the decaying vegetation on the forest floor, which nurture the growth of the next generation of plants. Long-term indigenous inhabitants have developed ways to use the plants, animals, and soils to eke out a living—but when migrant farmers from elsewhere remove the trees to utilize the soil, mistakenly assuming it will be able to support their crops, failure usually results. This reality notwithstanding, intensifying population pressure throughout the tropics continues to shrink what remains of the world's rainforests (a subject discussed further in Chapter 4A).

Relative Location and Biodiversity

Look again at Figure 10A-1, but in a more general way. The Malay Peninsula, adjacent Sumatera, Jawa, and the Lesser Sunda Islands east of Jawa seem to form a series of stepping stones toward New Guinea, and in the extreme lower-right corner of the map you can see Australia's Cape York Peninsula seeming to reach toward New Guinea. What would happen, you might ask, if sea level were to drop and those narrow bodies of water between the peninsulas and the islands dried up?

That is precisely what occurred—not just once, but repeatedly in the geologic history of Southeast Asia. Long ago, the orangutans whose descendants remain in Indonesia today were able to migrate from the mainland into warmer equatorial latitudes because sea levels dropped during periods of glaciation, and islands separated by water today were temporarily connected by land bridges. More recently, early human arrivals here also got assistance from nature: Australia's

Aborigines probably managed to cross what remained of the deeper trenches between islands by building rafts, but their epic journey was facilitated by land bridges as well.

Time and again, therefore, Southeast Asia was a receptacle for migrating species. Combine this with the realm's tropical environments, and it is no surprise that it is known for its **biodiversity** [4]. Scientists estimate that fully 10 percent of the Earth's plant and animal species are found in this comparatively small realm. Biogeographers can trace the progress of many of these species from the mainland across the archipelago toward Australia, but some were stopped by deeper trenches that remained filled with water even when sea levels dropped. An especially deep trench lies in the Lombok Strait between the islands of Bali (next to Jawa) and Lombok (Fig. 10A-1), a biogeographical boundary first recognized by the naturalist Alfred Russel Wallace, a contemporary of Charles Darwin.

As we shall presently observe, Southeast Asia's biodiversity had a fateful impact on its historical geography. Among the realm's specialized plants, it was the spices that attracted outsiders from India, China, and Europe, with consequences still visible on the map today.

Four Major Rivers

Water is the essence of life, and among world realms Southeast Asia is comparatively well endowed with moisture (Fig. G-7). Ample, occasionally even excessive, rainfall fills the rice-growing paddies of Indonesia and the Philippines. On the realm's mainland, where annual rainfall averages are somewhat lower and the precipitation is more seasonal, major rivers and their tributaries fill irrigation channels and form fertile deltas. The map of population distribution clearly highlights this spatial relationship between rivers and people, expressed in distinct coastal clustering (Fig. 10A-3).

FIGURE 10A-3

© H. J. de Blij, P. O. Muller, and John Wiley & Sons, Inc.

From the Field Notes . . .

"After a three-hour drive by bus southward from Hanoi, we reached the Van Long Nature Reserve in northeastern Vietnam's Red River Delta. It lies in a limestone terrain known as karst, part of the same geology that continues to the north through Halong Bay and into southern China beyond. Karst landscapes are formed through the slow but steady dissolution of limestone by (naturally) acidic water; the harder rock formations are more resistant to erosion and remain as towers, resulting in unique and sometimes whimsical landforms. The Van Long Nature Reserve is a biologically rich ecotourism site that was established in 2001 and involves local communities. Women from nearby villages guide visitors around in their basket boats, and they welcome the extra income in an area barely touched by Vietnam's emerging economy."

© Jan Nijman

www.conceptcaching.com

As Figure 10A-1 reminds us, rivers that are crucial to life in certain realms sometimes have their sources in neighboring realms, a geographic issue that can lead to serious regional discord. What right does an "upstream" state in one realm have to dam, or otherwise interfere with, a river that is vital to a state (or states) downstream? In Southeast Asia's case, three of its four major rivers originate in China: (1) the ***Mekong***, which crosses the mainland from north to south; (2) the ***Red*** River that reaches the sea via its course across northern Vietnam; and (3) the ***Irrawaddy***, which forms the lifeline of Myanmar (Burma). The fourth is an intra-realm river, the ***Chao Phraya***, that functions as the key artery of Thailand.

The Mighty Mekong

From its headwaters high in the snowy uplands of China's Qinghai-Xizang (Tibetan) Plateau, the Mekong River rushes and flows some 4200 kilometers (2600 mi) to its delta in southernmost Vietnam. This "Danube of Southeast Asia" traverses or borders five of the realm's countries, supporting rice farmers and fishing people, forming a transportation route where roads are few, and providing electricity from dams upstream. Tens of millions of people depend on the waters of the Mekong, from subsistence farmers in Cambodia to apartment dwellers in China. The Mekong Delta in southern Vietnam is one of the realm's most densely populated areas and produces enormous harvests of rice.

But problems loom. China is building a series of dams across the Lancang (as the Mekong is called there) to supply surrounding Yunnan Province with electricity. Although such hydroelectric dams should not interfere with water flow, countries downstream worry that a severe dry spell in the interior would impel the Chinese to slow the river's flow to keep their reservoirs full. Cambodia is especially concerned over the future of the Tonlé Sap, a large natural lake whose water is supplied by the Mekong (see Fig. 10B-2). In Vietnam, farmers worry about salt water invading the Delta's paddies should the Mekong's level drop. And the Chinese may not be the only dam builders in the future: Thailand has expressed an interest in building a dam on the Thai-Laos border where it is defined by the Mekong.

In such situations, the upstream states have an advantage over those downstream. Several international organizations have been formed to coordinate development in the Mekong Basin, including the Mekong River Commission (MRC), founded more than 50 years ago. China has offered to sell electricity from its dams to Thailand, Laos, and Myanmar. Coordinated efforts to reduce deforestation in the Mekong's drainage basin have had some effect. After consultations with the MRC, Australia built a bridge linking Laos and Thailand. There is even a plan to make the Mekong navigable from Yunnan to the coast, creating an alternative outlet for interior China. Sail the Mekong today, however, and you are struck by the slowness of development along this artery (see photo, p. 532). Wooden boats, thatch-roofed villages, and teeming paddies mark a river still crossed by antiquated ferries and flanked by few towns. Of modern infrastructure, one sees little. And yet the Mekong and its basin form the lifeline of mainland Southeast Asia's dominantly rural societies.

Rivers and States

Whereas the Mekong affects the lives of peoples in several Southeast Asian states, the other three major Southeast Asian rivers are essentially local. As the map shows, the Red River forms the focus for the heavily populated Tonkin Plain in northern Vietnam, where the capital,

Hanoi, lies on its banks. In Thailand, the relatively short but crucial Chao Phraya, on which Bangkok lies, is just one of a series of channels in that river's delta, formed by numerous streams rising in the country's interior. And Myanmar's Irrawaddy River (with some of whose headwaters rise in China) crosses that country from north to south, its wide valley one of the world's leading rice-producing areas and its largest city, Yangon, at the corner of its delta.

POPULATION GEOGRAPHY

Examine the map of Southeast Asia's population distribution (Fig. 10A-3), and you are immediately struck by the huge concentration of people on a relatively small island in Indonesia—a cluster larger than any other in the realm, the four mainland river deltas included. This is Jawa, and its population of almost 150 million is not only larger than any other island's, but also exceeds that of every country in the entire realm except Indonesia, of which it is a part. And even there, it contains more than half the country's population.

This population concentration is particularly noteworthy because Indonesia is not yet a highly urbanized country. In 2011, more than 50 percent of all Indonesians still lived in rural areas, and although Jawa, as is shown in Chapter 10B, is the most highly urbanized island of Indonesia, tens of millions of its inhabitants live off the land. What makes all this possible is a combination of fertile volcanic soil, ample water, and extremely warm temperatures that enable Jawa's farmers to grow three crops of rice in a single paddy during a single year, helping feed a national population still growing faster than the global average. But we should also be aware that the darkest red clusters on Jawa signify fast-growing urban areas, cornerstones in the building of a new economy with global linkages.

Within Indonesia, the contrasts between Jawa and the four other major islands—Sumatera, Borneo (Kalimantan), Sulawesi, and particularly Indonesian New Guinea—reflect the core-periphery relationship between these two sectors of the country. As the map suggests, such contrasts, less sharply defined, also mark other Southeast Asian countries, and the primate cities here (Bangkok, Manila, Yangon, Kuala Lumpur) are particularly dominant. Vietnam even contains two such anchors, which respectively represent its historic northern (Hanoi) and southern (Ho Chi Minh City) core areas.

As the table in Appendix B indicates and as the map confirms, Southeast Asia's states are not, by world standards, especially populous. Indonesia today is the world's fourth-ranking country in terms of population (242 million), but no other Southeast Asian country has even half that total. Three mainland countries contain between 50 and 100 million people. But take Laos, quite a large country territorially (about the size of the United Kingdom) and note that its

population only approaches 7 million; similarly, Cambodia, half the size of Germany, has about 16 million. In part, such modest numbers on the mainland reflect natural conditions less favorable to farming than those prevailing on volcanic soils or in fertile river basins, but more generally this realm did not grow as explosively during the last century as did neighboring realms. Indeed, Appendix B indicates that several countries in Southeast Asia today are growing at or below the global average of 1.2 percent annually.

The Ethnic Mosaic

Southeast Asia's peoples come from a common stock just as (Caucasian) Europeans do, but this has not prevented the emergence of regionally or locally discrete ethnic or cultural groups. Figure 10A-4 displays the broad distribution of ethnolinguistic groups in the realm, but be aware that this is a generalization. At the scale of this map, numerous smaller groups cannot be depicted.

Figure 10A-4 shows the rough spatial coincidence, on the mainland, between major ethnic group and modern political state. The Burman dominate in the country formerly called Burma (now Myanmar); the Thai occupy the state once known as Siam (now Thailand); the Khmer form the nation of Cambodia and extend northward into Laos; and the Vietnamese inhabit the long strip of territory facing the South China Sea.

Territorially, by far the largest population shown in Figure 10A-4 is classified as Indonesian, the inhabitants of the great island chain that extends from Sumatera west of the Malay Peninsula to the Malukus (Moluccas) in the east and from the Lesser Sunda Islands in the south to the Philippines in the north. Collectively, all these peoples shown on the map—the Filipinos, Malays, and Indonesians—are known as Indonesians, but they have been divided by history and politics. And note as well that the Indonesians in Indonesia itself include Javanese, Madurese, Sundanese, Balinese, and other major groups; again, hundreds of smaller ones cannot be mapped at this scale. In the Philippines, too, island isolation and contrasting ways of life are reflected in the cultural mosaic. Also part of this Indonesian ethnic-cultural complex are the Malays, whose heartland lies on the Malay Peninsula but who form minorities in other areas as well. Like most Indonesians, the Malays are Muslims, although Islam is a more powerful force within Malay society than it generally is in Indonesian culture.

In the northern part of the mainland region, numerous minorities inhabit remote corners of the countries in which the Burman (Burmese), Thai, and Vietnamese dominate. Those minorities tend to occupy areas on the peripheries of their countries, where the terrain is mountainous and the forest is dense, and where the governments of their national states do not have complete control. This remoteness and sense of detachment give rise to aspirations of secession, or at least resistance to governmental

FIGURE 10A-4

© H. J. de Blij, P. O. Muller, and John Wiley & Sons, Inc.

efforts to establish full authority, often resulting in bitter ethnic conflict.

Immigrants

Figure 10A-4 further reminds us that, again like eastern Europe, Southeast Asia is home to major ethnic minorities from outside the realm. On the Malay Peninsula, note the South Asian (Hindustani) cluster. Such Hindu communities

with Indian ancestries exist in many parts of the peninsula, but in the southwest they form the majority in a small area. In Singapore, too, South Asians form a significant minority. These communities arose during the European colonial period, but South Asians had arrived in this realm many centuries earlier, propagating Buddhism and leaving their architectural and cultural imprints on places as far removed as Jawa and Bali.

From the Field Notes . . .

"Like most major Southeast Asian cities, Bangkok's urban area includes a large and prosperous Chinese sector. No less than 14 percent of Thailand's population of 69 million is of Chinese ancestry, and the great majority of Chinese live in the cities. In Thailand, this large non-Thai population is well integrated into local society, and intermarriage is common. Still, Bangkok's "Chinatown" is a distinct and discrete part of the great city. There is no mistaking Chinatown's limits: Thai commercial signs change to Chinese, goods offered for sale also change (Chinatown contains a large cluster of shops selling gold, for example), and the urban atmosphere, from street markets to bookshops, is dominantly Chinese. This is a boisterous, noisy, energetic part of multicultural Bangkok, a vivid reminder of the Chinese commercial success in Southeast Asia."

www.conceptcaching.com

© H. J. de Blij

The Chinese

By far the largest immigrant minority in Southeast Asia, however, is Chinese. The Chinese began arriving here during the Ming and early Qing (Manchu) dynasties, and the largest exodus occurred during the late colonial period (1870–1940), when as many as 20 million immigrated. The European powers at first encouraged this influx, using the Chinese in administration and trade. But soon these **Overseas Chinese [5]** began to congregate in the major cities, where they established Chinatowns and gained control over much of the commerce (see photo above). By the time the Europeans tried to limit Chinese immigration, World War II was about to break out and the colonial era would end soon afterward.

Today Southeast Asia is home to about 32 million Overseas Chinese, more than 75 percent of the world total. Their lives have often been difficult. The Japanese relentlessly persecuted Chinese living in Malaya during World War II. Later, during the 1960s, Chinese in Indonesia were accused of communist sympathies, and hundreds of thousands were killed. More recently, in the late 1990s, Indonesian mobs again attacked Chinese and their property, this time because of their relative wealth and because many Chinese had become Christians during the colonial era and were now targeted by Islamic throngs. Resentment still continues, expressed by episodic flare-ups against Chinese in various parts of Southeast Asia.

Figure 10A-5 shows the migration routes and current concentrations of Chinese in Southeast Asia. Most originated in southern China's Fujian and Guangdong provinces, and a large number invested much of their wealth back in China when it opened up to foreign businesses three decades ago. Clearly, the Overseas Chinese of Southeast Asia have played a significant role in shaping the economic miracle in this sector of the Pacific Rim.

RELIGIONS OF SOUTHEAST ASIA

Southeast Asia, more than any other realm, is a historic crossroads of religions. With the migrants from the Indian subcontinent came their faiths: first Hinduism and Buddhism, later Islam. The Muslim religion, promoted by the growing number of Arab traders who appeared on the scene, became the dominant religion in Indonesia (where nearly 90 percent of the population adheres to Islam today). But in Myanmar, Thailand, and Cambodia, Buddhism remained supreme, and in all three countries the overwhelming majority of the people are now adherents. In culturally diverse Malaysia, the Malays are Muslims (to be a Malay *is* to be a Muslim), and almost all Chinese are Buddhists; but most Malaysians of Indian ancestry remain Hindus.

Although Southeast Asia has generated its own local cultural expressions, most of what remains in tangible form has resulted from the infusion of foreign elements. Take, for

Regional ISSUE   The Chinese Presence in Southeast Asia

THE CHINESE ARE TOO INFLUENTIAL!

"It's hard to imagine that there was a time when we didn't have Chinese minorities in our midst. I think I understand how Latin Americans [sic] feel about their 'giant to the north.' We've got one too, but the difference is that there are a lot more Chinese in Southeast Asia than there are Americans in Latin America. The Chinese even run one country because they are the majority there, Singapore. And if you want to go to a place where you can see what the Chinese would have in mind for this whole region, go there. They're knocking down all the old Malay and Hindu quarters, and they've got more rules and laws than we here in Indonesia could even think of. I'm a doctor here in Bandung and I admire their modernity, but I don't like their philosophy.

"We've had our problems with the Chinese here. The Dutch colonists brought them in to work for them, they got privileges that made them rich, they joined the Christian churches the Europeans built. I'm not sure that a single one of them ever joined a mosque. And then shortly after Sukarno led us to independence they even tried to collaborate with Mao's communists to take over the country. They failed and many of them were killed, but look at our towns and villages today. The richest merchants and the money lenders tend to be Chinese. And they stay aloof from the rest of us.

"We're not the only ones who had trouble with the Chinese. Ask them about it in Malaysia. There the Chinese started a full-scale revolution in the 1950s that took the combined efforts of the British and the Malays to put down. Or Vietnam, where the Chinese weren't any help against the enemy when the war happened there. Meanwhile they get richer and richer, but you'll see that they never forget where they came from. The Chinese in China boast about their coastal economy, but it's coastal because that's where our Chinese came from and where they sent their money when the opportunity arose.

"As a matter of fact, I don't think that China itself cares much for or about Southeast Asia. Have you heard what's happening in Yunnan Province? They're building a series of dams on the Mekong River, our major river, just across the border from Laos and Burma. They talk about the benefits and they offer to destroy the gorges downstream to facilitate navigation, but they won't join the Mekong River Commission. When the annual floods cease, what will happen to Tonlé Sap, to fish migration, to seasonal mudflat farming? The Chinese do what they want—they're the ones with the power and the money."

THE CHINESE ARE INDISPENSABLE!

"Minorities, especially successful minorities, have a hard time these days. The media portray them as exploiters who take advantage of the less fortunate in society, and blame them when things go wrong, even when it's clear that the government representing the majority is at fault. We Chinese arrived here long before the Europeans did and well before some other 'indigenous' groups showed up. True, our ancestors seized the opportunity when the European colonizers introduced their commercial economy, but we weren't the only ones to whom that opportunity was available. We banded together, helped each other, saved and shared our money, and established stable and productive communities. Which, by the way, employed millions of locals. I know: my family has owned this shop in Bangkok for five generations. It started as a shed. Now it's a six-floor department store. My family came from Fujian, and if you added it all up we've probably sent more than a million U.S. dollars back home. We're still in touch with our extended family and we've invested in the Xiamen SEZ.

"Here in Thailand we Chinese have done very well. It is sometimes said that we Chinese remain aloof from local society, but that depends on the nature of that society. Thailand has a distinguished history and a rich culture, and we see the Thais as equals. Forgive me, but you can't expect the same relationship in East Malaysia. Or, for that matter, in Indonesia, where we are resented because the 3 percent Chinese run about 60 percent of commerce and trade. But we're always prepared to accommodate and adjust. Look at Malaysia, where some of our misguided ancestors started a rebellion but where we're now appreciated as developers of high-tech industries, professionals, and ordinary workers. This in a Muslim country where Chinese are Christians or followers of their traditional religions, and where Chinese regularly vote for the Malay-dominated majority party. So when it comes to aloofness, it's not us.

"The truth is that the Chinese have made great contributions here, and that without the Chinese this place would resemble parts of South or Southwest Asia. Certainly there would be no Singapore, the richest and most stable of all Southeast Asian countries, where minorities live in peace and security and where incomes are higher than anywhere else. In fact, mainland Chinese officials come to Singapore to learn how so much was achieved there. Imagine a Southeast Asia without Singapore! It's no coincidence that the most Chinese country in Southeast Asia also has the highest standard of living."

Vote your opinion at www.wiley.com/college/deblij

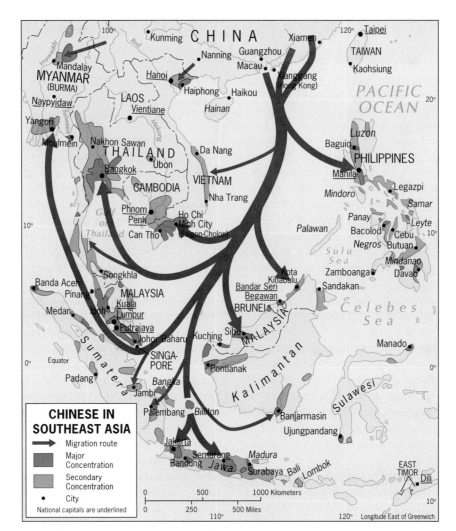

FIGURE 10A-5 © H. J. de Blij, P. O. Muller, and John Wiley & Sons, Inc.

Cambodia's renowned Angkor Wat offers a blend of Hindu and Buddhist iconographies: a statue of Vishnu (left) and nearby one of the Buddha (right).

© Jan Nijman

instance, Angkor Wat, the enormous complex of religious structures built in Cambodia during the twelfth century AD and today one of the world's most famous monuments. It was originally constructed as a Hindu temple, dedicated to the god Vishnu. The carvings on the walls of Angkor Wat tell the stories of the Hindu epics, and the temple's designs are closely associated with Hindu cosmology. But during the fourteenth century Buddhism took over this area and the temple complex became a place of worship for Buddhists.

Vietnam is perhaps the ultimate religious crossroads of the realm, where for many centuries there has been an almost casual blending of early Hinduism, then Buddhism, Daoism, and Confucianism—all mixing comfortably with age-old traditions of ancestor worship.

COLONIALISM'S HERITAGE: HOW THE POLITICAL MAP EVOLVED

When the European colonizers arrived in Southeast Asia, they encountered a patchwork of kingdoms, principalities, sultanates, and other traditional political entities whose leaders they tried to co-opt, overpower, or otherwise fold into their imperial schemes. There was no single, powerful center of indigenous culture as had developed in Han-dominated China. In the river basins and on the plains of the mainland, as well as on the islands offshore, a flowering of cultures had produced a diversity of societies whose languages, religions, arts, music, foods, and other achievements formed an almost infinitely varied mosaic—but none of those cultures had risen to imperial power when the Europeans arrived. Those European colonizers forged empires here, often by playing one state off against another; the Europeans divided and ruled. Out of this foreign intervention came the modern map of Southeast Asia, as only Thailand (formerly Siam) survived the colonial era as an independent entity. Thailand was useful to two competing powers, the French to the east and the British to the west: it served as a convenient buffer, and although the colonists carved pieces off Thailand's domain, the kingdom endured.

Indeed, the Europeans accomplished what local powers could not: the formation of comparatively large, multicultural states that encompassed diverse peoples and societies and welded them together. Were it not for the colonial intervention, it is unlikely that the 17,000 islands of far-flung Indonesia would today constitute the world's fourth-largest country in terms of population. Nor would the nine sultanates of Malaysia have been united, let alone with the peoples of northern Borneo across the South China Sea. For good or ill, the colonial intrusion consolidated a realm of few culture cores and numerous ministates into less than a dozen countries. The leading colonial competitors here in Southeast Asia were the Dutch, British, French, and Spanish (with the Spanish later replaced by the Americans in their stronghold, the Philippines). The Japanese had colonial objectives here as well (see Fig. 9A-10), but these aspirations came and went during the course of World War II.

The Dutch acquired the greatest prize: control over the vast archipelago now called Indonesia (formerly the Netherlands East Indies). France established itself on the eastern flank of the mainland, controlling all territory to the east of Thailand and the south of China. The British conquered the Malay Peninsula, gained power over the northern part of the island of Borneo, and established themselves in Burma (Myanmar) as well. Other colonial powers also gained footholds, but not for long. The exception was Portugal, which held on to its eastern half of the island of Timor (Indonesia) until well after the Dutch had been ousted from their East Indies. Figure 10A-6 shows the colonial framework in the late nineteenth century, before the United States assumed control over the Philippines in 1898. Note that while Thailand survived as an independent state, it lost territory to the British in Malaya and Burma and to the French in Cambodia and Laos.

The Colonial Imprint

The colonial powers divided their possessions into administrative units as they did in Africa and elsewhere. Some of these political entities became independent states when the colonial powers withdrew or were ousted by force (Fig. 10A-6).

French Indochina

France, one of the mainland's leading colonial powers, divided its Southeast Asian empire into five units. Three of these units lay along the east coast: Tonkin in the north next to China, centered on the basin of the Red River; Cochin China in the south, with the Mekong Delta as its focus; and in between these two, Annam. The other two French territories were Cambodia, which faces the Gulf of Thailand; and Laos, landlocked in the interior. Out of these five French dependencies there emerged three states: the three east-coast territories ultimately became a single state, Vietnam; the other two—Cambodia and Laos—each achieved separate independence.

The French had a name for their empire—*Indochina*. The *Indo* part of Indochina refers to cultural imprints received from South Asia: the Hindu presence; the importance of Buddhism, which came to Southeast Asia via Sri Lanka (Ceylon) and its seafaring merchants; the influences of Indian architecture and art (especially sculpture), writing and literature, and social structures and patterns. The *China* in the name *Indochina* signifies the role of the Chinese here. Chinese emperors coveted Southeast Asian lands, and China's power penetrated deep into the realm. Social and political upheavals in China, combined with the opportunities created by the European colonists, sent millions of Sinicized people southward. Chinese traders, pilgrims, seafarers, fishermen, and many others sailed from southeastern China to the coasts of Southeast Asia and established settlements there. Over time, these settlements attracted more Chinese immigrants, and Chinese influence in the realm grew (Fig. 10A-5). Relations between the Chinese settlers and the earlier inhabitants of Southeast Asia have at times been strained, even violent. The Chinese presence in Southeast Asia is long term, an invasion that has continued into modern times.

The name *Indochina* can only refer to a part of mainland Southeast Asia, however; it cannot be used to refer to the realm as a whole. Although the *Indo* segment of this regional name can also be taken to refer to the Buddhist influences that dominate here, it makes no reference to the momentous arrival of Islam, introduced by Arab seafarers in the twelfth and thirteenth centuries, and destined to transform the cultural geography of this realm.

British Imperialism

The British ruled a pair of major entities in Southeast Asia (Burma and Malaya) in addition to a large part of northern

FIGURE 10A-6

© H. J. de Blij, P. O. Muller, and John Wiley & Sons, Inc.

Borneo and many small islands in the South China Sea. Burma was attached to Britain's Indian Empire; from 1886 until 1937, it was governed from distant New Delhi. But when British India became independent in 1947 and split into several countries, Burma was not part of the grand design that created West and East Pakistan (the latter now Bangladesh), Ceylon (now Sri Lanka), and India. Instead, Burma (now Myanmar) was given the status of a sovereign republic in 1948.

In Malaya, the British developed a complicated system of colonies and protectorates that eventually gave rise to the equally complex, far-flung Malaysian Federation. Included were the former Straits Settlements (Singapore was one of these colonies), the nine protectorates on the Malay Peninsula (former sultanates of the Muslim era), the British dependencies of Sarawak and Sabah on the island of Borneo, and numerous islands in the Strait of Malacca and the South China Sea. The original Federation of

From the Field Notes . . .

"Walking along Tuanku Abdul Rahman Street in Kuala Lumpur, I had just passed the ultramodern Sultan Abdul Samad skyscraper when this remarkable view appeared: the old and the new in a country that seems to have few postcolonial hang-ups and in which Islam and democracy coexist. The

British colonists designed and effected the construction of the Moorish-Victorian buildings in the foreground (now the City Hall and Supreme Court); behind them rises the Bank of Commerce, one of many banks in the capital. Look left, and you see the Bank of Islam, not a contradiction here in economically diversified Malaysia. And just a few hundred yards away stands St. Mary's Cathedral, across the street from still another bank. Several members of the congregation told me that the church was thriving and that there was no sense of insecurity here. 'This is Malaysia, sir,' I was told. 'We're Muslims, Buddhists, Christians. We're Malays, Chinese, Indians. We have to live together. By the way, don't miss the action at the Hard Rock Café on Sultan Ismail Street.' Now there, I thought, was a contradiction as remarkable as this scene."

www.conceptcaching.com

© H. J. de Blij

Malaysia was created in 1963 by the political unification of recently independent mainland Malaya, Singapore, and the former British dependencies on the largely Indonesian island of Borneo. Singapore, however, left the Federation in 1965 to become a sovereign city-state, and the remaining units were later restructured into peninsular Malaysia and, on Borneo, Sarawak and Sabah. Thus the term *Malaya* properly refers to the geographic area of the Malay Peninsula, including Singapore and other nearby islands; the term *Malaysia* identifies the politico-geographical entity of which Kuala Lumpur is the capital city.

Netherlands "East Indies"

Following in the wake of the Portuguese, the first European colonizers in this realm, the Dutch came in search of what Southeast Asia had to offer—and now one aspect of this realm's biodiversity had fateful consequences. Among the plants domesticated by the local people on the islands of present-day Indonesia was a group collectively known as the *spices*. We know them today as black pepper, cloves, cinnamon, nutmeg, ginger, turmeric, and other condiments essential to flavorful meals. In Figure 10A-1 you can see, in eastern Indonesia between Sulawesi and New Guinea, a group of small islands called the Maluku Islands

(formerly Moluccas). The Dutch colonizers called these the **Spice Islands** because of the lucrative commerce in spices long carried on by Arab, Indian, and Chinese traders. What the Europeans wanted was control over this trade, and they were willing to go to war for it.

It may seem odd in today's world that spices could be important enough to be fought over, but at that time, in the pre-refrigeration era, spices not only conserved food but also added flavor to otherwise bland diets. Spices commanded sky-high prices on European markets. The Dutch East India Company, the commercial arm of the Dutch government, managed to take control of the Spice Islands, bringing untold wealth to the Netherlands and ushering in a period of enrichment through colonial exploitation known as the country's Golden Age.

From the mid-seventeenth through the late eighteenth century, the Dutch could develop their East Indies sphere of influence almost without challenge, for the British and French were preoccupied with the Indian subcontinent. By playing off the princes of Indonesia's states against one another in the search for economic concessions and political influence, by placing the Chinese in positions of responsibility, by imposing systems of forced labor in areas directly under its control, and by rearranging land ownership and power structures, the Company had a disastrous effect on the Indonesian societies it subjugated.

Java (Jawa), the most populous and productive island, became the focus of Dutch administration; from its capital at Batavia (now Jakarta), the Dutch East India Company extended its sphere of influence into Sumatra (Sumatera), Celebes (Sulawesi), and much of Borneo (Kalimantan) as well as the smaller islands of the East Indies. This was not accomplished overnight, and the struggle for territorial control was carried on long after the Company had yielded its administration to the Netherlands government. Dutch colonialism therefore threw a girdle around Indonesia's more than 17,000 islands, paving the way for the creation of the realm's largest and most populous state (home to almost a quarter-billion today).

From Spain to the United States

In the colonial tutelage of Southeast Asia, the Philippines, long under Spanish domination, had a unique experience. As early as 1571, the islands north of Indonesia were under Spain's control (they were named for Spain's King Philip II). Spanish rule began when Islam was penetrating the southern Philippines via northern Borneo. The Spaniards spread their Roman Catholic faith with great zeal, and between them the soldiers and priests consolidated Hispanic dominance over the predominantly Malay population. Manila, founded in 1571, became a profitable way-station on the route between southern China and western Mexico (Acapulco was the main trans-Pacific destination for the galleons leaving Manila's port). There was much profit to be made, but the indigenous people shared little in it. Great landholdings were awarded to loyal Spanish civil servants and to men of the church. Oppression eventually yielded revolution, and Spain was confronted with a major uprising in the Philippines when the Spanish-American War broke out elsewhere in 1898.

As part of the settlement of that brief war, the United States replaced Spain in Manila as colonial proprietor. That was not the end of the revolution, however. The Filipinos now took up arms against their new foreign ruler, and not until 1905, after terrible losses of life, did American forces manage to pacify their new dominion. Subsequently, U.S. administration in the Philippines was more progressive than Spain's had been. In 1934, Congress passed the Philippine Independence Law, providing for a ten-year transition to sovereignty. But before independence could be arranged, World War II intervened. In 1941, Japan conquered the islands, temporarily ousting the Americans; U.S. forces returned in 1944 and, with widespread Filipino support, defeated the Japanese in 1945. The agenda for independence was resumed, and in 1946 the sovereign Republic of the Philippines came into existence.

Today all of Southeast Asia's states are independent, but centuries of colonial rule have left strong cultural imprints. In their urban landscapes, their education systems, and countless other ways, this realm still carries the legacy of its colonial past.

SOUTHEAST ASIA'S EMERGING MARKETS

If Japan and the Asian Tigers set the early example in the 1960s and 1970s, with China following suit since the mid-1980s, it now seems to be the turn of several Southeast Asian countries to join the ranks of the world's rapidly emerging markets [6]. Vietnam, Indonesia, and Malaysia, in particular, have in recent years attracted substantial foreign investment and exhibited decidedly robust economic growth rates.

In the table in Appendix B, look at the column showing per capita income and note the very considerable disparities that mark this realm. The tiny oil state of Brunei has the highest income, although this figure is not indicative of what many of its immigrant workers earn. Singapore is in many ways the most comprehensively developed state in the realm, with an income level comparable to that of Switzerland and the U.S. Malaysia comes in a distant third and is followed by Thailand. Populous Indonesia and the Philippines exhibit per capita incomes less than one-tenth of Singapore's. Finally, Vietnam, Laos, Cambodia, and Myanmar, in that order, rank at the bottom of the list.

This table, however, only provides a snapshot and does not indicate the latest economic trends. According to many current observers, Vietnam has been growing the fastest in recent years. Of course, it remains a very poor country in urgent need of high growth rates to lift its masses to a better standard of living. Vietnam is still a communist state, and in certain ways its government seems to be more conservative than China's. But a stock exchange did open in Ho Chi Minh City (formerly Saigon) in 2000, and seven years later the country joined the World Trade Organization. If Vietnam takes further steps to liberalize its economy, observers say, it may well become Asia's next economic miracle.

Singapore's Centrality

There can be no doubt that Singapore is the economic heart of Southeast Asia. With just over 5 million people and only 619 square kilometers (240 sq mi) of territory, we are obviously not referring to size—this is all about the connections and centrality that enable Singapore to function as the leading node [7] in a realmwide economic network. Geographers generally rank Singapore as a top-tier world city because it has major international linkages and exerts global influence. Chapter 10B discusses the city-state itself in greater detail, but our task here is to emphasize its centrality in Southeast Asia's spatial economy. Singapore's container port (see photo at right) is not just the largest in the realm but in the entire world, underscoring its key role within and far beyond Southeast Asia. Moreover, Changi Airport Singapore was voted the world's best international airport in 2010.

Much of Singapore's exceptional regional position is based on its geography, especially its superb relative location. In the seventeenth century, it was Malacca further up the Malay Peninsula that was key to trade and shipping in Southeast Asia. The Strait of Malacca (Fig. 10A-6), named after that town, was already the most important sea route providing access to Southeast Asian waters for ships coming from the west. When the British sought to displace Dutch dominance in the region during the eighteenth century, they found themselves an even better local base of operations: Singapore possessed a larger and deeper natural harbor than Malacca and could therefore accommodate the larger steamships of the time. As British power steadily grew, Malacca became a backwater as Singapore rose to prominence.

Most of all, Singapore today is a symbol of modernity, a model of Southeast Asia's future. Throughout the realm, those who can afford it go there to shop, to connect to international flights, to transact business, to invest in real estate, or to send their children to one of the city-state's highly ranked universities. Imagine major parts of Southeast Asia becoming like Singapore over the next 25 years and how that would affect the world!

Prospects of Realmwide Integration: ASEAN

The overall development of Southeast Asia still has a long way to go. Political stability and increased regional integration will facilitate the process, and this is the long-term goal of **ASEAN [8]** or the Association of Southeast Asian Nations. Founded in the late 1960s, this supranational organization has primarily been concerned with security. But that has been a constantly challenging effort because a wide array of conditions mark its ten member-states. With the exception of minuscule East Timor, ASEAN encompasses all of the realm's countries; these include one influential city-state, an Islamic oil state (Brunei), two impoverished communist regimes (Vietnam and Laos), and a

Singapore has the biggest container port on Earth and is by far Southeast Asia's leading world city. This city-state not only lies at the heart of the realm's internal spatial economic system but also forms a hugely important gateway to the global economy.

© Justin Guariglia/Corbis Images

brutal military dictatorship whose population ranks among the most deprived on Earth (Myanmar).

One conflict that ASEAN has not been able to resolve is a long-running border dispute between Cambodia and Thailand, directly north of the Cambodian city of Siem Reap (Fig. 10A-1). At issue is the location of the ancient Preah Vihear Temple, which according to a 1962 ruling by the International Court of Justice belongs to Cambodia but continues to be aggressively claimed by Thailand. The temple was declared a UNESCO World Heritage Site in 2008, and that only exacerbated matters. As of press time, the latest exchange of gunfire across the border occurred in February 2011, reportedly killing ten soldiers. Unfortunately, ASEAN cannot play a role in the resolution of this and other conflicts because the organization is incapable of enforcing compliance.

Another problem that ASEAN has failed to resolve, one that literally affects the health of many in this realm, is the recurrent air pollution caused by Indonesia's massive, human-generated forest fires. Depending on weather conditions, thick plumes of smoke emanate from Sumatera (where most of the burning occurs, often to enlarge oil palm plantations) to Jawa, Singapore, Malaysia, and beyond. Repeatedly, Singapore's government has had to advise its residents with heart and respiratory conditions to avoid outdoor activities. Indonesia, the realm's most powerful country, has been painfully slow to address this problem and even refused to ratify the 2002 ASEAN agreement on Transboundary Haze Pollution.

In 1992, the organization was able to expand into the economic domain through **AFTA [9]**, the ASEAN Free Trade Agreement, and here the payoff has been far more substantial. AFTA has both triggered the lowering of tariffs and encouraged an upsurge in trade within Southeast Asia. As a result, intra-realm trade has boomed in recent years, an increasingly important development to avoid being completely overshadowed by China.

Of course, this would hardly be the first time that Southeast Asia found itself enmeshed in the spheres of influence of external powers. Over the past century, this realm was first dominated by European powers, and then it became part of Japan's empire; after independence, it was eventually integrated within the Japan-led *Yen bloc*, but nowadays it is increasingly drawn into the economic (and political) orbit of China. In 2010, ASEAN and China entered into a free-trade agreement, with the Chinese especially interested in acquiring raw materials from mainland Southeast Asia. Accordingly, the Chinese are investing heavily in state-of-the-art rail connections from its southern Yunnan Province all the way to Singapore. New rail lines are planned from Kunming (Yunnan's main city) into Laos, fanning out from there to Cambodia, Vietnam, Thailand, Malaysia, and Singapore; meanwhile, a direct line is under construction from China to Myanmar's Indian Ocean coast. Thus in order to avoid excessive dependence on China, Southeast Asia's emerging markets need to develop rapidly, diversify their trade partners, and strengthen intra-realm connections.

STATES AND BOUNDARIES

Although we tend to think of boundaries as lines on the map or fences on the ground, the legal definition of a boundary goes much farther than that. In fact, boundaries are actually invisible vertical planes extending above the ground into the air and below the ground into soil and rock (or water). Where these planes intersect with the ground, they form lines—the lines on the map.

A useful way to think about boundaries is to regard them as contracts between states. Such contracts take the form of treaties that contain the **definition** of every segment of the boundaries between them. These written definitions refer to actual landforms of the terrain through which the boundary lies—streams, larger water bodies, hills, ridges. Next, surveyors translate these descriptions into lines on large-scale maps that show every detail in the landscape. This process of **delimitation** creates the official boundary agreed to by the parties, and we see the results in generalized form on atlas maps. Sometimes neighboring states start arguing over the treaty language or the outcome of delimitation, which can result in seemingly pointless wars—such as the recent costly conflict between Ethiopia and Eritrea over a short stretch of their common border plus a few hundred square kilometers of mostly barren land. To avoid such problems, states mark certain stretches of their borders with fences, walls, or other barriers, a process referred to as **demarcation**. This does not always preclude disputes. For example, Israel's "Security Barrier" (see Chapter 7B) does not conform to any definition agreed to by its Palestinian neighbors.

Classifying Boundaries

Boundaries have many functions, and to understand this (and to learn why some boundaries tend to produce more trouble than others) it is helpful to view them in categorical perspective. Some boundaries conform to elongated features in the natural landscape (mountain ranges, rivers) and have **physiographic** origins. Others coincide with historic breaks or transitions in the cultural landscape and are sometimes referred to as *anthropogeographic* or more recently as **ethnocultural** boundaries. And as any world political map shows, many boundaries are simply straight lines, defined by endpoint coordinates with no reference to physical or cultural landscape features; these **geometric** boundaries often lead to problems when valuable natural resources are found to lie across or beneath them.

In general, the colonial powers and their successor governments defined the boundaries of Southeast Asia more judiciously than was the case in several other now-postcolonial parts of the world, notably Africa, the Arabian Peninsula, and Turkestan. The colonial powers that established the original treaties tried to create boundaries that lay in remote and/or sparsely peopled areas: for example, across interior Borneo. Nonetheless, a number of Southeast Asian boundaries have triggered problems, among them the geometric boundary between Papua, the portion of New Guinea ruled by Indonesia, and the country of Papua New Guinea, which occupies the eastern part of the island. The artificiality of this boundary is resulting in rising secessionist feelings among the population of Indonesian Papua.

From the Field Notes . . .

© Barbara A. Weightman

"I stood on the Laotian side of the great Mekong River which, during the dry season, did not look so great! On the opposite side was Thailand, and it was rather easy for people to cross here at this time of the year. But, the locals told me, it is quite another story in the wet season. Then the river inundates the rocks and banks you see here, it rushes past, and makes crossing difficult and even dangerous. The buildings where the canoes are docked are built on floats, and rise and fall with the seasons. The physiographic-political boundary between Thailand and Laos lies in the middle of the valley we see here."

www.conceptcaching.com

Even on a small-scale map of the kind we use in this chapter, we can categorize the boundaries in this realm. A comparison between Figures 10A-1 and 10A-4 reveals that the boundary between Thailand and Myanmar over long stretches is anthropogeographic (ethnocultural), most notably where the name *Kayin* (*Karen*), the Myanmar minority, appears in Figure 10A-4. And Figure 10A-1 shows that a lengthy segment of the Vietnam–Laos boundary is physiographic-political, in that it coincides with the Anna-mite Cordillera (Highlands).

Boundaries in Changing Times

A number of the world's boundaries are centuries—even many centuries—old, whereas others are of more recent origin. Hence, another way of interpreting their functions is to examine their evolution as part of the cultural landscape they partition. We distinguish four types of these genetic (evolutionary) boundaries; as it happens, you can find examples of all four in Southeast Asia.

Certain boundaries were defined and delimited before the present-day human landscape developed. In Figure 10A-7 (upper-left map), the border between Malaysia and Indonesia across the island of Borneo is an example of the first boundary type, an antecedent boundary [10]. Most of this border passes through very sparsely inhabited tropical rainforest, and the break in the settlement pattern can even be detected in the realm's population map (Fig. 10A-3).

A second category of boundaries evolved as the cultural landscape of an area took shape and became part of the ongoing process of accommodation between neighboring states. These subsequent boundaries [11] are represented in Southeast Asia by the map in the upper right of Figure 10A-7, which shows in some detail the border between Vietnam and China. This border is the result of a long process of adjustment and modification, the end of which may not yet have come.

The third category involves boundaries drawn forcibly across a unified or at least homogeneous cultural landscape. The colonial powers did this when they divided the island of New Guinea by delimiting a boundary in a nearly straight line (curved in only one place to allow for a bend in the Fly River), as shown in the lower-left map of Figure 10A-7. The superimposed boundary [12] they delimited gave the Netherlands the western half of New Guinea. When Indonesia became independent in 1949, the Dutch did not yield their part of New Guinea, which is peopled mostly by ethnic Papuans, not Indonesians. In 1962, the Indonesians invaded the territory by force of arms, and in 1969 the United Nations recognized its authority there.

GENETIC POLITICAL BOUNDARY TYPES

FIGURE 10A-7

© H. J. de Blij, P. O. Muller, and John Wiley & Sons, Inc.

CAMBODIA
Phnom Penh

COMPACT

THAILAND
Bangkok

PROTRUDED

VIETNAM
Hanoi

SOUTH AFRICA
Pretoria
LESOTHO
Cape Town

PERFORATED

ELONGATED

Manila

PHILIPPINES

Ho Chi Minh City
(Saigon-Cholon)

STATE TERRITORIAL
CONFIGURATIONS

FRAGMENTED

FIGURE 10A-8   © H. J. de Blij, P. O. Muller, and John Wiley & Sons, Inc.

This made the colonial, superimposed boundary the eastern border of Indonesia and had the effect of projecting Indonesia from the Southeast Asian realm into the adjoining Pacific realm. Geographically, all of New Guinea lies within the Pacific Realm.

The fourth genetic boundary type is the so-called **relict boundary [13]**—a border that has ceased to function but whose imprints (and sometimes influence) are still evident in the cultural landscape. The boundary between former North and South Vietnam (Fig. 10A-7, lower-right map) is a classic example: once demarcated militarily, it has had relict status since 1976 following the reunification of Vietnam in the aftermath of the Indochina War (1964–1975).

Southeast Asia's boundaries have colonial origins, but they have continued to influence the course of events in postcolonial times. Take one instance: the physiographic boundary that separates the main island of Singapore from the rest of the Malay Peninsula, the Johor Strait (see Fig.

10B-6). That physiographic-political boundary facilitated, perhaps crucially, Singapore's secession from the state of Malaysia in 1965. Without it, Malaysia might have been persuaded to halt the separation process; at the very least, territorial issues would have arisen to slow the sequence of events. As it was, no land boundary needed to be defined: the Johor Strait demarcated Singapore and left no question as to its limits.*

State and Territorial Morphology

Boundaries define and delimit states; they also create the mosaic of often interlocking territories that give individual countries their shape. This shape or ***territorial morphology***

---

*Except one: a tiny island at the eastern entrance to the Strait named Pedra Blanca (as Singapore calls it) or Pulau Batu Putih (the Malaysian version), which is still disputed today.

can affect a state's condition, even its survival. Vietnam's extreme elongation has influenced its existence since time immemorial. And, as is noted in Chapter 10B, Indonesia has tried to redress its fragmented nature (thousands of islands) by promoting unity through the "transmigration" of residents of Jawa from the most populous island to many of the others.

Political geographers identify five dominant state territorial configurations, all of which we have encountered in our world regional survey but which we have not categorized until now. All but one of these shapes is represented in Southeast Asia, and Figure 10A-8 to the left provides the terminology and examples:

- **Compact states [14]** have territories shaped somewhere between round and rectangular, without major indentations. This encloses a maximum amount of territory within a minimum length of boundary. Southeast Asian example: Cambodia.

- **Protruded states [15]** (sometimes called ***extended***) have a substantial, usually compact territory from which extends a peninsular corridor that may be landlocked or coastal. Southeast Asian examples: Thailand and Myanmar.

- **Elongated states [16]** (also known as ***attenuated***) have territorial dimensions in which the length is at least six times the average width, creating a state that lies astride environmental or cultural transitions. Southeast Asian example: Vietnam.

- **Fragmented states [17]** consist of two or more territorial units separated by foreign territory or a substantial body of water. Subtypes include mainland-mainland, mainland-island, and island-island. Southeast Asian examples: Malaysia, Indonesia, the Philippines, and East Timor.

- **Perforated states [18]** completely surround the territory of other states, so that they have a "hole" in them. No Southeast Asian example; the most illustrative case on the world political map is South Africa, perforated by Maryland-sized Lesotho.

In Chapter 10B, we will have frequent occasion to refer to the shapes of Southeast Asia's states. For so comparatively small a realm with so few countries, Southeast Asia displays a considerable variety of state morphologies. But one point of caution: states' territorial morphologies do not determine their viability, cohesion, unity, or lack thereof; they can, however, influence these qualities. Cambodia's compactness has not ameliorated its divisive political geography, for instance. But as we will find in our survey of the realm's regional geography, shape does play a key role in the still-evolving political and economic geography of Southeast Asia.

▶ **POINTS TO PONDER**

- Many countries in Southeast Asia have shown a remarkable capacity to accommodate and blend diverse religious traditions.

- The two smallest countries of the realm are also by far the wealthiest.

- How many products used in your daily life contain palm oil? Is there a way of knowing? If you use any, do you contribute to deforestation?

- Is mainland Southeast Asia on the way to becoming a Chinese periphery?

SOUTHEAST ASIA: REGIONS AND STATES

POPULATION

- Under 50,000
- 50,000–250,000
- 250,000–1,000,000
- 1,000,000–5,000,000
- Over 5,000,000

National capitals are underlined

— Railroad

Core area

Transition Zone

0   200   400   600   800 Kilometers
0 100 200 300 400 Miles

FIGURE 10B-1

© H. J. de Blij, P. O. Muller, and John Wiley & Sons, Inc.

REGIONS

Mainland Southeast Asia

Insular Southeast Asia

IN THIS CHAPTER

- Vietnam's cautious modernization
- Chinese railroads in Laos
- Thailand: A protest-plagued kingdom
- Myanmar plans a special economic zone
- Running out of space in Singapore
- Democratic Indonesia takes the lead

CONCEPTS, IDEAS, AND TERMS

The regional geography of Southeast Asia evolved from the former empires of competing colonial powers whose domination ensured compliance with their imperial designs. Subsequent postcolonial governments in this realm confronted strong centrifugal forces and generally succeeded in maintaining the stability of this regional framework. Neither outsiders (U.S. intervention in the 1960s aimed at dividing Vietnam into a communist North and a noncommunist South) nor locals (such as insurgents in Malaysia and secessionists in Indonesia) could fracture multicultural states, whose components did not begin their independent existence with much in common, either culturally or economically. There were times when it seemed that Malaysia might break up, but only Singapore left that federation, a negotiated secession without violence. Sprawling Indonesia dealt with separatists by force as well as through persuasion and accommodation. The troubled birth of East Timor in 2002, the realm's youngest state, is the exception rather than the rule. Meanwhile, Southeast Asia's governments have become more representative (with the glaring, tragic exception of Myanmar); its economies have matured and, in the majority of cases, prospered; and the realm's vital statistics (see Appendix B) reflect its overall development.

So much of the Southeast Asian realm consists of the waters between and among its land areas that, on the globe, it seems to be much larger territorially than it really is. Indeed, the realm's mainland component is actually smaller than its island entities combined: the mainland (including the peninsular part of Malaysia) consists of about 2.1 million square kilometers (800,000 sq mi) whereas the islands, even without including Indonesian Papua (which belongs to the Pacific Realm), account for some 2.4 million square

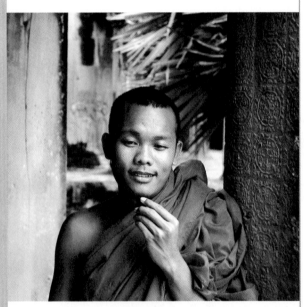

© Jan Nijman
Where were these photos taken?
Find out at www.wiley.com/college/deblij

© Jan Nijman

kilometers (930,000 sq mi)—a ratio of 46 to 54 percent. But, as we shall see, Malaysia overall exhibits stronger island than mainland properties, tilting the balance even further toward offshore Southeast Asia.

With eleven countries ranging from ministates Singapore and Brunei to regional giants Indonesia and Myanmar, Southeast Asia's states fall into every World Bank income category (see Fig. G-11). Singapore got rich from trade, Brunei from oil. Myanmar stayed poor when it did not need to because mismanagement prevented it from realizing its potential. Environmental, locational, and administrative problems bedeviled, and then restrained, Laos and Cambodia. But Malaysia forged ahead, Thailand was upward bound before recent political setbacks, and conservatively communist Vietnam is beginning to loosen its economic shackles. There was a time when this realm, in overall terms, was on a par with Subsaharan Africa. No longer. Southeast Asia's first-order regionalization recognizes the mainland–island dichotomy that is obvious from any map of this realm. But things are somewhat more complicated than that because the southern part of the Malay Peninsula, occupied by the states of Malaysia and Singapore, exhibits physiographic, historical, and cultural circumstances to justify its inclusion in the insular (island) rather than the mainland region. Using the political framework as our grid, we can map the broadest regional division of Southeast Asia in Figure 10B-1:

- **Mainland region:** Vietnam, Cambodia, Laos, Thailand, Myanmar (Burma)

- **Insular region:** Malaysia, Singapore, Indonesia, Brunei, the Philippines, East Timor (Timor-Leste)

Special note should also be taken of the discordance between the political grid and the cultural-geographic reality along the eastern periphery of the Insular region. Here, Indonesia controls the western half of the island of New Guinea—but all of New Guinea belongs in the neighboring Pacific Realm.

● MAJOR CITIES OF THE REALM

City	Population* (in millions)
Manila, Philippines	12.0
Jakarta, Indonesia	9.4
Bangkok, Thailand	7.2
Ho Chi Minh City (Saigon), Vietnam	6.5
Singapore, Singapore	5.2
Yangon, Myanmar (Burma)	4.6
Hanoi, Vietnam	3.1
Phnom Penh, Cambodia	1.7
Kuala Lumpur, Malaysia	1.6
Vientiane, Laos	0.9

*Based on 2012 estimates.

▪▪ MAINLAND SOUTHEAST ASIA

Five countries form the Mainland region of Southeast Asia (Fig. 10B-1): the three remnants of Indochina—Vietnam, Cambodia, and Laos—in the east; Thailand in the center; and Myanmar (formerly Burma) in the west. One religion, Buddhism, dominates cultural landscapes, but this is a multicultural, multiethnic region (see Fig. 10A-4). Although still one of the less urbanized regions of the world, the Mainland contains several major cities and the pace of urbanization is increasing. And as Figure 10B-1 shows, two countries (Vietnam and Myanmar) possess dual core areas.

We approach this region from the east, beginning our survey in Indochina where the United States fought and lost a disastrous war that ended in 1975, but whose impact on America continues to be felt today. After the Indochina War started formally in 1964 (U.S. involvement in Vietnam actually began earlier), some scholars warned that the conflict might spill over from Vietnam into Laos and Cambodia, and from there into Thailand, Malaysia, and even then-Burma. This view was based on the **domino theory [1]**, which holds that destabilization and conflict from any cause in one country can result in the collapse of order in one or more neighboring countries, triggering a chain of events that can affect a series of contiguous states in a region.

That did not quite happen, and some of the rhetoric of the time seems to have been infused with alarmist views from the West. Cambodia and Laos were indeed disastrously affected as the original Vietnam War became the wider Indochina War, but not the other "dominoes" (in fact, the last Laotian Hmong refugees who had been driven from their homes and who had found refuge in Thailand were forcibly ousted in late 2009).

Still, that does not necessarily invalidate the idea of the domino effect. Powerful states that intervene in weaker countries, as the United States did in Iraq in 2003 and Russia in Georgia in 2008, do so at the risk of destabilizing those countries' neighbors by setting in motion arms races and sowing the seeds of future intraregional conflict. Ethnic-cultural strife in one state can also engulf neighboring countries, as happened in central Africa during the 1990s when Rwanda's civil wars spread into the neighboring Congo (where a government fell as a result), and also drew in Burundi and Uganda.

The domino theory remains an intriguing device, and in our crowded world potential dominoes abound. Its most recent, dramatic manifestation, as shown in Chapters 7A and 7B, was the **diffusion [2]** of popular uprisings across North Africa and Southwest Asia in 2011.

Vietnam

Vietnam (population: 91 million) still carries the deep scars of the Indochina War, although the vast majority of Vietnamese have no personal memory of that terrible conflict. The more immediate concerns in Vietnam today are to reconnect the country with the outside world and to integrate

FIGURE 10B-2   © H. J. de Blij, P. O. Muller, and John Wiley & Sons, Inc.

its 2000-kilometer (1200-mi) strip of highly elongated territory through better infrastructure (Fig. 10B-2). This is also an emerging economy that is now making considerable progress. Poverty has dropped dramatically in the past decade as Vietnam has become a player in the global economy. In the early 2010s, it is regarded as a "hot" new location for assembly plants, which are attracted by wages lower than China's and more than a dozen economic development zones.

French Legacy

The French colonialists recognized that Vietnam, whose average width is less than 240 kilometers (150 mi), was not a homogeneous colony, so they divided it into three units: (1) *Tonkin*, land of the Red River Basin and Delta, centered on Hanoi in the north; (2) *Cochin China*, domain of the Mekong Delta, centered on Saigon in the south; and (3) *Annam*, focused on the ancient city of Hué in the middle. Today, the Vietnamese prefer to use *Bac Bo*, *Nam Bo*, and *Trung Bo*, respectively, to designate these areas.

The Vietnamese (or Annamese, also Annamites, after their cultural heartland) speak the same language, although the northerners can easily be distinguished from southerners by their accent. As elsewhere in their colonial empire, the French made their language the *lingua franca* of Indochina, but their tenure was cut short by the Japanese, who invaded Vietnam in 1940. During the Japanese occupation, Vietnamese nationalism became a powerful force, and after the Japanese defeat in 1945, the French could not regain control. In 1954, the French suffered a disastrous final trouncing on the battlefield at Dien Bien Phu in the far northwest and were ousted from the country.

North and South

But even after its forces routed the colonizers, Vietnam did not become a unified state. Separate regimes took control: a communist one in Hanoi and a noncommunist counterpart in Saigon. Vietnam's pronounced elongation had made things difficult for the French; now it played its role during the postcolonial period. Note, in Figure 10B-2, that Vietnam is widest in the north and south, with a slender "waist" in its middle zone. North and South Vietnam were worlds apart, and those worlds were represented in

Hanoi by communism and in Saigon by anticommunism. For more than a decade the United States tried to prop up the Saigon regime that controlled the south, but the communists prevailed, and like China, Vietnam still has a communist government today. After the end of hostilities, as many as 2 million Vietnamese refugees set out on often-flimsy boats onto the South China Sea; of those who survived, a majority settled in the United States.

Today, contrasts between north and south continue, although they are diminishing. The capital, Hanoi, has long lagged behind bustling Saigon, but now its skyline reflects modernization and the overland links to its port of Haiphong are vastly improved. With more than 3 million residents, Hanoi anchors the northern (Tonkin Plain) core area of Vietnam, the lower basin of the Red River (its agricultural hinterland). On rural roads, goods are still moved by human- or animal-drawn cart, but the roads themselves are being improved. The south, however, is experiencing even more profound change, with significant economic growth as the government slowly, and on its own terms, opens up the country to the global economy, thereby lifting many out of poverty.

Vietnam in Transition

In the first decades of communist rule, private enterprise was abolished and farmers were compelled to join collectives on

---

AMONG THE REALM'S GREAT CITIES . . .

SAIGON

OFFICIALLY, IT IS known as Ho Chi Minh City, renamed in 1975 after the ouster of American forces, but only bureaucrats and atlases call it that. To the people here, it is still *Saigon*, city on the Saigon River, served by, it seems, a thousand companies using the name. Saigon Taxi, Saigon Hotel, Saigon Pharmacy—there is hardly a city block without the name Saigon on some sign or billboard.

The phenomenon is significant. Hanoi's northerners ordained the change, but this still is South Vietnam, and many southerners see the North as peopled by go-slow, dogmatic communists obstructing the hopes and plans for their vibrant city of 6.5 million. According to UN population estimates, Saigon is more than twice as large as Hanoi, but like the capital it suffers from decades of deterioration, a broken-down infrastructure, and undependable public services. When, in the early 1990s, businesspeople could find no modern accommodations or working telephones, an Australian firm built an eight-story floating hotel with its own power plant, towed it up the Saigon River, and parked it on the CBD waterfront.

Therein lies Saigon's great advantage: large oceangoing vessels can reach the city in a three-hour sail up the meandering river. The Chinese developed its potential: in the 1960s, Chinese merchants controlled more than half of the South's exports and nearly all of its textile factories and foreign exchange. Saigon-Cholon (Cholon means "great

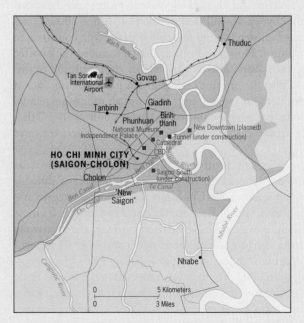

© H. J. de Blij, P. O. Muller, and John Wiley & Sons, Inc.

market") was a Chinese success story, but after the war the Hanoi regime nationalized all Chinese-owned firms. Tens of thousands of middle-class Chinese left the city, its economy devastated.

Two faces of changing Vietnam, three blocks apart in Ho Chi Minh City (Saigon): a statue of the revered former communist leader Ho Chi Minh in front of city hall, and the first Louis Vuitton store, catering to the emergent (upper) middle class.
© Jan Nijman

the communist Chinese model, resulting in near-famine conditions in the early 1980s. This coincided with China's economic opening, and Hanoi's leaders were quick to follow suit, though more timidly. By the mid-1990s, free enterprise was encouraged; farmers were allowed to cultivate for profit, and foreign investment was welcomed, although under far more restrictive terms than in China. The government has continued to provide adequate social services, resulting in health indicators better than those in China and rapidly rising attendance in its secondary schools.

A leading focus of Vietnam in recent years has been on increased agricultural production. Since 2000, it has ranked among the top five coffee-producing countries in the world. However, because mass-produced bulk coffee is grown plantation-style in the sun, the expansion of this crop into the interior highlands has produced significant deforestation, worsening the annual monsoon-related floods that plague this now-unstable countryside. Other important crops include rice, cotton, rubber, tea, and spices.

Vietnam continued to maintain a relatively fast-growing national economy even during recent difficult economic times. Ho Chi Minh City is booming again and now has a Special Economic Zone based on the Chinese model, as well as the New Saigon business and residential district. Saigon and environs now contribute 25 percent of Vietnam's industrial output and one-third of its tax revenues. In 2007, Vietnam was admitted to the World Trade Organization, and more such changes are in the offing. But reform seems to have slowed since 2009 and economists argue that the country is too dependent on state-run companies while it is in need of dynamic private initiative. Hanoi's communist planners have kept tighter economic control over the country than their Chinese counterparts and seem determined not to allow runaway capitalism of the Chinese variety—nor the wealth disparities that come with it.

Cambodia

Compact Cambodia is heir to the ancient Khmer Empire whose capital was Angkor and whose legacy is a vast landscape of imposing monuments including the great Buddhist-inspired temple complex, Angkor Wat. Today, 90 percent of Cambodia's 15.6 million inhabitants are ethnic Khmers, with the remainder divided between Vietnamese and Chinese. The present capital, Phnom Penh, lies on the Mekong River, which crosses Cambodia before it enters and forms its huge delta in Vietnam (Fig. 10B-2).

Geographically, Cambodia enjoys a number of advantages. As a compact state (see Fig. 10A-8), it encloses a maximum amount of territory within a minimum of boundary, and the favorable circumstances for cultural homogeneity tend to diminish centrifugal forces. But sometimes the most obvious geographic opportunities are undone by destructive politics. Neither spatial morphology nor ethnic uniformity could withstand the impact of the Indochina War and the subsequent murderous exploits of the *Khmer Rouge* regime in the late 1970s.

Postwar Problems

Cambodia's present social geography continues to suffer from the after-effects of that war. Its internal spatial advantages notwithstanding, this country could not overcome the liability of relative location adjacent to a conflict that was bound to spill across its borders (Fig. 10B-2). In 1970, internal disharmony fueled by external interference led to the ouster of the last king by the military; five years later, that regime was itself overthrown by communist revolutionaries, the so-called Khmer Rouge. These new rulers embarked on a vicious course of terror and destruction in order to reconstruct Kampuchea (as they called Cambodia) as a rural society. They drove townspeople into the countryside where they had no place to live or work, emptied hospitals and sent the sick and dying into the streets, outlawed religion and family, and in the process killed as many as 2 million Cambodians (out of a population of

From the Field Notes . . .

"Hanoi and most of the country's other urban areas look and function much as they have for generations. Main arteries throb with commerce and traffic (still mostly bicycles and mopeds); side streets are quieter and more residential. No high-rises here, but the problems of dense urban populations are nevertheless evident. Every time I turned into a side street, the need for street cleaning and refuse collection was obvious, and drains were clogged more often than not. 'We haven't allowed our older neighborhoods to be destroyed like they have in China,' said my colleague from Hanoi University, 'but we also don't have an economy growing so fast that we can afford to provide the services these people need.' Certainly the contrast between burgeoning coastal China and slower-growing Vietnam—a matter of communist-government policy—is evident in their cities. 'As you see, nothing much has changed here,' she said. 'Not much evidence of globalization. But we're basically self-sufficient, nobody goes hungry, and the gap between the richest and the poorest here in Vietnam is a fraction of what it is now in China. We decided to keep control.' You would find more agreement on these points here in Hanoi than in Saigon, but the Vietnamese state is stable and progressing."

www.conceptcaching.com

© H. J. de Blij

8 million). In the late 1970s, Vietnam, having won its own war, invaded Cambodia to drive the Khmer Rouge away. But this action sparked another reign of terror, and a stream of refugees crossed the border into Thailand. Eventually, remnants of the Khmer Rouge managed to establish a base in northwestern Cambodia, where their murderous leader, Pol Pot, who was never brought to justice for his actions, committed suicide in 1998.

Cambodia's postwar trauma continues today. Once self-sufficient and able to feed others, it now must import food. Rice and beans are the subsistence crops, but the dislocation in the farmlands set production back severely, and continuing strife in the countryside has disrupted supply routes for years. If there is hope for the national budget of the future, it may lie in recently discovered offshore oil reserves near Cambodia's relatively short coastline. Income from oil exports is set to begin in 2012 and could approach U.S. $2 billion annually by the end of this decade, a significant boost for an otherwise weak economy.

Cambodians hope that tourism, focused chiefly on the great temple complex at Angkor, will also become a mainstay of the economy. But political stability is a prerequisite for this industry, and it has not yet returned to this ancient kingdom (population today: 15.6 million).

Laos

Landlocked Laos has no fewer than five neighbors, one of which is East Asia's giant, China (Fig. 10B-2). The Mekong River forms a long stretch of its western boundary, and the important sensitive border with Vietnam to the east lies in mountainous terrain. With only 6.7 million people (over half of them ethnic Lao, related to the Thai of Thailand), Laos lies surrounded by comparatively powerful states. The country has no railroads, very few paved roads, and hardly any industry; moreover, less than 5 percent of the land is suitable for agriculture. Laos is only 27 percent urbanized; the capital, Vientiane, lies on the Mekong and has an oil pipeline to Vietnam's coast.

Laos has long included one small corner of the Golden Triangle of opium-poppy-cultivation fame, but under international (especially U.S. and European) pressure, the communist regime has forced the mainly hill-tribe people who produce most of it to abandon their crops. Because this was the only way they could make a living, these farmers were sent to resettlement villages in the lowlands, where they contracted malaria and other diseases not prevalent in their upland domain and were subject to cultural disintegration. The reward for the authorities was the continuation of foreign aid; the cost to the powerless hill people (especially the women) is incalculable. Here is an outstanding example of the power of the globalizing core reaching into the weakest of peripheries.

Communist Laos has been very slow to follow the examples from elsewhere in the realm to open up the economy. A much belated move in 2011 was the opening of the first stock market in Vientiane. A major new hydropower project on the Nam Theun River, a tributary of the Mekong, began operating in 2010 and now provides electricity to much of central Laos as well as selling surplus power to Thailand. The high-speed railway now under construction from Vientiane to Kunming in China is certain to pull the country more tightly into the Chinese orbit.

Thailand

In virtually every way, Thailand is the leading state of the Mainland region. In contrast to its neighbors, Thailand has been a strong participant in the realm's economic development. Its capital, Bangkok (population: 7.2 million), is the largest urban center in the Mainland region and one of the world's most prominent primate cities. The country's

population, 69 million in 2012, is growing at the slowest rate in this geographic realm.

Over the past few decades, only political instability and uncertainty have inhibited economic progress. Thailand is a constitutional monarchy with an elected parliament, but its progress toward a stable democracy has experienced a history of setbacks. In 2006, the armed forces ousted a controversial but popular prime minister, precipitating a power struggle that has severely impacted the national economy, notably the vitally important tourist industry that still languishes today. Mass demonstrations immobilized cities, closed down airports, and paralyzed surface transport at enormous cost to industry and commerce. The summer of 2010 saw renewed conflict on the streets of Bangkok (see photo below), with the armed forces clamping down on the so-called red shirts, a group of coalitions pressing for democratic reforms.

Thailand has a compact heartland that contains the core area, capital, and leading zones of productive capacity, while a 1000-kilometer (600-mi) protrusion [3] (a corridor of land extending away from the rest of the state), in places less than 32 kilometers (20 mi) wide, extends southward to the border with Malaysia (Fig. 10B-3). The boundary that defines this protrusion runs down the length of the upper Malay Peninsula to the Kra Isthmus, where neighboring Myanmar peters out and Thailand fronts the Andaman Sea (an arm of the Indian Ocean) as well as the Gulf of Thailand.

The Restive Peninsular South

In the entire country, no place lies farther from the capital than the southern end of this tenuous protrusion. Consider this spatial situation in the context of Figure 10A-4. The Malay ethnic population group extends northward from Malaysia more than 300 kilometers (200 mi) into Thai territory. In Thailand's five southernmost provinces (Fig. 10B-3, inset map), 85 percent of the inhabitants are Muslims (the figure for Thailand as a whole is less than 6 percent). The international border between Thailand and Malaysia is porous; in fact, you can cross it at will almost anywhere along the Kolok River by canoe, and inland it is a matter of walking a forest trail. For more than a century, these southern provinces have had closer ties with Malaysia across the border than with Bangkok 1000 kilometers (600 mi) away, and the Thai government has not tried to restrict movement or impose onerous rules on the Muslim population here.

In view of rising Islamic militancy and violence along this southern frontier, the south is now

During the spring and summer of 2010 a political crisis in Thailand turned into a social disaster when rebels supporting an ousted prime minister battled supporters of a discredited government in the streets of Bangkok, causing dozens of deaths, hundreds of injuries, extensive property damage, and economic havoc. By the time the army and police overpowered the "red shirts," parts of this globalized city looked like a war zone; as defeat loomed, the rebels torched one of the world's largest shopping malls, the Central World (below).

© AFP/Getty Images, Inc.

FIGURE 10B-3

© H. J. de Blij, P. O. Muller, and John Wiley & Sons, Inc.

coming to national attention. The Pattani United Liberation Organization, named after the functional capital of the Muslim south, was active in the 1960s and 1970s but had been dormant since. Then a series of bombings at a Pattani hotel, several schools, a Chinese shrine, and a Buddhist temple in 2001 and 2002 ushered in a new era of Muslim-inspired violence. Thailand's pronounced protrusion, its porous border, its dependence on tourism, and the vulnerable location of Phuket and other Andaman-coast resorts, combine to raise concern over this distant part of the national territory.

Bangkok on the Chao Phraya

As Figure 10B-1 shows, Thailand occupies the heart of the Mainland region of Southeast Asia. Even though Thailand has no Red, Mekong, or Irrawaddy Delta, its central lowland is watered by a set of streams that flow off the northern highlands and the Khorat Plateau to the east. One of these streams, the Chao Phraya, is the Rhine of Thailand. From the head of the Gulf of Thailand to Nakhon Sawan, this river is a highway of traffic. Barge trains loaded with rice head toward the coast, ferry boats head upstream, freighters

⦿ AMONG THE REALM'S GREAT CITIES . . .

BANGKOK

WITH A POPULATION of 7.2 million, Bangkok is mainland Southeast Asia's largest city, a sprawling metropolis on the banks of the Chao Phraya River, an urban agglomeration without a true center, an aggregation of neighborhoods ranging from the immaculate to the impoverished. In this city of great distances and sweltering tropical temperatures, getting around is often difficult because roadways are choked with traffic. A (diminishing) network of waterways affords the easiest way to travel, and life focuses on the busiest waterway of all, the Chao Phraya. Ferries and water taxis carry tens of thousands of commuters and shoppers across and along this bustling artery, flanked by a growing number of high-rise office buildings, luxury hotels, and ultramodern condominiums. Many of these modern structures reflect the Thais' fondness for domes, columns, and small-paned windows, creating a skyline unique in Asia.

Gold is Buddhist Thailand's symbol, adorning religious and nonreligious architecture alike. From a high vantage point in the city, you can see hundreds of golden spires, pagodas, and façades rising above the townscape. The Grand Palace, where royal, religious, and public buildings are crowded inside a white, crenellated wall 2 kilometers (more than 1 mi) long (see photo at right), is a gold-laminated city within a

© H. J. de Blij, P. O. Muller, and John Wiley & Sons, Inc.

city embellished by ornate gateways, dragons, and statuary. Across the mall in front of the Grand Palace are government buildings sometimes targeted by rioters in Bangkok's volatile political environment. And not far away are Chinatown (see photo, p. 524) and myriad markets, all components of Bangkok's throbbing commercial life.

transport tin and tungsten (of which Thailand is among the world's leading producers). Bangkok sprawls along both sides of the lower Chao Phraya floodplain, here flanked by skyscrapers, pagodas, factories, boatsheds, ferry landings, luxury hotels, and myriad modest dwellings all built in crowded confusion on very swampy ground. On the right bank of the Chao Phraya, Bangkok's west side, lie the city's remaining *klong* (canal) neighborhoods, where waterways and boats constitute the transport system. Bangkok still is known as the "Venice of Asia," although many *klongs* have been filled in and paved over to serve as roadways. Unfortunately, Bangkok's location on this soggy terrain has compounded its notorious traffic problems as a desperately needed underground transit system cannot be constructed.

Myanmar

Thailand's western neighbor, Myanmar (still referred to as Burma in anti-regime circles), is one of the world's poorest and most isolated countries where, it seems, time has stood still for centuries. Long oppressed by one of the world's most corrupt and brutal military dictatorships, it has potential as an oil producer, but its promise must await the regime change that would make multinationals more welcome. Myanmar shares with Thailand the narrow neck of the Malay Peninsula—but look again at Figure 10B-3 and note the contrast in surface communications.

Myanmar's problems are complicated by a shift in the Burmese core area that took place during colonial times. Prior to the colonial period, the core of embryonic Burma lay in the so-called dry zone between the Arakan Mountains (in the far west) and the Shan Plateau, which covers the country's triangular easternmost extension (Fig. 10B-4). The urban focus of the state was Mandalay, which had a central situation and relative proximity to the non-Burmese highlands all around. Then the British developed the agricultural potential of the Irrawaddy Delta farther south, and Rangoon (now called Yangon) became the hub of the colony. The Irrawaddy waterway links the old and the new core areas, but the center of gravity has shifted to the south.

As Figure 10A-4 indicates, the peripheral peoples (11 minorities) of Myanmar occupy a significant part of the state. A closer look at Figure 10B-4 shows that their domains hold the status of State, of which there are seven; the Burman-dominated areas (68 percent of a national population of just under 55 million) are designated as Divisions. The Shan of the northeast and far north, who are related to the neighboring Thai, account for about 9 percent of the population, or 4.9 million. The Kayin (Karen), who constitute more than 7 percent (4 million), live in the neck of Myanmar's protrusion and have proclaimed their desire to create an autonomous territory within a federal Myanmar. The Mon (2.2 percent, or 1.2 million) were in what is today Myanmar long before the Burmans and introduced

From the Field Notes . . .

"Thailand exhibits one of the world's most distinctive cultural landscapes, both in its urban and its rural areas. Graceful pagodas, stupas, and spirit houses of Buddhism adorn city and countryside alike. Gold-layered spires rise above, and beautify cities and towns that would otherwise be drab and featureless. The architecture of Buddhism has diffused into public architecture, so that many secular buildings, from businesses to skyscrapers, are embellished by something approaching a national style. The grandest expression of the form undoubtedly is a magnificent assemblage of structures within the walls of the Grand Palace in the capital, Bangkok, of which a small sample is shown here. Climb onto the roof of any tall building in the city's center, however, and the urban scene will display hundreds of such graceful structures, interspersed with the modern and the traditional townscape."

www.conceptcaching.com

© H. J. de Blij

Buddhism to the area; they want the return of ancestral lands from which they were ousted. Although the powerful military has dealt such aspirations a series of setbacks, centrifugal forces continue to bedevil the central regime. The military government's response has been to exert power by all available means rather than to accommodate these forces, even to the point of stifling political discourse—let alone opposition—among the Burmans themselves.

In 2007, the generals moved the capital 300 kilometers (200 mi) northward from Yangon to a newly constructed

From the Field Notes . . .

"Little did I realize, as I walked the central business district of Yangon (Myanmar) in mid-April 2008, that these streets would be turned into chaos by Cyclone Nargis just days later. This is Sule Pagoda Road, a few blocks north of the famed Octagonal Pagoda, the golden landmark in the center of town. Many signs in English remind you that this was once a British colony."

www.conceptcaching.com

© H. J. de Blij

FIGURE 10B-4

© H. J. de Blij, P. O. Muller, and John Wiley & Sons, Inc.

national-government center—ostensibly a city, but it can barely be called that as it is poorly connected to anywhere else and has few urban amenities. This center is located next to the interior town of Pyinmana and is known as Naypyidaw. The government asserts that its more central location will yield greater administrative efficiency as well as better protection against (unspecified) enemies.

Two recent events shook up Myanmar and exposed its failings in governance to the world in a way that its secretive junta does not like. In late 2007, a sizeable uprising of the people was led by Buddhist monks, and the world held its breath to see if this most repressive of states would finally open up. But that was not to be: yet again the junta clamped down, proclaiming that dissent would not be tolerated. Then in the spring of 2008, a powerful tropical cyclone slammed into the Irrawaddy Delta, killing at least 130,000 and displacing 2 to 3 million people. The world watched as the junta initially did not acknowledge the severity of the situation and only lackadaisically mobilized relief for storm victims. At first, foreign relief workers were prevented from bringing assistance that the government itself could not provide, and only weeks later was limited entry permitted for workers and supplies.

There were some tentative signs in 2010 of a possible turn for the better in Myanmar. The Dawei Development Project is a Thai-Burmese initiative to turn the small southern Burmese city of Dawei on the Andaman Sea into a major port. One of the first steps is the construction of a 160-kilometer (100-mi) highway from Dawei to the Thai town of Kanchanaburi, located about halfway between the port and Bangkok (Fig. 10B-4). Plans also call for what will become Myanmar's biggest industrial area by far, one resembling the special economic zones of China and Vietnam. This could mark a new economic beginning for Myanmar while far more advanced Thailand would gain valuable direct access to the nearby Andaman Sea and, beyond it, the Bay of Bengal and India.

Myanmar today ranks among the world's poorest, least developed, and most corrupt countries—not because of intrinsic indigence but because a rapacious military regime, tolerated and even commercially engaged by the international community, represses its peoples, crushes their aspirations, and repeatedly demonstrates the liabilities of life in the global periphery.

▪▪ INSULAR SOUTHEAST ASIA

On the peninsulas and islands of Southeast Asia's southern and eastern periphery lie six of the realm's 11 states (Fig. 10B-1). Few regions in the world contain so diverse a set of countries. Malaysia, the former British colony, consists of two major areas separated by hundreds of kilometers of South China Sea. The realm's southernmost state, Indonesia, sprawls across thousands of islands from Sumatera in the west to New Guinea in the east. North of the Indonesian archipelago lies the Philippines, a country that once was a U.S. colony. These are three of the most severely fragmented states on Earth, and each has faced the challenges that such politico-geographical division brings. This Insular region of Southeast Asia also contains two small but significant sovereign entities: a city-state and a sultanate. The city-state is Singapore, once a part of Malaysia (and one instance in which internal centrifugal forces were too great to be overcome). The sultanate is Brunei, an oil-rich Muslim territory on the island of Borneo that seems to be transplanted from the Persian Gulf. In addition, a third small entity, East Timor, only achieved independence a decade ago. Few parts of the world are more varied or interesting geographically.

Mainland–Island Malaysia

The state of Malaysia represents one of the three categories of fragmented states discussed in Chapter 10A: the mainland–island type, in which one part of the national territory lies on a continent and the other on one or more islands. Malaysia is a colonial political artifice that combines a pair of quite disparate components into a single state: the southern end of the Malay Peninsula and the northern sector of the island of Borneo. These are known, respectively, as *West Malaysia* and *East Malaysia* (Fig. 10B-1). The name "Malaysia" came into use in 1963, when the original Federation of Malaya, on the Malay Peninsula, was expanded to incorporate the territories of Sarawak and Sabah in Borneo. When the name *Malaya* is used, it refers to the peninsular part of the Federation, whereas *Malaysia* refers to the total entity.

Ethnic Components

The Malays of the peninsula, traditionally a rural people, displaced older aboriginal communities there and now constitute just over 50 percent of the country's population of 29.8 million. They possess a strong cultural identity expressed in adherence to the Muslim faith, a common language, and a sense of territoriality that arises from their perceived Malayan origins and their collective view of Chinese, Indian, European, and other foreign intruders.

The Chinese came to the Malay Peninsula and to northern Borneo in substantial numbers during the colonial period, and today they constitute 26 percent of Malaysia's population (they are the largest ethnic group in Sarawak). Hindu South Asians were in this area long before the Europeans, and for that matter before the Arabs and Islam arrived on these shores; today they still form a substantial minority of 7 percent of the population, clustered, like the Chinese, on the western side of the peninsula.

West Malaysia: The Dominant Peninsula

The populous peninsular part of Malaysia remains the country's dominant sector with 11 of its 13 States and over 75 percent of its population. Here, the Malay-dominated government has very tightly controlled economic and social policies while pushing the country's modernization. During the Asian economic boom of the 1990s, Malaysia's government embraced the notion of symbols: the capital, Kuala Lumpur, was endowed with (at that time) the world's tallest building; a space-age airport outpaced Malaysia's needs; a high-tech administrative capital was built at Putrajaya; and a nearby development was called Cyberjaya—all part of a so-called *Multimedia Supercorridor* to anchor Malaysia's core area (Fig. 10B-5).

The chief architect of this program was Malaysia's long-term and autocratic head of state, Mahathir bin Mohamad, leader of the Malay-dominated party that forms the majority in government. Mahathir had the support not only of the great majority of the country's Malays, but also the important and influential ethnic Chinese minority, which saw him as the only acceptable alternative to the more fundamentalist

FIGURE 10B-5   © H. J. de Blij, P. O. Muller, and John Wiley & Sons, Inc.

Islamic party challenging his rule. But Malaysia's headlong rush to modernize caused a backlash among more conservative Muslims, which in 2001 led to Islamist victories in two States, tin-producing Kelantan and energy-rich but socially poor Terengganu. As the fundamentalist governments in those eastern States imposed strict religious laws, Malaysians talked of two corridors marking their country: the Multimedia Supercorridor in the west and the Mecca Corridor in the east (Fig. 10B-5). But after Mahathir's resignation in 2003 and the appointment of a more moderate successor, Islamist fervor in the Mecca Corridor, which had featured calls for a *jihad* in Malaysia, has steadily subsided.

The Malay Peninsula's primacy began long ago, during colonial times, when the British created a substantial economy based on rubber plantations, palm-oil extraction, and mining (tin, bauxite, copper, iron). Singapore, located at the southern end of it, was the most prized possession until it seceded from the Malaysian Federation in 1965. The Strait of Malacca (Melaka) to the west continues to be one of the world's busiest and most strategic waterways, a crucial **choke point [4]** (a narrow waterway that constrains navigation) in the flow of resources and goods between major world realms. Piracy, long a significant concern in the strait, has been significantly curtailed through increased patrolling that began with the aid given to the area after the devastating 2004 Indian Ocean tsunami.

Malaysia, despite the loss of Singapore and notwithstanding its recurrent ethnic troubles, has become a leading player in Southeast Asia. The strong skills and modest wages of the local workforce have attracted many companies, and the government has capitalized on its opportunities, such as encouraging the creation of a high-technology manufacturing complex on the far northwestern island of Pinang, where Chinese outnumber Malays by two to one.

East Malaysian Borneo

The decision to combine the eleven Sultanates of Malaya with the States of Sabah and Sarawak on Borneo, creating the country now called Malaysia, had far-reaching consequences. Those two States constitute 60 percent of Malaysia's territory (although they are home to barely 20 percent of the population). They endowed Malaysia with major energy resources and huge stands of timber. They also complicated Malaysia's ethnic complexion because each State is home to more than two dozen indigenous groups (in fact, the immigrant Chinese form the largest single group in Sarawak). These locals complain that the federal government in Kuala Lumpur treats Malaysian Borneo as a colony, and politics here are contentious and fractious. It is likely that Malaysia will eventually confront devolutionary forces here in East Malaysia.

Brunei

Also located on Borneo—near where Sarawak and Sabah meet—is Brunei (Fig. 10B-1), a rich, oil-exporting Islamic sultanate far from the Persian Gulf. Brunei, the remnant of a former Islamic kingdom that once controlled all of Borneo and areas beyond, came under British control and was granted independence in 1984. Just slightly larger than

Delaware and with a population of 440,000, the sultanate is a mere ministate—except for the discovery of oil in 1929 and natural gas in 1965, which made this Southeast Asia's richest country. And there are indications that further discoveries will be made in the offshore zone owned by Brunei. The Sultan of Brunei rules as an absolute monarch; his palace in the capital, Bandar Seri Begawan, is reputed to be the world's largest. He will have no difficulty finding customers for his oil in energy-poor eastern Asia.

Brunei's territory is not only small but is also fragmented (see the inset map in Fig. 10B-7). A tiny sliver of Malaysian land separates the larger west from the east (where the capital lies), and Brunei Bay provides a water link between the two parts. It has been proposed that Brunei purchase this separating corridor from the Malaysians, but no action has yet been taken.

The country's population is growing apace through immigration: 66 percent of Brunei's residents are Malay and 11 percent Chinese. Most inhabitants live and work near the offshore oilfields in the northern district of Brunei Muara and in the capital city. Evidence of profligate spending is everywhere, ranging from sumptuous palaces to magnificent mosques to luxurious hotels. In this respect Brunei offers a stark contrast to neighboring East Malaysia, but even within Brunei there are significant differences. Brunei's interior remains an area of subsistence agriculture (a small minority of indigenous groups survive here) and rural isolation, its villages a world apart from the modern splendor of Bandar Seri Begawan.

Singapore

In 1965, a fateful event occurred in Southeast Asia. Singapore, the crown jewel of British colonialism in this realm, seceded from the recently independent (1963) Malaysian

From the Field Notes . . .

© H. J. de Blij

"Only when you rent a boat and see the reclamation work up close in the Jurong Islands (just off Singapore's southwestern shore) do you realize how massive a project this is to expand the city-state's space. Huge amounts of sand and earth are being repositioned to close channels, connect islands, and provide more space, primarily for industrial expansion. But this goes on in several other places along the Singaporean shore, including nearby Sentosa where locals and visitors go in ever-growing numbers for relaxation and entertainment, and where space is also at a premium."

www.conceptcaching.com

FIGURE 10B-6

© H. J. de Blij, P. O. Muller, and John Wiley & Sons, Inc.

Federation and became a sovereign state, albeit a ministate (Fig. 10B-6). With its magnificent relative location, its multiethnic and well-educated population, and its no-nonsense government, Singapore then overcame the severe limitations of space and the absence of raw materials to become one of the leading economic tigers of East and Southeast Asia.

With a mere 619 square kilometers (239 sq mi) of territory, space is at a premium in Singapore, and this is a constant worry for the government (see photo, p. 551). Singapore's only local spatial advantage over Hong Kong is that its small territory is less fragmented (there are just a few small islands in addition to the compact main island). With a population of just over 5 million and an expanding economy, Singapore must develop space-conserving high-technology and service industries. Benefiting from its relative location, the old port of Singapore had become one of the world's busiest (by numbers of ships served) even before its indepen-

dence. It thrived as an *entrepôt* [5] between the Malay Peninsula, Southeast Asia, Japan, and other emerging economic powers on the **Pacific Rim** [6]. Crude oil from Southeast Asia still is unloaded and refined at Singapore, then shipped to East Asian destinations. Raw rubber from the adjacent peninsula and from Indonesia's nearby island of Sumatera is shipped to China, Japan, the United States, and other countries. Timber from Malaysia, rice, spices, and other foodstuffs are processed and forwarded via Singapore. In return, automobiles, machinery, and equipment are imported into Southeast Asia and distributed through Singapore.

But that is the old pattern. Singapore's leaders today are redirecting the city-state's economy and moving toward an ever more technologically sophisticated industrial base. In Singapore the essentially one-party (borderline-authoritarian) government strictly controls business and many other aspects of life. Certain newspapers and magazines have been banned for criticizing the regime, and there are even fines for chewing gum, eating on the subway, failing to flush a public toilet, and wearing hair deemed longer than appropriate. However, its overall success following secession has tended to keep the critics quiet: while GNI per capita from 1965 to 2003 multiplied by a factor of more than 15—to almost U.S. $30,000—that of neighboring Malaysia reached just $10,320. Among other things, Singapore became (and for several years remained) the world's largest producer of disk drives for personal computers.

Singapore has moved in several different directions since 2000. First, it is concentrating on a triad of development themes: information technology, automation, and biotechnology. Second, it espouses notions of a **Growth Triangle [7]** involving Singapore's developing neighbors, Malaysia and Indonesia; those two countries would supply the raw materials and cheap labor, and Singapore the capital and technical expertise. Third, Singapore has opened its doors to capitalists of Chinese ancestry who left Hong Kong when China took it over in 1997 and who wanted to relocate their enterprises in Singapore. The population in Singapore is 75 percent Chinese, 14 percent Malay, and 9 percent South Asian. The government is Chinese-dominated, and its policies have served to sustain Chinese control. Indeed, China itself often cites Singapore's combination of authoritarianism and economic success as proving that communism and market economies can coexist.

Indonesia

The fourth-largest country in the world in terms of human numbers is also the globe's most expansive **archipelago [8]**. Spread across a chain of more than 17,000 mostly volcanic islands, Indonesia's 242 million people live both separated and clustered—separated by water and clustered on islands large and small.

The complicated map of Indonesia requires some close study (Fig. 10B-7). Five large islands dominate the archipelago territorially, but one of these, easternmost New Guinea, is not part of the Indonesian culture sphere, although its western half is under Indonesian control. The other four major islands are collectively known as the Greater Sunda Islands: **Jawa** (Java),* smallest but by

_____

*As in Africa and South Asia, names and spellings have changed with independence. In this chapter we will use contemporary spellings, except when we refer to the colonial era. Thus Indonesia's four major islands are Jawa, Sumatera, Kalimantan (the Indonesian part of Borneo), and Sulawesi. Prior to independence they were respectively known as Java, Sumatra, Dutch Borneo, and Celebes.

The skyline of Jakarta's CBD as seen from atop the towering Monas (National Monument) that rises at the center of the Indonesian capital's main ceremonial space, Merdeka (Freedom) Square. Air quality on this day is as good as it gets here in this incipient megacity usually shrouded in smog, produced by local factories and motor vehicles, and intensified by the region's frequent large-scale forest fires and eruptions of volcanic ash.

© Wolfgang Kaehler SuperStock

far the most populous and important; **Sumatera** (Sumatra) in the west, directly across the Strait of Malacca from Malaysia; **Kalimantan**, the southern, Indonesian sector of large, compact, mini-continent Borneo; and wishbone-shaped **Sulawesi** (Celebes) to the east. Extending eastward from Jawa are the Lesser Sunda Islands, including Bali and, near the eastern end, Timor. Another important island chain that lies within Indonesia is the Maluku (Molucca) Islands, between Sulawesi and New Guinea. The central water body of Indonesia is the Java Sea.

The Major Islands

Indonesia is a Dutch colonial creation, and Jawa was chosen as its colonial headquarters with Batavia (now Jakarta) as the capital. Today, **Jawa** remains the core of Indonesia. With close to 150 million inhabitants, Jawa is one of the world's most densely peopled places (see Figs. G-8 and 10A-3) as well as one of the most agriculturally productive, with its rice paddies rising up the highly fertile flanks of active volcanoes. This combination of extreme population density on top of a tectonically active crustal zone has the potential for disaster. As recently as November 2010, a two-week-long eruption of Mount Merapi near the city of Yogyakarta killed nearly 200 people (see Chapter 10A).

Jawa also is the most highly urbanized part of a country in which 57 percent of the people still live off the land. The primate city of Jakarta has now become the heart of a

FIGURE 10B-7

© H. J. de Blij, P. O. Muller, and John Wiley & Sons, Inc.

larger conurbation known as ***Jabotabek***, consisting of *Ja*-karta as well as *Bo*gor, *Ta*ngerang, and *Bek*asi. Since 1990, the population of this megalopolis has nearly doubled from 15 to 29 million. Already, Jabotabek is home to 12 percent of Indonesia's entire population and 28 percent of its urban population. Thousands of factories, their own-ers taking advantage of low prevailing wages, were built in this area, straining its infrastructure and overburdening the port of Jakarta. On an average day, hundreds of ships lie at anchor, awaiting docking space to offload raw mate-rials and take on finished products.

As has always been the case in Indonesia, Jawa is where the power lies. As a cultural group (though itself heterogeneous), the Jawanese consitute about 42 percent of the country's population. In the middle of the island, a politically powerful sultanate centers on Yogyakarta; Jawa also is the main base for the two national Islamic move-ments, one comparatively moderate and the other increas-ingly fundamentalist.

Sumatera, Indonesia's westernmost island, forms the western shore of the busy Strait of Malacca; Singapore lies across the Strait from approximately the middle of the island. Although much larger than Jawa, Sumatera has only about one-third as many people (50 million). In colo-nial times the island became a base for rubber and palm-oil plantations; its high relief makes possible the cultivation of a wide range of crops, and neighboring Bangka and Beli-tung yield petroleum and natural gas. Palembang is the key urban center in the south, but current attention focuses on the north. There, the Batak people accommodated

BRUNEI
- Oilfield
- Gasfield

colonialism and Westernization and made Medan one of Indonesia's booming cities.

In the far north, the Acehnese fought the Dutch into the twentieth century and, after Indonesia became a sovereign country, demanded autonomy and even outright independence for their State. Rebels fought the Indonesian military to a costly stalemate, and thousands died in the conflict—which would probably still be going on today but for a dramatic turn of events. The seafloor epicenter of the 2004 Indian Ocean tsunami lay near the far northern coast of Sumatera, and Aceh was directly in the path of the most powerful ocean waves. Entire towns and roads were swept away by tsunamis, and tens of thousands died; Banda Aceh, the State capital, was devastated. The international relief effort opened Aceh to foreigners in ways the Indonesians had long prevented, and the

Indonesian army as well as the rebels were engaged in rescue missions rather than warfare. This combination of circumstances facilitated a truce and negotiations resulting in a peace agreement under which the rebels agreed to drop their demand for independence and the Indonesian military withdrew.

Kalimantan is the Indonesian sector of the island of Borneo, a slab of the Earth's crystalline crust whose backbone of tall mountains is of erosional, not volcanic, origin. Larger than Texas, Borneo has a deep, densely rainforested interior that is a last refuge for some 35,000 orangutans as well as dwindling numbers of Asian elephants, rhinoceroses, and tigers. Along with a number of indigenous peoples, these wildlife species survive even as loggers and farmers relentlessly undermine their habitat. Borneo's Pleistocene heritage sustains a comparatively small human population

JAKARTA

JAKARTA, CAPITAL OF Indonesia and the realm's second-largest city, sometimes is called the Kolkata (Calcutta) of Southeast Asia. Stand on the elevated highway linking the port to the city center and see the villages built on top of garbage dumps by scavengers using what they can find in the refuse, and the metaphor seems to fit. There is poverty here unlike that in any other Southeast Asian metropolis.

But there are other sides to Jakarta (population: 9.4 million). Indonesia's economic progress has made its mark here, and the evidence is everywhere. Television antennas and satellite dishes rise like a forest from rusted, corrugated-iron rooftops. Cars (almost all, it seems, late-model), mopeds, and bicycles clog the streets, day and night. A meticulously manicured part of the city center contains a cluster of high-rise hotels, office buildings, and luxury-apartment complexes (see photo, p. 553). Billboards advertise planned communities on well-located, freshly cleared land.

Jakarta's population is a cross-section of Indonesia's, and the silver domes of Islam rise above the cityscape alongside Christian churches and Hindu temples. The city always was cosmopolitan, beginning as a cluster of villages at the mouth of the Ciliwung River under Islamic rule, becoming a Portuguese stronghold and later the capital of the Dutch East Indies (under the name of Batavia). Advantageously situated on the northwestern coast of Jawa, Indonesia's most populous island, Jakarta bursts at the seams with growth. Sail into the port, and hundreds of vessels, carrying flags from Russia to Argentina, await berths. Travel to the outskirts, and huge shantytowns are being expanded by a constant stream of new arrivals. So vast is the human agglomeration—nobody really knows how many people have descended on this near-megacity—that most live without adequate (or any) amenities.

© H. J. de Blij, P. O. Muller, and John Wiley & Sons, Inc.

Jakarta is the biggest city in the world without a metro transit system, but one is now finally under construction with a target date of 2016. In the meantime, public transport continues to rely exclusively on buses, and the city's traffic jams are notorious (transport delays are estimated to cost U.S. $3 billion per year). The gridlock has recently spurred discussions about moving the capital about 50 kilometers (30 mi) to the south. But it may well be that the government is just posturing in order to buy time with frustrated city dwellers as the new metro system takes shape.

(14 million on the Indonesian side, just under 6 percent of the country's population) on poor tropical soils. Indigenous peoples, principally the Dayak clans, have traditionally had less impact on the natural environment than the Indonesian and Chinese immigrants as well as multinational corporations that log the forests and clear woodland for farms. As Figure 10B-7 shows, the only towns of any size in Kalimantan lie on or near the coast; routes into the interior are still few and far between.

Sulawesi consists of a set of intersecting, volcanic mountain ranges rising above sea level; the 800-kilometer (500-mi) Minahasa Peninsula, propelled by volcanic activity, continues to build itself into the Celebes Sea. This northern peninsula, a favorite of the Dutch colonizers, remains the most developed part of an otherwise rugged and remote island, with Manado its relatively prosperous urban focus. Seven major ethnic groups inhabit the valleys and basins between the mountains, but the population of about 17 million

also includes many immigrants from Jawa, especially in and around the southern center of Ujungpandang. Subsistence farming is the leading mode of life, although logging, some mining, and fishing augment the economy. Clashes between Muslims and Christians occur intermittently in remote areas.

Papua, the Indonesian name for the western part of the island of New Guinea, has become an issue in Indonesian politics. Bordered on the east by a classic superimposed geometric boundary (Figs. 10B-7; 10A-7, lower-left map), it was taken over by Indonesia from the Dutch in 1969. Papua contains about 22 percent of Indonesia's territory, but its population is 3.1 million—just 1.3 percent of the national total. The indigenous inhabitants of this territory, which is in effect a colony, are Papuan, most living in the remote reaches of this mountainous and densely forested island. Papua is economically important to Indonesia, for it contains what is reputed to be the world's richest gold mine and its second-largest open-pit copper mine, but political consciousness has now

From the Field Notes . . .

© H. J. de Blij

"I drove from Manado on the Minahasa Peninsula in north-eastern Sulawesi to see the ecological crisis at Lake Tondano, where a fast-growing water hyacinth is clogging the water and endangering the local fishing industry. On the way, in the town of Tomolon, I noticed this side street lined with prefabricated stilt houses in various stages of completion. These, I was told, were not primarily for local sale. They were assembled from wood taken from the forests of Sulawesi's northern peninsula, then taken apart again and shipped from Manado to Japan. 'It's a very profitable business for us,' the foreman told me. 'The wood is nearby, the labor is cheap, and the market in Japan is insatiable. We sell as many as we can build, and we haven't even begun to try marketing these houses in Taiwan or China.' At least, I thought, this wood was being converted into a finished product, unlike the mounds of logs and planks I had seen piled up in the ports of Borneo awaiting shipment to East Asia."

www.conceptcaching.com

reached the Papuans. The Free Papua Movement has become increasingly active, holding small rallies in the capital, Jayapura, displaying a Papuan flag, and demanding recognition.

Diversity in Unity

Indonesia's survival as a unified state is as remarkable as India's or Nigeria's. With more than 300 discrete ethnic clusters, over 250 languages, and just about every religion practiced on Earth (although Islam dominates), actual and potential centrifugal forces are powerful here. Wide waters and high mountains perpetuate both cultural distinctions and differences. Indonesia's national motto is *bhinneka tunggal ika*: diversity in unity.

What Indonesia has achieved is etched against the country's continuing cultural heterogeneity. There are dozens of distinct aboriginal cultures; virtually every coastal community has its own roots and traditions. And the majority, the rice-growing Indonesians, include not only the numerous Jawanese—who have their own cultural identity—but also the Sundanese (who constitute 15.4 percent of Indonesia's population), the Madurese (3.3 percent), and others. Perhaps the best impression of this cultural diversity comes from the string of islands that extends eastward from Jawa to Timor, the Lesser Sunda Islands (Fig. 10B-7). The rice-growers of Bali adhere to a modified version of Hinduism, giving the island a unique cultural atmosphere; the population of Lombok is mainly Muslim, with some Balinese Hinduism; Sumbawa is a Muslim community; Flores is mostly Roman Catholic. In western Timor, Protestant groups dominate; in the now-independent east, where the Portuguese ruled, Roman Catholicism prevails.

Nevertheless, Indonesia nominally is the largest Muslim country in the world: overall, 88 percent of the people adhere to Islam, and in the cities the silver domes of neighborhood mosques rise densely above the townscape. Although until recently Indonesian Islam has been relatively moderate, lately more overt Islamization has been on the rise, with new laws banning public displays of affection and limiting the types of clothes women may wear in public.

And Indonesia has hardly been immune from terrorism. Serious terrorist attacks have occurred in the tourist areas of Bali and in central Jakarta, mounted by a terrorist organization named *Jemaah Islamiyah (JI)*, apparently headquartered in eastern Jawa. In addition to Indonesians, many Australians, for whom Indonesia—especially Bali—is a popular tourist destination, have lost their lives in these attacks.

Transmigration and the Outer Islands

As noted earlier, Indonesia's population of almost 250 million makes it the world's fourth most populous country. But Jawa contains no less than 57 percent of it on barely 7 percent of the national territory; with about 150 million people on an island the size of Louisiana, population pressure is staggering here. Moreover, Indonesia's annual rate of population growth remains above the realm's average of 1.3 percent. To deal with this chronic problem, and at the same time to strengthen the core's power over outlying areas, the Indonesian government for decades pursued a policy known as **transmigration [9]** (*transmigrasi* in Indonesian), inducing many from the densely populated inner islands (especially Jawa, Bali, and Madura) to relocate to such sparsely inhabited outer islands as Kalimantan and Sulawesi. During the last quarter of the twentieth century, millions moved to these peripheral locales as part of the government-sponsored program.

It was the Dutch colonialists who started the Transmigration Program, but it was abandoned when Indonesia achieved independence in the late 1940s. But the program was restarted in 1974, and by 2000 as many as 8 million Jawanese had been relocated to other islands. Yet surveys showed that half of these migrants experienced a decline in their standard of living; many were reduced to bare subsistence on tropical-forest land taken away from its indigenous inhabitants that turned out to be unsuitable for the farming methods used by the settlers. Cultural conflict, ecological havoc, and deforestation finally led the Indonesian government to cancel the program in 2001, but the damage it did will long outlive this ill-fated initiative.

Today, Indonesia's stature within Southeast Asia is on the rise. It is already by far the biggest country in the realm and is now making steady progress toward achieving a commensurate level of economic development and prosperity. But centrifugal forces are intensifying as well (growing Islamic militancy in Jawa is of particular concern), and ongoing decentralization and devolution throughout this immense archipelago will continue to challenge the central government.

East Timor

As Figure 10B-8 shows, Timor is the easternmost of the major Lesser Sunda Islands, and during the Dutch colonial

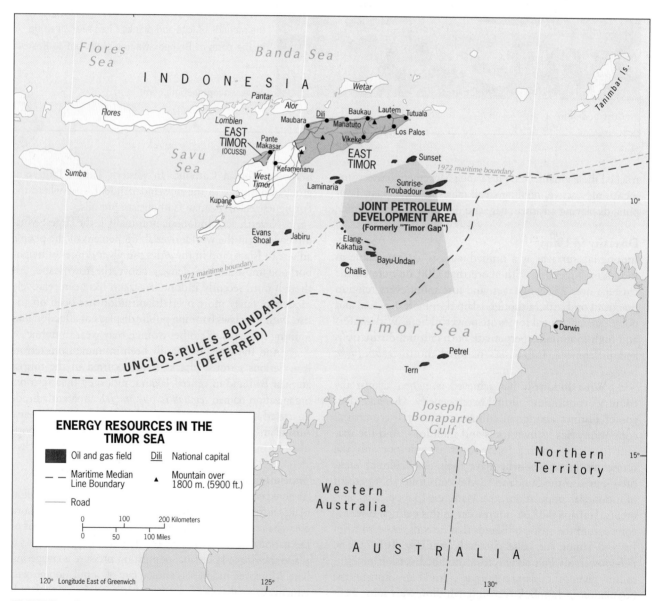

FIGURE 10B-8

© H. J. de Blij, P. O. Muller, and John Wiley & Sons, Inc.

period the Portuguese maintained a colony on the eastern half of it. When Indonesia shook off Dutch rule in the late 1940s, the Portuguese hung on in Timor, but in 1975 Indonesian forces overran that colony and annexed it formally in 1976. Indonesian rule, however, was even less benign than the Portuguese had been, and soon a bitter struggle for independence was under way. In 1999, after much death and destruction, East Timor's 800,000 inhabitants were allowed, under UN supervision, to vote on independence. They voted overwhelmingly in favor, provoking Indonesia's armed forces to redouble their ruthless repression. East Timor's already modest infrastructure was devastated, the death toll mounted daily, and only foreign intervention, led by Australia (which, among all the world's states, had been Indonesia's sole supporter in the matter), established a degree of order. East Timor became an independent state in 2002, the 198th member of the United Nations.

Subsequently, East Timor's leaders proclaimed the official name of their country to be *Timor-Leste*, the Portuguese version of "Timor East." As in the case of Côte d'Ivoire in West Africa, the French name for Ivory Coast, we use the two options interchangeably.

Nation-Building Nightmare

Nation-building in this Connecticut-sized country continues to be a difficult proposition. Figure 10B-8 helps explain why Australia supported Indonesia in its effort to absorb East Timor: by doing so, Australia would benefit from Indonesia's compliance in the delimitation of maritime boundaries in the Timor Sea, giving Australia a larger share of the known oil and gas reserves than an independent East Timor would be likely to relinquish. In the end, difficult negotiations between powerful and technologically capable Australia and weak and incapable East Timor produced a settlement that gave East Timor an acceptable share of future revenues—but not before Australia's international reputation for fairness was further damaged. On the map, the red maritime boundary is the one delimited under United Nations regulations; the blue boundary represents the bilateral agreement between Indonesia and Australia. At issue is the "gap" between them, most of it on East Timor's side but now a "joint petroleum development area."

But the billions of dollars in prospect cannot overcome the devastation and abject poverty that hinder the nation-building project here. As Figure 10B-8 shows, East Timor is a fragmented country, with a dominant east (where the coastal capital, Dili, is located) and a small exclave on the north coast of Indonesian West Timor called Ocussi (sometimes mapped as "Ocussi-Ambeno Province"). Although there is a road from this exclave to the main territory, relations between East Timor and Indonesia make its use impractical. Worse, the three-year United Nations buildup to independence in 2002 launched a society totally unprepared to go it alone. Competent governance has been in short supply. The freedom fighters who were demobilized upon independence were or-

ganized into police and army units that soon fell apart; widely available weapons empowered gangs and criminals, ethnic hostilities erupted, returned exiles and locals were at odds, and East Timor's still-fragile institutions collapsed.

East Timor is a predominantly agricultural country, but while United Nations experts were trying to erect political and social institutions (a judiciary, banking system, medical establishment, public education) farming was neglected and poverty worsened. Moreover, East Timor now has about 100,000 people concentrated in squalid refugee camps. To make matters worse, its population of 1.3 million continues to grow explosively, exhibiting a disastrous natural-increase rate of 3.1 percent annually (*double* that of the realm's second-fastest-growing country); not surprisingly, its fertility rate of 5.7 children per woman is one of the world's highest.

Given the limited dimensions of East Timor, its still relatively small population, its prospective oil income, and foreign assistance, the failure of nation-building here is a warning to the world that independence and freedom do not guarantee well-being and progress.

The Philippines

North of Indonesia, across the South China Sea from Vietnam, and south of Taiwan lies a lengthy archipelago of more than 7000 islands (only about 460 of them larger than 1 square kilometer [0.4 sq mi] in area) inhabited by 98 million people. The islands of the Philippines can be viewed as three groups: (1) Luzon, largest of all, and Mindoro in the north; (2) the Visayan group in the center; and (3) Mindanao, second-largest, located in the south (Fig. 10B-9). Southwest of Mindanao lies a small group of islands, the Sulu Archipelago, nearest to Malaysian Borneo, where Muslim-based insurgencies have kept this area in turmoil.

Few of the generalizations we have been able to make for Southeast Asia could apply in the Philippines without qualification. The country's location relative to the mainstream of change in this part of the world has had much to do with this situation. The islands, inhabited by peoples of Malay ancestry with Indonesian strains, shared with much of the rest of Southeast Asia an early period of Hindu cultural influence, which was strongest in the south and southwest and diminished northward. Next came the Chinese invasion, felt more strongly on the largest island of Luzon in the northern part of the Philippine archipelago. Islam's arrival was delayed somewhat by the position of the Philippines well to the east of the mainland and to the north of the Indonesian islands. The few southern Muslim beachheads were soon overwhelmed by the Spanish invasion during the sixteenth century. Today the Philippines, adjacent to the world's largest Muslim state (Indonesia), is 81 percent Roman Catholic, 9 percent Protestant, and only 5 percent Muslim.

FIGURE 10B-9   © H. J. de Blij, P. O. Muller, and John Wiley & Sons, Inc.

Muslim Insurgency

The Philippines' small Muslim population, concentrated in the southeastern flank of the archipelago (Fig. 10B-9), has long decried its marginalization in this predominantly Christian country. Over the past generation, a half-dozen Muslim organizations, including the Moro (Moor) National Liberation Front and the Moro Islamic Liberation Front, have promoted the Muslim cause through tactics ranging from peaceful negotiation with the government to violent insurgency. Densely forested Basilan Island became the base for an especially extreme group, Abu Sayyaf, which received support from the al-Qaeda network. International counter-terror operations were extended into this part of the Philippines when U.S. troops joined Filipino forces in pursuit. In the 2010s, the Muslim challenge continues to siphon off a disproportionate share of Manila's operating budget.

People and Culture

Out of the Philippines melting pot, where Malay, Arab, Chinese, Japanese, Spanish, and American elements have met and mixed, has emerged the distinctive Filipino culture. It is not a homogeneous or unified culture, as is reflected by the nearly 90 Malay languages in use in the islands, but it is in many ways unique. At independence in 1946, the largest of the Malay languages, Tagalog (also called Pilipino), became the country's official language. But English is widely learned as a second language, and a Tagalog-English hybrid, "Taglish," is increasingly heard today. The Chinese component of the

● AMONG THE REALM'S GREAT CITIES . . .

MANILA

MANILA, CAPITAL OF the Philippines and the realm's largest city, was founded by the Spanish invaders of Luzon more than four centuries ago. The colonists made a good choice in terms of site and situation. Manila sprawls at the mouth of the Pasig River where it enters one of Asia's finest natural harbors. To the north, east, and south a crescent of mountains encircles the city, which lies just 1000 kilometers (600 mi) southeast of China's Hong Kong.

Manila, named after a flowering shrub in the local marshlands, is bisected by the Pasig, which is bridged in numerous places. The old walled city, Intramuros, lies to the south. Despite heavy bombardment during World War II, some of the colonial heritage survives in the form of churches, monasteries, and convents. St. Augustine Church, completed in 1599, is one of the city's landmarks.

The CBD of Manila lies on the north side of the river. Although Manila has a well-defined commercial center with several avenues of luxury shops and modern buildings, the skyline does not reflect the high level of energy and activity common in Pacific Rim cities on the opposite side of the South China Sea. Neither is Manila a city of notable architectural achievements. Wide, long, and straight avenues flanked by palm, banyan, and acacia trees give it a look similar to San Juan, Puerto Rico.

In 1948, a newly built city immediately to the northeast of Manila was inaugurated as the de jure capital of the Philippines and named Quezon City. The new facilities were eventually to house all government offices, but many functions of the national government never made the move.

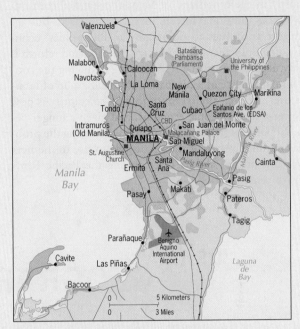

© H. J. de Blij, P. O. Muller, and John Wiley & Sons, Inc.

In the meantime, Manila's growth overtook Quezon's, so that it became part of the Greater Manila metropolis (which today is home to 12.0 million). Although the proclamation of Quezon City as the Philippines' official capital was never rescinded, Manila remains the de facto capital of the country today.

population is small (only 1.3 percent) but dominant in local business.

The Philippines population, concentrated where the good farmlands lie, is densest in three general areas: (1) the northwestern and south-central part of Luzon; (2) the southeastern extension of Luzon; and (3) the islands of the Visayan Sea between Luzon and Mindanao (see Fig. 10A-3). Luzon is the site of the capital, Manila-Quezon City (12 million—one-eighth of the entire national population), a megacity facing the South China Sea. Alluvial as well as volcanic soils, together with ample moisture in this tropical environment, enable self-sufficiency in rice and other staples and make the Philippines a net exporter of farm products despite a rather high population growth rate of 2.1 percent.

Perhaps more than any other people, Filipinos take jobs in foreign countries in massive numbers, proving their capacity to succeed in jobs they cannot find at home. The global merchant marine would not exist without Filipino

sailors, and Filipina nurses and domestic workers can be found from Dubai to Dubuque. Funds sent home to family members by the emigrants make the Philippines a world leader in monetary inflow known as **remittances [10]**.

Prospects

The Philippines seems to get little mention in discussions of Asia-Pacific developments, even though it would seem to be well positioned to share in the realm's economic growth. It has participated in offshore manufacturing in its numerous Export Processing Zones (which are similar to the Mexican *maquiladoras* discussed in Chapter 4B), but the necessary trade linkages have not been created and sustained. Moreover, governmental mismanagement as well as political instability have slowed the country's participation, but things have improved since 2000.

Despite a series of jarring events—the ouster of U.S. military bases, the damaging eruption of a volcano near the capital, the violence of Muslim insurgents, and a dispute over the nearby Spratly Islands in the South China Sea— the Philippines has made substantial economic progress over the past decade. Its electronics and textile industries (mostly in the hinterland of Manila) expanded continuously, and additional foreign investment arrived. But agriculture continues to dominate the Philippines' economy, unemployment remains high, further land reform is badly needed, and social restructuring (reducing the controlling influence of a comparatively small group of families over national affairs) must occur. The country is now a lower-middle-income economy, and given a longer period of political stability and success in reducing the problems outlined above, it should be able to rise to the next level and finally take its place among the Pacific Rim's growth centers.

● POINTS TO PONDER

- Thailand, the hub of Mainland Southeast Asia, has long-term, thorny border issues that involve refugees (Myanmar), river use (Laos), and land (Cambodia).
- Vietnam is asserting water rights in the South China Sea even as it disputes Chinese control over islands claimed by China.
- Under the constitution of Malaysia, a person of Malay ancestry must adhere to the Muslim faith.
- 90 percent of the population in East Timor lives off agriculture; in the United States, that figure is less than 3 percent.

using GEOGRAPHY

1   Thailand's collaboration with Myanmar in the Dawei Development Project may well be a stroke of genius by the government in Bangkok. Why? What are the potential geographic advantages to Thailand?

2   The Philippines seems rather removed from the dynamic, Pacific Rim economic developments of East and Southeast Asia. What are the main reasons for that? How would you factor in such geographic factors as location, colonial history, and ethnic networks?

INDONESIA

Jawa (Java) Bali Lombok Sumbawa Flores Savu Sea EAST TIMOR
Sumba Sawu Roti Timor

Timor Sea

Ashmore Islands (AUSTRALIA) Cartier Island (AUSTRALIA)

INDIAN OCEAN

Browse Island

Dampier Land

Broome

Port Hedland Roebourne
Barrow Island
North West Cape

Carnarvon
Cape Inscription
Dirk Hartog Island

Houtman Abrolhos

Geraldton

PERTH Fremantle Mandurah
Bunbury
Cape Naturaliste
Cape Leeuwin Northcliffe Albany

Bonaparte Arch.

Wyndham

KIMBERLEY PLATEAU

Halls Creek

GREAT SANDY DESERT

WESTERN

HAMMERSLEY RANGE

WESTERN PLATEAU AUSTRALIA

GIBSON DESERT

GREAT VICTORIA DESERT

Wiluna

Kalgoorlie

Hyden

Esperance

NULLARBOR PLAIN

EUCLA BASIN

Archipelago of the Recherche

Great Australian Bight

INDIAN OCEAN

Arafura Sea

Melville Island COBOURG PEN. Croker Island
Bathurst Island Van Diemen Gulf Jabiru
Darwin
Katherine

ARNHEM PLATEAU

Wessel Island

Cape Arnhem

Gulf of Carpentaria

Groote Eylandt
Numbulwar

Sir Edward Pellew Group
Mornington Island
Wellesley Islands

Kowanyama

Normanton

NORTHERN

TERRITORY

BARKLY TABLELAND

Tennant Creek

MACDONNELL RANGES Alice Springs

Uluru (Ayers Rock) (863 m, 2,831 ft.)

SIMPSON DESERT

Mount Isa

CARPENTARIA

GREAT

ARTESIAN

BASIN

LAKE EYRE BASIN
Lake Eyre

SOUTH

AUSTRALIA

Lake Torrens
Lake Gardiner

Lake Frome

Port Augusta Whyalla
Port Pirie
Port Lincoln Point Whidbey
Cape Carnot

Kangaroo Island

YORKE PEN.
EYRE PEN.
FLINDERS RANGES
ADELAIDE
Murray Bridge
Swan Hill
Horsham
Bendigo
Mount Gambier
VICTORIA Beton
Ballarat MELBOURNE
Geelong Sale
Warrnambool Cranbourne
Cape Otway
King Island

Bass Strait

PAPUA NEW GUINEA
Port Moresby New Guinea
Solomon Sea
Woodlark D'Entrecasteaux Islands
Louisiade Arch.

Moa Island Cape York Torres Strait
Prince of Wales Island Bamaga

CAPE YORK PEN.

Weipa

GREAT

DIVIDING

Coral Sea

CORAL SEA ISLANDS TERRITORY

Willis Islets
Coringa Islets Magdelaine Cays
Tregrosse Islets
Abington Reef

Cairns

Townsville

QUEENSLAND

GREAT DIVIDING RANGE

Longreach

Mackay

GREAT BARRIER REEF

Rockhampton
Gladstone

Bundaberg Sandy Cape
Hervey Bay Fraser Island

Maroochydore
BRISBANE
Gold Coast
Lismore Ballina

NEW SOUTH WALES

Broken Hill Darling

Tamworth

Coffs Harbour

Port Macquarie

Taree

Dubbo

Griffith Bathurst

Wagga Wagga Orange
Mildura
MURRAY-DARLING BASIN Murray
Albury Queanbeyan Newcastle
Wodonga AUSTRALIAN CAPITAL TERRITORY SYDNEY
Canberra Wollongong

Mt. Kosciuszko (2,230 m, 7,316 ft.)

Cape Howe

Wilsons Promontory

Furneaux Group

Cape Grim Hunter Is.
Marrawah Smithton
Devonport Cape Portland
Launceston
TASMANIA

South West Cape Hobart

Tasman Sea

INDIAN

OCEAN

Elevation (m)

3000
1500
600
300
150
0
below sea level
0
-150
-1500
-3000
-6000

Roads
Railroads

0 km 200 400 600 800 1000 1200
0 miles 200 400 600 800
Lazimuthal Equal-Area Projection
Scale 1:21,000,000

Te Hapua North Cape
Cape Reinga

PACIFIC OCEAN

0 km 200 400
0 miles 100 200 300

Transverse Cylindrical Equal-Area Projection
Scale 1:20,000,000

Tasman Sea

Great Barrier Island

Auckland Manukau
Hamilton Tauranga East Cape
North Island Rotorua
New Plymouth Gisborne
Wanganui Hastings
Palmerston North
Cape Farewell
Nelson Wellington
Blenheim Cape Palliser
Cook Strait
Greymouth

NEW ZEALAND

Mt. Cook (3,754 m, 12,316 ft.)

Christchurch

SOUTHERN ALPS

Milford Sound

South Island

Dunedin

Invercargill
Stewart Island

South West Cape

PACIFIC OCEAN

Foveaux

■■ REGIONS

Australia

New Zealand

IN THIS CHAPTER

- ◆ Australia's amazing biogeography
- ◆ Aboriginal claims to land and resources
- ◆ China covets Australian commodities
- ◆ Australia's Asian turn
- ◆ Foreign policy dilemmas Downunder
- ◆ New Zealand's matchless physiography

CONCEPTS, IDEAS, AND TERMS

FIGURE 11-1 © H. J. de Blij, P. O. Muller, and John Wiley & Sons, Inc.

Where were these photos taken?
Find out at www.wiley.com/college/deblij

The Austral Realm is geographically unique (Fig. 11-1). It is the only geographic realm that lies entirely in the Southern Hemisphere. It is also the only realm that has no land link of any kind to a neighboring realm and is thus completely surrounded by ocean and sea. It is second only to the Pacific as the world's least populous realm. Appropriately, its name refers to its location (the word **austral** [1] comes from the Latin for south)—a location far from the sources of its dominant cultural heritage, but close to its newfound economic partners on the Asian Pacific Rim.

● DEFINING THE REALM

Two countries constitute the Austral Realm: Australia, in every way the dominant one, and New Zealand, physiographically more varied but demographically much smaller than its giant partner (Fig. 11-2). Between them lies the Tasman Sea. To the west lies the Indian Ocean, to the east the Pacific, and to the south the frigid Southern Ocean.

This southern realm is at a crossroads. On the doorstep of populous Asia, its Anglo-European legacies are now infused by many other cultural strains. Polynesian Maori in New Zealand and Aboriginal communities in Australia are demanding greater rights and more acknowledgment of their cultural heritage. Pacific Rim markets are buying large quantities of raw materials. Chinese and other Asian tourists fill hotels and resorts. Queensland's tropical Gold Coast resembles Honolulu's Waikiki. The streets of Sydney and Melbourne display a multicultural panorama unimagined just two generations ago. All these changes have stirred political debate. Issues ranging from immigration quotas to indigenous land rights dominate, exposing social fault lines (city versus Outback in Australia; North versus South Island in

major geographic qualities of
AUSTRAL REALM

1. Australia and New Zealand constitute a geographic realm by virtue of territorial dimensions, relative location, and dominant cultural landscape.

2. Despite their inclusion in a single geographic realm, Australia and New Zealand differ physiographically. Australia has a vast, dry, low-relief interior; New Zealand is mountainous and has a temperate climate.

3. Australia and New Zealand are marked by peripheral development—Australia because of its aridity, New Zealand because of its topography.

4. The populations of Australia and New Zealand are not only peripherally distributed but also highly clustered in urban centers.

5. The economic geography of Australia and New Zealand is dominated by the export of livestock products and specialty goods such as wine. Australia also has significant wheat production and a rich, diverse base of mineral resources.

6. Australia and New Zealand are now integrated into the economic framework of the Asian Pacific Rim, principally as suppliers of raw materials.

7. As a result of heightened immigration from neighboring realms, Australia in a cultural sense is "returning" to Asia.

New Zealand). Aborigines and Maori first settled this realm, then the Europeans arrived, and now Asians are an increasingly significant economic and cultural element.

LAND AND ENVIRONMENT

Physiographic contrasts between massive, compact Australia and elongated, fragmented New Zealand are related to their locations with respect to the Earth's tectonic plates (consult Fig. G-4). Australia, with some of the geologically

most ancient rocks on the planet, lies at the center of its own plate, the Australian Plate. New Zealand, younger and less stable, lies at the convulsive convergence of the Australian and Pacific plates. Earthquakes are rare in Australia, and volcanic eruptions are unknown; New Zealand has plenty of both (the big Christchurch quakes of 2010–2011 being only the latest reminder). This locational contrast is also reflected by differences in relief (Fig. 11-1). Australia's highest relief occurs in what Australians call the Great Dividing Range, the mountains that line the east

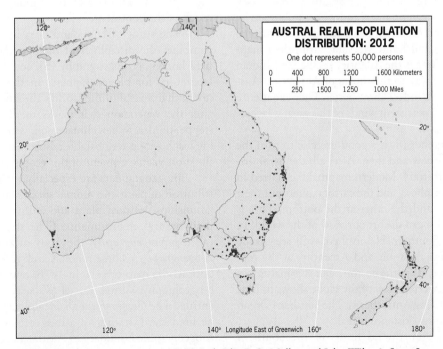

FIGURE 11-2   © H. J. de Blij, P. O. Muller, and John Wiley & Sons, Inc.

coast from the Cape York Peninsula to southern Victoria, with an outlier in Tasmania. The highest point along these old, now eroding mountains is Mount Kosciusko, 2230 meters (7316 ft) tall. In New Zealand, entire ranges are higher—Mount Cook in the Southern Alps, for example, reaches 3754 meters (12,316 ft). West of Australia's Great Dividing Range, the physical landscape mostly exhibits low relief, with some local exceptions such as the Macdonnell Ranges near the center; plateaus and plains dominate (Fig. 11-1). The Great Artesian Basin is a key physiographic region, providing underground water sources in what is otherwise desert country; to the south lies the continent's predominant river system, the Murray-Darling. The area mapped as *Western Plateau and Margins* in the lower map of Figure 11-3 contains much of Australia's mineral wealth.

Climates

Figure G-7 reveals the effects of latitudinal position and interior isolation on Australia's climatology. In this respect, Australia is far more varied than New Zealand, its climates ranging from tropical in the far north, where rainforests flourish, to Mediterranean in parts of the south. The interior is dominated by desert and steppe conditions, the semiarid steppes providing the grasslands that sustain tens of millions of livestock. Only in the east does Australia possess a zone of humid temperate climate, and here lies most of the country's economic core area. New Zealand, by contrast, is totally under the influence of the Southern and Pacific oceans, creating moderate, moist conditions, temperate in the north and colder in the south.

The Southern Ocean

Twice now we have referred to the Southern Ocean [2], but try to find this ocean on maps and globes published by famous cartographic organizations such as the National Geographic Society and Rand McNally. From their maps you would conclude that the Atlantic, Pacific, and Indian oceans reach all the way to the shores of Antarctica. Australians and New Zealanders know better. They experience the frigid waters and persistent winds of this great weather-maker on a daily basis.

For us geographers, it is a good exercise to turn the globe upside down now and then. After all, the usual orientation is quite arbitrary. Modern mapmaking started in the Northern Hemisphere, and the cartographers put their hemisphere on top and the other at the bottom. That is now the norm, and it can distort our view of the world. In bookstores in the Southern Hemisphere, you see upside down maps showing Australia and Argentina at the top, and Europe and Canada at the bottom. But this matter has a serious side. An inverted view of the globe shows us how vast the ocean encircling Antarctica is. The Southern Ocean may be remote, but its existence is real.

Where do the northward limits of the Southern Ocean lie? This ocean is bounded not by land but by a marine transition called the Subtropical Convergence [3]. Here the cold, extremely dense waters of the Southern Ocean meet the warmer waters of the Atlantic, Pacific, and Indian oceans. It is quite sharply defined by changes in temperature, chemistry, salinity, and marine fauna. Flying over it, you can actually observe it in the changing colors of the water: the Antarctic side is a deep gray, the northern side a greenish blue.

Although the Subtropical Convergence shifts seasonally, its position does not vary far from latitude 40° South, which also is the approximate northern limit of Antarctic icebergs. Defined this way, the great Southern Ocean is a huge body of water that circulates clockwise (from west to east) around Antarctica, which is why we also call it the West Wind Drift [4].

Biogeography

One of this realm's defining characteristics is its wildlife. Australia is the land of kangaroos and koalas, wallabies and wombats, possums and platypuses. These and numerous other *marsupials* (animals whose young are born very early in their development and then are carried in an abdominal pouch) owe their survival to Australia's early isolation during the breakup of Gondwana (see Fig. 6A-3). Before more advanced mammals could enter Australia and replace the marsupials, as happened in every other part of the world, this landmass was separated from Antarctica and India, and today it contains the world's largest assemblage of marsupial fauna.

Australia's vegetation has distinctive qualities as well, notably the hundreds of species of eucalyptus trees native to this geographic realm. Many other plants form part of Australia's unique flora, some with unusual adaptation to the high temperatures and low humidity that characterize much of the continent.

The study of fauna and flora in spatial perspective combines the disciplines of biology and geography in a field known as biogeography [5], and Australia is a giant laboratory for biogeographers. In the Introduction, we noted that several of the world's climatic zones are named after the vegetation that marks them: tropical savanna, steppe, tundra. When climate, soil, vegetation, and animal life reach a long-term, stable adjustment, vegetation forms the most visible element of this ecosystem.

Biogeographers are especially interested in the distribution of plant and animal species, as well as in the relationships between plant and animal communities and their natural environments. (The study of plant life is called *phytogeography*; the study of animal life is called *zoogeography*.) These scholars seek to explain the distributions the map reveals. In 1876, one of the founders of biogeography, Alfred Russel Wallace, published a book entitled *The Geographical Distribution of Animals* in which he fired the first shot in a long debate: where does the zoogeographic boundary of Australia's fauna lie? Wallace's fieldwork in the area revealed that Australian forms exist

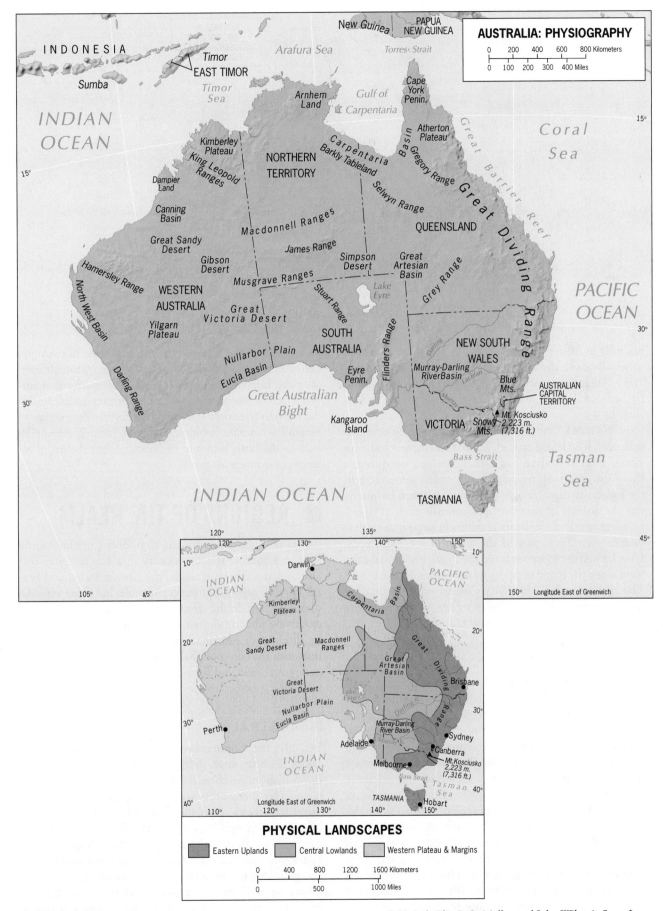

AUSTRALIA: PHYSIOGRAPHY

PHYSICAL LANDSCAPES

Eastern Uplands   Central Lowlands   Western Plateau & Margins

FIGURE 11-3

© H. J. de Blij, P. O. Muller, and John Wiley & Sons, Inc.

FIGURE 11-4   © H. J. de Blij, P. O. Muller, and John Wiley & Sons, Inc.

not only in Australia itself but also in New Guinea and in some islands to the northwest. As first noted in Chapter 10A, Wallace proposed that the faunal boundary should lie between Borneo and Sulawesi, and just east of Bali (Fig. 11-4).

Wallace's Line [6] soon was challenged by other researchers, who found species Wallace had missed and who visited islands Wallace had not. There was no question that Australia's zoogeographic realm ended somewhere in the Indonesian archipelago, but where? Western Indonesia was the habitat of nonmarsupial animals such as tigers, rhinoceroses, and elephants in addition to primates; New Guinea clearly was part of the realm of the marsupials. How far had the more advanced mammals progressed eastward along the island stepping stones toward New Guinea? The zoogeographer Max Weber found evidence that led him to postulate his own **Weber's Line**, which, as Figure 11-4 shows, lay very close to New Guinea.

The Human Impact

Not all research in zoogeography or phytogeography deals with such large questions. Much of it focuses on the relationships between particular species and their habitats—that is, the environment they normally occupy and of which they form a part. Such environments change, and these changes can spell disaster for the species. Climate change forces species to advance or retreat and is a topic often investigated by biogeographers. But it is not always clear whether environmental change or human intervention caused the problem. In Australia, the arrival of the **Aboriginal population [7]** (about 50,000 years ago) appears to have caused an ecosystem collapse because there is no evidence of significant climate change at the time. In an article in *Science*, Gifford H. Miller and his colleagues reported that widespread burning of the existing forest, shrub, and grassland vegetation across Australia appears

to have led to the spread of desert scrub and to the rapid extinction of most of the continent's large mammals soon after the human invasion occurred. The species that survived faced a second crisis when European colonizers introduced their livestock, leading to the further destruction of remaining wildlife habitats. Survivors include marsupials such as the koala bear and the wombat, but the list of extinctions is much longer.

● REGIONS OF THE REALM

Australia is the predominant component of the Austral Realm, a continent-scale country in a size category that also includes China, Canada, the United States, and Brazil. For two reasons, however, Australia has fewer regional divisions than do the aforementioned countries: the comparatively uncomplicated physiography of Australia and its diminutive human numbers. Our discussion, therefore, uses the core-periphery concept as a basis for investigating Australia and focuses on New Zealand as a region by itself.

■■ AUSTRALIA

In 2001, the Commonwealth of Australia celebrated its 100th birthday as an independent state. Still recognizing the British monarch as the head of state, the country entered its second century with a strong economy, stable political framework, high standard of living for most of its people, and favorable prospects ahead. Positioned on the Pacific Rim, almost as large as the 48 contiguous U.S. States, well endowed with farmlands and vast pastures, rivers, groundwater, mineral deposits, and energy resources, served by good natural harbors, and populated by 22.8 million mostly well-educated people, Australia is one of the most fortunate countries in the world.

Historical Geography

Aboriginal Australians reached this landmass as far back as 50,000 years ago, crossed the Bass Strait into Tasmania, and had developed a patchwork of indigenous cultures when Captain Arthur Phillip sailed into what is today Sydney Harbor (1788) to establish the beginnings of modern Australia. The Europeanization of Australia doomed the continent's Aboriginal societies. The first to suffer were those situated in the path of British settlement on the coasts, where penal colonies and free towns were founded.

Distance protected the Aboriginal communities of the northern interior longer than elsewhere; but in Tasmania, the indigenous Australians died off in just decades after having lived there for perhaps 45,000 years.

The Seven Colonies

Eventually, the major coastal settlements became the centers of seven different colonies, each with its own hinterland; by 1861, Australia was delimited by its now-familiar pattern of straight-line boundaries (Fig. 11-5). Sydney was

FIGURE 11-5

© H. J. de Blij, P. O. Muller, and John Wiley & Sons, Inc.

MAJOR CITIES OF THE REALM

City	Population* (in millions)
Sydney, Australia	4.7
Melbourne, Australia	3.9
Brisbane, Australia	2.0
Perth, Australia	1.6
Auckland, New Zealand	1.5
Adelaide, Australia	1.2
Canberra, Australia	0.4
Wellington, New Zealand	0.4

*Based on 2012 estimates.

the focus for New South Wales; Melbourne, Sydney's rival, anchored Victoria. Adelaide was the heart of South Australia, and Perth lay at the core of Western Australia. Brisbane was the nucleus of Queensland, and Hobart was the seat of government in Tasmania. The largest clusters of surviving Aboriginal people were in the so-called Northern Territory, with Darwin, on Australia's tropical north coast, its colonial city. Notwithstanding their shared cultural heritage, the Australian colonies were at odds not only with London over colonial policies but also with each other over economic and political issues. The building of an Australian nation during the late nineteenth century was a slow and difficult process.

Successful Federation

On January 1, 1901, following years of difficult negotiations, the Australia we know today finally emerged: the Commonwealth of Australia, consisting of six States and two Federal Territories (Fig. 11-5; Table 11-1). The two Federal Territories are the *Northern Territory*, assigned to protect the interests of the substantial Aboriginal population concentrated there and agitating for statehood, and the *Australian Capital Territory*, carved from southern New South Wales to accommodate the federal capital of Canberra, inaugurated in 1927.

Australia's six States are New South Wales (capital Sydney), at 7.5 million the most populous and politically powerful; Queensland (Brisbane), with the Great Barrier Reef offshore and tropical rainforests in its north; Victoria (Melbourne), small but populous by Australian standards with 5.6 million inhabitants; South Australia (Adelaide), where the Murray-Darling river system empties into the sea; Western Australia (Perth) with little more than 2 million people in an area of more than 2.5 million square kilometers (nearly 1 million sq mi); and Tasmania (Hobart), the island across the Bass Strait from the mainland's southeastern corner that lies in the path of the Southern Ocean's storms.

In earlier chapters, we have referred to the concept of federalism, a political idea with ancient Greek and Roman roots familiar to Americans, Canadians, South Asians, and, more recently, Russians. It is a notion of communal association whose name comes from the Latin *foederis*, implying alliance and coexistence, a union of consensus and common interest—a **federation [8]**. It stands in contrast to the idea that states should be centralized, or unitary. For this, too, the ancient Romans devised a term: *unitas*, meaning unity. Most European countries are **unitary states [9]**, including the United Kingdom of Great Britain and Northern Ireland. Although the majority of Australians came from that tradition (a kingdom, no less), they managed to overcome their differences and establish a Commonwealth that was, in effect, a federation of States with different viewpoints, economies, and objectives, separated by enormous distances along the rim of a remote island continent.

Sharing the Bounty

Despite the country's good fortunes, not everyone in Australia adequately shares in the national wealth. The indigenous (Aboriginal) population, although a small minority today of less than 600,000, remains disproportionately disadvantaged in almost every way, from lower life expectancies to higher

TABLE 11-1
States and Territories of Federal Australia, 2012

State	Area (1000 sq km [1000 sq mi])	Population (millions)	Capital	Population (millions)
New South Wales	801.6 (309.5)	7.5	Sydney	4.7
Queensland	1727.3 (666.9)	4.6	Brisbane	2.0
South Australia	983.9 (379.9)	1.7	Adelaide	1.2
Tasmania	67.9 (26.2)	0.5	Hobart	0.2
Victoria	227.7 (87.9)	5.6	Melbourne	3.9
Western Australia	2525.5 (975.1)	2.3	Perth	1.6
Territory				
Australian Capital Territory	2.3 (0.9)	0.4	Canberra	0.4
Northern Territory	1346.3 (519.8)	0.2	Darwin	0.1

unemployment levels than average, from lower high school graduation rates to much higher imprisonment ratios. But the nation is now embarked on a campaign to address these ills, with a formal apology issued by the government in 2008; its conciliatory actions range from enhanced social services for Aboriginals to favorable court decisions involving Aboriginal land claims.

When Australia was born as a federal state, its per capita GNI was the highest in the world. Australia's bounty fueled a massive flow of exports to Europe, and Australians prospered. That golden age could not last forever, and eventually the country's share of international trade declined. Nonetheless, Australia today ranks among the top 15 countries in the world in terms of GNI, and for the vast majority of Australians life is comfortable. In terms of the indicators of development discussed in Chapter 9B, Australia is far ahead of all its western Pacific Rim competitors except Japan and Singapore. As Australians celebrated their first century, they were, on average, earning far more than Thais, Malaysians, or Koreans. In terms of consumption of

The breathtakingly beautiful Great Barrier Reef, off the coast of Queensland in northeastern Australia (see Fig. 11-3). This is the largest coral reef on the planet, sometimes described as the only living thing on Earth that is visible from space. It was designated a World Heritage Site in 1981, and most of it is now lies within Great Barrier Reef Marine Park. Not surprisingly, it ranks as one of the world's most famous ecotourism destinations.

© Theo Allofs/Latitude/Corbis

From the Field Notes . . .

"My most vivid memory from my first visit to Alice Springs in the heart of the Outback is spotting vineyards and a winery in this parched, desert environment as the plane approached the airport. I asked a taxi driver to take me there, and got a lesson in economic geography. Drip irrigation from an underground water supply made viticulture possible; the tourist industry made it profitable. None of this, however, is evident from the view seen here: a spur of the Macdonnell Ranges overlooks a town of bare essentials under the hot sun of the Australian desert. What Alice Springs has is centrality: it is the largest settlement in a vast area Australians often call "the centre." Not far from the midpoint on the nearly 3200-kilometer (2000-mi) Stuart Highway from Darwin on the Northern Territory's north coast to Adelaide on the Southern Ocean, Alice Springs also was the northern terminus of the Central Australian Railway (before it was extended north to Darwin in 2003), seen in the middle distance. The shipping of cattle and minerals is a major industry here. You need a sense of humor to live here, and the locals have it: the town actually lies on a river, the intermittent Todd River. An annual boat race is held, and in the absence of water the racers carry their boats along the dry river bed. No exploration of Alice Springs would be complete without a visit to the base of the Royal Flying Doctor Service, which brings medical help to outlying villages and homesteads."

© H. J. de Blij

www.conceptcaching.com

energy per person, number of automobiles and kilometers of roads, levels of health, and literacy, Australia is decidedly a high-income developed country.

Distance

Australians often talk about distance. One of their leading historians, Geoffrey Blainey, labeled it a tyranny—an imposed remoteness from without and a divisive part of life within. Even today, Australia is far from nearly everywhere on Earth. A trans-Pacific jet flight from Los Angeles to Sydney takes about 14 hours nonstop and is correspondingly expensive. Freighters carrying products to European markets take ten days to two weeks to get there. Inside Australia, distances also are of continental proportions, and Australians pay the price—literally. Until some upstart private airlines started a price war, Australians paid more per mile for their domestic flights than air passengers anywhere else in the world.

But distance also was an ally, permitting Australians to ignore the obvious. Australia was a British progeny, a European outpost. Once you had arrived as an immigrant from Britain or Ireland, there were a wide range of environments, magnificent scenery, vast open spaces, and seemingly limitless opportunities. When the Japanese Empire expanded, Australia's remoteness saved the day. When boat people by the hundreds of thousands fled Vietnam in the mid-1970s aftermath of the Indochina War, almost none reached Australian shores. When immigration became an issue, its self-perceived comforts of isolation led Australia to adopt an all-white admission policy (only officially terminated in 1976) that was out of step with most of the rest of the world.

Immigrants

Today Australia is changing and rapidly so. Immigration policy now focuses on the would-be immigrants' qualifications, skills, financial status, age, and facility with the English language. With regard to skills, high-technology specialists, financial experts, and medical personnel are especially welcome. Relatives of earlier immigrants, as well as a quota of genuine asylum-seekers, are admitted as well. In recent years, total immigration has been between 120,000 and 180,000 annually, which keeps the country's population growing. According to the latest data, Australia's declining natural rate of increase today stands at 0.7 percent. The country is fast becoming a truly multicultural society. In Sydney, for instance, one in five residents is now of Asian ancestry. Overall, no less than a quarter of Australia's 2010 population was foreign-born, and another quarter consisted of first-generation Australians.

Core and Periphery

As Figure 11-2 shows, Australia is a large landmass, but its population is heavily concentrated in a core area that lies in the east and southeast, most of which faces the Pacific Ocean (locally named the Tasman Sea between Australia and New Zealand). Figure 11-5 shows that this crescent-like

Australian heartland extends from north of the city of Brisbane to the vicinity of Adelaide and includes the largest city, Sydney; the capital, Canberra; and the second-largest city, Melbourne. A secondary core area has developed in the far southwest, centered on Perth and its outport, Fremantle. Beyond lies the vast periphery, which the Australians call the **Outback [10]** (see Field Note box, p. 573).

To better understand the evolution of this spatial arrangement, it helps to refer again to the map of world climates (Fig. G-7). Environmentally, Australia's most favored strips face the Pacific and Southern oceans, and they are not large. We can describe the country as a coastal rimland with cities, towns, farms, and forested slopes giving way to the vast, arid, interior Outback. On the western flanks of the Great Dividing Range lie the extensive grassland pastures that catapulted Australia into its first commercial age—and on which still graze one of the largest sheep herds in the world (about 100 million sheep, producing more than one-fifth of all the wool sold in the world). Where it is moister, to the north and east, cattle by the millions graze on ranchlands. This is frontier Australia, over which livestock have ranged for nearly two centuries.

An Urban Culture

Despite the vast open spaces and romantic notions of frontier and Outback, Australia is an urban country, with 82 percent of all Australians living in cities and towns. As noted earlier, the primary core of Australia lies in the southeast of the continent (Fig. 11-5).

On the map, Australia's economic spatial organization is similar to Japan's: large cities lie along the coast, the centers of manufacturing complexes as well as the foci of agricultural areas. Contributing to this situation in Japan was mountainous topography; in Australia it was an arid interior. There, however, the similarity ends. Australia's territory is 20 times larger than Japan's, and Japan's population is almost six times that of Australia's. Japan's port cities are built to receive raw materials and to export finished products. Australia's cities forward minerals and farm products from the Outback to foreign markets and import manufactures from overseas.

Distances in Australia are much greater, and spatial interaction (which tends to decay with increasing distance) is reduced. In comparatively small, tightly organized Japan, you can travel from one end of the country to the other on superhighways, through tunnels, and over bridges with the utmost speed and efficiency. In Australia, the overland trip from Sydney to Perth, or from Darwin to Adelaide (using the new railway link via Alice Springs), is time-consuming and slow. Nothing in Australia compares to Japan's high-speed bullet trains.

The Cities

For all its vastness and youth, Australia has developed a remarkable cultural identity, a sameness of urban and rural landscapes that persists from one end of the continent to the other. Sydney, often called the New York of Australia, lies on a spectacular estuarine site, its compact, high-rise central business district overlooking a port bustling with ferry and freighter traffic. Sydney is a vast, sprawling metropolis of 4.7 million, with multiple outlying centers studding its far-flung suburbs; brash modernity and reserved British ways somehow blend here. Melbourne (3.9 million), sometimes regarded as the Boston of Australia, prides itself on its more interesting architecture and more cultured ways. Brisbane, the capital of Queensland, which also anchors Australia's Gold Coast and adjoins the Great Barrier Reef, is the Miami of Australia; unlike Miami, however, its residents can find nearby relief from the summer heat in the mountains of its immediate hinterland (as well as at its beaches). Perth, Australia's San Diego, is one of the world's most isolated cities, separated from its nearest Australian neighbor by two-thirds of a continent and from Southeast Asia and Africa by thousands of kilometers of ocean.

And yet, each of these cities—as well as the capitals of South Australia (Adelaide), Tasmania (Hobart), and,

to a lesser extent, the Northern Territory (Darwin)—exhibits an Australian character of unmistakable quality. Life is orderly and unhurried. Streets are clean, slums are few, graffiti rarely seen. By American and even European standards, violent crime (although rising) is uncommon. Standards of public transport, city schools, and health-care provision are high. Spacious parks, pleasing waterfronts, and plentiful sunshine make Australia's urban life more acceptable than just about anywhere else on Earth. Critics of Australia's way of life say that this very pleasant state of affairs has persuaded Australians that hard work is not all that necessary. But a few days' experience in the commercial centers of the major cities contradicts that assertion: the pace of life is quickening. The country's cultural geography evolved as that of a European outpost, prosperous and secure in its isolation. Now Australia has reinvented itself as a major component in an Australo-Asian chain, a Pacific partner in a transformed regional economic geography.

⊙ AMONG THE REALM'S GREAT CITIES . . .

SYDNEY

MORE THAN TWO centuries ago, Sydney was founded by Captain Arthur Phillip as a British outpost on one of the world's most magnificent natural harbors. The free town and penal colony that struggled to survive evolved into Australia's largest city. Today, metropolitan Sydney (4.7 million) is home to more than one-fifth of the country's entire population. An early start, the safe harbor, fertile nearby farmlands, and productive pastures in its hinterland combined to propel Sydney's growth. Later, as road and railroad links made Sydney the focus of Australia's growing core area, industrial development and political power augmented its primacy.

With its incomparable setting (photo, p. 566) and mild sunny climate, its many urban beaches, and its easy reach to the cool Blue Mountains of the Great Dividing Range, Sydney is one of the world's most liveable cities. Good public transportation (including an extensive cross-harbor ferry system from the doorstep of the waterfront CBD), fine cultural facilities headed by the multi-theatre Opera House complex, and many public parks and other recreational facilities make Sydney attractive to visitors as well. A healthy tourist trade, much of it from East Asian countries, bolsters the city's economy. Sydney's hosting of the 2000 Olympic Games was further testimony to its rising global visibility.

Increasingly, Sydney also is a multicultural city. Its small Aboriginal sector is being overwhelmed by the arrival of large numbers of Asian immigrants. The Sydney suburb of Cabramatta symbolizes the impact: more than half of its nearly 100,000 residents were born elsewhere, mostly in Vietnam. Unemployment is high, drug use is a problem, and crime and gang violence persist. Yet, despite the deviant be-

© H. J. de Blij, P. O. Muller, and John Wiley & Sons, Inc.

havior of a small minority, tens of thousands of Asian immigrants have established themselves in some profession. As the photo and caption on page 576 suggest, today things are definitely improving in this dynamic community.

These developments herald Sydney's coming of age. The end of Australia's isolation has brought Asia across the country's threshold, and again the leading metropolis is showing the way.

The name Cabramatta conjures up varied reactions among Australians. During the 1950s and 1960s, many immigrants from southern Europe settled in this western suburb of Sydney, attracted by affordable housing. In the 1970s and 1980s, Southeast Asians arrived here in large numbers, and during this period Cabramatta, in the eyes of many, became synonymous with gang violence and drug dealing. More recently, however, Cabramatta's ethnic diversity has come to be viewed in a more favorable light, and is now seen as the "multicultural capital of Australia," a tourist attraction and proof of Australia's capacity to accommodate non-Europeans. Meanwhile, Cabramatta has been spruced up with Oriental motifs of various kinds. This "Freedom Gate" in the Vietnamese community is flanked by a Ming horse and a replica of a Forbidden City lion—all reflecting better times for an old gateway for immigrants as well as refugees.

© Robert Francis/Alamy

Economic Geography

Because of its early history as a treasure trove of raw materials, Australia to this day is seen worldwide as a country whose economy depends primarily on its exportable natural resources. In fact, however, Australia's economy depends overwhelmingly on services, not commodity exports. Tourism alone contributes around 5 percent—about the same as the value of mineral exports. The action is in those bustling coastal cities far more than in the Outback.

When Australia became established as a state, however, it needed goods from overseas, and here the tyranny of distance played a key role. Imports from Britain (and later the United States) were expensive mainly because of transport costs. This encouraged local entrepreneurs to establish their own industries to produce these goods more cheaply. Economic geographers call such industries **import-substitution industries [11]**, and this is how local industrialization got its start.

When the prices of foreign goods became lower because transportation had become more efficient and therefore cheaper, local businesses demanded protection from the colonial governments, and high tariffs were erected against imported goods. Local products could now continue to be

made inefficiently because their market was guaranteed. If Japan could not afford this, how could Australia? We can see the answer on the map (Fig. 11-6). Even before federation in 1901, all the colonies could export valuable minerals whose earnings shored up those inefficient, uncompetitive local industries. By the time the colonies unified, pastoral industries were also contributing income. So the miners and the farmers paid for those imports Australians could not produce themselves, plus the products made in the cities. No wonder the cities grew: here were secure manufacturing jobs, jobs in state-run service enterprises, and jobs in the growing government bureaucracy. When we noted earlier that Australians once had the highest per capita GNI in the world, this was initially achieved in the mines and on the farms, not in the cities.

Agricultural Abundance

Australia has material assets of which other countries on the Pacific Rim can only dream. In agriculture, sheep-raising was the earliest commercial venture, but it was the technology of refrigeration that brought world markets within reach of Australian beef producers. Wool, meat, and wheat have long been the country's big three income earners; Figure 11-6 displays the immense pastures in the east, north, and west that constitute the ranges of Australia's huge herds. The zone of commercial grain farming forms a broad crescent extending from northeastern New South Wales through Victoria into South Australia, and covers much of the hinterland of Perth. And keep in mind the scale of this map: Australia is only slightly smaller than the 48 contiguous States of the United States! Commercial grain farming in Australia is big business. As the climatic map (Fig. G-7) would suggest, sugarcane grows along most of the humid, subtropical coastal strip of Queensland, and Mediterranean crops (including grapes for Australia's highly successful and expanding wine industry) cluster in the hinterlands of Adelaide and Perth. Mixed horticulture concentrates in the basin of the Murray River system, including rice, grapes, and citrus fruits, all under irrigation. In addition, as elsewhere in the world, dairying has developed near the large metropolitan areas. With its considerable range of environments, Australia yields a great diversity of crops.

Mineral Wealth

Australia's mineral resources, as Figure 11-6 shows, also are diverse. Major gold discoveries in Victoria and New South Wales produced a ten-year gold rush starting in 1851 and ushered in a new economic era. By the middle years of that decade, Australia was producing 40 percent of the world's gold. Subsequently, the search for more gold led to the

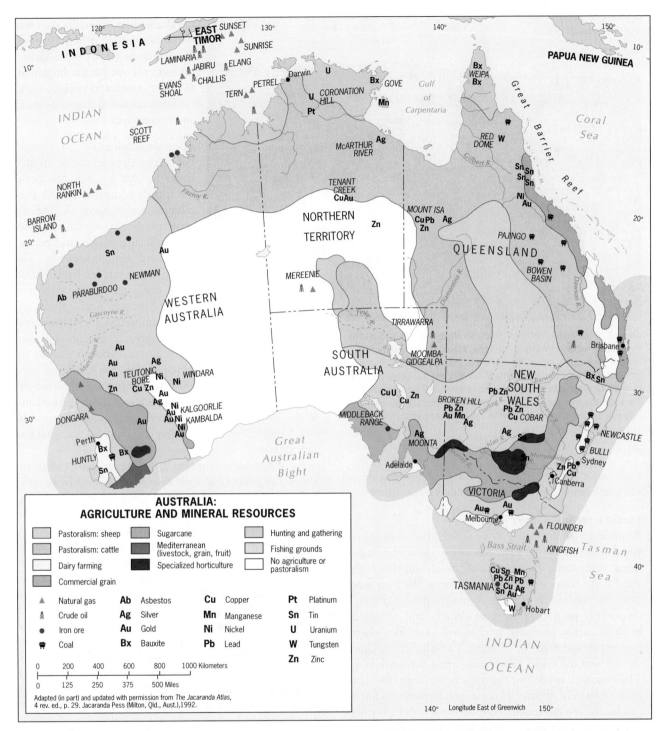

AUSTRALIA:
AGRICULTURE AND MINERAL RESOURCES

Pastoralism: sheep
Pastoralism: cattle
Dairy farming
Commercial grain
Sugarcane
Mediterranean (livestock, grain, fruit)
Specialized horticulture
Hunting and gathering
Fishing grounds
No agriculture or pastoralism

▲ Natural gas
🛢 Crude oil
● Iron ore
🛒 Coal

Ab Asbestos
Ag Silver
Au Gold
Bx Bauxite

Cu Copper
Mn Manganese
Ni Nickel
Pb Lead

Pt Platinum
Sn Tin
U   Uranium
W   Tungsten
Zn Zinc

0   200   400   600   800   1000 Kilometers
0   125   250   375   500 Miles

Adapted (in part) and updated with permission from *The Jacaranda Atlas*, 4 rev. ed., p. 29. Jacaranda Pess (Milton, Qld., Aust.),1992.

FIGURE 11-6

© H. J. de Blij, P. O. Muller, and John Wiley & Sons, Inc.

discoveries of additional minerals. New finds are still being made today, and even oil and natural gas have been found both inland and offshore (see the symbols in Fig. 11-6 in the Bass Strait between Tasmania and the mainland, and off the northwestern coast of the continent). Coal is mined at numerous locations, notably in the east near Sydney and Brisbane but also in Western Australia and even in Tasmania; before coal prices fell, this was a valuable export. Substantial deposits of metallic and nonmetallic minerals abound—from the complex at Broken Hill and the mix of minerals at Mount Isa to the huge nickel deposits at Kalgoorlie and Kambalda, the copper of Tasmania, the tungsten and bauxite of northern Queensland, and the asbestos of Western Australia. A closer look at the map reveals the wide distribution of iron ore (photo, p. 578), and for this raw material as for many others, Japan was Australia's best customer for several years (until it was recently surpassed by China). In fact, the enormous demand for raw materials

About 320 kilometers (200 mi) south-southwest of Port Hedland in the Pilbara region of Western Australia, Mount Whaleback near the town of Newman is one of the world's largest sources of high-grade iron ore. China's insatiable demand for this ore has generated the planet's largest "open-cut" iron mine, a huge and growing gash in the natural landscape—just one such impact resulting from Australia's role as raw-material supplier to industrializing economies.

© Photoshot Holdings Ltd/Alamy

from India as well as China by 2011 marked the longest running commodity boom in Australia's recent history.

Manufacturing's Limits

Australian manufacturing, as we noted earlier, remains oriented to domestic markets. Australian automobiles, electronic equipment, and cameras are clearly not challenging the Pacific Rim's economic tigers for a place on world markets—not yet, at any rate. Australian manufacturing is diversified, producing some machinery and equipment made of locally produced steel as well as textiles, chemicals, paper, and many other items. These industries cluster in and near the major urban areas where the markets are. The domestic market in Australia is not large, but it remains relatively affluent. This makes it attractive to foreign producers, and Australia's shops overflow with high-priced goods from Japan, South Korea, Taiwan, the United States, and Europe. Indeed, despite its long-term protectionist practices, Australia still does not produce many goods that could be manufactured at home. Overall, the continuing prominence of the primary sector [12] indicates that further economic development is still needed.

Australia's Challenges

The Commonwealth of Australia may be changing, but its neighbors in Southeast and South Asia are changing even faster. Australia's European bonds are weakening as its Asian ties are strengthening, and it plays a bigger role in

the Pacific Rim. Australia does face certain challenges at home as well. These include: (1) Aboriginal claims, (2) concerns involving immigration, (3) environmental degradation, and (4) issues related to Australia's status and regional role.

Aboriginal Issues

For several decades the Aboriginal issues focused on two questions: the formal admission by the government and majority of mistreatment of the Aboriginal minority with official apologies and reparations, and land ownership. The first question was resolved in 2008 when newly elected (now former) Prime Minister Kevin Rudd offered a formal apology for the historic mistreatment of the Aborigines. The second question has major geographic implications. Although making up only 2 percent of the total population, the Aboriginal population of almost 600,000 (including many of mixed ancestry) has been gaining influence in national affairs, and in the 1980s Aboriginal leaders began a campaign to obstruct exploration on what they designated as ancestral and sacred lands.

When the Europeans arrived as colonists, they invoked the doctrine of *terra nullis* (land owned by no one) and occupied land that had been used communally by the indigenous people. The situation is similar to what happened in the Americas when Amerindian concepts of land ownership conflicted with the private ownership ideas of the Europeans. Until 1992, Australians had taken it for granted that Aboriginals had no right to land ownership, but in that year the Australian High Court made the first of a series of rulings in favor of Aboriginal claimants. A subsequent court decision implied that vast areas (probably as much as 78 percent of the whole continent) could potentially be subject to Aboriginal claims (Fig. 11-7). Today the Aboriginal land issue [13] remains mainly (though not exclusively) an Outback issue, but it has the potential to overwhelm Australia's court system and to inhibit economic growth.

Lately, Australians have initiated debate on ways to bring so-called market-driven incentives to Aboriginal areas. At the moment, much Aboriginal land is administered communally by Aboriginal Land Councils, which makes the people land-rich but keeps them dirt-poor because the system in effect prevents private enterprise, including even the construction of family-owned housing. Some Aboriginal leaders are arguing that the system perpetuates an undesirable dependence on federal handouts, keeps jobs from becoming available, and restrains economic incentive. But not all Aboriginal leaders agree, worrying that privatization of any kind will undermine the Land Councils and

FIGURE 11-7 © H. J. de Blij, P. O. Muller, and John Wiley & Sons, Inc.

create a small minority of better-off businesspeople, leaving the great majority of tradition-bound Aborigines even worse off. Still, the fact that the idea has some Aboriginal support is a sign of changing times in the relationships among Australian communities.

In 2007, Australia was shaken by a report on conditions in Aboriginal desert camps in the Northern Territory that chronicled rampant alcoholism, domestic violence, unemployment, school truancy, and a general breakdown of Aboriginal culture. The then-prime minister, calling this a national emergency, ordered the federal government to take control of 60 Aboriginal communities in the Northern Territory through the use of police and army patrols, health checks, and other measures. While there were accusations of political grandstanding and questions about sincerity, there could be no doubt that Australia's conscience was once again troubled by the fate of the country's first inhabitants left behind in the era of modernization.

Immigration Issues

The immigration issue is older than Australia itself. Fifty years ago, when Australia had less than half the population it has today, 95 percent of the people were of European ancestry, and more than three-quarters of them came from the British Isles. Eugenic (race-specific) immigration policies maintained this situation until the 1970s. Today, the picture is dramatically different: of 22.8 million Australians, only

VOICE FROM THE
Region

© Julie Tregale
Julie Tregale, Sydney, Australia

BEING ASIAN IN AUSTRALIA

"I live in a suburb on the north side of Sydney, Australia's number-one city. Born and raised in Indonesia on the island of Jawa, I am one of many Asians who have migrated to Australia's cities in recent times. Most immigrants come here to look for work or to get an education. At the workplace and in the universities, there are always Asian faces. The government encourages immigration because Australia has an aging population and they need foreign workers. As a result, the country is changing fast. It's good that Australia has opened up to the rest of Asia and that it has become a much more diverse culture. But I can also see that people are worried about everybody getting along, about integration. Some ethnics turn their area into their little country where no one speaks English. That makes it difficult for Australians to go there. Also, it seems that gang fights between different ethnicities are happening more and more. Being a mother of two, I also see some issues at schools. The students who excel academically are usually Asians. Sometimes, Australian parents complain that Asian students have been coached and how unfair it is, etc. But I think it's about the different priorities that each family has on how they want to prepare their child and that comes from values they grew up with. I see that more Asians are geared into high academic achievement whereas the Australians are more into sports (rugby, soccer, swimming). A friend of mine who is Vietnamese-born but has lived here nearly all her life said that it's not always an advantage to have high academic achievements because in reality, in the workplace, you also need to be accepted, have the social skills, and fit in. In all, I am not surprised that the growing number of immigrants has caused some discrimination and tension. Government should take an active role to balance things out and make Australia a great place to live for all."

about one-third are of British-Irish origin, and Asian immigrants outnumber both European immigrants and the natural increase each year. During the early 1990s, about 150,000 legal immigrants arrived in Australia annually, most from Hong Kong, Vietnam, China, the Philippines, India, and Sri Lanka. Annual immigration quotas have

Regional ISSUE   Indigenous Rights and Wrongs

THE LEAST WE SHOULD DO IS APOLOGIZE

"I can't believe this. Here we are more than two dozen years beyond our two-hundredth-year anniversary, the evidence of our mistreatment of the Aborigines is everywhere, and we can't, as a nation, say we're sorry? How could we get ourselves into a public debate over this? What does it take to offer sincere apologies to the original Australians and to tell them (and not only to tell them) that we'll do better?

"We should remind ourselves more often than we do of what happened here. When our ancestors arrived on these shores there were more than a million Aboriginal people in Australia, with numerous clans and cultures. They had many virtues, but one in particular: they didn't appropriate land. Land was assigned to them by the creator (what they call the Dreaming), and their relationship to it was spiritual, not commercial. They didn't build fences or walls. Neither had they adopted some bureaucratic religion. Just think, a world where land was open and free and religion was local and personal. Wouldn't that do something for the Middle East and Northern Ireland! Well, we showed up and started claiming and fencing off the land that, under British rules, was there for the taking. If the Aborigines got in the way, they were pushed out, and if they resisted, they got killed. You don't even want to think about what happened in Tasmania. A campaign of calculated extermination. Between 1800 and 1900 the Aboriginal population dropped from 1 million to around 50,000. And when we became a nation, they weren't even accorded citizenship.

"None of this stopped some white Australian men from getting Aboriginal women pregnant. But from 1910 on church and state managed to make things even worse. They took the young children away from their mothers and put them in institutions, where they would be "Europeanized" and then married off to white partners, so that they would lose their Aboriginal inheritance. This, if you'll believe it, went on into the 1960s! Think of the scenes, these kids being kidnapped by armed officials from their mothers. And we have to argue about saying we're sorry?

"Well, at least we're discussing it in the open, and the misdeeds of the past are coming out. Aboriginal Australians finally got the vote in 1962, and they're able to get elected to parliament. I work in a State government office that assists Asian immigrants here in Sydney, and I only wish that we'd done for the Aborigines what we've done for the immigrants we admit."

ENOUGH IS ENOUGH

"Former Prime Minister John Howard was right when he said that apologizing for history should be a private matter. I was born on this Outback sheep station 40 years ago, and we employ a dozen Aboriginal workers, most of whom were born in this area. I had nothing to do with what happened more than a century ago, or even the first 75 years of the past century. What I would have done is irrelevant, and what my great-great-grandparents may have done has nothing to do with me. All over the world people are born into situations not of their making, and they should all go around apologizing?

"And as a matter of fact I believe that this country has bent over backwards, in my time at least, to undo the alleged wrongs of the past. Look, the Aboriginal minority counts about 550,000 or 2.5 percent of the population. Compare this to the map: they've got the Northern Territory and other parts of the country too, and that's close to 15 percent of Australia. And now we're going to give them more? I could see disaster in the making when that guy Eddie Mabo got the High Court to agree to his so-called 'native title' claim to land with which he and his people hadn't done a thing. Since when do Aboriginal customs take precedence over the law of the land? But even worse was that decision supporting the Wiks up there in Cape York. That was unbelievable: native title on land already held in pastoral lease? That affects all of us. Pretty soon the whole of Australia will be targeted by these Aboriginal claimants and their lawyers. And by the way, in the old days those people moved around all the time. Who's to say what clans owned what land when it comes to claiming native title? These court cases are going to tie us up in legal knots for generations to come.

"And now I gather that some Aboriginal people are calling for a treaty between themselves and the government, for what purpose I'm not sure—I guess it'll be like the one the Kiwis made with the Maori there. Is this really what we need if we're going to put the past behind us? Look, I like the fellows working here, but you've got to realize that no laws or treaties are going to change all of the problems they have. They're getting preference for many kinds of government employment, remedial help in all kinds of areas, but still they wind up in jail, leaving school, giving up their jobs. Yes, I'm sure wrongs were done in the past. But apologizing isn't going to make any difference. They have their opportunity now. In a lot of countries they would have never gotten it: Aussies are pretty decent people. It's up to them to make the best of it."

Vote your opinion at www.wiley.com/college/deblij

since been reduced, but were then allowed to rise again—reaching about 185,000 at the end of the 2000s—with Asian immigrants continuing to outnumber those from Western sources. This situation upsets some in the long-dominant Anglo-Australian society, and restrictions on foreign ownership of Australian real estate reflect fears of what extremist groups call the Asianization of the country. But the obvious success of many Asian settlers in Australia underscores the opportunities still available in this free and open society. Sydney, the leading recipient of the Asian influx, has become a mosaic of ethnic neighborhoods, some of which have gone through periods of gang violence and drug dealing, but have stabilized and even prospered over time. Nevertheless, multiculturalism will remain a long-term challenge for Australia in the twenty-first century.

Environmental Issues

Environmental degradation [14] is practically synonymous with Australia. First the Aborigines, then the Europeans and their livestock, inflicted heavy damage on Australia's natural environments and ecologies. Great stands of magnificent forest were destroyed. In Western Australia, centuries-old trees were simply ringed and left to die so that the sun could penetrate through their leafless crowns to nurture the grass below for pasture for introduced livestock. In island Tasmania, where Australia's native eucalyptus tree reaches its greatest concentrations (comparable to North American redwood stands), tens of thousands of hectares of this irreplaceable treasure have been lost to chain saws and pulp mills. Many of Australia's unique marsupial species have been driven to extinction, and many more are endangered or threatened. "Never have so few people wreaked so much havoc on the ecology of so large an area in so short a time," observed a geographer in Australia

recently. But awareness of this environmental degradation is growing. In Tasmania, the Green environmentalist political party has now become a force in State affairs, and its activism has slowed deforestation, dam-building, and other development projects. Still, many Australians fear the environmentalist movement as an obstacle to economic growth at a time when the economy needs stimulation. This, too, is an issue for the future.

Another environmental problem involves Australia's wide and long-term climatic variability. In a dominantly arid continent, droughts in the moister fringes are the worst enemy, and Australia's history is replete with devastating dry spells. Australia is vulnerable to El Niño events (see Chapter 5B), but recent global warming may also be playing a role in the process. One such serious drought, regarded as the worst in living memory, lasted throughout the 2000s and imperils the entire region watered by the Murray-Darling river system, Australia's breadbasket (Fig. 11-6). By some estimates, an area that accounts for 40 percent of Australia's agriculture and contains 85 percent of its irrigated land is on the verge of ruin (see photos below). Part of this is caused by nature, but other factors have to do with the increasing demand for water in Australia's burgeoning urban areas and by excessive damming, well-drilling, and water diversion in the Murray-Darling Basin's upstream tributaries. The Australian government wants to put in place a coordinated drainage-basin-control program for the Murray-Darling region, but this requires cooperation from above (the State governments involved) and below (local farmers who survive by diverting water to raise their crops). So far, the plan has not been adopted, but if the drought continues to deepen, Australia's environmental crisis will leave no alternative—and even then it may be too little too late.

During the first decade of the twenty-first century Australia lay in the grip of its worst drought on record, with calamitous consequences ranging from deadly forest fires to parched farmlands. One of Australia's most profitable agricultural industries, winemaking, suffered severely as many winegrowers were driven off their land. The photo on the left shows a once-mature vineyard near Yarrawonga, on the New South Wales-Victoria border north of Melbourne along the Murray River, desiccated and abandoned during the summer of 2009. When the rains finally came, they caused disastrous floods, washing away topsoil and eroding the heat-baked countryside. The photo on the right, taken in January 2010, shows the results at Coonamble, New South Wales along a tributary of the Darling River.

(Left) © Global Warming Images/Alamy; (Right) © Photo by James Croucher/Newspix/Getty Images, Inc.

Australia's Place in the World

Several issues involving Australia's status at home, relations with neighbors, and position in the world are also stirring up national debate. A persistent domestic question is whether Australia should become a republic, ending the status of the British monarch as the head of state, or whether it should continue its status quo in the British Commonwealth.

Relations with neighboring *Indonesia* and *East Timor* have become more complex. For many years, Australia had what may be called a special relationship with Indonesia, whose help it needs in curbing illegal seaborne immigration. It was also profitable for Australia to counter international (UN) opinion and recognize Indonesia's 1976 annexation of Portuguese East Timor, for in doing so Australia could deal directly with Jakarta for the oil reserves beneath the Timor Sea (see Fig. 10B-8). Thus Australia gave neither recognition nor support to the rebel movement that fought for independence in East Timor. But this story had a surprisingly happy ending. When the East Timorese campaign for independence succeeded in 1999 and Indonesian troops began an orgy of murder and destruction, Australia sent an effective peacekeeping force and spearheaded the United Nations effort to stabilize the situation (for more see Chapter 10B). Today, a new chapter has opened in Australia's relations with these northern neighbors.

Australia also maintains a long-term relationship with *Papua New Guinea (PNG)* (see Chapter 12). This association, too, has gone through difficult times. In recent years, the inhabitants of Papua New Guinea have strongly resisted privatization, World Bank involvement, and globalization generally. Australia assists PNG in several spheres, but its motives are sometimes questioned. Not long ago, the construction of a projected gas pipeline from PNG to the Australian State of Queensland precipitated fighting among tribes over land rights, resulting in dozens of casualties, and Australian public opinion reflected doubts concerning the appropriateness of this venture. When political violence and chaos overtook the *Solomon Islands* east of Papua New Guinea in 2003, Australian forces intervened: a failing state in Australia's neighborhood could become a base for terrorist activity.

Immediately after the 9/11 attack on U.S. targets in New York and Washington, Australian leaders expressed strong support for the American campaign against terrorism, but they made it a point also to assure the Indonesian government that the War on Terror was not a war on Islam. Although the great majority of Australians supported this stance and troops were sent to Iraq, their will was severely tested in 2002 when a terrorist attack on a nightclub on the Indonesian island of Bali killed 88 vacationing Australians—and again in 2005 when another attack in Bali was followed by the apprehension of a terrorist cell preparing for an assault within Australia itself. The elections of 2007 brought in a new Labour Party-led government with a different orientation, especially toward the United States, and Australia's troops were pulled out of Iraq.

In all, Australia's growing stature in the world and its closer ties with neighboring realms has prompted a reconsideration of the country's very identity and where it best fits. Australia has at once assumed a wider global presence, and it has in some ways (re-)turned to Asia and the Pacific Rim. The exclusive relationship with Britain now belongs to the past; trade and investment with (relatively) nearby countries are on the increase; political sensitivity to issues in neighboring realms has grown; and the growing number of Asian faces on the streets of Sydney and elsewhere tells the story.

Territorial dimensions, relative location, and raw-material wealth have helped determine Australia's place in the world and, more specifically, on the Pacific Rim. Australia's population, still less than 25 million early in the second decade of the twenty-first century, is smaller than Taiwan's and only slightly larger than that of the Caribbean island of Hispaniola. However, Australia's importance in the international community far exceeds its human numbers.

▪▪ NEW ZEALAND

Fifteen hundred miles east-southeast of Australia, in the Pacific Ocean across the Tasman Sea, lies New Zealand, also known as *Aotearoa* in Maori (meaning "land of the long white cloud"). In an earlier age, New Zealand would have been part of the Pacific geographic realm because its population was all Maori, a people with Polynesian roots. But New Zealand, like Australia, was invaded and occupied by Europeans. Today, its population of 4.5 million is about 70 percent European, and the Maori form a substantial minority of about 660,000, with many of mixed Euro-Polynesian ancestry (including Pacific Islanders).

New Zealand consists of two large mountainous islands and many scattered smaller islands (Fig. 11-8). The two large islands, with the South Island somewhat larger than the North Island, look diminutive in the great Pacific Ocean, but together they are larger than Britain. In contrast to Australia, the two main islands are mostly mountainous or hilly, with several peaks rising far higher than any on the Australian landmass. The South Island has a spectacular snowcapped range appropriately called the Southern Alps, with numerous peaks reaching beyond 3300 meters (10,000 ft). The smaller North Island has proportionately more land under low relief, but it also has an area of central highlands along whose lower slopes lie the pastures of New Zealand's chief dairying district. Hence, while Australia's land lies relatively low in elevation and exhibits much low relief, New Zealand's is on the average high and is dominated by rugged terrain.

As we noted at the beginning of this chapter, the convergence of the Australian and Pacific tectonic plates underlies much of New Zealand, and it renders the country prone to volcanic eruptions and earthquakes. On February 22, 2011, a magnitude 6.3 quake struck very close to the center of Christchurch, the South Island's biggest city (essentially an aftershock of an even bigger earthquake five months earlier, whose epicenter was located at the western edge of the urban region). Structural damage to the CBD and its immediate surroundings was severe, and the death toll reached 182, making it New Zealand's deadliest quake in 80 years.

Human Spatial Organization

The most promising areas for habitation, therefore, are the lower-lying slopes and lowland fringes on both islands. On the North Island, the largest urban area, Auckland, occupies a comparatively low-lying peninsula. On the South Island, the largest lowland is the agricultural Canterbury Plain, centered on Christchurch. What makes these lower areas so attractive, apart from their availability as cropland, is their magnificent pastures. The range of soils and pasture plants allows both summer and winter grazing. Moreover, the Canterbury Plain, the chief farming region, also produces a wide variety of vegetables, cereals, and fruits. About half of all New Zealand is pasture land, and much of the farming provides fodder for the pastoral industry. Forty million sheep and four million cattle dominate these livestock-raising activities, with wool, meat, and dairy products providing about two-thirds of the islands' export revenues.

Despite their contrasts in size, shape, physiography, and history, New Zealand and Australia share a number of characteristics. Apart from their joint British heritage, they

FIGURE 11-8   © H. J. de Blij, P. O. Muller, and John Wiley & Sons, Inc.

From the Field Notes . . .

"The drive from Christchurch to Arthur's Pass on the South Island of New Zealand was a lesson in physiography and biogeography. Here, on the east side of the Southern Alps, you leave the Canterbury Plain and its agriculture and climb into the rugged topography of the glacier-cut, snowcapped mountains. A last pasture lies on a patch of flatland in the foreground; in the background is the unmistakable wall of a U-shaped valley sculpted by ice. Natural vegetation ranges from pines to ferns, becoming even more luxuriant as you approach the moister western side of the island."

www.conceptcaching.com

© H. J. de Blij

share a sizeable pastoral economy with growth in specialty goods such as wines, a small local market, the problem of great distances to world markets, and a desire to stimulate (through protection) domestic manufacturing. The very high degree of urbanization in New Zealand (86 percent of the total population) once again resembles Australia: substantial employment in city-based industries, mostly the processing and packing of livestock and farm products, as well as government jobs.

Spatially, New Zealand further shares with Australia its pattern of **peripheral development [15]** (Fig. 11-2), imposed not by deserts but by high rugged mountains and the fragmented nature of the country. The country's major cities—Auckland and the capital of Wellington on the North Island; Christchurch and Dunedin on the South Island—are all located on the coast, and the entire rail and highway network is thereby peripheral in its configuration. Moreover, the two main islands are separated by Cook Strait, a windswept waterway that can only be crossed by ferry or air (Fig. 11-8). On the South Island, the Southern Alps are New Zealand's most formidable barrier to surface communications.

The Maori Factor and New Zealand's Future

Like Australia, New Zealand has had a history of difficult relations with its indigenous population. The Maori, who account for 15 percent of the country's population today, appear to have reached the islands during the tenth century AD. By the time the European colonists arrived, the Maori had had a tremendous impact on the islands' ecosystems, especially on the North Island where most of them lived. In 1840, the Maori and the British signed a treaty at Waitangi that granted the colonists sovereignty over New Zealand but guaranteed the Maori rights covering established tribal

lands. Although the British abrogated parts of the treaty in 1862, the Maori had reason to believe that vast reaches of New Zealand, as well as offshore waters, were theirs.

As in Australia, judicial rulings during the 1990s supported the Maori position, which led to expanded claims and growing demands. Culturally, the declaration of Maori as an official language in New Zealand and its teaching throughout the school system are seen as significant progress toward an acceptance of the Maori cultural heritage. But one of the Maori's persistent complaints is the slow pace of integration of the minority into modern New Zealand society. Although Maori claims encompass much of rural New Zealand, they also cover prominent sites in the major cities. Today, the Maori question is the leading national issue in the country.

The Green Factor

New Zealand is well known for its progressive politics and high quality of life. Among the factors that contribute to superior environmental progressiveness is its status as one of the leading "green" societies in the world, with a long-active Green Party and an established program of environmental conservation. Although the Maori and then the European colonists degraded New Zealand's landscapes, their descendants have been exemplary in observing environmentally friendly and sustainable policies.

Environmental scientists recently ranked New Zealand first in the world (the United States, was 28th) in a report that examined a range of environmental indices such as clean water, air pollution, renewable energy, and biodiversity conservation. Approximately 30 percent of its land area is protected from development, and in 2007 New Zealand declared that it would become the first carbon-neutral country in the world by 2020. More than 70 percent of its

From the Field Notes . . .

© H. J. de Blij

"It was Sunday morning in Christchurch on New Zealand's South Island, and the city center was quiet. As I walked along Linwood Street I heard a familiar sound, but on an unfamiliar instrument: the Bach sonata for unaccompanied violin in G minor—played magnificently on a guitar. I followed the sound to the artist, a Maori musician of such technical and interpretive capacity that there was something new in every phrase, every line, every tempo. I was his only listener; there were a few coins in his open guitar case. Shouldn't he be playing before thousands, in schools, maybe abroad? No, he said, he was happy here, he did all right. A world-class talent, a street musician playing Bach on a Christchurch side street, where tourists from around the world were his main source of income. Talk about globalization."

www.conceptcaching.com

energy is already derived from renewable sources (hydro and geothermal), compared to eight percent in the United States. New Zealand is also a nuclear-free country that does not permit even visiting naval vessels with nuclear capabilities to anchor at its ports. And the country has even established Environmental Courts to hear cases involving environmental management decisions. With its pioneering green initiatives, New Zealand demonstrates that a country, albeit one with a small population, can successfully work to improve its natural environment if its leaders possess the necessary political will.

Dominant cultural heritage and prevailing cultural landscape form two criteria on which the delimitation of the Austral Realm is based. But in both Australia and New Zealand, the cultural mosaic is changing, and the convergence with neighboring realms is proceeding.

POINTS TO PONDER

- Australia's population growth is marked by immigration (nearly 200,000 per year) exceeding natural increase (about 150,000 annually).

- New Zealand experiences between 10,000 and 15,000 earthquakes every year; of these, 100–150 can be felt, and about 20 are of magnitude 5 or higher.

- Britain was Australia's most important trading partner until the early 1980s, when it was displaced by Japan. In this century, Australia's biggest trading partner is China.

- Over the past decade, Australians have killed about 50 million kangaroos. Almost all the meat has gone into pet food.

 using GEOGRAPHY

1. Both Japan and New Zealand were struck by major earthquakes in 2011 (although Japan's was far more catastrophic). Consider the geography of the cities and surrounding areas of Sendai (Japan) and Christchurch (New Zealand) that were hit the hardest. What are the similarities and differences between these two urban regions in terms of exposure to seismic risk?

2. Australia may have a modest population, but its territorial dimensions, resource base, economic wealth, military capacity, and relative location make it a regional power. Its relationships with its neighbors, however, have at times been problematic. To what extent does relative location affect Australia's interaction with Indonesia? How has Australia's relationship with East Timor, before and after the latter's independence in 2002, been affected by its links with Indonesia, and what was the key issue in this context? What interests does Australia maintain in Papua New Guinea? Why would Australia be involved in the problems that plague the Solomon Islands?

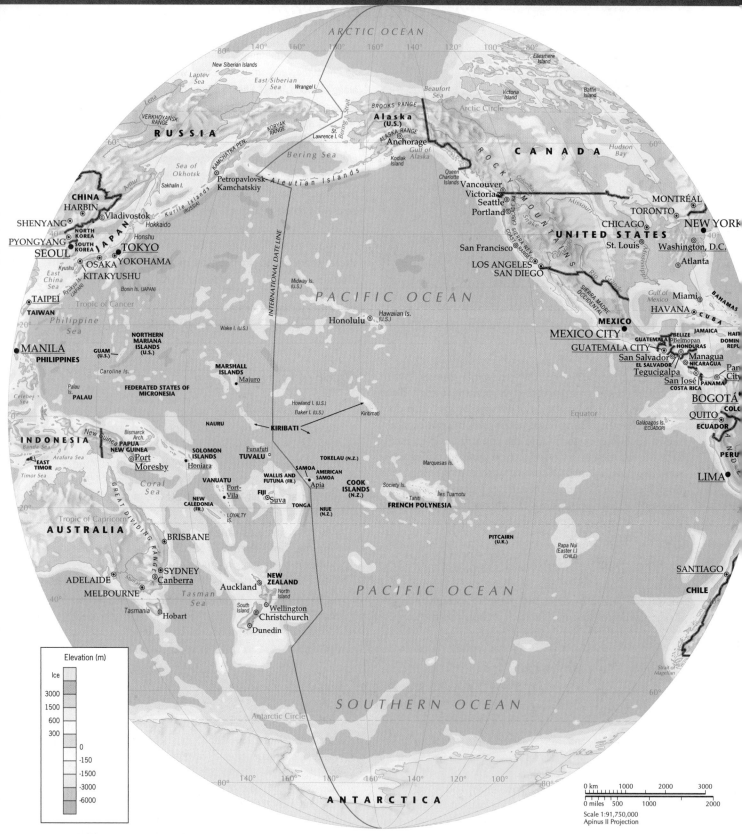

▪▪ REGIONS

Melanesia

Micronesia

Polynesia

IN THIS CHAPTER

- ◆ Water, water everywhere—but who owns it?
- ◆ Islands high and low
- ◆ The conundrum of divided Papua
- ◆ Colonists and foreigners: Still there, still dominant
- ◆ The partitioning of Antarctica
- ◆ Geopolitics in the warming Arctic Basin

CONCEPTS, IDEAS, AND TERMS

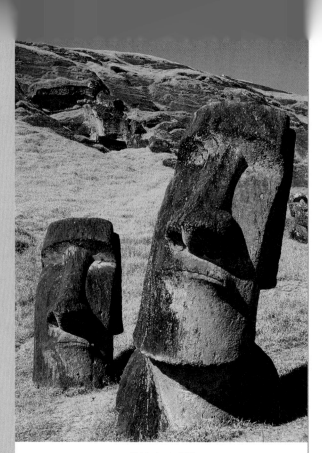

© H. J. de Blij
Where were these photos taken?
Find out at www.wiley.com/college/deblij

In this final chapter, we focus on three of the world's vast expanses so different from those discussed in previous chapters that we need an entirely new perspective. Water and ice, not land and soil, dominate the physiography. Environmental changes occurring here have consequences for the entire planet.

The largest of these expanses, the *Pacific Realm*, covers almost half the globe from the Bering Sea to the Southern Ocean and from Ecuador to Indonesia. The Pacific Ocean, larger than all the world's land areas combined, is studded with tens of thousands of islands of which New Guinea is by far the largest. The two polar areas to which it is linked could hardly be more different: the *Arctic* is water, frozen at the surface, but you can reach the North Pole in a submarine; the *Antarctic* is land, weighed down by the planet's largest accumulation of permanent ice. The surface journey to the South Pole, first accomplished just over a century ago, has cost many an explorer's life.

Climate change is altering geography everywhere, as is the enhanced technology of mineral (especially oil and gas) exploration and exploitation. Rising sea levels would imperil thousands of populated low-elevation islands. Areas of ocean floor once beyond the scope of human intervention are now coming within reach. Melting ice is clearing new maritime shipping routes. But, as we will see, international rules to govern who owns what in these remote frontiers are subject to dispute. When land boundaries jut outward into the sea, it depends on who draws the maps. We should be aware of the potential for competition, even conflict.

© H. J. de Blij

major geographic qualities of
THE PACIFIC REALM

1. The Pacific Realm's total area is the largest of all geographic realms. Its land area, however, is the smallest, as is its population.

2. The island of New Guinea, with 10.7 million people, alone contains almost 80 percent of the Pacific Realm's population.

3. The Pacific Realm, with its wide expanses of water and numerous islands, has been strongly affected by United Nations Law of the Sea provisions regarding states' rights over economic assets in their adjacent waters.

4. The highly fragmented Pacific Realm consists of three regions: Melanesia (including New Guinea), Micronesia, and Polynesia.

5. Melanesia forms the link between Papuan and Melanesian cultures in the Pacific.

6. The Pacific Realm's islands and cultures may be divided into volcanic high-island cultures and coral-based low-island cultures.

7. In Micronesia, U.S. influence has been particularly strong and continues to affect local societies.

8. In Polynesia, indigenous culture exhibits remarkable consistency and uniformity throughout the region, its enormous dimensions and dispersal notwithstanding. Yet at the same time, local cultures are nearly everywhere severely strained by external influences. In Hawai'i, as in New Zealand, indigenous culture has been largely submerged by Westernization.

● DEFINING THE REALM

Our survey begins with the Pacific geographic realm that covers just about an entire hemisphere, the one commonly called the Sea Hemisphere (Fig. 12-1). This Sea Hemisphere meets the Russian and North American realms in the far north and merges into the Southern Ocean in the south. Despite the preponderance of water, this fragmented, culturally complex realm does possess regional identities. It includes the Hawaiian Islands, Tahiti, Tonga, and Samoa—fabled names in a world apart.

In terms of modern cultural and political geography, Indonesia and the Philippines are not part of the Pacific Realm, though Indonesia's political system reaches into it; nor do Australia and New Zealand belong to it. Before the European invasion and colonization, Australia would have been included (because of its Aboriginal population) as well as New Zealand (because of its Maori population's Polynesian affinities). But the Europeanization of their countries has engulfed indigenous Australians and Maori New Zealanders, and the regional geography of Australia and New Zealand today is decidedly not Pacific. In New Guinea, on the other hand, Pacific peoples, numerically and culturally, remain the dominant element.

In examining the Pacific Realm, it is important to keep in mind the dimensional contrasts this part of the world presents. The realm is enormous, but its total land area is a mere 975,000 square kilometers (377,000 sq mi),

about the size of Texas and New Mexico, and over 90 percent of this lies in New Guinea.* The population is so widely scattered that a distribution map would be impractical, and it only totals a bit less than 14 million (about as many people as there are in Beijing, China).

COLONIZATION AND INDEPENDENCE

The Pacific islands were colonized by the French, British, and Americans; an indigenous Polynesian kingdom in the Hawaiian Islands was annexed by the United States, and since 1959 has been the fiftieth State. Yet today's map is still an assemblage of independent and colonial territories (Fig. 12-1). France controls New Caledonia and French Polynesia. The United States administers Guam as well as American Samoa, the Line Islands, Wake Island, Midway Islands, and several smaller islands; the United States also has special relationships with other territories, former dependencies that are now nominally independent. The British, through New Zealand, have responsibility for the Pitcairn group of islands, and New Zealand administers and supports the Cook, Tokelau, and Niue Islands. Easter Island, that storied speck of land in the southeastern

*The numbers in the Data Table in Appendix B do not match these totals because only the political entity of Papua New Guinea is listed, not the Indonesian province of Papua that occupies the western part of the island. Here, as Figure 10B-1 shows, the political and the realm boundaries do not coincide.

Pacific, is part of Chile. Indonesia rules Papua, the western half of the island of New Guinea.

Other island groups have now become independent states. The largest are Fiji, once a British dependency, the Solomon Islands (also formerly British), and Vanuatu (until 1980 ruled jointly by France and Britain). Also on the current map, however, are such microstates as Tuvalu, Kiribati, Nauru, and Palau. Foreign aid is crucial to the survival of most of these countries. Tuvalu, for example, has a total area of around 25 square kilometers (10 sq mi), a population of some 10,000, and a per capita GNI of about U.S. $1500, derived from fishing, sales of copra (coconut meat used to make oil), and some tourism. But what really keeps Tuvalu going is an international trust fund set up by Australia, New Zealand, the United Kingdom, Japan, and South Korea. Annual grants from that fund, as well as remittances sent back to families by workers who have left for New Zealand and elsewhere, allow Tuvalu to survive.

THE PACIFIC REALM AND ITS MARINE GEOGRAPHY

Certain land areas may not be part of the Pacific Realm (we mentioned the Philippines and New Zealand), but the Pacific Ocean extends from the shores of North and South America to mainland East and Southeast Asia and from the Bering Sea southward to the Subtropical Convergence. This means that several seas, including the Sea of Japan (East Sea), the East China Sea, and the South China Sea, are part of the Pacific Ocean. As we will see below, this relationship matters. Pacific coastal countries, large and small, mainland and island, compete for jurisdiction over the waters that bound them.

The Pacific Realm and its ocean, therefore, form an ideal place to focus on **marine geography [1]**. This field encompasses a variety of approaches to the study of oceans and seas; some marine geographers focus on the biogeography of coral formations, others on the geomorphology of beaches, and still others on the movement of currents. A particularly interesting branch of marine geography has to do with the definition and delimitation of political boundaries at sea. Here geography intersects political science and maritime law.

The State at Sea

Littoral (coastal) states do not end where atlas maps suggest they do. States have claimed various forms of jurisdiction over coastal waters for centuries, closing off bays and estuaries and ordering foreign fishing fleets to stay away from nearby fishing grounds. Thereby arose the notion of the **territorial sea [2]**, where all the rights of a coastal state would prevail. Beyond lay the **high seas [3]**, free, open, and unfettered by national interests.

It was in the interest of colonizing, mercantile states to keep territorial seas narrow and high seas wide, thereby interfering as little as possible with their commercial fleets. In the seventeenth and eighteenth centuries the territorial sea was 3, 4, or at most 6 nautical miles wide, and the colonizing powers claimed the identical widths for their colonies (1 nautical mile = 1.85 kilometers or 1.15 statute miles).*

During the twentieth century, these constraints weakened. States without trading fleets saw no reason to limit their territorial seas. States with nearby fishing grounds traditionally exploited by their own fleets wanted to keep the increasing number of foreign trawlers away. States with shallow **continental shelves [4]** (offshore continuations of coastal plains) wished to control their resources on and below the seafloor, made more accessible by improved technology. States also disagreed on the methods by which offshore boundaries, whatever their width, should be defined. Early efforts by the League of Nations (the precursor to the United Nations) during the 1920s to resolve these issues met with only partial success, mainly in the technical sphere of boundary delimitation.

Scramble for the Oceans

In 1945, the United States helped precipitate what has become known as the scramble for the oceans. President Harry S Truman issued a proclamation that claimed U.S. jurisdiction and control over all the resources "in and on" the continental shelf down to its margin, around 100 fathoms (183 meters/600 ft) deep. In some areas, the shallow continental shelf of the United States extends more than 300 kilometers (200 mi) offshore, and Washington did not want foreign countries drilling for oil just beyond the 3-mile territorial sea.

Few observers foresaw the impact the Truman Proclamation would have, not only on U.S. waters but on the oceans everywhere, including the Pacific. It set off a rush of additional claims. In 1952, a group of South American countries, some with little continental shelf to claim, issued the Declaration of Santiago, claiming exclusive fishing rights up to a distance of 200 nautical miles off their coasts. Meanwhile, as part of the Cold War competition, the Soviet Union claimed a 12-mile territorial sea and urged its allies to do likewise.

---

*Here and in the rest of this section all distances involving maritime boundaries are given in nautical miles only. Throughout the modern history of maritime law, the latter unit has been the only one used for this kind of boundary-making. Any distance stated in nautical miles can be converted to its metric equivalent by multiplying it by 1.85. The conversion table in Appendix A will aid in this computation.

UNCLOS Intervention

At this point the United Nations intervened, and a series of UNCLOS (United Nations Conference on the Law of the Sea) meetings began. These meetings addressed issues ranging from the closure of bays to the width and delimitation of the territorial sea, and after three decades of negotiations they achieved a convention that changed the political and economic geography of the world's oceans forever. Among its key provisions were the authorization of a 12-mile territorial sea for all countries as well as the establishment of a 200-mile (230-statute-mi/370-km)-wide **Exclusive Economic Zone (EEZ)** [5] over which a coastal state would have total economic rights. Resources in and under this EEZ (fish, oil, minerals) belong to the coastal state, which could either exploit them or lease, sell, or share them as it saw fit. These zones were noted in Chapter 2A in regard to the new competition for the Arctic Ocean, but as we will see EEZs take on even greater importance in this Pacific Realm.

These new provisions had a far-reaching impact on the world's oceans and seas (Fig. 12-2), especially the Pacific. Unlike the Atlantic Ocean, the Pacific is studded with islands large and small, and a microstate consisting of a single small island suddenly acquired an EEZ covering 166,000 square nautical miles. European colonial powers still controlling minor Pacific possessions (most notably France) saw their maritime jurisdictions vastly expanded. Small, low-income archipelagos could now bargain with large, rich fishing nations over fishing rights in their EEZs. And for all the UNCLOS Convention's provisions for the "right of innocent passage" of shipping through EEZs and via narrow straits, the extent of the world's high seas has been diminished accordingly.

Maritime Boundaries

The extension of the territorial sea to 12 nautical miles and the EEZ to an additional 188 nautical miles created new **maritime boundary** [6] problems. Waters less than 24 nautical miles wide separate many countries all over the world, so that **median lines** [7], equidistant from opposite shores, have been delimited to establish their territorial seas. And even more countries lie closer than 400 nautical miles apart, requiring further maritime-boundary delimitation to determine their EEZs. In such maritime regions as the North Sea, the Caribbean Sea, and the East and South China seas (and more recently the Arctic Ocean), a maze of maritime boundaries emerged, some of them subject to dispute. Political changes on land can also lead to significant modifications at sea.

A case in point involves newly independent East Timor and its neighbor across the Timor Sea, Australia (see Fig. 10B-8). Australia had divided the waters and seafloor of the Timor Sea with Indonesia while recognizing Indonesia's 1976 annexation of East Timor. Thus when East Timor achieved independence in 2002, the so-called *Timor Gap* became an issue: where was the median line that would divide Timorese and Australian claims to the oil and gas reserves in this zone? The Australians argued that since the line defined with Indonesia predated East Timor's independence, it should continue in effect, giving Australia the bulk of the energy resources. But UNCLOS regulations, to which Australia subscribed, required a re-delimitation, giving East Timor a much larger share. This led Australia to withdraw from UNCLOS in 2002 so that it would not be bound by its rules. After difficult negotiations, the Australians in 2005 offered East Timor a half share of the energy revenues from the natural gas reserve that was in dispute, in addition to income already flowing from another major gasfield in the so-called Joint Petroleum Development Area. In return, East Timor's government agreed to defer the maritime-boundary issue for 50 years, leaving a future generation to solve this question. It was estimated that, over the next 30 years, East Timor might receive U.S. $13 billion from these sources, and, depending on the life and capacity of the reserves, perhaps even more. This will go a long way toward meeting the fledgling state's most urgent needs.

EEZ Implications

The UNCLOS provisions created new opportunities for some states to expand their spheres of influence. Wider territorial-sea and EEZ allocations raised the stakes: claiming an island now entailed potential control over a huge maritime area. In Chapter 2A we highlighted the need to carve up the Arctic as the ice melts and resources beckon. In Chapters 9B and 10B, we referred to several island disputes off mainland East Asia, which involve Japan and Russia, Japan and South Korea, Japan and China, China and Vietnam, and China and the Philippines. Ownership of many islands there is uncertain, and small specks of island territory have become large stakes in the scramble for the oceans. For instance, in the case of the Spratly Islands (Fig. 10B-1), six countries claim ownership, including both Taiwan and China. China's island claims in the South China Sea support Beijing's contention that this body of water is part and parcel of the Chinese state—a position that worries other states with coasts facing it.

Figure 12-2 reveals what EEZ regulations have meant to Pacific Realm countries such as Tuvalu, Kiribati, and Fiji, with nearly circular EEZs now surrounding the clusters of islands in this vast ocean space. Japan, Taiwan, and other nations with fisheries have purchased fishing rights in these EEZs from the island governments. Nonetheless, violations of EEZ rights do occur; recently, Vanuatu and the Philippines were at odds over unauthorized Filipino fishing in Vanuatu's EEZ.

The process of boundary delimitation continues. In an earlier edition of this book, we included a map of the South China Sea and nearby waters off East and Southeast Asia showing the median-line boundaries delimited according to

FIGURE 12-2

© H. J. de Blij, P. O. Muller, and John Wiley & Sons, Inc.

UNCLOS specifications. But that map changed when coastal states engaged in bilateral and multilateral negotiations—and argued over island ownership with major boundary implications. Indeed, the Pacific (and world) map of maritime boundaries remains very much a work in progress.

This point is underscored by a development that occurred at the end of the 1990s, when the implications of provisions in Article 76 of the 1982 UNCLOS Convention were reexamined. Whenever a continental shelf extends beyond the 200-mile limit of the EEZ, the Convention apparently allows a coastal state to claim that extension as a natural prolongation of its landmass. Although there is as yet no regulation that would also allow the state to extend its EEZ to the edge of the continental shelf, it is not difficult to foresee such an amendment to

the Convention. As it stands, states are now delimiting their proposed natural prolongations under a current deadline, which puts poorer states at a disadvantage since the required marine surveys are quite expensive. Lately, this issue has become quite heated, especially in the Arctic Ocean as countries scramble to claim prolongations. A recent article in *Science* reported that the big winners are likely to include the United States, Canada, Australia, New Zealand, Russia, and India—although a total of 60 coastal countries could ultimately benefit to some extent from the provision.

Thus the scramble for the oceans continues, and with it the constriction of the world's high seas and open waters. As we shall see, the scramble may be entering a new stage and geographic arena: the Arctic Ocean, where oil,

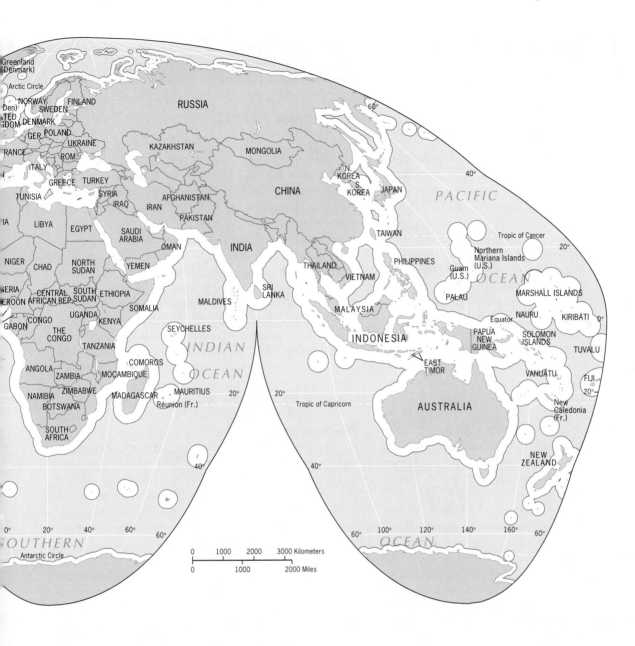

natural gas, fishing grounds, and maritime routes all figure in the coming competition.

● REGIONS OF THE PACIFIC

Sail across the Pacific Ocean, and one spectacular vista follows another. Dormant and extinct volcanoes, sculpted by erosion into basalt spires draped by luxuriant tropical vegetation and encircled by reefs and lagoons, tower over azure waters. Low atolls with nearly snow-white beaches, crowned by stands of palm trees, seem to float in the water. Pacific islanders, where foreign influences have not overtaken them, appear to take life with enviable ease.

More intensive investigations, however, reveal that such Pacific sameness is more apparent than real. Even the

Pacific Realm, with its long sailing traditions, its still-diffusing populations, and its historic migrations, has a durable regional framework. Figure 12-3 outlines the three regions that constitute this realm:

Melanesia: Papua (Indonesia), Papua New Guinea, Solomon Islands, Vanuatu, New Caledonia (France), Fiji

Micronesia: Palau, Federated States of Micronesia, Northern Mariana Islands, Republic of the Marshall Islands, Nauru, western Kiribati, Guam (United States)

Polynesia: Hawaiian Islands (U.S.), Samoa, American Samoa, Tuvalu, Tonga, eastern Kiribati, Cook and other New Zealand-administered islands, French Polynesia, Easter Island (Chile)

Regional
ISSUE   Who Should Own the Oceans?

WE WHO LIVE HERE SHOULD OWN THE WATERS

"Without exclusive fishing rights in our waters up to 200 nautical miles offshore, our economy would be dead in the water, no pun intended, sir. As it is, it's bad enough. I'm an official in the Fisheries Department of the Ministry of Natural Resources Development here in Tuvalu, and this modest building you're visiting is the most modern in all of the capital, Funafuti. That's because we're the only ones making some money for the state other than our Internet domain (*tv*), which has been the big story around here since we commercialized it. We don't have much in the way of natural resources and we export some products from our coconut and breadfruit trees, but otherwise we need donors to help us and major fishing nations to pay us for the use of our 750,000 square kilometers of ocean.

"Which brings me to one of our big concerns. Three-quarters of a million square kilometers sound like a lot, but look at the map. It's a mere speck in the vast Pacific Ocean. Enormous stretches of our ocean seem to belong to nobody. That is, they're open to fishing by nations that have the technology to exploit them, but we don't have those fast boats and factory ships they use nowadays. Still, the fish they take there might have come to our exclusive grounds, but we don't get paid for them. It seems to us that the Pacific Ocean should be divided among the countries that exist here, not just on the basis of nearby Exclusive Economic Zones but on some other grounds. For example, islands that are surrounded by others, like we are, ought to get additional maritime territory someplace else. Islands that aren't surrounded should get much more than just 200 nautical miles. To us the waters of the Pacific mean much more than they do to other countries, not only economically but historically too. I don't like Japanese and American and Norwegian fishing fleets on these nearby 'high seas.' They don't obey international regulations on fishing, and they overfish wherever they go. When there's nothing left, they'll simply end their agreement with us, and we'll be left without income from a natural resource over which we never had control.

"I don't expect anything to change, but we feel that the era of the 'high seas' should end and all waters and subsoil should be assigned to the nearest states. The world doesn't pay much attention to our small countries, I realize. But we've put up with colonialism, world wars, nuclear weapons testing, pollution. It's time we got a break."

THE OCEAN SHOULD BE OPEN AND RESTRICTED

"Ours is the Blue Planet, the planet of oceans and seas, and the world has fared best when its waters were free and open for the use of all. Encroachment from land in the form of 'territorial' seas (what a misnomer!) and other jurisdictions only interfered with international trade and commercial exploitation and caused far more problems than they solved. Already much of the open ocean is assigned to coastal countries that have no maritime tradition and don't know what to do with it other than to lease it. And when coastal countries lie closer than 400 nautical miles to each other, they divide the waters between them for EEZs and don't even leave a channel for international passage. Yes, I realize that there are guarantees of uninhibited passage through EEZs. But mark my words: the time will come when small coastal states will start charging vessels, for example cruise ships, for passage through 'their' waters. That's why, as the first officer of a passenger ship, I am against any further allocation of maritime territory to coastal states. We've gone too far already.

"In the 1960s, when those UNCLOS meetings were going on, a representative from Malta named Arvid Pardo came up with the notion that the whole ocean and seafloor beyond existing jurisdictions should be declared a 'common heritage of mankind' to be governed by an international agency that would control all activities there. The idea was that all economic activity on and beneath the high seas would be controlled by the UN, which would ensure that a portion of the proceeds would be given to 'Geographically Disadvantaged States' like landlocked countries, which can't claim any seas at all. Fortunately nothing came of this—can you imagine a UN agency administering half the world for the benefit of humanity?

"So you can expect us mariners to oppose any further extensions of land-based jurisdiction over the seas. The maritime world is complicated enough already, what with choke points, straits, narrows, and various kinds of prohibitions. To assign additional maritime territory to states on historic or economic grounds is to complicate the future unnecessarily and to invite further claims. From time to time I see references to something called the 'World Lake Concept,' the preliminary assignment of all the waters of the world to the coastal states through the delimitation of median lines. Such an idea may be the natural progression from the EEZ concept, but it gives me nightmares."

Vote your opinion at www.wiley.com/college/deblij

MAJOR CITIES OF THE REALM

City	Population* (in millions)
Honolulu, Hawai'i (U.S.)	1.0
Port Moresby, Papua New Guinea	0.3
Nouméa, New Caledonia	0.2
Suva, Fiji	0.2

*Based on 2012 estimates.

Ethnic, linguistic, and physiographic criteria are among the foundations for this regionalization of the Pacific Realm, but we should not lose sight of the dimensions. Not only is the land area small, but as we also noted earlier the current population of this entire realm (including Indonesia's Papua Province) was only slightly more than 13 million—10.1 million without Papua—about the same as one very large city. Even fewer people live in this realm than in another vast area of far-flung settlements—the oases of North Africa's Sahara.

■■ MELANESIA

The large island of New Guinea lies at the western end of a Pacific region that extends eastward to Fiji and includes the Solomon Islands, Vanuatu, and New Caledonia (Fig. 12-3). The human mosaic here is complex, both ethnically and culturally. Most of the 10.7 million people of New Guinea (including the Indonesian part—the province of Papua—and the independent state of Papua New Guinea) are Papuans, and a large minority is Melanesian. Altogether there are as many as 700 communities speaking different languages; the Papuans are most numerous in the densely forested highland interior and in the lowland south, whereas the Melanesians inhabit the north and east. The region as a whole has more than 8 million inhabitants, making this the most populous Pacific region by far.

With 7.1 million people today, **Papua New Guinea (PNG)** became a sovereign state in 1975 after nearly a century of British and Australian administration. Almost all of PNG's limited development is taking place along its coasts, whereas most of the interior remains hardly touched by the changes that transformed neighboring Australia. Perhaps four-fifths of the population is part of the self-sufficient subsistence economy, growing root crops, raising pigs, hunting wildlife, and gathering forest products. Old traditions of the kind lost in Australia continue on here, protected by remoteness and the rugged terrain.

Welding this disparate population into a nation is a task hardly begun, and PNG faces numerous obstacles in addition to its cultural complexity. Not only are hundreds of languages in use, but more than 40 percent of the population is illiterate. English, the official language, is used by the educated minority but is of little use beyond the coastal zone and its towns. The capital, Port Moresby, has just over 300,000 inhabitants, reflecting the very low level of urbanization (13 percent) in this developing economy.

From the Field Notes . . .

"Arriving in the capital of New Caledonia, Nouméa, in 1996, was an experience reminiscent of French Africa 40 years earlier. The French tricolor was much in evidence, as were uniformed French soldiers. European French [Caldoche] residents occupied hillside villas overlooking palm-lined beaches, giving the place a Mediterranean cultural landscape. And New Caledonia, like Africa, is a source of valuable minerals. It is one of the world's largest nickel producers, and from this vantage point you could see the huge treatment plants, complete with concentrate ready to be shipped (left, under conveyor).

What you cannot see here is how southern New Caledonia has been ravaged by the mining operations, which have denuded whole mountainsides. Working in the mines and in this facility are the local Kanaks, Melanesians who make up about 43 percent of the population of about 240,000. Violent clashes between Kanaks and Caldoche French have obstructed government efforts to change New Caledonia's political status in such a way as to accommodate pressures for independence as well as continued French administration."

www.conceptcaching.com

© H. J. de Blij

FIGURE 12-3

© H. J. de Blij, P. O. Muller, and John Wiley & Sons, Inc.

Yet Papua New Guinea is not without economic opportunities. Oil was discovered in the 1980s, and by 2000 crude oil was PNG's largest export by value. Gold now ranks second, followed by copper, silver, timber, and several agricultural products that include coffee and cocoa, reflecting the country's diversity of environments and resources. Pacific Rim developments have affected even PNG: most exports go to its nearest neighbor, Australia, but Japan ranks close behind.

Turning eastward, it is a measure of Melanesia's cultural fragmentation that as many as 120 languages are spoken in the approximately 1000 islands that make up the **Solomon Islands** (about 80 of these islands support almost all the people, who number about 525,000). Inter-island, historic animosities among the islanders were worsened by the events of World War II, when U.S. forces moved thousands of Malaitans to Guadalcanal. This started a postwar cycle of violence that led Australia to intervene in 2003.

New Caledonia, still under French rule, is in a very different situation. Only around 43 percent of the population of about 300,000 are Melanesian; close behind are

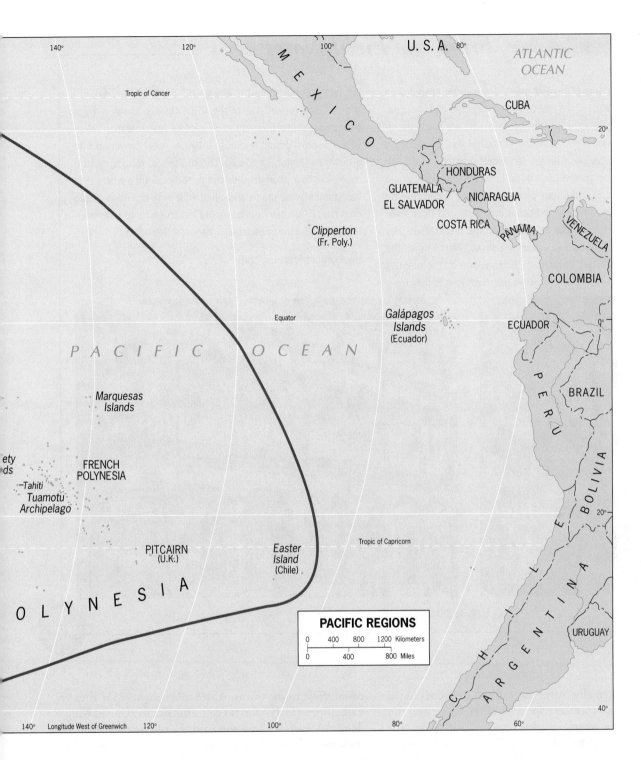

the 37 percent of European ancestry (known as the Caldoche), many of them descended from the inhabitants of the penal colony France established here during the nineteenth century. Nickel mines, based on reserves that rank among the world's largest, dominate New Caledonia's export economy. The mining industry attracted additional French settlers, and social problems soon arose. Most of the French population lives in or near the capital city of Nouméa, steeped in French cultural landscapes, in the southeastern quadrant of the island. Melanesian demands for an end to colonial rule have led to violence, and the two

communities remain engaged in a long-ongoing process of coming to terms.

On its eastern margins, Melanesia includes one of the Pacific Realm's most interesting countries, **Fiji**. On two larger and over 100 smaller islands live more than 900,000 Fijians, of whom 51 percent are Melanesians and 44 percent South Asians, the latter brought to Fiji from India during the British colonial occupation to work on the sugar plantations. When Fiji achieved independence in 1970, the indigenous Fijians owned most of the land and held political control, while the Indians were concentrated

From the Field Notes . . .

"Returning in Suva (Fiji) more than 20 years after I had done a study of its CBD, I noted that comparatively little had changed, although somehow the city seemed more orderly and prosperous than it was in 1978. At the Central Market I watched a Fijian woman bargain with the seller over a batch of taro, the staple starch-provider in local diets. I asked her how she would serve it. 'Well, it's like your potato,' she said. 'I can make a porridge, I can cut it up and put it in a stew, and I can even fry pieces of it and make them look like the French fries they give you at the McDonald's down the street.' Next I asked the seller where his taro came from. 'It grows over all the islands,' he said. 'Sometimes, when there's too much rain, it may rot—but this year the harvest is very good.' . . . Next I walked into the crowded Indian part of the CBD. On a side street I got a reminder of the geographic concept of agglomeration. Here was a colonial-period building, a former hotel converted into a business center. I counted 15 enterprises, ranging from a shoe store to a photographer and including a shop where rubber stamps were made and another selling diverse tobacco products. Of course (this being the Indian sector of downtown Suva) tailors outnumbered all other establishments."

www.conceptcaching.com

© H. J. de Blij

in the towns (chiefly Suva, the capital) and dominated commercial life. It was a recipe for trouble, and it was not long in coming when, in a later election, the politically active Indians outvoted the Fijians for seats in the parliament. A coup by the Fijian military in 1987 was followed by a revision of the constitution, which awarded a majority of seats to ethnic Fijians. Before long, however, the Fijian majority splintered and instability returned. By 2006, the country witnessed its fourth military takeover in 20 years.

This political malfunctioning has cast a long shadow across the country. In 2009, Fiji was suspended from the **British Commonwealth [8]** for its refusal to call democratic elections. The sugar industry was damaged by the nonrenewal of Indian-held leases by Fijian landowners, which also resulted in a substantial movement of Indians from the countryside to the towns, where unemployment is already high. Interestingly, whereas countries like the United Kingdom and Australia have shunned the regime they consider undemocratic, China has stepped in to fill the void. Chinese foreign aid to Fiji has increased sevenfold since the last coup, and Chinese investments in infrastructural projects have increased as well. This is but another sign that China's influence is now being projected on a global scale, and not always in a manner that existing powers find reassuring.

Melanesia, the most populous region in the Pacific Realm, is bedeviled by centrifugal forces of many kinds. No two countries present the same form of multiculturalism; each has its own challenges to confront, and some of these spill over into neighboring (or more distant foreign) islands.

▪▪ MICRONESIA

North of Melanesia and to the east of the Philippines lie the islands that constitute the region known as Micronesia (Fig. 12-3). The name (*micro* means small) refers to the size of the islands: the 2000-plus islands of Micronesia are not only tiny (many of them no larger than one square kilometer [half a square mile]), but they are also much lower-lying, on average, than those of Melanesia. Some are volcanic islands (**high islands [9]**, as the people call them), but they are outnumbered by islands composed of coral, the **low islands [10]** that barely protrude above sea level (photo below). Guam, with 550 square kilometers (210 sq mi), is Micronesia's largest island, and no island elevation anywhere in Micronesia reaches 1000 meters (3300 ft).

Take, for example, the country of **Kiribati**, consisting of three main island groups that lie across Micronesia and Polynesia (Fig. 12-3). The average elevation of Kiribati's 33 islands is 2 meters (6.5 ft) and, like all other low islands in this realm, they are highly vulnerable to rising sea levels at a time of global warming.

The high-island/low-island dichotomy is useful not only in Micronesia, but also throughout the realm. Both the physiographies of these islands and the economies they support differ in crucial ways. High islands wrest substantial orographic rainfall from the moist maritime air and tend to be well watered with good volcanic soils. As a result, agricultural products show some diversity, and life is reasonably secure. Populations tend to be larger on these high islands than on the low islands, where drought is the rule and fishing and the coconut palm are the mainstays of life. Small communities cluster on the low islands, and

A classic example of a *low island* in the archipelago that forms Kiribati, which lies near where the equator and International Date Line intersect—as well as the realm's three regions.

© H. J. de Blij

over time many of these have died out. The major migrations, which sent fleets to populate islands from Hawai'i to New Zealand, tended to originate in the high islands.

Until the mid-1980s, Micronesia was largely a United States Trust Territory (the last of the post–World War II trusteeships supervised by the United Nations). But, as Figure 12-3 shows, Micronesia is now divided into countries bearing the names of independent states. Today the **Marshall Islands**, where the United States tested nuclear weapons (giving prominence to the name *Bikini*), is a republic in "free association" with the United States, having the same status as Palau and the Federated States of Micronesia. The **Northern Mariana Islands** are a commonwealth "in political union" with the United States, which in effect provides billions of dollars in assistance to these countries, in return for which they commit themselves to avoid foreign policy actions that are contrary to American interests.

The Republic of **Palau** is another fascinating example of how some of these small Pacific nations make a living. This group of islands some 800 kilometers (500 mi) east of the Philippines became officially independent in 1994 and has a total population of about 25,000. Palau has no military but entered an agreement with the United States to provide security in return for a 50-year lease for a base. It is highly dependent on American foreign assistance. Palau also gets funds and investments from Taiwan, a reward for Palau's official recognition of Taiwan's sovereign status as the Republic of China (and for withholding recognition of the People's Republic of China).

Also part of Micronesia are the U.S. territory of **Guam**, where independence is not in sight and where U.S. military installations and tourism provide most of its income, and the remarkable Republic of **Nauru**. With a population of less than 20,000 and only 20 square kilometers (8 sq mi) of land, Nauru got rich by selling its phosphate deposits to Australia and New Zealand, where they are used as fertilizer. Per capita incomes had risen above U.S. $12,000, making Nauru one of the realm's high-income societies. But the main phosphate deposits have run out; now uncertainty prevails as mining of a secondary layer proceeds.

In this region of tiny islands, most people subsist on farming or fishing, and virtually all the countries need infusions of foreign aid to survive. The natural economic complementarity between the high-island farming cultures and the low-island fishing communities all too often is negated by distance, spatial as well as cultural. Life here may seem idyllic to the casual visitor, but for the Micronesians it is usually a daily challenge.

▪▪ POLYNESIA

To the east of Micronesia and Melanesia lies the heart of the Pacific, enclosed by a great triangle stretching from the Hawaiian Islands to Chile's Easter Island to New Zealand. This is Polynesia (Fig. 12-3), a region of numerous islands (*poly* means many), ranging from volcanic mountains rising above the Pacific's waters (Mauna Kea on Hawai'i reaches over 4200 meters [nearly 13,800 ft]), clothed by luxuriant tropical forests and drenched by well over 250 centimeters (100 in) of rainfall each year, to low coral atolls where a few palm trees form the only vegetation and where drought is a persistent problem. The Polynesians have somewhat lighter-colored skin and wavier hair than do the other peoples of the Pacific Realm. Anthropologists differentiate between these original Polynesians and a second group, the Neo-Hawaiians, who are a blend of Polynesian, European, and Asian ancestries. In the U.S. State of Hawai'i—actually an archipelago of more than 130 islands—Polynesian culture has been transformed by infusions from both Europe and Asia.

Its vastness and the diversity of its natural environments notwithstanding, Polynesia clearly constitutes a geographic region within the Pacific Realm. Polynesian culture, though spatially fragmented, exhibits a remarkable consistency and uniformity from one island to the next, from one end of this widely dispersed region to the other. This consistency is particularly expressed in vocabularies, technologies, housing, and art forms. The Polynesians are uniquely adapted to their maritime environment, and long before European sailing ships began to arrive in their waters, Polynesian seafarers had learned to navigate their wide expanses of ocean in huge double canoes as long as 45 meters (150 ft). They traveled hundreds of kilometers to favorite fishing zones and engaged in inter-island barter trade, using maps constructed from bamboo sticks and cowrie shells and navigating by the stars. However, modern descriptions of a Pacific Polynesian paradise of emerald seas, lush landscapes, and gentle people distort harsh realities. Polynesian society was forced to accept much loss of life at sea when storms claimed their boats; families were ripped apart by accident as well as migration; hunger and starvation afflicted the inhabitants of smaller islands; and the island communities were often embroiled in violent conflicts and cruel retributions.

The political geography of Polynesia is complex. In 1959, the **Hawaiian Islands** became the fiftieth State to join the United States. The State's population is now 1.5 million, with over 80 percent living on the island of Oahu. There, the superimposition of cultures is symbolized by the panorama of Honolulu's skyscrapers against the famous extinct volcano at nearby Diamond Head. The Kingdom of **Tonga** became an independent country in 1970 after seven decades as a British protectorate; the British-administered Ellice Islands were renamed **Tuvalu** and received independence from Britain in 1978. Other islands continue under French control (including the Marquesas Islands and Tahiti), under New Zealand's administration (Cook Islands), and under British, U.S., and Chilean flags.

In the process of politico-geographical fragmentation, Polynesian culture has suffered severe blows. Land developers, hotel builders, and tourist dollars have set **Tahiti** on a course along which Hawai'i has already traveled far. The Americanization of eastern **Samoa** has created a new society different from the old. Polynesia has lost much of its ancient cultural consistency; today, the region is a patchwork of new and old—the new often bleak and barren, with the old under intensifying pressure.

VOICE FROM THE
Region

©Fatima Sauafea-Le'au
Fatima Sauafea-Le'au,
fisheries biologist,
American Samoa

FA'A SAMOA—THE SAMOAN WAY

"I was born and raised in American Samoa in the small coastal village of Paepaeulupo'o. Legends and stories are passed down from several generations and form the foundation of our cultural and traditional values, known as the *Fa'a Samoa*. One cultural tradition that has been carried on for many years is the management of the reef areas next to village communities. For centuries we Samoans have lived off the ocean reefs that support a rich fish fauna. Management of the reefs depended greatly on how well organized the village was in regard to its different institutions: the matai (village chief) system; the women's group; and the men's group. These are the different systems that make up a village community, and they each utilize the reefs and resources in different ways. But in modern times American Samoa has gone through many changes that have altered our social fabric and that have resulted in poor management practices in many places. As a fisheries biologist, I think modern scientific approaches are very useful and effective in demonstrating key processes to understand the status and health of our resources and environment, and how human activities damage them. At the same time, as a Samoan I truly believe that the incorporation of traditional knowledge with modern science will enhance effective management of our resources. Traditional ways are important in sustaining the reefs in the long run, and they also provide our people with a sense of ownership. They possess knowledge about local resources and their problems, as well as information based on their daily observations of changes occurring within their habitat. We can be modern and at the same time live in harmony with nature. As a scientist and as a Samoan, my hope is to see more village communities on my island engaged in managing our reefs and resources, and keeping *Fa'a Samoa* alive."

The countries and cultures of the Pacific Realm lie in an enormous ocean basin along whose rim a great drama of economic and political transformation will play itself out during the twenty-first century. Already, the realm's own former margins—in Hawai'i in the north and in New Zealand in the south—have been so recast by foreign intervention that little remains of the kingdoms and cultures that once prevailed. Now the Pacific Basin faces changes far greater even than those brought here by European colonizers. Once upon a time the shores and waters of the Mediterranean Sea formed an arena of regional transformation that changed the world. Then it was the Atlantic, avenue of the Industrial Revolution and stage of fateful war. Now it seems to be the turn of the Pacific as the world's largest country and next superpower (China) faces the richest and most powerful (the United States). Giants will jostle for advantage in this arena; how will the microstates of the Pacific Realm fare?

● POLAR FUTURES

PARTITIONING THE ANTARCTIC

South of the Pacific Realm lies Antarctica and its encircling Southern Ocean. The combined area of these two gargantuan geographic expanses constitutes 40 percent of the entire planet—two-fifths of the Earth's surface containing a mere one one-thousandth of the world's population.

Is Antarctica a geographic realm? In physiographic terms, yes—but not on the basis of the criteria we use in this book. Antarctica is a continent, practically twice the size of Australia, but virtually all of it is covered by a dome-shaped ice sheet nearly 3.2 kilometers (2 mi) thick near its center. The continent frequently is referred to as "the white desert" because, despite all of its ice and snow, annual precipitation is extremely low—averaging less than 15 centimeters (6 in) per year. Temperatures are frigid, with winds so strong that Antarctica also is called the home of the blizzard. For all its size, no functional regions have developed here, nor have any towns or transport networks except the supply lines of research stations. Antarctica still is a frontier, even a scientific frontier still slowly giving up its secrets. Underneath all that ice lie some 70 lakes of which Lake Vostok is the largest at over 14,000 square kilometers (5400 sq mi). It may be as much as 600 meters (2000 ft) deep. No one has yet seen a sample of its water.

Like virtually all frontiers, Antarctica has always attracted pioneers and explorers. Whale and seal hunters destroyed huge populations of Southern Ocean fauna during the eighteenth and nineteenth centuries, and explorers planted the flags of their countries on Antarctic shores. By the first decade of the twentieth century, the quest for the South Pole had become an international obsession; Roald Amundsen, the Norwegian, reached it first in 1911. All

From the Field Notes . . .

"The Antarctic Peninsula is geologically an extension of South America's Andes Mountains, and this vantage point in the Gerlach Strait leaves you in no doubt: the mountains rise straight out of the frigid waters of the Southern Ocean. We have been passing large, flat-topped icebergs, but these do not come from the peninsula's shores; rather, they form on the leading edges of ice shelves or on the margins of the mainland where continental glaciers slide into the sea. Here along the peninsula, the high-relief topography tends to produce jagged, irregular icebergs. The mountain range that forms the Antarctic Peninsula continues across Antarctica under thousands of feet of ice and is known as the Transantarctic Mountains. Looking at this place you are reminded that beneath all this ice lies an entire continent, still little-known, with fossils, minerals, even lakes yet to be discovered and studied."

www.conceptcaching.com

© H. J. de Blij

this soon led to national claims in Antarctica during the interwar period (1918–1939).

The geographic effect was the partitioning of Antarctica into pie-shaped sectors centered on the South Pole (Fig. 12-4). In the least frigid area of the continent, the Antarctic Peninsula, British, Argentinian, and Chilean claims overlapped—and they still do. One sector, Marie Byrd Land (shown in neutral beige on the main map), was never formally claimed by any country.

Why should states be interested in territorial claims in so remote and difficult an area? As we described in Chapter 2A in the section on the Arctic, these remote areas near the poles contain substantial resources and are attractive for possible future use, despite the difficulties of access. Antarctica is no different as both land and sea contain raw materials that may some day become crucial: proteins in the waters, and fuels and minerals beneath the ice. As noted above Antarctica (14.2 million square kilometers [5.5 million sq mi]) is almost twice the size of Australia, and the Southern Ocean is nearly as large as the North and South Atlantic combined. However distant actual future exploitation may be, countries want to keep their stakes here.

But the claimant states (those with territorial claims) recognize the need for cooperation. During the late 1950s, they joined in the International Geophysical Year (IGY) that launched major scientific programs and established a number of permanent research stations throughout the continent. This spirit of cooperation led to the 1961 signing of the **Antarctic Treaty [11]**, which ensures continued scientific collaboration, prohibits military activities, protects the environment, and holds national claims in abeyance. In 1991, when the treaty was extended under the terms of the Wellington Agreement, concerns were raised that it does not do enough to control future resource exploitation.

Notice, in Figure 12-4, that the map shows no maritime claims off the pie-shaped sectors that blanket most of Antarctica. In fact, some of the claimant states did draw maps that extended their claims into the Southern Ocean, but the 1991 treaty extension terminated that initiative and restricted the existing claims to the landmass. It is of course a possibility that those claims could be reinstated should the Wellington Agreement collapse: national self-interest has abrogated many an international treaty in the past. But delimiting maritime claims off Antarctica is particularly challenging for practical reasons. As Figure 12-4 shows, Antarctica in a number of places is flanked by *ice shelves* attached to its coast, permanent slabs of floating ice (such as the Ronne and the Ross). These ice shelves are replenished on the landward side and break off (*calve*) on the seaward side, where huge icebergs float northward into the Southern Ocean to become part of the wide belt of *pack ice* that encircles the continent. From where would

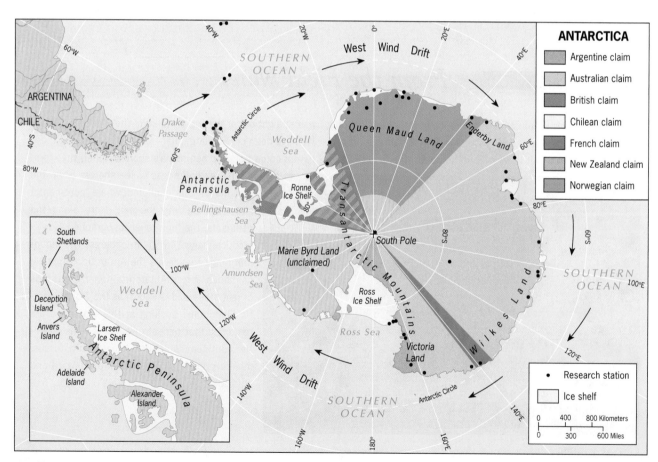

FIGURE 12-4

© H. J. de Blij, P. O. Muller, and John Wiley & Sons, Inc.

any territorial sea or EEZ be measured? From the inner edge of the ice shelf? That would put the territorial "sea" boundary on the ice shelf! From the ever-changing outer edge of the ice shelf? Measuring a territorial sea or EEZ from such an unstable perimeter would not work either. No UNCLOS regulations would be applicable in situations like this, so no matter what the maps showed, such claims are legally unsupportable. That is just as well. The last thing the world needs is a scramble for Antarctica.

In an age of growing national self-interest and increasing raw material consumption, the possibility nonetheless exists that Antarctica and its offshore waters may yet become an arena for international rivalry. Until now, its remoteness and its forbidding environments have saved it from that fate. The entire world benefits from this because evidence is mounting that Antarctica plays a critical role in the global environmental system, so that human modifications may have worldwide (and unpredictable) consequences.

GEOPOLITICS IN THE ARCTIC BASIN

To observe how different the physiographic as well as the political situation is in the Arctic, consider the implications of Figure 12-5. Not only does the North Pole lie on the floor of a relatively small body of water grandiosely labeled the Arctic Ocean, but the entire Arctic is ringed by countries whose EEZs, delimited under UNCLOS rules, would allocate much of the ocean floor (the *subsoil*, to use its legal designation) to those states. Moreover, again under UNCLOS rules, states can expand their rights to the seafloor even farther than the 200-mile EEZ if they are able to prove their continental shelves continue beyond that limit—in fact, up to 350 nautical miles offshore.

As the map shows, the Siberian Continental Shelf is by far the largest in the Arctic; indeed, it is the largest in the world. It extends from Russia's north coast beneath the waters of the Arctic Ocean and under the floating ice cap at the ocean's center. And because several island groups off the Siberian mainland (such as Franz Josef Land, North Land, and the New Siberian Islands) belong to Russia, the Russians can claim virtually all of Eurasia's northern continental shelf under the existing regulations. But even that is not enough: the Russians want to extend their claim all the way to the North Pole itself, where in 2007 they sent a submarine to plant a Russian flag on the seafloor nearly 4000 meters (13,000 ft) below the Pole (see photo at right).

To bolster their claim, the Russians assert that the Lomonosov Ridge, a submerged mountain range, is a "natural extension" of their Siberian Continental Shelf. Few if any neutral observers would agree that this ridge, or several others rising from the deep floor of the Arctic Ocean, is a continental-shelf landform. But this is the stuff of international disputes over maritime boundaries, and the legal wrangling is sure to go on for decades.

Another look at Figure 12-5 reveals the other states with frontage on the Arctic Ocean: Norway, through its Svalbard Islands (including Spitsbergen); Denmark, because

it still has ultimate authority over now-autonomous Greenland; Canada, with tens of thousands of kilometers of Arctic coastline but proportionately less continental shelf; and the United States, with Alaska's limited but important continental shelves to its north and west. The United States stands alone among these candidates by not having ratified the UNCLOS treaties, so that the Americans do not feel bound by the regulations to which everyone else adheres. In any case, you can quickly see that an Antarctic-type, pie-shaped-sector partitioning would not work here; but it is clear as well that the evolving contest among claimant states is fraught with risk. And the stakes are high: estimates of the quantity of oil and natural gas reserves to be found below Arctic waters and ice run as high as 25 percent of the world's remaining total. No matter what method is used to calculate which country gets how much, the biggest winner will be Russia, regardless of what the resolution of its Lomonosov claim will be.

Disputation and Navigation

Even as a dispute over Arctic maritime boundaries looms, another issue is emerging, this one related to the effects of recently sustained global warming. The Greenland Ice Sheet, in many ways a smaller version of Antarctica's, is now in a melting phase, and the zone of seasonally expanding

On August 3, 2007, tens of millions of Russian television viewers witnessed live the scene pictured here: the operator of a Russian mini-submarine planting a metal Russian flag at the North Pole, on the seafloor of the Arctic Ocean beneath the polar ice cap. As Figure 12-5 shows, the North Pole lies far beyond the Russian continental shelf, but in the upcoming international negotiations relating to national claims in the Arctic Basin, the Russians will claim that the Lomonosov "Ridge" actually constitutes an extension of their continental shelf, thereby entitling Russia to draw international boundaries up to, and even beyond, the North Pole. Other countries with actual and pending Arctic claims make less dramatic but equally assertive moves in the run-up to negotiations that have the potential to transform the map of the Arctic region. With new sea routes and newly exploited oil and gas reserves in prospect, diplomacy will be difficult but crucial.
© AP/Wide World Photos

COMPETITION IN THE ARCTIC

- —··— EEZ Boundary
- —·— Treaty Boundaries
- ····· Disputed Boundaries
- ☐ Continental Shelf

MELTING ICE ON ARCTIC WATERS, 1980–2008

SHIPPING ROUTES

- ☐ 1980–2000 average extent
- ☐ 2008 extent
- — North West (Canadian) Passages
- — Northern (Russian) Maritime Passage

FIGURE 12-5

© H. J. de Blij, P. O. Muller, and John Wiley & Sons, Inc.

and contracting Arctic sea ice has also shown significant losses in both surface extent and thickness. Researchers with the Greenland Ice Core Project reported in a 2007 issue of *Science* that during a previous warming, more than 450,000 years ago, the southernmost part of Greenland was covered by boreal forest—suggesting that the ice sheet as well as the sea ice may disappear entirely if current global warming continues.

The ecological consequences would be far-reaching, threatening the habitats of polar bears, whales, walruses, seals, and other species. But the geographic implications would be significant as well. When the sea ice melts in summer and does not recover in winter, once-blocked waterways open up and accessibilities can be transformed (Fig. 12-5, subsidiary maps). The Northwest Passage, that fabled high-latitude route between the Atlantic and Pacific oceans, has been the object of hope and despair for centuries, causing much loss of human life. But in 2007, for the first time in recorded history, it was completely ice-free for a brief period. Consider what that would mean if vessels bound for East Asia from western Europe or the U.S. eastern seaboard could sail year-round from the Labrador Sea (between Greenland and Canada) to the Bering Sea (between Alaska and northeasternmost Russia) via this waterway!

But who owns the waterways vacated by the ice? Canada calls the Northwest Passage a domestic waterway, giving it the right to control all shipping. The United States holds that the Northwest Passage is an international waterway, and American and Russian ships have sailed through it without Canada's permission. The Canadians have affirmed their position by starting work on a deepwater port at Nanisivik on Nunavut's Baffin Island at the eastern entrance to the Northwest Passage, and by building a military base at Resolute Bay to the west; they also announced the construction of eight Arctic patrol vessels to police the route. In the meantime, the United States is opening a new Coast Guard station to patrol the narrow Bering Strait. The lines, if not yet the battle lines, are being drawn in the warming Arctic.

Humanity's growing numbers, escalating demands, environmental impacts, and technological capacities are transforming even the most remote recesses of our resilient planet. No international resolution is more critical than the avoidance of destructive conflict in fragile polar environments.

POINTS TO PONDER

- Governments of low-lying island-states in the Pacific and Indian oceans are demanding international help as they are confronted by the threat of rising sea level.

- A vast raft of garbage, much of it consisting of plastics, floats in the waters of the Pacific Ocean and continues to grow.

- Fiji receives support from Beijing for officially recognizing the People's Republic of China (PRC) but not Taiwan; Palau gets support from the Taiwanese for officially recognizing Taiwan but not the PRC.

- As national claims in Antarctica lie dormant, states with coastlines on the Arctic Ocean—including Russia, Canada, and the United States—are sketching maritime boundaries as well as charting claims in the far north.

using GEOGRAPHY

1. One of the most important achievements of the United Nations has been a series of conferences out of which grew the UN Convention on the Law of the Sea (UNCLOS). This Convention redrew the maritime map of the world, stabilizing what had been a scramble for the oceans, limiting the width of the territorial sea, and introducing the concept of the Exclusive Economic Zone (EEZ). How has the EEZ changed the economic geography of the Pacific geographic realm? To what extent has the EEZ functioned to expand as well as limit states' offshore activities? Considering the map of the Pacific Realm, what countries are favored over others in terms of maritime holdings?

2. Review the polar projections in Figures 12-4 and 12-5. They remind us that all world maps are by definition two-dimensional distortions of the three-dimensional realities on the globe. Describe what the polar projections reveal about the upper Northern and Southern Hemispheres that world maps using other projections cannot.

METRIC (STANDARD INTERNATIONAL [SI]) AND CUSTOMARY UNITS AND THEIR CONVERSIONS

NOTE: Appendices C, D, and E may be found on the book's website at www.wiley.com/college/deblij

Length

Metric Measure

1 kilometer (km)	= 1000 meters (m)
1 meter (m)	= 100 centimeters (cm)
1 centimeter (cm)	= 10 millimeters (mm)

Nonmetric Measure

1 mile (mi)	= 5280 feet (ft)
	= 1760 yards (yd)
1 yard (yd)	= 3 feet (ft)
1 foot (ft)	= 12 inches (in)
1 fathom (fath)	= 6 feet (ft)

Conversions

1 kilometer (km)	= 0.6214 mile (mi)
1 meter (m)	= 3.281 feet (ft)
	= 1.094 yards (yd)
1 centimeter (cm)	= 0.3937 inch (in)
1 millimeter (mm)	= 0.0394 inch (in)
1 mile (mi)	= 1.609 kilometers (km)
1 foot (ft)	= 0.3048 meter (m)
1 inch (in)	= 2.54 centimeters (cm)
	= 25.4 millimeters (mm)

Area

Metric Measure

1 square kilometer (km^2)	= 1,000,000 square meters (m^2)
	= 100 hectares (ha)
1 square meter (m^2)	= 10,000 square centimeters (cm^2)
1 hectare (ha)	= 10,000 square meters (m^2)

Nonmetric Measure

1 square mile (mi^2)	= 640 acres (ac)
1 acre (ac)	= 4840 square yards (yd^2)
1 square foot (ft^2)	= 144 square inches (in^2)

Conversions

1 square kilometer (km^2)	= 0.386 square mile (mi^2)
1 hectare (ha)	= 2.471 acres (ac)
1 square meter (m^2)	= 10.764 square feet (ft^2)
	= 1.196 square yards (yd^2)
1 square centimeter (cm^2)	= 0.155 square inch (in^2)
1 square mile (mi^2)	= 2.59 square kilometers (km^2)
1 acre (ac)	= 0.4047 hectare (ha)
1 square foot (ft^2)	= 0.0929 square meter (m^2)
1 square inch (in^2)	= 6.4516 square centimeters (cm^2)

Temperature

To change from Fahrenheit (F) to Celsius (C)

$$°C = \frac{°F - 32}{1.8}$$

To change from Celsius (C) to Fahrenheit (F)

$$°F = °C \times 1.8 + 32$$

Note that Appendix B is found on pages DT-1 to DT-7 that run backward from the inside cover

GLOSSARY*

Aboriginal land issue The legal campaign in which Australia's **indigenous peoples** have claimed title to traditional land in several parts of that country. The courts have upheld certain claims, fueling Aboriginal activism that has raised broader issues of indigenous rights.

Aboriginal population See **indigenous peoples**.

Absolute location The position or place of a certain item on the surface of the Earth as expressed in degrees, minutes, and seconds of **latitude**, 0° to 90° north or south of the equator, and **longitude**, 0° to 180° east or west of the *prime meridian* passing through Greenwich, England (a suburb of London).

Accessibility The degree of ease with which it is possible to reach a certain location from other locations. *Inaccessibility* is the opposite of this concept.

Acculturation Cultural modification resulting from intercultural borrowing. In **cultural geography**, the term refers to the change that occurs in the **culture** of **indigenous peoples** when contact is made with a society that is technologically superior.

Advantage The most meaningful distinction that can now be made to classify a country's level of economic **development**. Takes into account geographic **location, natural resources**, government, political stability, productive skills, and much more.

AFTA The ASEAN Free Trade Agreement that since 1992, through lowered tariffs and other incentives, has fostered increased trade within Southeast Asia. This is the economic centerpiece of the **Association of Southeast Asian Nations (ASEAN)**, a **supranational** organization whose members include 10 of that realm's 11 states (only East Timor does not participate).

Agglomeration **Process** involving the clustering or concentrating of people or activities.

Agrarian Relating to the use of land in rural communities or to agricultural societies in general.

Agriculture The purposeful tending of crops and livestock in order to produce food and fiber.

Al-Qaeda The terrorist organization that evolved into an expanding global network under the directorship of Usama bin Laden between the mid-1990s and his elimination by the U.S. in 2011. It sought to coordinate the efforts of once loosely allied Muslim revolutionary movements, and unleash a *jihad* aimed at what it perceived to be Islam's enemies in the West. Afghanistan became al-Qaeda's headquarters; its exploits were funded by numerous Islamic sources and ranged from terrorist training in local camps to lethal attacks on American targets, most notably the September 11, 2001 attacks on New York and Washington.

Alluvial Referring to the mud, silt, and sand (collectively *alluvium*) deposited by rivers and streams. *Alluvial plains* adjoin many larger rivers; they consist of the renewable deposits that are laid down during floods, creating fertile and productive soils. Alluvial **deltas** mark the mouths of rivers such as the Nile (Egypt) and the Ganges (Bangladesh).

Altiplano High-elevation plateau, basin, or valley between even higher mountain ranges, especially in the Andes of South America.

Altitudinal zonation Vertical regions defined by physical-environmental zones at various elevations (see Fig. 4A-4), particularly in the highlands of South and Middle America. See *tierra caliente, tierra templada, tierra fría, tierra helada,* and *tierra nevada*.

American Manufacturing Belt North America's near-rectangular Core Region, whose corners are Boston, Milwaukee, St. Louis, and Baltimore. This region dominated the industrial geography of the United States and Canada during the industrial age; still a formidable economic powerhouse that remains the North American realm's geographic heart.

Animistic religion The belief that inanimate objects, such as hills, trees, rocks, rivers, and other elements of the natural landscape, possess souls and can help or hinder human efforts on Earth.

Antarctic Treaty International cooperative agreement on the use of Antarctic territory.

Antecedent boundary A political boundary that existed before the **cultural landscape** emerged and stayed in place while people moved in to occupy the surrounding area.

Anthracite coal Hardest and highest carbon-content coal, and therefore of the highest quality.

Apartheid Literally, *apartness*. The Afrikaans term for South Africa's pre-1994 policies of racial separation, a system that produced highly segregated socio-geographical patterns.

Aquaculture The use of a river segment or an artificial pond for the raising and harvesting of food products, including fish, shellfish, and even seaweed. The Japanese pioneered the practice, which is now spreading globally, and is already the dominant source of seafood as the oceans become fished out.

Aquifer An underground reservoir of water contained within a porous, water-bearing rock layer.

Arable Land fit for cultivation by one farming method or another. See also **physiologic density**.

Archipelago A set of islands grouped closely together, usually elongated into a *chain*.

Area A term that refers to a part of the Earth's surface with less specificity than **region**. For example, *urban area* alludes generally to a place where urban development has occurred, whereas *urban region* requires certain specific criteria on which such a designation is based (e.g., the spatial extent of commuting or the built townscape).

Areal interdependence A term related to **functional specialization**. When one area produces certain goods or has certain raw materials and another area has a different set of raw materials and produces different goods, their needs may be *complementary*; by exchanging raw materials and products, they can satisfy each other's requirements.

Arithmetic density A country's population, expressed as an average per unit area, without regard for its **distribution** or the limits of **arable** land. See also **physiologic density**.

Aryan From the Sanskrit *Arya* (meaning "noble"), a name applied to an ancient people who spoke an **Indo-European language** and who moved into northern India from the northwest.

ASEAN The **Association of Southeast Asian Nations (ASEAN)**, a **supranational** organization whose members include 10 of that realm's 11 states: Brunei, Cambodia, Indonesia, Laos, Malaysia, Myanmar (Burma), the Philippines, Singapore, Thailand, and Vietnam (only minuscule East Timor does not participate).

Asian Tiger See **economic tiger**.

Atmosphere The Earth's envelope of gases that rests on the oceans and land surface and penetrates open spaces within soils. This layer of nitrogen (78 percent), oxygen (21 percent), and traces of other gases is densest at the Earth's surface and thins with altitude.

Atoll A ring-like coral reef surrounding an empty lagoon that probably formed around the rim of a now-completely-eroded volcanic cone standing on the seafloor. They are common in certain tropical areas of the Pacific Ocean where they are classified among that realm's **low islands**.

Austral South.

---

*Words in **boldface type** within an entry are defined elsewhere in this Glossary

Autocratic A government that holds absolute power, often ruled by one person or a small group of persons who control the country by despotic means.

Balkanization The fragmentation of a **region** into smaller, often hostile political units. Named after the historically contentious Balkan Peninsula of southeastern Europe.

Barrio Term meaning "neighborhood" in Spanish. Usually refers to an urban community in a Middle or South American city; also applied to low-income, inner-city concentrations of Hispanics in such western U.S. cities as Los Angeles.

Bauxite Aluminum ore; usually deposited at shallow depths in the wet tropics.

Biodiversity Shorthand for *biological diversity*; the total variety of plant and animal species that exists in a given area.

Biodiversity hotspot A much higher than usual, world-class geographic concentration of natural plant and/or animal species. Tropical rainforest environments have dominated, but their recent ravaging by **deforestation** has had catastrophic results. The example given in the book is the southern (Costa Rica-Panama) segment of the **land bridge** that forms Central America—over recent geologic time a biogeographical highway for evolutionary change.

Biogeography The study of *flora* (plant life) and *fauna* (animal life) in spatial perspective.

Biome One of the broadest justifiable (geographic) subdivisions of the plant and animal world (less than a dozen exist overall). An assemblage and association of plants and animals that forms a regional ecological unit of subcontinental dimensions.

Birth rate The *crude birth rate* is expressed as the annual number of births per 1000 individuals within a given population.

Bituminous coal Softer coal of lesser quality than **anthracite**, but of higher grade than **lignite**. When heated and converted to coking coal or *coke*, it is used to make steel.

Boreal forest See **taiga**.

Break-of-bulk point A location along a transport route where goods must be transferred from one carrier to another. In a port, the cargoes of oceangoing ships are unloaded and put on trains, trucks, or perhaps smaller river boats for inland distribution. An *entrepôt*.

BRIC Acronym for the four biggest emerging national markets in the world today—**B**razil, **R**ussia, **I**ndia, and **C**hina.

British Commonwealth Now renamed *The Commonwealth of Nations* (and known simply as *The Commonwealth*), this nonpolitical inter-governmental organization is constituted by 54 member-states. All but two (Rwanda and Moçambique) formerly belonged to the British Empire, from which the organization originated. Its members cooperate in free association to promote, among other things, democracy, human rights, free trade, and world peace.

Buffer state See **buffer zone**.

Buffer zone A country or set of countries separating ideological or political adversaries. In southern Asia, Afghanistan, Nepal, and Bhutan were parts of a buffer zone set up between British and Russian-Chinese imperial spheres. Thailand was a *buffer state* between British and French colonial domains in mainland Southeast Asia.

Caliente See *tierra caliente*.

Cartogram A specially transformed map not based on traditional representations of **scale** or area.

Cartography The art and science of making maps, including data compilation, layout, and design. Also concerned with the interpretation of mapped patterns.

Caste system The strict **social stratification** and segregation of people—specifically in India's Hindu society—on the basis of ancestry and occupation.

Cay A low-lying small island usually composed of coral and sand. Pronounced *kee* and often spelled "key."

Cell phone revolution Applied specifically to much of Subsaharan Africa and certain other parts of the developing world, the linking of farmers with centers of information for weather and market conditions via cell-phone text messaging—allowing them to negotiate higher prices for their products.

Central business district (CBD) The downtown heart of a central city; marked by high land values, a concentration of business and commerce, and the clustering of the tallest buildings.

Centrality The strength of an urban center in its capacity to attract producers and consumers to its facilities; a city's "reach" into the surrounding region.

Centrifugal forces A term employed to designate forces that tend to divide a country—such as internal religious, linguistic, ethnic, or ideological differences.

Centripetal forces Forces that unite and bind a country together—such as a strong national culture, shared ideological objectives, and a common faith.

Cerrado Regional term referring to the fertile savannas of Brazil's interior Central-West that make it one of the world's most promising agricultural frontiers (mapped in Fig. 5A-5). Soybeans are the leading crop, and other grains and cotton are expanding. Inadequate transport links to the outside world remain a problem.

Choke point A narrowing of an international waterway causing marine traffic congestion, requiring reduced speeds and/or sharp turns, and increasing the risk of collision as well as vulnerability to attack. When the waterway narrows to a distance of less than 38 kilometers (24 mi), this necessitates the drawing of a **median line (maritime) boundary**. Examples are the Hormuz Strait between Oman and Iran at the entrance to the Persian Gulf, and the Strait of Malacca between Malaysia and Indonesia.

City-state An independent political entity consisting of a single city with (and sometimes without) an immediate **hinterland**. The ancient city-states of Greece have their modern equivalent in Southeast Asia's Singapore.

Climate The long-term conditions (over at least 30 years) of aggregate **weather** over a region, summarized by averages and measures of variability; a synthesis of the succession of weather events we have learned to expect at any given location.

Climate change theory An alternative to the **hydraulic civilization theory**; holds that changing **climate** (rather than a monopoly over **irrigation** methods) could have provided certain cities within the ancient Fertile Crescent with advantages over other cities.

Climate region A **formal region** characterized by the uniformity of the **climate** type within it. Figure G-7 maps the global distribution of such regions.

Climatology The geographic study of **climates**. Includes not only the classification of climates and the analysis of their regional distribution, but also broader environmental questions that concern climate change, interrelationships with soil and vegetation, and human–climate interaction.

Coal See **anthracite coal, bituminous coal, fossil fuels**, and **lignite**.

Collectivization The reorganization of a country's **agriculture** under communism that involves the expropriation of private holdings and their incorporation into relatively large-scale units, which are farmed and administered cooperatively by those who live there.

Colonialism Rule by an autonomous power over a subordinate and an alien people and place. Though often established and maintained through political structures, colonialism also creates unequal cultural and economic relations. Because of the magnitude and impact of the European colonial thrust of the last few centuries, the term is generally understood to refer to that particular colonial endeavor. Also see **imperialism**.

Command economy The tightly controlled economic system of the former Soviet Union, whereby central planners in Moscow assigned the production of particular goods to particular places, often guided more by socialist ideology than the principles of **economic geography**.

Commercial agriculture For-profit **agriculture**.

Common market A **free-trade area** that not only has created a **customs union** (a set of common tariffs on all imports from outside the area) but also has eliminated restrictions on the movement of capital, labor, and enterprise among its member countries.

Communal tension Persistent stress among a country's sociocultural groups that can often erupt into communal violence. India is particularly noteworthy in this regard: frequent conflicts among its highly diverse cultural communities are generally based on (politicized) religion, but they can also be **caste**-based.

Compact state A politico-geographical term to describe a state that possesses a roughly circular, oval, or rectangular territory in which the distance from the geometric center to any point on the boundary exhibits little variance. Poland and Cambodia are good examples of this shape category.

Complementarity Exists when two regions, through an exchange of raw materials and/or finished products, can specifically satisfy each other's demands. See also **areal interdependence**.

Confucianism A philosophy of ethics, education, and public service based on the writings of Confucius (*Kongfuzi*); traditionally regarded as one of the cornerstones of Chinese **culture**.

Congo Two countries in Africa have the same short-form name, *Congo*. In this book, we use ***The Congo*** for the larger Democratic Republic of the Congo, and ***Congo*** for the smaller Republic of Congo.

Coniferous forest A forest of cone-bearing, needleleaf evergreen trees with straight trunks and short branches, including spruce, fir, and pine. See also **taiga** and **boreal forest**.

Connectivity The degree of direct linkage between a particular location and other locations within a regional, national, or global transportation **network**.

Contagious diffusion The distance-controlled spreading of an idea, innovation, or some other item through a local population by contact from person to person—analogous to the communication of a contagious illness.

Conterminous United States The 48 **contiguous** or adjacent States that occupy the southern half of the North American realm. Alaska is not contiguous to these States because western Canada lies in between; neither is Hawai'i, separated from the mainland by over 3000 kilometers (2000 mi) of ocean.

Contiguous Adjoining; adjacent.

Continental drift The slow movement of continents controlled by the processes associated with **plate tectonics**.

Continental shelf Beyond the coastlines of many landmasses, the ocean floor declines very gently until the depth of about 660 feet (200 m). Beyond the 660-foot line the sea bottom usually drops off sharply, along the *continental slope*, toward the much deeper mid-oceanic basin. The submerged continental margin is called the continental shelf, and it extends from the shoreline to the upper edge of the continental slope.

Continentality The variation of the continental effect on air temperatures in the interior portions of the world's landmasses. The greater the distance from the moderating influence of an ocean, the greater the extreme in summer and winter temperatures. Continental interiors also tend to be dry when the distance from oceanic moisture sources becomes considerable.

Conurbation General term used to identify a large multimetropolitan complex formed by the coalescence of two or more major **urban areas**. The Atlantic Seaboard **Megalopolis**, extending along the northeastern U.S. coast from southern Maine to Virginia, is a classic example.

Copra The dried-out, fleshy interior of a coconut that is used to produce coconut oil.

Cordillera Mountain chain consisting of sets of parallel ranges, especially the Andes in northwestern South America.

Core See **core area**; **core-periphery relationships**.

Core area In geography, a term with several connotations. *Core* refers to the center, heart, or focus. The core area of a **nation-state** is constituted by the national heartland, the largest population cluster, the most productive region, and the part of the country with the greatest **centrality** and **accessibility**—probably containing the capital city as well.

Core-periphery relationships The contrasting spatial characteristics of, and linkages between, the *have* (core) and *have-not* (periphery) components of a national or regional **system**.

Corridor In general, refers to a spatial entity in which human activity is organized in a linear manner, as along a major transport route or in a valley confined by highlands. More specifically, the politico-geographical term for a land extension that connects an otherwise **landlocked state** to the sea.

Cross-border linkages The ties between two closely-connected localities or regions that face each other across an international boundary. These relationships are often longstanding, and intensify further as **supranationalism** proceeds (especially among the EU countries of western Europe). In North America, neighboring southwestern Ontario and southeastern Michigan are an example of such ties, which propel movements of people and goods across that stretch of the U.S.-Canada border.

Cultural diffusion The **process** of spreading and adopting a cultural element, from its place of origin across a wider area.

Cultural ecology The myriad interactions and relationships between a **culture** and its **natural environment**.

Cultural environment See **cultural ecology**.

Cultural geography The wide-ranging and comprehensive field of geography that studies spatial aspects of human **cultures**.

Cultural landscape The forms and artifacts sequentially placed on the **natural landscape** by the activities of various human occupants. By this progressive imprinting of the human presence, the physical (natural) landscape is modified into the cultural landscape, forming an interacting unity between the two.

Cultural pluralism See **plural(istic) society**.

Cultural revival The regeneration of a long-dormant **culture** through internal renewal and external infusion.

Culture The sum total of the knowledge, attitudes, and habitual behavior patterns shared and transmitted by the members of a society. This is anthropologist Ralph Linton's definition; hundreds of others exist.

Culture hearth Heartland, source area, innovation center; place of origin of a major **culture**.

Culture region A distinct, culturally discrete spatial unit; a **region** within which certain cultural norms prevail.

Customs union A **free-trade area** in which member countries set common tariff rates on imports from outside the area.

Death rate The *crude death rate* is expressed as the annual number of deaths per 1000 individuals within a given population.

Deciduous A deciduous tree loses its leaves at the beginning of winter or the onset of the dry season.

Definition In **political geography**, the written legal description (in a treaty-like document) of a boundary between two countries or territories. See also **delimitation**.

Deforestation The clearing and destruction of forests (especially tropical rainforests) to make way for expanding settlement frontiers and the exploitation of new economic opportunities.

Deglomeration Deconcentration.

Deindustrialization **Process** by which companies relocate manufacturing jobs to other regions or countries with cheaper labor, leaving the newly-deindustrialized region to convert to a service economy while struggling with the accompanying effects of increased unemployment and meeting the retraining needs of its workforce.

Delimitation In **political geography**, the translation of the written terms of a boundary treaty (the **definition**) into an official cartographic representation (map).

Delta **Alluvial** lowland at the mouth of a river, formed when the river deposits its alluvial load on reaching the sea. Often triangular in shape, hence the use of the Greek letter whose symbol is Δ.

Demarcation In **political geography**, the actual placing of a political boundary on the **cultural landscape** by means of barriers, fences, walls, or other markers.

Demographic burden The proportion of a national population that is either too old or too young to be productive and that must be cared for by the productive population.

Demographic transition model Multi-stage **model**, based on western Europe's experience, of changes in population growth exhibited by countries undergoing industrialization. High **birth rates** and **death rates** are followed by plunging death rates, producing a huge net population gain; birth and death rates then converge at a low overall level. See Figure 8A-10.

Demography   The interdisciplinary study of population—especially **birth rates** and **death rates**, growth patterns, longevity, **migration**, and related characteristics.

Dependencia **theory**   Originating in South America during the 1960s, it was a new way of thinking about economic development and underdevelopment that explained the persistent poverty of certain countries in terms of their unequal relations with other (i.e., rich) countries.

Desert   An arid expanse supporting sparse vegetation, receiving less than 25 centimeters (10 in) of precipitation per year. Usually exhibits extremes of heat and cold because the moderating influence of moisture is absent.

Desertification   **Process** of **desert** expansion into neighboring **steppelands** as a result of human degradation of fragile semiarid environments.

Development   The economic, social, and institutional growth of national **states**.

Devolution   The **process** whereby regions within a **state** demand and gain political strength and growing autonomy at the expense of the central government.

Dhows   Wooden boats with characteristic triangular sails, plying the seas between Arabian and East African coasts.

Dialect   Regional or local variation in the use of a major language, such as the distinctive accents of many residents of the U.S. South.

Diffusion   The spatial spreading or dissemination of a **culture** element (such as a technological innovation) or some other phenomenon (e.g., a disease outbreak). For the various channels of outward geographic spread from a source area, see **contagious, expansion, hierarchical,** and **relocation diffusion**.

Distance decay   The various degenerative effects of distance on human spatial structures and interactions.

Diurnal   Daily.

Divided capital   In **political geography**, a country whose central administrative functions are carried on in more than one city is said to have divided capitals. The Netherlands and South Africa are examples.

Domestication   The transformation of a wild animal or wild plant into a domesticated animal or a cultivated crop to gain control over food production. A necessary evolutionary step in the development of humankind: the invention of **agriculture**.

Domino theory   The belief that political destabilization in one **state** can result in the collapse of order in a neighboring state, triggering a chain of events that, in turn, can affect a series of **contiguous** states.

Double cropping   The planting, cultivation, and harvesting of two crops successively within a single year on the same plot of farmland.

Double Delta   South Asia's combined **delta** formed by the Ganges and Brahmaputra rivers. All of Bangladesh lies on this enormous deltaic plain, which also encompasses surrounding parts of eastern India. Well over 200 million people live here, attracted by the fertility of its soils that are constantly replenished by the **alluvium** transported and deposited by these two of Asia's largest river systems. Natural hazards abound here as well, ranging from the flooding caused by excessive **monsoonal** rains to the intermittent storm surges of powerful cyclones (**hurricanes**) that come from the Bay of Bengal to the south.

Dravidian languages   The language family, indigenous to the South Asian realm, that dominates southern India today; as opposed to the Indo-European languages, whose tongues dominate northern India.

Dry canal   An overland rail and/or road **corridor** across an **isthmus** dedicated to performing the transit functions of a canalized waterway. Best adapted to the movement of containerized cargo, there must be a port at each end to handle the necessary **break-of-bulk** unloading and reloading.

Dynasty   A succession of Chinese rulers that came from the same line of male descent, sometimes enduring for centuries. Dynastic rule in China lasted for thousands of years, only coming to an end just a bit more than a century ago in 1911.

Ecology   The study of the many interrelationships between all forms of life and the natural environments in which they have evolved and continue to develop. The study of *ecosystems* focuses on the interactions between specific organisms and their environments. See also **cultural ecology, biome,** and **biodiversity**.

Economic geography   The field of geography that focuses on the diverse ways in which people earn a living and on how the goods and services they produce are expressed and organized spatially.

Economic restructuring   The transformation of China into a market-driven economy in the post-Mao era, beginning in the late 1970s.

Economic tiger   One of the burgeoning beehive countries of the western **Pacific Rim**. Following Japan's route since 1945, these countries have experienced significant modernization, industrialization, and Western-style economic growth since 1980. Three leading economic tigers are South Korea, Taiwan, and Singapore. The term is increasingly used more generally to describe any fast-developing economy.

Economies of scale   The savings that accrue from large-scale production wherein the unit cost of manufacturing decreases as the level of operation enlarges. Supermarkets operate on this principle and are able to charge lower prices than small grocery stores.

Ecosystem   See **ecology**.

Ecumene   The habitable portions of the Earth's surface where permanent human settlements have arisen.

Ejidos   Mexican farmlands redistributed to peasant communities after the Revolution of 1910–1917. The government holds title to the land, but user rights are parceled out to village communities and then to individuals for cultivation.

Elite   A small but influential upper-echelon social class whose power and privilege give it control over a country's political, economic, and cultural life.

El Niño-Southern Oscillation (ENSO)   A periodic, large-scale, abnormal warming of the sea surface in the tropical latitudes of the eastern Pacific Ocean that has global implications, disturbing normal weather patterns in many parts of the world, especially South America.

Elongated state   A **state** whose territory is decidedly long and narrow in that its length is at least six times greater than its average width. Chile and Vietnam are two classic examples.

Emerging market   The world's fastest growing national market economies as measured by economic growth rates, attraction of **foreign direct investment**, and other key indicators. Led by the **BRICs** (Brazil, Russia, India, and China), but this club is now expanding to include many other countries.

Emigrant   A person **migrating** away from a country or area; an out-migrant.

Empirical   Relating to the real world, as opposed to theoretical abstraction.

Enclave   A piece of territory that is surrounded by another political unit of which it is not a part.

Endemism   Referring to a disease in a host population that affects many people in a kind of equilibrium without causing rapid and widespread deaths.

Entrepôt   A place, usually a port city, where goods are imported, stored, and transshipped; a **break-of-bulk point**.

Environmental degradation   The accumulated human abuse of a region's **natural landscape** that, among other things, can involve air and water pollution, threats to plant and animal **ecosystems**, misuse of **natural resources**, and generally upsetting the balance between people and their habitat.

Epidemic   A local or regional outbreak of a disease.

Escarpment   A cliff or very steep slope; often marks the edge of a plateau. The Great Escarpment that lines much of Africa's east coast is a classic example.

Estuary   The widening mouth of a river as it reaches the sea; land subsidence or a rise in sea level has overcome the tendency to form a **delta**.

Ethanol   The leading U.S. biofuel that is essentially alcohol distilled from corn mash. Much of it is produced in the historic Corn Belt centered on Iowa and Illinois. Many risks and problems accompany this energy source, which overall is not an efficient replacement for **fossil fuels**.

Ethnic cleansing The slaughter and/or forced removal of one **ethnic** group from its homes and lands by another, more powerful ethnic group bent on taking that territory.

Ethnicity The combination of a people's **culture** (traditions, customs, language, and religion) and racial ancestry.

European state model A **state** consisting of a legally defined territory inhabited by a population governed from a capital city by a representative government.

European Union (EU) **Supranational** organization constituted by 27 European countries to further their common economic interests. In alphabetical order, these countries are: Austria, Belgium, Bulgaria, Cyprus, the Czech Republic, Denmark, Estonia, Finland, France, Germany, Greece, Hungary, Ireland, Italy, Latvia, Lithuania, Luxembourg, Malta, the Netherlands, Poland, Portugal, Romania, Slovakia, Slovenia, Spain, Sweden, and the United Kingdom.

Exclave A bounded (non-island) piece of territory that is part of a particular **state** but lies separated from it by the territory of another state. Alaska is an exclave of the United States.

Exclusive Economic Zone (EEZ) An oceanic zone extending up to 200 **nautical miles** (370 km) from a shoreline, within which the coastal **state** can control fishing, mineral exploitation, and additional activities by all other countries.

Expansion diffusion The spreading of an innovation or an idea through a fixed population in such a way that the number of those adopting grows continuously larger, resulting in an expanding area of dissemination.

Extraterritoriality The politico-geographical concept suggesting that the property of one **state** lying within the boundaries of another actually forms an extension of the first state.

Failed state A country whose institutions have collapsed and in which anarchy prevails. Somalia is a current example.

Fatwa Literally, a legal opinion or proclamation issued by an Islamic cleric, based on the holy texts of Islam, long applicable only in the *Umma*, the realm ruled by the laws of Islam. In 1989, the Iranian Ayatollah Khomeini extended the reach of the *fatwa* by condemning to death author Salman Rushdie, a British citizen living in the United Kingdom.

Favela Shantytown on the outskirts or even well within an urban area in Brazil.

Fazenda Coffee plantation in Brazil.

Federal state A political framework wherein a central government represents the various subnational entities within a **nation-state** where they have common interests—defense, foreign affairs, and the like—yet allows these various entities to retain their own identities and to have their own laws, policies, and customs in certain spheres.

Federation See **federal state**.

Fertile Crescent Crescent-shaped zone of productive lands extending from near the southeastern Mediterranean coast through Lebanon and Syria to the **alluvial** lowlands of Mesopotamia (in Iraq). Once more fertile than today, this is one of the world's great source areas of **agricultural** and other innovations.

First Nations Name given Canada's **indigenous peoples** of American descent, whose U.S. counterparts are called Native Americans.

Fjord Narrow, steep-sided, elongated, and inundated coastal valley deepened by glacier ice that has since melted away, leaving the sea to penetrate.

Floating population China's huge mass of mobile workers who respond to shifting employment needs within the country. Most are temporary urban dwellers with restricted residency rights, whose movements are controlled by the *hukou* system.

Floodplain Low-lying area adjacent to a mature river, often covered by **alluvial** deposits and subject to the river's floods.

Forced migration Human **migration** flows in which the movers have no choice but to relocate.

Foreign direct investment (FDI) A key indicator of the success of an **emerging market** economy, whose growth is accelerated by the infusion of foreign funds to supplement domestic sources of investment capital.

Formal region A type of **region** marked by a certain degree of homogeneity in one or more phenomena; also called *uniform region* or *homogeneous region*.

Formal sector The total activities of a country's legal economy that is taxed and monitored by the government, whose **gross domestic product (GDP)** and **gross national product (GNP)** are based on it; as opposed to an **informal economy**.

Forward capital Capital city positioned in actually or potentially contested territory, usually near an international border; it confirms the **state's** determination to maintain its presence in the region in contention.

Fossil fuels The energy resources of **coal**, natural gas, and petroleum (oil), so named collectively because they were formed by the geologic compression and transformation of tiny plant and animal organisms.

Four Motors of Europe *Rhône-Alpes* (France), *Baden-Württemberg* (Germany), *Catalonia* (Spain), and *Lombardy* (Italy). Each is a high-technology-driven region marked by exceptional industrial vitality and economic success not only within Europe but on the global scene as well.

Fragmented modernization A checkerboard-like spatial pattern of **modernization** in an **emerging-market** economy wherein a few localized regions of a country experience most of the development while the rest are largely unaffected.

Fragmented state A **state** whose territory consists of several separated parts, not a **contiguous** whole. The individual parts may be isolated from each other by the land area of other states or by international waters. The United States and Indonesia are examples.

Francophone French-speaking. Quebec constitutes the heart of Francophone Canada.

Free-trade area A form of economic integration, usually consisting of two or more **states**, in which members agree to remove tariffs on trade among themselves. Frequently accompanied by a **customs union** that establishes common tariffs on imports from outside the trade area, and sometimes by a **common market** that also removes internal restrictions on the movement of capital, labor, and enterprise.

Free Trade Area of the Americas (FTAA) The ultimate goal of **supranational** economic integration in North, Middle, and South America: the creation of a single-market trading bloc that would involve every country in the Western Hemisphere between the Arctic shore of Canada and Cape Horn at the southern tip of South America.

Fría See **tierra fría**.

Frontier Zone of advance penetration, usually of contention; an area not yet fully integrated into a national **state**.

Functional region A **region** marked less by its sameness than by its dynamic internal structure; because it usually focuses on a central **node**, also called *nodal region* or *focal region*.

Functional specialization The production of particular goods or services as a dominant activity in a particular location. See also **local functional specialization**.

Fundamentalism See **revivalism (religious)**.

Gentrification The upgrading of an older residential area through private reinvestment, usually in the downtown area of a central city. Frequently, this involves the displacement of established lower-income residents, who cannot afford the heightened costs of living, and conflicts are not uncommon as such neighborhood change takes place.

Geographic change Evolution of **spatial** patterns over time.

Geographic realm The basic spatial unit in our world regionalization scheme. Each realm is defined in terms of a synthesis of its total human geography—a composite of its leading cultural, economic, historical, political, and appropriate environmental features.

Geography of development The subfield of economic geography concerned with spatial aspects and regional expressions of **development**.

Geometric boundaries Political boundaries **defined** and **delimited** (and occasionally **demarcated**) as straight lines or arcs.

Geomorphology The geographic study of the configuration of the Earth's solid surface—the world's landscapes and their constituent landforms.

Ghetto An intraurban region marked by a particular **ethnic** character. Often an inner-city poverty zone, such as the black ghetto in U.S. central cities. Ghetto residents are involuntarily segregated from other income and racial groups.

Glaciation See **Pleistocene Epoch**.

Globalization The gradual reduction of regional differences at the world **scale**, resulting from increasing international cultural, economic, and political exchanges.

Green Revolution The successful recent development of higher-yield, fast-growing varieties of rice and other cereals in certain developing countries.

Gross domestic product (GDP) The total value of all goods and services produced in a country by that state's economy during a given year.

Gross national product (GNP) The total value of all goods and services produced in a country by that state's economy during a given year, plus all citizens' income from foreign investment and other external sources.

Growth pole An urban center with a number of attributes that, if augmented by investment support, will stimulate regional economic **development** in its **hinterland**.

Growth triangle An increasingly popular economic **development** concept along the western **Pacific Rim**, especially in Southeast Asia. It involves the linking of production in growth centers of three countries to achieve benefits for all. For example, Singapore would supply capital and technical know-how to lead high-technology manufacturing projects based on raw materials and inexpensive labor supplied by adjacent areas of Malaysia (Johor State) and Indonesia (Riau Islands).

Hacienda Literally, a large estate in a Spanish-speaking country. Sometimes equated with the **plantation**, but there are important differences between these two types of agricultural enterprise.

Hanification Imparting a cultural imprint by the ethnic Chinese (the "people of Han"). Within China often refers to the steadily increasing migration of Han Chinese into the country's **periphery**, especially Xinjiang and Xizang (Tibet). **Overseas Chinese** imprints, more generally referred to as **Sinicization**, have been significant as well, most importantly in the Southeast Asian realm.

Heartland theory The hypothesis, proposed by British geographer Halford Mackinder during the early twentieth century, that any political power based in the heart of Eurasia could gain sufficient strength to eventually dominate the world. Furthermore, since Eastern Europe controlled access to the Eurasian interior, its ruler would command the vast "heartland" to the east.

Hegemony The political dominance of a country (or even a region) by another country. The former Soviet Union's postwar grip on Eastern Europe, which lasted from 1945 to 1990, was a classic example.

Helada See *tierra helada*.

Hierarchical diffusion A form of **diffusion** in which an idea or innovation spreads by trickling down from larger to smaller adoption units. An urban **hierarchy** is usually involved, encouraging the leapfrogging of innovations over wide areas, with geographic distance a less important influence.

Hierarchy An order or gradation of phenomena, with each level or rank subordinate to the one above it and superior to the one below. The levels in a national urban hierarchy—or **urban** system—are constituted by hamlets, villages, towns, cities, and (frequently) the **primate city**.

High island Volcanic islands of the Pacific Realm that are high enough in elevation to wrest substantial moisture from the tropical ocean air (see **orographic precipitation**). They tend to be well watered, their volcanic soils enable productive agriculture, and they support larger populations than **low islands**—which possess none of these advantages and must rely on fishing and the coconut palm for survival.

High seas Areas of the oceans away from land, beyond national jurisdiction, open and free for all to use.

High value-added goods Products of improved net worth.

Highveld A term used in South Africa to identify the high, grass-covered plateau that dominates much of the country. The lowest-lying areas (mainly along the narrow coastlands) in South Africa are called *lowveld*; areas that lie at intermediate elevations form the *middleveld*.

Hindutva "Hinduness" as expressed through Hindu nationalism, Hindu heritage, and/or Hindu patriotism. The cornerstone of a fundamentalist movement that has been gaining strength since the late twentieth century that seeks to remake India as a society dominated by Hindu principles prevail. It has been the guiding agenda of the Bharatiya Janata Party (BJP), which has emerged a powerful force in national politics and in big States like Maharashtra, Gujarat, and Madhya Pradesh.

Hinterland Literally, "country behind," a term that applies to a surrounding area served by an urban center. That center is the focus of goods and services produced for its hinterland and is its dominant urban influence as well. In the case of a port city, the hinterland also includes the inland area whose trade flows through that port.

Historical inertia A term from manufacturing geography that refers to the need to continue using the factories, machinery, and equipment of heavy industries for their full, multiple-decade lifetimes to cover major initial investments—even though these facilities may be increasingly obsolete.

Holocene The current *interglacial* epoch (the warm period of glacial contraction between the glacial expansions of an **ice age**); extends from 10,000 years ago to the present. Also known as the *Recent Epoch*.

Human evolution Long-term biological maturation of the human species. Geographically, all evidence points toward East Africa as the source of humankind. Our species, *Homo sapiens*, emigrated from this hearth to eventually populate the rest of the **ecumene**.

Hukou A longstanding Chinese system whereby all inhabitants must obtain and carry with them residency permits that indicate where an individual is from and where they may exercise particular rights such as education, health care, housing, and the like. Under communism this practice has been used by central planners to rigidly control the internal movement of population.

Hurricane A tightly-wound, tropical cyclonic storm capable of inflicting great wind and water damage in low-lying coastal zones. Originates at sea in the hot, moist atmosphere of the lower latitudes, and can reach Category-5 wind speeds in excess of 249 kph (155 mph). Minimal hurricane wind speed is 119 kph (74 mph); but even below that intensity, tropical storms (63-119 kph [39-74 mph]) can wreak significant damage. The name *hurricane* is confined to North and Middle America; in the western North Pacific Ocean, such storms are called *typhoons*; in the Indian Ocean Basin, they are known as (*tropical*) *cyclones*.

Hurricane Alley The most frequent pathway followed by tropical storms and **hurricanes** over the past 150 years in their generally westward movement across the Caribbean Basin. Historically, hurricane tracks have bundled most tightly in the center of this route, most often affecting the Lesser Antilles between Antigua and the Virgin Islands, Puerto Rico, Hispaniola (Haiti/Dominican Republic), Jamaica, Cuba, southernmost Florida, Mexico's Yucatán, and the Gulf of Mexico.

Hydraulic civilization theory The theory that cities which managed to control **irrigated** farming over large **hinterlands** held political power over other cities. Particularly applies to early Asian civilizations based in such river valleys as the Chang (Yangzi), the Indus, and those of Mesopotamia.

Hydrologic cycle The **system** of exchange involving water in its various forms as it continually circulates between the **atmosphere**, the oceans, and above and below the land surface.

Ice age A stretch of geologic time during which the Earth's average atmospheric temperature is lowered; causes the equatorward expansion of continental ice sheets in the higher latitudes and the growth of mountain glaciers in and around the highlands of the lower latitudes.

Immigrant A person **migrating** into a particular country or area; an in-migrant.

Imperialism The drive toward the creation and expansion of a **colonial** empire and, once established, its perpetuation.

Import-substitution industries The industries local entrepreneurs establish to serve populations of remote areas when transport costs from distant sources make these goods too expensive to import.

Inaccessibility See **accessibility**.

Indentured workers Contract laborers who sell their services for a stipulated period of time.

Indigenous peoples Native or *aboriginal* peoples; often used to designate the inhabitants of areas that were conquered and subsequently colonized by the **imperial** powers of Europe.

Indo-European languages The major world language family that dominates the European **geographic realm** (Fig. 1A-7). This language family is also the most widely dispersed globally (Fig. G-9), and about half of humankind speaks one of its languages.

Industrial Revolution The term applied to the social and economic changes in agriculture, commerce, and especially manufacturing and urbanization that resulted from technological innovations and greater specialization in late-eighteenth-century Europe.

Informal sector Dominated by unlicensed sellers of homemade goods and services, the primitive form of capitalism found in many developing countries that takes place beyond the control of government. The complement to a country's **formal sector**.

Infrastructure The foundations of a society: urban centers, transport networks, communications, energy distribution systems, farms, factories, mines, and such facilities as schools, hospitals, postal services, and police and armed forces.

Insular Having the qualities and properties of an island. Real islands are not alone in possessing such properties of **isolation**: an **oasis** in the middle of a **desert** also has qualities of insularity.

Insurgent state Territorial embodiment of a successful guerrilla movement. The establishment by antigovernment insurgents of a territorial base in which they exercise full control; thus a **state** within a state.

Intercropping The planting of several types of crops in the same field; commonly used by **shifting cultivators**.

Interglacial See **Pleistocene Epoch**.

Intermontane Literally, between the mountains. Such a location can bestow certain qualities of natural protection or **isolation** to a community.

Internal migration **Migration** flow within a country, such as ongoing westward and southward movements toward the **Sunbelt** in the United States.

International migration **Migration** flow involving movement across an international boundary.

Intervening opportunity In trade or **migration** flows, the presence of a nearer opportunity that greatly diminishes the attractiveness of sites farther away.

Inuit **Indigenous** peoples of North America's Arctic zone, formerly known as Eskimos.

Irredentism A policy of cultural extension and potential political expansion by a **state** aimed at a community of its nationals living in a neighboring state.

Irrigation The artificial watering of croplands.

Islamic Front The southern border of the African Transition Zone that marks the religious **frontier** of the **Muslim** faith in its southward penetration of Subsaharan Africa (see Fig. 6B-9).

Islamization Introduction and establishment of the **Muslim** religion. A **process** still under way, most notably along the **Islamic Front**, that marks the southern border of the African Transition Zone.

Isohyet A line connecting points of equal rainfall total.

Isolation The condition of being geographically cut off or far removed from mainstreams of thought and action. It also denotes a lack of receptivity to outside influences, caused at least partially by poor **accessibility**.

Isoline A line connecting points of equal value; see **isohyet**, **isotherm**.

Isotherm A line connecting points of equal temperature.

Isthmus A **land bridge**; a comparatively narrow link between larger bodies of land. Central America forms such a link between Mexico and South America.

Jihad A doctrine within Islam. Commonly translated as *holy war*, it entails a personal or collective struggle on the part of **Muslims** to live up to the religious standards prescribed by the *Quran* (Koran).

Juxtaposition Contrasting places in close proximity to one another.

Karst The distinctive natural landscape associated with the chemical erosion of soluble limestone rock.

Land alienation One society or culture group taking land from another. In Subsaharan Africa, for example, European **colonialists** took land from **indigenous** Africans and put it to new uses.

Land bridge A narrow **isthmian** link between two large landmasses. They are temporary features—at least when measured in geologic time—subject to appearance and disappearance as the land or sea level rises and falls.

Land Hemisphere The half of the globe containing the greatest amount of land surface, centered on western Europe (Fig. 1A-4). In **geomorphology**, this can also refer to the position of the African continent, which lies central to the world's landmasses.

Land reform The spatial reorganization of **agriculture** through the allocation of farmland (often expropriated from landlords) to **peasants** and tenants who never owned land.

Land tenure The way people own, occupy, and use land.

Landlocked An interior **state** surrounded by land. Without coasts, such a country is disadvantaged in terms of **accessibility** to international trade routes, and in the scramble for possession of areas of the **continental shelf** and control of the **exclusive economic zone** beyond.

Language family Group of languages with a shared but usually distant origin.

"Latin" American city model The Griffin-Ford model of intraurban spatial structure in the Middle American and South American realms, diagrammed in Figure 5A-8.

Latitude Lines of latitude are **parallels** that are aligned east-west across the globe, from 0° latitude at the equator to 90° North and South latitude at the poles.

Leached soil Infertile, reddish-appearing, tropical soil whose surface consists of oxides of iron and aluminum; all other soil nutrients have been dissolved and transported downward into the subsoil by percolating water associated with the heavy rainfall of moist, low-latitude climates.

League of Nations The international organization that emerged after the First World War (1914–1918) whose purpose was to maintain international peace and promote cooperation in solving international economic, social, and humanitarian problems. It consisted of as many as 58 member-countries by the mid-1930s, but it failed to prevent World War II (1939–1945) and is best remembered as the flawed predecessor to the United Nations. The latter was formed in 1945, has grown since then from 51 to 193 member-countries today, and continues its primary mission of preventing international wars, serving as a forum for dialogue and conflict resolution.

Leeward The protected or downwind side of a **topographic** barrier with respect to the winds that flow across it.

Liberation Theology A powerful religious movement that arose in South America during the 1950s, and subsequently gained followers all over the global **periphery**. At its heart is a belief system, based on a blend of Christian faith and socialist thinking, that interprets the teachings of Christ as a quest to liberate the impoverished masses from oppression. It particularly appealed to the indigenous Amerindian populations that dominate the **demographic** mosaic of South America's Andean West region.

Lignite Low-grade, brown-colored variety of **coal**.

Lingua franca A "common language" prevalent in a given area; a second language that can be spoken and understood by many people, although they speak other languages at home.

Littoral Coastal or coastland.

Llanos The interspersed **savanna** grasslands and scrub woodlands of the Orinoco River's wide basin that covers most of interior Venezuela and Colombia.

Local functional specialization A hallmark of Europe's **economic geography** that later spread to many other parts of the world, whereby particular people in particular places concentrate on the production of particular goods and services.

Location Position on the Earth's surface; see also **absolute location** and **relative location**.

Location theory A logical attempt to explain the locational pattern of an economic activity and the manner in which its producing areas are interrelated.

Loess Deposit of very fine silt or dust that is laid down after having been windborne for a considerable distance. Notable for its fertility under **irrigation** and its ability to stand in steep vertical walls.

Longitude Angular distance (0° to 180°) east or west as measured from the *prime meridian* (0°) that passes through the Greenwich Observatory in suburban London, England. For much of its length across the mid-Pacific Ocean, the 180th meridian functions as the *international date line*.

Low island Low-lying coral islands of the Pacific Realm that—unlike **high islands**—cannot wrest sufficient moisture from the tropical maritime air to avoid chronic drought. Thus productive agriculture is impossible, and their modest populations must rely on fishing and the coconut palm for survival.

Lusitanian The Portuguese sphere, which by extension includes Brazil.

Madrassa **Revivalist** (**fundamentalist**) religious school in which the curriculum focuses on Islamic religion and law and requires rote memorization of the *Quran* (Koran), Islam's holy book. Founded in former British India, these schools were most numerous in present-day Pakistan but have **diffused** as far as Turkey in the west and Indonesia in the east. Afghanistan's **Taliban** emerged from these institutions in Pakistan

Maghreb The region occupying the northwestern corner of Africa, consisting of Morocco, Algeria, and Tunisia.

Main Street Canada's dominant **conurbation** that is home to nearly two-thirds of the country's inhabitants; extends southwestward from Quebec City in the mid-St. Lawrence Valley to Windsor on the Detroit River.

Mainland-Rimland framework The twofold regionalization of the Middle American realm based on its modern cultural history. The Euro-Amerindian *Mainland*, stretching from Mexico to Panama (minus the Caribbean coastal strip), was a self-sufficient zone dominated by **hacienda land tenure**. The Euro-African *Rimland*, consisting of that Caribbean coastal zone plus all of the Caribbean islands to the east, was the zone of the **plantation** that relied heavily on trade with Europe. See Figure 4A-7.

Maquiladora The term given to modern industrial plants in Mexico's U.S. border zone. These foreign-owned factories assemble imported components and/or raw materials, and then export finished manufactures, mainly to the United States. Import duties are disappearing under **NAFTA**, bringing jobs to Mexico and the advantages of low wage rates to the foreign entrepreneurs.

Marchland An area or **frontier** of uncertain boundaries that is subject to various national claims and an unstable political history. Refers specifically to the movement of various armies, refugees, and migrants across such zones.

Marine geography The geographic study of oceans and seas. Its practitioners investigate both the physical (e.g., coral-reef **biogeography**, ocean–**atmosphere** interactions, coastal **geomorphology**) as well as human (e.g., **maritime boundary**-making, fisheries, beachside development) aspects of oceanic environments.

Maritime boundary An international boundary that lies in the ocean. Like all boundaries, it is a vertical plane, extending from the seafloor to the upper limit of the air space in the atmosphere above the water.

Median-line boundary An international **maritime boundary** drawn where the width of a sea is less than 400 **nautical miles**. Because the **states** on either side of that sea claim **exclusive economic zones** of 200 nautical miles, it is necessary to reduce those claims to a (median) distance equidistant from each shoreline. **Delimitation** on the map almost always appears as a set of straight-line segments that reflect the configurations of the coastlines involved.

Medical geography The study of health and disease within a geographic context and from a spatial perspective. Among other things, this geographic field examines the sources, **diffusion** routes, and distributions of diseases.

Megacity Informal term referring to the world's most heavily populated cities; in this book, the term refers to a **metropolis** containing a population of greater than 10 million.

Megalopolis When spelled with a lower-case *m*, a synonym for **conurbation**, one of the large coalescing supercities forming in diverse parts of the world. When capitalized, refers specifically to the multimetropolitan (*Bosnywash*) corridor that extends along the northeastern U.S. seaboard from north of Boston to south of Washington, D.C.

Melting pot Traditional characterization of American society as a blend of numerous **immigrant ethnic** groups that over time were assimilated into a single societal mainstream. This notion always had its challengers among social scientists, and is now increasingly difficult to sustain given the increasing complexity and sheer scale of the U.S. ethnic mosaic in the twenty-first century.

Mercantilism Protectionist policy of European **states** during the sixteenth to the eighteenth centuries that promoted a **state**'s economic position in the contest with rival powers. Acquiring gold and silver and maintaining a favorable trade balance (more exports than imports) were central to the policy.

Meridian Line of **longitude**, aligned north-south across the globe, that together with **parallels** of **latitude** forms the global grid system. All meridians converge at both poles and are at their maximum distances from each other at the equator.

Mestizo Derived from the Latin word for *mixed*, refers to a person of mixed European (white) and Amerindian ancestry.

Métis **Indigenous** Canadian people of mixed native (**First Nations**) and European ancestry.

Metropolis Urban **agglomeration** consisting of a (central) city and its suburban ring. See also **urban (metropolitan) area**.

Metropolitan area See **urban (metropolitan) area**.

Microstate A sovereign **state** that contains a minuscule land area and population. They do not have the attributes of "complete" states, but are on the map as tiny yet independent entities nonetheless. Europe contains many examples, including Monaco, San Marino, Andorra, and Liechtenstein.

Migration A change in residence intended to be permanent. See also **forced, internal, international**, and **voluntary migration**.

Migratory movement Human relocation movement from a source to a destination without a return journey, as opposed to cyclical movement (see also **nomadism**).

Model An idealized representation of reality built to demonstrate its most important properties. A **spatial** model focuses on a geographical dimension of the real world.

Modernization In the eyes of the Western world, the **Westernization process** that involves the establishment of **urbanization**, a market (money) economy, improved circulation, formal schooling, the adoption of foreign innovations, and the breakdown of traditional society. Non-Westerners mostly see "modernization" as an outgrowth of **colonialism** and often argue that traditional societies can be modernized without being Westernized.

Monsoon Refers to the seasonal reversal of wind and moisture flows in certain parts of the subtropics and lower-middle latitudes. The *dry monsoon* occurs during the cool season when dry offshore winds prevail. The *wet monsoon* occurs in the hot summer months, which produce onshore winds that bring large amounts of rainfall. The air-pressure differential over land and sea is the triggering mechanism, with windflows always moving from areas of relatively higher pressure toward areas of relatively lower pressure. Monsoons make their greatest regional impact in the coastal and near-coastal zones of South Asia (Fig. 8A-3), Southeast Asia, and East Asia.

Mosaic culture Emerging cultural-geographic framework of the United States, dominated by the fragmentation of specialized social groups into homogeneous communities of interest marked not only by income, race, and **ethnicity** but also by age, occupational status, and lifestyle. The result is an increasingly heterogeneous sociospatial complex, which resembles an intricate mosaic composed of myriad uniform—but separate—tiles.

Mulatto A person of mixed African (black) and European (white) ancestry.

Multilingualism A society marked by a mosaic of local languages. Constitutes a **centrifugal force** because it impedes communication within the larger population. Often a *lingua franca* is used as a "common language," as in many countries of Subsaharan Africa.

Multinationals Internationally active corporations capable of strongly influencing the economic and political affairs of many countries in which they operate.

Muslim An adherent of the Islamic faith.

NAFTA (North American Free Trade Agreement) The **free-trade area** launched in 1994 involving the United States, Canada, and Mexico.

Nation Legally a term encompassing all the citizens of a **state**, it also has other connotations. Most definitions now tend to refer to a group of tightly knit people possessing bonds of language, **ethnicity**, religion, and other shared **cultural** attributes. Such homogeneity actually prevails within very few states.

Nation-state A country whose population possesses a substantial degree of **cultural** homogeneity and unity. The ideal form to which most **nations** and **states** aspire—a political unit wherein the territorial state coincides with the area settled by a certain national group or people.

NATO (North Atlantic Treaty Organization) Established in 1950 at the height of the Cold War as a U.S.-led **supranational** defense pact to shield postwar Europe against the Soviet military threat. NATO is now in transition, expanding its membership while modifying its objectives in the post-Soviet era. Its 28 member-states (as of late 2011) are: Albania, Belgium, Bulgaria, Canada, Croatia, Czech Republic, Denmark, Estonia, France, Germany, Greece, Hungary, Iceland, Italy, Latvia, Lithuania, Luxembourg, the Netherlands, Norway, Poland, Portugal, Romania, Slovakia, Slovenia, Spain, Turkey, the United Kingdom, and the United States.

Natural hazard A natural event that endangers human life and/or the contents of a **cultural landscape**.

Natural increase rate Population growth measured as the excess of live births over deaths per 1000 individuals per year. Natural increase of a population does not reflect either **emigrant** or **immigrant** movements.

Natural landscape The array of landforms that constitutes the Earth's surface (mountains, hills, plains, and plateaus) and the physical features that mark them (such as water bodies, soils, and vegetation). Each **geographic realm** has its distinctive combination of natural landscapes.

Natural resource Any valued element of (or means to an end using) the environment; includes minerals, water, vegetation, and soil.

Nautical mile By international agreement, the nautical mile—the standard measure at sea—is 6076.12 feet in length, equivalent to approximately 1.15 statute miles (1.85 km).

Near Abroad The 14 former Soviet republics that, in combination with the dominant Russian Republic, constituted the USSR. Since the 1991 breakup of the Soviet Union, Russia has asserted a sphere of influence in these now-independent countries, based on its proclaimed right to protect the interests of ethnic Russians who were settled there in substantial numbers during Soviet times. These 14 countries include Armenia, Azerbaijan, Belarus, Estonia, Georgia, Kazakhstan, Kyrgyzstan, Latvia, Lithuania, Moldova, Tajikistan, Turkmenistan, Ukraine, and Uzbekistan.

Neocolonialism The term used by developing countries to underscore that the entrenched **colonial** system of international exchange and capital flow has not changed in the postcolonial era—thereby perpetuating the huge economic advantages of the developed world.

Neoliberalism A national or regional development strategy based on the privatization of state-run companies, lowering of international trade tariffs, reduction of government subsidies, cutting of corporate taxes, and overall deregulation to business activity.

Neolithic The "New Stone Age" (ca. 9500-4500 BC), which was the final stage of cultural evolution and technological development among prehistoric humans. Among the numerous accomplishments that took place, two of the most important for human geography were: (1) the rise of farming through the domestication of plants and animals, and (2) the first permanent village settlements that mark the origin of urbanization.

Network (transport) The entire regional **system** of transportation connections and **nodes** through which movement can occur.

Nevada See *tierra nevada*.

New World Order A description of the international system resulting from the 1991 collapse of the Soviet Union whereby the balance of nuclear power theoretically no longer determines the destinies of **states**.

Node A center that functions as a point of **connectivity** within a regional **network** or **system**. All urban settlements possess this function, and the

higher the position of a settlement in its **urban system** or **hierarchy**, the greater its nodality.

Nomadism Cyclical movement among a definite set of places. Nomadic peoples mostly are **pastoralists**.

North American Free Trade Agreement See **NAFTA**.

Nucleation Cluster; **agglomeration**.

Oasis An area, small or large, where the supply of water (from an **aquifer** or a major river such as the Nile) permits the transformation of the immediately surrounding **desert** into productive cropland.

Occidental Western. Also see *Oriental*.

Offshore banking Term referring to financial havens for foreign companies and individuals, who channel their earnings to accounts in such a country (usually an "offshore" island-state) to avoid paying taxes in their home countries.

Oligarchs Opportunists in post-Soviet Russia who used their ties to government to enrich themselves.

OPEC (Organization of Petroleum Exporting Countries) The international oil *cartel* or syndicate formed by a number of producing countries to promote their common economic interests through the formulation of joint pricing policies and the limitation of market options for consumers. The 12 member-states (as of late 2011) are: Algeria, Angola, Ecuador, Iran, Iraq, Kuwait, Libya, Nigeria, Qatar, Saudi Arabia, United Arab Emirates (UAE), and Venezuela.

Oriental The root of the word "oriental" is from the Latin for *rise*. Thus it has to do with the direction in which one sees the sun "rise"—the east; *oriental* therefore means Eastern. *Occidental* originates from the Latin for fall, or the "setting" of the sun in the west; *occidental* therefore means Western.

Orographic precipitation Mountain-induced precipitation, especially when air masses are forced to cross **topographic** barriers. Downwind areas beyond such a mountain range experience the relative dryness known as the **rain shadow effect**.

Outback The name given by Australians to the vast, peripheral, sparsely settled interior of their country.

Outer city The non-central-city portion of the American **metropolis**; no longer "sub" to the "urb," this outer ring was transformed into a full-fledged city during the late twentieth century.

Outsourcing Turning over the partial or complete production of **a** good or service to another party. In **economic geography**, this usually refers to a company arranging to have its products manufactured in a foreign country where labor and other costs are significantly lower than in the home country (which experiences a commensurate loss in jobs)

Overseas Chinese The more than 50 million ethnic Chinese who live outside China. About half live in Southeast Asia, and many have become quite successful. A large number maintain links to China and as investors played a major economic role in stimulating the growth of **SEZs** and Open Cities in China's **Pacific Rim**.

Pacific Rim A far-flung group of countries and components of countries (extending clockwise on the map from New Zealand to Chile) sharing the following criteria: they face the Pacific Ocean; they exhibit relatively high levels of economic development, industrialization, and urbanization; their imports and exports mainly move across Pacific waters.

Pacific Ring of Fire Zone of crustal instability along **tectonic plate** boundaries, marked by earthquakes and volcanic activity, that ring the Pacific Ocean Basin (see Fig. G-5).

Paddies (paddyfields) Ricefields.

Pandemic An outbreak of a disease that spreads worldwide.

Pangaea A vast, singular landmass consisting of most of the areas of the present-day continents. This *supercontinent* began to break up more than 200 million years ago when still-ongoing **plate** divergence and **continental drift** became dominant **processes** (see Fig. 6A-3).

Parallel An east-west line of **latitude** that is intersected at right angles by **meridians** of **longitude**.

Partition The subdivision of the British Indian Empire into India and Pakistan at the end of colonial rule on August 15, 1947. Shortly before their departure from what is now the South Asian realm, the British were persuaded to create separate countries for South Asia's massive Hindu (India) and Muslim (West and East Pakistan) populations. Boundaries were drawn hurriedly and ineffectively through thousands of kilometers of highly complicated cultural terrain, triggering the biggest mass migration in human history as millions fled their homes to be sure they were not on the "wrong" side of the new border. Both India and Pakistan (the successor to West Pakistan; East Pakistan seceded in 1971 to become independent Bangladesh) continue to suffer the consequences of the Partition that occurred 65 years ago, including the scars of four subsequent armed conflicts, the dangerous unresolved impasse in Kashmir, and persistent tensions magnified by the nuclear capabilities of both.

Pastoralism A form of **agricultural** activity that involves the raising of livestock.

Peasants In a **stratified** society, peasants are the lowest class of people who depend on **agriculture** for a living. But they often own no land at all and must survive as tenants or day workers.

Peninsula A comparatively narrow, finger-like stretch of land extending from the main landmass into the sea. Florida and Korea are examples.

Peon (*peone*) Term used in Middle and South America to identify people who often live in serfdom to a wealthy landowner; landless **peasants** in continuous indebtedness.

Per capita Capita means *individual*. Income, production, or some other measure is often given per individual.

Perforated state A **state** whose territory completely surrounds that of another state. South Africa, which encloses Lesotho and is perforated by it, is a classic example.

Periodic market Village market that is open every third day or at some other regular interval. Part of a regional network of similar markets in a preindustrial, rural setting where goods are brought to market on foot and barter remains a leading mode of exchange.

Peripheral development Spatial pattern in which a country's or region's development (and population) is most heavily concentrated along its outer edges rather than in its interior. Australia, with its peripheral population distribution, is a classic example (note its **core area** in Fig. 11-5), and nearby New Zealand exhibits a similar configuration of people and activities.

Periphery See **core-periphery relationships**.

Permafrost Permanently frozen water in the near-surface soil and bedrock of cold environments, producing the effect of completely frozen ground. Surface can thaw during brief warm season.

Physical geography The study of the geography of the physical (natural) world. Its subfields encompass **climatology, geomorphology, biogeography, soil geography, marine geography**, and water **resources**.

Physical landscape Synonym for **natural landscape**.

Physiographic political boundaries Political boundaries that coincide with prominent physical features in the **natural landscape**—such as rivers or the crest ridges of mountain ranges.

Physiographic region (province) A **region** within which there prevails substantial **natural-landscape** homogeneity, expressed by a certain degree of uniformity in surface **relief, climate**, vegetation, and soils.

Physiography Literally means *landscape description*, but commonly refers to the total **physical geography** of a place; includes all of the natural features on the Earth's surface, including landforms, **climate**, soils, vegetation, and water bodies.

Physiologic density The number of people per unit area of **arable** land.

Pilgrimage A journey to a place of great religious significance by an individual or by a group of people (such as a pilgrimage [*hajj*] to Mecca for **Muslims**).

Plantation A large estate owned by an individual, family, or corporation and organized to produce a cash crop. Almost all plantations were established within the tropics; in recent decades, many have been divided into smaller holdings or reorganized as cooperatives.

Plate tectonics Plates are bonded portions of the Earth's mantle and crust, averaging 100 kilometers (60 mi) in thickness. More than a dozen such plates exist (see Fig. G-4), most of continental proportions, and they are in motion. Where they meet one slides under the other (**subduction**), crumpling the surface crust and producing significant volcanic and earthquake activity; a major mountain-building force.

Pleistocene Epoch Recent period of geologic time that spans the rise of humankind, beginning about 2 million years ago. Marked by *glaciations* (repeated advances of continental ice sheets) and more moderate *interglacials* (ice sheet contractions). Although the last 10,000 years are known as the **Holocene** Epoch, Pleistocene-like conditions seem to be continuing and we are most probably now living through another Pleistocene interglacial; thus the glaciers likely will return.

Plural(istic) society A society in which two or more population groups, each practicing its own **culture**, live adjacent to one another without mixing inside a single **state**.

Polder Land reclaimed from the sea adjacent to the shore of the Netherlands by constructing dikes and then pumping out the water trapped behind them.

Police state A **state** in which the government (usually marked by dictatorial leadership) exercises totalitarian control over the political, social, and economic life of its citizens. Repression of the people includes rigid restrictions on their movements, freedom to communicate and express their views, and aggressive monitoring and enforcement by means of a secret-police force. Among the most blatant examples in mid-2011 were Syria, Myanmar (Burma), and especially North Korea.

Political geography The study of the interaction of geographic space and political **process**; the spatial analysis of political phenomena and processes.

Pollution The release of a substance, through human activity, which chemically, physically, or biologically alters the air or water it is discharged into. Such a discharge negatively impacts the environment, with possible harmful effects on living organisms—including humans.

Population decline A decreasing national population. Russia, which now loses about half a million people per year, is the best example. Also see **population implosion**.

Population density The number of people per unit area. Also see **arithmetic density** and **physiologic density** measures.

Population distribution The way people have arranged themselves in geographic space. One of human geography's most essential expressions because it represents the sum total of the adjustments that a population has made to its natural, cultural, and economic environments. A population distribution map is included in every chapter in this book.

Population expansion (explosion) The rapid growth of the world's human population over the past century, attended by accelerating *rates* of increase.

Population geography The field of geography that focuses on the spatial aspects of **demography** and the influences of demographic change on particular countries and regions.

Population implosion The opposite of **population explosion**; refers to the declining populations of many European countries and Russia in which the **death rate** exceeds the **birth rate** and **immigration** rate.

Population movement See **migration** and **migratory movement**.

Population projection The future population total that demographers forecast for a particular country. For example, in the Data Table in Appendix B such projections are given for all the world's countries for 2025.

Population pyramid Graphic representation or *profile* of a national population according to age and gender. Such a diagram of age-sex structure typically displays the percentage of each age group (commonly in five-year increments) as a horizontal bar, whose length represents its relationship to the total population. Males in each five-year age cohort are shown to the left of the center line, females to the right. See Figure 8A-12.

Postindustrial economy Emerging economy, in the United States and a number of other highly advanced countries, as traditional industry is increasingly eclipsed by a higher-technology productive complex dominated by services, information-related, and managerial activities.

Primary economic activity Activities engaged in the direct extraction of **natural resources** from the environment such as mining, fishing, lumbering, and especially **agriculture**.

Primate city A country's largest city—ranking atop its urban **hierarchy**—most expressive of the national culture and usually (but not in every case) the capital city as well.

Process Causal force that shapes a spatial pattern as it unfolds over time.

Productive activities The major components of the spatial economy. For individual components see: **primary economic activity, secondary economic activity, tertiary economic activity, quaternary economic activity,** and **quinary economic activity**.

Protruded state Territorial shape of a **state** that exhibits a narrow, elongated land extension (or *protrusion*) leading away from the main body of territory. Thailand is a leading example.

Push-pull concept The idea that **migration** flows are simultaneously stimulated by conditions in the source area, which tend to drive people away, and by the perceived attractiveness of the destination.

Qanat In **desert** zones, particularly in Iran and western China, an underground tunnel built to carry **irrigation** water by gravity flow from surrounding mountains (where **orographic precipitation** occurs) to the arid flatlands below.

Quaternary economic activity Activities engaged in the collection, processing, and manipulation of *information*.

Quinary economic activity Managerial or control-function activity associated with decision-making in large organizations.

Racial profiling The use, by police and other security personnel, of an individual's race or ethnicity in the decision to engage in law enforcement.

Rain shadow effect The relative dryness in areas downwind of mountain ranges resulting from **orographic precipitation**, wherein moist air masses are forced to deposit most of their water content as they cross the highlands.

Rate of natural population increase See **natural increase rate**.

Realm See **geographic realm**.

Refugees People who have been dislocated involuntarily from their original place of settlement.

Region A commonly used term and a geographic concept of paramount importance. An **area** on the Earth's surface marked by specific criteria, which are discussed in the Introduction (see pp. 11–12).

Regional boundary In theory, the line that circumscribes a **region**. But razor-sharp lines are seldom encountered, even in nature (e.g., a coastline constantly changes depending on the tide). In the **cultural landscape**, not only are regional boundaries rarely self-evident, but when they are ascertained by geographers they most often turn out to be **transitional** borderlands.

Regional complementarity See **complementarity**.

Regional concept The geographic study of **regions** and regional distinctions, as discussed in the Introduction (see pp. 11–12).

Regional disparity The spatial unevenness in standard of living that occurs within a country, whose "average" overall income statistics invariably mask the differences that exist between the extremes of the affluent **core** and the poorer, disadvantaged **periphery**.

Regional geography Approach to geographic study based on the spatial unit of the **region**. Allows for an all-encompassing view of the world, because it utilizes and integrates information from geography's topical (**systematic**) fields, which are diagrammed in Figure G-15.

Regional state A "natural economic zone" that defies political boundaries and is shaped by the global economy of which it is a part; its leaders deal directly with foreign partners and negotiate the best terms they can with the national governments under which they operate.

Regionalism The consciousness of and loyalty to a **region** considered distinct and different from the **state** as a whole by those who occupy it.

Relative location The regional position or **situation** of a place relative to the position of other places. Distance, **accessibility**, and **connectivity** affect relative location.

Relict boundary A political boundary that has ceased to function, but the imprint of which can still be detected on the **cultural landscape**.

Relief Vertical difference between the highest and lowest elevations within a particular area.

Religious revivalism See **revivalism** (**religious**).

Relocation diffusion Sequential **diffusion process** in which the items being diffused are transmitted by their carrier agents as they relocate to new areas. The most common form of relocation diffusion involves the spreading of innovations by a **migrating** population.

Remittances Money earned by **emigrants** that is sent back to family and friends in their home country, mostly in cash; forms an important part of the economy in poorer countries.

Restrictive population policies Government policy designed to reduce the **rate of natural population increase**. China's one-child policy, instituted in 1979 after Mao's death, is a classic example.

Revivalism (religious) Religious movement whose objectives are to return to the foundations of that faith and to influence state policy. Often called *religious fundamentalism*; but in the case of Islam, **Muslims** prefer the term *revivalism*.

Rift valley The trough or trench that forms when a thinning strip of the Earth's crust sinks between two parallel faults (surface fractures).

Rural-to-urban migration The dominant **migration** flow from countryside to city that continues to transform the world's population, most notably in the less advantaged geographic realms.

Russification Demographic resettlement policies pursued by the central planners of the Soviet Empire (1924–1991), whereby ethnic Russians were encouraged to **emigrate** from the Russian Republic to the 14 non-Russian republics of the USSR.

Sahel Semiarid **steppeland** zone extending across most of Africa between the southern margins of the arid Sahara and the moister tropical **savanna** and forest zone to the south. Chronic drought, **desertification**, and overgrazing have contributed to severe famines in this area since 1970.

Satellite states The countries of eastern Europe under Soviet **hegemony** between 1945 and 1989. This tier of countries—the "satellites" captured in Moscow's "orbit" following World War II—was bordered on the west by the Iron Curtain and on the east by the USSR. Using the names then in force, they included Bulgaria, Czechoslovakia, East Germany, Hungary, Poland, and Romania. The Balkan states of former Yugoslavia and Albania to the south were also under tight communist control, but their systems differed from Moscow's to the extent that neither was generally considered to be a Soviet satellite.

Savanna Tropical grassland containing widely spaced trees; also the name given to the tropical wet-and-dry climate type (*Aw*).

Scale Representation of a real-world phenomenon at a certain level of reduction or generalization. In **cartography**, the ratio of map distance to ground distance; indicated on a map as a bar graph, representative fraction, and/or verbal statement. *Macroscale* refers to a large area of national proportions; *microscale* refers to a local area no bigger than a county.

Scale economies See **economies of scale**.

Secession The act of withdrawing from a political entity, usually a **state,** as when the U.S. South tried unsuccessfully to secede from the United States in 1861 and sparked the Civil War. Successful secessions in more recent times include Singapore from Malaysia in 1965 and Slovakia from former Czechoslovakia in 1993. As **devolution** has become a force to be reckoned with around the world since the late twentieth century, the threat of secession can sometimes become a bargaining tool to gain increased local autonomy; examples include Canada (Quebec), Spain (Catalonia), and the United Kingdom (Scotland).

Secondary economic activity Activities that process raw materials and transform them into finished industrial products; the *manufacturing* sector.

Sedentary Permanently attached to a particular area; a population fixed in its location; the opposite of **nomadic**.

Separate development The spatial expression of South Africa's "grand" **apartheid** scheme, whereby nonwhite groups were required to settle in segregated "homelands." The policy was dismantled when white-minority rule collapsed in the early 1990s.

Sex ratio A **demographic** indicator showing the ratio of males to females in a given population.

Shantytown Unplanned slum development on the margins of cities in disadvantaged countries, dominated by crude dwellings and shelters mostly made of scrap wood and iron, and even pieces of cardboard.

Sharecropping Relationship between a large landowner and farmers on the land wherein the farmers pay rent for the land they farm by giving the landlord a share of the annual harvest.

Sharia The strict criminal code based in Islamic law that prescribes corporal punishment, amputations, stonings, and lashing for both major and minor offenses. Its occurrence today is associated with the spread of **religious revivalism** in **Muslim** societies.

Shatter belt **Region** caught between stronger, colliding external cultural-political forces, under persistent stress, and often fragmented by aggressive rivals. Eastern Europe and mainland Southeast Asia are classic examples.

Shifting agriculture Cultivation of crops in recently cut-down and burned tropical-forest clearings, soon to be abandoned in favor of newly cleared nearby forest land. Also known as *slash-and-burn agriculture*.

Sinicization Giving a Chinese cultural imprint; Chinese **acculturation**. See also **Hanification**.

Site The internal locational attributes of an urban center, including its local spatial organization and physical setting.

Situation The external locational attributes of an urban center; its **relative location** or regional position with reference to other non-local places.

Small-island developing economies The additional disadvantages faced by lower-income island-states because of their often small territorial size and populations as well as overland **inaccessibility**. Limited resources require expensive importing of many goods and services; the cost of government operations per capita are higher; and local production is unable to benefit from **economies of scale**. The eastern Caribbean islands of the Lesser Antilles are a good example.

Social stratification See **stratification** (**social**).

Southern Ocean The ocean that surrounds Antarctica (discussed on page 568 in Chapter 11).

Spatial Pertaining to space on the Earth's surface. Synonym for *geographic(al)*.

Spatial diffusion See **diffusion**.

Spatial interaction See **complementarity, intervening opportunity**, and **transferability**.

Spatial model See **model**.

Spatial process See **process**.

Spatial system The components and interactions of a **functional region**, which is defined by the areal extent of those interactions. See also **system**.

Special Administrative Region (SAR) Status accorded the former dependencies of Hong Kong and Macau that were taken over by China, respectively, from the United Kingdom in 1997 and Portugal in 1999. Both SARs received guarantees that their existing social and economic systems could continue unchanged for 50 years following their return to China.

Special Economic Zone (SEZ) Manufacturing and export center in China, created since 1980 to attract foreign investment and technology transfers. Seven SEZs—all located on China's Pacific coast—currently operate: Shenzhen, adjacent to Hong Kong; Zhuhai; Shantou; Xiamen; Hainan Island, in the far south; Pudong, across the river from Shanghai; and Binhai New Area, next to the port of Tianjin.

Squatter settlement See **shantytown**.

State A politically organized territory that is administered by a sovereign government and is recognized by a significant portion of the international community. A state must also contain a permanent resident population, an organized economy, and a functioning internal circulation system.

State boundaries The borders that surround **states** which, in effect, are derived through contracts with neighboring states negotiated by treaty. See **definition, delimitation**, and **demarcation**.

State capitalism Government-controlled corporations competing under free-market conditions, usually in a tightly regimented society. South Korea in the postwar era was a classic example.

State formation The creation of a **state**, exemplifying traditions of human **territoriality** that go back thousands of years.

State planning Involves highly centralized control of the national planning process, a hallmark of communist economic systems. Soviet central planners mainly pursued a grand political design in assigning production to particular places; their frequent disregard of the principles of **economic geography** contributed to the eventual collapse of the USSR.

State territorial morphology A **state's** geographical shape, which can have a decisive impact on its spatial cohesion and political viability. A **compact** shape is most desirable; among the less efficient shapes are those exhibited by **elongated, fragmented, perforated**, and **protruded** states.

Stateless nation A national group that aspires to become a **nation-state** but lacks the territorial means to do so; the Palestinians and Kurds of Southwest Asia are leading examples.

Steppe Semiarid grassland; short-grass prairie. Also the name given to the semiarid climate type (*BS*).

Stratification (social) In a layered or stratified society, the population is divided into a **hierarchy** of social classes. In an industrialized society, the working class is at the lower end; **elites** that possess capital and control the means of production are at the upper level. In the traditional **caste system** of Hindu India, the "untouchables" form the lowest class or caste, whereas the still-wealthy remnants of the princely class are at the top.

Subduction In **plate tectonics**, the **process** that occurs when an oceanic plate converges head-on with a plate carrying a continental landmass at its leading edge. The lighter continental plate overrides the denser oceanic plate and pushes it downward.

Subsequent boundary A political boundary that developed contemporaneously with the evolution of the major elements of the **cultural landscape** through which it passes.

Subsistence Existing on the minimum necessities to sustain life; spending most of one's time in pursuit of survival.

Subtropical Convergence A narrow marine **transition zone**, girdling the globe at approximately latitude 40°S, that marks the equatorward limit of the frigid **Southern Ocean** and the poleward limits of the warmer Atlantic, Pacific, and Indian oceans to the north.

Suburban downtown In the United States (and increasingly in other high-income countries), a significant concentration of major urban activities around a highly accessible suburban location, including retailing, light industry, and a variety of leading corporate and commercial operations. The largest are now coequal to the American central city's **central business district (CBD)**. A leading feature of the **outer city**.

Sunbelt The popular name given to the southern tier of the United States, which is anchored by the mega-States of California, Texas, and Florida. Its warmer climate, superior recreational opportunities, and other amenities have been attracting large numbers of relocating people and activities since the 1960s; broader definitions of the Sunbelt also include much of the western United States, even Colorado and the coastal Pacific Northwest.

Superimposed boundary A political boundary emplaced by powerful outsiders on a developed human landscape. Frequently ignores preexisting cultural-spatial patterns, such as the border that still divides North and South Korea.

Supranational A venture involving three or more **states**—political, economic, and/or cultural cooperation to promote shared objectives. The **European Union** is one such organization.

System Any group of objects or institutions and their mutual interactions. Geography treats systems that are expressed spatially, such as in **functional regions**.

Systematic geography Topical geography: **cultural, political, economic geography**, and the like.

Taiga The subarctic, mostly **coniferous** snowforest that blankets northern Russia and Canada south of the **tundra** that lines the Arctic shore. Known as the **boreal forest** in North America.

Takeoff Economic concept to identify a stage in a country's **development** when conditions are set for a domestic Industrial Revolution.

Taliban The term means "students" or "seekers of religion." Specifically, refers to the Islamist militia group that emerged from *madrassas* in Pakistan and ruled neighboring Afghanistan between 1996 and 2001; it has been trying to regain control of that country in its continuing conflict with U.S.-led NATO troops. Taliban rule, in adherence with an extremist interpretation of *Sharia* law, was marked by one of the most virulent forms of militant Islam ever seen.

Tar sands The main source of oil from non-liquid petroleum reserves. The oil is mixed with sand and requires massive open-pit mining as well as a costly, complicated process to extract it. The largest known deposits are located in the northeast of Canada's province of Alberta, and by most estimates these Athabasca Tar Sands constitute one of the largest oil reserves in the world. The high oil prices of recent years have led to greatly expanded production here, but the accompanying environmental degradation caused by strip-mining and waste disposal (see photo, p. 189) has triggered a widening protest movement that may limit the exploitation of this resource.

Taxonomy A **system** of scientific classification.

Technopole A planned techno-industrial complex (such as California's Silicon Valley) that innovates, promotes, and manufactures the products of the **postindustrial** informational economy.

Tectonics See **plate tectonics**.

Templada See *tierra templada*.

Terms of trade In international economics refers to an agreement between trading partners that stipulates the quantity of imports that can be purchased by one country through the sale of a fixed quantity of exports to the other.

Terracing The transformation of a hillside or mountain slope into a step-like sequence of horizontal fields for intensive cultivation.

Territoriality A country's or more local community's sense of property and attachment toward its territory, as expressed by its determination to keep it inviolable and strongly defended.

Territorial sea Zone of seawater adjacent to a country's coast, held to be part of the national territory and treated as a component of the sovereign **state**.

Tertiary economic activity Activities that engage in *services*—such as transportation, banking, retailing, education, and routine office-based jobs.

Tierra caliente The lowest of the **altitudinal zones** into which the human settlement of Middle and South America is classified according to elevation. The *caliente* is the hot humid coastal plain and adjacent slopes up to 750 meters (2500 ft) above sea level. The natural vegetation is the dense and luxuriant tropical rainforest; the crops include sugar and bananas in the lower areas, and coffee, tobacco, and corn along the higher slopes.

Tierra fría Cold, high-lying **altitudinal zone** of settlement in Andean South America, extending from about 1800 meters (6000 ft) in elevation up to nearly 3600 meters (12,000 ft). **Coniferous** trees stand here; upward they change into scrub and grassland. There are also important pastures within the *fría*, and wheat can be cultivated.

Tierra helada In Andean South America, the highest-lying habitable **altitudinal zone**—ca. 3600 to 4500 meters (12,000 to 15,000 ft)—between the tree line (upper limit of the *tierra fría*) and the snow line (lower limit of the *tierra nevada*). Too cold and barren to support anything but the grazing of sheep and other hardy livestock.

Tierra nevada The highest and coldest **altitudinal zone** in Andean South America (lying above 4500 meters [15,000 ft]), an uninhabitable environment of permanent snow and ice that extends upward to the Andes' highest peaks of more than 6000 meters (20,000 ft).

Tierra templada The intermediate **altitudinal zone** of settlement in Middle and South America, lying between 750 meters (2500 ft) and 1800 meters (6000 ft) in elevation. This is the "temperate" zone, with moderate temperatures compared to the *tierra caliente* below. Crops include coffee, tobacco, corn, and some wheat.

Topography The surface configuration of any segment of **natural landscape**.

Toponym Place name.

Transculturation Cultural borrowing and two-way exchanges that occur when different **cultures** of approximately equal complexity and technological level come into close contact.

Transferability The capacity to move a good from one place to another at a bearable cost; the ease with which a commodity may be transported.

Transition zone An area of spatial change where the **peripheries** of two adjacent **realms** or **regions** join; marked by a gradual shift (rather than a sharp break) in the characteristics that distinguish these neighboring geographic entities from one another.

Transmigration The now-ended policy of the Indonesian government to induce residents of the overcrowded, **core-area** island of Jawa to move to the country's other islands.

Treaty ports **Extraterritorial enclaves** in China's coastal cities, established by European colonial invaders under unequal treaties enforced by gunboat diplomacy.

Triple Frontier The turbulent and chaotic area in southern South America that surrounds the convergence of Brazil, Argentina, and Paraguay. Lawlessness pervades this haven for criminal elements, which is notorious for money laundering, arms and other smuggling, drug trafficking, and links to terrorist organizations, including money flows to the Middle East.

Tropical deforestation See **deforestation**.

Tropical savanna See **savanna**.

Tsunami A seismic (earthquake-generated) sea wave that can attain gigantic proportions and cause coastal devastation. The tsunami of December 26, 2004, centered in the Indian Ocean near the Indonesian island of Sumatera (Sumatra), produced the first great natural disaster of the twenty-first century. Our new century's second major tsunami disaster occurred along the coast of Japan's northeastern Honshu Island on March 11, 2011.

Tundra The treeless plain that lies along the Arctic shore in northernmost Russia and Canada, whose vegetation consists of mosses, lichens, and certain hardy grasses.

Turkestan Northeasternmost region of the North Africa/Southwest Asia realm. Known as Soviet Central Asia before 1992, its five (dominantly Islamic) former Soviet Socialist Republics have become the independent countries of Kazakhstan, Uzbekistan, Turkmenistan, Kyrgyzstan, and Tajikistan. Today Turkestan has expanded to include a sixth state, Afghanistan.

Turkish Model In the wake of the regime changes in the North Africa/Southwest Asia realm brought about by the "Arab Spring" of 2011, moderates have cited Turkey as the best model of democratic governance for this part of the world. Specifically, this involves a multi-party democracy that has a place for, but is not dominated by, Islamic political parties.

Uneven development The notion that economic development varies spatially, a central tenet of **core-periphery relationships** in realms, regions, and lesser geographic entities.

Unitary state A **nation-state** that has a centralized government and administration that exercises power equally over all parts of the **state**.

United Nations See discussion under **League of Nations**.

Unity of place The great German natural scientist Alexander von Humboldt's notion that in a particular locale or region intricate connections exist among climate, geology, biology, and human cultures. This laid the foundation for modern geography as an *integrative discipline* marked by a spatial perspective.

Urbanization A term with a variety of connotations. The proportion of a country's population living in urban places is its level of urbanization. The **process** of urbanization involves the movement to, and the clustering of, people in towns and cities—a major force in every geographic realm today. Another kind of urbanization occurs when an expanding city absorbs rural countryside and transforms it into suburbs; in the case of cities in disadvantaged countries, this also generates peripheral **shantytowns**.

Urban (metropolitan) area The entire built-up, non-rural area and its population, including the most recently constructed suburban appendages. Provides a better picture of the dimensions and population of such an area than the delimited municipality (central city) that forms its heart.

Urban realms model A spatial generalization of the contemporary large American city. It is shown to be a widely dispersed, multinodal **metropolis** consisting of increasingly independent zones or *urban realms*, each focused on its own **suburban downtown**; the only exception is the shrunken central realm, which is focused on the central city's **central business district** (see Fig. 3A-11).

Urban system A **hierarchical** network or grouping of urban areas within a finite geographic area, such as a country.

Veld See **highveld**.

Voluntary migration Population movement in which people relocate in response to perceived opportunity, not because they are forced to **migrate**.

Wahhabism A particularly virulent form of (Sunni) **Muslim revivalism** that was made the official faith when the modern **state** of Saudi Arabia was founded in 1932. Adherents call themselves "Unitarians" to signify the strict fundamentalist nature of their beliefs.

Wallace's Line As shown in Figure 11-4, the zoogeographical boundary proposed by Alfred Russel Wallace that separates the marsupial fauna of Australia and New Guinea from the non-marsupial fauna of Indonesia.

Weather The immediate and short-term conditions of the **atmosphere** that impinge on daily human activities.

West Wind Drift The clockwise movement of water as a current that circles around Antarctica in the **Southern Ocean**.

Westernization The Western view of the **process** of **modernization** that involves the establishment of **urbanization**, a market (money) economy, improved circulation, formal schooling, adoption of foreign innovations, and the breakdown of traditional society. Non-Westerners mostly see "modernization" as an outgrowth of **colonialism** and often argue that traditional societies can be modernized without being Westernized

Wet monsoon See **monsoon**.

Windward The exposed, upwind side of a **topographic** barrier that faces the winds that flow across it.

World city Either London, New York, or Tokyo. The highest-ranking urban centers of **globalization** with financial, high-technology, communications, engineering, and related industries reflecting the momentum of their long-term growth and **agglomeration**.

World geographic realm See **geographic realm**.

A page number followed by "f" indicates the entry is within a figure; a page number followed by "t" indicates the entry is within a table.

This Data Table is a valuable resource, and should be consulted throughout your reading. Like all else in this book, this table is subject to continuous revision and modification. Compared to previous editions, we have deleted some indices, elaborated others, and introduced new ones. For example, in a world with ever-slower population growth, the so-called Doubling Time index—the number of years it will take for a population to double in size based on its current rate of natural increase—has lost most of its utility. On the other hand, when it comes to **Life Expectancy** and **Literacy**, general averages conceal significant differences between male and female rates, which in turn reflect conditions in individual countries, so we report these by gender. Also in this edition, we continue to use the **Corruption Index**, not available for all countries but an important reflection of a global problem. The **Big Mac Price** index, a measure introduced by the journal *The Economist*, tells you much more than what a hamburger with all the trimmings would cost in real dollars in various countries of the world—it also reflects whether those countries' currencies are overvalued or undervalued. And the final column, in which we formerly used to reported the per-capita GNP (Gross National Product), now reveals the **GNI**, that is, the Gross

National Income per person and what this would buy in each country. In the language of economic geographers, this is called the GNI-PPP, the per-capita Gross National Income in terms of its Purchasing Power Parity.

Indexes that may not be immediately obvious to you include **Arithmetic Population Density**, the number of people per square kilometer in each country; **Physiologic Population Density**, the number of people per square kilometer of agriculturally productive land; **Birth** and **Death Rates** per thousand in the population, resulting in the national population's rate of **Natural Increase**; a population's **Infant Mortality**, the number of deaths per thousand in the first year of life, thus largely reflecting the number of deaths at birth; **Child Mortality**, the deaths per thousand of children in their first five years; the **Corruption Index**, based on Transparency International data in which 10.0 is perfect and 0.1 is the worst; the **Big Mac Price** index, which tells you why Argentina in 2011 was the best place to buy a hamburger in U.S. dollars; and the **Per Capita GNI ($U.S.)**, the GNI-PPP index referred to above, which tells you how spendable income varies around the globe. For additional details on sources and data, please consult the Data Sources section of the Preface.

	Land Area Sq. km.	Sq. mi.	Population 2012 (Millions)	Population 2025 (Millions)	Population Density Arithmetic	Physiologic	Birth Rate	Death Rate	Natural Increase %	Infant Mortality per 1,000	Child Mortality per 1,000	Life Expectancy Male (years)	Female (years)	Percent Urban Pop. %	Literacy Male %	Literacy Female %	Corruption Index	Big Mac Price ($US)	Per Capita GNI ($US)
Korea, North	120,541	46,541	23.0	23.6	191	1194	15	10	0.5	26	33	61	66	60	99.0	99.0			—
Korea, South	99,259	38,324	49.3	49.1	497	2614	9	5	0.4	5	7	77	83	82	99.2	96.4	5.4	2.82	$28,120
Mongolia	1,566,492	604,826	2.9	3.3	2	186	25	6	1.9	24	29	63	70	61	99.2	99.3	2.7		$3,480
Taiwan	36,180	13,969	23.3	23.9	644	2575	8	6	0.2			75	82	78	97.6	90.2	5.8	2.34	—
REALM	**4,494,790**	**1,735,448**	**613.4**	**698.6**	**136**	**655**	**20**	**7**	**1.3**	**28**	**35**	**68**	**72**	**42**	**93.0**	**86.1**			**$4,510**
SOUTHEAST ASIA																			
Brunei	5,770	2,228	0.4	0.5	71	3557	16	3	1.3	5	7	75	80	72	94.7	88.2	5.5		$50,200
Cambodia	181,040	69,900	15.6	19.0	86	392	25	8	1.6	68	88	59	63	20	79.7	53.4	2.1		$1,820
East Timor	14,869	5,741	1.3	1.9	86	715	41	10	3.1	48	56	60	62	22			2.5		$4,690
Indonesia	1,904,561	735,355	242.1	273.2	127	748	20	6	1.4	30	39	69	73	43	91.9	82.1	2.8	2.51	$3,830
Laos	236,800	91,429	6.7	8.3	28	939	28	7	2.1	46	59	63	66	27	73.6	50.5	2.1		$2,060
Malaysia	329,750	127,317	29.8	34.9	90	377	21	5	1.6	6	6	72	77	63	91.5	83.6	4.4	2.19	$13,740
Myanmar (Burma)	676,577	261,228	54.4	61.7	80	502	20	11	0.9	54	71	56	60	31	89.0	80.6	1.4		$1,290
Philippines	299,998	115,830	98.0	117.6	327	990	26	5	2.1	26	33	70	74	63	95.5	95.2	2.4	2.19	$3,900
Singapore	619	239	5.2	5.7	8338	416,908	10	4	0.6	2	3	79	84	100	96.4	88.5	9.3	3.08	$47,940
Thailand	513,118	198,116	68.9	72.6	134	336	15	9	0.6	12	14	66	72	31	97.2	94.0	3.5	2.17	$5,990
Vietnam	331,689	128,066	91.0	103.2	274	1248	17	5	1.2	20	24	72	76	28	95.7	91.0	2.7		$2,700
REALM	**975,341**	**376,804**	**10.0**	**13.1**	**18**	**841**	**29**	**9**	**2.0**	**42**	**55**	**60**	**65**	**22**	**65.9**	**52.1**			**$2,114**
AUSTRAL REALM																			
Australia	7,741,184	2,988,888	22.8	26.9	3	42	14	6	0.7	4	5	79	84	82	100.0	100.0	8.7	3.84	$34,040
New Zealand	270,529	104,452	4.5	5.0	16	137	14	7	0.8	5	6	78	82	86	100.0	100.0	9.3	3.59	$25,090
REALM	**8,012,942**	**3,093,814**	**27.2**	**31.9**	**3**	**47**	**14**	**6**	**0.8**	**4**	**5**	**79**	**84**	**83**	**100.0**	**100.0**			**$32,850**
PACIFIC REALM																			
Federated States of Micronesia	699	270	0.1	0.1	148	286	25	6	1.9	32	39	67	68	22	67.0	87.2			$3,000
Fiji	18,270	7,054	0.9	0.9	51	318	24	7	1.7	15	18	66	71	51	95.0	90.9			$4,270
French Polynesia	3,999	1,544	0.3	0.3	77	962	18	5	1.3			72	77	53	94.9	95.0			—
Guam	549	212	0.2	0.2	375	1706	19	4	1.5			76	82	93	99.0	99.0			—
Marshall Islands	179	69	0.1	0.1	591	3478	34	6	2.8	29	35	64	67	68	92.4	90.0			—
New Caledonia	18,581	7,174	0.3	0.3	17	1650	16	5	1.2			72	80	58	57.4	58.3			—
Papua New Guinea	462,839	178,703	7.1	9.1	15	1532	31	10	2.2	52	68	57	62	13	63.4	50.9	2.1		$2,000
Samoa	2,841	1,097	0.2	0.2	73	171	26	5	2.0	21	25	72	74	22	100.0	100.0	4.1		$4,340
Solomon Islands	28,899	11,158	0.5	0.8	18	606	33	8	2.5	30	36	62	63	17	62.4	44.9	2.8		$2,580
Vanuatu	12,191	4,707	0.2	0.3	17	172	31	6	2.5	14	16	66	69	24	57.3	47.8	3.6		$3,940

	Land Area Sq. km.	Land Area Sq. mi.	Population 2012 (Millions)	Population 2025 (Millions)	Population Density Arithmetic	Population Density Physiologic	Birth Rate	Death Rate	Natural Increase %	Infant Mortality per 1,000	Child Mortality per 1,000	Life Expectancy Male (years)	Life Expectancy Female (years)	Percent Urban Pop.	Literacy Male %	Literacy Female %	Corruption Index	Big Mac Price ($US)	Per Capita GNI ($US)
Egypt	1,001,445	386,660	83.8	103.6	84	2790	27	6	2.1	18	21	69	75	43	66.6	43.7	3.1	2.28	$5,460
Iran	1,633,182	630,575	77.1	87.1	47	429	19	6	1.3	26	31	70	73	69			2.2		$10,840
Iraq	438,319	169,236	33.2	44.7	76	582	32	6	2.6	35	44	64	72	67	70.7	45.0	1.5		—
Israel	21,059	8,131	7.9	9.4	373	1777	22	5	1.6	3	4	79	83	92	97.9	94.3	6.1	3.86	$27,450
Jordan	89,210	34,444	6.9	8.5	77	1537	31	4	2.6	22	25	72	74	83	94.9	84.4	4.7		$5,530
Kazakhstan	2,717,289	1,049,151	16.8	18.4	6	51	23	9	1.4	26	29	63	74	54	99.1	96.1	2.9		$9,690
Kuwait	17,819	6,880	3.2	4.1	181	18,100	22	2	2.0	8	10	76	80	98	84.3	79.9	4.5		$52,610
Kyrgyzstan	198,499	76,641	5.5	6.5	28	395	24	7	1.6	32	37	64	72	35	98.6	95.5	2.0		$2,130
Lebanon	10,399	4,015	4.4	4.7	426	1374	20	5	1.5	11	12	70	74	87	92.3	80.4	2.5		$10,880
Libya	1,759,532	679,359	6.7	8.1	8	384	23	4	1.9	17	19	72	77	77	90.9	67.6	2.2		$15,630
Morocco	446,548	172,413	32.9	36.6	74	335	21	6	1.5	33	38	69	73	57	61.9	36.0	3.4		$4,330
Oman	212,459	82,031	3.2	4.2	15	1509	20	3	1.8	9	12	70	74	72	80.4	61.7	5.3		$20,650
Palestinian Territories	6,260	2,417	4.2	6.0	677	33,829	33	4	2.8	25	30	70	73	83	96.3	87.4			—
Qatar	11,000	4,247	1.7	2.1	157	15,703	9	1	0.8	10	11	75	77	100	80.5	83.2	7.7		—
Saudi Arabia	2,149,680	829,996	30.7	35.7	14	715	28	2	2.6	18	21	74	78	81	84.1	67.2	4.7	2.67	$22,950
Sudan (North)	1,877,063	724,735	33.1		17.6		31	10	2.1	69	115	57	60	46	44.0	18.0	2.5		$2,145
Syria	185,179	71,498	23.6	28.6	128	426	28	3	2.5	14	16	72	76	54	88.3	60.4	2.5		$4,350
Tajikistan	143,099	55,251	8.0	9.8	56	796	28	4	2.4	52	61	64	69	26	99.6	98.9	2.1		$1,860
Tunisia	163,610	63,170	10.8	12.1	66	205	18	6	1.2	18	21	72	76	66	81.4	60.1	4.3		$7,070
Turkey	774,816	299,158	75.4	85.0	97	256	18	6	1.2	19	20	69	74	76	93.6	76.7	4.4	3.89	$13,770
Turkmenistan	488,099	188,456	5.3	6.1	11	274	22	8	1.4	42	45	61	69	47	98.8	96.6	1.6		$6,210
United Arab Emirates	83,600	32,278	5.5	7.0	66	6628	15	2	1.4	7	7	77	79	83	75.5	79.5	6.3	2.99	—
Uzbekistan	447,397	172,741	29.1	34.4	65	542	23	5	1.8	32	36	65	71	36	98.5	96.0	1.6		$2,660
Western Sahara	252,120	97,344	0.5	0.7	2	5	34	9	2.5			58	62	81					—
Yemen	527,966	203,849	25.0	34.5	47	1581	38	8	3.0	51	66	62	64	29	67.4	25.0	2.2		$2,210
SOUTH ASIA																			
REALM	**4,487,762**	**1,732,734**	**1641.3**	**1946.0**	**152**	**304**	**24**	**7**	**1.7**	**51**	**66**	**64**	**66**	**29**	**66.1**	**39.9**			**$2,882**
Bangladesh	143,998	55,598	169.4	195.0	1867	1867	22	7	1.5	41	52	65	67	25	51.7	29.5	2.4		$1,440
Bhutan	47,001	18,147	0.7	0.9	15	513	25	8	1.7	52	79	67	68	32	61.1	33.6	5.7		$4,880
India	3,287,576	1,269,340	1227.1	1444.5	373	655	23	7	1.5	50	66	63	65	29	68.6	42.1	3.3		$2,960
Maldives	300	116	0.3	0.4	1037	10,368	22	3	1.9	11	13	72	74	35	96.3	96.4	2.3		$5,280
Nepal	147,179	56,826	29.1	35.7	198	943	28	8	1.9	39	48	64	65	17	59.1	21.8	2.2		$1,120
Pakistan	796,098	307,375	193.4	246.3	243	838	30	7	2.3	71	87	66	67	35	57.6	27.8	2.3	2.46	$2,700
Sri Lanka	65,610	25,332	21.2	23.2	323	1114	19	7	1.2	13	15	72	76	15	94.5	88.9	3.2	1.86	$4,480
EAST ASIA																			
REALM	**11,774,215**	**4,546,050**	**1585.0**	**1703.9**	**135**	**1091**	**12**	**7**	**0.5**	**16**	**17**	**73**	**77**	**52**	**94.9**	**84.7**	**3.5**		**$9,447**
China	9,572,855	3,696,100	1359.1	1484.7	142	1014	12	7	0.5	17	19	72	76	47	92.3	77.4	3.5	1.95	$6,020
Japan	377,799	145,869	127.4	119.3	337	2594	9	9	0.0	2	3	79	86	86	100.0	100.0	7.8	3.67	$35,220

	Land Area Sq. km.	Land Area Sq. mi.	Population 2012 (Millions)	Population 2025 (Millions)	Population Density Arithmetic	Population Density Physiologic	Birth Rate	Death Rate	Natural Increase %	Infant Mortality per 1,000	Child Mortality per 1,000	Life Expectancy Male (years)	Life Expectancy Female (years)	Percent Urban Pop.	Literacy Male %	Literacy Female %	Corruption Index	Big Mac Price ($US)	Per Capita GNI ($US)
Gabon	267,668	103,347	1.6	2.0	6	291	29	10	1.9	52	69	58	61	84	79.8	62.2	2.8		$12,270
Gambia	11,300	4,363	1.9	2.5	168	798	38	12	2.7	78	103	54	57	54	43.8	29.6	3.2		$1,280
Ghana	238,538	92,100	25.1	31.8	105	457	31	9	2.2	47	69	58	61	48	79.5	61.2	4.1		$1,430
Guinea	245,860	94,927	11.5	15.9	47	777	41	11	3.0	88	142	55	58	28	55.1	27.0	2.0		$1,190
Guinea–Bissau	36,120	13,946	1.7	2.3	47	358	43	18	2.4	115	193	45	48	30	53.0	21.4	2.1		$530
Ivory Coast	322,459	124,502	23.0	30.8	71	310	37	14	2.4	83	119	51	54	50	54.6	38.5	2.2		$1,580
Kenya	580,367	224,081	42.2	51.3	73	909	37	10	2.7	55	84	56	57	18	89.0	76.0	2.1		$1,580
Lesotho	30,349	11,718	1.9	2.0	64	579	28	19	0.9	61	84	40	43	23	73.6	93.6	3.5		$2,000
Liberia	111,369	43,000	4.4	6.1	39	980	43	11	3.3	80	112	54	57	58	69.9	36.8	3.3		$300
Madagascar	587,036	226,656	21.2	28.6	36	724	37	9	2.7	41	58	59	62	31	87.7	72.9	2.6		$1,040
Malawi	118,479	45,745	16.3	22.9	138	655	44	15	2.9	69	110	48	50	14	74.5	46.7	3.4		$830
Mali	1,240,185	478,838	16.2	22.3	13	326	46	15	3.1	101	191	50	53	33	47.9	33.2	2.7		$1,090
Mauritania	1,025,516	395,954	3.6	4.4	3	347	34	11	2.3	74	117	55	59	40	50.6	29.5	2.3		$2,000
Mauritius	2,041	788	1.3	1.4	643	1237	12	7	0.5	15	17	69	76	42	87.7	81.0	5.4		$12,480
Moçambique	801,586	309,494	24.5	31.2	31	765	40	16	2.3	96	142	47	49	31	59.9	28.4	2.7		$770
Namibia	824,287	318,259	2.3	2.8	3	277	28	9	1.9	34	48	60	62	35	82.9	81.2	4.4		$6,270
Niger	1,266,994	489,189	17.0	27.4	13	336	52	17	3.5	76	160	48	49	20	23.5	8.3	2.6		$680
Nigeria	923,766	356,668	166.3	217.4	180	530	42	17	2.4	86	138	47	48	47	72.3	56.2	2.4		$1,940
Réunion	2,510	969	0.8	1.0	327	2181	18	5	1.3			75	82	92	84.8	89.2			—
Rwanda	26,340	10,170	11.0	15.8	417	993	42	14	2.9	70	111	50	53	17	73.7	60.6	4.0		$1,010
São Tomé and Principe	961	371	0.2	0.2	221	514	37	7	2.9	52	78	64	69	58	70.2	39.1	3.0		$1,780
Senegal	196,720	75,954	13.2	17.4	67	560	39	11	2.8	51	93	54	57	41	47.2	27.6	2.9		$1,760
Seychelles	451	174	0.1	0.1	227	1512	18	7	1.0	11	12	68	78	53	82.9	85.7	4.8		$19,770
Sierra Leone	71,740	27,699	6.1	8.1	85	1060	40	16	2.4	123	192	46	49	36	50.7	22.6	2.4		$750
Somalia	637,658	246,201	10.0	13.9	16	782	46	16	3.0	109	180	48	51	34	85.5	84.5	1.1		—
South Africa	1,221,034	471,444	50.8	54.4	42	320	21	12	0.9	43	62	54	57	52	68.3	46.0	4.5	2.45	$9,780
South Sudan	628,755	242,763	8.5	13.5	45		45	20	2.5			43	46	12	21.0	11.0			$490
Swaziland	17,361	6,703	1.2	1.5	71	647	31	16	1.5	52	73	46	45	22	80.9	78.7	3.2		$5,010
Tanzania	945,087	364,900	47.7	67.4	51	1010	42	12	3.0	68	108	55	56	25	84.1	66.6	2.7		$1,230
Togo	56,791	21,927	7.1	9.3	126	293	33	8	2.5	64	98	60	63	40	72.2	42.6	2.4		$820
Uganda	241,040	93,066	36.1	53.4	150	441	47	13	3.4	79	128	52	53	17	77.7	57.1	2.5		$1,140
Zambia	752,607	290,583	14.0	20.3	19	265	45	20	2.5	86	141	41	42	37	85.2	71.2	3.0		$1,230
Zimbabwe	390,759	150,873	12.9	16.8	33	368	30	17	1.3	56	90	41	44	37	95.5	89.9	2.4		—
REALM	**19,318,887**	**7,459,064**	**615.6**	**739.2**	**65**	**714**	**25**	**7**	**1.9**	**33**	**43**	**67**	**71**	**57**	**83.1**	**64.5**			**$8,387**
NORTH AFRICA/SOUTHWEST ASIA																			
Afghanistan	652,086	251,772	30.3	39.4	47	388	39	18	2.1	134	199	44	44	22	51.0	20.8	1.4		—
Algeria	2,381,730	919,591	37.3	43.6	16	522	23	5	1.8	29	32	71	74	63	75.1	51.3	2.9		$7,940
Bahrain	689	266	1.3	1.6	1936	21,515	15	2	1.3	10	12	73	77	100	91.0	82.7	4.9		—

Continued

	Land Area		Population		Population Density		Birth Rate	Death Rate	Natural Increase %	Infant Mortality per 1,000	Child Mortality per 1,000	Life Expectancy		Percent Urban Pop. %	Literacy		Corruption Index	Big Mac Price ($US)	Per Capita GNI ($US)
	Sq. km.	Sq. mi.	2012 (Millions)	2025 (Millions)	Arithmetic	Physiologic						Male (years)	Female (years)		Male %	Female %			
Puerto Rico	8,951	3,456	4.0	4.1	451	5015	12	7	0.4	9		75	82	94	93.7	94.0	5.8		—
Saint Lucia	619	239	0.2	0.2	328	1170	14	7	0.7	19	20	71	76	28					$9,190
St. Vincent & the Grenadines	391	151	0.1	0.1	260	930	17	8	0.9	18	12	70	74	40					$8,770
Trinidad and Tobago	5,131	1,981	1.3	1.4	256	1068	14	8	0.6	31	35	66	73	12	99.0	97.5	3.6		$23,950
REALM (SOUTH AMERICA)	17,867,238	6,898,579	400.4	444.0	22	358	18	6	1.2	17	20	70	77	82	90.1	89.0			$10,150
Venezuela	912,046	352,143	29.7	35.1	33	815	21	5	1.6	15	18	71	77	88	93.3	92.7	2.0		$12,830
Uruguay	177,409	68,498	3.4	3.5	19	277	14	9	0.5	11	13	72	80	94	97.4	98.2	6.9	3.74	$12,540
Suriname	163,270	63,039	0.5	0.6	3	314	19	7	1.3	24	26	65	73	67	95.9	92.6			$7,130
Peru	1,285,214	496,224	30.4	34.5	24	788	21	6	1.6	19	21	71	76	76	94.7	85.4	3.5	3.54	$7,980
Paraguay	406,747	157,046	6.7	8.1	17	277	25	6	1.9	19	23	70	74	58	94.4	92.2	2.2		$4,820
Guyana	214,969	83,000	0.8	0.8	4	192	23	7	1.6	29	35	62	70	28	99.0	98.1	2.7		$2,510
French Guiana	89,999	34,749	0.2	0.3	2	233	28	3	2.4	14	14	75	81	81					—
Ecuador	283,560	109,483	14.7	16.6	52	470	21	5	1.6	20	24	72	78	65	83.6	82.3	2.5		$7,760
Colombia	1,138,906	439,734	46.8	53.5	41	1027	20	6	1.4	16	19	71	78	75	93.6	90.2	3.5	4.39	$8,510
Chile	756,626	292,135	17.4	19.1	23	767	15	6	0.9	7	9	76	82	87	95.9	95.5	7.2	3.34	$13,270
Brazil	8,547,360	3,300,154	197.6	212.4	23	330	17	6	1.0	21	21	69	77	84	85.5	85.4	3.7	4.91	$10,070
Bolivia	1,098,575	424,162	10.8	13.3	10	492	27	7	2.0	40	51	64	68	65	92.1	79.4	2.8		$4,140
Argentina	2,780,388	1,073,514	41.3	46.2	15	149	18	8	1.0	13	14	72	79	91	96.9	96.9	2.9	1.78	$14,020
REALM (SUBSAHARAN AFRICA)	24,292,983	9,379,573	863.9	1150.7	36	472	40	14	2.6	79	127	51	53	34	73.1	59.9			$1,974
Angola	1,246,693	481,351	20.0	27.4	16	534	42	17	2.5	98	161	45	49	57	55.6	28.5	1.9		$5,020
Benin	112,620	43,483	10.4	13.6	92	577	40	10	3.0	75	118	57	60	41	47.8	23.6	2.8		$1,460
Botswana	581,727	224,606	1.9	2.2	3	321	30	11	1.9	43	57	55	55	60	74.4	79.8	5.8		$13,100
Burkina Faso	274,000	105,792	17.3	25.4	63	527	46	12	3.4	91	166	51	54	23	31.2	13.1	3.1		$1,160
Burundi	27,829	10,745	8.9	11.6	318	740	36	15	2.1	101	166	49	52	10	56.3	40.5	1.8		$380
Cameroon	475,439	183,568	20.9	26.5	44	275	37	14	2.3	95	154	50	52	53	81.8	69.2	2.2		$2,180
Cape Verde	4,030	1,556	0.5	0.7	129	1171	25	6	2.0	23	28	69	76	61	84.3	65.3	5.1		$3,450
Central African Rep.	622,978	240,533	5.0	6.6	8	268	38	16	2.2	112	171	47	50	38	59.6	34.5	2.1		$730
Chad	1,283,994	495,753	12.2	16.9	9	316	46	17	2.9	124	209	47	50	27	66.9	40.8	1.7		$1,160
Comoros	2,230	861	0.7	0.9	330	623	33	7	2.6	75	104	62	66	28	63.5	49.1	2.1		$1,170
Congo	341,998	132,046	4.1	5.5	12	1198	38	13	2.5	81	128	52	54	60	87.5	74.4	2.0		$3,090
Congo, The	2,344,848	905,351	71.9	101.4	31	767	47	17	2.9	126	199	46	49	33	86.6	67.7	2.0		$290
Djibouti	23,201	8,958	0.9	1.1	40	4020	29	11	1.8	75	94	54	57	76	65.0	38.4	3.2		$2,330
Equatorial Guinea	28,050	10,830	0.7	1.0	26	290	39	16	2.3	88	145	48	51	39	92.5	74.5	1.9		$21,700
Eritrea	117,598	45,405	5.5	7.4	47	1168	37	9	2.9	39	55	57	62	21	43.9	33.4	2.6		$630
Ethiopia	1,104,296	426,371	89.7	119.8	81	738	39	12	2.7	67	104	54	57	16	83.7	70.0	2.7		$870

	Land Area		Population		Population Density		Birth Rate	Death Rate	Natural Increase %	Infant Mortality per 1,000	Child Mortality per 1,000	Life Expectancy		Percent Urban Pop.	Literacy		Corruption Index	Big Mac Price ($US)	Per Capita GNI ($US)
	Sq. km.	Sq. mi.	2012 (Millions)	2025 (Millions)	Arithmetic	Physiologic						Male (years)	Female (years)		Male %	Female %			
Slovenia	20,251	7,819	2.1	2.0	104	744	11	9	0.2	2	3	76	82	50	98.6	96.8	6.4		$26,910
Spain	505,988	195,363	47.4	48.4	94	240	11	8	0.3	4	4	78	84	77	100.0	100.0	6.1		$31,130
Sweden	449,959	173,730	9.4	10.2	21	300	12	10	0.2	2	3	79	83	84	100.0	100.0	9.2	6.56	$38,180
Switzerland	41,290	15,942	7.8	8.3	190	1581	10	8	0.2	4	4	80	84	73	99.5	97.4	8.7	6.19	$46,460
Ukraine	603,698	233,089	45.5	41.9	75	128	11	15	−0.4	13	15	63	74	69	100.0	100.0	2.4	1.84	$7,210
United Kingdom	244,878	94,548	62.7	68.6	256	985	13	9	0.4	5	6	77	82	80	97.6	89.2	7.6	3.48	$36,130
REALM	**17,261,421**	**6,664,672**	**158.3**	**158.7**	**9**	**149**	**12**	**14**	**−0.2**	**13**	**14**	**62**	**74**	**73**	**99.1**	**99.1**			**$15,531**
RUSSIAN REALM																			
Russia	17,075,323	6,592,819	141.3	140.8	8	138	12	14	−0.2	11	12	62	74	73	99.7	99.2	2.1	2.33	$15,630
Armenia	29,800	11,506	3.1	3.3	105	525	15	10	0.6	20	22	68	75	64	99.4	98.1	2.6		$6,310
Azerbaijan	86,599	33,436	9.2	10.3	106	483	17	6	1.1	30	34	70	75	54	98.9	95.9	2.4		$7,770
Georgia	69,699	26,911	4.6	4.3	66	443	13	10	0.3	26	29	69	79	53	99.7	99.4	3.8		$4,850
REALM	**19,941,634**	**7,699,508**	**347.7**	**391.1**	**17**	**141**	**13**	**8**	**0.6**	**7**	**8**	**76**	**81**	**79**	**97.6**	**97.4**			**$45,890**
NORTH AMERICA																			
Canada	9,970,600	3,849,670	34.4	39.7	3	69	11	7	0.4	5	6	78	83	80	95.7	95.3	8.9	4.00	$36,220
United States	9,629,047	3,717,796	313.3	351.4	33	163	14	8	0.6	7	8	75	80	79	99.0	99.0	7.1	3.73	$46,970
REALM	**2,714,579**	**1,048,105**	**200.2**	**223.3**	**74**	**456**	**21**	**5**	**1.6**	**19**	**23**	**72**	**78**	**71**	**90.3**	**87.1**			**$11,910**
MIDDLE AMERICA																			
Antigua and Barbuda	440	170	0.1	0.1	231	1282	14	6	0.9	11	12	73	77	31					$20,570
Bahamas	13,880	5,359	0.3	0.4	22	2201	15	6	0.9	9	12	71	77	83	95.4	96.8	7.8		—
Barbados	430	166	0.3	0.3	705	1807	13	8	0.5	10	11	71	76	38	98.0	96.8			—
Belize	22,960	8,865	0.3	0.4	14	342	27	4	2.3	16	18	71	74	51					$6,040
Costa Rica	51,100	19,730	4.7	5.3	92	1026	17	4	1.3	10	11	77	82	59	95.5	95.7	5.3	3.83	$10,950
Cuba	110,859	42,803	11.3	11.1	102	248	11	8	0.3	4	6	76	80	75	99.8	99.8	3.7		—
Dominica	751	290	0.1	0.1	135	675	15	8	0.7	8	10	72	78	73			5.2		$8,300
Dominican Republic	48,731	18,815	10.2	11.6	210	678	23	6	1.7	27	32	69	75	67	84.0	83.7	3.0		$7,890
El Salvador	21,041	8,124	6.4	6.9	302	775	20	7	1.4	15	17	67	76	63	82.5	76.1	3.6		$6,670
Grenada	339	131	0.1	0.1	299	936	17	9	0.8	13	15	68	72	31					$8,060
Guadeloupe	1,709	660	0.4	0.5	237	1582	14	7	0.7	7		76	83	100	90.5	90.5			—
Guatemala	108,888	42,042	15.2	20.0	140	776	34	6	2.8	33	40	66	73	47	76.2	61.1	3.2		$4,690
Haiti	27,749	10,714	10.2	12.2	367	1111	28	9	1.8	64	87	59	62	48	51.0	46.5	2.2		$1,180
Honduras	112,090	43,278	8.0	9.8	71	394	28	5	2.3	25	30	70	75	50	72.5	72.0	2.4		$3,870
Jamaica	10,989	4,243	2.8	2.9	252	1008	20	7	1.2	26	31	68	75	52	82.5	90.7	3.3		$7,360
Martinique	1,101	425	0.4	0.4	368	1839	13	7	0.6	8		77	83	89	96.0	97.1			—
Mexico	1,958,192	756,062	113.7	123.4	58	415	19	5	1.4	15	17	74	79	77	93.1	89.1	3.1	2.50	$14,270
Netherlands Antilles	800	309	0.2	0.2	253	2534	14	7	0.7	11		73	80	92	96.6	96.6			—
Nicaragua	129,999	50,193	6.2	7.5	48	218	23	4	1.8	22	26	69	73	56	64.2	64.4	2.5		$2,620
Panama	75,519	29,158	3.6	4.3	48	531	20	5	1.6	16	23	73	78	64	92.6	91.3	3.6		$11,650